International Conference on Recent Advancements and Effectual Researches in Engineering Science and Technology (RAEREST 2018)

IOP Conference Series: Materials Science and Engineering
Volume 396

Kerala State, India
20 – 21 April 2018

ISBN: 978-1-5108-7026-0
ISSN: 1757-8981

Printed from e-media with permission by:

Curran Associates, Inc.
57 Morehouse Lane
Red Hook, NY 12571

Some format issues inherent in the e-media version may also appear in this print version.

Printed by Curran Associates, Inc. (2018)

For permission requests, please contact the Institute of Physics at the address below.

Institute of Physics
Dirac House, Temple Back
Bristol BS1 6BE UK

Phone: 44 1 17 929 7481
Fax: 44 1 17 920 0979

techtracking@iop.org

Additional copies of this publication are available from:

Curran Associates, Inc.
57 Morehouse Lane
Red Hook, NY 12571 USA
Phone: 845-758-0400
Fax: 845-758-2633
Email: curran@proceedings.com
Web: www.proceedings.com

Table of contents

Volume 396

International Conference on Recent Advancements and Effectual Researches in Engineering Science and Technology (RAEREST)

20-21 April 2018, Kerala State, India

Accepted papers received: 19 July 2018
Published online: 29 August 2018

Preface

Papers - Civil Engineering

Structural and Concrete Technology 1

Computer Science Engineering

Computational Intelligence

Electronics and Communication Engineering

Signal Processing

IOP Conf. Series: Materials Science and Engineering **396** (2018) 011001 doi:10.1088/1757-899X/396/1/011001

Editorial – RAEREST 2018

This volume is a collection of the papers presented at the second Global Colloquium on *Recent Advancements and Effectual Researches in Engineering, Science and Technology (RAEREST – 2018),* which was organized and hosted at St. Joseph's College of Engineering & Technology, Palai, Kerala during April 20th and 21st, 2018.

St. Joseph's College of Engineering & Technology, Palai (SJCET) is owned and managed by the diocese of Palai, and was the fulfilment of a long-cherished dream the diocese of providing facilities for higher education with international stature. SJCET's objective is to develop itself into a college with a distinct identity and character, where knowledge and training are imparted for life-long learnability along with developing moral, professional and ethical values in engineers of the future. The campus is well equipped so as to create an environment fully conducive to realize the objectives. Discipline, hard work, positive thinking, commitment to excellence and abiding faith in the Almighty are the guiding principles that propel this institution towards its vision of emerging as a Centre of Excellence in technical education in the country. SJCET meets almost all national and international benchmarks stipulated for a world class engineering institution by undertaking consultancy works, promoting several in-house projects, conducting exhibitions for public, expert talks and conferences.

RAEREST 2018 is the third international conference organized and hosted by St. Joseph's College of Engineering and Technology, Palai, in successful succession to the International Multi-Conference on Automation, Communication, Control and Compressed Sensing (iMac4s) and RAEREST- 2016. These international conferences focuses on providing an opportunity to global participants to share ideas, knowledge and information with eminent and experienced professionals from industries, research organizations, faculty, research scholars from academic institutions, working in their area of interest. The theme of RAEREST – 2018 is "Human Empowerment through Technology". The program of the conference constitutes of plenary & keynote lectures from eminent scientists or researchers working in almost all areas of

IOP Conf. Series: Materials Science and Engineering **396** (2018) 011001 doi:10.1088/1757-899X/396/1/011001

engineering and its related applications. Dr. P V Aravind from Technical University of Delft, Netherlands who himself is a pioneer in applications of technology for human empowerment delivered two lectures to the participants of this conference in the broad area of Fuel Cells. Dr. A K Ashraff from the Indian Space Research Organization, Dr. Tapan Kumar Hotta from Vellore Institute of Technology, Vellore, Dr. M J Jalaja & Dr. Bino I Thomas, both from Rajiv Gandhi Institute of Technology, Dr. Vincent George from Govt. Engineering College, Palakkad and Dr. Varghese Paul from Cochin University of Science and Technology were other noted plenary speakers to the conference.

Close to 200 research papers were received from which around 75 papers were selected for oral presentation through a double blind peer review. The conference is a culmination of almost a year's hard work of many faculties at SJCET, right from the planning phase, through the publicity rounds, communications with publishers, the peer review process in the double blind system, discussions with sponsors, reverting with authors on finalization of publication papers and a smooth execution of the conference.

Dr. Madhukumar S

Dr. Rajesh Baby

Prof. Sabarinath G

IOP Conf. Series: Materials Science and Engineering **396** (2018) 011002 doi:10.1088/1757-899X/396/1/011002

Peer review statement

All papers published in this volume of *IOP Conference Series: Materials Science and Engineering* have been peer reviewed through processes administered by the proceedings Editors. Reviews were conducted by expert referees to the professional and scientific standards expected of a proceedings journal published by IOP Publishing.

IOP Conf. Series: Materials Science and Engineering **396** (2018) 012001 doi:10.1088/1757-899X/396/1/012001

Effect of Glass Fibers, Fly Ash and Quarry Dust on Strength and Durability Aspects of Concrete – *An Experimental Study*

PL Meyyappan[1], K Kumaran [1] M Gopalakrishnan[1] and E Harikrishnan[1]

[1] Kalasalingam Academy of Research and Education, Krishnnakoil, 626126, India

E-mail: meyyappan9881@gmail.com

Abstract. In this study, partial replacement of cement by fly ash and quarry dust has been made in order to limiting the usage of cement and thereby it leads a way to reduce the cost of construction. Moreover it is suitable for handling the waste products from thermal station and crushers and reduces the impact of environmental pollution issues. In order to compensate the strength and durability related aspects, glass fibers are incorporated in the range of 1 %, 2 % and 3 % of volume fraction along with replacement materials. Replacement of cement is achieved by various replacement levels of 10 %, 20 % and 30 % fly ash and quarry dust. The replacement is made in both individual and combined manner. For evaluating strength and durability characteristics on M20 concrete, compressive strength and porosity tests were carried out. The experimental investigation revealed that the addition of glass fibers up to 2 % along with replacement materials in 20 % will be the optimum in improving the strength and durability aspects. Regression analysis is also carried out to validate the experimental results.

1. Introduction

Most of the concrete research studies emphasized that the utilization of fibers in to concrete has gained a lot in improving the properties in both precast and in situ *concrete* construction [3], [8],[9]. Among all the kinds of fibers, glass fibers in to the concrete have an advantage of being light weight, high strength and more sustainable [9], [10]. In the recent days, the utilization of waste products such as fly ash, quarry dust, marble dust, rice husk, blast furnace slag, peels of fruits and vegetables, coconut and palm shells, plastics, rubber and e-waste etc is used as a replacement material for cement. By utilizing the waste products as the replacement material, the cost of construction is reduced to some desirable extent and reduces the Co_2 emission by limiting in the production of concrete [1], [2], [4]. Further it helps a lot in disposing the waste and thereby the pollution control issues can also be addressed in sequential way. Though a lot of research works have been reported on the effect of glass fibers on concrete, not much work has been carried out in a combined way of utilizing waste materials along with it. Hence, in this study an attempt has been made to investigate the effect of addition of glass fibers, fly ash and quarry dust on the strength and durability aspects of concrete.

2. Materials Used

The following materials were used in this investigation:

2.1. Cement

Cement 43 grade Ordinary Portland Cement (OPC) confirming to the requirements of IS:8112 code

2.2. Sand
River sand having a specific gravity of 2.68 and confirming to zone II of IS:383

2.3. Coarse Aggregate
Crushed granite coarse aggregate of 20 mm maximum size with specific gravity of 2.64

2.4. Glass fibers
Alkali resistant glass fibers having aspect ratio 1693, a diameter of 15 microns and a length of 25 mm.

2.5. Fly ash
Class C fly ash having specific gravity of 2.35 obtained from Thermal power station which is located at Tuticorin in Tamil Nadu is used for the partial replacement for cement

2.6. Quarry dust
Quarry dust passing through 4.75 mm retained on 150 micron IS sieves obtained from quarry near by Srivilliputtur in Tamil Nadu is used for the partial replacement for cement as a filler material

2.7. Water
Ordinary portable water from single source used throughout the experimental work

3. Mix Design and Testing
The designed mix used in this experimental study is 1:1.43:2.72:0.5 (M20). The mix design for the above grade was done by using Indian Standard Code 10262-2009. Table 1 shows the mix proportion details for the above grade of concrete. Using the mix design, a total 240 numbers of standard concrete cube specimens 150 mm x 150 mm x 150 mm were used for compressive strength test and porosity test. The cement is replaced at 10 %, 20% and 30% by flyash and quarry dust in individual and combined composition additionally with incorporating 0 %, 1 %, 2% and 3 % of volume fraction of glass fibers.

Table 1. Mix Proportions Details

Concrete Mix Designations	Cement (Kg/m^3)	Fly Ash (Kg/m^3)	Quarry Dust (Kg/m^3)	Glass Fibers	Fine aggregate (Kg/m^3)	Coarse aggregate (Kg/m^3)	Water (litres)
M01, M02, M03, M04	412	-	-	0%,1%, 2%, 3%	590	1123.5	206
MF0, MF1, MF2, MF3	370.8	41.2	-	0%,1%, 2%, 3%	590	1123.5	206
2MF0, 2MF1, 2MF2, 2MF3	329.6	82.4	-	0%,1%, 2%, 3%	590	1123.5	206
3MF0, 3MF1, 3MF2, 3MF3	288.4	123.6	-	0%,1%, 2%, 3%	590	1123.5	206
MQ0, MQ1, MQ2, MQ3	370.8	-	41.2	0%,1%, 2%, 3%	590	1123.5	206
2MQ0, 2MQ1, 2MQ2, 2MQ3	329.6	-	82.4	0%,1%, 2%, 3%	590	1123.5	206
3MQ0, 3MQ1, 3MQ2, 3MQ3	288.4	-	123.6	0%,1%, 2%, 3%	590	1123.5	206
MFQ0, MFQ1, MFQ2, MFQ3	370.8	20.6	20.6	0%,1%, 2%, 3%	590	1123.5	206
2MFQ0, 2MFQ1, 2MFQ2, 2MFQ3	329.6	41.2	41.2	0%,1%, 2%, 3%	590	1123.5	206
3MFQ0, 3MFQ1, 3MFQ2, 3MFQ3	288.4	61.8	61.8	0%,1%, 2%, 3%	590	1123.5	206

Cement, fine aggregates, coarse aggregates, fly ash and quarry dust were first mixed and glass fibers were added in the dry state through sieving and water was added at the end. This was done to achieve uniform distribution of fibers. The specimens were casted and the moulds were removed after 24 hours of casting. Then it was allowed for water curing for the period of 28 days. Compression strength test was carried out at a uniform stress of 14 N/mm^2 in the compressive testing machine. For porosity test, the specimens were placed in hot oven at 100°C temperature for a period of 48 hours. The volume of pores can be estimated by the weight difference of specimen before and after placing in oven and it is expressed in terms of percentage.

4. Results and Discussions

The values of compressive strength and porosity obtained for the above grade of concrete containing 0%, 1%, 2% and 3% of volume fraction of glass fibers with the replacement of cement by flyash and quarry dust in individual and combined composition have been presented (Table 3 – 5). The each test result is the average test result of three specimens.

4.1. Effect of flyash and glass fibers

From the Table 1, it is understood that if the flyash replacement is increased upto 20 % the compressive strength increases with a gain factor of 1.3% for without incorporation of glass fibers and if the replacement increased to 30%, the value of compressive strength decreases with a loss factor of 1.21% for without incorporating glass fibers. When the glass fibers were incorporated upto 2% of volume fraction, the compressive strength increases to a extend of 26.81N/mm^2 with a gain factor of 1.26 % in condition of without replacement of flyash. In the experimental result it is understood that, the combined effect of flyash (20 % replacement) and glass fiber (2% of volume fraction), the compressive strength achieves a maximum of 34.95 % with a gain factor of 1.65 %. It is also noted that, when replacement of flyash extended beyond 20 % and volume fraction of glass fibers beyond 2 %, there seems to be fall down in the strength characteristics (figure 1). It may be due of punching effect of glass fibers and absence of bonding materials. From figure 2, it is understood that, the fineness of flyash occupying the pores of the concrete and thereby the percentage of porosity (volume of pores) decreases from 6.39 % to 5.25 %, when the replacement percentage is around 20 % without adding glass fibers.

Table 2. Test results for replacement of cement by flyash

Concrete Mix Designations	Cement %	Fly ash %	Glass fibers %	Compressive strength N/mm^2	Porosity %
M01	100	0	0	21.24	6.39
M02	100	0	1	24.18	5.81
M03	100	0	2	26.81	5.30
M04	100	0	3	23.87	5.95
MF0	90	10	0	23.64	5.88
MF1	90	10	1	29.45	5.42
MF2	90	10	2	31.74	5.21
MF3	90	10	3	25.45	5.72
2MF0	80	20	0	26.81	5.25
2MF1	80	20	1	32.64	4.97
2MF2	80	20	2	34.95	4.88
2MF3	80	20	3	27.14	5.49
3MF0	70	30	0	22.18	5.88
3MF1	70	30	1	24.14	5.79
3MF2	70	30	2	27.65	5.21
3MF3	70	30	3	22.42	5.91

When the replacement is increased to 30 %, the volume of pores is around 5.88 % which is lesser than normal concrete. When the glass fibers incorporated in to the concrete, the volume of pores reduced from 6.39 % to 5.95 % in without replacement condition. In a combined way, the flyash and glass fibres presence, the volume of pores were drastically reduced to 4.88 % (reduction by 1.31%) because of immersion of fibers (2 % in volume fraction) and fine particles of flyash (20 % replacement of cement) filled up the pores. If the fiber and flyash content increased to 3% and 30 % respectively, volume of pores were almost same as normal concrete due to non uniform mixing and improper compaction.

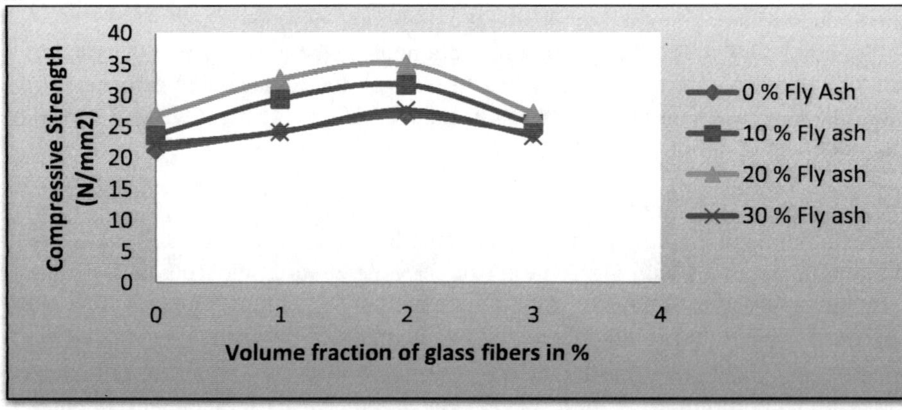

Figure 1. Compressive strength Vs Volume fraction of glass fibers by Flyash replacement

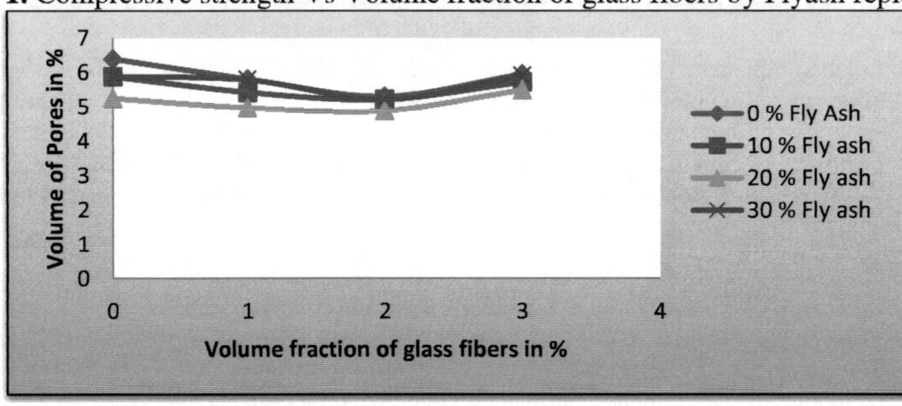

Figure 2. Volume of pores Vs Volume fraction of glass fibers by Flyash replacement

4.2. Effect of quarry dust and glass fibers

From the Table 3, it is understood that the replacement of cement by quarry dust is effective upto 20% in terms of compressive strength and porosity aspects with the gain factor around 1.13%. If it is further increased to 30 %, the strength reduced to 0.81% for without glass fibers. If the glass fibers are added upto 2 % in terms of volume fraction, the compressive strength reaches to a maximum of 29.71N/mm^2 with a gain factor of 1.4%.

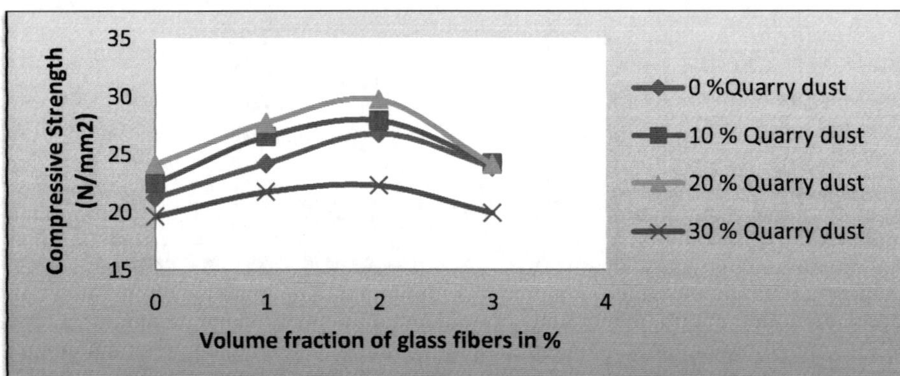

Figure 3. Compressive strength Vs Volume fraction of glass fibers by Flyash replacement

Table 3. Test results for replacement of cement by quarry dust

Concrete Mix Designations	Cement %	Quarry dust %	Glass fibers %	Compressive strength N/mm^2	Porosity %
M01	100	0	0	21.24	6.39
M02	100	0	1	24.18	5.81
M03	100	0	2	26.81	5.30
M04	100	0	3	23.87	5.92
MQ0	90	10	0	22.45	6.14
MQ1	90	10	1	26.51	5.38
MQ2	90	10	2	27.93	5.21
MQ3	90	10	3	24.17	5.88
2MQ0	80	20	0	24.13	5.54
2MQ1	80	20	1	27.74	5.24
2MQ2	80	20	2	29.71	4.91
2MQ3	80	20	3	24.06	5.64
3MQ0	70	30	0	19.62	6.93
3MQ1	70	30	1	21.73	6.30
3MQ2	70	30	2	22.27	6.14
3MQ3	70	30	3	19.91	6.76

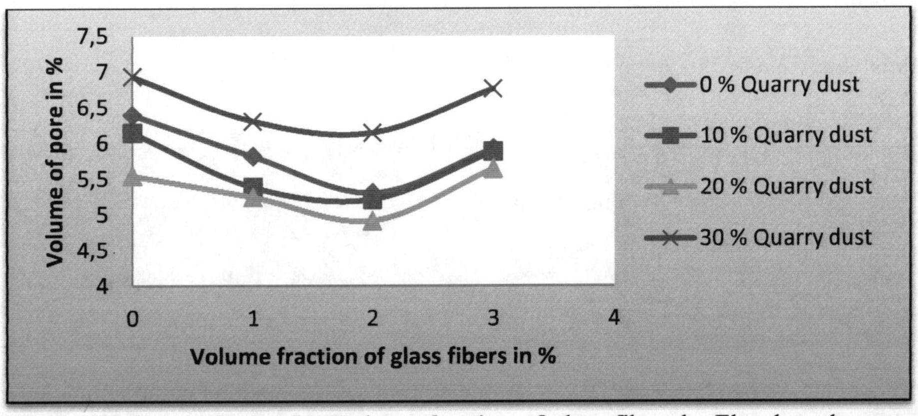

Figure 4. Volume of pores Vs Volume fraction of glass fibers by Flyash replacement

Figure 3 clearly shows that, the quarry dust can found to be optimum upto 20 % replacement in addition of 2 % volume fraction of steel fibers. It is also noticed that, compressive strength decreases with a reduction factor of 0.74 % as compared with normal concrete. This may be due to the missing component of cementatious material and poor bonding with the constituents. In line with figure 4, the 30 % replacement of cement by quarry dust has pores upto 6.14 % and 6.93 % for with (2 % of volume fraction) and without addition of glass fibers respectively whereas the normal concrete is having 6.39 % of pores. Due to fineness nature of quarry dust, it occupying the pores in the concrete and thereby the pores are getting reducing. In comparing with the flyash, the results were not showing significance.

4.3. Combined effect of fly ash, quarry dust and glass fibers

The compressive strength of concrete with the replacement of fly ash and quarry dust in combined way of 10% and 20% is 24.82 N/mm^2 and 28.95 N/mm^2 respectively for without glass fibers (Table 4). When the combined replacement is increased to 30%, the compressive strength is reduced to 23.73 N/mm^2. If the glass fibers are incorporated as 1%, and 2% in terms of volume fraction and also with the replacement of 10% and 20% of combined fly ash and quarry dust, the compressive strength value is in the range of 31.81 N/mm^2 to 38.79 N/mm^2. When the percentage of glass fibers increased to 3%, the compressive strength reduced to 23.31 N/mm^2 with 30% combined replacement. It is found that, the optimum percentage of fly ash and quarry dust can be significantly effective for the replacement of cement in the combined way is 20 % (Each 10%) as seen in figure 5.

Table 4. Test results for replacement of cement by flyash and quarry dust

Concrete Mix Designations	Cement %	Fly ash %	Quarry dust %	Glass fibers %	Compressive strength N/mm^2	Porosity %
M01	100	0	0	0	21.24	6.39
M02	100	0	0	1	24.18	5.81
M03	100	0	0	2	26.81	5.3
M04	100	0	0	3	23.87	5.92
MFQ0	90	5	5	0	24.82	5.58
MFQ1	90	5	5	1	31.81	4.98
MFQ2	90	5	5	2	34.91	4.58
MFQ3	90	5	5	3	26.72	5.31
2MFQ0	80	10	10	0	28.95	4.83
2MFQ1	80	10	10	1	35.25	4.72
2MFQ2	80	10	10	2	38.79	4.34
2MFQ3	80	10	10	3	28.22	5.16
3MFQ0	70	15	15	0	23.73	5.64
3MFQ1	70	15	15	1	25.34	5.38
3MFQ2	70	15	15	2	29.03	5.15
3MFQ3	70	15	15	3	23.31	5.67

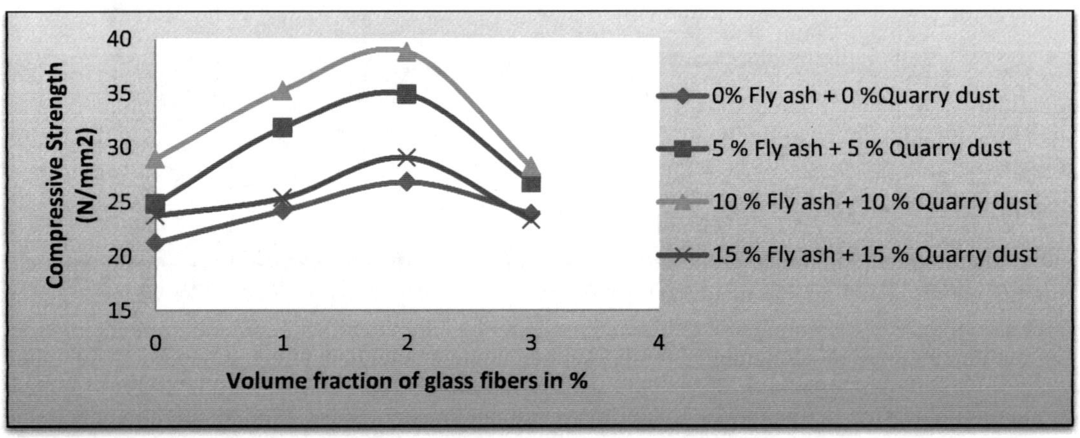

Figure 5. Compressive strength Vs Volume fraction of glass fibers by F + Q replacement

In overall, the strength gain factor is around 1.82% for the combined action of fly ash, quarry dust and glass fibers may be due to proper bonding and incorporation of the constituents. In figure 6, it is understood that, volume of pores is maximum of 6.39 % for without fibers and replacement materials. The volumes of pores were effectively reduced up to 4.34 % when glass fibers (2% of volume fraction) and replacement materials (10% of flyash and 10% of quarry dust) are introduced. Due to bunching of glass fibers and more percentage of non-cementatious materials, the volumes of pores were increasing and it is almost same for fibers (1% and 3% in volume fraction) along with replacement materials (5% and 15% respectively). It is also understood the performance of quarry dust in a combined manner is more effective than in a individual manner.

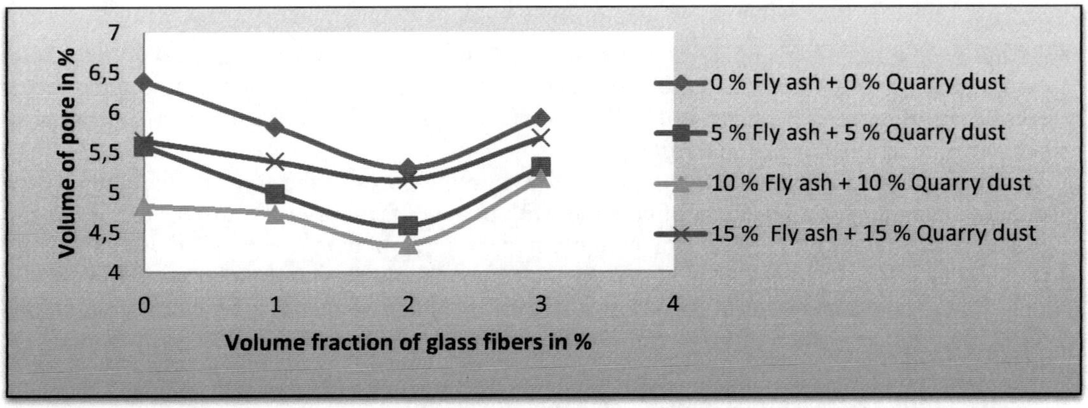

Figure 6. Volume of pores Vs Volume fraction of glass fibers by F + Q replacement

4.4. Regression analysis

A simple regression analysis is a type of projecting modeling technique which examines the correlation between a dependent (target) and independent variable (s) (predictor). This technique is used for forecasting, time series modelling and finding the fundamental effect relationship between the variables. By using regression analysis, the best fit equations (Eq 1. to Eq.3) were obtained between Compressive strength, volume fraction of glass fibers and replacement of materials are given below:

$$CS = 1.052\, V_{gf} + 0.057\, R_f + 24.44 \quad (1)$$

$$CS = 0.176\, V_{gf} - 0.118\, R_q + 25.44 \quad (2)$$

$$CS = 0.135\, V_{gf} + 0.105\, R_{fq} + 24.44 \quad (3)$$

Similarly, the best fit equations (Eq. 4 to Eq. 6) were obtained between Porosity, volume fraction of glass fibers and replacement of materials are given below:

$$P = 0.063\, V_{gf} + 0.018\, R_f + 4.932 \quad (4)$$

$$P = -0.095\, V_{gf} + 0.021\, R_q + 5.795 \quad (5)$$

$$P = -0.121\, V_{gf} + 0.028\, R_{fq} + 4.062 \quad (6)$$

Based on the regression analysis, the relationship between compressive strength and porosity of concrete is also obtained. Using Eq.7 the compressive strength of glass fiber reinforced concrete (GFRC) can be estimated once it its porosity is known.

$$CS = -0.585\, P^2 + 1.783\, P + 33.78 \quad (7)$$

Where, *CS = Compressive strength of concrete at 28 days*
P = Porosity (Volume of pores)
V_{gf} = *Volume fraction of glass fibers*
R_f = *Replacement levels of cement by fly ash*
R_q = *Replacement levels of cement by quarry ash*
R_{fq} = *Replacement levels of cement by fly ash and quarry dust in a combined manner.*

There was a strong correlation between the *experimental data* and the *predictive values* with plus or minus 10% to 15%.

5. Conclusions

(a) Addition of glass fibers up to 2 %, in to the concrete resulted in significant increase in compressive strength and decrease in porosity without any replacements. When glass fibers extended further to 3 %, there is a falling trend in compressive strength and rising trend in porosity due to bunching of glass fibers.

(b) Addition of pozzolanic materials of fly ash filled up the pores in the concrete and thereby the increase in compressive strength of concrete with the gain factor of 1.2% to 1.3 %. Further if it is added along with glass fibers, the compressive strength is increased with the gain factor up to 1.6 % to 1.7 %.

(c) The individual replacement of quarry dust does not contribute much in terms of strength and durability characteristics of concrete. However, in the experimental results proved that, the combined effect of glass fibers, fly ash and quarry dust improved the compressive strength of concrete up to 1.82% and also drastic reduction in porosity.

(d) The optimum percentage of fiber content and replacement materials is found to be 2 % of volume fraction and 20 % of replacement materials.

(e) Empirical equations based on the regression analysis have been established to estimate the compressive strength of glass fiber reinforced concrete.

References

[1] Baboo Rai, Sanjay Kumar and Kumar Satish 2014 Effect of Fly Ash on Mortar Mixes with Quarry Dust as Fine Aggregate *Journal of Adavnces in Material Science and Engineering* **1** pp 1-7.

[2] Krishnamoorthi A and Mohan Kumar G 2013 Properties of Green Concrete Mix by Concurrent use of Fly ash and Quarry dust *IOSR Journal of Engineering* **3** pp 48-54.

[3] Anjali H Jamale and Urmila R Kaw 2015 Effect of Quarry Dust and Fly Ash Mix on strength properties of M40 grade Concrete *International Journal of Engineering Research and General Science* **3** pp 391-397.

[4] Chandana Sukesh, Katakam Bala Krishna, Sri Lakshmi Sai Teja P and S.Kanakambara Rao 2013 Partial Replacement of Sand with Quarry Dust in Concrete *International Journal of Innovative Technology and Exploring Engineering* **2** pp 254-258.

[5] Md.Abid Alam, Imran Ahmad and Fazlur Rehman 2015 Experimental Study on Properties of Glass Fibre Reinforced Concrete *International Journal of Engineering Trends and Technology* **24** pp 297-301.

[6] Frederick T. Wallenberger, James C. Watson, and Hong Li 2001 Glass Fibers, *ASM International Hand Book* **21** pp 21-30.

[7] Chaitanya kumar JD, Abhilash GVS, Khasim Khan P, Manikanta sai G and Tarakaram V 2016 Experimental Studies on Glass Fiber Concrete *American Journal of Engineering Research* **5** pp 100-104.

[8] Xiao-quan Shi, Zhi-qiang Zhang, Zhi-ye Li 2012 Experimental study of the shear capacity of glass fiber reinforced polymer reinforced concrete beam with circular cross section *Journal of Shanghai Jiaotong University (Science)* **17** pp 408-414.

[9] Sathish Kumar. V, Blessen Skariah Thomas and Alex Christopher 2012 An Experimental Study on the Properties of Glass Fibre Reinforced Geopolymer Concrete *International Journal of Engineering Research and Applications* **2** pp 722-726

[10] Kumar Shantveerayya and Vikas Nikkam 2016 An Experimental Study on the Properties of Glass Fibre Reinforced and Ground Granulated Blast Furnace Slag Concrete *SSRG International Journal of Civil Engineering* **3** pp 14-19.

Use of Clay Tile Chips as Coarse Aggregate in Concrete

Aswathy Mohan[1,b],Job Thomas[2,a]and Neethu Joseph[3]

[1] M-Tech Student, Department of Civil Engineering, SJCET, Palai, India
[2]Associate Professor, Department of Civil Engineering, School of Engineering, Cochin University of Science and Technology, Cochin, India
[3] Assistant Professor, Department of Civil Engineering, SJCET, Palai, India

[a]E-mail: job_thomas@cusat.ac.in
[b]E-mail: amnetmr@gmail.com

Abstract. Roof of majority of old buildings in Kerala is constructed with Mangalore pattern clay tiles. Hence, a large amount of broken clay tiles are produced during the demolition of these old buildings. In this study, the potential of using the broken clay tiles as aggregate in concrete is studied. The crushed natural stone aggregate in concrete is replaced by 25, 50, 75 and 100% by volume. The control mix containing natural aggregate and mix with clay tile aggregates partially or fully replaced prepared in the laboratory and tested. The compressive strength, modulus of elasticity, modulus of rupture, tensile strength were determined. The test results indicated that replacement of natural aggregate by 25% by volume with the broken clay tile can be recommended. Thereafter, the reduction in the strength is found to be significant.

Keywords: Clay Tile, Modulus of Rupture, Young's Modulus, Split Tensile Strength, Compressive Strength

1. Introduction

Use of waste material in construction is a green approach. The alternate materials to replace constituent materials of concrete are a hot research topic today. It provides the value addition to the waste materials and helps to achieve the sustainability goals in the construction. Manglore pattern clay Roof tiles were used for roof construction in majority of the old buildings in Kerala. Hence, large amount of broken roof tile chips are available in the construction and demolition (C&D) waste generated during the renovation and reconstruction work of these old buildings. Hence, it is important to identify a method to utilize the broken clay chips. In this study, potential of the use of these broken clay tile chips in concrete production as coarse aggregate is studied.

Aruna*et al*[1] studied the potential of use of broken tiles for the part replacement to coarse aggregates in concrete. The percentage replacement used was 0 to 25 percent. The results showed that the compressive strength is reduced by 3 to 13% corresponding to replacement ratios between 5 and 25 %.Elci [2] studied the use of floor and wall tiles as aggregate in concrete production. The wall tile is reported to have lower density and high water absorption compared to floor tile. Both tile aggregate sowed good adhesive properties. Prasad *et al.* [3] used clay tile aggregate to replace crushed stone aggregate in 25 MPa concrete. The replacement ratios of 10, 20, 30 and 40 % were tried and the test results indicated that up to 30 % replacement can be recommended. Saravanakumar and Maruthuchalam [4] used waste clay tile chips for replacing coarse aggregate and fly ash for replacing cement. The test result showed that compressive strength of clay tile aggregate concrete is 82% and 75% of control mix when the aggregate replacement ratio is 20 and 30% respectively. Ibrahim et al. [5] studied the use of clay brick chips for the production of light weight concrete. The specimens were prepared with varying the percentage of clay bricks aggregate content. The result showed that replacement ratio of 25% yields a compressive strength of 25MPa. Khalaf [6] studied the use of crushed brick as coarse aggregate in concrete. The test results showed that 28-days compressive strength of concrete containing brick aggregate is greater than that of concrete containing crushed stone aggregate

Yang et al. [7] studied the properties of concrete containing used recycled crushed aggregate (RCA) and crushed clay bricks (CCB). The proportion of CCB: RCA of 0:100, 20:80, 50:50 and 100:0 was used. The strength of the mix containing CCB: RCA of 20:80 was found to be greater than other mixes. Anderson et al. [8] used waste ceramic tiles as coarse aggregate. The aggregate replacement ratios of 20-100% were used. A slight increase in the strength was reported when the percentage replacement ratio of coarse aggregate is 50%. Awoyera et al. [9] used ceramic tile waste as coarse aggregate in concrete. The results showed that the increase in strength is about 36.1 % corresponding to a replacement ratio of 100%. Kavitha and Sundar [10] carried out experimental study to use ceramic tile waste as partial replacement of coarse aggregate in concrete. It was suggested that the optimum dosage of the ceramic tile aggregate is 25% of the total aggregate in concrete. Isah [11] suggested that the optimum replacement ratio of the waste ceramic tile chips as aggregate in concrete is 10%.Sekar [12] studied the influence of partial replacement of coarse aggregate with ceramic tile waste in concrete. The replacement percentage of 0, 15, 30 and 45 % was used. The test result showed that there is a significant increase in strength up to 15% replacement and beyond which there is no such improvement in the strength. McGinnis13] studied the strength of concrete with recycled aggregate. The strength was found to decrease by 16.6 and 26.4% corresponding to a replacement ratio of 50 and 100% respectively.Mohammed *et al.* [14] studied the use of recycled brick aggregate as coarse aggregate in concrete. The replacement percentage between 10-50% was considered in this study. The result showed that there is no significant reduction in the compressive strength due to the addition of brick chip aggregates. Minapu*et al.* [15] experimented to use light weight porous pumice stone pieces as the coarse aggregate in 30 MPa concrete. Out of the various coarse aggregate replacement ratios tried in this study, 20 % replacement of pumice aggregate was found to yield best strength.Rajeswari and George [16] studied the use of pumice stone as aggregate in concrete. The crushed stone aggregate was replaced with pumice stone in the ratios of 0, 50, 60 and 70%. The maximum strength was reported for a mix with 60% replacement of coarse aggregate. Abdulla [17] prepared concrete containing construction and demolition waste. The aggregate replacement ratio of 100 % was used. The test results indicated that the use of demolition waste aggregate in concrete does not influence the strength properties of concrete.

Mangalore pattern roof tiles are burned pressed clay tiles having relatively better properties when compared to the brick aggregates. The use of waste pieces of the clay roof tiles has become practical, since they are cheap and it will reduce environmental pollution. Concrete is one of the most important building materials used in construction industry. Also replacement of raw materials in concrete is also a practical case. Concrete is the mixture of cement, coarse aggregate, fine aggregate and water in a definite proportion. The use of many waste materials as coarse aggregate in concrete are reported in the literature [1-17]. The potential of the use of waste pieces of burned pressed clay roof tiles as coarse aggregate in concrete are not studied in detail. In this study, partial and full replacement of conventional crushed stone aggregate in concrete with clay roof tile chips is examined. The properties such as compressive strength, tensile strength, modulus of rupture, modulus of elasticity were examined.

2. Material Properties

The main constituents of concrete mix are water, cement, coarse aggregate and fine aggregate. Chemical admixture is also used in order to increase the workability of the mix. The properties of the constituent materials are determined.

2.1 Cement

Portland pozzolana cement conforming to IS:1489[18] was used. Portland pozzolana cement is manufactured by mixing75 % of ordinary Portland cement (OPC) with 25% of fly ash pozzolana. The specific gravity and standard consistency of cement is found to be 2.96 and 32% respectively.

2.2 Fine Aggregate

Fine aggregate fills the voids between the coarse aggregate in the concrete. The specific gravity of the fine aggregate is found to be 2.6. Particle size distribution of the fine aggregate is given in Figure 1 and it indicates that the fine aggregate is conforming to zone II of IS:383 [19].

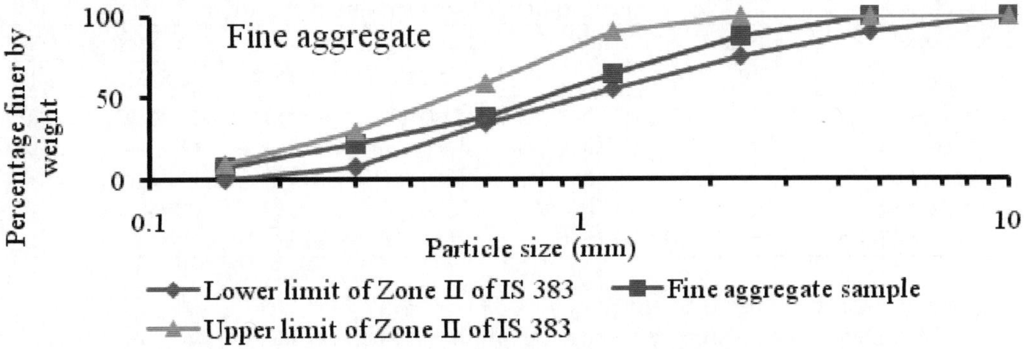

Figure 1. Particle size distribution of fine aggregate

2.3Crushed stone aggregate (CSA)

The coarse aggregates occupy about 60% of the total volume of the concrete. Crushed stone aggregate was used as the coarse aggregate in control mix. The specific gravity of the CSA was found to be 2.73. The well graded CSA was used and the particle size distribution curve is given in Figure 2.

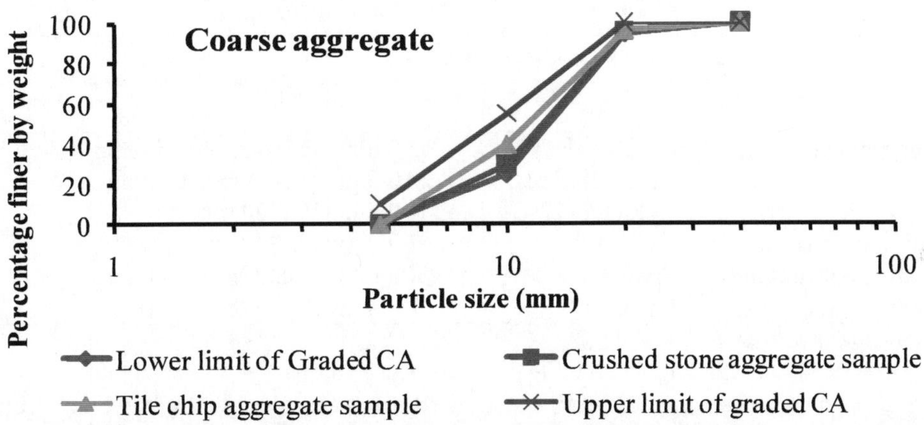

Figure 2. Particle size distribution of coarse aggregate(CA)

2.4Tile chips aggregate (TCA)

Broken Mangalore Pattern roof tiles were collected from the building demolition site. These tile pieces are further broken into 20 mm down size and used as aggregate in concrete. The specific gravity of the clay tile chip aggregate was found to be 2.1. The particle size distribution curve of the TCA indicates that the aggregate is grades as per IS:383[19].

2.5Water and chemical admixture

Potable water was used as mixing water in concrete. Sulphonated naphthalene based super plasticiser was used as the chemical admixture to improve the workability of the concrete.

3. Mix Proportion

The weight of the constituent materials were arrived as based on absolute volume method given by IS: 10262[20]. The mixes were designated to indicate the proportion of the crushed stone aggregate

(CSA) and tile chip aggregate (TCA). For example, designation of 75C25T indicates the mix containing CSA of 75% by volume of total aggregate and TCA of 25%. The weight of the materials per cubic meter of concrete is given in Table 1

Table 1. Details of concrete mix

Mix designation	Quantity in kg for one cubic meter of concrete					
	Admixture	Water	Cement	Fine aggregate	Coarse aggregate	
					CSA	TCA
100C0T	2.8	175	350	723	1192	--
75C25T	2.8	175	350	723	894	229
50C50T	2.8	175	350	723	596	458
25C75T	2.8	175	350	723	298	688
0C100T	2.8	175	350	723	--	917

4. Casting and Curing of Specimens

The fresh concrete was prepared using a pan mixture. Materials were weigh-batched. The coarse aggregates were taken in saturated surface dry (SSD) condition. The tile aggregates were soaked for 24 hours and then spread on non-absorbing platform for 1 hour to dry the surface. The dry mixture of the materials was prepared initially in the mixing machine. The 90% of the total water was added to the dry mixture and thoroughly mixed. The balance 10% of the water was mixed with chemical admixture. The admixture-water solution was then added and mixing is continued. The homogeneous fresh mix is then filled in the clean oiled moulds. The details of specimens are given inTable 2.

Table 2. Details of test specimens

Sl No.	Test	Shape of specimen	Size of specimen L x B x H	Number of specimens
1	Compressive strength	Cube (7-days)	150 mm x 150 mm x 150 mm	3
		Cube (28-days)	150 mm x 150 mm x 150 mm	3
		Cylinder (28-days)	150 mm Ø x 300 mm (H)	3
2	Split tensile strength	Cylinder*	150 mm Ø x 300 mm (H)	3
3	Modulus of rupture	Prism*	500 mm x 100 mm x 100 mm	3
4	Modulus of Elasticity	Cylinder*	150 mm Ø x 300 mm (H)	3

*Tested at an age of 28 days

A total of 90 specimens were cast in this study. All the specimens were demoulded after 24 hours of casting and cured by immersing in water for required period.

5. Testing of Specimens

The specimens were tested as per IS:516[21]. The compression test was conducted using a compression testing machine of capacity 2000kN. The modulus of elasticity test was conducted by attaching a compresso-meter of least count of **0.001**mm. The modulus of rupture test was carried out using flexure testing machine having a capacity of 100kN. The modulus of rupture specimens were tested over a span of 400mm. Three specimens were tested and the average was reported in this paper.

6. Results and Discussion

The results of the tests of compressive and tensile strength, modulus of rupture and modulus of elasticity test are given in Table 3. The strength loss due to the addition of tile aggregate is computed and is given in Figure 3 and 4. . The strength loss is the ratio of difference between the strength of the mix and the control mix to the strength of control mix.

Table 3. Experimental test results

Mix Designation	7-dayscube compressive strength (N/mm^2)	28-dayscube compressive strength (N/mm^2)	28-dayscylinder compressive strength (N/mm^2)	Split tensile strength (N/mm^2)	Modulus of rupture (N/mm^2)	Modulus of Elasticity (kN/mm^2)
100C0T	15.1	26.3	21.2	3.6	4.6	25.4
75C25T	14.2	25.6	20.3	3.3	4.4	22.8
50C50T	13.7	22.7	18.2	2.9	3.6	21.1
25C75T	12.5	18.4	15.2	2.6	3.0	17.0
0C100T	10.6	15.3	12.7	2.2	2.8	10.5

6.1 Compressive Strength

The compressive strength for all specimens is given in Table 3. With the addition of TCA increases the compressive strength decreases. The strength loss in compressive strength due to the addition of TCAs in concrete is shown in Figure 3. The strength loss in compression was observed for full (100%) replacement of CSA with TCA, which was found to be 41% and 40% for 28-days compressive strength of cylinder and cube respectively. The decrease for 7-days cube compressive strength was found to be 30%. The specific gravity of the tile aggregate was 2.1, which was about 77% of the CSA. This indicates that the tile aggregate contains more voids when compared to CSA. Hence, TCAs are weaker aggregates than the CSAs. This may be reason for having lower strength for TCA concrete when compared to CSA concrete. However, the reduction in the compressive strength due to the replacement of 25% CSA with TCA is found to be nominal (3 to 5 %). Hence, the replacement of CSA with TCA up to 25% can be recommended without modification of the mix design procedure. The higher replacement ratios can be considered after incorporating necessary modifications to compensate the strength loss in the concrete.

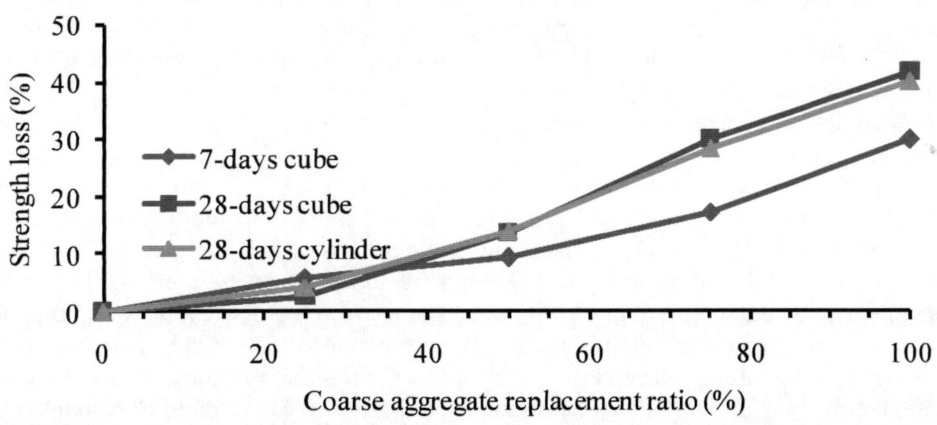

Figure 3. Compressive strength loss due to the addition of tile aggregate

6.2 Split Tensile Strength

The average split tensile strength of concrete containing TCA is given in Table 3. The split tensile strength is found to decrease with increase in replacement ratio of the coarse aggregate. The strength loss in tension is given in Figure 4. The loss in spit tensile strength due to 100% replacement of CSA with TCA is found to be 38%. The reduction in the split tensile strength due to the replacement of 25% CSA with TCA is found to be 8 - 9%. In the cracked surface of the specimen, both splitting of the aggregate and failure of the aggregate interface was visible. The water absorbed on the surface of the tile aggregate cause the reduction in the alkalinity of the cementatious compound at the aggregate interface. Hence, the aggregate interface of the tile aggregate is weaker than CSA. This is the reason for failure of the aggregate-mortar interface in the concrete containing TCA. The weak aggregate and weak interface fail during the splitting and this resulted in lower magnitude of splitting tensile strength of the concrete containing TCA.

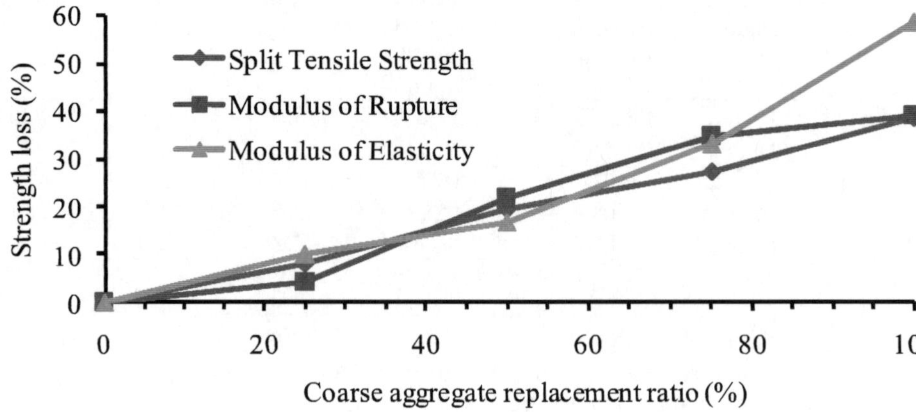

Figure 4. Strength loss due to the addition of tile aggregate

6.3 Modulus of Rupture

The modulus of rupture test result is given in Table 3. The modulus of rupture was found to decrease with the increase in the replacement ratio of CSA with TCA. The strength loss of the modulus of rupture due to replacement of CSA with TCA is given in Figure 4. The maximum strength loss of modulus of rupture is found to be 39% and it was found for the mix with aggregate replacement ratio of 100%. The reduction in the modulus of rupture due to the replacement of 25% CSA with TCA is found to be 5 %. The adhesion of TCA with the surrounding mortar is relatively low due to the dispersion of absorbed water in TCA in pre-soaking stage. This leads to the easy braking of the interface when subjected to tensile force in the bottom fibres of the prism specimen, which may be attributed to the strength loss in modulus of rupture in concrete containing TCA.

6.4 Modulus of Elasticity

The modulus of elasticity of all specimens is given in Table 3. The modulus of elasticity was found to decrease with increase in the replacement ratio of the coarse aggregate. The strength loss corresponding to modulus of elasticity was determined and is given in Figure 4. The strength loss in modulus of elasticity due to the full (100%) replacement of CSA with TCA was found to be 58%. The reduction in the modulus of elasticity due to the replacement of 25% CSA with TCA is found to be around 10%. The modulus of elasticity is recorded in the re-cracking stage. The CSA being a strong aggregate, the modulus of elasticity of specimen containing CSA was found to be relatively greater in magnitude when compared to the TCA. The deformation of the weak tile aggregate and the aggregate-mortar interface collectively responsible for the reduction in the modulus of elasticity of the concrete.

7. Prediction Model

The models for predicting the strength properties of TCA concrete is proposed by carrying out the regression analysis. The independent variable is taken as the 28-days cube compressive strength (f_{cu-28d}) and coarse aggregate replacement ratio(R). R is taken as the independent variable. The magnitude of f_{cu-28d} to be substituted in MPa and R is equal to 0, 0.25, 0.5, 0.75 or 1.0. The prediction models are given in Table 7.The proposed prediction models can be used to estimate the various strength properties of TCA concrete, if the cube strength test data and replacement ratio are known.

Table 4. Prediction of strength property of TCA concrete

Property	Proposed Model	Predicted/ experimental	
		Mean	SD*
7-days cube compressive strength (MPa)	$f_{cu-7d} = 0.62 f_{cu-28d}$	0.89	0.09
28-days cylinder compressive strength (MPa)	$f_{cy-28d} = 0.81 f_{cu-28d}$	0.97	0.01
Split tensile strength (MPa)	$f_{sp} = (0.71 - 0.13R)\sqrt{f_{cu-28d}}$	0.98	0.02
Modulus of rupture (MPa)	$f_r = (0.89 - 0.21R)\sqrt{f_{cu-28d}}$	0.94	0.04
Modulus of elasticity (GPa)	$E_c = (5.12 - 2.03R)\sqrt{f_{cu-28d}}$	1.16	0.19

*SD=Standard Deviation

8. Conclusion.

Clay tile chips can be used as a satisfactory aggregate in concrete to get good quality concrete and will give a satisfactory strength. The replacement of crushed stone aggregate (CSA) with clay tile chip aggregate (TCA) yielded a 7th day compressive strength of up to 14.2 N/mm^2. The test results indicated that the replacement ratio up to 25% can be adopted without modifying the mix design procedure because the reduction in compressive strength would be as 3 to 5%. Higher replacement ratios are also permitted provided appropriate adjustments in the mix design procedure may be incorporated. It is expected that the test results of this study would help the concrete engineers to arrive at the appropriate replacement ratio of TCA in concrete for making sustainable concrete. The mix design procedure using tile aggregate can be developed in future. It is expected that the test data will be useful for formulating code provisions to permit the use of tile aggregate.

References

[1] Aruna D, Prabhu R, Yaragal S C and Venkataramana K2015Studies on usage potential of broken tiles as part replacement to coarse aggregates in concretes*Int. J. Res. Eng. Techno.* **4**(7) 110-114.

[2] Elci H 2015 Utilisation of crushed floor and wall tile wastes as aggregate in concrete production. *J. Clean. Prod.* 112(1) 775-783

[3] Prasad N N, Hanitha P and Anil N C 2016 Partial replacement of coarse aggregate by crusher tiles and fine aggregate by garanite powder to improve the concrete properties. IOSR J. Mech. Civil Eng. **13**(6) 168-176.

[4] Saravanakumar P and Maruthachalam D 2017 Studies on structural behaviour of recycled aggregate concrete with waste clay tiles and fly ash*Int. J.Civil Eng. Techno.* **8**(7) 545-552.

[5] Ibrahim N M, Salehuddin S, Amat R C, Rahim N U and Izhar T N T 2013 Performance of light weight foamed concrete with waste clay bricks as coarse aggregate. *APCBEE Procedia***5**(2013) 497-501.

[6] Khalaf F M 2006Using crushed clay brick as coarse aggregate in concrete. *J. Mater. Civil Eng.* **18**(4)518-526.

[7] Yang J, Du Q and Bao Y 2011Concrete with recycle concrete aggregate and crushed clay bricks. *Cons, Build. Mater.***25** (4)1935-1945.

[8] Anderson D J, Smith S T and Au F T K 2016 Mechanical properties of concrete utilising waste ceramic as coarse aggregate. *Cons. Build. Mater.* **117**(2016) 20-28

[9] Awoyera P O, Ndambuki J M Akinmusuru J O and Omole D O 2016Characterization of ceramic waste aggregate concrete. *HBRC J.***11**(3)1-6.

[10] Kavitha B and Sundar M L 2017 Experimental study on partial replacement of coarse aggregate with ceramic tile wastes and cement with glass powder. *Int. J. ChemTech. Res.* 10(8) 74-80.
[11] Isah R J R 2017 Experimental investigation on ceramic tile wastes as a replacement of coarse aggregate in concrete using acrylic polymer as an admixture. *Int. Res. J. Eng.Tech.* 4(10) 416-422.
[12] Sekar M 2017 Partial replacement of coarse aggregate by waste ceramic tile in concrete. *Int. J. Res. Appl. Sci. Eng.* 5(3). 473-479.
[13] McGinnis M J, Davis M, Rosa A, Weldon B D and Kurama Y C 2015. Strength and stiffness of concrete with recycled concrete aggregtes. *Cons. Build. Mater*. **154**(2017) 258-269.
[14] Mohammed T U,Hasnat A, Awal M A and Bosunia S Z 2014 Recycling of brick aggregate concrete as coarse aggregate*J. Mater. Civil Eng.***27**(7)1-10.
[15] Minapu L K, Ratnam M K M V and Rangaraju U 2014 Experimental study on light weight aggregate concrete with pumice stone, silica fumes and fly ash as a partial replacement of coarse aggregate. *Int. J. Inn. Res. Sci*, **3**(12) 18131-18138.
[16] Rajeswari S and George S 2016 Experimental study of light weight concrete by partial replacement of coarse aggregate using pumice aggregate. *Int. J. Sci. Eng. Res.* 4(5) 50-53.
[17] Abdulla N A 2015 Effect of recycled coarse aggregate type on concrete*J. Mater. Civil Eng.* 1-9.

[18] IS; 1489:1991 Portland Pozzolana Cement – Specification *Bureau of Indian Standards, New Delhi.*
[19] IS: 383:1970Indian Standard Code of Coarse and Fine Aggregate from Natural Sources for Concrete, *Bureau of Indian Standards*, New Delhi
[20] IS: 10262:2009 Concrete Mix Proportioning- Guidelines, *Bureau of Indian Standards, New Delhi.*
[21] IS: 516:1959Methods of tests for strength of concrete, *Bureau of Indian Standards, New Delhi*

Mechanical Properties of Concrete Containing Brick Chips

Christy Francis Lonth[1,b], Job Thomas[2,a] and Neethu Joseph[3]

[1]M-Tech Student, Department of Civil Engineering, SJCET, Palai, India
[2]Associate Professor, Department of Civil Engineering, Cochin University of Science and Technology, Cochin, India
[3]Assistant Professor, Department of Civil Engineering, SJCET, Palai, India

[a]E-mail: job_thomas@cusat.ac.in
[b]E-mail: christyflonth@gmail.com

Abstract. Non-availability of coarse aggregate affects the progress of the construction industry. This is the peak time to find alternate materials for the concrete construction. In this study, the potential of using brick chips as coarse aggregate is studied. The mechanical properties of concrete such as compressive and tensile strength, modulus of rupture and modulus of elasticity were determined. The variable of the study was coarse aggregate replacement ratio by volume. The virgin crushed stone aggregate was used as coarse aggregate in control mix. The coarse aggregate replacement ratio such as 0, 25, 50, 75 and 100 was used. The test results indicated that 25% of the virgin aggregate in concrete can be replaced without any modification in the mix design procedure.

Keywords: Brick chips, Coarse aggregate, Compressive strength.

1. Introduction

Concrete is the commonly used material for various constructions. The utilization of concrete is rapidly increasing due to development in infrastructure and construction activities all around the world. However, the greater production of concrete causes continuous and extensive extraction of aggregate, which lead to its depletion and ecological imbalance. Recently, restriction on mining in the quarries led to increase in cost and non-availability of the aggregate for concrete production. The use of alternate materials such as wood chips, plastic waste, polyethylene, textile waste, rice husk ash, rubber tyres, paper, vegetable fibers and pulp industry waste, waste glass, broken bricks, groundnut shell etc. for replacing aggregates in concrete is required to be investigated. In this study, the potential of using brick chips prepared from broken brick pieces in the construction and demolition waste is investigated.

Generally, the lack of knowledge of the performance of concrete made with brick as its aggregates is a major obstacle for reuse of brick waste in concrete. The test data of Adamson et al.[1] showed that the use of brick aggregate for replacing virgin granite aggregates results in a loss of compressive strength of concrete. Cachim [2] concluded that a 15% replacement of the coarse aggregate can be done without compromising concrete strength. An increase of 10% in flexural and tensile strengths was also reported when crushed brick was used as the aggregate in concrete compared with normal aggregates. Debieb and Kenai [3] reported a decrease in the strength, but it is possible to produce concrete containing crushed bricks with characteristics similar to stone aggregates concrete provided the percentage replacement is limited to 25% and 50% for coarse and fine aggregate respectively. Dey and Pal [4] studied the strength parameters, workability and fire resistance of brick aggregate concrete. The outcome of the test performed by them indicated that concrete produced with crushed brick aggregate have very good heat resistance up to a temperature of 600°C. Emon et al. [5] suggested to use low cost steel wire fibre in brick chip aggregate concrete to improve its strength characteristics. The study by Ibrahim et al. [6] evaluated the performance of lightweight foamed concrete using waste clay brick as coarse aggregate. The compressive strength of lightweight concrete for 25% substitution of brick showed highest compressive strength of 25MPa with increase in the quantity of brick aggregate. Islam et al. [7] replaced both crushed stone and brick chips with polypropylene and reported that a partial replacement of 10% by volume of coarse aggregate can be suggested without modification of mix design procedure. Kanchidurai et al. [8] concluded 25-50% of over burnt brick bat wastes can be replaced with the normal concrete. Also, the poor workability of replaced concrete can be easily overcome by adding water reducing admixtures. Khalaf [9] reported that the 28-day compressive strength of concrete made with crushed brick increases when compared to that of concrete containing granite aggregate. Mansur et al. [10] proposed the use of crushed brick aggregate as coarse aggregate in concrete way back in 1999. The study was done by selecting four basic mixes for crushed aggregate to achieve concrete with grades from 30 to 60 MPa. The maximum 28-day strength achieved for brick-concrete was 72 MPa with a water cement ratio of 0.3 and that of control concrete was 64 MPa. McGinnis et al. [11] determined the strength and stiffness of concrete containing recycled concrete aggregates. The decrease in compressive strength was found to be 16.6% at 50% replacement and 26.4% at 100% replacement. Rashid et al. [12] proposed that high strength concrete of $f_c^{'}$ equals to 31.02 MPa to 45.5 MPa which can be prepared with brick aggregate made from high strength brick. Subramani and Kumaran [13] observed that the brick waste in concrete can be used for the production of lightweight concrete. It was suggested that use of waste materials will not only reduce the cost of construction but will also contribute to the safe disposal of waste materials. The use of brick aggregate instead of stone aggregate in various components of a building structure can result in a significant reduction of dead load on column as well as foundation. Hence, the replacement of stone aggregate by brick chips may yield cost-effectiveness in making concrete, which will be economically and environmentally beneficial.

This project aims to evaluate the potential of brick chips to as coarse aggregate in concrete. The broken brick pieces collected from the building demolition site were further crushed to 20 mm size and used. The strength properties of concrete such as compressive strength, tensile strength, modulus of rupture and modulus of elasticity were determined and compared with the corresponding mix containing virgin crushed stone aggregate. The coarse aggregate occupies about 60 to 70 percent of the volume of the concrete and hence, the properties of the coarse aggregate influence the strength characteristics of the concrete significantly. The influence of strength properties on various replacement ratio of the coarse aggregate is reported in this paper. In India, guidelines for the use of brick chips for the production of concrete are not being available. It is expected that the results of this study will be useful for developing the standard code practices.

2. Material Properties

The materials used are cement, manufactured sand, crushed stone aggregate, brick chip aggregate, water and admixture. The properties of the material are determined by conducting laboratory test.

2.1. Cement

The Fly ash based Portland Pozzolana Cement was used for this study. The specific gravity of the cement was found to be 2.96 and other physical properties of the cement are given in Table 1.

2.2. Crushed stone aggregate

The coarse aggregate used in the control mix was crushed stone aggregate (CSA) of nominal size of 20 mm. The specific gravity of the CSA was found to be 2.73 and the other physical properties are given in Table 2.

2.3. Brick chip aggregate

Broken brick pieces are further crushed to 20 mm size and are used as coarse aggregate in concrete. The partial or full replacement of CSA with brick chip aggregate (BCA) was considered in this study. The specific gravity of the brick aggregate was found to be 1.90. Other physical properties of BCA are given in Table 2.

Table 1. Physical properties of Cement

Property	Test results
Standard Consistency	32%
Specific gravity	2.96
Initial setting time	140 minutes
Final setting time	300 minutes
3-days compressive strength	17.5 N/mm^2
7-days compressive strength	24.1 N/mm^2
28-days compressive strength	36.2 N/mm^2

Table 2. Physical properties of Aggregate

Property	Fine Aggregate	Coarse Aggregates	
		Crushed stone aggregate (20 mm)	Brick chip aggregate (20 mm)
Specific Gravity	2.7	2.73	1.90
Loose Bulk Density (kg/m^3)	1525	1435	1090
Water Absorption	1.7	1.2	12.2
Particle size gradation	Zone II*	Graded aggregate*	Graded aggregate *

* conforming to IS:383 [14]

2.4. Water and Admixture

Potable water was used as the mixing water in concrete. Sulphonated naphthalene polymer was used as the super plasticizer. The admixture used in this study is a chloride free brown coloured liquid, which can instantly disperse in water. The dosage of the super plasticizer was considered as 0.8% by weight of cement.

3. Details of Concrete Mix

The raw materials were tested and the properties were determined. The proportion for the concrete mix of strength of 25 MPa was arrived at by absolute volume method given by IS: 10262 [15].

Table 3. Details of concrete mix

Sl No.	Mix designation	Quantity in kg for one cubic meter of concrete					
		Admixture	Water	Cement	Fine aggregate	Coarse aggregate	
						CSA	BCA
1	100C0B	2.8	175	350	723	1192	--
2	75C25B	2.8	175	350	723	895	208
3	50C50B	2.8	175	350	723	596	415
4	25C75B	2.8	175	350	723	298	623
5	0C100B	2.8	175	350	723	--	830

In the control mix, crushed stone aggregate (CSA) was used as the coarse aggregate. In other mixes, the CSA is partly or fully replaced with brick chip aggregate (BCA). The aggregate replacement ratios of 25, 50, 75 and 100 % were used. The mix was designated to indicate the replacement ratio and coarse aggregate type. For example, 25C75B indicates a mix containing 25 percent of CSA (C) and 75 percent of BCA (B) by volume. The weight of the constituent materials per cubic meter of concrete corresponding to various mixes is given in Table 3.

4. Casting and Testing

A total of 90 specimens were cast and tested in this work. The constituent materials were batched by weight and mixed using a pan mixer machine. The coarse aggregates were weighed in saturated surface dry condition. They are pre-wetted for one day and allowed to dry the surface in air prior to casting. The standard specimens given by IS:516 [16] were cast and vibrated using a table top vibrator. The details of the specimen are given in Table 4.

Table 4. Details of test specimens

Sl No.	Test	Shape of specimen	Size of specimen L x B x H	Number of specimens
1	Compressive Strength	Cube (7-days)	150 mm x 150 mm x 150 mm	3
		Cube (28-days)	150 mm x 150 mm x 150 mm	3
		Cylinder (28-days)	150 mm Ø x 300 mm (H)	3
2	Split tensile Strength	Cylinder*	150 mm Ø x 300 mm (H)	3
3	Modulus of Rupture	Prism*	500 mm x 100 mm x 100 mm	3
4	Modulus of Elasticity	Cylinder*	150 mm Ø x 300 mm (H)	3

*Tested at an age of 28 days

The specimens were cured for the required period by immersing in water in a tank. All specimens were taken out from tank about one hour before the test. The surface is cleaned and wiped with cotton. The cylinder specimens for determining the strength and modulus of elasticity were capped with neat paste and levelled the surface. The modulus of rupture test specimens was tested over a span of 400 mm and by applying two-point load in the middle third region. All the specimens are tested as per the details given in IS:516 [16]. The details of testing of the specimens are given in Figure 1.

(a) Cube compression test (b) Cylinder compression test (c) Split Tensile strength test (d) Modulus of rupture test (e) Modulus of elasticity test

Figure 1. Test set-up of specimens

5. Results and Discussion

The compressive and tensile strength, modulus of rupture and modulus of elasticity of concrete containing with and without brick chip aggregate were determined. The test results of control concrete mix containing crushed stone aggregate (CSA) is compared with the corresponding results of concrete containing brick chip aggregate (BCA) in partial or full. The influence of addition of BCA is discussed.

5.1. Compressive Strength

The average compressive strength of three specimens was determined and is given in Table 5. The addition of brick chip aggregate (BCA) resulted in reduction in the compressive strength. The maximum decrease in cube compressive strength was found to be for concrete mix of 0C100B and is equal to 41 and 43 % at an age of 7 and 28 days respectively. The maximum decrease in the cylinder compressive strength was found to be 46 %. The BCAs are weaker aggregates when compared to the CSA and hence, the crack easily propagates through the BCA. This leads to the early failure of the BCA concrete specimens when compared to the CSA concrete specimens.

Table 5. Compressive strength of concrete

Sl No.	Mix designation	Compressive strength in N/mm^2		
		Cube		Cylinder
		7-days	28-days	28-days
1	100C0B	15.1	26.3	21.2
2	75C25B	14.2	25.3	20.3
3	50C50B	13.3	22.3	17.8
4	25C75B	10.2	17.1	14.1
5	0C100B	8.8	14.9	11.3

The strength ratio is the ratio of the strength of the mix to the corresponding strength of the respective control mix. For example, the strength ratio of 7-days cube strength of 25C75B is 10.2/15.1=0.68, where 10.2 and 15.1 are the 7-day strength of mix 25C75B and 100C0B. The variation of the strength ratio of concrete in compression corresponding to coarse aggregate replacement ratio is computed and is shown in Figure 2. The reduction in compressive strength of concrete of cubes at 7-days and 28-days and cylinder strength at 28-days are almost equal corresponding to replacement ratio of coarse aggregate.

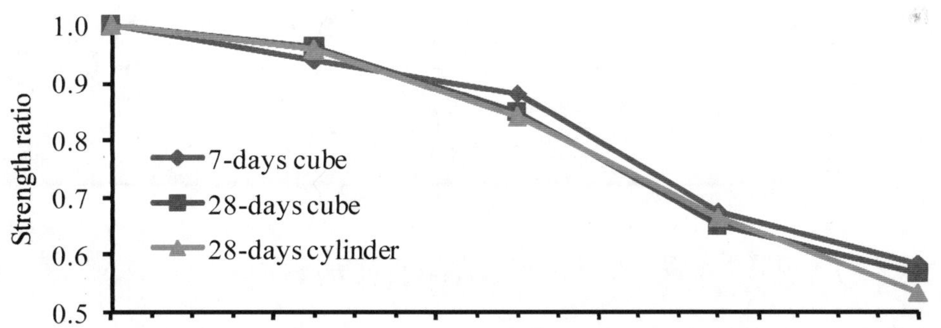

Figure 2. Variation of strength ratio in compression due to the addition of BCAs in concrete

The strength ratio corresponding to various compressive strength tests of mix 75C25B was found to be between 94 and 96%. Similarly, the compressive strength ratios of mix 50C50B were found to be between 84 and 88%. The reduction of strength is found to be nominal (6-4 %) for mix containing BCAs about 25% of the volume of the total coarse aggregate. Hence, the 25% replacement of CSA with BCA in concrete can be recommended without any modification in the mix design procedure.

The reduction in the strength of mix containing BCAs about 50% by volume of total coarse aggregate is found to be 12-16%. Hence, the 50% replacement of CSA with BCA in concrete can be recommended with appropriate modification in the mix design procedure.

5.2. Split Tensile Strength

The average split tensile strength of three specimens is determined and given in Table 6. The maximum decrease in the split tensile strength was found to be 49% for the mix designated by 0C100B when compared to that of corresponding control mix (100C0B). In the mix CSA, the cracks are tortuous because the splitting failure occurs through the mortar phase and interface between aggregate and mortar. The failure of aggregate is not likely in the case of mix containing CSA. However, the brick chip aggregate (BCA) are relatively weak and hence the cracks propagates through the aggregate. This is the reason for the lower strength of concrete containing BCA when compared to that of corresponding control mix. The variation of split tensile strength ratio of the concrete containing BCA is given in Figure 3. The lowest split tensile strength ratio was found to be equal to 51% for mix 0C100B. The decrease in the split tensile strength of concrete mix 75C25B is found to be nominal (10 percent) when compared to the control mix.

Table 6. Tensile and elasticity properties of concrete

Sl. No.	Mix Designation	Split Tensile strength (MPa)	Modulus of rupture (MPa)	Modulus of Elasticity (GPa)
1	100C0B	3.69	4.60	25.40
2	75C25B	3.33	4.30	22.20
3	50C50B	2.62	3.73	20.38
4	25C75B	2.19	3.20	16.97
5	0C100B	1.88	2.57	10.11

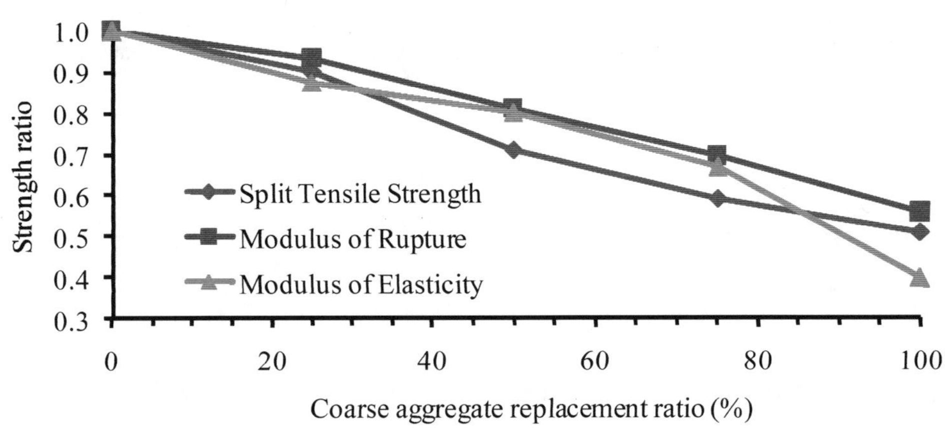

Figure 3. Variation of strength ratio in tension and elasticity due to the addition of BCAs in concrete

5.3. Modulus of Rupture

Modulus of rupture is the tensile strength when subjected to flexure or bending load and is given in Table 6. In this, the top fibres in the prism specimens are subjected to compression and bottom fibres are subjected to tension and this aspect is discussed in Thomas [17]. Hence, the tensile strength observed in modulus of rupture test is higher than the tensile strength observed in the split tensile strength test. The modulus of rupture of control mix (100C0B) is found to be 25% greater than its split tensile strength. Similarly, modulus of rupture of mix 0C100B is 36% higher in magnitude than its split tensile strength.

The modulus of rupture strength ratio of concrete containing BCAs is shown in Figure 3. The lowest strength ratio of modulus of rupture was found to be 56% for mix 0C100B. The decrease in the modulus of rupture of concrete mix 75C25B is found to be nominal (7 percent) when compared to the control mix.

5.4. Modulus of Elasticity

Modulus of elasticity is measured in the Hookean zone of the stress-strain deformation characteristic curve. The deformation of the concrete composite within the elastic stage is directly proportional to the deformation characteristics of constituent materials and their interface [17]. The brick aggregates are manmade material and contain inherent pores within when compared to the crushed stone aggregate (CSA). Hence, the deformation of brick chip aggregate (BCA) is relatively higher than the CSA, which resulted in the lower magnitude of modulus of elasticity for concrete containing BCA when compared to concrete containing CSA. The strength ratio of modulus of elasticity is computed and given in Figure 3. The lowest strength ratio of modulus of elasticity was found to be 40% for mix 0C100B. The modulus of elasticity strength ratio of the mix 75C25B was found to 87%.

6. Prediction Model

The models for predicting the strength properties of BCA concrete is proposed by carrying out the regression analysis. The 28-days cube compressive strength (f_{cu-28d}) and coarse aggregate replacement ratio(R) is taken as the independent variable. The magnitude of f_{cu-28d} to be substituted in MPa and R is equal to 0, 0.25, 0.5, 0.75 or 1.0. The prediction models are given in Table 7.

Table 7. Prediction of strength property of BCA concrete

Property	Proposed Model	Predicted/ Experimental	
		Mean	SD*
7-days cube compressive strength (MPa)	$f_{cu-7d} = 0.58 f_{cu-28d}$	0.98	0.02
28-days cylinder compressive strength (MPa)	$f_{cy-28d} = 0.80 f_{cu-28d}$	1.05	0.03
Split tensile strength (MPa)	$f_{sp} = (0.71 - 0.24R)\sqrt{f_{cu-28d}}$	0.96	0.03
Modulus of rupture (MPa)	$f_r = (0.91 - 0.22R)\sqrt{f_{cu-28d}}$	1.03	0.02
Modulus of elasticity (GPa)	$E_c = (5.08 - 2.0R)\sqrt{f_{cu-28d}}$	1.17	0.11

*SD=Standard Deviation

The proposed prediction models can be used to estimate the various strength properties of BCA concrete, if the cube strength test data and replacement ratio are known.

7. Cost analysis

Today, the approximate cost of 100 cubic feet of crushed stone aggregate is INR. 5000/-. The brick bats are waste material and hence initial cost of purchase is zero. However, processing or recycling charges to break into chips will be involved before its use. If crushing machineries are used, it is expected that the total cost of processing will be comparable with the cost of the crushed stone. Moreover, the cost of the crushed stone aggregate will increase in future if quarrying is banned due to environmental protection reasons. In that situation, the cost escalation of natural aggregates will be huge due to its scarcity. The use of recycled materials will help to mitigate the possible decline in the growth of construction industry.

8. Conclusion

Based on the experimental study on the mechanical properties of concrete containing brick chip aggregate (BCA), following conclusions are arrived at.

- The decrease in the strength properties of concrete is nominal when BCA replacement of 25% by volume of total coarse aggregate is considered.
- BCA replacement of 25% by volume of total coarse aggregate in concrete is recommended without any modification in the mix design procedure.
- 100% replacement of crushed stone aggregate (CSA) with BCA can be considered for the preparation of non-structural concrete.

Brick chip aggregates (BCA) are sustainable alternate to the conventional virgin crushed stone aggregate (CSA). The prediction models proposed in this study will be useful for the designer to arrive at the various strength properties of BCA concrete. The durability properties of concrete containing BCA may be investigated in the future.

References

[1] Adamson M, Razmjoo A and Poursaee A 2015 Durability of concrete incorporating crushed brick as coarse aggregate *Constr. Build. Mater.* **94** 426–432.

[2] Cachim P B 2009 Mechanical properties of brick aggregate concrete *Constr. Build. Materi.* **23**(3) 1292-1297.

[3] Debieb F and Kenai S 2008 The use of coarse and fine crushed bricks as aggregate in concrete *Constr. Build. Mater.* **22** 886-893.

[4] Dey G and Pal J 2013 Use of brick aggregate in standard concrete and its performance in elevated temperature *Int. J. Eng. Technol.* **5**(4) 523-526.

[5] Emon A B, Manzur T and Yazdani N 2016 Improving performance of lightweight concrete with brick chips using low cost steel wire fiber *Constr. Build. Mater.* **106** 575–583.

[6] Ibrahim N M , Salehuddin S, Amat R C, Rahim N L and Izhar T N T 2013 Performance of lightweight foamed concrete with waste clay brick as coarse aggregate *APCBEE Procedia* **5** 497 – 501.

[7] Islam M J, Shafian S A and Sarwar N 2016 Replacement of traditional coarse aggregate *Proc. 3rd Int. Conf. Adv. Civ. Eng.* 321-326.

[8] Kanchidurai S, Bharani G and Mohan K S R 2017 Strength and durability studies on concrete with partial replacement over burnt brick bat waste *IOP Conf. Series: Earth Envir. Sci.* **80** 1-9.

[9] Khalaf F M 2006 Using crushed clay brick as coarse aggregate in concrete *J. Mater. Civ. Eng.* **18** 518-526.

[10] Mansur M A, Wee T H and Lee S C 1999 Crushed bricks as coarse aggregate for concrete *ACI Mater. J.* **96** 478-484.

[11] McGinnis M J, Davis M, Rosa A, Weldon B D and Kurama Y C 2017 Strength and stiffness of concrete with recycled concrete aggregates *Constr. Build. Mater.* **154** 258–269.

[12] Rashid M A, Hossain T and Islam M A 2009 Properties of higher strength concrete made with crushed brick as coarse aggregate *J. Civ. Eng.* **37**(1), 43-52.

[13] Subramani T and Kumaran S 2015 Experimental investigation of using concrete waste and brick waste as a coarse aggregate *Int. J. Appl. Innov. Eng. Manage.* **4**(5) 294-303.

[14] IS 383:1970, Indian Standard Code of Coarse and Fine Aggregate from Natural Sources for Concrete, *Bureau of Indian Standards*, New Delhi.

[15] IS 10262:2009, Indian Standard Recommended Guidelines for Concrete Mix Design, *Bureau of Indian Standards,* New Delhi.

[16] IS 516:1959, Method of tests for Strength of Concrete, *Bureau of Indian Standard,* New Delhi.

[17] Thomas J 2015 Concrete Technology *Cengage Learning India Pvt Ltd*, New Delhi.

IOP Conf. Series: Materials Science and Engineering **396** (2018) 012004 doi:10.1088/1757-899X/396/1/012004

Study of hardened properties of hybrid fibre reinforced concrete using steel and glass fibre

Athira Anand[1, a], **Manish Jose**[2]

[1] M-Tech Student, Department of Civil Engineering, SJCET, Palai, Kerala, India
[2] Assistant Professor ,Department of Civil Engineering , SJCET, Palai, Kerala,India

[a] E-mail: athiradanand@gmail.com

Abstract. In the present study the effect of using hybrid microfiber reinforcement on the mechanical properties of concrete is studied. Concrete is an important construction material which has low tensile strength and flexural strength. It is proved that addition of fibers improve different properties of concrete like tensile strength and ductility. Improvement in performance by the use of more than one fibre in same concrete is studied here. Hooked steel fibers and glass microfibers (6mm and 12 mm) are used in this experimental study. Influence of each fibre and their combined effect is studied. Fibre is added by volume fraction and it is fixed as 0.5% of the total volume. Mechanical properties are tested by Indian standard methods. From the results it is found that the properties are improved by the use of hybrid fibre reinforcement in concrete

keywords: Fibre, Compressive strength, Spilt tensile strength, Modulus of rupture.

1. Introduction

Concrete is an important heterogeneous building material which is brittle in nature. It is very user friendly so that it can be moulded to any shape. It has a very long life span and economical too. Also fresh concrete can be moulded very easily. According to our need it can be cast into any strength and specification. Now a days even it is really easy to place concrete by pumping and all. Mechanized mixing and placing helps to reduce manpower also. All such reasons make concrete such a popular construction composite. Along with lot of advantages, like any other material it has some disadvantages too. Concrete has low tensile strength and is sometimes permeable to moisture and other chemicals. So it is vulnerable to corrosion. Cementitious mash up like concrete is made by the complex arrangement of different inclusions like pores, aggregates and micro cracks. Inclusions can be of different size and shapes. When subjected to load concrete behaviour depend on these inclusions i.e., their distribution and continuity. Occurrence of micro cracks at the aggregate mortar junction cause weakness in concrete. When exposed to load these micro cracks spread and result in brittle fracture in conventional concrete. It happens due to due to low tensile strength and ductility of concrete. These cracks occur during solidification due to different types of shrinkage. The width of these initial cracks seldom exceeds a few microns, but their other two dimensions may be of higher magnitude. When loaded these micro cracks spread due to stress concentration and gets expanded, also new cracks are formed. Failure is initiated by cracks in concrete which is governed by tensile properties. Experiment and numerical evaluation are the main methods of studying resisting cracks. Different methods have been used to resolve these problems. It include reinforcing, prestressing etc. Fibre reinforced concrete is one such method for improving the tensile strength of concrete. It has been recognised that the addition of small, closely spaced and uniformly dispersed fibres to concrete would act as crack arrester and would substantially improve its static and dynamic properties. Fibres obstruct the weakness plane and prevent the propagation of cracks under the application of load and in turn failure or structure or member. Steel, glass, nylon, asbestos etc. are commonly used fibres. Fibres with high modulus of elasticity improve flexural and impact resistance. While reinforcements are laid according to design and standard construction practise, fibres are randomly distributed in matrix. It improve the properties of the matrix itself. It also improve the ductility and post cracking behaviour .They not only reduce the permeability of concrete .Thus reduce the intrusion of chemicals or moisture from external environment. Concrete in which two or more than fibres are added is called hybrid fibre reinforced concrete. Correct combination of different fibres improve the properties of concrete and also result in performance synergy. The objective of this study is to make use of the merits of two fibres, short steel fibres and glass micro fibres. The amount of fibre added is expressed by volume fraction. Performance of concrete before cracking is improved by the use of hybrid fibres .This happens due to the bridging action of fibres. The smaller fibre bridges micro cracks and controls their growth. Larger fibre is used to arrest the spreading of macro cracks. Bridging of macro cracks which increase later the ductility of the concrete. In present study 7 and 28 day mechanical properties are found out and compared

Banthia and Guptha [1] used steel polypropylene and carbon fibres to prepare hybrid fibre reinforced concrete and study their synergy effect. Hybrid concrete made with steel and polypropylene had better properties. Bariceivic *et al.*[2] by finding the appropriate amount and proportion of fibres, made sustainable hybrid fibre reinforced concrete using recycled tire steel and factory made fibres. SHFRC with superior behaviour than usual FRC with factory made fibres was made. Use of fibres had no effect on workability when fibre volume was limited to 1.2%.It was also concluded that the properties of concrete in post cracking area is determined by the proportion of fibres. Chen and Qiao [3] studied influence of hybrid microfiber reinforcement in fracture and crack propagation. Steel fibres were used as macrofiber and polyvinyl alcohol was used as microfiber. Use of hybrid fibre increased the ability to withstand both formation and propagation of cracks. Deng and Li [4] conducted a study on the mechanical properties and fracture properties of concrete using steel and synthetic fibres. Results showed that better behaviour is obtained for hybrid concrete and had synergic effects. Lawler et al. [5] used poly vinyl alcohol fibre steel micro fibres and steel macro fibres to find the influence of fibre combinations in concrete .Hardened properties, water permeability and shrinkage crack resistance of concrete were studied. PVA hybrid had grater ultimate strength than steel hybrid by a reasonable margin. Mastali and Dalvand [6] studied the fresh and hardened properties of self-compacting concrete made with poly propylene and recycled steel fibre. They found out that there is reasonable increase in impact resistance and mechanical properties of concrete. Wang et al. [7] studied the influence of using different type and size of steel and polypropylene fibre to find out for which combination mechanical properties are better. Compared to conventional concrete hybrid concrete split tensile strength is improved by 44.34 % bending strength by 49.05 % and impact resistance 6.8 times. Yao et al. [8] conducted experiment on the mechanical properties of concrete after keeping the volume fraction of concrete constant as 0.5%.Three different hybrid concrete were made using carbon and polypropylene ,carbon and steel ,and polypropylene and steel. From the results concluded that carbon and steel had better performance because of similar modulus.

2. Objectives and scope

Different fibres have different effects on the properties of concrete. Some improve ductility, while some improve durability. Improvement depend on the fibre added, its proportion, orientation etc. In the present study two type of fibres are introduced in concrete and their impact on mechanical properties are compared with conventional plain concrete properties.

- To find out the design mix for
    - Conventional concrete
- Find the compressive strength for specimens of each individual mix

IOP Conf. Series: Materials Science and Engineering **396** (2018) 012004 doi:10.1088/1757-899X/396/1/012004

- Study the flexural behaviour of beam specimen made with each mix
- Study the split tensile strength of cylinder specimen
- Study the influence of addition of hybrid fibres from the test results

3. Materials and specimen

3.1 Materials

Study is performed using 53 grade Portland Pozzolana Cement. M sand used had a specific gravity of 2.63 and sieve analysis indicated that it is confirming to zone II. 20 mm coarse aggregate with specific gravity 2.75 is used. Materials were proportioned as per mix design. To improve workability and other properties admixture FOSROC SP 430 IS added. Two type of micro fibres were used. Glass fibre of length 12mm and 6 mm. Alkali resistant glass fibre is used which provide good durability. Hooked steel fibre 50mm length were used as macro fibre. Their physical properties are given in Table 1.The total volume fraction of fibres was kept to a maximum of 0.1 percent to maintain economical and workable mix.

Table 1 Properties of fibres

Fibre	Length	Tensile strength(MPa)	Youngs modulus(GPa)	Elongation (%)	Combustibility
Steel	50	2500	210	4	Non combustible
Glass fibre 6mm	6	1700	73	2.4	Non combustible
Glass fibre 12 mm	12	1700	73	2.4	Non combustible

3.1.1Fibres

Glass fibre is available in different forms such as E glass fibre, Alkali resistant fibre and ECR fibre for different needs. Alkali resistant glass fibre used is shown in figure below. Steel fibres are available in different forms like plain, corrugated, crimped, hooked, flattened etc. Hooked steel fibre used is shown in figure below.

Figure 1 Glass fibre

IOP Conf. Series: Materials Science and Engineering **396** (2018) 012004 doi:10.1088/1757-899X/396/1/012004

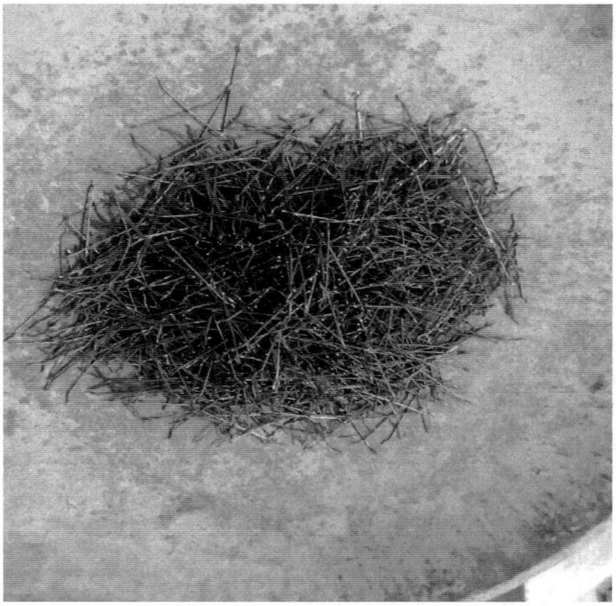

Figure 2 Steel fibre

3.2 Mix and specimen

Each individual mix were prepared in 6 separate batches to determine the test results for individual properties. There mix design for testing program is given below. The cement, M Sand, aggregate and fibres where dry mixed for 2 minutes .Water and admixture were added and mixed for another 2 minutes in a concrete mixer.

3.2.1 Mix design

The strength and durability of concrete will depends on the materials and methods chosen with adequate mixing. By selecting the right mix proportion for the concrete, the concrete will good workability can be achieved. The mix design was according to IS: 10262 – 1982 guidelines for M25 mix including replacements are designed for the investigation. The mix design data and results are given below. Fibres are added by volume percentage. Their volume fraction in each mix is also given below in the following table

Table 2 Mix design

Target mean strength	31.6 N/mm^2
Water Content	140 liters
Cement Content	350 kg/ M3
Fine aggregate content	758.727 kg/M3
Coarse aggregate content	1245.5 kg/M3

3.2.2 Volume of fibres

Fibres are added by volume percentage. Their volume fraction in each mix is given below.

Table 3 Volume fraction

Test series	Steel	Glass(6)	Glass(12)
C mix	0	0	0
G1	0	0.5	0
G2	0	0	0.5

IOP Conf. Series: Materials Science and Engineering **396** (2018) 012004 doi:10.1088/1757-899X/396/1/012004

S	0.5	0	0
SG1	0.5	0.5	0
SG2	0.5	0	0.5

4. Experimental program and testing

In order to find the mechanical properties of concrete cube specimens of size 150mm x150mm x 150mm, cylinder specimens of size 150mm diameter and 300mm height, and prism of size 500mmx 10mmx100mm were casted using cast iron moulds. Concrete was mixed as per design were filled in the specimen moulds layer by layer and compacted by vibration. Specimens were cured for 28 days and surface dried before testing.

Figure 3 specimens

5. Results and discussion

As specified earlier hardened properties of conventional concrete i.e., control mix is compared with fibre reinforced and hybrid fibre reinforced concrete. Results obtained are discussed here

5.1 Compression testing

Mixing fibre in concrete can increase porosity at same time can limit crack growth. Addition of fibres have two opposite effects on concrete. Considering the results here due to hybrid fibre compressive strength is increased. 3 specimens were tested for each mix to find 28 day strength in CTM.

Table 4 Compressive strength of concrete

Mix	Compressive strength N/mm^2
C mix	27.3
G1	28.02
G2	27.91
S	29.43
SG1	32.8
SG2	32.1

IOP Conf. Series: Materials Science and Engineering **396** (2018) 012004 doi:10.1088/1757-899X/396/1/012004

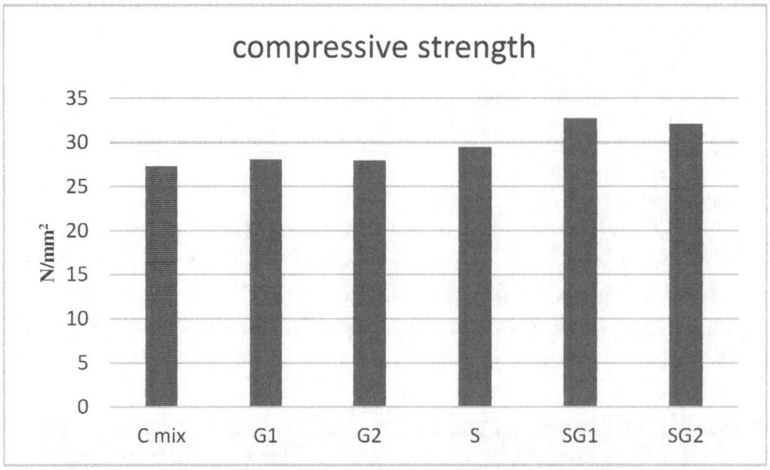

Figure 4 Average compressive strength of mixes

5.2 Split tensile strength

Split tensile strength of the cylinder specimens was obtained by the test in CTM. From the results it is clear that tensile strength is increased by the addition of fibre

Table 5 Split tensile strength of concrete

Mix	Split tensile strength N/mm^2
C mix	3.2
G1	3.58
G2	3.39
S	3.09
SG1	4.24
SG2	4.12

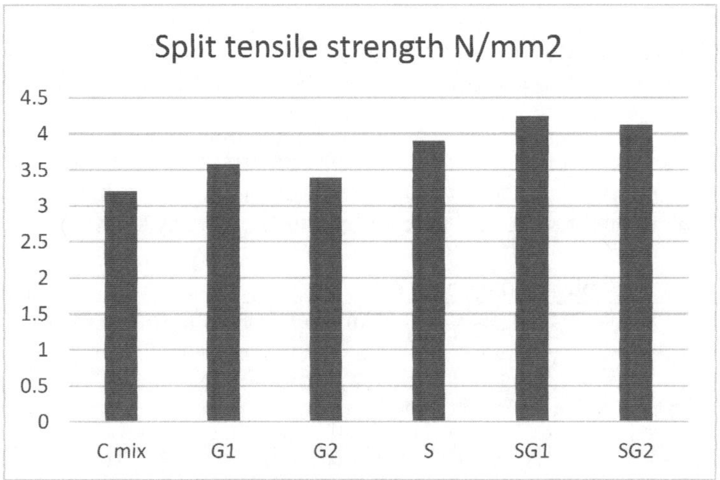

Figure 5 Split tensile strength of different mixes

5.3 Flexural strength

3 prism specimens were tested for each mix in flexural testing machine. Results obtained from the test show that flexural strength is increased by the addition of fibre.

Table 6 Flexural strength of concrete

Mix	Flexural strength N/mm^2
C mix	5
G1	6.4
G2	6.6
S	6.7
SG1	6.96
SG2	6.88

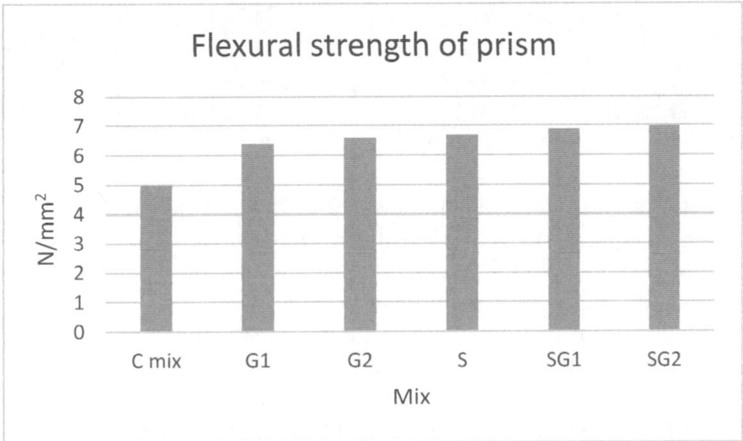

Figure 6 Flexural strength of different mixes

6. Mode of failure

Crack and failure pattern of PCC and fibre reinforced concrete have some difference. In PCC crack formed initially enlarges and result in failure while in hybrid fibre reinforced concrete several cracks are formed. Due to the addition of fibre the failure mode is transformed from brittle to a little ductile.

Figure 7 Failure pattern for PCC

Figure 8 Failure pattern for HyFRC

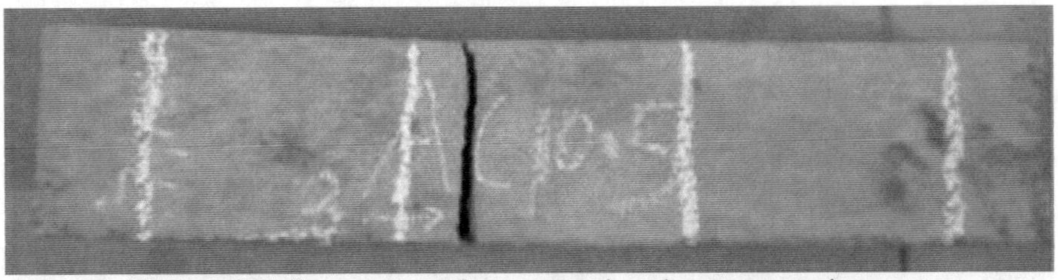

Figure 9 Failure pattern for conventional concrete prism

Figure 10 Failure pattern for hybrid concrete prism

7. Conclusion

This paper presents the results of experimental studies conducted on the properties of different concrete specimens. Mix compositions were reinforced with single and hybrid fibre. Volume fraction of each fibre was fixed as 0.5 % to obtain economical and workable mix. Based on the test carried on cube cylinder and prism specimen the following conclusions are highlighted.

- Compressive strength is improved by the addition of fibres .Maximum increase in compressive strength is 20.14% for SG1 .It may be due to the influence of glass fibres .Glass fibres brought down the permeability and shrinkage.
- Split tensile strength is also improved by the addition of fibres and maximum increase is for SG1 which is 32.5%
- Flexural strength is also increased by the addition of fibres and maximum increase is 39.2% obtained for SG1
- Hybrid fibre reinforced mix with 6mm short fibre is found effective, because of the short length they can bridge the cracks without being ruptured and hence carry higher load.

References

[1] Banthia N and Guptha R 2004 Hybrid fibre reinforced concrete (HyFRC): fibre synergy in high strength matrices *J. Materials and structures* **37** 717-716

[2] Barcevic A, Bjegovic D, Skazlic M 2017 Hybrid reinforced concrete with unsorted recycled tire steel fibers *J. Materials in civil engineering* **29**

[3] Chen Y, Qiao P 2011 Fracture of hybrid microfiber reinforced cement composites *J. Aerospace Engg* **24** (154-161)

[4] Deng Z and Li J 2006 Mechanical behaviours of concrete combined with steel and synthetic Macro-fibers *International Journal of Physical Sciences* **1**(2) 057-066

[5] Lawler J S,Zampini D and Shah S P 2005 Microfiber and Macrofiber Hybrid Fiber-Reinforced Concrete *J. Mater. Civ. Engg* **17** 595-604

[6] Mastali M, Dalvind A 2017 Fresh and Hardened Properties of Self Compacting Concrete Reinforced with Hybrid Recycled Steel – Polypropylene Fiber *J. Mater. Civ. Engg* **29**

[7] Wang P,Huang Z,Jiang J,Yongjun Wu 2012 Performances of Hybrid Fiber Reinforced Concrete With Steel Fibers and Polypropylene Fibers*J civil Engg& urban planning* 458-461

[8] Yao W,Li J ,Wu K 2003 Mechanical properties of hybrid fiber-reinforced concrete at low fiber volume fraction *Cement and Concrete Research* **33** 27–30

Numerical Study of Dual Web Steel Plate Girders Under Buckling, Monotonic Load Response And Hysteresis Behaviour

Merin Maria Alex[1],Nikhil Mohanan[2],Asha Alice Kulavattom[3]
[1] M-tech Student, SJCET Palai, India
[2] Research Engineer, FEM CFD Research and Development, Kerala, India
[3] Assistant professor, SJCET Palai, India

Abstract. Steel Plate Girders are generally I shaped steel beams made of two or more individual steel plates which are bolted, welded or riveted to form the girder's flanges and web. Plate girders are usually used in railway bridges. Predominantly, these girders contain only a single central web plate. In this paper, we propose and study the behavior of dual parallel web girders under buckling, monotonic loading, and hysteresis response. The control beam is first modeled and validated based on experimental results and a comprehensive numerical investigation has been discussed in this paper. From the analysis results, model with 20 mm web spacing shows better performance The finite element code of ANSYS APDL 16.2 is used for modeling the test cases.

Keywords: Plate girder, Buckling, Hysteresis

1. Introduction

The plate girders are generally I-beams made from separate steel plates which may be bolted or welded or riveted to form the horizontal flanges and vertical web of the beam. Stresses on the flanges are larger near the centre of the span than near the end of the span, hence the top and bottom flange plates are reinforced in the mid portion of the span. Vertical stiffeners prevent the buckling of web due to shear stresses. Vertical stiffeners are uniformly spaced along the girder with additional stiffeners over the supports and wherever concentrated loads occurs. Plate girders are mainly used as beams under railway bridges.

Behaviour of plate girders under load is important. Previous studies have been made to improve the method for predicting the buckling resistance and cyclic behavior of steel plate girders. The design and buckling behaviour of stainless steel plate girders subjected to shear failure have attracted attentions in recent year[2-4].Olsson[1] conducted eight tests on austenitic and duplex stainless steel plate girders and developed a new design expressions for calculating the shear resistance. Real et al.[5] conducted a test setup of nine plate girders which are made using austenitic stainless steels. The test results showed the effect of material non-linearity on the shear strength. Shanmugam et.al[14]conducted experimental and numerical study on ultimate load bearing capacity of curved plate girders. It was seen that the load carrying capacity of girders decreased with increase in curvature. Saliba and Gardner[12-13] conducted nine tests on duplex stainless steel plate girders with rigid end condition and a design approach was presented to determine ultimate shear resistance of plate girders. Hassanein[10-11] carried out numerical studies on stainless steel plate girders to determine the influence of imperfections on the buckling behavior. New design methods for predicting the ultimate buckling resistance of stainless steel plate girders have been proposed[6-9].In monotonic loading, an increasing load is applied to the structure, to determine its yield and ultimate strength. The effect of cyclic loading is also studied. The main objective of this paper is to investigate buckling behaviour, monotonic load response and hysteresis behaviour of dual web steel plate girders.

IOP Conf. Series: Materials Science and Engineering **396** (2018) 012005 doi:10.1088/1757-899X/396/1/012005

2. Methodology

2.1 Experimental Tests

The test set up is given in figure 1, in which the plate girders were subjected to a concentrated load at mid span. Two bearings, supported over a rigid steel pedestal is used to provide the simply supported condition.60 mm diameter cylinder was used to form the end bearing. The pin support on one side was provided by a bearing that was welded on a flat plate, while the other was a roller support, allowing longitudinal displacements and rotations. The small clamping frames were provided at each end, allowing restraints on plate girders with varying flange widths. A thin layer of lubricating oil was applied between the two plates and specimen, to avoid friction. Two lateral supports were provided by using a pair of triangular frames, with aspect ratio of 1.5 at each side of the plate girder, to prevent lateral-torsional buckling failure. The test set up is given in figure 1. A 5000 kN capacity hydraulic testing machine is used for performing the shear buckling tests. The experiment was done by X.W.Chen *et.al* (2017) Wuhan University, China.

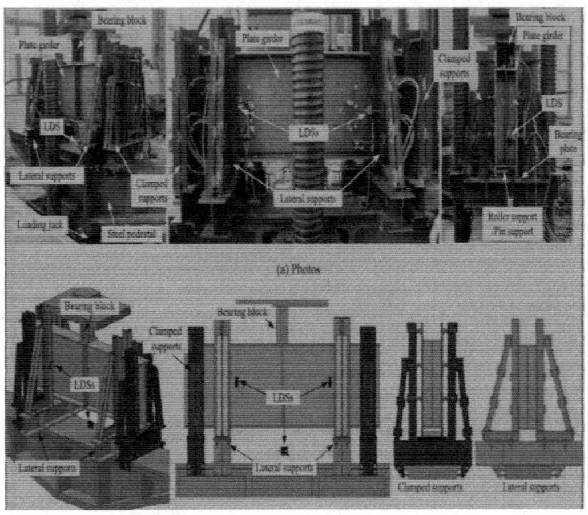

Figure 1 Test setup[15]

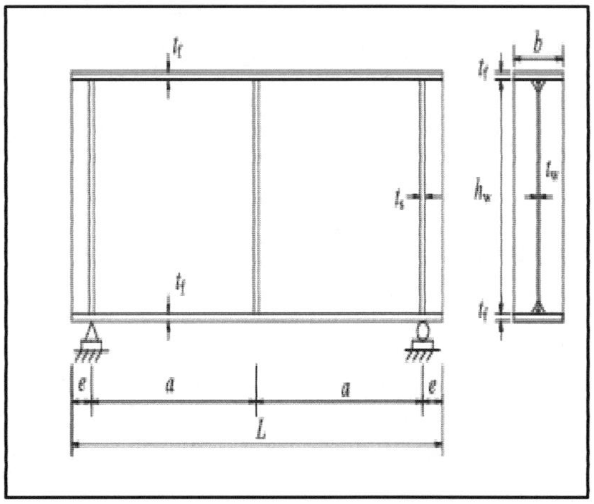

Figure 2 Geometry of test specimen[15]

2.2 Finite Element Modeling and Calibration of Test Specimens

The test specimen consists of a stainless steel plate girder with three transverse stiffeners as given in figure 2. The geometry of test specimen was modeled as shell element due to the less complexity involved, and is shown in Figure 3 and the geometric dimensions are given in Table 1. The material properties of the test specimen are shown in Table 2.

Table 1. Geometric dimensions for test plate girder

Plate girder	L (mm)	a (mm)	e (mm)	h_w (mm)	b (mm)	t_w (mm)	t_f (mm)	t_s (mm)
Specimen	798.8	299.2	100.2	299.5	134	3.82	11.85	11.85

IOP Conf. Series: Materials Science and Engineering **396** (2018) 012005 doi:10.1088/1757-899X/396/1/012005

Table 2. Material Properties

Grade	E(GPa)	υ
1.4301	200	0.3

2.2.1. Geometric Modeling

The modeling is done using ANSYS Static Structural. The geometric model is shown in figure 3.

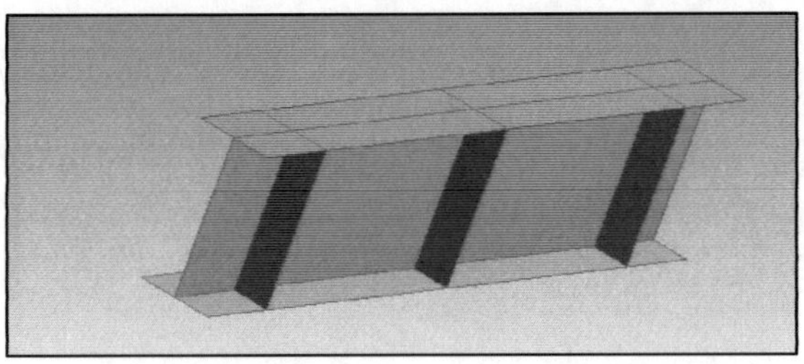

Figure 3 Model of test steel plate girder

Figure 4 Meshed test model

2.2.2.Meshing and Discretization

Mesh convergence study was conducted. The meshed model is shown in figure 4. It was found that the results tend to remain constant beyond 25000 elements under a mesh size of 17mm Quadratic elements and hence the choice for the study.

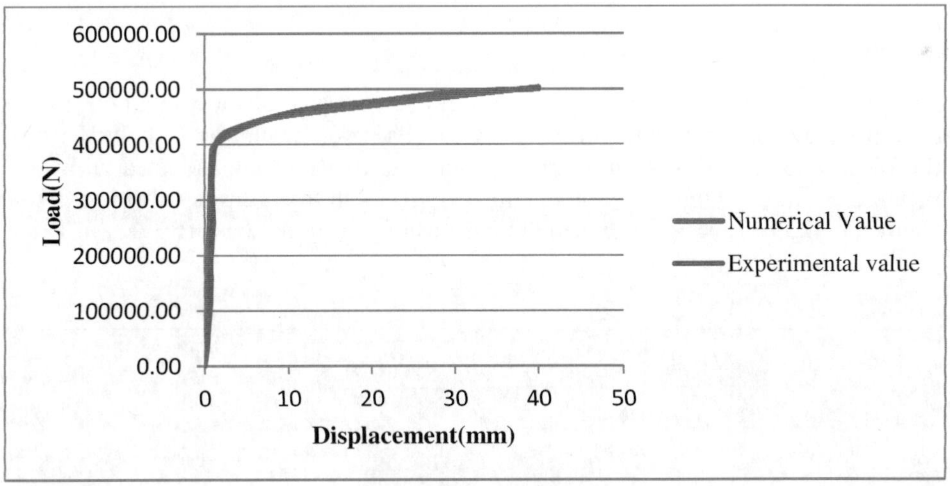

Figure 5 Calibration Curve

Table 3. Calibration result

	Load(N)	Error %
Experimental result	500100.00	
		0.583
Numerical result	503020.00	

3. Calibration of Finite Element Model

The result obtained from the base journal is similar to that obtained from software analysis. The result is obtained by plotting displacement versus load graph and it is seen that experimental value obtained from the base journal is approximately equal to the value that we obtained during our validation procedure. The obtained result is given in Table 3 and the graph is shown in figure 5.The deformed shapes of tested plate girder obtained experimentally from journal and that from the FE model is shown in figure 6. The obtained error percentage is 0.583% which is less than 5% hence numerical validation is considered to be acceptable.

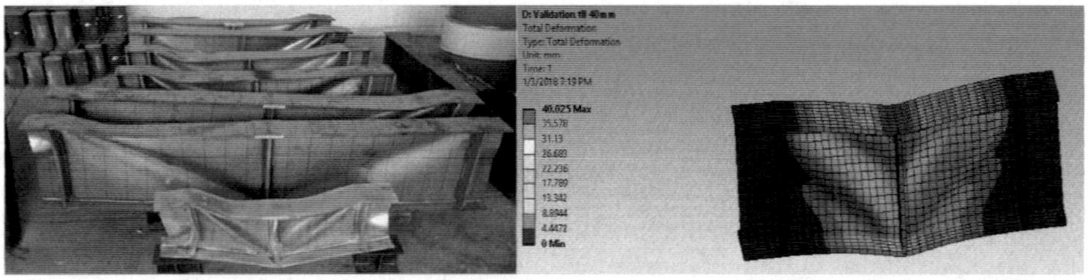

Figure 6 Deformation pattern

4. Project Models

Here, all the support and load conditions used here are same as that of the validation geometry. The single web of the plate girder is split into dual web. The web thickness is halved from original geometry thickness to study the iso-volumetric strength. Six models were modeled with web spacing of 10 mm, 20 mm, 30 mm, 40 mm, 50 mm and 60 mm. The web thickness is halved to 1.91 mm. The models are shown in figure 7.Designation and dimensions of project models are given in Table 4.

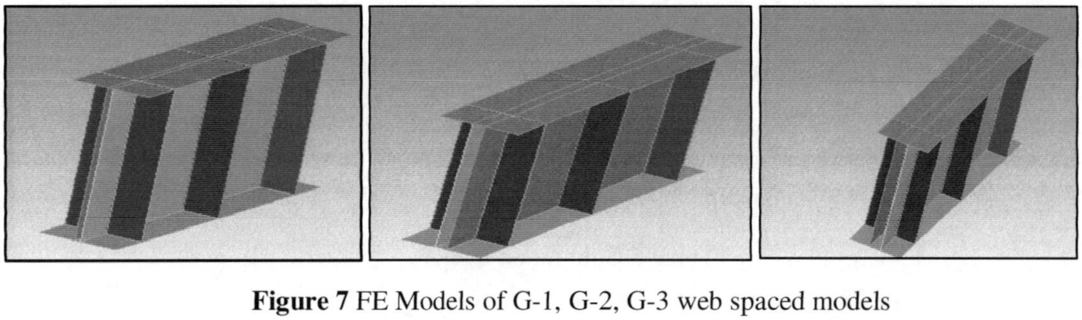

Figure 7 FE Models of G-1, G-2, G-3 web spaced models

IOP Conf. Series: Materials Science and Engineering **396** (2018) 012005 doi:10.1088/1757-899X/396/1/012005

Table 4. Dimensions of project models

No.	Designation	L (mm)	A (mm)	e (mm)	h_w (mm)	b (mm)	t_w (mm)	t_f (mm)	t_s (mm)	Spacing of web (mm)
1	G-1	798.8	299.2	100.2	299.5	134	1.91	11.85	11.85	10
2	G-2	798.8	299.2	100.2	299.5	134	1.91	11.85	11.85	20
3	G-3	798.8	299.2	100.2	299.5	134	1.91	11.85	11.85	30
4	G-4	798.8	299.2	100.2	299.5	134	1.91	11.85	11.85	40
5	G-5	798.8	299.2	100.2	299.5	134	1.91	11.85	11.85	50
6	G-6	798.8	299.2	100.2	299.5	134	1.91	11.85	11.85	60

4.1. Boundary conditions of project models

4.1.1 Post Buckling

Linear buckling: An initial perturbation load of 1 kN was applied to generate the linear stresses for the Eigen value buckling and is shown in figure 8. A fixed support is provided on two edges on either side of the plate girder. A displacement is provided on the top and bottom edges with x axis restrained. The boundary conditions are shown in figure 9.

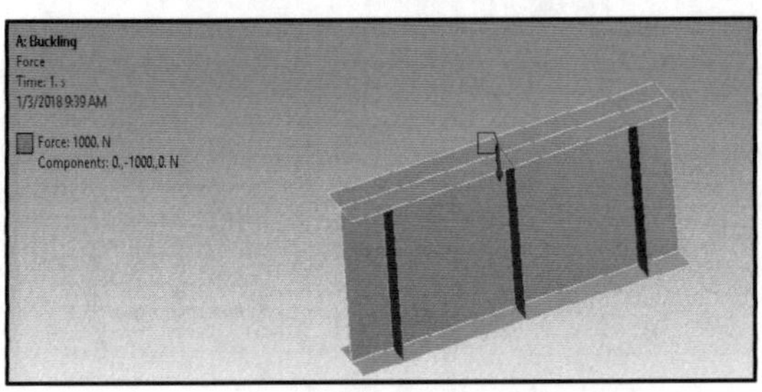

Figure 8 Initial loading

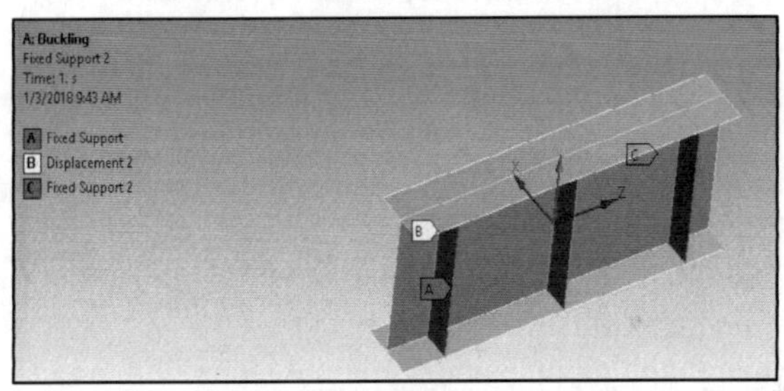

Figure 9 Initial Boundary condition

Nonlinear Buckling: Eigen value buckling mode shape is selected as the initial geometry for the Nonlinear Post Buckling Analysis. Finite Element Modeler was used for transferring the imperfect mode shape geometry. The boundary conditions are shown in figure 10. A fixed support is provided on one end and a displacement constraint on other end to attain simply supported conditon. A lateral displacement constraint is also provided on top and bottom edges. A displacement controlled load of 40 mm is provided from top to make the structure buckle.The load deflection curves were plotted after solving.

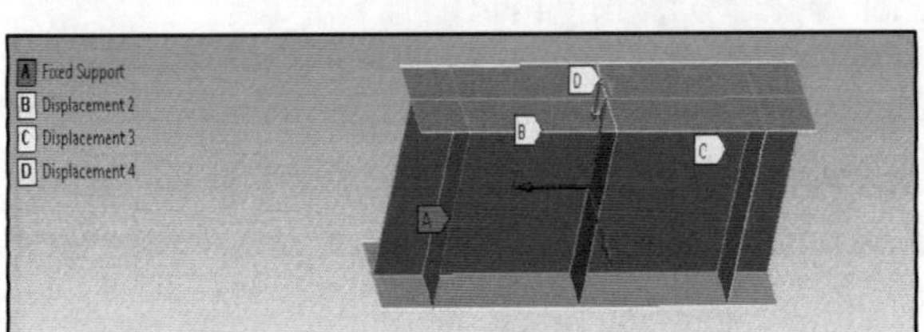

Figure 10 Boundary conditions of axial buckling

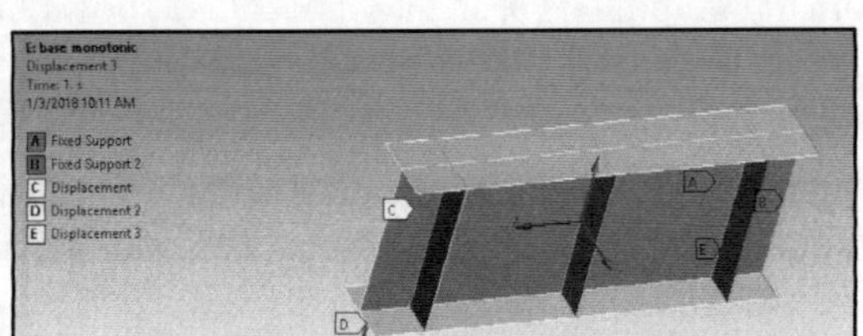

Figure 11 Boundary conditions of monotonic load

4.1.2 Monotonic Load Performance and Testing

The boundary conditions used in monotonic loading are shown in figure 11.A fixed support is provided on two edges on one end and a displacement constraint was given on other end to suit simply supported condition. A displacement constrained load of 100 mm is provided on five edges on one end and a fixed support is provided on five edges on other end. A displacement constraint in the lateral X is provided on sixteen edges on top and bottom. The load deflection curves were plotted after solving.

4.1.3 Hysteresis Behaviour

The boundary conditions are similar to monotonic loading. The programme is based on ATC 24 Cyclic Testing Guidelines The strain controlled load was increment in the order of $0.25d_y$, $0.5d_y$, $0.75d_y$, d_y, $2d_y$, $3d_y$, and $4d_y$, to understand the hysteresis behaviour. Though it is recommended to continue the test until complete failure, we have chosen to reduce the cycles and test the comparative behaviour. The d_y value was calculated from the load obtained in monotonic test.

5. Results And Discussions

5.1 Buckling Analysis

The load- displacement graph obtained from the buckling analysis is shown in figure 12 and critical buckling load values in figure 13.

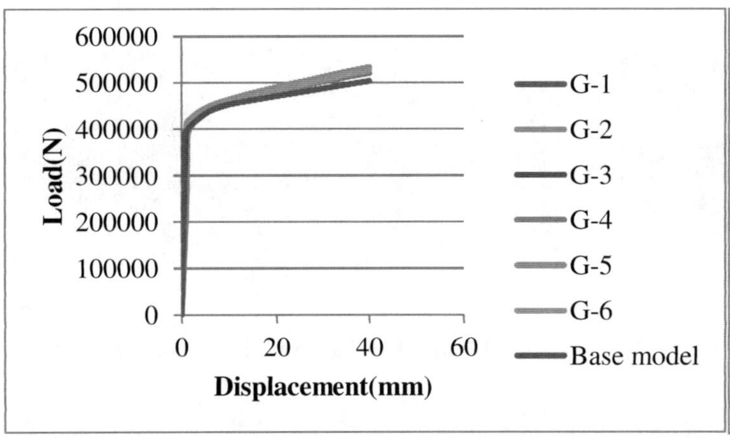

Figure 12 Load- displacement graph in buckling analysis

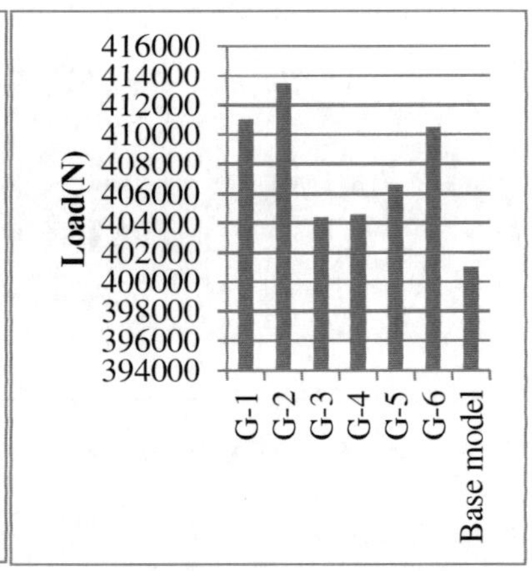

Figure 13 Critical buckling load values

From the graph, it is seen that model G-2 shows higher critical buckling load.

5.2 Monotonic Loading

The load – displacement graph obtained from monotonic loading analysis is shown in figure 14. The model G-2 with 20 mm spaced web plate girder shows maximum yielding load value and is given in figure 15.

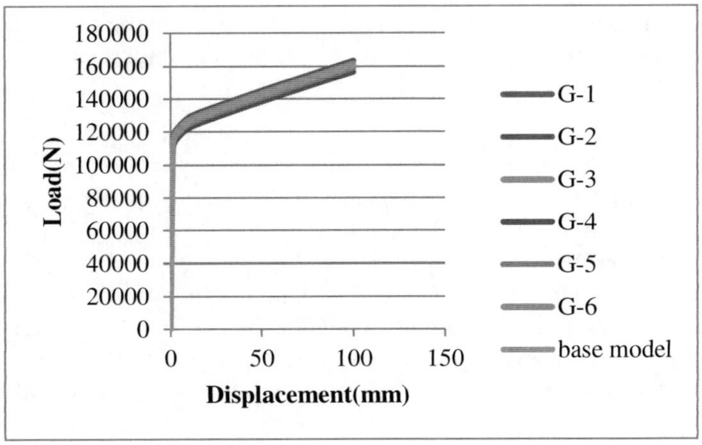

Figure 14 Load- displacement graph

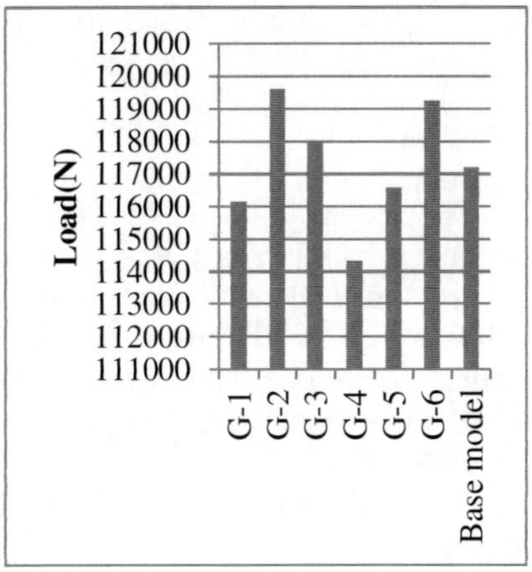

Figure 15 Yield load values

5.3 Hysteresis behaviour

The hysteresis curve corresponding to the load displacement for each model is shown in figure 16.

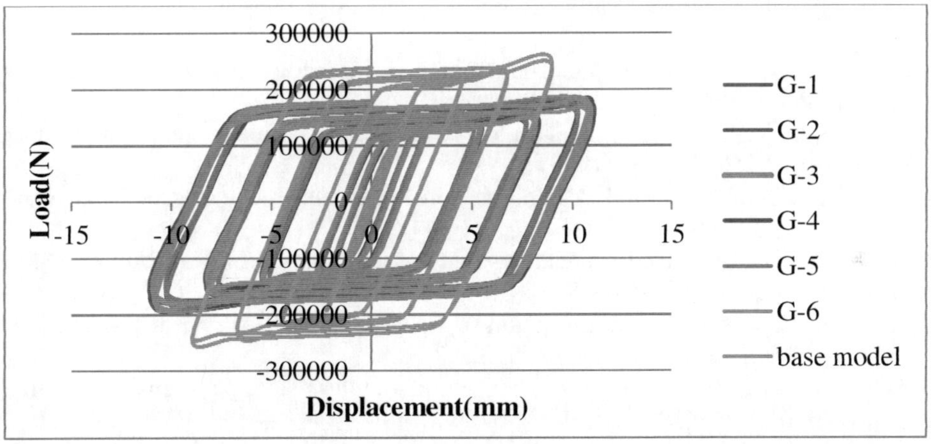

Figure 16 Hysteresis behaviour of models

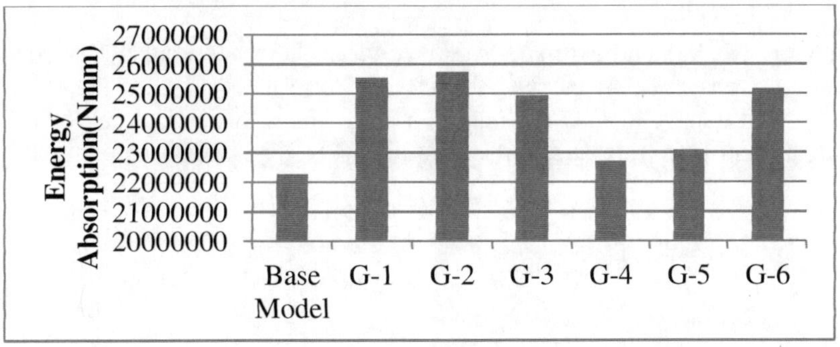

Figure 17 Energy absorption

The energy absorption capability is calculated as the area of the cyclic curve obtained for each model as given in figure 17.Model G-2 with 20 mm web spacing has more area which indicates more energy absorption.

6. Conclusion

From the buckling analysis, it is found that model G-2 with 20 mm web spacing has greatest critical buckling load and hence more load carrying capacity. In the monotonic load response we find that that model G-2 with 20 mm web spacing has better load carrying capacity. Lastly, for the hysteresis response, model G-2 tends to show better energy absorption capabilities which gradually decreases and then increases when it approaches the box girder configuration. Conclusively, we find that model G-2 with 20 mm web spacing shows better load carrying capacity and energy absorption and hence can be adopted as the optimum model which may be experimentally verified and set for industrial applications.

Reference

[1] A. Olsson, Stainless Steel Plasticity-material Modeling and Structural Application(2001).
[2] B. Young, W.M. Lui, Tests of cold-formed high-strength stainless steel compression members, Thin-Walled Struct. (2006)
[3] C.H. Yoo, S.C. Lee, Mechanics of web panel post buckling behavior in shear, J. Struct. Eng(2006)
[4] D.W. White, M.G. Barker, Shear resistance of transversely stiffened steel I-girders, J. Struct. Eng. ASCE (2008)
[5] E. Real, E. Mirambell, I. Estrada, Shear response of stainless steel plate girders, Eng.Struct. 29 (7) (2007).
[6] J. Becque, M. Lecce, K.J.R. Rasmussen, The direct strength method for stainless steel compression members, J. Constr. Steel (2008).
[7] K.J.R. Rasmussen, G.J. Hancock, Design of cold-formed stainless steel tubular members. I: columns, J. Struct. Eng. ASCE 119 (8) (1993)
[8] L. Gardner, M. Ashraf, Structural design for non-linear metallic material, Eng. Struct. 28 (6) (2006).
[9]M. Theofanous, L. Gardner, Experimental and numerical studies of lean duplex stainless steel beams, J. Constr. Steel Res. 66 (6) (2010) 816–825.
[10] M.F. Hassanein, Finite element investigation of shear failure of lean duplex stainless steel plate girders, Thin-Walled Struct. 49 (8) (2011) 964–973.
[11] M.F. Hassanein, Imperfection analysis of austenitic stainless steel plate girders failing by shear, Eng. Struct. 32 (3) (2010) 704–713.
[12]N. Saliba, L. Gardner, Experimental study of shear response of lean duplex stainless steel plate girders, Eng. Struct. 46 (2013) 375–391.
[13] N. Saliba, E. Real, L. Gardner, Shear design recommendations for stainless steel plate girders, Eng. Struct. 59 (2014).
[14] N.E. Shanmugam, M. Mahendrakumar, V. Thevendran, Ultimate load behaviour of horizontally curved plate girders, J. Constr. Steel Res. 59 (2003)
[15] X.W. Chen, H.X. Yuan, X.X. Du, Y. Zhao, J. Ye, L. Yang, Shear buckling behaviour of welded stainless steel plate girders with transverse stiffeners(2017)

Effect of multi-layers in Aluminium corrugated sandwich panels under blast loading.

ALKA GEORGE[1] · NIKHIL MOHANAN[2] ,ASHA ALICE KULAVATTOM[3]

[1]MTech Student, SJCET, Palai,India

[2]Research Engineer, FEM CFD Research and Development, Kerala, India

[3]Assistant Professor, SJCET, Palai, India

Abstract: Composites and sandwiched aluminium structures are of great importance when considering various structures of aerospace, mechanical and civil structural significance. The failure of these structures due to shockwaves and explosions are of critical importance when designing structures and hence in this study we investigate the dynamic response of aluminium corrugated sandwich panels based on the objective criteria composite layering. The result of these experiments are studied based on numerical explicit modelling of the panels and conclusions have been drawn to analyse the significance of parameters under such dynamic loads.

Keywords: Corrugated sandwich panels,TNT blast,layering,dynamic response.

1.Introduction

Terrorism is becoming a crisis that world is facing and recent attacks highlights the importance of building safety. Hence it is paramount to design blast resistant structures. On the basis of various studies, it is found that structures built using aluminium sandwich panels, either pre-cast or cast in-situ, shows great resistance against the blast attacks. Aluminium sandwich panels are extensively used in different fields such as construction works, industries, railway systems and marine works. It can be efficiently used in these fields, since it is having easier fabrication methods, low density and higher strength compared to monolithic plates. Also it can be used in various structures were weight savings are necessary.

Various studies have been conducted on static and dynamic response of aluminium sandwich panels. X Li etal[1]conducted an experimental study on dynamic response of aluminium corrugated sandwich panels using four-cable ballistic pendulum system. Numerical simulation were also performed on two configurations of specimens considered. Xiaochao et.al[2] proposed an auxetic re-entrant cell honeycomb cores and their dynamic response and blast resistance is numerically investigated. Shiqiang Li et.al[3] experimented on blast response of metallic sandwich panels with stepwise graded aluminium honeycomb cores. It was found that for graded panels with relative density descending core arrangement, transmitted force attenuation and energy dissipation were larger than ungraded panels. M.D.Goelet.etal[4]studied about the blast resistance of stiffened sandwich panels with aluminium cenosphere syntactic foam core. Quantitative assessment is used for the dynamic response study, focussing maximum central point deflection of back sheet of sandwich structure. The effect of foam thickness, stiffness configuration and strain rate are considered in the study .Zhang etal[5]conducted an experimental study on dynamic response of metallic trapezoidal corrugated-core sandwich panels subjected to air blast loading. Sigit.P.Santhosha etal[6] done similar study on response analysis of blast impact loading of metal-foam sandwich panels. Structural responses in terms of displacement, velocity and acceleration is taken into account.The experimental and numerical investigation of the blast response of flexible sandwich panels are also reported [7-13]

2.Experiment

2.1. Experimental setup

The corrugated sandwich panels for the experiment were collected and the face sheet and core were attached by hot melt adhesive members. The face sheet and core was made using AL-1200H18,and the mechanical properties are specified as follows: density $\rho=2.71\times10^{3}kg/m^{3}$,Youngs modulus

IOP Conf. Series: Materials Science and Engineering **396** (2018) 012006 doi:10.1088/1757-899X/396/1/012006

E=70GPa,Poissons ratio λ=0.33,tensile strength=210MPa,yield stress=140MPa.[1].A four cable ballistic pendulum system was used for the experimental process.It was used to measure the impulse applied on the specimen.Tthe entire experimental setup consists of clamp frames, a steel beam and counterweight balance, and the system was suspended using four wire ropes shown in Fig.1.The specimen was mounted onto the frame by 16 screws on the front of the pendulum. The explosive charge or the detonator was mounted in front of the specimen. A laser displacement sensor was mounded behind the pendulum inorder to measure its translation. From the recorded oscillated amplitude, the impulse exerted on the front face of the pendulum was calculated and from the exposed area of specimen effective impulse of the specimen was further calculated. The dimensions of specimen was 300mm×300mm and the blast exposed region had the dimension 250mm×250mm.A total of eight experiments was conducted for two different geometries. The experiment was conducted by Xin Li, in Shenzhen, China 2014.

Figure.1.Four cable ballistic pendulum system[1]

Table1: Dimensions of corrugated sheets used in the experiment(X, Li etal

Configuration	H_f/Face sheet thickness	H_c/Core height	t/Core thickness	Length/a	Length/b	θ	Cell length/s
1	0.8	4	0.2	4	8	63.4	12
2	0.8	8	0.2	7	14	66.4	21

Table1: Dimensions of corrugated sheets used in the experiment(X, Li etal)

Eight test were conducted for two different specimens of varying face sheet thickness under the above experimental conditions.

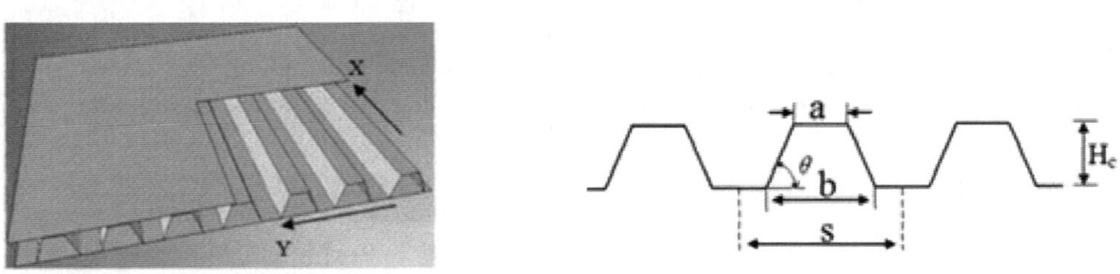

Figure2.Sketch of the specimen [1]

2.2. Test results

The residual back deflection of the back face sheet and failure modes of face sheets and core are analysed. Deformation o of the specimen is obtained by the time-displacement curve plotted by sensor. With the increase of TNT mass and decreasing the face off distance, the impulse exerted on the specimen increases which is listed in Table 2.

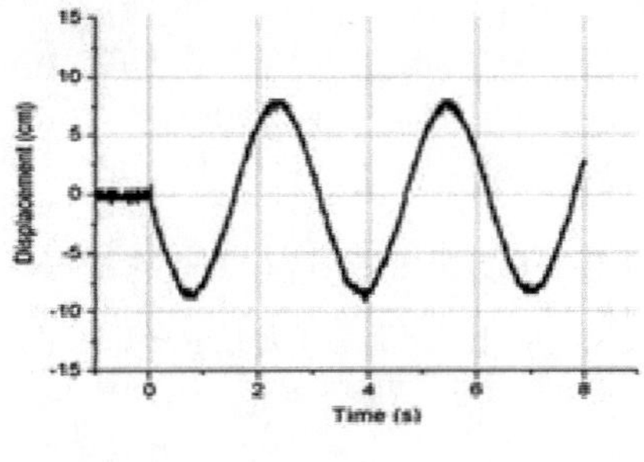

Figure3.Time-Displacement curve[1]

Table2. Impulse and maximum residual back deflection

Test number	1	2	3	4	5	6	7	8
Configuration	1	1	1	1	2	2	2	2
m/Mass of TNT(g)	10	10	15	30	15	15	30	30
R/Distance between panel and TNT(mm)	120	100	100	210	120	100	240	190
I/Impulse(NS)	11.8	13.7	14.3	23.8	13.2	14.1	19	26.7
δ/Maximum residual back deflection(mm)	13.5	19	23.6	43.4	21.7	24.5	41.4	61.5

Plastic deformations are only formed and no pitting failures are observed in the specimen.The core is divided into three regions, full-folded, partially-folded and clamped region.The plastic deformation is larger in the full-folded region, since it is located at the centre. The deformation or the impulse produced in the clamped region is small due to presence of clamped frame.

3.Methodology

3.1. Finite element model

The dynamic response study of sandwich panels are analysed using ANSYS AUTODYN software.150mmx150mmaluminium corrugated panels with core and face sheet dimensions as in Table.1, configuration 2 is modelled. A clamp frame 20mm×5mm is provided at the eddges of the ANSYS model in order to provide constrains to the panel.

Table.3. Mechanical properties of material

Density	2700 kg/m^3
Bulk modulus	6.862745×10^7
Poisson's ratio,λ	0.33
Yeild stress,σ	140MPa
Elastic modulus,*E*	70GPa
Shear modulus	2.631579×10^7
Yeild stress	1.4×10^{5}MPa
Tangent modulus	5×10^5

IOP Conf. Series: Materials Science and Engineering **396** (2018) 012006 doi:10.1088/1757-899X/396/1/012006

Meshing is done after mesh convergence for 0.5mm mesh size with 1.9 lakh nodes.

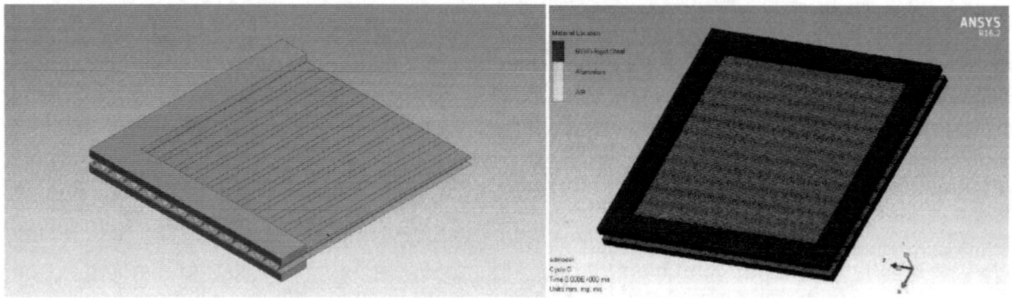

Figure.4 a)Quarter model of Sandwich panel b)Meshed model of Sandwich panel

Mass of 15gm TNT is placed at a stand-off distance 120mm vertical the middle point of the model as in test number 5 given in Table.2.and the FEM model of air is shown in Fig.5.

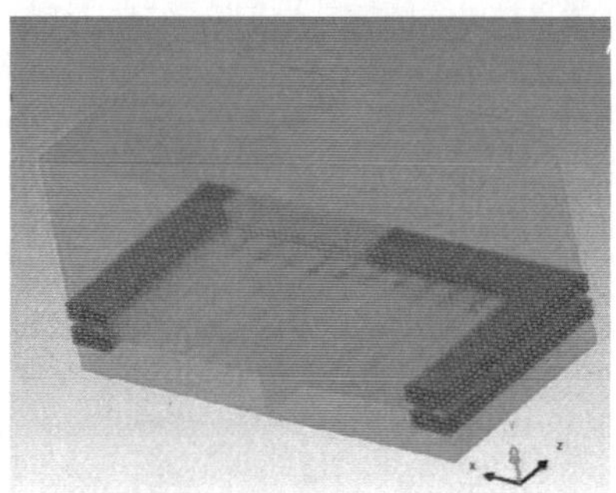

Figure.5.FEM model of air

3.2. Validation

The results obtained from the validation model shows acceptable agreement with the base journal. Displacement-Time graph is plotted and the results shows that the average error percentage is 4.73%, hence the numerical validation is considered acceptable shown in Table.4.

Table.4. Validation results

	Experimental result	Numerical result	Error
Maximum deflection	30.53	28.8	5.67%
Plastic deformation	27.3	26.3	3.8%

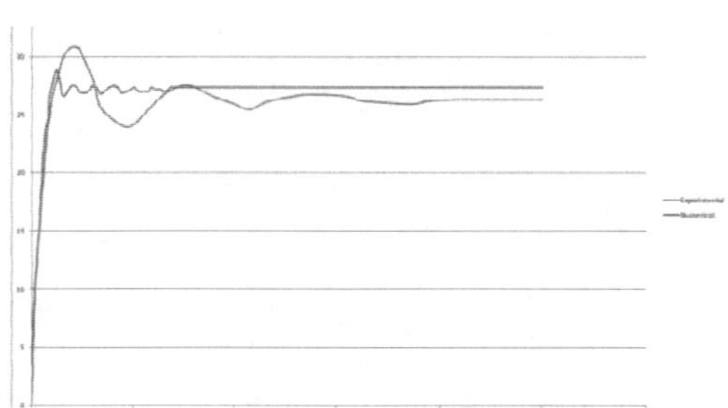

IOP Conf. Series: Materials Science and Engineering **396** (2018) 012006 doi:10.1088/1757-899X/396/1/012006

Figure.6. Validation curve

3.3. Project model

a)Geometric models

Eight geometric multi-layered models are done under the same volume and core height criteria that of the validation model. The corresponding face sheet and core thickness is adjusted using the acceptable gauge sizes of aluminium. The alignment and orientation of layers are altered in the various geometries. The geometric details of model is shown in Table.5.

Table.5. Details of project models

No.	Designation	Core height(mm)	Plate thickness(mm)	Core thickness(mm)
1	L2-SS-0-0	8	0.511	0.18
2	L2-SS-0-90	8	0.511	0.18
3	L2-SI-0-0	8	0.511	0.18
4	L2-SI-0-90	8	0.511	0.18
5	L3-SSS-0-0-0	8	0.361	0.16
6	L3-SSS-0-90-0	8	0.361	0.16
7	L3-SIS-0-0-0	8	0.361	0.16
8	L3-SIS-0-0-90	8	0.361	0.16

Four models are done for both double layer,L2 and triple layer,L3.The alignment of face sheets in each model are varied either straight,S or inverted,I accordingly. Also the orientation of selected plates along the vertical axis is varied by 90^0.Sample quarter model of L2-SI-0-90 and L3-SIS-0-0-90 are shown in Fig.7. 15gm of TNT is added at a standoff distance 120mm for 5millisececonds.

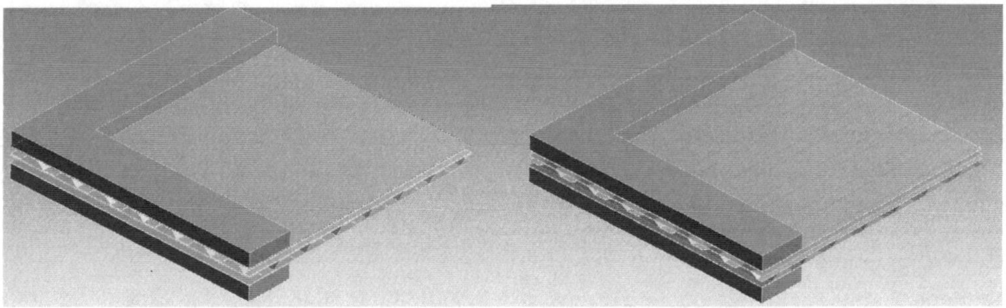

Figure.7.Quarter models of a)L2-SI-0-90 b)L3-SIS-0-0-90

b) Results and discussions

The deformation behaviour of double layer and triple layer is different in models. In double layer models(model no.1,2,3,4), it is deformed downwards and after maximum deflection point, the deformation is proceeded almost constantly.In triple layer models(model no.5,6,7,8),during the initial period, the sheet is deflected in the downward direction and after 0.8milliseconds it is found to be deflected back due to the vaccum created at the point of blast.Maximum deflection and minimum deflection of various models are shown in Table.6.From the results obtained it is observed that deflection values of triple layer models are less compared to that of double layer models.The maximum deflection is observed in L2-SI-0-0 and minimum deflection is observed in L3-SIS-0-0-0 and L3-SSS-0-90-0.The deformation behaviour of L2-SI-0-0 and L3-SIS-0-0-0 and the comparison curve of all the models are shown in Fig.8. ,negative deflection in Y- axis and time in X-axis.

Table.6.Maximum and minimum deflection values of models

Model	Maximum deflection(mm)	Minimum deflection(mm)
L1-S-0(validation model)	28.8	24.35
L2-SS-0-0	29.35	23.93
L2-SS-0-90	28.61	24.61
L2-SI-0-0	29.19	29.063
L2-SI-0-90	29.43	28.23
L3-SSS-0-0-0	29.43	15.47
L3-SSS-0-90-0	29.07	11.19
L3-SIS-0-0-0	29.08	11.10
L3-SIS-0-0-90	27.76	11.43

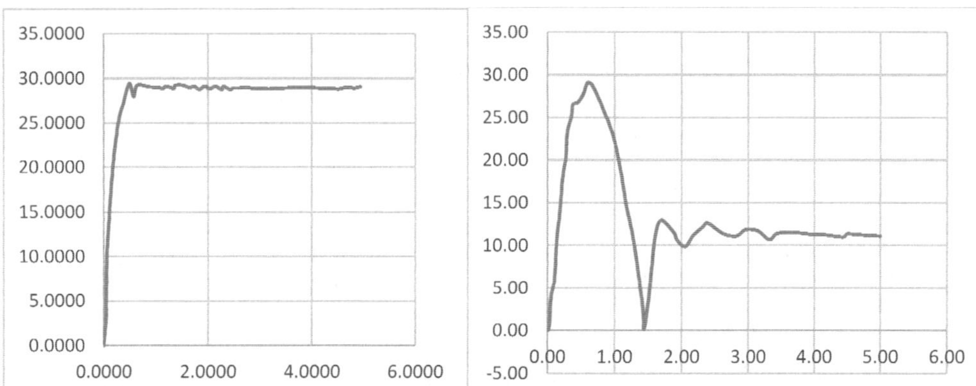

Figure.8. a)Deflection-time curve of L2-SI-0-0 and L3-SIS-0-0-0.

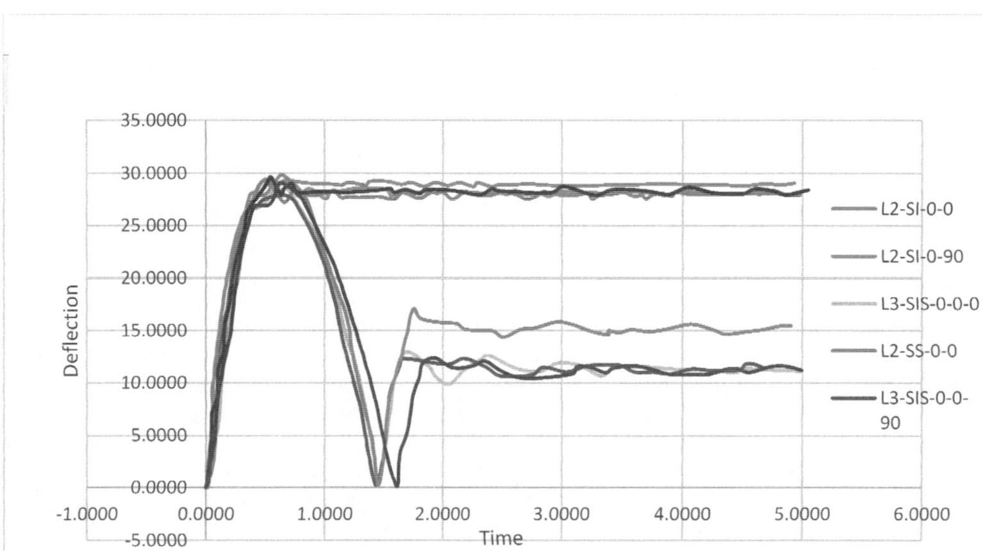

Figure.8.b) Comparison graph of models.

Conclusion

The dynamic response of multi-layered trapezoidal panels under blast loading is studied and it is observed that, as number of layer of face sheet is increased, the panel deflection is decreased. It is found that triple layered, L3-SIS-0-0-0 and L3-SSS-0-90-0 models shows satisfactory results under the blast load.

Reference

1.XinLi,Zhihua Wang (2013),Response of aluminium corrugated sandwich panels under air blast loadings: Experiment and numerical simulation

2.D.Karagiozova,G.N.Nurick,G.S.Langdon(2015), Response of flexible sandwich panels to blast loading.

3.XiaochaoJin , Zhihua Wang (2014),Dynamic response of sandwich structures with graded auxetic honeycomb cores under blast loading

4.Da-LinXiang, Ji-LiRong,and XuanHe(2015),Experimental Investigation of Dynamic Response and Deformation of Aluminium Honeycomb Sandwich Panels Subjected to Underwater Impulsive Loads

5.Shiqiang Li,Xin Li (2016),finite element analysis of sandwich panels with stepwise graded aluminium honeycomb cores under blast loading

6.Manmohan Das Goel,Vasant.A.Matsagar,Anil.K.Gupta(2015), Blast resistance of stiffened sandwich apnels with aluminium cenosphere syntatic foam.

7. Pan Zhang,Jun Lui,Yuansheng Cheng etal(2016),Dynamic response of metallictrapezoidal corrugated-core sandwich panels subjected to air blast loading- An experimental study.

8.Dong Yongxiang, Feng Shunshan, Xia Changjing, and Gui Lele(2015),Dynamic Behaviour of Concrete Sandwich Panel under Blast Loading

9.Cho-Chung Liang, Ming-Fang Yang, Pin-Wen Wu(2015), Optimum design of metallic corrugated core sandwich panels subjected to blast loads. Ocean Eng 2001;28: 825e61

10.Rubino V, Deshpande VS, Fleck NA(2009), The collapse response of sandwich beams with a Y-frame core subjected to distributed and local loading. Int J MechSci 2008;50:233e46

11.MarkellosAndreou, AnastasiosKotsoglou and StavroulaPantazopoulou(2016), Modelling Blast Effects on a Reinforced Concrete Bridge

12.Sigith.P.Santhosha,Faizal Arrifurrahman, MohHaffidzIzzudin (2017),Response Analysis of Blast Impact Loading of Metal-foam Sandwich Panels.

13.Kaurin, Christian Mæhle, Varslot, Magnus Olaf (2010),Blast loading on square steel plates; A coparative study of numerical methods

14.KazemahvaziSohrab, Zenkert Dan. Corrugated all-composite sandwich structures. Part 1: modeling. Compos SciTechnol 2009;69

Seismic assessment of irregular reinforced concrete frame structures using triple friction pendulum bearing

Suni Susan Babu[1] and Manish Jose[2]

[1]Mtech Student, Dept. of Civil Engineering, St Joseph College Of Engineering Palai, Kottayam, Kerala, India
[2]Assistant Professor, Dept.of Civil Engineering, St Joseph College Of Engineering Palai, Kottayam Kerala, India

E-mail : manishjose@yahoo.com

Abstract. In the present scenario due to the fluctuating demand and need for the recent generation has made the designers and engineers plan the building have irregularity. The current study copes with the irregular building having the asymmetric plan as per code IS 1893:2002(Part1). If these buildings are constructed in the earthquake-vulnerable area then they should be designed to resist earthquake forces. Base isolation is the one of the most prominent and widely accepted passive control technique developed to prevent or minimize these forces by isolating the superstructure from the substructure. The Triple Friction Pendulum Bearing (TFPB) is picked as an isolator for the current study.TFPB has offered better seismic performance, lower bearing costs and lower construction cost and has adaptive behaviour. The response of a G+10 and G+20 base isolated and fixed base building are evaluated. For the analysis, symmetrical and different asymmetrical plan shapes like L, T,+, C are considered. Response Spectrum Analysis is carried out for the study by using the software SAP 2000 version 19 for two cases. The results show that in the base isolated building there is a grow in storey displacement, time period and greatly reduced the base shear, storey acceleration compared to base fixed structures.

Keywords: Triple Friction Pendulum Bearing, Base Isolation, Plan irregular building, Time period, Base Shear

1. Introduction

During the last decades, modern architectural designs have made plenty of challenges for civil engineers. Studies of previously earthquakes demonstrate that the irregular buildings in plan and /or height are usually vulnerable than the regular ones. The prime explanation for the collapse in asymmetric is attributed to the non-monotonous lateral deformation of the resisting plane.[9]Now a day the base isolation techniques represent an interesting design strategy for decoupling the structure from the damaging effects of the earthquake by increasing the structure flexibility and providing a convenient damping ratio together with energy dissipation.[1,2] Mainly two types of seismic isolators namely elastomeric and frictional one, used for isolation. Single, Double Concave Friction Pendulum is two different types of frictional isolators which are recently used in numerous buildings and bridges. From this type of isolators, the Triple Concave Friction Pendulum (TCFP) isolator is the newest generation and the earthquake behaviour of it is being examined by researchers.[9]For the present study Triple Friction Pendulum Bearing is used as the base isolator. It was proposed by Victor Zayas.

The TFPB associates with three pendulums in one bearing and each having the properties to select and accumulate the structure's response to different seismic strengths and frequencies. This approach is based on the assertions that lower the coefficient of friction, less the shear transmitted. The TFPB differs from the single, that there is 3 friction pendulum mechanism are achieved by using 4 concave surfaces with sliding occurring on two of the surfaces at a given time, existing in each bearing instead of one pendulum mechanism. These mechanisms are actuated at different phases based on the strong seismic demand. An image of the TFP's disassembled parts and a cross-section of a TFP bearing are shown in figures 1, 2 below, respectively. [4]

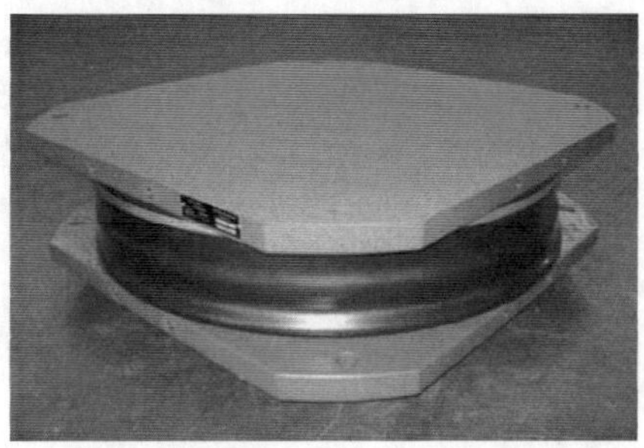

Figure 1. Triple friction pendulum bearing

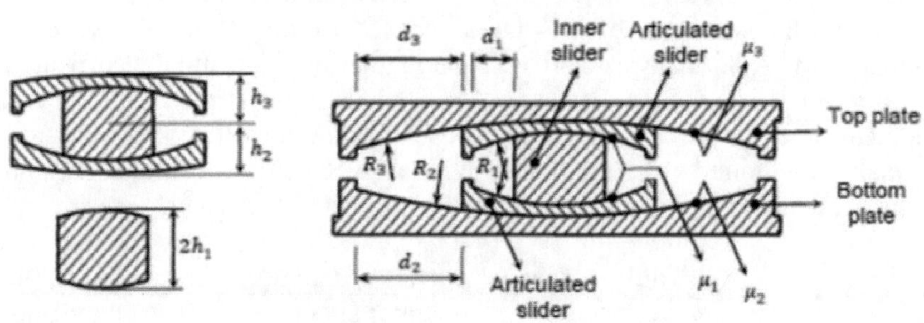

Figure 2. Sectional view and basic parameters

2. Objective

The objectives of the current study are,

- The work is focussed on the study of the effects of seismic forces on multistoried buildings with base fixed and isolated in SAP 2000.
- To assess the effect of the asymmetric plan on the seismic response of different multistoried building and to find more acceptable building among them.
- To design and study the effectiveness of Triple Friction Pendulum Bearing used as a base isolator, and analyze its performance when increasing the height of the buildings.

IOP Conf. Series: Materials Science and Engineering **396** (2018) 012007 doi:10.1088/1757-899X/396/1/012007

3. Methodology

In the existent study, G+10 and G+20 reinforced concrete building having the symmetric and asymmetric plan are considered. Four geometric plan irregular models are considered. The building is situated in zone V, having the medium soil of India is considered.

Table 1. Loadings and properties on the building

Loadings	
Live load on floor	3kN/m^2
Live load on roof	1.5kN/m^2
Floor finish	1.5kN/m^2
Geometric Properties	
Plan dimension in X direction	36m
Plan dimension in Y direction	20m
Building Type	SMRF
Storey height	3.1
Beam size	230X400mm
Column Size(Bottom 5 Storey)	250X750mm
Remaining 6 Storey	230X500mm
Material Properties	
Grade of Concrete	M25
Grade of Steel	Fe415
Soil Type	Medium(Type-II)
Zone Factor: Z	0.36
Importance factor: I	1
Special Moment Resisting Frame: R	5

The work started with the modelling and analysis of G+10 and G+20 building for two cases. The initial one is fixed and the second is base isolated. Triple Friction Pendulum isolator is used as the base isolator in this study. After analysis of fixed base regular model using the software SAP 2000, the maximum vertical reaction is obtained. By using ASCE-Code 7 the Triple Friction Pendulum Bearing is designed manually and the same is used as a base isolator for all models considered. After the modelling, the analysis is done by the method of Response Spectrum for the fixed and isolated base and the observations are compared.

Table 2. Link properties of Isolator

Property	Inner slider	Outer Slider
Element height(mm)	121	161
Elastic Stiffness(kN/m)	50175.029	42717.503
Effective Stiffness(kN/m)	36085384	36085384
friction Coefficient-fast	0.159	0.187
friction Coefficient-slow	0.08	0.093
Radius(mm)	526	3395
Rate parameter(sec/mm)	0.0006	0.0007

4. Modelling and Analysis

The Design Software SAP 2000 has been used for the modelling and analysis of the buildings. For the present study, Linear Dynamic Analysis is performed.

4.1. Models considered

The following models are considered for the analysis and it is build up using the Sap 2000 software.

Table 3. Models Considered For the Study

Notation	Shape	Base Condition
1A	Regular	Fixed Base
1B	Regular	TFPB
2A	T-shaped	Fixed Base
2B	T-shaped	TFPB
3A	L-shaped	Fixed Base
3B	L-shaped	TFPB
4A	C-shaped	Fixed Base
4B	C-shaped	TFPB
5A	+ shaped	Fixed Base
5B	+ shaped	TFPB

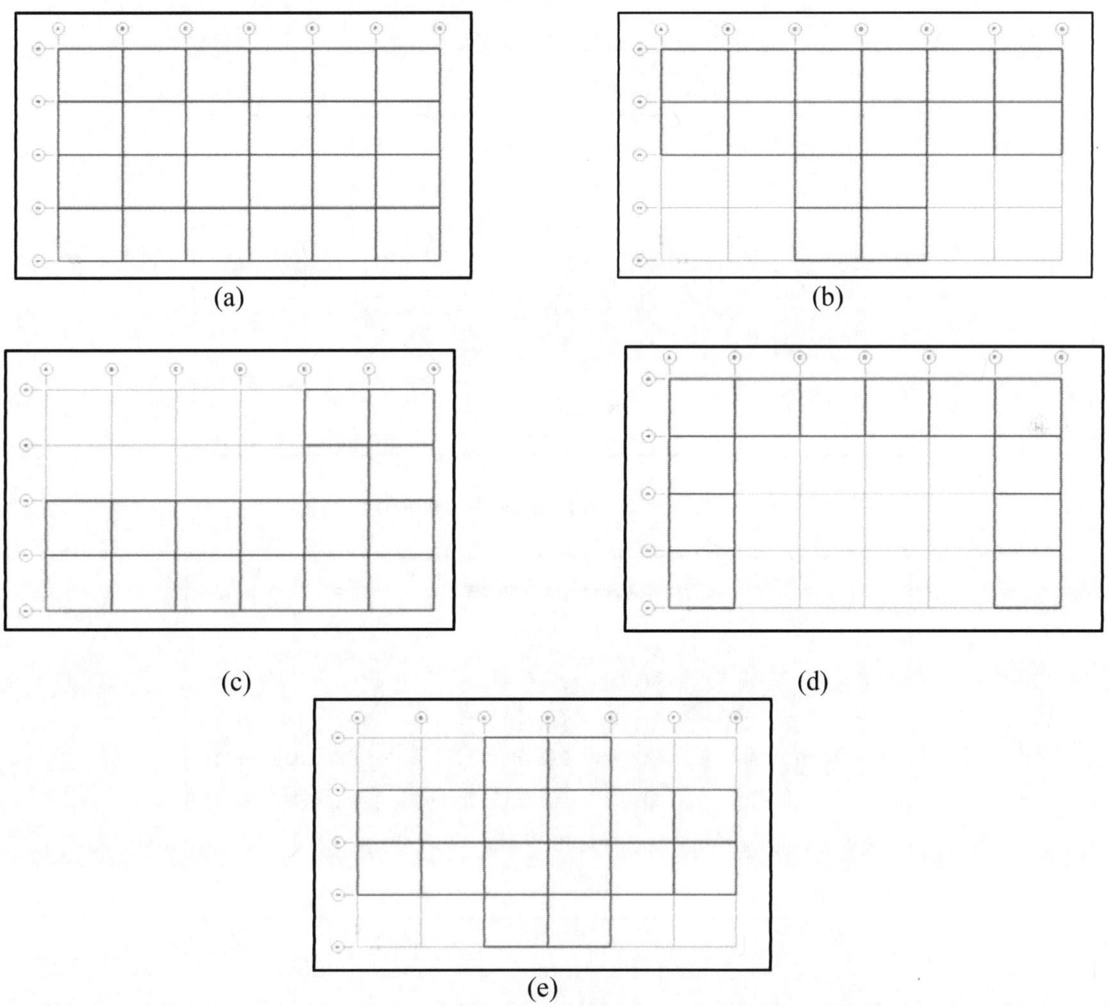

Figure 3. Plan view of (a) Regular,(b)T,(c) L,(d) C and (e) + shaped G+10 and G+20 storied RC Building

IOP Conf. Series: Materials Science and Engineering **396** (2018) 012007 doi:10.1088/1757-899X/396/1/012007

5. Results and Discussions

5.1. Time period

Table 4. Comparison of mode period for G+10 and G+20 of different models

Mode Period(sec)						
	G+10			G+20		
Models	Mode-1	Mode-2	Mode-3	Mode-1	Mode-2	Mode-3
1A	2.014	1.598	1.391	4.151	3.249	2.864
1B	2.457	1.993	1.759	4.557	3.622	3.215
2A	2.487	2.058	1.876	5.053	4.178	3.821
2B	2.972	2.505	2.332	5.487	4.59	4.253
3A	1.954	1.619	1.349	4.065	3.338	2.795
3B	2.376	2.003	1.7032	4.446	3.695	3.129
4A	1.801	1.589	1.328	3.749	3.378	2.779
4B	2.184	1.949	1.658	4.102	3.668	3.098
5A	1.952	1.587	1.349	4.049	3.245	2.793
5B	2.374	1.964	1.703	4.433	3.599	3.127

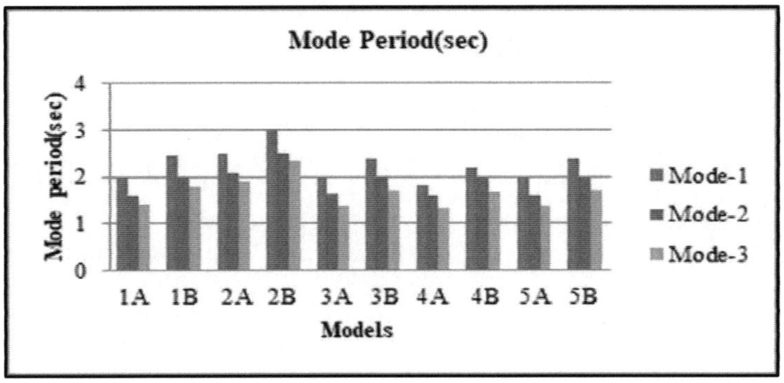

Figure 4. Mode period for G+10 storied building

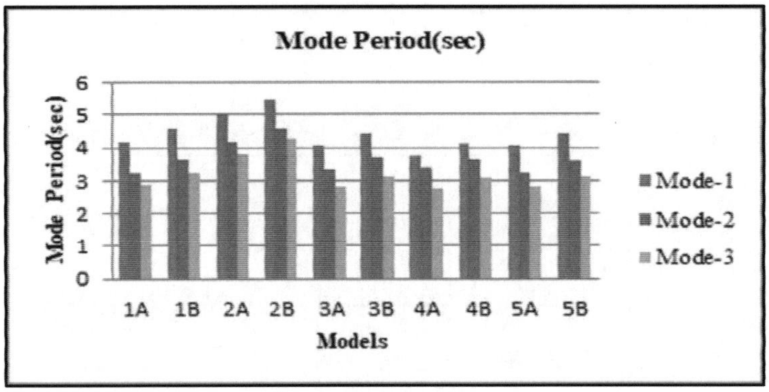

Figure 5. Mode period for G+20 storied building

The table 4 and figure 4 & 5 shows the mode period corresponding to Mode-1, Mode-2 and Mode-3 for different models. The values clearly show that the mode period of TFPB is increased by around 22% for regular, L, C, + shaped G+10 compared to fixed and it is around by 10% for G+20.This is due to the reason of flexibility of the isolator. From that T-shape shows a low increase in mode period compared to other models.

5.2. Base Shear

Table 5. Comparability of base shear for G+10 and G+20 of different models

	Base Shear (kN)			
	G+10		G+20	
Models	RSX	RSY	RSX	RSY
1A	810.808	748.032	974.149	720.792
1B	727.246	652.199	810.808	728.608
2A	666.951	527.1	620.835	619.579
2B	591.769	464.863	612.306	594.278
3A	579.028	536.491	690.675	506.989
3B	522.419	471.404	656.387	503.988
4A	466.835	473.832	534.814	444.185
4B	427.851	420.233	528.470	420.233
5A	580.521	537.87	696.548	556.724
5B	523.401	472.333	661.075	508.982

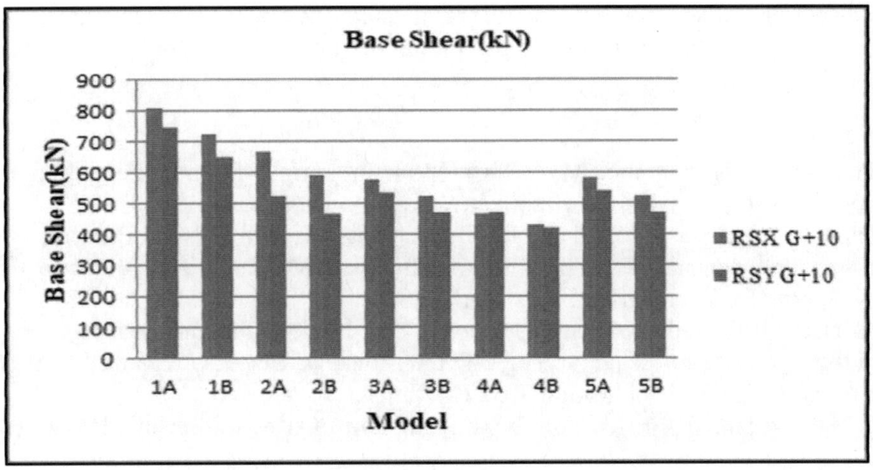

Figure 6. Base shear for G+10 storied building

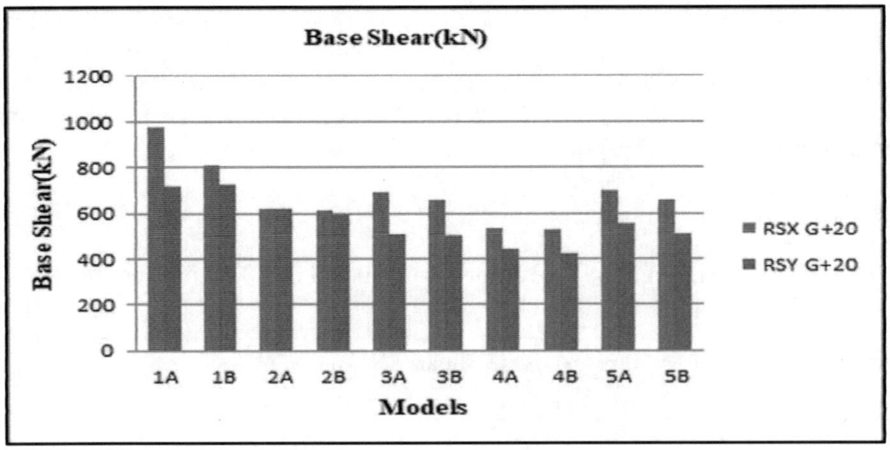

Figure 7. Mode period for G+20 storied building

The table 5 and figure 6,7 shows the base shear in X and Y- direction for different models of G+10 and G+20.For G+10 RC frame building, the outcome reveals that the ratio of base shear for TFPB decreases by around 10% than the fixed base. The model 4B shows the vast reduction in base shear compared to others. For G+20 RC frame building shows that base shear decreases by around 20%. TFPB is more suitable for G+20 multi-storey building.

6. Conclusion

It is observed that the base isolation increases the flexibility at the base level of the buildings. The natural period of a building increases with the height. When considering the fixed base and isolated building, the base shear is reduced considerably for TFPB.C-Shaped building gives a maximum performance and T-shape building is major exposure to the earthquake than the other models.TFPB is more suitable for G+20 multi-storey building and it is found that the time period increases for isolated building due to the presence of flexible isolator. The TFPB reduces the base shear, acceleration and increases the displacement and mode period. From the study, it has been concluded that by using the isolators for plan irregular building gives an outstanding performance at higher seismic zone area.

References

[1] Fenz D M and Constantinou, M C 2008 Modeling Triple Friction Pendulum Bearings for Response-History Analysis, *Earthquake Spectra*, **vol. 24**, No. 4, pp.1011-28

[2] Constantinou, M C, et al. 2011 LRFD-Based Analysis and Design Procedures for Bridge Bearings and Seismic Isolation, (Technical Report MCEER-11-0004, State University of New York at Buffalo, Buffalo, New York, USA)

[3] Nhan D Dao, 2013 Predicting the Displacement of Triple Pendulum Bearings in a Full- Scale Shaking Experiment using a Three-Dimensional Element, *Earthquake Engineering & Structural Dynamics*, Publication 7, 32 Citations

[4] Pejman Namiranian,2014 Effects of Triple Pendulum Bearing on Seismic Response of Isolated Buildings under Near-Field Excitation, *Earthquake Engineering & Structural Dynamics*, Publication 5, 0 Citations

[5] Khloud El Bayoumi,2015 Modelling of Triple Friction Pendulum Bearing in Sap 2000, *International Journal Of Advances In Engineering & Technology*, **vol. 8**, issue 1, pp.1964-71

[6] Ryan K L, Dao N D 2015 Influence of vertical ground shaking on Horizontal Response of Seismically Isolated Buildings with Friction Bearings, *ASCE Journal of Structural*

Engineering,pp-04015089-1-12
[7] Khoshnoudian F, Fallahian M, and Loghman V 2015 Torsionally Seismic Behaviour of Triple Friction Pendulum Bearing, *Advances in Structural Engineering*, **vol. 18**, pp.2151-66
[8] Mazza F and Mazza M 2016 Nonlinear seismic analysis of irregular r.c.framed buildings base-isolated with friction pendulum system under near-fault excitations, *Soil Dynamics and Earthquake Engineering*,pp.299-312
[9] Loghman V, Tajammolian H and Khoshnoudian F, 2017 Rotational Components of Near-Fault Earthquakes Effects on Triple Friction Pendulum Base Isolated asymmetric Structures,pp-110-127
[10] Weber F, Distl J, Braun C,2017 Isolation Performance Assessment Of Adaptive Behaviour of Triple Friction Pendulum, *Journal Of Civil Engineering Research*,pp.17-33
[11] IS 1893:2000(part 1), Earthquake Resistant Criteria Design Of Structures, Indian Seismic Code
[12] IS 456:2000 Code For Plain And Reinforced Concrete
[13] ASCE-Code 7:2010 Minimum Design Loads For Buildings And Other Structures

Numerical Study on Axial Behaviour of Concrete Filled Double Skin Steel Tubular (CFDST) Column with Cross Helical FRP Wrappings

Anna Skaria [1] and Mariamol Kuriakose [2]
[1]PG student, Department of Civil Engineering, St. Joseph's College of Engineering and Technology, Palai, Kottayam, Kerala, India.
[2]Assistant professor, Department of Civil Engineering, St. Joseph's College of Engineering and Technology, Palai, Kottayam, Kerala, India.
Email: annaskaria16@gmail.com

Abstract. Composite columns offers many structural benefits and widely used in civil engineering structures. CFDST consist of inner and outer steel tubes, and concrete sandwiched in between tubes. The steel tubes and concrete work together well and integrity of steel concrete interface is maintained. Steel tubes in inner and outer can acts as permanent formwork and primary reinforcement. Steel tubes provides a confinement effect to concrete and in turn concrete prevents the inward buckling of tubes. As there is deteriorations due to ageing and corrosion strengthening techniques are used. Strengthening can be done by using fibre reinforced polymer (FRP) as an external reinforcement. This paper aims to study about the load carrying capacity of fibre reinforced CFDST columns under axial load using ANSYS 16.2 (Workbench) software. The modeling of CFDST columns were studied with variation in angle (40°, 45°,50°,60°), material change and thickness change of FRP sheets.

Keywords- CFDST, strengthening, axial load, external reinforcement.

1. Introduction

Composite columns have the potential of becoming a part of structural members in low rise and high rise buildings, as it utilize the advantages of both concrete and steel. Due to their excellent static and earthquake resistant properties, such as high strength, high durability, large energy absorption capacity, bending stiffness, fire performance along with favorable construction ability concrete filled steel tubular (CFST) columns are increasingly used in construction industry. The composite action of steel and concrete makes the concrete filled tubular (CFT) columns better than the standard steel or reinforced concrete columns. These potential benefits of composite action largely depends on the steel-concrete interface.

CFDST columns are another advancement of CFT structures in which concrete is sandwiched between steel tubes. The in-filled concrete prevents the inward buckling of concrete while steel tube prevents the lateral deformation of concrete. CFDST have properties similar to that of CFT columns but in addition they are lighter, stronger and posses more energy absorption capacity. From the environmental point of view, CFDST uses less concrete, and therefore more sustainable. Steel tubes provided inside can acts as reinforcement as well as formwork, thereby reducing the cost of construction. The inner cavity can be used for placing power cables, drainage pipes, telecommunication lines etc.

In recent years, it is observed that many deteriorations are occurring on the CFDST structures due to ageing, cracking, yielding, large deformation. Therefore, necessary strengthening techniques such as use of fibre reinforced polymer (FRP) are applied. FRP have high tensile strength, more ductility, corrosion resistant and in addition they will not cause any alteration to the appearance of the members. From the past researchers, the use fibres as strengthening material will significantly improves the strength and stiffness of steel tubular members. A numerical study has been carried out to investigate the suitability of cross helical fibre reinforced polymer fabrics in strengthening of CFDST members under axial load.

IOP Conf. Series: Materials Science and Engineering **396** (2018) 012008 doi:10.1088/1757-899X/396/1/012008

2. Scope

In this paper, the behavior of CFDST columns surrounded by FRP layers under axial load were studied. The FRP is provided as cross helical strips circumscribed about the column. The parameters used in this study is change in fibre material (glass fibre reinforced polymer(GFRP),carbon fibre reinforced polymer(CFRP), basalt fibre reinforced polymer(BFRP), aramid fibre reinforced polymer(AFRP), change in helix angle(40°,45°,50°,60°)and change in thickness (0.3mm, 0.6 mm, 0.9 mm).The non linear static analysis is carried out to determine the deformation, axial stress and strain of different models.

3. Validation

3.1 Experimental test

CFDST column specimens with outer diameter 205.3 mm, inner diameter 140.3 mm and a height of 400 mm were casted. These specimens were then tested under monotonic axial compression. LVDTs were used to obtain the axial deformation. All compression tests were carried out with an MTS machine of displacement control rate of 0.24 mm/min. All tests data, including loads, strains and displacement were recorded simultaneously by a data logger [1].

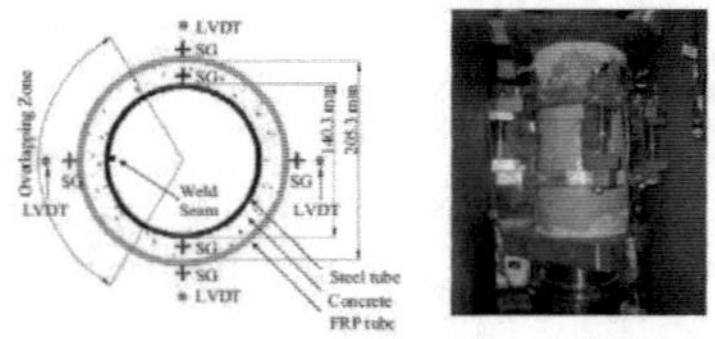

Fig 1. Experimental setup

3.2 Finite element model

Finite element model was developed using finite element analysis software ANSYS 16.2. The cross section was modeled by using circle and column geometry is obtained by extruding it. Bottom end is fixed with no degrees of freedom and load is applied axially at top.

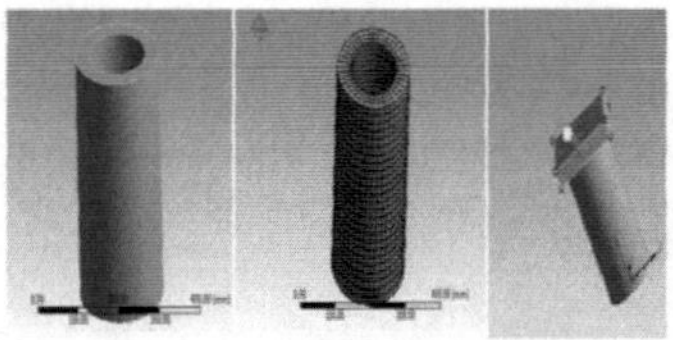

Fig 2. Numerical model

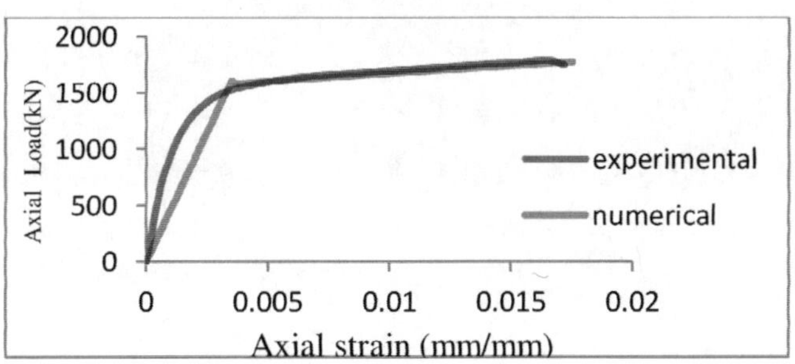

Fig 3. Validation curve

IOP Conf. Series: Materials Science and Engineering **396** (2018) 012008 doi:10.1088/1757-899X/396/1/012008

Fig 3 shows the validation curve. Both the experimental and numerical curves were found to be in good agreement. Therefore, numerical validation is considered to be acceptable.

4. Finite element analysis

4.1 Material specification

The material properties of concrete, steel and fibre reinforced polymer (FRP) are shown in table 1, table 2 and table 3 respectively.

Table 1. Material properties of concrete

Density	2400 kg/m^3
Poisson's ratio	0.18
Elastic modulus	31622.7 N/mm^2
Characteristic compressive strength	40N/mm^2

Table 2. Material property of structural steel

Density	7850 kg/m^3
Poisson's ratio	0.30
Elastic modulus	1.9x10^5 N/mm^2
Yield strength	250 N/mm^2

Table 3. Material properties of FRP

Properties	GFRP	CFRP	BFRP	AFRP
Density	2000 kg/m^3	1700 kg/m^3	2650 kg/m^3	1400 kg/m^3
Elastic modulus N/mm^2	Ex-45000	Ex-240000	Ex-37700	Ex-13600
	Ey-10000	Ey-4830	Ey-5237	Ey-1482
	Ez-10000	Ez-4830	Ez-5237	Ez-1482

4.2 Meshing

Different grid sizes are considered to reach to appropriate mesh size. The adopted element meshing is shown in fig 4.

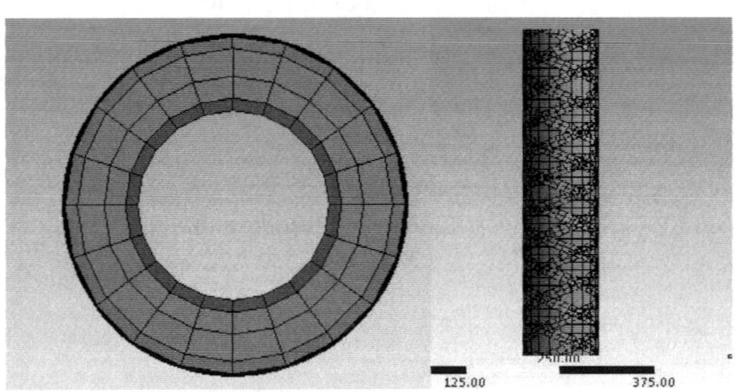

Fig 4. Meshed specimen

4.3 Geometry

All CFDST columns selected are 600 mm in height, outer diameter of 91.5 mm and inner diameter of 62.543 mm, thickness of steel tube is 3.6 mm. Hard contact is used in both normal and tangential direction of the contact between steel tube and concrete, and outer steel tube and FRP.

4.4 Boundary conditions

The end plate is assumed to be an elastic rigid plate. An axial load is applied to the top end plate along the respective axis and other end is provided as fixed support.

4.5 Modeling

About 6 numbers of CFDST columns are modeled. Out of this one model is considered as control specimen, without the FRP wrapping and 4 other models are prepared with FRP wrapping cross helically provided with an angle of 40° , 45°, 50°, 60° and are shown in fig 5, fig 6, fig 7, fig 8, fig 9. Table 4 shows the details of the model created.

Table 4. Model details

Model no	Designation	FRP wrappings	Helix angle	Pitch length
1	Control specimen	-	-	-
3	Model-40	Cross helical	40	241.207
4	Model-45	Cross helical	45	287.46
5	Model-50	Cross helical	50	342.58
6	Model-60	Cross helical	60	497.89

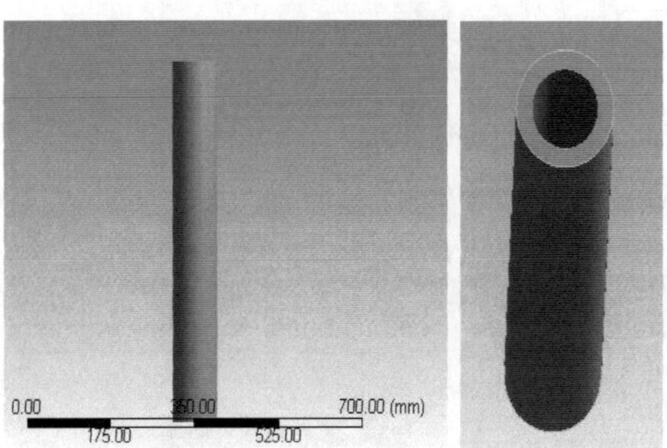

Fig 5. Control specimen

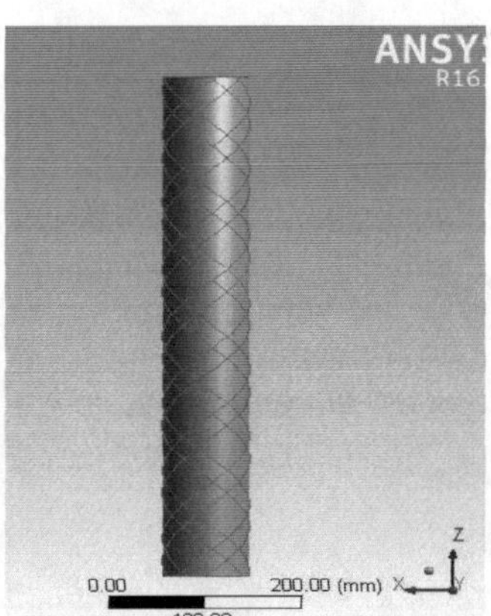

Fig 6. Model- 40

IOP Conf. Series: Materials Science and Engineering **396** (2018) 012008 doi:10.1088/1757-899X/396/1/012008

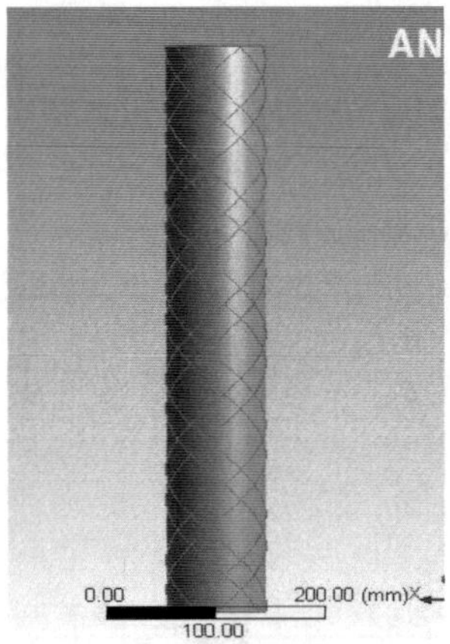

Fig 7. Model- 45

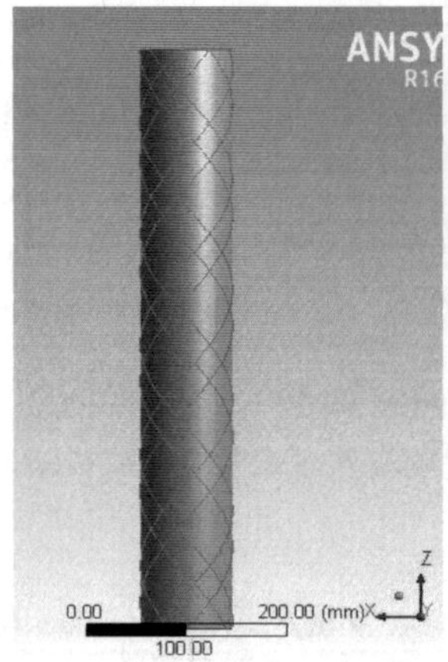

Fig 8. Model-50

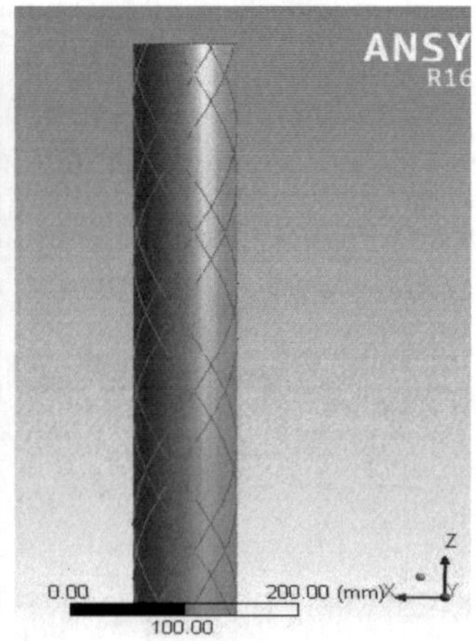

Fig 9. Model-60

5. Analysis

Static structural analysis in ANSYS gives the deformation, force reaction and stresses. Fig 10, fig 11, fig 12, fig 13, fig 14 shows the deformation of control specimen and specimen with FRP wrapping with various helix angle.

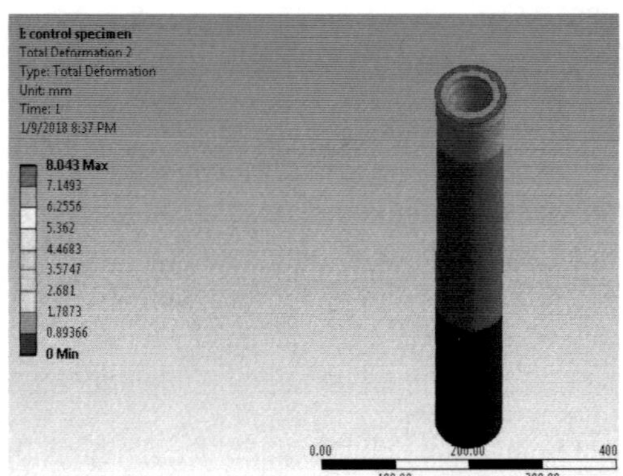

Fig 10. Deformation diagram without FRP

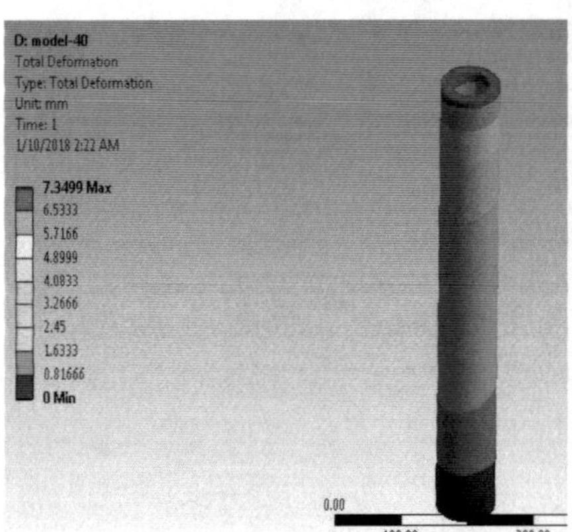

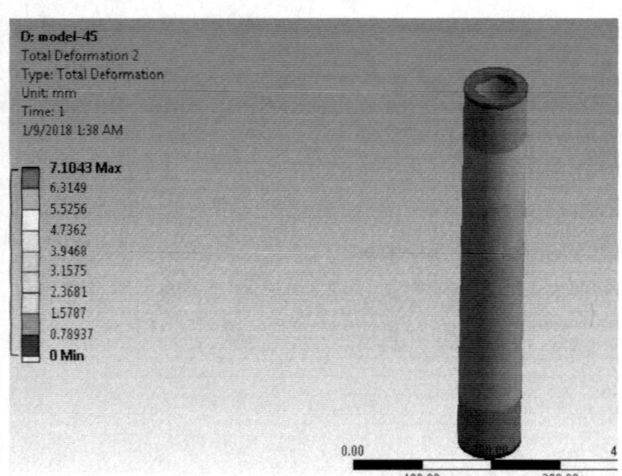

Fig 11. Deformation diagram of model-40 **Fig 12.**Deformation diagram of model-45

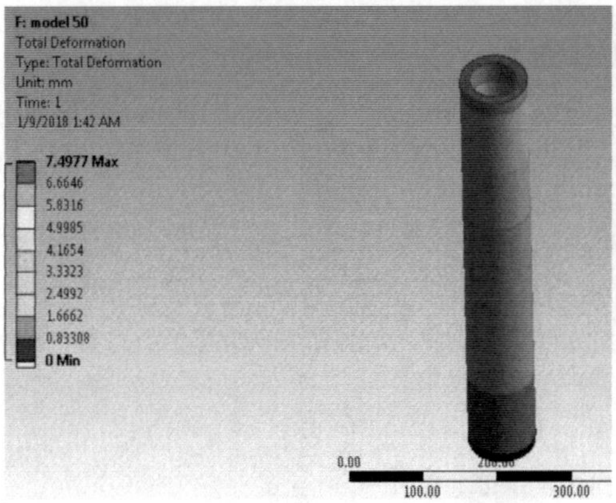

Fig 13. Deformation diagram of model-50

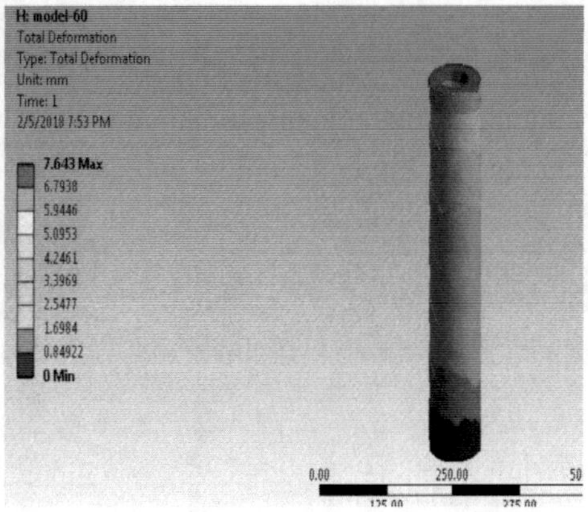

Fig 14. Deformation diagram of model-60

6. Results and discussions

6.1 Effect of change in helix angle

The FRP is provided with varying the helix angle 40°, 45°, 50° and 60°. The yielding load and ultimate load carrying capacity is studied.

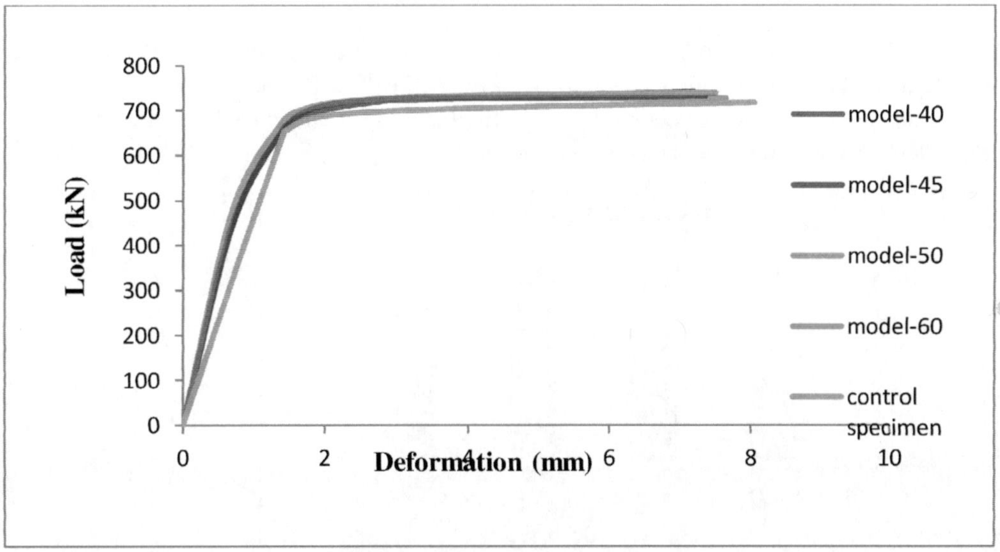

Fig 15. Load-deformation graph

Fig 15 shows the load deformation graph for CFDST columns with different helix angle of FRP strips. Graph shows that the model with FRP with helix angle 45° have a higher yield point compared with the control specimen. The CFDST column with FRP strip at an angle of 45° shows lesser deformation and a higher stiffness. This is mainly due to the confinement effect by the FRP strip.

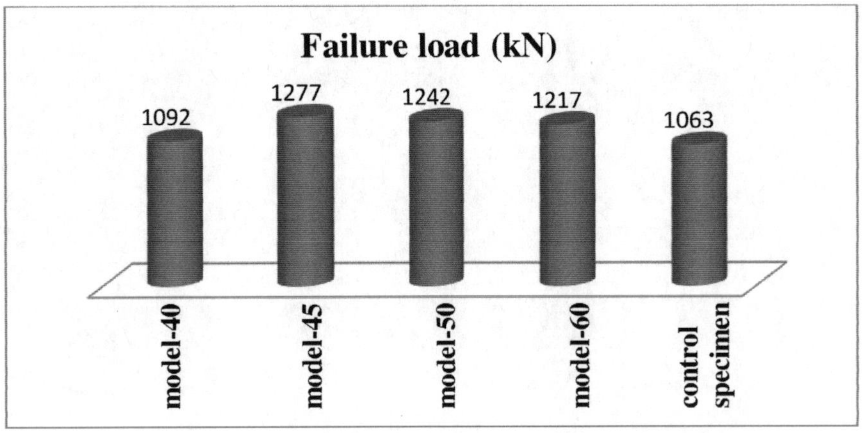

Fig 16. Failure load for different specimens

Fig 16 shows the failure load of CFDST columns determined by using failure criteria. The load carrying capacity increases with an increase in the helix angle up to 45° and then decreases. The specimen with FRP of helix angle 45° has higher load carrying capacity. The external wrapping of FRP strips provides external confinement pressure effectively and enhances the load carrying capacity.

6.2 Effect of change in material

The material of FRP is changed and the materials used are glass fibre reinforced polymer(GFRP), carbon fibre reinforced polymer(CFRP), basalt fibre reinforced polymer(BFRP) and aramid fibre reinforced polymer (AFRP). For study helix angle is kept constant as 45°. The failure load determined by using equivalent stress failure criteria for different material is shown in Fig. 17.

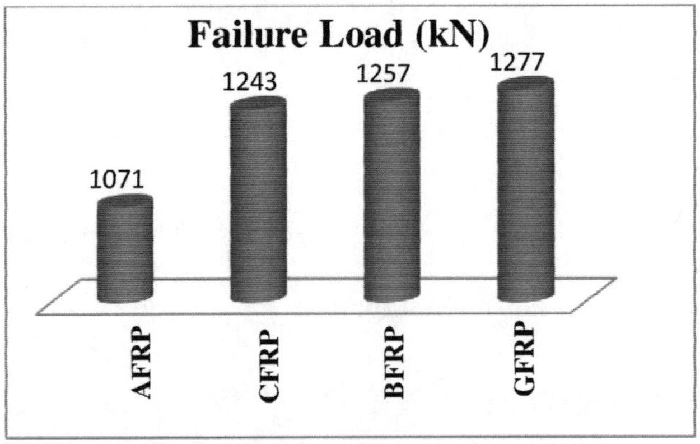

Fig 17. Failure load for different material

6.3 Effect of change in thickness of FRP

The thickness of FRP is varied and is as 0.3 mm 0.6mm and 0.9mm. The helix angle is kept constant (45°) for the study. Fig 18 shows the failure load by different thickness.

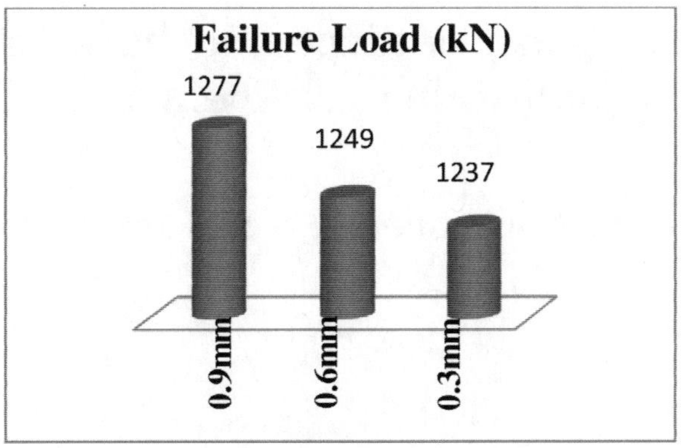

Fig 18. Failure load for different thickness

7. Conclusion

The present study is an investigation of the axial behavior of concrete filled double steel tubular columns with FRP wrappings. Based on the results obtained from study, following conclusions can be obtained.

- The load bearing capacity of columns with FRP wrapping with helix angle 45° is greater than the specimen without wrappings. This is mainly due to the proper confinement.
- The load carrying capacity increases up to 45° and then decreases.
- Load–deformation graph indicates that the stiffness will be higher than control specimen.
- The application of GFRP strips increases axial load carrying of CFDST columns compare with BFRP, CFRP and AFRP.
- With the increase in the thickness the ultimate load carrying capacity also increases.
- Stress concentration is more at the interface between steel and concrete and it does not have any adverse change with change in helix angle.
- It can be concluded that the strength of the CFDST column can be increased by providing FRP wrappings.

8. References

[1] T. Yu , B.Zhang , Y.B.Cao , J.G.Teng "Behavior of hybrid FRP-concrete-steel double-skin tubular columns subjected to cyclic axial compression, *Thin-Walled Structures* 87 (2012), 196-203)

[2] Lin-Hai Han, Qing-Xin Ren, Wei Li, "Tests on the stub stainless steel–concrete–carbon steel double-skin tubular (DST) Columns", *Journal of Constructional Steel Research* 67 (2011) 437–452

[3] Jun wang, ding zhou," Mechanical behaviour of concrete filled double skin steel tubular (CFDST) stub columns confined by FRP under axial compression", *Steel and Composite Structures* (2014)17(4):431-452

[4] J.C.M. Ho ,C.X. Dong, "Improving the strength, stiffness and ductility of CFDST columns by external confinement ",*Thin-Walled Structures* 75 (2014),18–29

[5] Jin-Kook Kim, Hyo- Gyoung Kwak ,"Behavior of Hybrid Double Skin Concrete Filled Circular Steel Tube Columns" Steel and Composite Structures 14(2) (2013)

[6] Y. Essopjee, M. Dundu, "Performance of concrete-filled double-skin circular tubes in compression", *Composite Structures* 133 (2015) 1276–1283

[7] G. Ganesh Prabhu , M. C. Sundarraja , Yun Yon, "Compressive behavior of circular CFST columns externally reinforced using CFRp composites", *Thin-Walled Structures* 87 (2015), 139–148

[8] Aamir Hassan, Ms. S. Sivakamasundari, "Study of Structural Strength and Behavior of Concrete Filled Double Skin Columns", *International Journal of Engineering Trends and Technology (IJETT)* – Volume 18 , Number 6 – Dec (2014)

IOP Conf. Series: Materials Science and Engineering **396** (2018) 012009 doi:10.1088/1757-899X/396/1/012009

Dynamic blast response of diagonally biased stiffened steel plate under confinement

Anju Alexander [1], **Nikhil Mohanan** [2], **Bincy S** [3,a]
[1] M-tech Student, SJCET Palai, India
[2] Research engineer, FEM CFD research and development, Kerala, India
[3] Assistant professor, SJCET Palai, India

[a] E mail: bincy@sjcetpalai.ac.in

Abstract: The explosion rate is increasing in marine and industrial field .This paper investigates the confined blast response of steel square plates stiffened with diagonally biased plate stiffeners. ANSYS Autodyn was used to conduct the parametric study and to analyze the effect of these diagonally biased stiffener plates on structural response and failure of these plates. Furthermore, a study on the plastic behavior and failure of these high strength steel plates due to the impact of blast waves were conducted. Among the explosives available, TNT is used as the charge mass.

Keywords: TNT, Stiffeners, Finite Element, dynamic response,

1. Introduction

Due to the increase in different accidental or the intentional events like terrorist attack or gas leakage etc. hence the behavior of the structural components that are subjected to different kind of blast loadings has become an important subject for the research in recent years. This structures occasionally are subjected to different kinds of blast loading such as air blast, surface blast, gas explosion, confined blast during terrorist attack or during combat environment. The blast hazards in this structures will lead to the collapse of entire structure or will reduce the structural safety. The research in this field are mainly focused on two important aspects that are to determine the blast load from the explosions and to predict the dynamic response of stiffened steel plates using experimental, analytical, numerical approaches [1-8].

The rigid perfectly elastic methods which are used to estimate the residual deformation of the metal plate were first developed by jones [9-11]. The yield line which are formed in the plate plays an important role in the deflection of the plate. Yield lines are considered for the analytical model of stiffened plates.[10,14] The rectangular plate where modeled with segments interconnected by yield line. It yield a roof shaped displacement in most of the analytical models. This method gives accurate results when initial kinetic energy to the maximum possible elastic strain energy was about to be larger than about 10 and the load pulse duration was short with respected to the fundamental natural period of elastic vibration[15,16].

When an explosion occur in a confined environment the characteristics of the blast loads are different from open blast. A quasi static pressure load with a large duration is generated after the initial and reflected shock wave. [5] when the pulse duration was large or the structure is relatively stiff the rigid plastic solution is non conservative compared with the elastic plastic solution [7]. The dynamic response of various stiffener configuration, considering the mesh dependency, loading duration, strain- rate sensitivity by means of numerical analysis it is seen that the time duration and the stiffener configuration has greater effect on the overall Behaviour of the plate, stiffener response with the strain rate effect have higher stiffener response with increase in the value of t_d [3,2]. Under accidental fire the initial behavior of the stiffener plate alters, Repeated buckling caused by the travelling of fire causes alternating tension and compression which has grater implication on the integrity of welded connections [12]. The obtained result is valid only wen used with specified mesh, the mesh near the crack should be refined to ensure acceptable resolution and the broad range of mesh should satisfies the criteria and the yield acceptable predictive capability [17] .

2. Methodology

2.1 Experimental setup

The experimental setup was designed to carry out confined blast acting in a stiffened plate. The set up has a structural chamber made from high strength steel as shown in figure 1 (a). The dimensions and the model is shown in figure 1 (b-c).

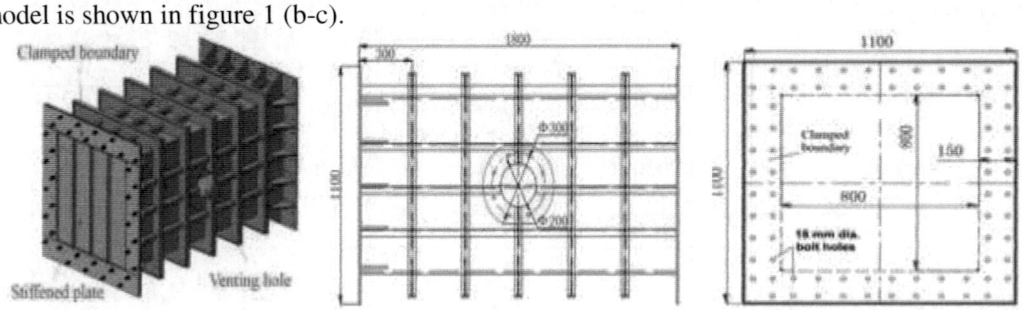

(a) Test setup for stiffened plate (b) Front view (c) Side view

Figure 1: The dimensions of experimental setup. [1]

The length of the chamber is 11800mm and the square plate is 1100 x 1100 mm. plates were made with low carbon steel. The chamber was designed to be strong to withstand the load and to minimize the deformation of walls under confined blasting. A vent hole is provided at the center of the front wall as shown in figure 2(a). The area of the vent hole can be changed by installing different hollow flange plate to change the blast load acting on the test specimen. The specimens were fixed at both end on the support plate by means of 36M16 bolt. After the test plates were clamped the TNT charge in the middle of the test ring were initiated to produce blast load as shown in figure 2(b)

The plate which is used in the experiment are one way stiffened with two or three flats as shown in figure 1. The effective loading area is 800 x 800 mm as shown in Figure 2. The clamping frame was provided with rotational and translational restraints through the support plate which is located against the test plate by bolt. The diameter of the bolt hole in the test plate and the clamp frame are 2mm larger than the bolt. The stiffened plates were manufactured by welding steel flats to the plates using carbon dioxide welding method. Stitch welding technique was used for the manufacturing of the stiffened plates .This confined explosion test was funded as a part of National Defense Fundamental Research Project (B1420133057),National Natural Science Foundation of China (51409202 and 11502180) and the Fundamental Research Funds for the Central Universities (2015-yb-005) and experiment was conducted by Cheng Zheng, Xiang-shao Kong et al(2016) Wuhan University of Technology, Wuhan 430063, China

IOP Conf. Series: Materials Science and Engineering **396** (2018) 012009 doi:10.1088/1757-899X/396/1/012009

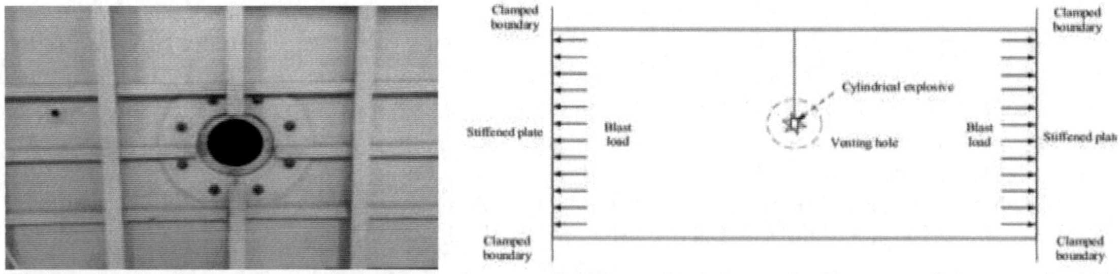

Figure 2 (a) Adjustable vent hole in the front wall [1] (b) schematic diagram of the test ring [1]

The square stiffened plates were tested with different thickness and loading conditions. Before the dynamic test ultrasonic thickness gauge was used for the accurate measurement of the plate and the stiffeners. a cases was studied the corresponding geometry of the stiffened plate and the loading condition and the TNT mass were shown in the table 1.

Table 1: dimensions of stiffener plates and loading conditions

late ref.	Plate thickness δ (mm)	Number of ribs	Length of stiffeners (mm)	Thickness Of stiffeners b (mm)	Height of stiffeners h (mm)	Radius of Venting hole (mm)	Mass of TNT
5	2.3	3	800	2.3	30	50	55

2.2 Finite element modeling

Stiffened plate of size 1100 x 1100 were made with low carbon steel with different thickness. The effective loading area is 800 x 800 mm. The diameter of the bolt hole in the test plate and the clamp frame are 2 mm larger than the bolt. The thickness of the plate is 2.3mm. the square steel plate is stiffened using stiffener plates which are placed diagonal to the effective area .thickness 1.6 mm and the mechanical properties of low carbon steel Q235 is shown in table 2, 0.5mm mesh with hard behaviour is used the mesh convergence is performed.

The numerical modeling is performed with ANSYS AUTODYN. In the numerical model, the stiffener, the plate are discretized by bilinear four-node quadrature shell elements having one quadrature point and fully coupled Eulerian lagrangian approaches are used to simulate the coupling interaction between the blast load and the stiffened plate. The structural chamber is treated as a rigid boundary of air fixed support is provided along the boundary that is the rotational and translation of all nodes of the clamp are restrained and flow out boundary is applied to the zone where vent hole is provide and also to the area behind the plate. The FE model of air is shown in Figure 3 (d). bonded body interaction is provided between the stiffeners and the steel plate. The square plate is stiffened using diagonally biased stiffeners 10 models were modeled the dimensional details is shown in table 3. Geometrical models are shown in figure 3 the finite element model for air is shown in figure 8

Table 2: mechanical properties of material

Mass density	= 7850 kg/m3	Strain hardening modulus B	= 275 × 106 Pa
Poisson ratio	= 0.28	Strain hardening index n	= 0.36
Young's modulus E	=2.06 × 1011 Pa	Strain rate hardening index C	= 0.022
Yielding stress A	= 270 × 106 Pa	Temperature softening parameter m	= 1.03
Johnson–Cook constitutive relation		Melting temp 800 'C	

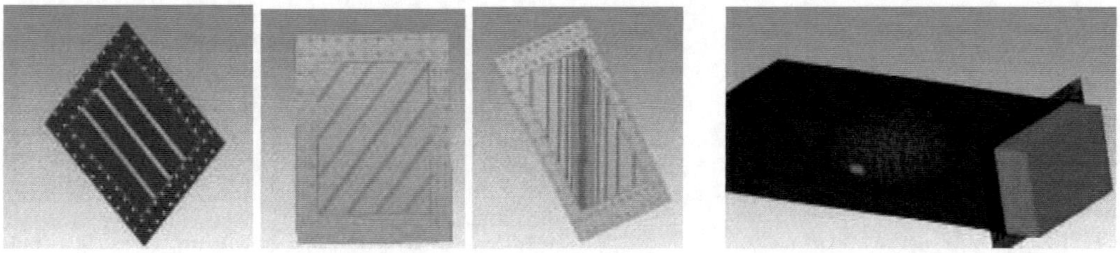

(a) VG (b) DB-B 1.0 (c) DB-B 0.75 (d) FE model of air

Figure 3: FE Model of stiffened plate and fluid

Table 3:Dimensions of stiffened plates

Plate ref.	Designation	Plate thickness δ (mm)	Biase ratio	Thickness of stiffeners b (mm)	Height of stiffeners h (mm)	Radius of venting hole (mm)	Mass of TNT (g)
1	WS	2.3	0	0	0	50	55
2	DB-B 1.0	2.3	1.0	1.6	25	50	55
3	DB-B 0.9	2.3	0.9	1.6	20	50	55
4	DB-B 0.8	2.3	0.8	1.6	16	50	55
5	DB-B 0.75	2.3	0.75	1.6	10	50	55
6	DB-B 0.7	2.3	0.7	1.6	8	50	55
7	DBC-B 1.0	2.3	1.0	1.6	30	50	55
8	DBC-B 0.9	2.3	0.9	1.6	30	50	55
9	DBC-B 0.8	2.3	0.8	1.6	30	50	55
10	DBC-B 0.75	2.3	0.75	1.6	30	50	55
10	DBC-B 0.7	2.3	0.7	1.6	30	50	55
12	VG	2.3	0	2.3	30	50	55

2.3 Calibration

The result obtained from the base journal is similar to that obtained from software analysis. The result is obtained by plotting displacement versus time graph it is seen that the end moment and the maximum deflection of the experimental value and also the analytical value obtained from the base journal is approximately equal to the value that we obtained during our validation procedure. the obtained result is plot on table 4 and the graph is shown in figure 4(a). Mesh convergence is shown in figure 4(b) The obtained error percentage is 3.4% which is less than 5% hence numerical validation is considered to be acceptable

Table 4: validation result

	Analytical	Experimental	Numerical	Analytical Error	Experimental Error
Maximum Deflection	25.47		26.57	4.1%	
End Deflection	23.27	23.502	22.7	2.4%	3.4%

IOP Conf. Series: Materials Science and Engineering **396** (2018) 012009 doi:10.1088/1757-899X/396/1/012009

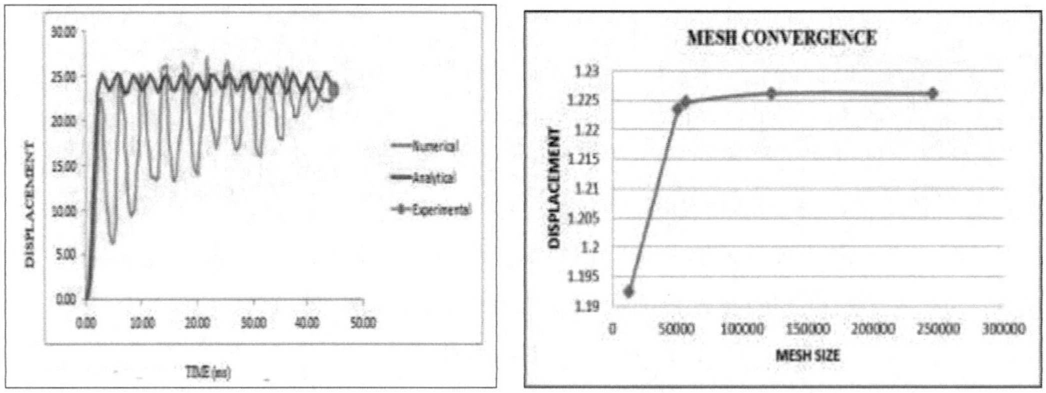

Figure 4 : (a) Validation curve (b) Mesh convergence

3. Result and discussion

The displacement and strain history at the center of the plate is taken as the output in the numerical solution the numerical result of the parameters are shown in table 4 and the bar chart showing the deflection is shown in figure 5

Table 5: Displacement at the center of the plate

Plate no	Designation	Maximum deflection	End deflection
1	DB – B 1.0	25	21
2	DB – B 0.9	28	23
3	DB – B 0.8	26	22
4	DB – B 0.75	25	20
5	DB – B 0.7	24	13
6	DBC – B 1.0	26	20
7	DBC – B 0.9	26	22
8	DBC – B 0.8	25	19
9	DBC – B 0.75	20	17
10	DBC – B 0.7	22	17
11	WS	36	30
12	VG	26.57	22.7

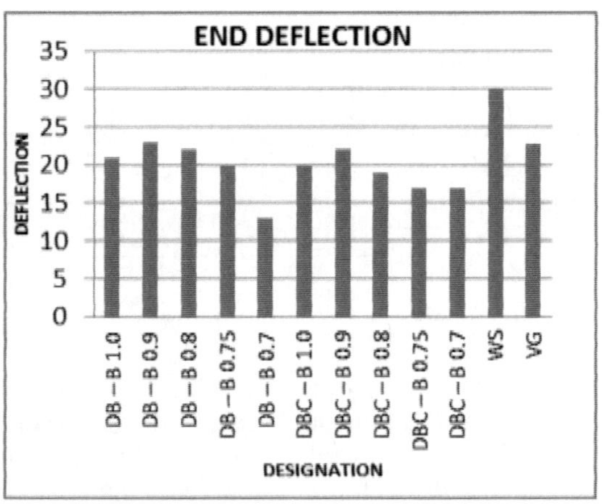

Figure 5: Displacement bar chart

In dynamic test the confined blast wave have a long duration and relatively high peak pressure. The time displacement history graph in comparison with square plates without stiffeners is shown in figure 6.From the result, much variation in the vibration characteristics exist between the end deflections under biased specimen. The end deflection was reduced from about 30mm in the test specimen to 13mm in the biased specimen. The contour plot of DB–B-0.8 is shown in figure 6 (f)

IOP Conf. Series: Materials Science and Engineering **396** (2018) 012009 doi:10.1088/1757-899X/396/1/012009

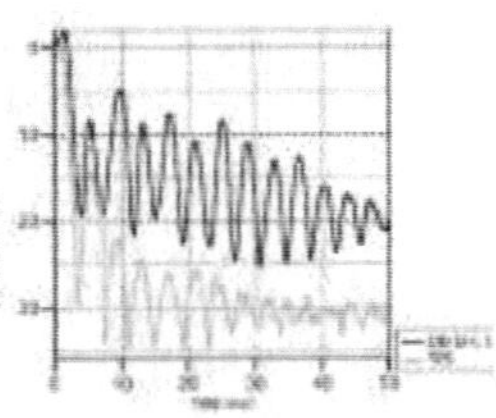
(a) DB-B 1.0

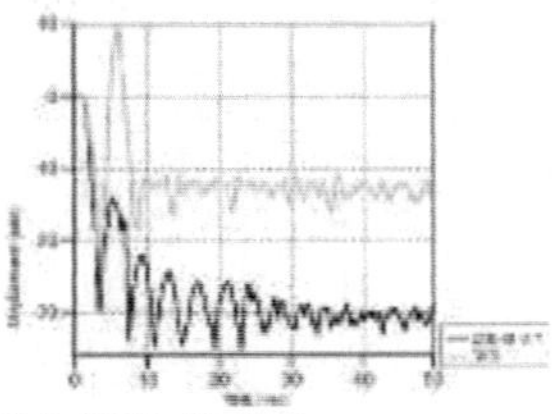
(b) DB-B 0.7

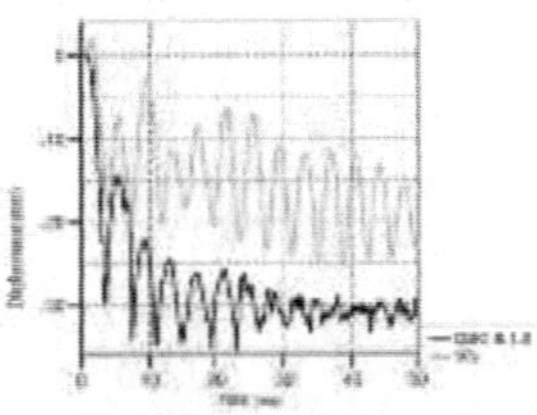
(c) DBC – B 1.0

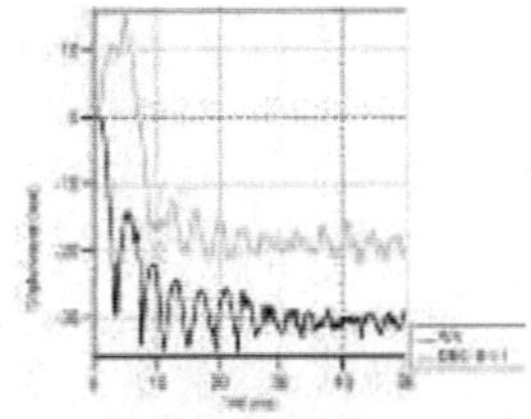
(d) DBC - B 0.7

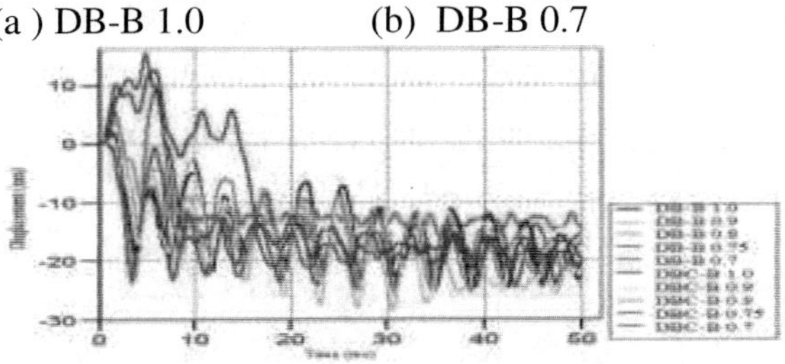
(e) Comparison of the time displacement graph

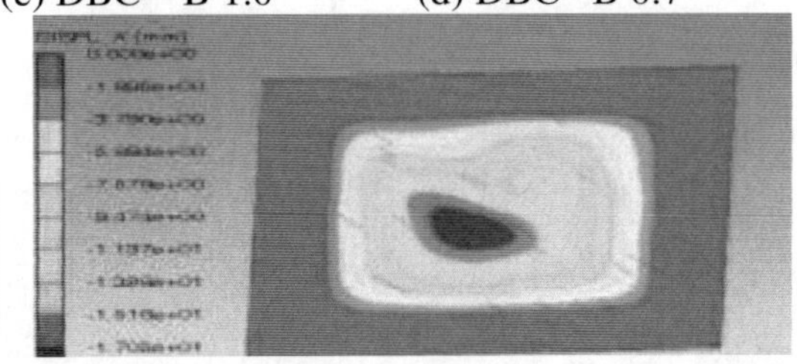
(f)Deformation profile of stiffened plate in DB-B 0.8

Figure 6: displacement time graph

The stress acting on the plate varies with the change in the position of the yield line. With the change in the peak stress values the deformation varies. The plate without stiffeners has maximum stress concentration at the center leading to maximum deformation causing fracture. These could initiate the crack propagation along the yield lines, causing failure. Within our stiffened models, the CB-B 0.9 has maximum deformation while the stress concentration is maximum towards the center, this is due to the low bias ratio. Moreover, specimen CBC-B 0.75 has minimum deformation, and as seen from stress contours, the normalized stresses are distributed along the stiffened plates and along the edge of the plate. It has been seen that about 10% decrease in stress concentration was achieved with diagonally biased stiffeners plates compares with WS plates.

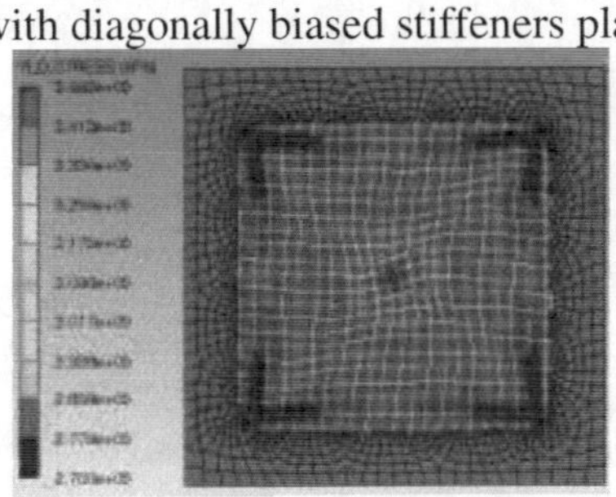
(a)

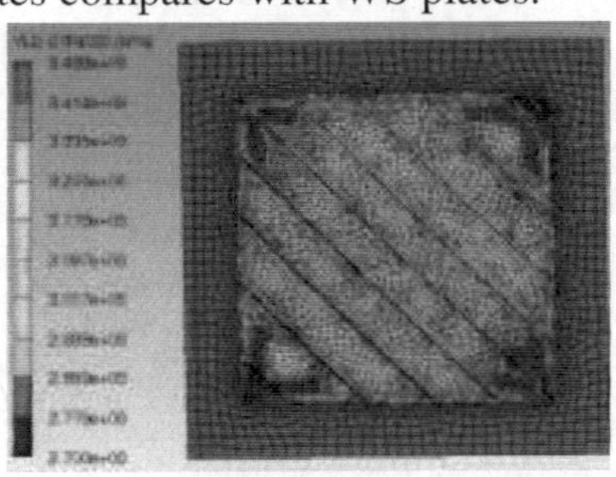
(b)

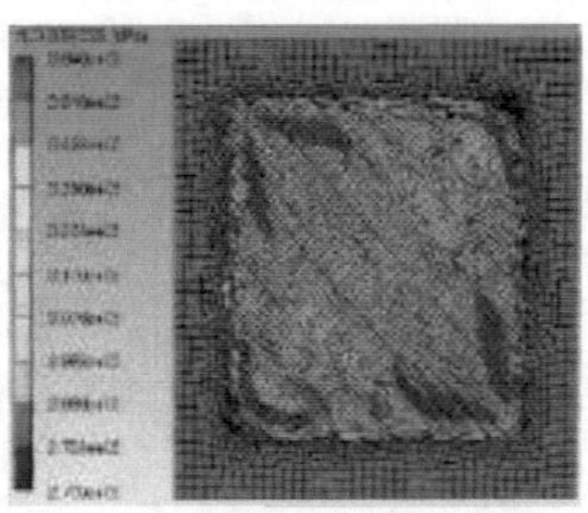
(c)

Figure 7: (a) Stress profile of stiffened plate in WS (b) Stress profile of stiffened plate in DB-B 0.9 (c) Stress profile of stiffened plate in CBC-B 0.7.

4. Conclusion

This paper aims to investigate the behaviour of the square stiffened steel plate under confined blast loading through numerical formulation. From the results it was observed that with the increase in stiffener bias or with the reduction in the spacing between the stiffener-plates for the same volume of steel, the deflection can be reduced. In general for an unstiffened plate, the deflection is maximum at the center; by increasing the stiffeners at the center the deflection can be reduced drastically. From the results it is seen that maximum variation in the residual deflection for various bias, reduces to less than 10%, However, with reduced spacing the vibration on the plates can be damped faster and hence can give more fatigue life to the specimen plates. Conclusively, the optimum bias ratio is 0.7 and the

IOP Conf. Series: Materials Science and Engineering **396** (2018) 012009 doi:10.1088/1757-899X/396/1/012009

stresses decrease by about 10% for the diagonally biased stiffeners leading to a reduced deformation and increased vibration damping for these plates.

Reference

[1] C Zheng , X Kong , W Wu , F Liu 2016 The elastic-plastic dynamic response of stiffened plates under confined blast load, International Journal of Impact Engineering 141–153

[2] Olson M D ,Nurick G N , Fagnan J R, Levin A 1995 Deformation and tearing of blast loaded stiffened square plates. Int J Impact Eng,273–91.

[3] C Yuen S, Nurick G N 2005 Experimental and numerical studies on the response of quadrangular stiffened plates. Part I: subjected to uniform blast load. Int J Impact Eng;31(1):55–83.

[4] Yuen S C K ,Langdon G S , Nurick G N 2005 Experimental and numerical studies on the response of quadrangular stiffened plates. Part II: localised blast loading. Int J Impact Eng 2005:85–111.

[5] Peles S, Neuberger A, Rittel D 2009 Spring back of circular clamped armor steel plates subject to spherical air-blast loading. Int J Impact Eng:53–60.

[6] Kong X, Wu W, Li J, Chen P, Liu F 2014 Experimental and numerical investigation on a multi-layer protective structure under the synergistic effect of blast and fragment loadings. Int J Impact Eng :146–62.

[7] Schleyer G K, Hsu S S 2000 A modelling scheme for predicting the response of elastic–plastic structures to pulse pressure loading. Int J Impact Eng:759–77.

[8] Schleyer GK, Hsu SS, White MD, Birch RS 2003 Pulse pressure loading of clamped mild steel plates. Int J Impact Eng :223–47.

[9] Jones N 1997 Structural impact. Cambridge: Cambridge University Press; 1989 (Paper back edition 1997).

[10] Jones N. 2012 Impact loading of ductile rectangular plates. Thin Wall Struct:68–75.

[11] Jones N 2014 Dynamic inelastic response of strain rate sensitive ductile plates due to large impact, dynamic pressure and explosive loadings. Int J Impact Eng;74:3–15.

[12] Jiang J, Olson M D. 1995 Rigid-plastic analysis of underwater blast loaded stiffened plates. Int J Mech Sci:843–59.

[13] Yankelevsky DZ 1985 Elasto-plastic blast response of rectangular plates. Int J Impact Eng:107–19.

[14] Imam B M, Collins J 2013 Assessment of flat deck metallic plates-yield line and membrane analyses. J Construct Steel Res;82:131–41.

[15] Nurick G N, Martin J B 1989 Deformation of thin plates subjected to impulsive loading – a review, part II: experimental studies. Int J Impact Eng;8:171– 86.

[16] Nurick G N, Martin J B 1989 Deformation of thin plates subjected to impulsive loading – a review, part I: theoretical considerations. Int J Impact Eng:159–70.

[17] K Nahshon, M G Pontin, A G Evans, J W Hutchinson , W Zok 2007 dynamic shear rupture of steel plates, journal of mechanics of materials and structures Vol. 2, no. 10

Falling weight impact response of prestressed concrete slabs with manhole

Sharoan Merin Mathews [1], Nikhil Mohanan [2] and Manish Jose [3]

[1]MTech Student, Department of Civil Engineering, SJCET, Palai, India
[2]Research Engineer, FEM CFD Research and Development, Kerala, India
[3]Assistant Professor, Department of Civil Engineering, SJCET, Palai, India

E-mail: manishjose@yahoo.com

Abstract. Blast and impact resistance structures have become an indispensable requirement for the design of concrete structures due to the prominent danger imposed. The increasing number of accidents and global terrorism on public civil structures pose a serious threat to the infrastructure and human life. It has been seen that prestressing of concrete can improve the impact resistance of the structures. The objective of this paper is to investigate the response of reinforced concrete slabs, prestressed in its transverse direction by steel wires of 7 mm diameter against impact load imposed by a hammer falling from 5 m height tested and calibrated using ANSYS Autodyn software package. The control slab is compared with slabs having opening/ cut-out of different aspect ratio and at different positions. Three different sizes for the hole/opening is considered here provided the area of the opening/hole is kept constant. Post analysis, the most optimum size of the hole was found out to be 500 mm x 500 mm and its position was at the edge of the slab.

Keywords: Prestressing, reinforced concrete slab, impact load, ANSYS Autodyn

1. Introduction

The major threat posed on our built environment is by impact loads which give potentially dangerous consequences. The damaging impact loads usually include vegetation, automobiles, or other objects hurled toward built structures during hurricanes, large vehicular impacts on bridge-piers, impact loads from explosions, ship impact, aeroplane impact, rock-fall impact, or missile impact etc. The major challenge faced by modern engineers is to mitigate the detrimental effects of impact loads on the structural integrity of the infrastructure and thereby making it more resilient [12]. Many studies have proven that the load carrying capacity and toughness of concrete specimens can be increased by increasing the percentage of steel fibre, by providing a high percentage of reinforcement [11], and also by changing reinforcement configuration [3,7,10]. In contrast to the alteration in reinforcement arrangement, the energy absorption capacity and damage resistance of reinforced concrete (RC) slabs are not much influenced by the increase in reinforcement ratio [3,10] despite the increase in the drop height [8].

IOP Conf. Series: Materials Science and Engineering **396** (2018) 012010 doi:10.1088/1757-899X/396/1/012010

In another study, two prestressed concrete rock-shed frames were given a heavy impact by dropping a mass of up to 5000 kg to investigate it's the impact resistance capacity [2]. The rigid connection at the column junction of 'T' shaped fully-rigid frame resulted in its higher energy absorption capacity when compared to the inverted 'L' type frame ('Γ' shaped).

The examination of available literature revealed that the attempts to investigate the structural behaviour of prestressed concrete are very rare. A few studies reported on prestressed concrete are limited to the dynamic response of prestressed concrete sleepers under single and repeated impacts [1,4,5]. In the present study, the effect of impact loading on reinforced concrete slabs with openings of different aspect ratio at different positions and which is prestressed by steel wires of 7 mm diameter in the transverse direction is studied here.

2. Methodology

2.1. Experimental Study

A prestressed concrete square plate of size 800 mm and thickness 100 mm was cast using concrete mix proportion 1:1.74:1.68 (C: FA: CA). The expected characteristic compressive strength of concrete was 40 N/mm^2.The water-cement ratio used was 0.35. The final strength achieved after 28 days of curing was 48.4 N/mm^2 with a standard deviation of 1.60. High Yield Strength Deformed bars of 8 mm diameter at 140 mm spacing was used in both main and transverse spans of the concrete plates. Also,10 numbers of 4 mm diameter prestressing wires were provided at a spacing of 80 mm center to center with an eccentricity of 25 mm from the center of the plate thickness (figure 1). These prestressing wires were pre-tensioned by pulling the wires with the help of a 50t capacity hydraulic jack before casting the concrete plates to obtain a stress of 10% of unconfined compressive strength (48.4 N/mm^2). Impact test was performed on the plates using a heavy steel mass of 243 kg (figure 2) by dropping it from a height of 1 m. Displacement sensors were placed in the center and quarter span of the bottom surface of the plate to measure the displacement. This experiment was conducted by Vimal Kumar *et.al.* funded at Department of Civil Engineering, Indian Institute of Technology, Roorkee [11].

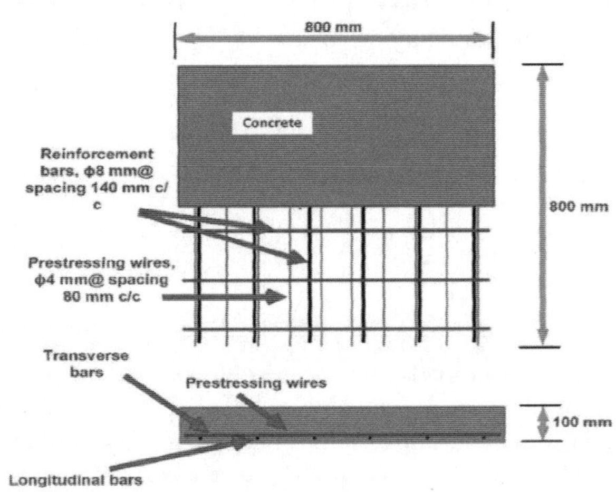

Figure 1. Detailing of reinforcement bars in prestressed concrete plate [11]

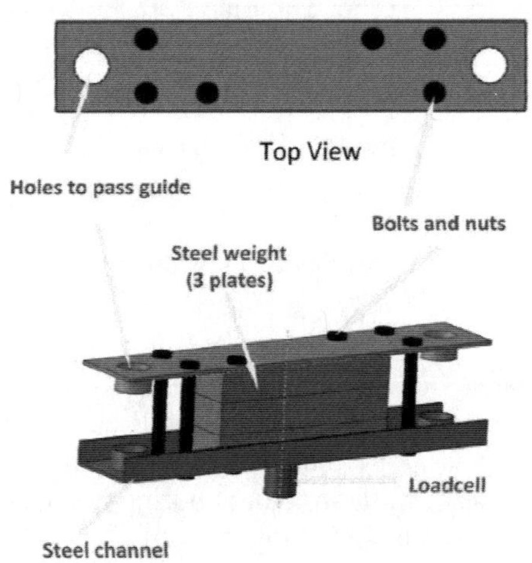

Figure 2. The detail of weight assembly used in the experiment [11]

The maximum displacement obtained at the center span of the plate is 16.09 mm (figure 3).

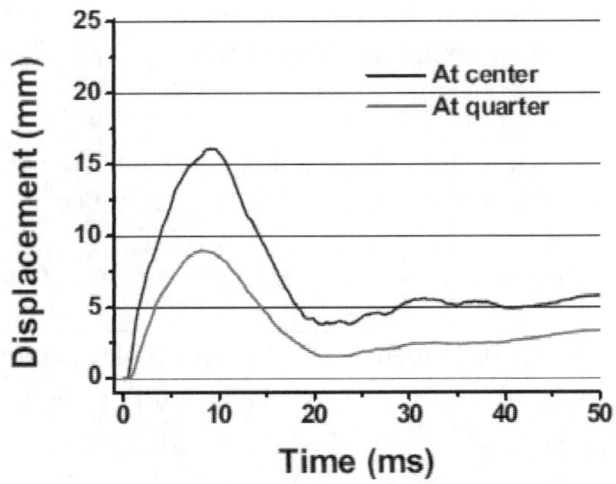

Figure 3. Displacement Vs Time graph obtained from experiment [11]

2.2 Finite element modelling for experimental calibration

A prestressed concrete square plate of 800 mm size and 100 mm thickness was modelled using design modeler in ANSYS workbench (figure 4). The reinforcement details were given as per the experiment (figure 5). Material properties of concrete such as compressive strength 48.4 N/mm^2, Young's modulus of 34785.05 N/mm^2 and Poisson's ratio of 0.18 is given.

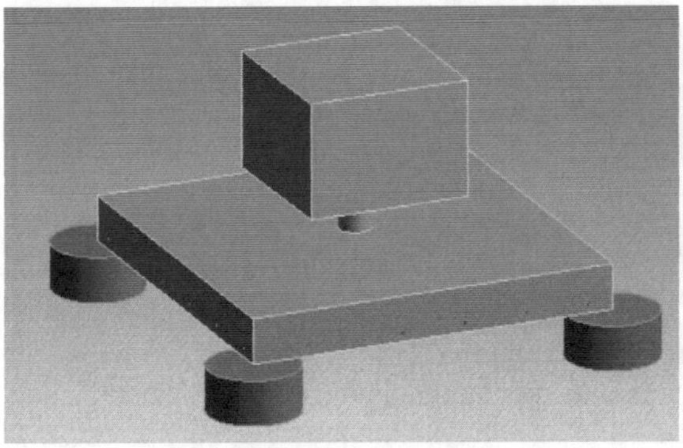

Figure 4. Model Geometry

Figure 5. Reinforcement details

Mesh convergence study was performed and the optimum mesh size obtained was 15 mm. Contacts between reinforcements and concrete were given. Prestressing of concrete was done in static structural by bolt pre-tensioning method. The load given to obtain 10% prestress was 59769 N. The boundary conditions given for prestressing are (i) Rotational constraint for prestressing wires and main reinforcements (ii) the sides of the concrete plate along the direction of prestressing wires were free to move along the direction of prestressing and in vertical direction (iii) the other two sides are free to move only in the direction of prestressing (iv) standard earth gravity is given. Average normal stress for the whole structure after prestressing was obtained as 4.84 MPa approximately 10% prestressed.

The impact analysis is done using ANSYS Autodyn. To reduce the time for solving, the hammer was moved to a position just above the concrete slab and an initial velocity of 4.4274 m/s is given for the hammer to generate the required momentum. The analysis end time was set to 20 ms. Material model used for concrete is Standard RHT Concrete model. Steel used for reinforcements are modeled using Johnson Cook strength model. The supports along with the hammer are modeled with Rigid EOS with the density equal to that of structural steel. The major boundary conditions for this analysis are that all fixed condition to the supports, lateral constraints on the Hammer and the standard earth gravity to all bodies. Subsequently, the hammer is free to move only in vertical direction. Gauges are set at center and quarter span to measure the displacements and strain rates. In addition, the top edges are constrained not to move in the vertical direction as well according to the experimental setup. The results from Autodyn showed the maximum deformation of the plates at the center span as 16.908 mm.

The values of displacement obtained from experiment and from the software were compared. The comparison data is shown in the table below (table 1) and the calibration results are shown in figure 6.

Table 1. Comparison of experimental to numerical model.

Method	Maximum Displacement	Maximum deflection Error %
Experiment	16.090 mm	Error %= [(16.908-16.09)/16.09]x 100
Autodyn Software	16.908 mm	= 5.0839%

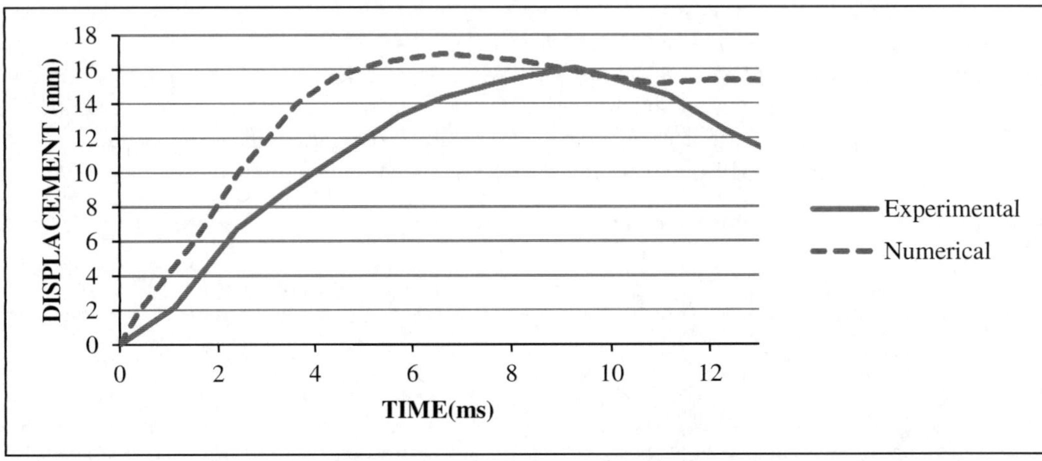

Figure 6. Comparison of experimental and numerical time deflection curves

From the FE model, we see that there isn't much variation while comparing experimental and numerical results and the error margin obtain was 5.0839%. Hence the mesh convergence size and the material properties used in the calibration model can be used for the proposed study.

2.3 Numerical modelling of prestressed slabs with manholes

The mesh size, material properties and boundary conditions used in the proposed study is the same as that used in the calibration model. A one-way concrete slab with dimensions of 3000 mm length, 1000 mm width and 150 mm thickness, was modelled. 10 mm diameter bars of 11 numbers are provided as the longitudinal reinforcements and 10 mm diameter bars at 250 mm center to center spacing are provided as the transverse reinforcements [6]. The prestress wires are given according to the ratio of the volume of main reinforcements to the volume of prestressing wires as per the experimental setup. Therefore 18 numbers of 7 mm diameter wires along the transverse direction, i.e., each one has a length of 1000 mm are used as prestressing wires. A total of 10 number of models of the slab were modelled, which include a control slab without opening, and the rest 9 models having an opening of different aspect ratio at three different positions. The area of the opening is kept constant. The three different sizes are 500 mm x 500 mm, 555.556 mm x 450 mm and 625 mm x 400 mm. The positions considered are [i] center of the slab [ii] with 3 prestress wires to the shorter span [iii] with 6 prestress wires to the shorter span. The prestressing analysis was carried out for all the models in the same way as in the calibration process. The designation, prestress wire spacing, hole size, position, and average prestress value at midplane is given in the table below (table 2).

Table 2. Model details.

Model No.	Designation	Prestress wires spacing (mm)	Hole Size (mm)	Position of hole	Average Prestress value at midplane of concrete (N/mm^2)
1	T174	174.118	No hole		4.84
2	T151-500X500M	151.25	500 x 500	Middle	4.84
3	T151-500X500Q	151.25	500 x 500	Quarter	4.84
4	T151-500X500E	151.25	500 x 500	Edge	4.84
5	T147.5-555.556X450M	147.5	555.556 x 450	Middle	4.84
6	T147.5-555.556X450Q	147.5	555.556 x 450	Quarter	4.84
7	T147.5-555.556X450E	147.5	555.556 x 450	Edge	4.84
8	T143-625X400M	143.375	625 x 400	Middle	4.84
9	T143-625X400Q	143.375	625 x 400	Quarter	4.84
10	T143-625X400E	143.375	625 x 400	Edge	4.84

Rigid supports are given on opposite sides of the slab. The hammer has dimensions of 0.15 x 1.25 x 0.75 m [9]. It is dropped from a height of 5 m. The impacting surface of the hammer is hemispherical. The material used is rigid structural steel. The figure below shows some of the models and reinforcement details (figure 7).

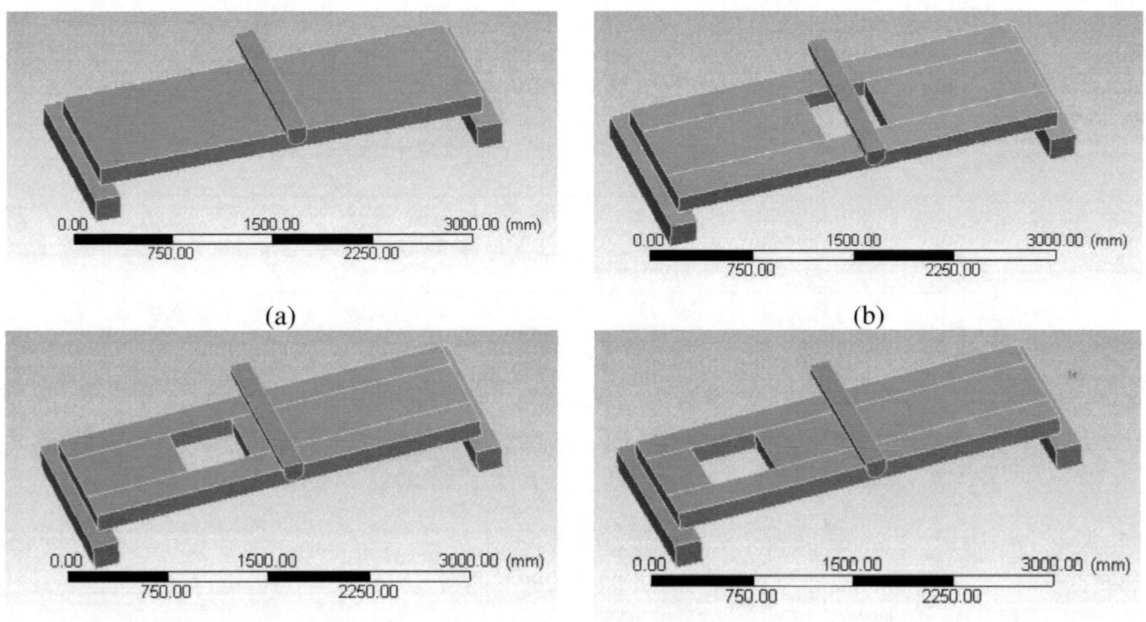

Figure 7. (a) T174 (b) T151-500X500 M (c) T151-500X500Q (d) T151-500X500 E

The impact analysis conditions are also same as that of the calibration model. The analysis end time is set as 40 ms. The initial velocity of the hammer for 5m is given as 9.9028 m/s. All the material properties are the same. The Autodyn model of T174 is shown below (figure 8).

IOP Conf. Series: Materials Science and Engineering **396** (2018) 012010 doi:10.1088/1757-899X/396/1/012010

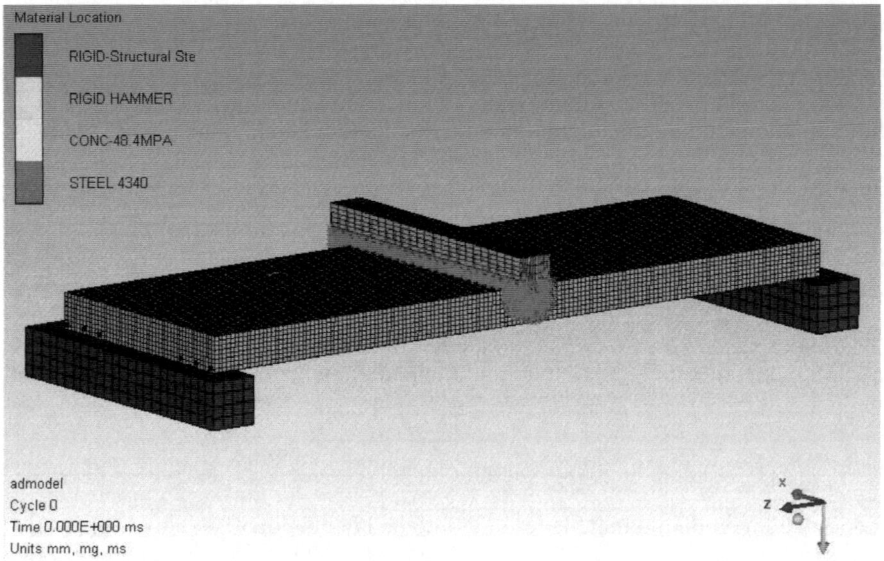

Figure 8. Autodyn model of T174

3. Results and Discussions

The displacement versus time history of all the models is obtained from the output of Autodyn. The table below gives the maximum displacement obtained from the gauge lying below and in line with the hammer (table 3).

Table 3. Maximum Displacement obtained for models.

Model No.	Designation	Maximum Displacement (mm)
1	T174	33.049
2	T151-500X500M	53.332
3	T151-500X500Q	54.647
4	T151-500X500E	31.640
5	T147.5-555.556X450M	57.833
6	T147.5-555.556X450Q	54.088
7	T147.5-555.556X450E	31.702
8	T143-625X400M	46.203
9	T143-625X400Q	47.998
10	T143-625X400E	31.705

The obtained output, displacement Vs time graphs of each hole size are plotted in excel and compared with the base model (figure 9 (a-c)).

IOP Conf. Series: Materials Science and Engineering **396** (2018) 012010 doi:10.1088/1757-899X/396/1/012010

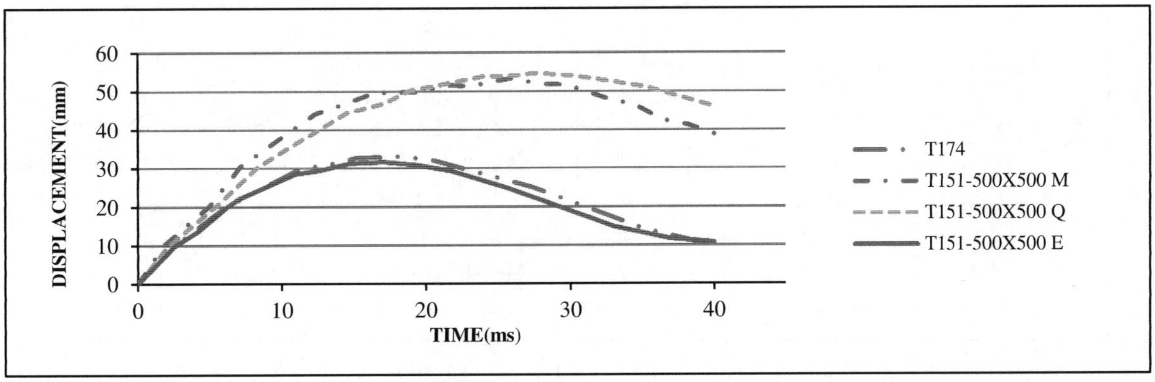

(a) Comparison of T174 with T151-500x500 at different positions

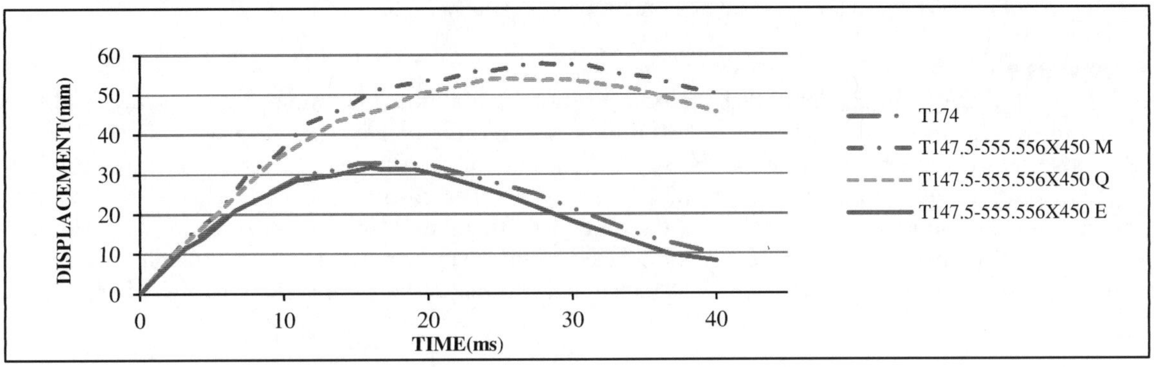

(b) Comparison of T174 with T147.5-555.556 x450 at different positions

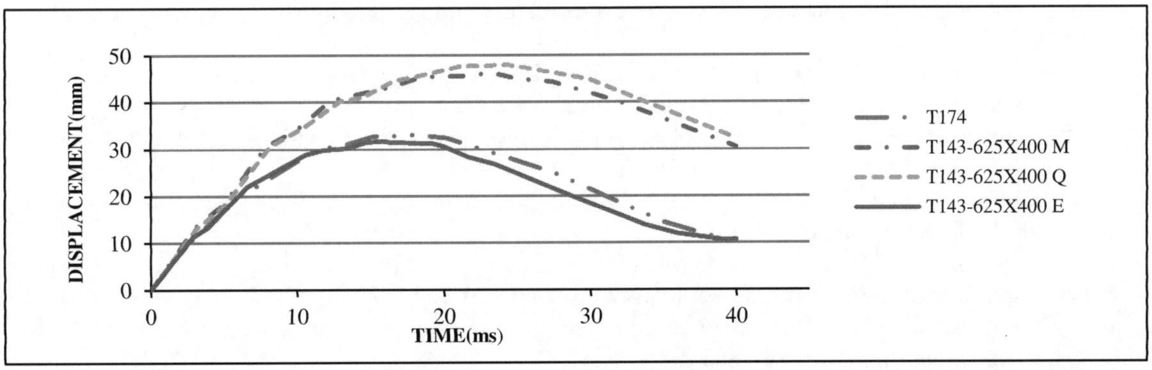

(c) Comparison of T174 with T143-625x400 at different positions

Figure 9. Comparison of displacement Vs Time graph of different models at different position (a-c)

From the comparison, we can understand that in all cases edge position is best suited for providing a hole, because it showed minimum deflection in all cases. In order to find out the suitable size for the hole, all the edge hole graphs of various sizes are compared (figure 10).

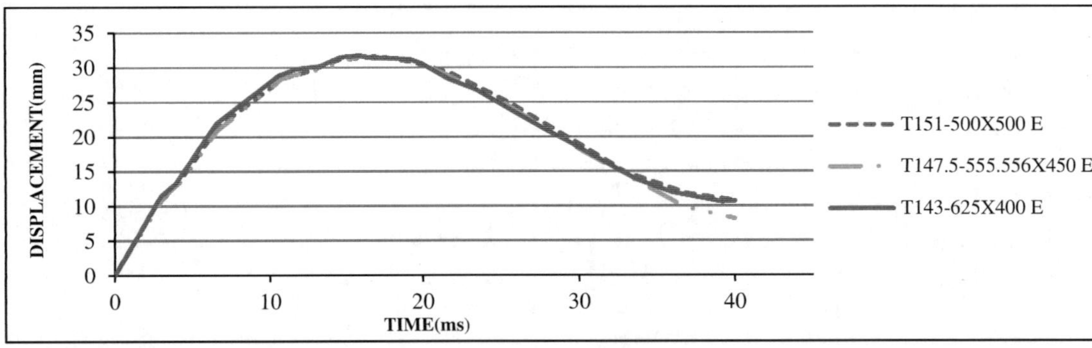

Figure 10. Comparison of Displacement Vs Time graph of edge hole for various sizes

From the comparison, we find that the results are nearly the same, however, the hole size of 500 mm x 500 mm showed the least displacement progressively.

4. Conclusion

From this study, we can conclude that the best position for the hole/opening is edge position and the optimum size for the hole is 500 mm x 500 mm or square hole of any size. The possible reason for edge position to be the best position is that, in this case the prestress energy is used up entirely by the impactor with more central cross section, while in the other cases the cross-sectional area where prestress is concentrated is less and hence they are less effective. Also, the entire energy is absorbed by the concrete surrounding the hole when impacted head-on, whereas, in the case of the hole near the support, most of the energy is transferred to the support instead of being absorbed by the concrete near the hole.

5. References

[1] Ye X, Wang N, and Mindess S 1994 Effect of loading rate and support conditions on the mode of failure of prestressed concrete railroad ties subjected to impact loading *Cem. Concr. Res.* **24** (7) 1286–98

[2] Kishi N, Konno H, Ikeda K, and Matsuoka K G 2002 Prototype impact tests on ultimate impact resistance of PC rock-sheds *Int. J. Impact Eng.***27** 969–85

[3] Zineddin M and Krauthammer T 2007 Dynamic response and behavior of reinforced concrete slabs under impact loading *Int. J. Impact Eng.* **34** 1517–34

[4] Kaewunruen S and Remennikov A M 2009 Impact capacity of railway prestressed concrete sleepers *Eng. Fail. Anal.***16** 1520–32

[5] Kaewunruen S and Remennikov A M 2009 Progressive failure of prestressed concrete sleepers under multiple high-intensity impact loads *Eng. Struct.***31** 2460–73

[6] Anil O, Kaya N, and Arslan O 2013 Strengthening of one-way RC slab with opening using CFRP strips *Const. and Build. Mat.***48** 883–93

[7] Adhikary SD, Li B, and Fujikake K 2014 Effects of high loading rate on reinforced concrete beams ACI Struct. J. **111** (3) 651–60

[8] Zhan T, Wang Z, and Ning J 2015 Failure behaviors of reinforced concrete beams subjected to high impact loading *Eng. Fail. Anal.* **56** 233–43

[9] Qasrawi Y, Heffernan P J, and Fam A 2015 Numerical modeling of concrete-filled FRP tube dynamic behavior under blast and impact loading *J. Struct. Eng* 0733-9445/04015106(1-13)

[10] Othman H and Marzouk H 2016 An experimental investigation on the effect of steel reinforcement on impact response of reinforced concrete plates *Int. J. Impact Eng.***88** 12–21

[11] Kumar V, Iqbal M A and Mittal A K 2017 Experimental investigation of prestressed and reinforced concrete plates under falling weight impactor Article in *Thin-Walled Structures* 1-11

[12] Ranade R, Li V C, Heard W F, and Williams B A 2017, Impact resistance of high strength-high ductility concrete *Cem. Concr. Res.***98** 24–35

Parametric Study of Longitudinal Hollow Steel Fibre Reinforced Concrete (SFRC)Beams

Bibi M Jacob[1], Bincy S[2]

1 PG Student, Structural Engineering and Construction Management, St Josephs College of Engineering and Technology

2 Assistant Professor, Dept. of Civil Engineering, St. Joseph´s College of Engineering and Technology, Palai, Kerala, India

E-mail: bibijacob01@gmail.com, bincy@sjcetpalai.ac.in

Abstract. The main aim of this study is to investigate on the crack patterns and flexural strength of simply supported RC rectangular hollow beams and SFRC rectangular hollow beams (with a circular hole in longitudinal direction) under a loading frame with a gradual increasing load. The experimental program consists of casting and testing of 4RC and 4 SFRC beams of size 1400 x 150 x150. M 25 grade of concrete was used for the casting of all the beams. Hollow beams were made using pipes of diameters 32mm, 40mm and 50mm and the position of placing the pipes are at a distance of 3 to 4 cm from bottom reinforcement. All other properties of the beam were kept constant such as the cross section of the beam and the amount of longitudinal steel. In this study we also focus on the material management by the introduction of a hollow core. By this material optimization we can decrease the dead loads which contributes to the seismic effect in high rise buildings. The influence of area of hole is taken as reference for the study. A comparison of experimental results of RC and SFRC were made to predict the crack behaviour and the flexural load carrying capacity of hollow beams.

Keywords: Hollow beams, Circular hole, Steel fibre reinforced concrete, Crack pattern

1. Introduction

Concrete is a versatile building material that it is only second to that of water according to the per capita consumption in the world. Understanding the basics and fundamentals of concrete is necessary to produce a good quality concrete. The term concrete relate to a mixture of cement, sand, aggregate held together into a pasting substance. The paste is made up of substances such as cement and water and can also contain substantial materials like

IOP Conf. Series: Materials Science and Engineering **396** (2018) 012011 doi:10.1088/1757-899X/396/1/012011

superplasticizers. In this experimental study we have used FOSROC SP 430 as the superplasticizer. Every year billions of buildings are being built up such as industrial buildings, commercial buildings, dwelling units, bridged etc. In these buildings another billions of RC beams are being built. Although a good information is available about the strength of normal RC beams, no valid datas are available about the strength of hollow RC and hollow SFRC beams.

Introduction of a longitudinal hole in the beam helps in reducing the self weight of the beams. As a result of this, it helps in reducing the size of the structural members which supports the beams such as columns and foundations, thus leading to the material saving in large amounts. In addition to these advantages other utility services can be laid through these without much difficulty such as telephone lines, electrical lines, sewage ducts and water service lines whenever necessary. In the RC beams, the hollow section can be made in a shapes such as circular, trapezoidal, triangular, diamond and many irregular shapes. But circular and rectangular holes are more commonly seen.

The properties of the normal concrete can be further be improved in terms of workability, durability and strength characteristics from reducing porosity by compaction, improved characteristics of cement paste and by increasing the aggregate paste bond. In addition to this, the strength can be further improved by the introduction of fibres into the normal mix. The fibres used can be steel, carbon or glass fibres. These type of concrete made are named as fibre reinforced concrete. It has being revealed that the reinforced concrete with a acceptable amount of fibre acquires large strength and better performance in terms of their flexure, toughness, compression, in which the degree of increased performance lies in the type of fibre used. Steel fibre is a versatile fibre used in constructions by the majority of architects and builders.

Advantages of hollow cross section

- Reduced the weight, which helps in reducing the cost of transportation, ease of handling and assembling of the specimen.
- Substantial reduction in material quantities, the materials required are much less than that used for conventional systems.

2. Objectives and Scope

The main aim of this experimental study is to:

- To study the flexural behavior of hollow RC and SFRC beams of different diametrical holes.
- To determine the ultimate load carrying capacity of the hollow RC and SFRC beams.
- To study the crack patterns of the different beams.
- To make an investigation on the relation between the diameters of the holes to that of the occurrence of the cracks.

3. Methodology

To study the behaviour of hollow reinforced concrete beams and hollow steel fibre reinforced concrete beams. This research will help us to focus on the material optimization which results in the overall economy of a structure. The methodology which should be followed is discussed in the next paragraph.

Properties of materials such as sand, aggregate, cement are being found. It is very important to find out the properties as they are the ones which impart strength and durability to the concrete. Then the mix design is formulated. It is being formulated using IS – 10262. The average compressive strength of concrete at 7 days and 28 days are found. This provides us with the information whether the mix designed gave us with an expected strength as that provided by M 30 concrete. If this strength is upto the limit, then the design mix can be used for the beam. If otherwise, redesign the mix. Next reinforced beams and the fibre reinforced beams are casted as simply supported beams. The curing of the beams are being done for 28 days. The composite beams are tested using two-point loading. Atlast the formulation and interpretation of results is done.

4. Materials

The fundamental components of concrete mix are water, cement, coarse aggregate and fine aggregate. Addition of admixture with concrete mix will helps to increase its properties. For this thesis work superplasticizer is used as admixture, which helps to maintain workability without adding more water

4.1 Cement

Cement is the binder material which starts to set when mixed with water which causes a series of hydration chemical reactions. The constituents will slowly hydrate and the mineral hydrates solidify; the interlocking of the hydrates gives cements its strength. Cement used in the work was PPC (Portland Pozzolana Cement). 53 grade Portland pozzolana Ambuja cement is adopting for experimental programme.

4.2 Fine Aggregate

In this study M sand which belongs to zone II is free from impurities is used. The size of it is lesser than 4.75 mm. The specific gravity and fineness modulus of this fine aggregate where found to be 2.88 and 2.87 respectively. The percentage of passing is within the limits as per IS:383-1970.

4.3 Coarse Aggregate

The coarse aggregate used here is 20mm in size belongs to zone II, crushed angular shape and is free from dust particles. The specific gravity and fineness modulus was found to be 2.79 and 6.93 respectively and the impact value was found to be 9.44 %. The percentage of passing is within the limits as per IS: 383-1970.

IOP Conf. Series: Materials Science and Engineering **396** (2018) 012011 doi:10.1088/1757-899X/396/1/012011

4.4 Fibre

Steel fibres of 30mm length and 0.5mm diameter is being used for the thesis work Fig:1. These fibres provide greater flexibility with various types of Portland cements and admixtures. Its loose and unglued characteristics provides even distribution mix in a fast and convenient manner thus effective in avoiding balling.

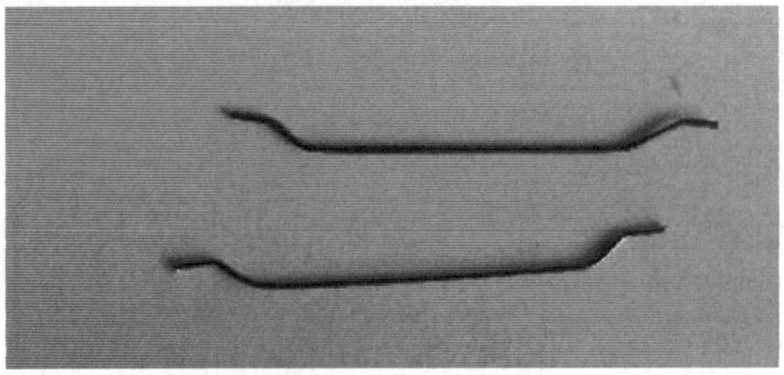

Fig: 1: Hooked end fibre

4.5 Admixture

Admixture used for this project is superplastizer, FOSROC SP430 which is free of chloride. It will help to reduce the amount of water demand and also it will help in early strength gain of concrete. It also increases the permeability and durability of the concrete

5. Experimental Investigation

5.1 The Details of Test Specimens

A total of 8 beams of the dimensions of 1400x150x150mm were casted and tested. The details of the beams are are given in table 1. The beams are having longitudinal holes of diameters 32, 40 and 50 mm. For making the longitudinal holes with ease, frictionless PVC pipes are used and these pipes are tied to the reinforcements as given in Fig:2. The reinforcements were given a cover of 20mm.

Table 1: Test specimen

NO.	BEAM DESIGNATION	DIAMETER OF HOLE(mm)	TYPE OF FAILURE
1	NO PIPE NC	0	FLEXURE
2	32mm PIPE NC	32	FLEXURE
3	40mm PIPE NC	40	FLEXURE
4	50mm PIPE NC	50	FLEXURE
5	NO PIPE FRC	0	FLEXURE
6	32 mm PIPE FRC	32	FLEXURE
7	40 mm PIPE FRC	40	FLEXURE
8	50 mm PIPE FRC	50	FLEXURE

IOP Conf. Series: Materials Science and Engineering **396** (2018) 012011 doi:10.1088/1757-899X/396/1/012011

Fig:2: Reinforcement cage of the beam

5.2 Concrete Saving

Volume of 1 beam, v1 = 1400x150x150 = 31500000 mm^3

Volume of 50mm pipe, v2= π x 25^2 x1400 = 2748893.572 mm^3

% reduction of concrete = (v2/v1) x100 = 8.72%

Since we have assumed for a small beam, the percentage of reduction of concrete is small. But in actual cases, that is for a huge building the percentage of saving will also be more. Hence making the project economical.

5.3 Experimental Setup

All the beams casted were tested in a loading frame of capacity 500 kN and were loaded until the failure of the specimen occurred using a two-point loading system. Also the load was given gradually using a hydraulic jack and were distributed uniformly to the two points using a steel joist. The experimental setup can be seen in the Fig:3.

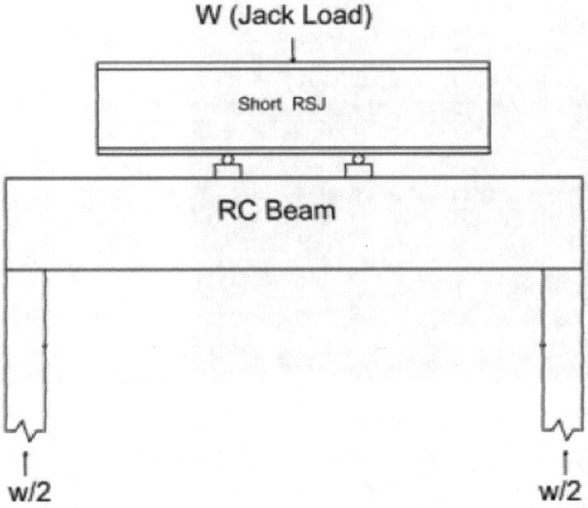

Fig:3 Experimental Setup

IOP Conf. Series: Materials Science and Engineering **396** (2018) 012011 doi:10.1088/1757-899X/396/1/012011

6. Result

6.1 Load Carrying Capacity

The presence of hollow recess in reinforced concrete beams was found to decrease the load carrying capacity by only a small value. Thus these hollow sections can be successively used in structures where relatively smaller load carrying capacity is needed. The decrease in load carrying capacity w.r.t no pipe is shown in Table 2.

Table 2: Decrease in Load Carrying Capacity

PIPE DIAMETER	NC(%)	FRC(%)
32mm	2.5	1.05
40mm	4	2.95
50mm	7.85	8

6.2 Load- Deflection Responses

The load-deflection curves for SFRC are shown in Fig:4. All the beams followed the same pattern of load-deflection response. After yielding of steel reinforcement we can see there an enormous rate of increase in deflection for subsequent loads.

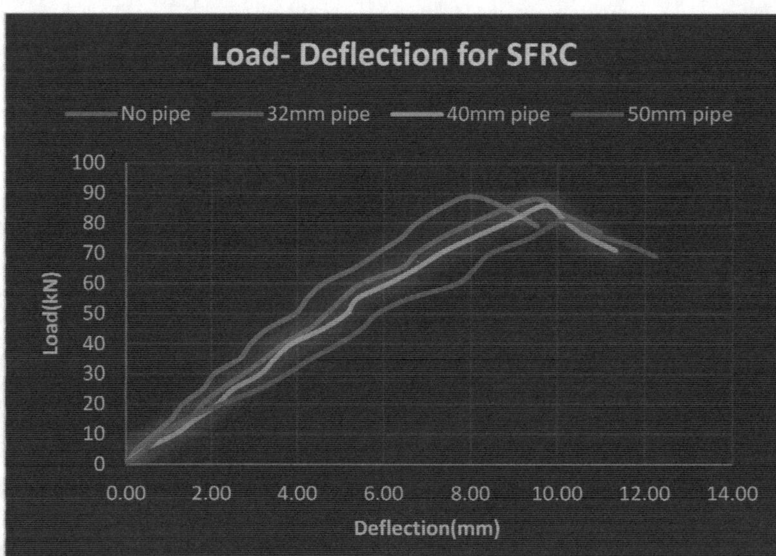

Fig:4 : Load - Deflection Graph for SFRC

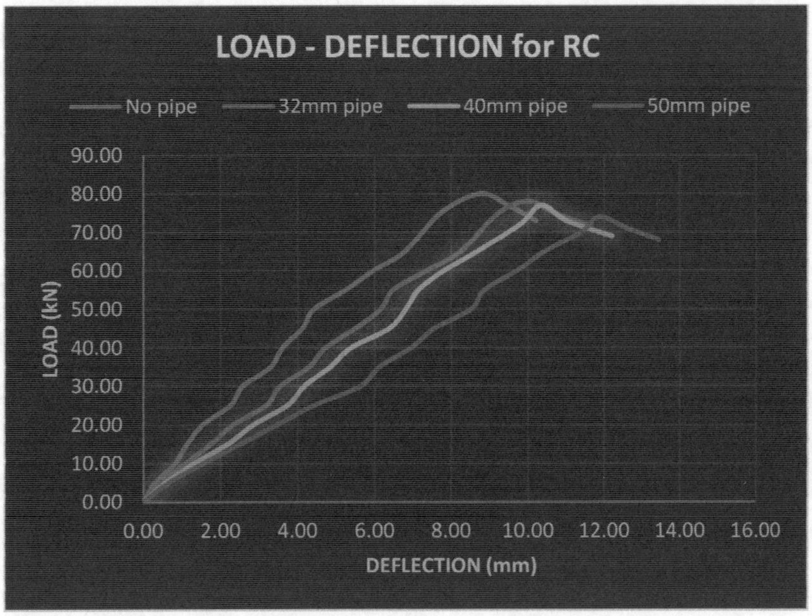

Fig:5: Load - Deflection Graph for RC

6.3 Crack Pattern

Initial stages of loading, all beams were un-cracked beam. When the applied load reached to the rupture strength of the concrete on specimens, the concrete started to crack. The failure pattern in the all the tested beams was observed as a flexure failure. The beams showed initial cracking in the constant bending moment region and then the cracks patterns in the vertical direction as the load was increased further. All the beams showed the same pattern of failure and the failure modes are shown in Fig:5.

Fig:6: Formation of Cracks

7. Conclusion

- In general, the no. of crack is larger in the hollow section than solid one.
- But the addition of steel fibre can reduce the cracks and the load carrying capacity can be tolerated.

- The RC beams and the SFRC beams having a longitudinal hole, failed under flexural mode.
- The beams failed as a result of the initiation of yielding of reinforcing steel and the subsequent crushing of concrete in the compression zone.
- There is scope for taking essential utility services, such as water supply, sewage, air-conditioning ducts, and electrical and telephone cables, through the longitudinal holes of these beams whenever necessary.
- The propagation of crack in hollow beams was relatively slower than RC solid beams. Also the number of cracks is less when compared with cracks in RC solid beams. This may be caused due to the fact in bending effect in tensile zone is restricted by introducing hollow longitudinal hole.
- The presence of hollow section in reinforced concrete beams does not result in any appreciable decrease in the load carrying capacity. And thus these hollow sections can be successively used in structures where relatively smaller load carrying capacity is needed.

References

[1] A.Murugesan, A. Narayanan 2018 *Deflection of Reinforced Concrete Beams with Longitudinal Circular Hole* American Society of Civil Engineers, Volume **23**, Issue 1

[2] N. Parthiban, M.Neelamegam 2017 *Flexural Behavior Of Reinforced Concrete Beam With Hollow Core In Shear Section* International Research Journal of Engineering and Technology (IRJET) ISSN: 2395 -0056 Volume: **04** Issue: 04 | Apr -2017

[3] A.Murugesan, A. Narayanan 2016 *Influence of a Longitudinal Circular Hole on Flexural Strength of Reinforced Concrete Beams* American Society of Civil Engineers, Volume **22** Issue 2

[3] M. Kumar, S. Mangalagi, R. Harsoor 2013 Behavior *of R.C.C. Beam with Circular Opening Strengthened by CFRP and GFRP Sheets* ISSN: 2319-1163 | ISSN: 2321-7308

[4] V.S Satheesh, B. Nyodu 2017 Flexural *Behavior of Hollow RC beam using Glass Fiber* IJSTE/ Volume **3** / Issue 09 / 073

[5] S.Manikandan, S.Dharmar, S. Robertravi 2015 *Experimental Study On Flexural Behaviour Of Reinforced Concrete Hollow Core Sandwich Beams* International Journal of Advance Research In Science And Engineering IJARSE, Vol. No.**4**, Special Issue (01), March 2015

[6] W.A Warvosh, H.N Al- Maliki, M. Munna 2015 Torsional *Behavior of Reinforced Concrete Hollow Core Beams* Journal of Engineering and Development, Vol.**20**, ISSN 1813-7822

[7] A.J.H Alshimmeri, H.N.G Al-Maliki 2014 *Structural Behavior of Reinforced Concrete Hollow Beams under Partial Uniformly Distributed Load* Journal of Engineering, Volume **20**

[8] Jain Joy, Rajesh Rajeev 2014 *Effect of Reinforced Concrete Beam with Hollow Neutral Axis* (IJSRD/Vol. **2**/Issue 10/2014/099)

IOP Conf. Series: Materials Science and Engineering **396** (2018) 012015 doi:10.1088/1757-899X/396/1/012015

Study of Traffic Flow at a Three Legged Busy Intersection in Vellore, India

S Vasantha Kumar, P. Priyadharshini, B. Dhanalakshmi and K. Settu

School of Civil and Chemical Engineering, Vellore Institute of Technology (VIT), Vellore – 632014, Tamil Nadu, India.
Email: svasanthakumar@vit.ac.in

Abstract. The rapid growth in economy has resulted in an intense increase of personal vehicle usage which resulted in road traffic congestion in all metropolitan cities like Chennai, Bengaluru, etc. As Vellore has a good transportation network, it experienced a substantial urban growth especially after it becomes a city corporation in 2008. One of the ill or side effects of urbanization is traffic bottlenecks and before suggesting remedies to address the traffic problems, understanding the current situation is very important. Analysis of traffic flows or volumes at busy intersections in order to find the mode share percentage, directional distribution and peak period traffic volume is the first and foremost step to understand the present situation of traffic. Hence in the present study, a busy intersection in Vellore called Chittor bus stand intersection near VIT was selected and volume data was collected through video surveys. The collected flow or volume data was carefully examined to identify the modal share, directional distribution and morning & evening peak hour traffic volume. The results showed that the two wheelers and three wheelers comprise 82% of the total volume and the proportion of cars is only 11%. The traffic data analysis showed that the light commercial vehicles (LCV) and heavy motorized vehicles (HMV) constitute 7% of the total traffic. A highest hourly volume of 7573 PCU's was observed between 9.30 am and 10.30 am during morning and 7336 PCU's between 5.40 pm and 6.40 pm in evening. By taking a 10% growth in traffic every year, then in the year 2021, the traffic volume at the intersection may reach 10,000 PCU's and may warrant for a flyover or grade separator at the intersection.

1. Introduction

Migration of people to urban centres due to economy growth in India has resulted in traffic problems in almost all the cities in our country. People who live in cities are expected to increase from 3500 million in the year 2010 to 6300 million in 2050 [1]. This twofold increase will lead to many problems in all the major cities of the world and similar situation exists in India too. Tamil Nadu stands next to Gujarat and Maharashtra as 50% of the state population are living in cities. The major reason for traffic congestion in cities is the lesser use of public transportation and increase in usage of personal vehicles such as cars and two-wheelers. In Chennai alone, private or personal vehicles have experienced a significant increase from 1 million in 1999 to 4.2 million in 2017. Similar to Chennai, other cities in Tamil Nadu like Trichy, Salem, Madurai, Coimbatore, Vellore, etc. were experiencing considerable urbanization in recent years which resulted in personal vehicular growth. As Vellore is located close to the capital city of Chennai and good transportation network, it experienced a substantial urban growth. For any city, it is essential to have a circular ring road or bypass road around the city in order to separate the bypassable traffic from city traffic. However in Vellore, still now there is no such ring or circular road which can connect the important national highways that passes through the Vellore city. Hence all trucks and trailers are using the roads which are within the city limits and hence create traffic congestion during peak hours in one of the arterial road in Vellore, namely, Vellore – Katpadi road.

In order to address the traffic congestion problem in Vellore – Katpadi road, it is very essential to first understand the present context, i.e., first we need to study the traffic volumes at intersections along the route to get an idea of modal share of different vehicles, peak hour traffic volume and direction-wise distribution. This traffic volume study also will help us to check whether a traffic signal or a flyover (grade separator) is required at the intersection. Lal et al. [2] collected the traffic and pedestrian data at one of the busy intersections in Ettumanoor in Kottayam, Kerala and proposed various traffic improvement plans/measures to increase the vehicular flow at the junction. Errampalli and Kayitha [3] conducted traffic volume studies at busy intersections and proposed traffic management plan for the city of Port Blair in India. Many such studies were carried out in India and abroad. In the present work, one of the important intersections in Vellore – Katpadi road, namely Chittor bus stand intersection near VIT was took and traffic volume data was collected through videographic surveys. The collected video data was analyzed to get the modal share of different vehicles, traffic volume in peak hour and directional distribution. More details about the study location are given in the following section.

2. Details of study area

The Chittor bus stand junction is one of the important and busy intersections in Vellore city and is located 500m from Katpadi railway junction and 2 km from VIT University. In Fig.1, Google map of the intersection with some important landmarks in the city such as old and new bus stand, Katpadi railway junction, VIT University and CMC hospital are shown.

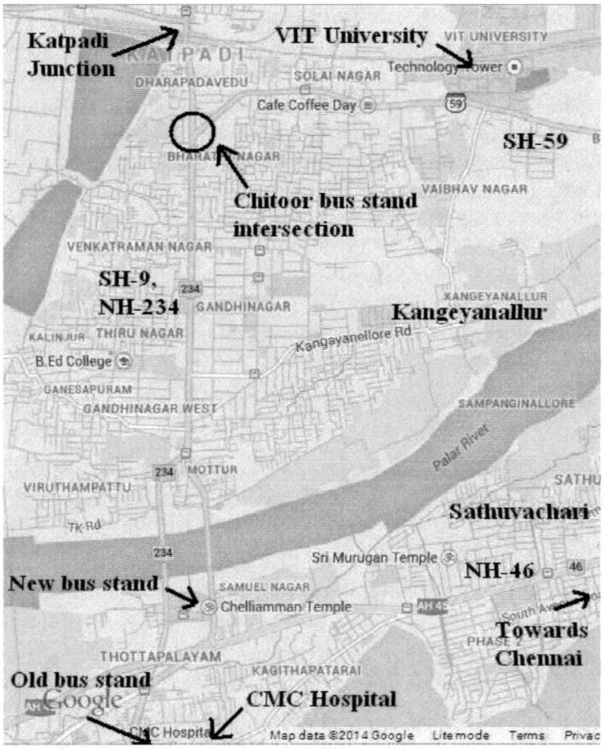

Fig.1 Google map showing the study location

The SH-59 (Thiruvalam-Venkatagirikottah Road) intersects SH-9 (Cuddalore-Chittor road) at this three-legged T-type intersection. NH-234 (Mangalore to Villupuram) also passes through this intersection. The SH-9/NH-234 is a four lane road with two lanes in each direction. The SH-9 is a two lane road with one lane in each direction. As there is no ring road connecting these national and state highways, all the bypassable traffic are passing through this particular intersection which causes traffic congestion in peak hours. In addition to the bypassable vehicles, all the local vehicles which come from Vellore fort and Sathuvachari towards Katpadi railway station and VIT also pass through this intersection only. Hence the vehicles experiences high delay during peak hours and this is a major motivation for carrying out the present study. At present, the traffic is controlled by traffic police manually that too only during peak hours. Traffic data collection using video survey is discussed in the following section.

3. Data Collection and Extraction

The conventional way in which the traffic data collection is done is to put traffic enumerators who will stand on the roadsides and will note down the class-wise number of vehicles in a sheet of paper. There are two problems with this manual data collection. One is, it is not economical especially if more number of enumerators was employed and the other one is these enumerators sometimes may forget to note down the vehicles especially if the traffic is high. In order to overcome these problems, video data collection is generally preferred. The main advantage is we can extract the traffic flow data at any desired time interval by simple playing the collected video. In the present study also, the same has been adopted. A high rise building having a wide area coverage at the intersection was selected and traffic recording was done using SONY Handycam. Eight hrs. of video recording from 7 to 11 am and 3 to 7 pm was carried out on 8th December, 2017. A snapshot from the taken video is shown in Fig.2.

Fig.2 Snapshot from the video showing Chitoor bus stand intersection

For volume extraction, 6 directions of traffic movement were considered and were listed here. (1). From Vellore towards Katpadi (direction number 1 in Fig.3); (2). From Vellore towards VIT University (direction number 2 in Fig.3); (3). From Katpadi towards Vellore (direction number 3); (4). From Katpadi towards VIT (direction number 4); (5). From VIT towards Katpadi (direction number 5); (6). From VIT towards Vellore (direction number 6). To bring the heterogeneity in vehicle classes, five vehicle classes or modes were considered and they are two-wheelers, three-wheelers (autos and share autos), passenger cars, light commercial vehicles (LCV) and heavy motorized vehicles (HMV) which includes buses and trucks. The traffic volume in 5 min. intervals from 7 am to 11 am and 3 to 7 pm in all the six directions of traffic was manually counted using the video data collected. The next step of converting the class-wise traffic volumes to equivalent passenger cars (PCU) was done using the PCU factors given in IRC 106 [4]. The PCU factors used are 0.75, 2, 1, 1.4 and 2.2 for two-wheelers, three-wheelers, cars, LCVs and HMVs respectively. Both the class-wise volumes and PCU converted traffic volumes were analyzed thoroughly to study the traffic conditions at the selected intersection and the analysis results are given in the following section.

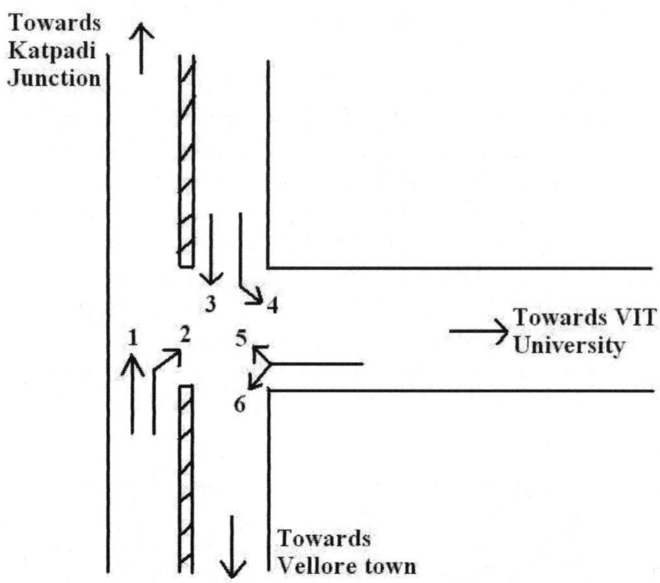

Fig.3 Direction-wise traffic movement diagram at Chittor bus stand intersection

4. Analysis of Traffic Flow

The modal share of various vehicle types was calculated using the extracted traffic volume data and the results are shown in Fig.4. It can be seen from Fig.4 that 32,348 two-wheelers have observed during the analysis period of eight hours, which accounts to 66% of the total traffic. Next to motor cycles, 8113 autos were observed, which amounts to 16% of the total traffic. Adding both, two wheelers and three wheelers comprise 82% of the total traffic volume at the selected intersection. The reason for this high proportion is that, for office/educational trips, the mode choice is two-wheelers or autos/share autos. A traffic study conducted in Chennai revealed that two-wheeled vehicles constitute about 45%, three wheelers such as autos constitute 6%, cars and SUVs constitute 47% and HMVs counts to only 2% [5]. The proportion of cars in cities like Chennai is about 50% of the total traffic as the length of the trips are usually in the range of 20 to 40 km and hence people preference is mostly cars or SUVs for such longer trips. However, the proportion of cars is only 11% of the total traffic in the present study. The possible reason is that, for office trips people prefer two-wheelers or motor cycles than cars as the trip lengths are generally small in the range of 5 to 10 kms in medium sized cities like Vellore. But as the city grows, the trip length will increase substantially and people may prefer to use cars than two-wheelers. The LCVs and HMVs proportion is 7% of the total traffic. In cities like Chennai, Delhi, Bengaluru, the entry of heavy vehicles into the city during peak hours were prohibited in order to ensure smooth and efficient flow of traffic within city limits. However, it may not be possible to implement such restriction in Vellore because there is no circular ring road exist at present for these heavy vehicles to move.

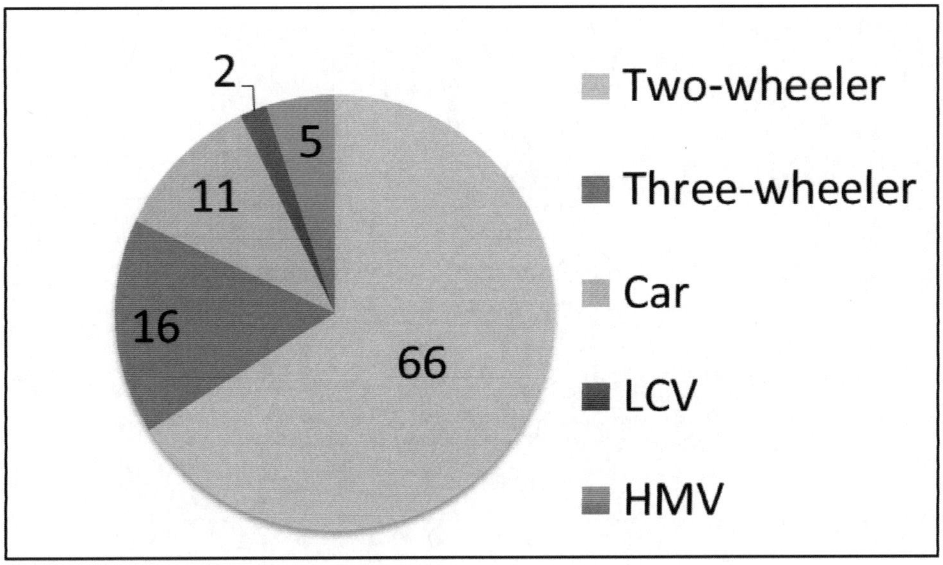

Fig.4 Percentage composition of various vehicle types

The direction-wise flow of traffic in terms of total PCU's is shown in Fig.5. As anticipated, the directions-1 & 3 have exhibited higher traffic flow when we compare it to other directions as it mainly contains through or direct traffic. Next to directions-1 & 3, directions-2 (Vellore to VIT University) & 6 (VIT to Vellore) were showing higher traffic flow. In morning, direction-2 was showing higher flow as it mainly consists of office and educational trips towards VIT and other institutions in Katpadi-Tiruvalam road. In evening, direction-6 was comparatively higher as it indicates the return travel from office to home.

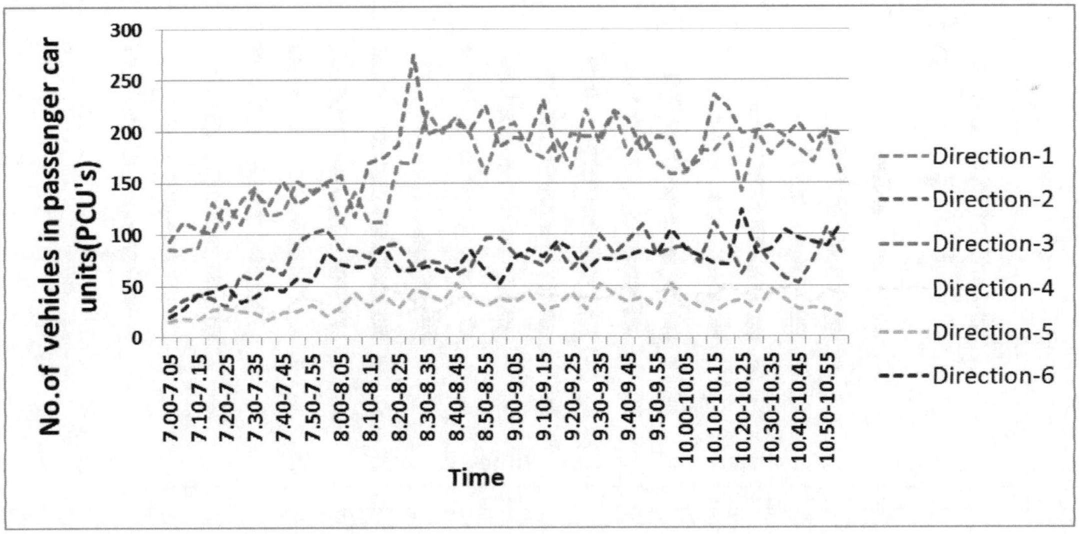

Fig.5a Direction-wise traffic flow in morning

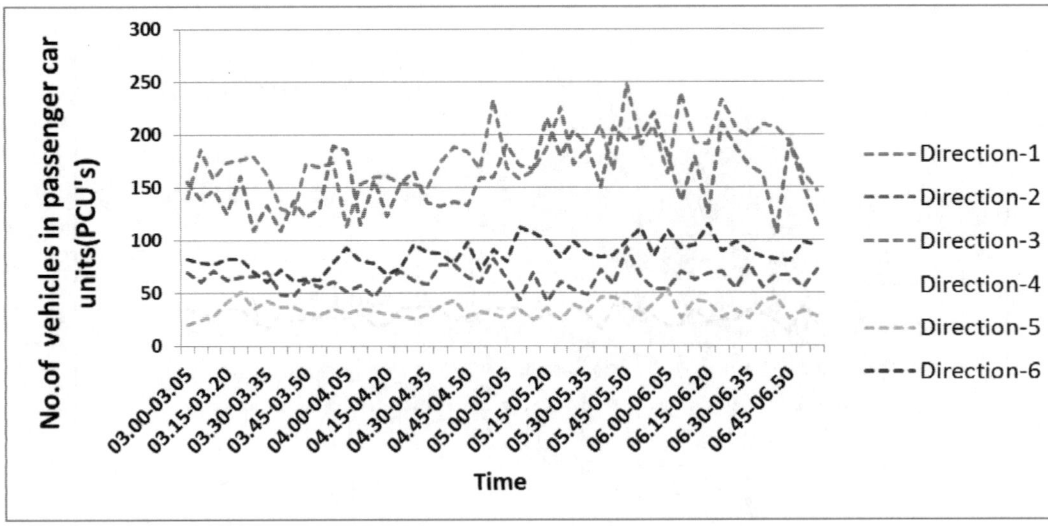

Fig.5b Direction-wise traffic flow in evening

The total number of vehicles in PCU's is shown in Fig.6. It was observed that, the highest hourly volume is 7573 PCU's between 9.30 am and 10.30 am during morning and 7336 PCU's between 5.40 pm and 6.40 pm in evening. As per IRC-92 [6], when the total traffic is greater than 10,000 PCU's in one hour, then the flyover may be justified at the intersection. But in the present case, the total volume is only 7573 PCU's, which is less than the limit of 10,000 PCU's and hence a flyover or interchange is not required at present. By considering a 10% growth in traffic every year [7], in 2021, the traffic volume may go beyond 10,000 PCU's and may warrant or require a flyover at the selected intersection.

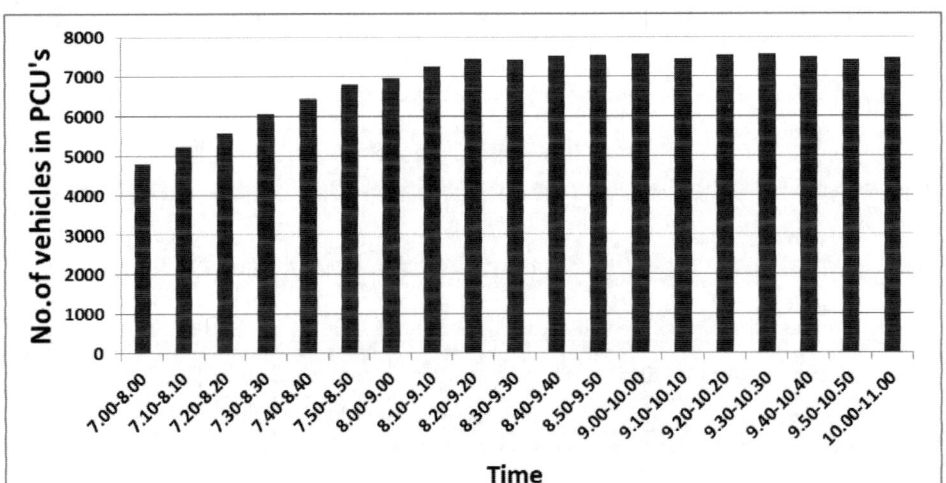

Fig.6a Total number of vehicles expressed in PCU's during morning

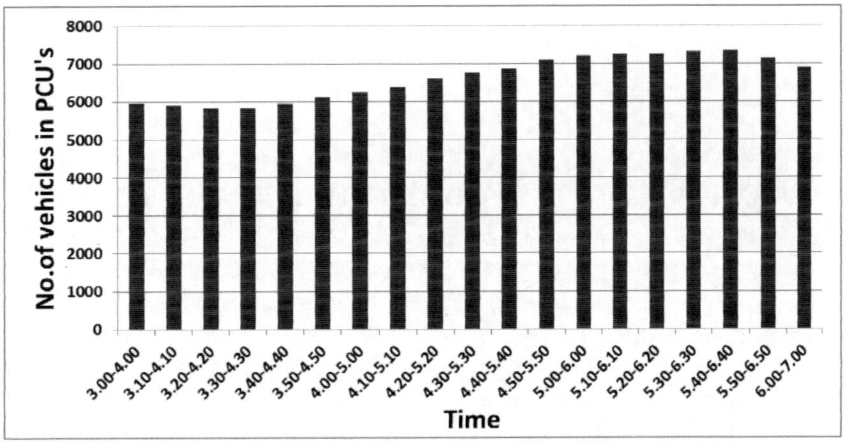

Fig.6b Total number of vehicles expressed in PCU's during evening

5. Concluding Remarks

Urbanization in cities is a serious issue in developing countries like India. More people are migrating to metro cities and this phenomenon of urbanization results in the vehicular growth in the cities. As Vellore city is on the Chennai - Bengaluru national highway and is close to Chennai, the city faced a tremendous urban growth in recent years. It is gradually changing from a town to a metropolitan city. As a first step to solve the traffic problems in Vellore, a busy intersection namely Chittor bus stand intersection near VIT was selected and traffic study was carried out through video surveys. The survey data was carefully analyzed to find the percentage share of different vehicle types, direction-wise distribution and morning & evening peak hour traffic volume. The results would be very useful to traffic police authorities to identify suitable traffic control measures and implement it.

6. References

[1] M.L. Moss and H.O. Neill 2012 *Urban Mobility in the 21st Century* NYU Rudin Center for Transportation Policy, Newyork, USA.

[2] G. Lal, L.G. Divya, K.J. Nithin, S. Mathew and B. Kuriakose 2016 Sustainable traffic improvement for urban road intersections of developing countries: a case study of Ettumanoor, India *Procedia Technology* 25 115 – 121.

[3] M. Errampalli and R. Kayitha 2016 Traffic management plan for Port Blair city, India *Transportation Research Procedia* 17 548 – 557.

[4] IRC: 106 1990 *Guidelines for capacity of urban roads in plain areas* Indian Roads Congress Code of Practice, Indian Roads Congress, New Delhi, India.

[5] S. Vasantha Kumar and L. Vanajakshi 2013 Mode-wise travel time estimation on urban arterials using transit buses as probes *Proc. of the 92nd Annual Meeting of Transportation Research Board (CD-ROM)*, Washington, D.C.

[6] IRC: 92 1985 *Guidelines for the design of interchanges in urban areas* Indian Roads Congress Code of Practice, Indian Roads Congress, New Delhi, India.

[7] IRC: 108 1996 *Guidelines for traffic prediction on rural highways* Indian Roads Congress Code of Practice, Indian Roads Congress, New Delhi, India.

IOP Conf. Series: Materials Science and Engineering **396** (2018) 012016 doi:10.1088/1757-899X/396/1/012016

A Review on Various Techniques used in Predicting Pollutants

Aiswarya Baby

P.G Scholar, Department of Computer Science and Engineering, St Joseph's College of Engineering and Technology,Palai, India

E-mail: aiswaryapbaby@gmail.com

Aneena Ann Alexander

Assistant Professor, Department of Computer Science and Engineering, St Joseph's College of Engineering and Technology,Palai,India

E-mail: aneenaalex@gmail.com

Abstract. Pollution is now becoming a biggest problem in today's age. In numerous nations, because of the high number of vehicles that are accessible on the streets, air contamination has progressed toward becoming a difficult issue. This eventually leads to emission of different gases. So there is a need to predict the amount of emission beforehand. There have been various research commitments in the region of anticipating toxins. Artificial intelligence (AI) are often employed where a problem requires estimation or prediction. Most of the methods used are based on neural network, fuzzy logic, hybrid methods and others. This paper discusses findings and limitations of existing contributions in the field of pollution prediction.
keywords:Artificial Intelligence, pollution prediction, Neural Network, Multilayer Perceptron.

1. Introduction

Artificial Intelligence(AI) is playing an increasing role in todays life. Intelligence is considered as the ability to collect reason and knowledge about knowledge to solve complex problems. In the near future intelligent machines will replace human capabilities in many areas. Artificial intelligence is the study and developments of intelligent machines and software that can learn, reason, communicate, gather knowledge, perceive and manipulate the objects. It makes machines more helpful and more brilliant. It works with the help of scientific theorems (if then statements and logics) and artificial neurons (artificial neural network). AI technologies have matured to the point in offering real practical benefits in many of their applications. Major Areas in Artificial Intelligence include Natural Language Processing ,Expert Systems, Neural Computing ,Robotics and Sensory Systems, Computer Vision and Scene Recognition, Speech Understanding, Intelligent Computer Aided Instruction.

A large number of tools are used to solve the most difficult problems in computer science. Some of the methods are Search and optimization, Logic , Probabilistic methods for uncertain reasoning , Classifiers and statistical learning methods, Neural networks, Deep feed forward neural networks, Deep recurrent neural networks, Control theory, Languages, Evaluating progress.

2. Literature Review

2.1. Greenhouse Gas Estimation

Environmental pollution is the major challenge that the world faces today. The transportation system is one of the greatest contributors to the deterioration of the environmental situation because vehicles are the major contributors to greenhouse gas emissions. In this manner, transportation is one of the real sources of the greenhouse gas emissions. Two major approaches that are mainly used to estimate the GHGs emissions, are bottom-up models and top-down models proposed by the IPCC [1]. In top-down. approach the entire carbon emission is calculated from the total fuel consumption in the city. Later this total carbon emission is divided among each economic sector, such as transportation sector. Whereas in bottom-up approach from each economic sector the fuel consumption is calculated. Then according to the carbon emission parameters of the fuel, the carbon emission amount is calculated.

However the two models have many uncertainties because of uncertainties in the empirical assumptions and the uncertainty in the primary data such as fuel consumption, carbon content, and oxidation factors[2]. The top-down approach mainly focuses on market interactions. It doesnt give much importance to the technological detail in the energy sector. While the major focuses of bottom-up approach is on substitutability of individual energy technologies and it lack macro-economic feedbacks between other economic sectors and the energy.

2.2. Greenhouse Gas Prediction

2.2.1. Neuro-Fuzzy InferenceTechniques : An neuro-fuzzy inference system (ANFIS) consolidates both neural networks and fuzzy systems. A neuro-fuzzy system depends on a fuzzy system that is trained by a learning algorithm derived from neural network theory. The learning technique works on nearby data, and results in narrow modifications in the underlying fuzzy system. Using neuro-fuzzy inference system an environmental monitoring system is designed to predict the pollutant levels in Italy [3]. Input data are those collected from sensors environmental parameters. The model efficiently predicted the pollutant levels. The criteria used for evaluation was Root Mean Square Error (RMSE). This evaluation criterion can stress the models performance.

2.2.2. Ordered Weighted Averaging based Time Series Model : OWA is a multicriteria evaluation procedure. The OWA methodology relies upon a few parameters, which can be specified by fuzzy quantifiers. OWA based time series model are used to predict future occasions in light of past occasions. The amount of Ozone emitted in Taiwan is predicted using this method [4]. Here the data from air monitoring stations are used. Ozone is selected as the attribute since it is the air pollutant species of concern. This strategy combines the merits of both time series method (TSM) and ordered weighted averaging (OWA). The criteria used for evaluation was RMSE and Mean Square Error (MSE) . This can stress the models performance and takes a lot of computational time.

2.2.3. Grey Dynamic Model : The grey dynamic model performs well in modelling and forecasting. The Grey Dynamic Model was utilized to forecast the oil pollutants in China [5]. Since the oil utilization in China is expanding step by step, the development of oil stations is in the ascendant. Concentrations of CH,CO, NOx are used as input data. These information are then pre-processed and are utilized to train the Grey model. The problem here is that the dataset used are not diverse.

2.2.4. Response Surface Methodology : Response surface methods (RSM) are information based techniques. They perform well when there is a rich tapestry of accurate information provided.

Due to their ease of use as well as dexterity of being applied across the board of industrial applications They have gained huge fame. This technique was utilized to foresee the NOx emission [6] and input data like coal quality, total air supply, oxygen concentration and air velocity are considered. Four different Response surface models were used and the model with quadratic terms had the best prediction. The challenge for RSM is to demonstrate the continuum and to capture precisely the rich collection of physical phenomena.

2.2.5. Matern Function Based Extended Fractional Kalman Filtering : This method is used to addresses the problem of predicting air pollution emissions [7]. Uncertainty in data involves measurement error or error related to instrument precision .Here the input data are the observed station data. But Kalman Filtering techniques suffer from a few impediments, for example, that identified with the need of exact stochastic models of sensor random errors and its immunity to noise, observability.

2.2.6. Digital Imaging, Contourlet Transform and Radial Basis Function Network Techniques : Digital images are electronic snapshots taken of a scene or scanned from documents, for example, photographs, printed texts, and artwork. Through the combination of flame radical imaging, contourlets transform, and radial basis function network techniques, NOx emissions in a biomass ignition process is predicted [8]. Using a spectroscopic imaging system the images of four flame radicals (OH*, CN*, CH* and C2*) are captured. Based on contourlet coefficients the features of the images are then identified .These features are used to train the radial basis function network techniques(RBF). The The fundamental disservice is that this is time consuming.

2.2.7. Regression Model : The economy of a country can be improved by rapid industry development. But this leads to the consequences of heavy CO2 emissions. So there is a need for the CO2 emission prediction. Regression model was used for the CO2 emission prediction purpose [9]. The relationships among variables can be estimated by this method.

Another Regression model application is in the estimation and forecast of pollutants in road tunnels [10]. The sensors collect data on traffic parameters, such as traffic flow , speed , truck ratio , and others as well as environmental variables, like atmospheric pressure , air flow velocity , air flow direction. The data collected are pre-processed and is then applied to the Regression model. The relationship that is established between variables based on some limited data may not hold great if more and more data are taken into consideration.

2.2.8. FCM-HMM Clustering and TS Fuzzy Inference Algorithm : A multi model frame is constructed based on FCM-HMM clustering and TS fuzzy inference multi-model to predict the air pollution index [11]. Data used are meteorological and climate factor. But HMM frequently have various unstructured parameters and dependencies between hidden state cannot be expressed.

2.2.9. Support Vector Machine : Support vector machine (SVM) is a discriminative classifier and is used for the prediction of the co2 level [12]. Input data is the co2 amount which is emitted by burning coal and when generating electricity. The data is normalized, cross validated and then applied to the support vector machine. The major drawback of the support vector approach lies in choice of the kernel.

2.2.10. Neural Network Technique : Various gases have been predicted using Neural Network technique (NN). Neural networks are parallel computing devices. They are essentially an attempt to make a computer model of the brain. Its fundamental goal is to build up a system that perform various computational tasks faster than the traditional systems. A multilayer perceptron neural network based model for short prediction of SO concentrations was developed for the stations around the SoStanj Thermal Power Plant [13]. The input features are fed into the model and backpropagation algorithm is used to train the model. One major drawback observed is that forecast of the so2 gas is for a brief timeframe because of the limited amount of data.

In order to achieve good generalizing capabilities of the model, one of the most important tasks that should be solved is the selection of the patterns used for neural network-based model training [14]. Two types of pattern selection procedures were developed: a meteorological knowledge - based strategy and a Kohonen neural network-based strategy. Patterns selection reduce the computational cost. It also improve the models generalizing capabilities and hence enhance the quality of prediction. The result of these two strategies showed a high increase of the models generalizing capabilities. The strategy based on Kohonen neural network is certainly more suitable for use in other research fields such as speech recognition, or otherwise where huge data bases are available.

Semi-supervised learning is a class of machine learning techniques that make use of both unlabeled and labeled data for training - normally a small amount of labeled data with a large amount of unlabeled data. Semi-supervised learning (SSL) approach was used to predict air quality information using input data such as traffic flow, POIs , meteorology, structure of road networks, and human mobility [15]. This method provides accurate air quality information. But verifying SSL assumptions or mathematically formalizing them is very difficult.

Feed-forward ANNs enable signals to travel one way only that is from input to output. There are no loops. Feed-forward ANNs tend to be straightforward networks that associate inputs with outputs. Feedback networks can have signals travelling in both directions by introducing loops in the network. Feedback networks are powerful and can get extremely complicated. Using a neural network model based on feed-forward neural networks and recurrent neural network accurate forecasting of fine particulate matter concentration is done [16]. Here the dataset used are not diverse. So there occurs a problem of over fitting.

Multi Layer perceptron (MLP) is a feedforward neural network with one or more layers between input and output layer. Backpropagation learning algorithm is used to train this type of network. MLPs are widely used for recognition, pattern classification, prediction and approximation. Such a network is used for the transportation carbon emission prediction in the cities [17]. The model showed improved accuracy. Multi Layer Perceptron can solve problems which are not linearly separable.

Table 1. Comparison Table

Techniques	Pros	Cons	Performance metrics
Neuro-Fuzzy InferenceTechniques	• ability to change the qualitative aspects of human knowledge into the process of precise quantitative analysis	• time consuming	• efficiency
Grey Dynamic Model	• useful for dealing with poor, incomplete, and uncertain information	• low prediction precision	• accuracy

Techniques	Pros	Cons	Performance metrics
Ordered Weighted Averaging based Time Series Model	• allows decision-makers to change the form of attribute combinations from a minimum-type combination through all intermediate types to a maximum-type combination.	• takes a lot of computational time	• efficiency and accuracy
Response Surface Methodology	• simplicity and easiness	• difficult to model the continuum and to capture accurately the rich repertoire of physical phenomena.	• predictive accuracy
Matern Function Based Extended Fractional Kalman Filtering	• results to a faster convergence in terms of iterations compared to traditional methods	• Space and computation complexity	• accuracy
Digital Imaging, Contourlet Transform and RBF Network Techniques	• RBF is quick to train • the hidden layer of RBF is easier to interpret than the hidden layer in an MLP	• Although the RBF is quick to train,after training is finished , when it is used it is slower than a MLP	• stability
Regression Model	• Easy to implement and interpret	• involves very lengthy and complicated procedure of calculations and analysis.	• accuracy
FCM-HMM Clustering and TS Fuzzy Inference Algorithm	• fairly readable • HMM generally provides better compression than a simple Markov model	• training can be quite slow • suffers heavy computational burdens thereby becoming very time-consuming	• improved prediction performance
Support Vector Machine	• prediction accuracy is generally high • robust	• long training time • difficult to understand the learned function	• accuracy

IOP Conf. Series: Materials Science and Engineering **396** (2018) 012016 doi:10.1088/1757-899X/396/1/012016

Techniques	Pros	Cons	Performance metrics
NN Technique	• no prior knowledge of the data generating process is needed for implementing NN • problem of model misspecification does not occur	• addition of too many hidden units incites the problem of over fitting the data	• accuracy

3. Conclusion

In this paper, various methods that are used to predict the greenhouse gas emission were discussed. Artificial Intelligence plays a very important role in forecasting various greenhouse gas emission. In forecasting models, the major concern is with accuracy. More and more complex models have been introduced to gain accuracy. It is found that Multi Layer perceptron is more efficient than other techniques in terms of accuracy. In future, various efficient hybrid algorithms/model should be substituted in place of complex models to decrease the execution time of the model.

4. References

[1] I. P. on Climate Change, 2006 IPCC Guidelines for National Greenhouse Gas Inventories. Intergovernmental Panel on Climate Change, 2016.

[2] Detlef P.vanVuuren , MoniqueHoogwijk , TerryBarker , KeywanRiahi , StefanBoeters , *Comparison of top-down and bottom-up estimates of sectoral and regional greenhouse gas emission reduction potentials*,Science Direct on 2009.

[3] F.CM. orabito, M.Versaci, *Prediction and Estimation of Atmospheric Pollutant Levels by Soft Computing Approach* ,Proceedings. IJCNN 01. International Joint on 2001.

[4] Sue-Fen Huang,Ching-Hsue Cheng *Forecasting The Air Quality Using OWA Based Time Series Model*,International Conference on Machine Learning and Cybemetics on 2008.

[5] Zhao Xiao-gang, Zhou Yi, Zhao Jian-yu *Study on the Prediction of Air Pollutants in Urban Oil Depot Based on Grey Dynamic Model* ,International Symposium On Water Resource and Environmental Protection on 2011.

[6] Ligang Zheng, Shuijun Yu, Minggao Yu , *Prediction of Nitrogen Oxides from Coal Combustion by Using Response Surface Methodology* ,International Conference on Computer Distributed Control and Intelligent Environmental Monitoring on 2012.

[7] S. Metia, S. D. Oduro and Q. P. Ha *Air Pollution Prediction Using Matern Function Based Extended Fractional Kalman Filtering* ,2014 13th International Conference on Control Automation Robotics Vision (ICARCV) on 2014.

[8] Nan Li, Gang Lu, Xinli Li, Yong Yan,*Prediction of Pollutant Emissions of Biomass Flames Using Digital Imaging, Contourlet Transform and Radial Basis Function Network Techniques* ,IEEE International Instrumentation and Measurement Technology Conference Proceedings on 2014.

[9] Tan Chun Ho, Sim Chong Keat Mohd Zubir Mat Jafri and Lim Hwee San ,*A Prediction Model for CO2 Emission From Manufacturing Industry and Construction in Malaysia*,International Conference on Space Science and Communication on 2015.

[10] Ondrej Pribyl, Pavel Pribyl , *On the estimation of pollutants in road tunnels based on traffic data* ,ELEKTRO on 2016.

[11] Ling Wang , Xi-yuan Xiao , Jian-yao Meng, *Prediction of Air Pollution Based on FCM-HMM Multimodel* , 35th Chinese Conference(CCC) on 2016.

[12] Chairul Saleh, Nur Rachman Dzakiyullah, Jonathan Bayu Nugroho ,*Carbon dioxide emission prediction using support vector machine*, Materials Science and Engineering 114 on 2016 .

[13] Primoi Mlakar and Marija Boinar ,*Perceptron Neural Network - Based Model Predicts Air Pollution*,Intelligent Information System, 1997. IIS 97. Proceedings on 1997.

[14] Marija Boinar, Jamova , *Pattern Selection Strategies for a Neural Network -Based Short Term Air Pollution Prediction Model* ,Intelligent Information System, 1997. IIS 97. Proceedings on 1997.

IOP Conf. Series: Materials Science and Engineering **396** (2018) 012016 doi:10.1088/1757-899X/396/1/012016

[15] Y. Zheng, F. Liu, and H.-P. Hsieh ,*U-air: when urban air quality inference meets big data*, in Proceedings of the 19th ACM SIGKDD international conference on Knowledge discovery and data mining.ACM, 2013.
[16] Mihaela Opreal, Marian Popescu, Sanda Florentina Mihalache, *A Neural Network Based Model for PM2.5 AirPollutant Forecasting* ,20th International Conference on System Theory, Control and Computing (ICSTCC) on 2016.
[17] Xiangyong,Karoru,Dong,Chen and Hai Ji; *Predicting Transportation Carbon Emission with Urban Big Data"* IEEE International Journal on 2017.

Authenticating Communication of Autonomous Vehicles With Artificial Intelligence

Neetha George[1] and Jeena Thomas[2]

1 P G Scholar, Department of Computer Science and Engineering, St.Joseph's College of Engineering and Technology, Palai, Kerala, India.

2 Assistant Professor,Department of Computer Science and Engineering, St.Joseph's College of Engineering and Technology, Palai, Kerala, India.

E-mail: neethageorge06@gmail.com

E-mail: jeena.thomas@sjcetpalai.ac.in

Abstract. Autonomous vehicles in Vehicular Ad-hoc Networks(VANET)-the spontaneous creation of a wireless network for data exchange to the domain of vehicles may pave the way for future systems where computers take over the art of driving. It aids in transferring secure messages for proper communication between the vehicles. Unauthorized access like injecting spoofed messages to VANET can arise as an extreme threat. So remedies should be taken at early degree of architectonic process.Denial-of Service(DoS) attack takes place while validating each message in VANET ,so an equity between message authentication and DoS prevention should be maintained.This paper discusses on diverse security demanding situations and distinct techniques to secure message transfer between the vehicles that are connected wirelessly.
Keywords-Artificial Intelligence,Vehicular Ad-hoc Network(VANET),wireless communication,message authentication, security issues

1. Introduction

The Internet of Things (IOT)[1][2],is one of the most flourishing technology which makes vehicles connected each other. It connects devices, machines and tools to the internet by means of wireless technologies. VANET[3][4] is a type of network in which the interaction of vehicles occur by exchanging messages between them. The safe exchange of messages can lead to collision avoidance, traffic congestion control, thus ensuring the driver a smooth driving experience. But there can be also many attacks faced by VANET[5] like monitoring attacks, social attacks, timing attacks, application attacks, and network attacks. So the connected vehicles should be secured properly. There are several existing methods for securing them.

Regarding the colossal gains expected from vehicular communications and massive vehicles(hundreds of millions worldwide), it is understood that the intra-communication of vehicles can turn into the greatest admissible recognition of mobile ad hoc networks.GPS acceptors in addition to communication capabilities,which are the most suitable assimilation of on-board computers and spotting devices, open immense trading chances, hence boosts astounding research challenges. One of the major issues is security; very little attention has been dedicated untill now to the safety of vehicles in the network. Probably, the size of the network, the speed of the vehicles, the relevance of their geographic location, the very

IOP Conf. Series: Materials Science and Engineering **396** (2018) 012017 doi:10.1088/1757-899X/396/1/012017

intermittent interconnection, and the inevitable gradual distribution of effective resources shape the threat very peculiar and challenging.

2. Overview of VANET

2.1. Architecture of VANET

Vehicles are connected to each other wirelessly with Vehicular Ad-hoc Network (VANET)[5], an advanced technology that takes moving cars as communication nodes to form a instinctive network. Routers, deliberately placed along the road, insure constant coverage for vehicular communications. VANET. Vehicles communicate to each other with On-Board Units(OBU) rigged on vehicles. Road-Side Units (RSU) expand communications with both geographical inclusion and tremendous speed of data.Dedicated Short-Range Communication (DSRC)[6][7] - a two way wireless communication that is short- to- medium-range ensures enormous transmission of data crucial in communications-based valid security functions is entrusted for VANETs. Vehicles also work with OBUs armed with sensors, such as RAdio Detection And Ranging (RADAR) and LIght Detection And Ranging (LIDAR), to avoid collisions.

The Central Control Unit acts as a heart of the whole system. The GPS[8] unit and the sensors provides one-way communication which is necessary positioning, velocity and time, to empower interdependent services which are position based. The GSM unit is a two-way communication which ensures the mobile wireless communication among vehicle to vehicle and vehicle to infrastructure communication. The Central Control unit of OBU[11] works on and decode the data from top layer and store it in flash memory to record all the neighbouring vehicles behaviour. Information Sharing broadcasts the position of vehicles, their acceleration and speed several times per second, which is got from GPS module[9]

The Judgement and Decision Making block facilitates to take an appropriate decision to assure the protection thereby averting the intrusions by false drivers. Information regarding vehicle speed, radio control, map view, emergency or collision warnings and traffic status is being shared with the user. User can get entire details displayed to his smartphone or a tablet via Bluetooth wireless link. The alerting messages are being sent and controlled by application running on the user interface device when any risk is encountered. A Road-Side Unit (RSU)[10][11] works as a static OBU powered by more computing properties and have frequent wired connection to the Internet that is considered to be the backbone. RSUs are usually installed at each 100-200 meters across a road to provide networking infrastructure for enhanced performance and enforced security in transmitting messages in vehicular network.Figure 1 illustrates various components of an On-Board Unit.

2.2. Security Attacks in VANET

VANET security has evolved to be a major concern in the society. All the current stations should be authenticated before they contact the available services in the network. The attack enlists the process of identification which represents the whole network to serious aftereffects. In a Vehicular network the authentication safeguards to protect the secured nodes from the outside or inside attackers who tricks the network using a fake identification.The securing scheme process will take place whenever a vehicle tends to link to the network or any service.

Attacks on confidentiality will be a key security mechanism in VANET communications, which ensures the data should accessed by authorized user. within the absence of mechanism, the VANET have to ensure the exchanged message confidentiality, while it's far drastically prone to assaults. Attacks on availability is a most crucial role in VANET which assure the network has been an functional one and also offer needed information during functioning time. Attacks on authentication and identification has emerged as an important discussion in security of VANET. All the actual stations need to be secured in the network before they come in association with the available services. The attack employs the process of identification which represents the

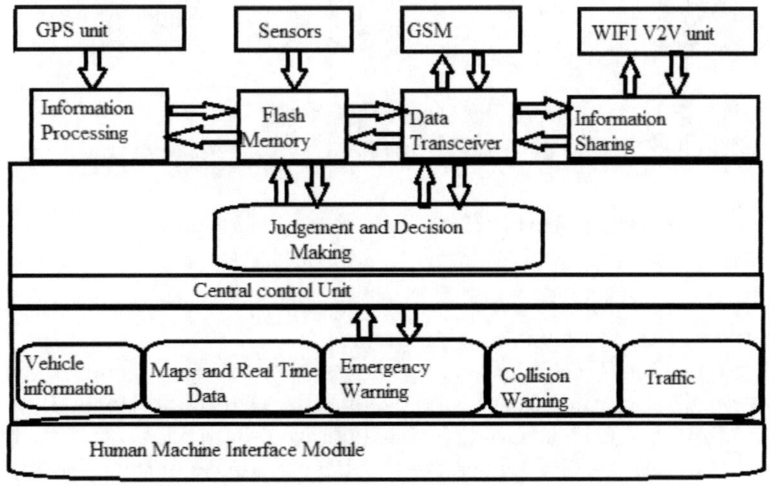

Figure 1. On-Board Unit

whole network to serious consequences. Whenever a vehicle tends to link to the network or any service, the authentication process will takes place. Attacks on integrity and data trust in a vehicle has to ensure that swapped data should not been amended during transmission. These mechanisms aids to safeguard information against deletion, alteration or supplementary attack. Attacks on non-repudiation means, the ability to endorse that the provider and the acceptor are the individuals who can send and receive the messages. The non-abrogation of data origin establish that the information is send to respective sender and the non-repudiation of advent have to make sure that data has been received only authenticated receiver.[12]

3. Authentication Mechanisms in VANET

3.1. Authentication with Digital Signatures

Authentication with Digital Signatures[13][14] is one the excellent techniques for securing the VANET messages since they sign each of the messages before passing them to the acceptor side.It uses a Public key Infrastructure(PKI) to provide security due to huge number of individuals in the network.Each of the vehicle in the network will be contributed with a pair of public/private key.A private key signs the message before it is transferred and includes Certification Authority(CA)certificate.A tamper-proof device along with private keys is necessary in each vehicle where the confidential messages are tracked and out-gone data are signed. Little time later, the sender simulcasts the key and instructs all that this revealed key is not to be used in the future. Receivers cache the actual message until the key is received and then verify the signature.Symmetric cryptographic primitives are used by this verification.Digital signature working is shown in figure 2

Digital Signatures provide speed,security,authenticity,tracking,improper prevention facilities. The existing broadcast authentication standard of Digital Signatures in VANETs is susceptible to signature flooding.Here the excessive signature verification requests exhaust the computable functions of targets. To overcome this challenge of Digital signatures, two efficient broadcast authentication schemes, Fast Authentication (FastAuth) and Selective Authentication (SelAuth) can be used.

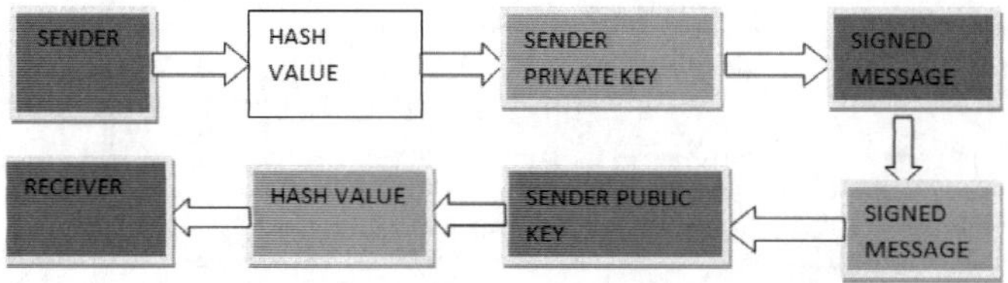

Figure 2. Authentication with Digital Signature

3.2. Authentication with Fast Authentication scheme

FastAuth,is an adequate One-Time Signature(OTS) scheme to secure beacon messages. The innovation in FastAuth[15] is the layout of a sequential Huffman Hash Tree (CHT),which provides the predictability in vehicle mobility to generate small signatures. Every beacon has to be checked when they are received because the beacon may contain definitive and crucial safety information. The recent VANET signature standard is computationally costly, but traditional OTS provides instant verification with increased communication overhead.Inspired by this exclusive challenge,the target of FastAuth is to gain fast authentication with short signatures.

Each vehicle in FastAuth divides its timeline into a sequence of prediction intervals at higher levels. In each prediction interval, a vehicle performs three steps: Beacon Prediction, Key Pair Construction, and Signature Broadcast. In Beacon Prediction at the starting of a prediction interval, all vehicles forcast their beacon messages for the next I beacons. For this vehicles use the probability distribution of the distance vector between two successive beacons based on data of the past trajectory.In Key Pair Construction before transferring any beacon message in this interval, the vehicle needs to build an OTS public key and one interval worth of OTS private keys. A chained Huffman hash tree (CHT), links these pre-computed keys together in a fashion that reduces the size of signatures and generates a single public key.In Signature Broadcast after beacon prediction and construction of the key, a vehicle signs its OTS public key using ECDSA signatures and then simulcasts this ECDSA signed public key PKots along with the first beacon in this prediction interval. Moreover, to retain a high beacon update frequency during severe packet loss, we assimilate Forward Error Correction into FastAuth to recover missed beacons.[16]

3.3. Authentication With Selective Authentication Scheme

To overcome the challenge for creating a decisive verification scheme for multi-hop applications, SelAuth[15], a signature verification protocol which can quickly block the escalation of invalid signatures without verifying all receiving signatures at every hop can be effectively used. In particular, SelAuth uses neighbor identification to avoid enactment and per neighbor verification probability, adjusted dynamically as wrong signatures are received, to achieve isolation. Neighbors of a vehicle communicates directly with the vehicle. L SelAuth also uses warning pushback to accelerate the desolation of pernicious vehicles. As in many other probabilistic verification schemes, vehicles running SelAuth verify an incoming message with a certain probability in order to help identify invalid signatures.However, an important difference is that in preceding work such a probability depends entirely on the local status of the receiver, disregarding where this message is from or whether other vehicles have checked this message. SelAuth leverages ancillary information shared between neighbours to facilitate the probability adjustment for fast and efficient isolation.

The core components of SelAuth are forwarder identification and warning pushback.This enables SelAuth to concentrate faster than other probabilistic verification schemes and be more resource-efficient than the Verify-All approach. Forwarder (or neighbor) identification enables the receiver of a message to effectively find which of its neighbors sent or forwarded this message. In SelAuth, vehicles detecting an invalid signature will initiate a Complaint message to warn vehicles at the previous hop.It is referred to be the Pushback message as it is pushed towards the producer.

4. Securing autonomous vehicles With Artificial Intelligence

An artificial-intelligence system continuously learns from its past incidents and by its pertinent feature to discern and recognize its surroundings. It can think like humans and can save lives,thus improve traffic safety and autonomous driving. An agent is said to behave intelligently when has to decide what actions are best for a particular situation and its end results. It is adaptable to the changing goals and environment and can learn from its past.[19]

4.1. Artificial Intelligence For Positioning

With the help of sensors which collects the information about its environment,agents can take decision about the correct state of its locations and surroundings . In case of VANET, when the communication is guided by beaconing,[17] a vehicle is unaware about its actual and upcoming location and other vehicles speed. The single data available to a vehicle is the sequential transfer rate of the beacons. To realize the assimilation of DSRC and GPS findings, the sensor data fusion logic is a serious issue. Considering the computation efficiency in a dynamic vehicular environment, the widely used techniques like Kalman and particle filters[18] for a sub optimal solution to Bayesian filter may suffer from the collusive computational burden, and thus a local approach with pre-defined assumptions to the posterior density is a suitable choice[20]

4.1.1. Authentication With Particle Filter Two basic checks are provided by Message Authentication: integrity check and identification check. Message authentication is to differentiate the malicious vehicles from the valid ones Broadcasting the beacon messages is one of the relevant research area because an effective number of message transmitted in VANETs are broadcast messages. Significant algorithms are required to more reduce broadcast storms which emerges due to packet flooding. To integrate message authentication with particle filter there is a security scheme based on Schoch;s concept called context adaptive beacon verification (CABV)[18] which aims at decrementing the computational overhead in authenticating beacon messages for protected vehicular message transfer. Here to validate the signatures each initial to the n beacons coming from vehicles are verified . To avoid the in-between intruded beacons, a linear and non linear estimators are taken for upcoming location forecast. If the predicted positions and that are tracked in the beacon works change immensely, then a signature is provoked.

5. Modified Approach of Context Adaptive Beacon Verification Method With Particle Filter

In the modified approach,at first particle filter will track the location of the vehicle and it is initialized with beacon messages.The vehicle will update its current location frequently to the network.The distance of the vehicle tracked from the particle filter is compared with this distance.If there is wide difference,the vehicle considered as malicious one and the beacon counter is incremented.

Figure 3 illustrates the working of the modified approach of Context Adaptive Beacon Verification method along with particle filter.

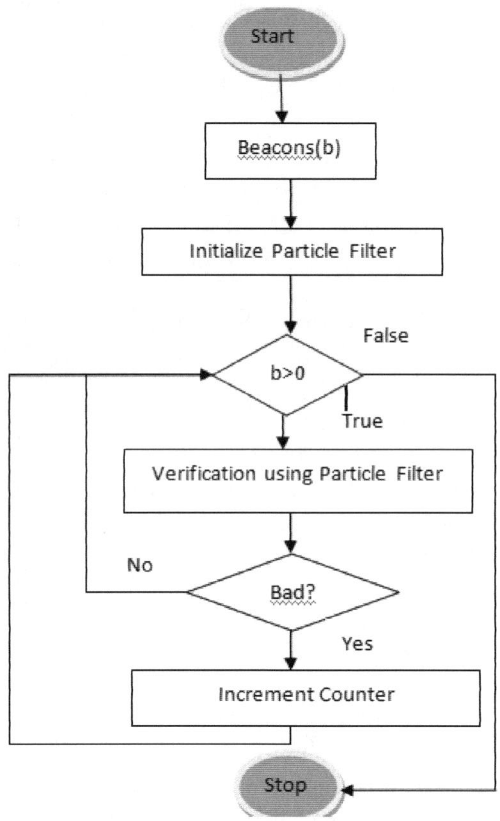

Figure 3. : Modified Approach for Working of CABV method With Particle Filter

6. Conclusion

Since autonomous vehicles are paving a great future to self driving,large amount of vulnerable attacks take place in vehicular networks.This review shows promising results of securing the wireless communication of autonomous vehicles including various methods.We first analyzed Authentication of VANETs with Digital Signatures ,which uses signed messages.Eventhough it secured the network,resulted in computational burden.We later discussed on Fast Authentication and Selective Authentication schemes which were able to overcome some disadvantages of Digital Signatures,but had computational overhead leaving spoofed beacons undetected.Then we saw Particle filter that particularly reduces communication overhead and detection level of spoofed messages will be kept same. Context adaptive beacon verification(CABV) with particle filter proved that it can detect and prevent spoofed attacks and can reduce the computational overhead.But this method also leaves certain number of spoofed beacons undetected.So the study of modified approach of CABV with particle filter is considered in future which can increase the efficiency .The detailed study of further possible vulnerable attacks with this approach will remain for the future scope of development.

7. References

[1] Floerkemeier, C., Langheinrich, M., Fleisch, E., Mattern, F., Sarma, S.E.: 2008 The *Internet of Things.* First International Conference, IOT 2008, LNCS 4952, Springer

[2] Shen Subin, Mao Yanqin, Fan Quli, Zongping, Huang Wei. 2010 *Conceptual model and architecture of Internet of Things [J].* Journal Of Nanjing University Of Posts And Telecommunications (Natural Science Edition)

[3] S. Zeadally, R. Hunt, Y. S. Chen, A. Irwin and A. Hassan,2012 *Vehicular Ad Hoc Networks (VANETS): Status, Results and Challenges.,*
[4] M. Raya and J. P. Hubaux,2005 *The Security of Vehicular ad hoc networks.* in 3rd ACM workshop on Security of ad hoc and sensor networks., Alexandria, VA, USA
[5] Stampoulis, A. and Chai, Z. (2007). *A survey of security in vehicular networks.*
[6] Bai, F., and Krishnan, H. 2006. Reliability analysis of DSRC *wireless communication for vehicle safety applications.* IEEE intelligent transportation systems conference , Toronto
[7] Kenney J B. 2011 *Dedicated Short-Range Communications (DSRC) Standards in the United States.* Proceedings of the IEEE, Vol. 99(7) pp.
[8] Alfred Leick, John Wiley and Sons, 1995 *The Global Positioning System* GPS Satellite Surveying 2nd Edition,
[9] Thong, S.T.S.; Chua Tien Han; Rahman, T.A., *Intelligent Fleet Management System with Concurrent GPS and GSM Real-Time Positioning Technology,2007.* 7th International Conference on ITS
[10] J. Lee and C. Kim, 2010 *A roadside unit placement scheme for Vehicular Telematics networks,* in AST10
[11] Qiong Yang, Lin Wang, Weiwei Xia, Yi Wu, Lianfeng Shen 2014 International Conference on Connected Vehicles and Expo (ICCVE)
[12] I.A. Sumra, I. Ahmad, H. Hasbullah, J.-L. bin, Ab Manan,2011 *Behavior of attacker and some new possible attacks in vehicular ad hoc network (VANET),* in: 3rd International Congress on Ultra Modern Telecommunications and Control Sys- tems and Workshops (ICUMT) IEEE
[13] Igor Furgel and Kerstin Lemke 2004*A review of the digital tachograph system.* In Workshop on Embedded IT-Security in Cars (escar)
[14] Lutz Gollan and Christoph Meinel. 2002*Digital signatures for automobiles.* In Systemics, Cybernetics and Informatics (SCI)
[15] Hsu-Chun Hsiao, Ahren Studer,Chen Chen, Adrian Perrig 2011 *Flooding-Resilient Broadcast Authentication for VANETs* Las Vegas, Nevada, USA.
[16] Kiho Lim, D. Manivannan 2016 *An efficient protocol for authenticated and secure message delivery in vehicular ad hoc networks* at Science Direct vehicular communications
[17] Kargl, E. Schoch, B. Wiedersheim and T. Leinmuller,2008 *Secure and efficient beaconing for vehicular networks,* in 5TH ACM International Workshop on Vehicular Inter-Networking, San Francisco, California, USA
[18] F. Gustafsson, F. Gunnarsson, N. Bergman, U. Forssell, J. Jansson, R. Karlsson and P. J. Nordlund,2002 *Particle Filters for Positioning, Navigation and Tracking.,* in IEEE Transactions on Signal Processing S. Eichler, C. Schroth, T. Kosch and M. Strassberger,2006 *Strategies for Context-Adaptive Message Dissemination in Vehicular Ad Hoc Networks.,* in IEEE
[19] Y. J. Abueh and H. Liu, *Message Authentication in Driverless Cars,*2016 in IEEE Symposium on Technologies for Homeland Security (HST)
[20] L. D. Ambroggi,2016 *Artifical Intelligence Systems for Autonomous Driving On the Rise,* IHS Markit,Available: https://technology.ihs.com/579746/artificial-intelligence-systems-forautonomous-driving-on-the-rise-ihs-says.

IOP Conf. Series: Materials Science and Engineering **396** (2018) 012018 doi:10.1088/1757-899X/396/1/012018

A Study On The Impact Of Machine Learning Tools For Detecting Anxiety Disorders In Preschool Children

Anugraha K R[1] and Vineetha S[2]

1 Student of Computer Science &Engineering , RIT ,Kottayam.

2 Faculty of Computer Science &Engineering , RIT ,Kottayam.

E-mail: anugrahakr@gmail.com

Abstract. Nowadays anxiety disorders in preschool children is an interesting area for researchers. There are different types of anxiety disorders in which generalized anxiety disorder and separation anxiety disorders has high chance of occurrence among preschool children. Early identification of anxiety disorders in childhood can often prevent the persistence of these difficulties and appropriate treatments can improve the quality of their life. In light of the high prevalence of these disorders and the associated short and long-term impairment, there is a critical need for developing reliable and valid methods for measuring and monitoring the severity of these disorders and their symptoms. This survey discusses different studies in preschool anxiety disorders and different methods that are used as scales for measuring anxiety disorders. This paper also describes a machine learning technique using ADTree classifier for developing a screening tool for detecting anxiety disorders in preschool children. It was found that ADScore generated by ADTree can be used to measure the severity of disorder.

1. Introduction

Childhood plays an important role in structuring the social activities of a person. The social and emotional competencies are recognized as critical factors for children's success in later stages of adulthood [1]. Also the interaction of children among their primary caretakers and children set the stage for future academic and personal growth [2]. Even though a child is given utmost care, some may experience anxiety disorders due to some stress full events, which may develop into serious psychological disorders in their future.

Anxiety is normal and common in childhood. In most cases, anxiety in children is temporary and may be triggered by a despairing event. For example, a young child may experience separation anxiety when starting preschool or kindergarten. In some cases, however, this anxiety can be persistent and intense and can interfere with a child's daily routines and activities such as going to school, making friends, or sleeping. Such constant and serious anxieties that doesn't go away with reassurance and comfort, are called anxiety disorders.

According to Diagnostic and Statistical Manual of Mental Disorders, anxiety disorders include disorders that share features of excessive fear and anxiety and related behavioral disturbances. The most common types of anxiety disorders are Generalized anxiety disorder (GAD), Separation anxiety disorder (SAD),

Social Phobia and post-traumatic stress disorder. Early detection and diagnosis of such psychiatric disorders is crucial for their subsequent development and quality of life.

This paper discusses about the various anxiety disorders and their frequency in children. This paper also discusses about the different methods which are followed till date for finding the anxiety disorder.

2. Related works

Jessie C. Anderson et al [3] conducted study among preschoolers about anxiety disorders. In this study they found out that the most prevalent disorders were attention deficit, oppositional, and separation anxiety disorders(SAD), and the least prevalent were depression and social phobia. Conduct disorder, overanxious disorder, and simple phobia had intermediate prevalence. They were also able to find that the children identified at 11 years as having multiple disorders had a history of behavior problems since 5 years of age. Fifty-five percent of the disorders occurred in combination with one or more other disorders, and 45% as a single disorder. Orvaschel et al [4] also studied about the anxiety disorders among children. They reported that these anxiety syndromes were found in about 1-5% of children. They also pointed out that many of these anxiety disorders will be found in association with each other.

Lauren Franz et al. found that GAD, SAD and social phobia are the most common anxiety disorders found in preschool children. They conducted the study among reasonable representative samples of preschool aged children, using structured psychiatric assessments with known psychometric properties.In their study, they were able to find that about 1% of the children suffered from anxiety disorders. The range of prevalence estimates for generalized anxiety disorder reported in these studies is 0.6% to 6.5%, whereas prevalence estimates of separation anxiety disorder range from 0.3% to 5.4%, and social phobia range from 0.5% to 4.4%. They also found out that these anxiety disorders are highly comorbid at all ages. Comorbidity data from the preschool period remains limited with studies showing the co-occurrence between broad categories of emotional and behavioral disorders. Through this study they also reported that, comorbidity, even in the preschool years, is perhaps expected, as population-based twin studies indicate that the genetic liability for specific anxiety disorders partly overlaps.

The Duke Preschool Anxiety Study[9] aimed to do the following: establish prevalence rates for generalized anxiety disorder, separation anxiety disorder, and social phobia according to the DSM-IV criteria; detail patterns of comorbidity among generalized anxiety disorder, separation anxiety disorder, and social phobia; and examine the patterns of comorbidity among generalized anxiety disorder, separation anxiety disorder, and social phobia and other anxious or non-anxious disorders in preschool-aged children in pediatric primary care. The Duke Preschool Anxiety Study was done in three phases over a long period of time. The different phases include 1). The primary care questionnaire screening phase, 2). An in-home parent interview phase and 3). Lab based case control phase. From this study, they pointed out that among the preschoolers with anxiety disorders, only one fourth of them contains symptoms of more than one of the anxiety disorders. They also found that the Generalized Anxiety Disorder displayed the greatest degree of comorbidity with other anxiety disorders.

These anxiety disorders can be more commonly found in the early stages of life itself which may lead to serious psychological disorders if left undiagnosed. Appropriate measures should be found out for finding these disorders in the early stage itself. Some of the different practices or scales that are used for finding the anxiety disorders are discussed below.

2.1. Pediatric Anxiety Rating Scale (PARS)

The Pediatric Anxiety Rating Scale (PARS)[5] was developed as a clinically rated instrument for assessing both impairments and overall anxiety severity. It can be used to examine the course of anxiety symptoms and for judging the impact of treatments on the extremity and intervention of anxiety symptoms in youths. In this work they created a checklist for screening the children and youths. The checklist was derived mostly from the *DSM-IV* criteria for anxiety disorders including social phobia (SoP), separation anxiety disorder (SAD), and generalized anxiety disorder (GAD) and by collecting details from a group of expert child and adolescent psychiatrists and psychologists at the participating sites of the Research Units on Pediatric Psychopharmacology (RUPP) who were asked to comment on: 1). Face validity, 2). Comprehensiveness, 3).Overlap with other constructs, and 4).Readability and clarity. This 50-item symptom checklist organizes items into the following categories: Social Interactions or Performance Situations (9 items), Separation (10 items), Generalized (8 items), Specific Phobia (4 items), Physical Signs and Symptoms (13 items), and other (6 items).

The PARS scale assessment is done by collecting information about symptoms from interviews with both child and parent(s).Endorsed symptoms are then collectively rated by the clinician on seven dimensions of severity, using a 6-point scale (0 for none, and 1–5 for minimal to extreme) for each dimension. These dimensions are (1) number of symptoms (none to more than 10), (2) frequency (none to several hours per day), (3) severity of distress associated with anxiety symptoms (none to extreme), (4) severity of physical symptoms (none to extreme), (5) avoidance (none to extreme), (6) interference at home (no interference to totally or almost totally unable to function at home), and (7) interference out of home (no interference to totally or almost totally unable to function out of home such as with peers or at school). The PARS Total Score is then calculated by adding the above scales of dimensions excluding number of symptoms and severity of physical symptoms. A score of 3 on each of these 5-point scales indicates a clinically significant level of severity, avoidance, or interference.

2.2 Multidimensional Anxiety Scale for Children (MASC)

Johns S. March and his colleagues developed the Multidimensional Anxiety Scale for Children (MASC)[6]. The MASC was developed to evaluate common anxiety symptoms in children across the elementary junior and senior high school age range. So when collecting the details for the anxiety symptom assessment procedure,they opt for not to assume anything about the normative clustering of pediatric anxiety symptoms other than to hypothesize that specific descriptors should fall within emotional, cognitive, physical or behavioral symptom domains .
Several steps were involved in the creation of MASC scale:

1. Available self-report anxiety scales covering general and specific symptom domains as well as the *DSM-lll-Rcriterion* items were inspected again.
2. Each item/question from these measures was then arranged into the one of the four symptom domains specified above.
3. The number of items in each pool were reduced by:
    - Retaining items that can be easily understood, covered the desired age range, and closely reflected one of the four chosen anxiety dimensions.
    - Eliminating duplicates and rewording.

4. Expert clinicians, members of an anxiety disorders support group, and lay non-experts classified the selected 60 items(15 per group) into the four selected domains. A Q sort procedure was used for this, which increased the item-content validity.
5. Based on the selected experts' comments and the pattern of classification, a 41 item, 4-point Likert scale-having approximately 10 items perhypothesized symptom domain-was developed.

During further study, they were able to find out that the MASC structure, which reflects the in vivo structure of pediatric anxiety symptoms, is invariant across gender and age and shows excellent internal reliability. Shared variance with scales sampling symptom domains of interest was highest for anxiety, intermediate for depression, and lowest for externalizing symptoms, indicating adequate convergent and divergent validity.

2.3. Screen for Child Anxiety Related Emotional Disorders (SCARED)

The Screen for Child Anxiety Related Emotional Disorders (SCARED)[7], is a self-report instrument that was developed as a screening tool for childhood anxiety disorders. Several steps were involved in the creation of final SCARED items. At first an85 item questionnaire was generated by experienced clinicians, identifying symptoms of separation anxiety disorder (SAD), generalized anxiety disorder (GAD), social phobia, and school phobia (a simple phobia) based on *DSM-IV* classification of anxiety disorders. The questionnaire was then administered to children between the ages of 9 and 18 years and to their parents who attended the Child Mood and Anxiety Outpatient Clinic at Western Psychiatric Institute and Clinic in Pittsburgh. The severity of symptoms for 3 months was rated using a 0 to 2point rating scale, with 0 meaning not true or hardly ever true, 1 meaning sometimes true, and 2 meaning true or often true and the collected data was used for preliminary analysis. Finally, a38-item SCARED was derived after the factor analysis and deletion of items that overlapped with depressive symptoms.

2.4. Child behaviour checklist(CBCL)

The CBCL[11] is a standardized form which parents could use to describe their children's behavioral and emotional problems. The CBCL consists of a set of questions and for each question the parents could mark their response as 0 if the item is not true for their children, 1 if the item is somewhat or sometimes true and 2 if the item is very true or often true. There are two versions for CBCL form, the first version for children of age 2-3 and the second version for children of age 4-18. The CBCL/2-3 syndromes are designated in six areas: anxious/depressed, withdrawn, sleep problem, aggressive behavior and destructive behavior. The CBCL/4-18 consists of additional competence items to assess the child's activities, social relation and school functioning. The symptoms for these syndromes are rated by the parents based on their child's behavior for 2 months in CBCL/2-3 and for 6 months in CBCL/4-18.

The data obtained with the CBCL are summarized on a profile that displays the parent's rating of each item. The child's score on each syndrome is the sum of numbers that their parents marked on the individual items that comprise the syndrome. The profile could be used to analyze the child's standing on syndromes of problem that were derived from statistical analyses of CBCLs filled out for large number of clinically referred children. On the profile there are 2 broken lines which indicate a border line range between the normal and clinical ranges, the bottom line through 95% and top line through 98%. The score that are below the bottom broken line are considered normal and those that are above the top broken line are in clinical range. Scores between the broken line are high enough to be of concern but not high enough to be considered very deviant.

2.5. Preschool Age Psychiatric Assessment (PAPA)

Preschool Age Psychiatric Assessment (PAPA) [12] by Helen Link Egger is another effective screening tool for anxiety disorders. The PAPA is a parent interview for diagnosing psychiatric disorders in preschool children based on the parent version of the Child and Adolescent Psychiatric Assessment (CAPA).The trained PAPA interviewers interview the parent or other primary caregiver. A 3-month primary period is given to the parents for the assessment. There are 2 versions for PAPA. One is paper version and other is ePAPA. If the paper version is used then the individual PAPA variables are then entered into a database. If the ePAPA or electronic version of PAPA is used then the interview and coding are completed on a PC, making separate data entry unnecessary.

PAPA separately assesses the presence of the symptoms and the disabilities resulting from these symptoms. The symptoms are assessed mainly in 4 domains. They are: DSM-1V-TR and ICD-10 diagnostic criteria, Research Diagnostic criteria, and relevant behaviors and symptoms experienced by preschoolers and their families. 30 areas are covered while assessing the disabilities which includes the

IOP Conf. Series: Materials Science and Engineering **396** (2018) 012018 doi:10.1088/1757-899X/396/1/012018

child's relationship with his or her parents, other adults, siblings and peers as well as child's functioning in home, at school and out of the home.

One of the main advantages of PAPA is that, it combines the characteristics of both interviewer-based and respondent-based interview. Like respondent- based interviews, the PAPA uses a highly structured protocol, with required questions and probes. However, as trained interviewers are used, they could provide clear examples about behaviors or feelings relevant to the symptom, and have the symptom at a prespecified level of severity which helps the subjects understand the question clearly.

Egger et al. conducted the test-retest reliability of Preschool Age Psychiatric Assessment[12]. Three phases were used for performing the test, the screening phase, Test Re- Test phase (TRT) and data analysis phase. In the screening phase equal number of parents of boys and girls were selected based on some selection criteria. In the TRT phase, selected parents were subjected to interviews. The interviewers were blind to the parent's screen status. After the first interview a second interview was scheduled within a period of about 1 week. The second interview was by an interviewer who was blind to the results of the first. Cohen's k method was used to assess agreement on categorical variables, and the intra class correlation coefficient (ICC) was used to assess agreement between syndrome scale scores. Computed weighted reliability statistics from the whole sample was used to produce unbiased estimates of reliability for pediatric primary care. These weights were inversely proportional to the probability of selection into the test-retest sample. They also computed non weighted estimates for the screen high group alone to approximate expected reliability in psychiatric clinic samples. They achieved diagnostic reliability similar to those obtained from interviews with k ranging from 0.36 to 0.79. They also achieved good reliability in the assessment of disability and SED which proves that PAPA could be helpful for both researchers and clinicians. As the study included assessment of children of different ages and from different backgrounds, they suggest thatPAPA could be used with various population of children.

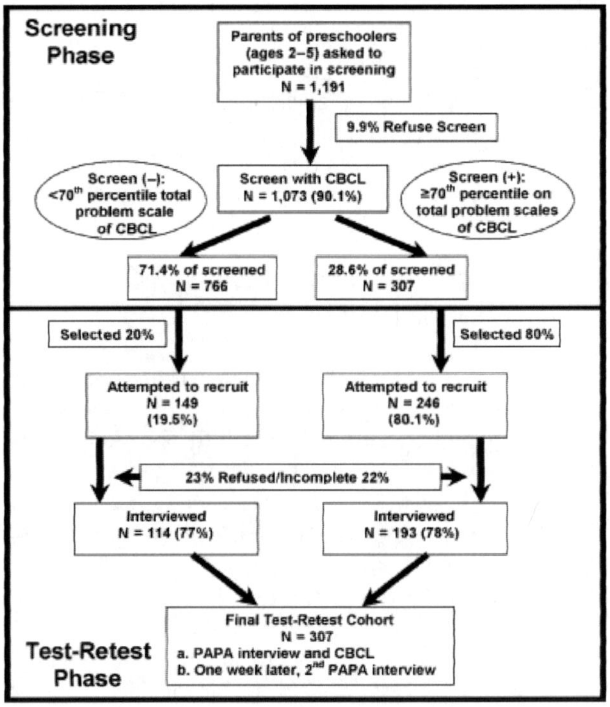

Figure 2.5.1 : Design of PAPA test-retest study[12]

Thus the assessment of psychopathology in childhood is commonly carried out by performing long comprehensive interviews such as Preschool Age Psychiatric Assessment (PAPA). Unfortunately, the time required to complete a full interview is too long to apply it at the scale of the actual population at risk, and most of the population goes undiagnosed or is diagnosed significantly later than desired. So numerous works are being done to find the relationship among the different factors/items and combine them, thereby reducing time required for completing these different assessment scales and methods

One such notable work was done by Birhameret. al [8]. In this work they tried to reduce the number of items in SCARED and provide same reliability as that of the full version SCARED. Factor analysis was conducted by administering the SCARED to a group of children and their parents. In the preliminary analysis, they found that the social factor did not discriminate well between patients with social phobia and other anxiety disorders. So 3 items were added to the 38-item version for the conducting the test. Based on the factor analysis on the 41 item version, 5 factors were obtained: panic/somatic, generalized anxiety, separation anxiety, social phobia, and school phobia. The 5 factor SCARED demonstrated good internal consistency and discriminant validity.

The recent study about preschool anxiety disorders was held by Kimberly Carpenter [13] and his colleagues. In this work, they aim to learn from unique and very rich previously collected PAPA examples the inter-correlations between different questions in order to provide a reliable risk analysis in the form of a much shorter interview. They use for this purpose the alternating decision trees algorithm, which combines decision trees with boosting to produce small and interpretable decision rules. Rather than a binary prediction, the algorithm provides a measure of confidence in the classification outcome. This is highly desirable from a clinical perspective, where it is preferable to abstain a decision on the low-confidence cases and recommend further screening. In order to prevent over-fitting, they propose to use network inference analysis to predefine a set of candidate question with consistent high correlation with the diagnosis.

Among the different measurement techniques for anxiety disorders, PAPA is more reliable and most widely used. PAPA include a large number of questions and therefore it may take hours to complete one PAPA interview. So it is important to reduce the number of questions to detect anxiety disorders without compromising the reliability of the test. Even though works are done regarding reducing the complexity of the PAPA assessment, a more reliable and shorter version of PAPA to assess the most commonly found anxiety disorders would be highly beneficial.

Since the dataset provided by PAPA is high dimensional a screening tool is very essential to detect the disorder. After PAPA interview the collected data must be clinically analyzed to detect the disorder. This whole procedure is very time consuming. Machine learning technique using ADTree classifier is found to be very effective to develop screening tool to detect anxiety disorder[14]. In order to reduce the complexity of constructing ADTree different machine learning techniques can be used to select most informative features from PAPA dataset .

3. ADTree Classifier

An alternating decision tree(ADTree) is a decision tree consists of alternating layers of prediction nodes and splitter nodes. The values in a root node represent the initial probability for assigning the target class according to the training dataset. Alternating decision tree classifiers are then built according to a particular structure using boosting wherein simple rules are successively added to the alternating decision tree classifier until the unit classifier of the tree exhibits satisfactory performance. Since boosting iteration adds three nodes (one splitter node and two prediction nodes) to the tree, more boosting iterations will result in larger and potentially more accurate trees.

IOP Conf. Series: Materials Science and Engineering **396** (2018) 012018 doi:10.1088/1757-899X/396/1/012018

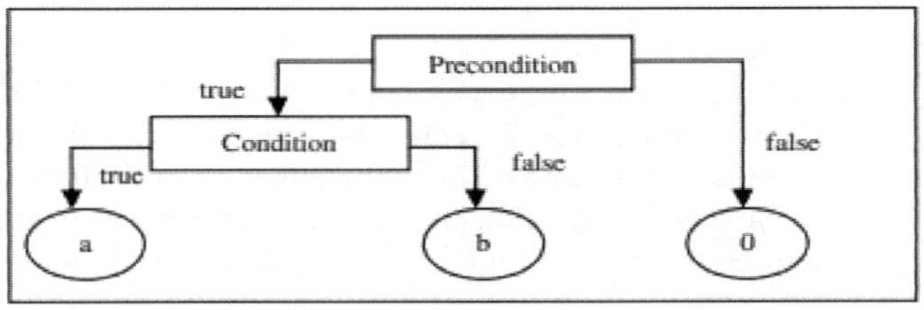

Figure 3.1: Graphical representation of single decision rule of ADTree[14]

Unlike original decision trees: an instance is mapped into a path along the tree from the root to one of the leaves and output is the label of the leaf, the classification result of an alternating decision tree became the sign of the sum of the predictions along the multi-path associated with the given instances. When some feature values are unknown, the alternating decision tree algorithm only considers the reachable decision nodes. Also, the alternating decision tree algorithm has shown competitive performances and has produced smaller and intuitive classification rules than general decision tree algorithms. The figure below is an example of ADTree.

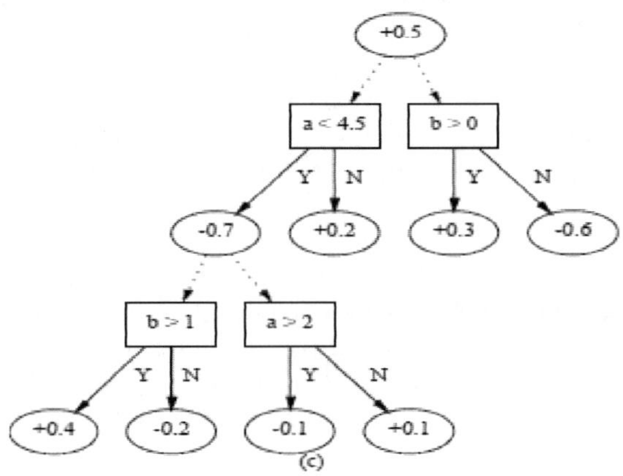

Figure 3.2 : Simple ADTree[14]

4. Conclusion

In this paper different methods such as PARS, MASC, SCARED, CBCL and PAPA for measuring anxiety disorders are analyzed. From that PAPA is found to be reliable and valid structured diagnostic parent report interview for detecting anxiety disorders. But PAPA is very time consuming as it involve a long set of questions. So the advantage of PAPA assessment can be utilized by extracting the essential questions that would identify the SAD and GAD. The most recent approach to develop a screening tool for detecting GAD and SAD is a machine learning based method. . ADTree classifier can be used to develop a screening tool for detecting anxiety disorder. Instead of making a binary decision, ADTree classifier produces an ADScore, which can be used to calculate risk score for GAD and SAD. Risk score value lies between zero and one. A score near to zero indicates less chance of disorder and score near to one indicates high chance of disorder.

5. References

IOP Conf. Series: Materials Science and Engineering **396** (2018) 012018 doi:10.1088/1757-899X/396/1/012018

[1]Kagan, S.L., Moore, E. and Bredekamp, S. eds., 1998. *Reconsidering children's early development and learning toward common views and vocabulary*: National Education Goals Panel. DIANE Publishing..
[2] Denham, S.A., 2006. *Social-emotional competence as support for school readiness: What is it and how do we assess it?*. Early education and development, 17(1), pp.57-89.
[3] Anderson, J.C., Williams, S., McGee, R. and Silva, P.A., 1987. *DSM-III disorders in preadolescent children: Prevalence in a large sample from the general population*. Archives of general psychiatry, 44(1), pp.69-76.
[4] Orvaschel, H. and Weissman, M.M., 1986. *Epidemiology of anxiety disorders in children: A review.* Anxiety disorders of childhood, pp.58-72.
[5] Research Units on Pediatric Psychopharmacology Anxiety Study Group, 2002. *The pediatric anxiety rating scale (PARS): Development and psychometric properties*. Journal of the American Academy of Child & Adolescent Psychiatry ,1061-1069.
[6]March, J.S., Parker, J.D., Sullivan, K., Stallings, P. and Conners, C.K., 1997. *The Multidimensional Anxiety Scale for Children (MASC): factor structure, reliability, and validity*. Journal of the American academy of child & adolescent psychiatry, 36(4), pp.554-565.
[7] Birmaher, B., Khetarpal, S., Brent, D., Cully, M., Balach, L., Kaufman, J. and Neer, S.M., 1997. *The screen for child anxiety related emotional disorders (SCARED): Scale construction and psychometric characteristics*. Journal of the American Academy of Child & Adolescent Psychiatry, 36(4), pp.545-553.
[8]Birmaher, Boris, et al. 1999. *Psychometric properties of the Screen for Child Anxiety Related Emotional Disorders (SCARED): a replication study*.Journal of the American Academy of Child & Adolescent Psychiatry 38.10 , 1230-1236.
[9]Franz, L., Angold, A., Copeland, W., Costello, E.J., Towe-Goodman, N. and Egger, H., 2013. *Preschool anxiety disorders in pediatric primary care: prevalence and comorbidity*. Journal of the American Academy of Child & Adolescent Psychiatry, 52(12), pp.1294-1303..
[10] Biederman, J., Wozniak, J., Kiely, K., Ablon, S., Faraone, S., Mick, E., Mundy, E. and Kraus, I., 1995. *CBCL clinical scales discriminate prepubertal children with structured interview—derived diagnosis of mania from those with ADHD*. Journal of the American Academy of Child & Adolescent Psychiatry, 34(4), pp.464-471.
[11]Egger, H.L. and Angold, A., 2004. *The Preschool Age Psychiatric Assessment (PAPA): A structured parent interview for diagnosing psychiatric disorders in preschool children.* Handbook of infant, toddler, and preschool mental health assessment, pp.223-243.
[12]Egger, H.L. and Angold, A., 2006. *Common emotional and behavioral disorders in preschool children:* presentation, nosology, and epidemiology. Journal of child psychology and psychiatry, 47(3-4), pp.313-337.
[13] Carpenter, K., Sprechmann, P., Fiori, M., Calderbank, R., Egger, H. and Sapiro, G., 2014, May. *Questionnaire simplification for fast risk analysis of children's mental health.* In Acoustics, Speech and Signal Processing (ICASSP), 2014 IEEE International Conference on (pp. 6009-6013). IEEE.
[14] Sok, H.K., Chowdhury, M.S., Ooi, M.P.L., Kuang, Y.C. and Demidenko, S., 2013, May. *Using the ADTree for feature reduction through knowledge discovery*. In Instrumentation and Measurement Technology Conference (I2MTC), 2013 IEEE International (pp. 1040-1044). IEEE.

IOP Conf. Series: Materials Science and Engineering **396** (2018) 012019 doi:10.1088/1757-899X/396/1/012019

Smart Car Parking With Reservation System Using QR Generator

Aswathy James
Department of Computer Science, St.Josephs College Of Engineering And Technology,Palai,India
E-mail: aswathy.james3@gmail.com

Prince Abraham
Asst.Professor Department of Computer Science, St.Josephs College Of Engineering And Technology,Palai,India
E-mail: prince.abraham@sjcetpalai.ac.in

Abstract. Nowadays parking has become an expensive resource in the almost any major city's in the world, and its limited availability is the concurrent cause of urban traffic congestion and air pollution. The common method of finding a parking space is manual where the driver usually finds a space on the street through luck and experience. This process takes time and effort and may lead to the worst case of failing to find any parking space if the driver is driving in a city with high vehicle density.In this paper, an attempt has been done to automate the car as well as the car parking system with a Smart Parking System (SPS) which is based on the integration of an Android app and QR Code reader.The introduction of a novel algorithm that increases the efficiency of the current smart-parking system and develops an android app to collect information about the occupancy state of parking spaces, and to inform the drivers to the nearest vacant parking spot. .The entering into or leaving the parking slot is controlled by an Android-based application. .The algorithm helps improve the probability of successful parking and minimizes the user waiting time.

1. Introduction

Major cities all around the world are facing a new problem nowadays lack of sufficient parking space. With families getting smaller and the total number of motor vehicles is more than the total number of heads per family, the parking scenario is woefully falling short of the current requirements in the country. The situation is such that on any given working day approximately 40% of the roads in urban India are taken up for just parking the cars. The problem has been further exacerbated by the fact that nowadays even people from the low-income group are able to own cars. Particularly in the cities and the big towns, there is a problem where the supply-demand ratio makes parking a problem for parking space providers, the motorists or both. Here comes the significance of a smart car parking system(SPS) technology.

In the development of traffic management systems, an intelligent parking system was created to minimize the cost of hiring people and for optimal use of resources for car-park owners. Currently, the commonly used method of finding a parking space is manual where the driver usually struggles a lot to get a parking slot and sometimes fails to get one parking space if the driver is driving in a city with high vehicle density. The alternative is to find a predefined car park with high capacity. However, this is not an optimal solution because the car park could usually be far away from the user destination.

Here with the project, we aim to develop an autonomous car parking system which is controlled by an Android application and thus to provide an efficient car parking system. The proposed system ensures the driver to reserve a particular parking space in a given parking area, which will be intimated according to the availability of free space. Although the application is capable of suggesting the most optimal parking space for the user based on the destination provided, it has to be further modified in such a way that it can notify the queuing time required when a given slot is already been occupied by someone else.

2. Existing Systems

Various methods are used to improve the intelligent parking mechanisms.Study of these existing systems shows that these systems need little or more human intervention for the functioning. One of the existing systems is [01] Intelligent Systems For Car Parking With Image Processing.In this paper, a brown rounded image on parking slot is captured using the camera and it is used to detect the free parking slots. The currently available parking spaces are displayed on the seven segment display. First, the image of the parking slot with the brown rounded image is taken. Then create the binary images according to the brown rounded images. Due to this, we have to remove the noise of the images and identifies the object boundaries. The image detection module determines which objects are round, by determining each objects area and perimeter. Accordingly, the free parking slots are allocated

Another Existing method is Integration of [02] RFID and WSN Technologies in a Smart Parking System.Basically, the SPS provides innovative services for the automatic supervision of paid parking spaces through the deployment of an IEEE 802.15.4-based WSN able to collect and deliver the data to the central server. A customized application on the server analyzes the received information and also sends an alert message to the mobile application of the traffic cop in case of unauthorized use of a reserved space or expiration of a parking receipt. Drivers can also use the system to pay the fee. The framework of the system consists of WSNs, Smart Gateway (SG), Central Server (CS) and two different mobile applications, called Parking App and Policeman App, designed for vehicle drivers and traffic cops, respectively. The main peripherals of the deployed Zigbee network are Router (R) and Coordinator (C) nodes. The R nodes provide forwarding and routing capabilities, where the c node collects the received data and forwards them to the Central Server. In the RFID-WSN integrated system, the Router Reader (RR) node typology has been introduced, which identifies an R node interfaced with a UHF RFID reader.The designed system consists of a WSN with some R node and RR nodes are spread out in the parking area. In particular, R nodes, equipped with a light sensor, are placed on each parking slot to monitor their state, while the RR nodes are on poles located their neighboring reserved parking space.

The information retrieved by the nodes is delivered, in a multi-hop manner, to the C node, which delivers them to the Smart Gateway. This last one, in turn, tests the collected data and sends them, together with the position of the parking zone, to the CS. the SG provides also an NFC way to finalize users payment for their parking fee. The main function of the RR nodes includes control the reserved parking space and fill that space by using only authorized cars, labeled by UHF RFID tags. More specifically, when the CS receives the information that a reserved parking space has been occupied, it checks if a new RFID tag has been read by the RR node responsible for controlling that specific reserved space, and, in such a case, it verifies its authorizations. The CS maintains a database handling a lot of information about parking spaces available and users payments. In case of improperly use of a reserved space or expiration

IOP Conf. Series: Materials Science and Engineering **396** (2018) 012019 doi:10.1088/1757-899X/396/1/012019

of parking receipt, a parking monitoring application on the CS informs the traffic staff, exploiting the Google Cloud Messaging (GCM)

[03] A vision based car parking system includes 2 type of images.The positive image and negative image.The positive image contains images of cars from various angles but the negative image does not contain the images of cars.based on these two images parking slots are reserved

3. Proposed System

The public parking system developed within the SPS project will partially remodel the public off-street parking system so that only the authorized users can use these facilities and only these users can be reserved the parking space.First, the users can register at the SPS Android App and also book the vacant parking slot. In the rest of the paper, we will refer to an on-street public space managed by SPS as SPS parking spot/space. To establish the SPS architecture, the following business rules have been defined.

First, Reservation confirmation with a QR Code can be sent only to the authorized customers. Second, A reservation is held for a grace period (e.g. 15 min) after the start of the reserved interval in order to account for customers who do not show up in time. If the customer arrives within that period he will park his vehicle to his reserved area. If the customer does not reach within that particular period a message will inform him/her about the expiration of his/her reservation. Third, If the customer can arrive anytime between the grace period, a vacant and unreserved spot will be offered for the remaining period if any free space is available although he has to reserve a new space from his app. The customer will be billed from the start to the end of his/her original reservation. Fourth, No-show customers will not be billed for the canceled reservation. On a further modification of the project, If the customer fails to clear their parking slot at the scheduled time then he can pay the additional charge.A message (e.g. SMS or email) will be sent to the customer to notify these events.SPS business policies do not allow overbooking since IPA is thought for customers who are willing to pay for an on-street parking stall in order to have the absolute certainty to find a parking spot at their arrival in a high-density trafficked area of the city. If SPS allowed overbooking, there will be no difference between a regular parking spot and an SPS stall. Furthermore, SPS parking stalls must be paid independently at the time of exit from the slot, so there is no loss of revenues.

The SPS architecture consists of two modules user and server(admin).This subsystem is a core module of the SPS system. It manages the communication with the customer, for example, accounting, reservation, cancellation, and billing. This module communicates with the user.First user login to the system then enters the destination then choose the suitable parking space.i. The user can see the available slot in the parking space and book the slot before arriving at the venue.Book the slot.The booking is confirmed by the generation of unique QR code which is used in future for verification purpose.The user can cancel the booking.

Next module is Server(admin).The administrator is the person who manages the domain. This module provides the facilities for the admin for managing users. This module could update the details of a new parking area to the existing system.The booking will expire if the user does not turn up within 15mins from the time specified.The QR code is generated when the slot is booked. Once the user leaves, the system updates the database.

4. IMPLEMENTATION

Project implementation(or project execution) is the phase where visions and plans become reality. This is the logical conclusion, after evaluating, deciding, visioning, planning, applying for funds and finding the financial resources of a project. Technical implementation is one part of executing a project, the Implementation process is simply a translation of the design into physical realization, using the language of the target architecture.

- User login.

IOP Conf. Series: Materials Science and Engineering **396** (2018) 012019 doi:10.1088/1757-899X/396/1/012019

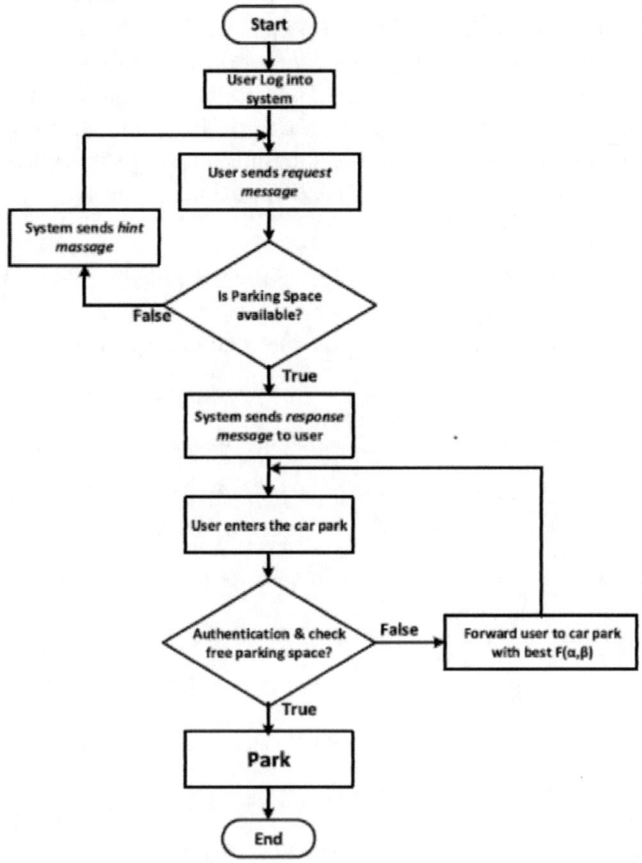

Figure 1. Proposed Architecture of the Smart Parking System with Reservation

- Enter the destination.
- Choose the suitable parking space.
- The user can see the available slot in the parking space and book the slot before arriving at the venue.
- Book the slot
- The booking is confirmed by the generation of unique QR code which is used in future for verification purpose.
- User can cancel the booking.
- Admin can add parking area and slots.
- The booking will expire if the user does not turn up within 15mins from the time specified.
- Slot QR.is being generated
- Once the user leaves, the system updates the database

5. Result And Discussions

In this project, we have presented a novel parking reservation system, called SPS with Reservation, for the management of the off-street parking spots in consolidated cities. SPS puts the management of parking spots into a different perspective that goes over the simple

IOP Conf. Series: Materials Science and Engineering **396** (2018) 012019 doi:10.1088/1757-899X/396/1/012019

automation of parking system through the use of advanced technological solutions, such as wireless networks and sensor communication. In fact, SPS is concerned with (i) the quality of life in modern cities, in terms of the amount of pollution and effects of the urban traffic congestion on the abilities of the drivers and (ii) the quality of mobility in urban areas.

On further modification, we are looking forward to implementing an intelligent parking assistant (IPA) architecture aims at overcoming current public parking management solutions. Further considerations would be made about the evaluation of the revenue of the parking site. Future modifications the SPS system will include:

- a module for the Smartphone App to navigate the users to the assigned parking spots using a vocal guide.
- a module to use Bluetooth technology to complete the check-in process.

6. References

[1] R. Yusnita, Fariza Norbaya, and Norazwinawati Basharuddin "IntelligentSystems For Car Parking With Image Processing "

[2] i. L. Mainetti, L. Palano, L. Patrono, M. L. Stefanizzi, and R. Vergallo,"Integration of RFID and WSN technologies in a smart parking system,"in Proc. 22nd Int. Conf. Softw., Telecommun. Comput. Netw. (SoftCOM),2014, pp. 104110.

[3] Hamada R. H.Al-Absi,Justin Dinesh Daniel Devaraj,Patrick Sebastian,Yap Vooi Voon"A vision based car parking system"

[4] A Cloud-Based Smart-Parking System Based on Internet-of-Things Technologies.

[5] C. W. Hsu, M. H. Shih, H. Y. Huang, Y. C. Shiue, and S. C. Huang,"Verication of smart guiding system to search for parking space viaDSRC communication," in Proc. 12th Int. Conf. ITS Telecommun. (ITST),2012, pp. 7781.

[6] Wigmore, Internet of Things (IoT). Newton, MA, USA: TechTarget, Jun. 2014

IOP Conf. Series: Materials Science and Engineering **396** (2018) 012020 doi:10.1088/1757-899X/396/1/012020

SENTIMENT ANALYSIS ON TEXTUAL REVIEWS

Mirsa Karim
PG Scholar,Department of Computer Science and Engineering St.Josephs College of Engineering and Technology ,Pala,Kottayam,India
E-mail: mirsakarim@gmail.com

Smija Das
Assistant professor, Department of Computer Science and Engineering St.Josephs College of Engineering and Technology,Pala, Kottayam,India
E-mail: smija.das@sjcetpalai.ac.in

Abstract. These days there is an expansion in review websites.It has turned out to be considerably more intricate to mine fundamental data from survey sites and take proper choice .Using Natural Language Processing, there is need to identify sentiment of content or document.In this paper Sentiment Analysis is done in view of Rule based mechanism and machine learning approach.Both of these strategies are analyzed and discovered that machine learning is most appropriate for Sentiment Analysis in light of the exactness measurement.Sentiment Vader and Senti word net are the Rule based algorithms utilized and LDA analysis on Naive bayes is the machine learning strategy used.

1. INTRODUCTION

Sentiment Analysis is a forthcoming exploration progressing field that is becoming important because of utilization of different applications. Supposition Analysis is additionally called as sentiment mining. Audits are given by individuals in an unstructured way in type of forums,blogs etc.Then preprocessing of surveys is done and seen if the survey is certain, negative or nonpartisan. Order approaches like vocabulary and machine learning based methodologies are utilized for Sentiment Analysis. Vocabulary based approach is of word reference based approach and corpus based approach. Machine learning strategies are most broadly used to group and anticipate supposition as either positive or negative conclusion. Machine learning calculations are for the most part named directed or unsupervised approach. Regulated approach takes named dataset where each preparation set has effectively alloted its supposition. Unsupervised approach takes unlabelled dataset where audit isn't characterized with its mark . Assessment investigation alludes to the undertaking of recognizing supposition from surveys.

1.1. Applications and challenges of sentiment analysis

The primary utilization of assumption investigation is in this manner giving the clients the possibility and suggestion in selection of products.A client is normally pulled in to certain part of item if there should be an occurrence of picking an item. A solitary worldwide rating could be exact in decision making process. Estimation investigation can team up the conclusions of the analysis and assess evaluations on specific parts of the item. Another utility of feeling investigation is for organizations that need to know the interest of clients on their products.If client is unsatisfied with specific part of item organization can change the angle and furthermore help to discover what all viewpoints are more pulled in by clients. At long last, Sentiment Analysis has been proposed as a part of different innovations. Sentiment Analysis for discovering conclusion or assessment of client about specific item.

Key challenges on opinion examination are tread precisely on exactness numbers,utliize both machine learning and human knowledge,adopt a multi-technique look into plan,Stop regarding assessment investigation as a hobby,keep a receptive outlook about the discoveries.

2. RELATED WORK

Bhumika M. Jadav[4], performed Sentiment Analysis to order surveys in light of its assessment as either positive or negative category.Here unstructured information was changed over to organized information utilizing preprocessing and later to numeric score esteem. RajatBhat, SwapnilGaonkar[15] utilized Hadoop Distributed File framework (HDFS) to store information set and keep running on MapReduce outline for performing Sentiment Analysis. ArunaSathish[8] centers around the fundamental utilizations of Sentiment Analysis in web based business and in addition data security. By following the client remarks a reasonable thought of brand observation and consumer loyalty was acquired. Xiaojiang Lei[15], XuemingQian propose a social client's audits assessment estimation approach and compute every client's supposition score on things/benefits and considered administration notoriety, which mirrors the clients' thorough assessment. Finally, benefit notoriety factor is combined into suggestion framework to make a precise rating forecast.

XuemingQian, He Feng[1]considered three necessary factors, personal attraction, interpersonal similary consideration, and interpersonal impact, and combined them into a systemic accurate recommendation model probabilistic matrix factorization technique was used for recommendation model. ShengxiangGao, Zhengtao Yu, Linbin Shi, Xin Yan, and Haixia Song[6] proposed a method to find expert rating by using previous rating histories,based on some association rules and result was found by creating problablistic matrix factorisation for review experts.

Evangelos PsomakelisKonstantinos Tserpes[14] contemplated approaches like sack of words,n-grams. Order calculations and dictionary based methodologies were utilized for execution evaluation.Main arrangement calculations utilized for assumption investigation was Multilayer Perceptrons, Nave Bayes,SVM C4.5as well as their mixes. The outcomes demonstrated that machine learning approach was most appropriate for anticipating assumption of tweet. Cataldo Musto, Giovanni Semeraro, Marco Polignano[9] proposed a vocabulary based approach for forecast of estimation in Twitter posts. This investigation depends on the lexical sources like SentiWordNet, WordNet-Affect, MPQA and SentiNet.Anas Collomb[10] proposed procedures for correlation of techniques used to assess the notoriety of things utilizing slant examination. The machine learning approaches utilize calculations to prepare dataset to anticipate the notion. The vocabulary based technique utilizes introduction of sentences for estimation examination. The administer based method characterize given sentence into positive and negative by feeling thought. Shreya Banker1 and Rupal Patel[11] played out a near report to discover the assumption related with audit. Measurable methods,like Naive Bayes are better order strategies for supposition analysis,Machine learning methodology can take the valence of word for estimation in view of recurrence. Bruno Ohana[12] suggested that for conclusion investigation positive and negative word scores can be checked utilizing sentword net by assessment introduction.

3. METHADOLOGY

In this Paper Sentiment Analysis is done based on Rule based mechanism and machine learning approach.Both of these methods are compared and found that machine learning is best suited for Sentiment Analysis based on the accuracy measurement.Sentiment Vader and Senti word net are the Rule based mechanisms used and LDA analysis on naive bayes is the machine learning technique used.

3.1. DATA SET

Moview review dataset (http://www.cs.cornell.edu/people/pabo/movie-review-data/) from cornell university labelled as positive and negative is taken.

3.2. RULE BASED MECHANISMS:

Rule based Sentiment Analysis effectively use rule mining algorithms to discover the features of a product and to find opinion associated with a particular product.

3.2.1. SENTIMENT VADER

VADER(Valence Aware Dictionary and Sentiment Reasoner) is a vocabulary and control based Sentiment Analysis instrument that is particularly receptive to assumptions communicated in social media.This calculation is utilized to group notion into four classes:positive,negative neutral and compound. The compound score, is the entirety of all the dictionary evaluations which have been standardized to go between - 1 and 1.

norm score = score/math.sqrt((score*score) + alpha)

The review is considered as positive, if compound score is $>=$ 0.5,is considered as neutral if score $>$ -0.5 and$<$ 0.5 and is considered as negative if score is $<=$-0.5

3.2.2. SENTIWORD NET

Sentiword net is Natural language processing tool for calculating score of word and to find opinion nature.

Sentiword algorithm is as follows:
Input file is created
Pos tagger is used to parse each sentence
Tag is allocated to each word ie,whether noun,adjective,verb etc
Tag is then checked and assigned to sentiword net to calculate score
Sentiment type of word is then returned
Total no of positive and negative adjective words are counted for each sentence
If count is odd number then sentence is negative ,else positive

3.3. MACHINE LEARNING APPROACHES

Machine learning approaches learns the sentiment associated with the corresponding sentence but they require training data which is somewhat difficult to obtain and they are often much more computationally costly in associated with time taken for classification , cpu processing etc.

IOP Conf. Series: Materials Science and Engineering **396** (2018) 012020 doi:10.1088/1757-899X/396/1/012020

3.3.1. LDA ANALYSIS ON NAIVE BAYES:

3.3.2. Data Preprocessing for LDA

All the unique words are constructed in the vocabulary V , each word has a label wi 1, 2, . . . , Nd .For each document dj , we choose a dimensional Dirichlet random variable $\theta m Dirichlet$ $(a) Foreachtopiczk, wherek_[1, \gamma], wechoose\theta kDirichlet$ $(b) Foreachtopiczk, theinferenceschemeisbasedupontheobservationthat$:

$$p\left(\Theta,\Phi|D^{\mathrm{train}},\alpha,b\right)=\sum_{z}p\left(\Theta,\Phi|z,D^{\mathrm{train}},\alpha,b\right) \times P\left(z,|D^{\mathrm{train}},\alpha,b\right).$$

3.4. Extracting Product Features

Tags are added ie:, the symbol / before product features to distinguish other words in reviews.

LDA is a Bayesian model, which is utilized to model the relationship of reviews, topics and words.

The terminologies used in LDA model is described as:
1) v : vocabulary consisting of unique words
2) wi 1, 2, . . .,Nd : unique word in review
3) dm: document associated with user
4) T: total number of topics
5) θm: the multinomial distribution of topics
6) $_{\phi}$k : the component for each topic
7)a,b : Dirrichlet priors associated with the multinomial distributionm and k

3.5. User Sentimental Measurement:

Then for a review 'r' that user'u'posts for the item 'i', Sentiment score is calculated as follows:

s(r)= 1 /Nc (Q *Rw *Dw)

where c denotes the clause.Nc denotes the number of clauses.Q denotes the negation check coefficient. Dw is determined by the empirical rule.The Naive bayes classifier works on Bayesian theorem and they perform classification using multinomial logistic regression.

The Nave bayes classifier is a simple classifier that relies on Bayesian probability and the nave assumption that feature probabilities are independent of one another. Training data give information about conditional and priori probabilities which feed to Naive bayes classifcation which form MachineLearning based classifier. In conditional probabilities look for words that appear more in positive and negative reviews.when a review is given it is split to words and look conditional probability whether they are likely to occur in positive or negative reviews.

3.6. Extract reviews
Extract reviews get reviews from positive and negative files and return 2 list one for positive reviews and other for negative reviews.For every review return tuple and label of review.

3.7. Build Vocabulary
For building Vocabulary iterate to all words in reviews and extract words .Then will get A unique collection of words.Add all words to a set.Then create a dictionary of words by iterating through all words in vocabualary.If a word is Present in vocabulary dictionary hold true for value else show false.

3.8. Training
Extraction of features will give all the features needed for training data.It is then given to nltk classifier.once we have trained classifier we will use classifier for individual reviews and then apply to test data, which individualy classify positive and negative review and store it in list .Once prediction is complete see how many predicted labels are corect.

4. PERFORMANCE EVALUATION
SentiWord Net,Sentiment Vader and LDA Analysis on naive Bayes was used inorder to predict sentiment associated with review and found that machine learning technique is best suited for Sentiment Analysis.In Sentiword net and Vader ,label 1 is assigned to positive review and 0 for negative review.Summing all values of positive and negative reviews with the total length of reviews is used for performance measurement.While in Naive bayes label 1 for positive reviews and -1 for negative reviews is used.

Based on the Accuracy measurement obtained from the same data set given it is shown that Sentiment Vader has 54.7 percent Accuracy which is low .Naive bayes has Accuracy of 75.2 which is high as compared to rule based mechanisms which shows that machine learning techniques is more accurate in calculating Sentiment of reviews.

Algorithm Used	SentiWord Net	Sentiment Vader	LDA on Naive Bayes
Accuracy	59.17	54.76	75.2

5. ARCHITECTURE
In the proposed system architecture Text Reviews ie, sequential data is given as input.Then Sentiment Analysis is done by Sentiwordnet,Sentiment Vader and LDA analysis on Naive bayes.Sentiword net use POS Tagging ie. assigning a unique tag to each word in review whether they are verb,adverb,noun or adjective.Then synset score is evaluated for Sentiment Analysis.ynsets are set of synonyms.

Vader uses compound score to calculate sentiment associatd with review.For training Naive bayes build a Vocabualary and extract features from the review. HDInsight is a completely overseen cloud benefit that makes it simple, quick, and savvy to process enormous measures of information. Utilize prevalent open-source systems, for example, Hadoop, Spark, Hive, LLAP, Kafka, Storm, R and more. Sky blue HDInsight empowers an expansive scope of situations, for example, ETL, Data Warehousing, Machine Learning, IoT and more.The IPython Notebook is presently known as the Jupyter Notebook. It is an intelligent computational condition, in which we can consolidate code execution, rich content, arithmetic, plots and rich media.

IOP Conf. Series: Materials Science and Engineering **396** (2018) 012020 doi:10.1088/1757-899X/396/1/012020

Power BI is a suite of business investigation instruments that convey bits of knowledge all through your association. Interface with several information sources, disentangle information prepare, and drive specially appointed investigation. Deliver lovely reports, at that point distribute them for your association to devour on the web and crosswise over cell phones.

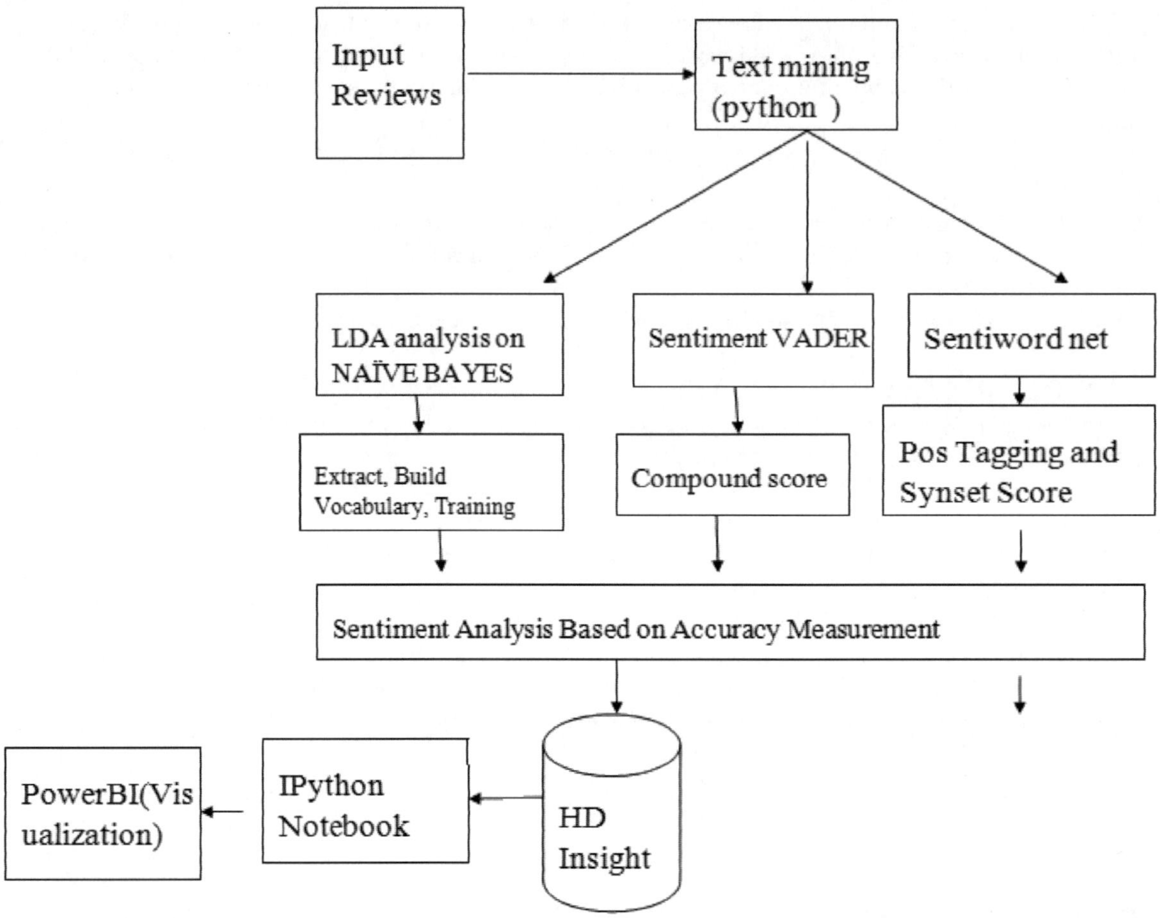

Figure 1. Proposed System Architecture

6. CONCLUSION

Sentiment Analysis ,otherwise called assessment mining is utilized to distinguish estimation related with a sentence. This examination would help the client to perform fitting rating. Sentiment examination dependably manages extremity of sentence.In this paper feeling examination is finished by utilizing Rule based and Machine learning based systems and found that machine learning procedure is more precise and accurate in anticipating the conclusion of a sentence or finding sentiment associated with sentence. Future work can be improved as research in joining a classifier with different methodologies, for example, word vectors may create preferable outcomes over every individual classifier can deliver without anyone else.

IOP Conf. Series: Materials Science and Engineering **396** (2018) 012020 doi:10.1088/1757-899X/396/1/012020

References

[1] XuemingQian, He Feng, Guoshuai Zhao, Tao Mei, Personalized Recommendation Combining User Interest and Social Circle, IEEE TRANSACTIONS ON KNOWLEDGE AND DATA ENGINEERING,2007

[2] Yan-Ying Chen,An-Jung Cheng, and Winston ,Travel Recommendation by Mining People Attributes and Travel Group Types From Community ContributedPhotos,IEEETRANSACTIONSONMULTIMEDIA,VOL.15,NO.6,OCTOBER2013

[3] Wenjuan Luo1,,FuzhenZhuang, Ratable Aspects over Sentiments: Predicting Ratings for Unrated Reviews ,2014 IEEE International Conference on Data Mining

[4] Bhumika M. Jadav,Sentiment Analysis using Support Vector Machine based on Feature Selection and Semantic AnalysisInternational Journal of Computer Applications 2015

[5] Xiaojiang Lei, XuemingQian,Rating Prediction via Exploring Service Reputation,2015 IEEE 17th International Workshop on Multimedia Signal processing

[6] ShengxiangGao, Zhengtao Yu, Linbin Shi, Xin Yan, and Haixia Song Review Expert Collaborative Recommendation Algorithm Based on Topic RelationshipIEEE/CAA JOURNAL OF AUTOMATICA SINICA, VOL. 2, NO. 4, OCTOBER 2015

[7] Hao Ma, Haixuan Yang, Michael R. Lyu, Irwin King,SoRec: Social Recommendation Using Probabilistic Matrix Factorization 2015 IEEE International Conference on Data Mining

[8] ArunaSathish,Sentiment Analysis in E-Commerce and Information Security,International Journal of Innovative Research in Computer and Communication Engineering,2016

[9] Cataldo Musto,Giovanni Semeraro, A comparison of Lexicon-based approaches for Sentiment Analysis of microblog posts, IEEE International Conference on Data Mining,2016

[10] Anas Collomb, A Study and Comparison of Sentiment Analysis Methods for Reputation Evaluation,IEEETRANSACTIONSONMULTIMEDIA,2016

[11]] Shreya Banker1 and Rupal Patel, A BRIEF REVIEW OF SENTIMENT ANALYSIS METHODS International Journal of Information Sciences and Techniques (IJIST) Vol.6, No.1/2, March 2016

[12]] Bruno Ohana , Brendan Tierney, Sentiment Classification of Reviews Using SentiWordNet2016

[13] Xiaojiang Lei, Xueming Qian, Rating Prediction Based on SocialSentiment From Textual Reviews, IEEE TRANSACTIONS ON MULTIMEDIA, VOL. 18, NO. 9, SEPTEMBER 2016

[14] Evangelos Psomakelis,Konstantinos Tserpes, COMPARING METHODS FOR TWITTER SENTIMENT ANALYSISConference: Conference:International Conference on Knowledge Discovery,2017

[15] RajatBhat, SwapnilGaonkar,Sentiment Analysis of Product Reviews using Hadoop,IJSRD - International Journal for Scientific Research Development— Vol. 4, Issue 12, 2017

IOP Conf. Series: Materials Science and Engineering **396** (2018) 012021 doi:10.1088/1757-899X/396/1/012021

A Performance Improvement Inference Method For Link Prediction in Social Graphs

Ashly Ann Jo

PG Scholar,Department of Computer Science,St.Joseph's College of Engineering and Technology,Pala,India;

E-mail: appuachi02@gmail.com

Prince.V.Jose

Assistant Professor,Department of Computer Science,St.Joseph's College of Engineering and Technology,Pala,India;

E-mail: pvj.prince@gmail.com

Abstract. Social network analysis has turned into a conspicuous field in link prediction.The precise social network models are additionally address a few downsides. Since the link forecast in Social network analysis confront a few threats, for example, the in partially correct rules. An ideal inference mechanism should scale up towards vast scale information. The inference methods uses probabilistic evidence data since it can easily predict the vulnerabilities. There are diverse responses for Social network analysis have suggested over years. In this approach develop a model to predict the nearness of associations among nodes in broad scale casual groups, for example, informal organizations, which are exhibited by Markov Logic Networks (MLNs) and Bayesian Networks.This show gives a successful inference model which can deal with complex conditions and somewhat partially correct rules. The proposed system predicts the accuracy and efficiency of link prediction utilizing MAP Markov Logic induction technique and MPE Bayesian derivation strategy.

1. Introduction

Social network analysis is a mainstream approach to display the collaboration among the general population in a gathering or group. It can be envisioned as a diagram, where a vertex compares to a man in that gathering and an edge speaks to some type of relationship between the relating people. The affiliations are generally determined by shared interests that are normal in that gathering. Be that as it may, informal communities are extremely powerful protests, since new edges and vertices are added to the chart over the time. Understanding the progression that drives the advancement of interpersonal organization is a mind boggling issue because of countless parameters.In any case, a similarly less demanding issue is to comprehend the relationship between two particular nodes. This paper talk about the elements that drive the affiliations and to make the relationship between two hubs influenced by different nodes. The particular issue case that address in this exploration is to foresee the probability of a future relationship between two nodes, realizing that there is no relationship between the nodes in the present condition of the graph[1]. This issue is normally known as the Link Prediction issue.

IOP Conf. Series: Materials Science and Engineering **396** (2018) 012021 doi:10.1088/1757-899X/396/1/012021

This link prediction can be reduced by improving the efficiency and accuracy of prediction model using inference method.This paper proposed an inference method for link prediction model which derive a subgraph from social network graph.Thus the proposed system can achieve maximum expectation result from the given query in social network graph edges.

2. Related Works

2.1. Statistical Relational Learning

Statistical relational learning is an emerging area that combines statistics and artificial intelligence. The impact of number of samples on inference accuracy in machine learning for addressing uncertainty with complex inherent structural constraints. In statistical relational learning the uncertainty is solved by using statistical methods. Structural constraints can be represented using first-order logic. The surmising in measurable social learning is ordinarily determined by utilizing probabilistic graphical models, e.g., Bayesian networks and Markov networks. Bayesian systems speak to a joint appropriation of irregular factors as a coordinated non-cyclic chart. Its nodes are the irregular factors while the edges relate to guide impact starting with one node then onto the next. The BayesStore project [18] is one recent work based on Bayesian networks for probabilistic inference. Compared to Bayesian networks, Markov networks represent a joint probability distribution of random variables as an undirected graph, where the nodes represent the variables and the edges correspond to the direct probabilistic interaction between neighbouring variables. Oliveira and Gomes [4] developed Markov logic networks for web reasoning over partially correct rules. However, applying Markov Logic Networks in prediction has its drawbacks such ad its low efficiency in domains involving many objects. Some approaches are proposed to minimize this problem. Singla and Domingos [19] proposed a lazy implementation of Markov logic networks, named as LazySAT. Mihalkova and Richardson [20] exhibited a meta-inference calculation that can accelerate the derivation by maintaining a strategic distance from excess calculation. Shavlik and Natarajan [6] provided a pre-processing technique to reduce the effective size of inference networks. However, all the above approaches experimented only with very small-scale data sets, involving only a very limited number of objects.

2.2. Link Prediction on Social Networks

Liben-Nowell [21] formalized the link prediction problem and developed approaches to link prediction based on measures for analysing the proximity of nodes in a network. Unfortunately, they considered only the features that are based on the link structure of the network itself. Leroy [22] acquainted a system with anticipate the structure of an interpersonal organization when the system itself is thoroughly absent while some other data in regards to the nodes are accessible. They first generated a bootstrap probabilistic graph using any available feature and then applied the link prediction algorithms in [21] to the probabilistic graph. Backstrom [23] proposed supervised random walks based method for link prediction that performs a PageRank-like random walk on a social network.The proposed approach utilizes Markov Logic Networks to capture complex and structural relation and interaction among social entities.

2.3. Sampling in Probabilistic Databases

Sampling based methodologies [13], [14], [15], [19] are proposed for overseeing inadequate and indeterminate information in probabilistic databases [16], [17],[24], [25]. The thought is basic furthermore develop an arbitrary examples while watching the earlier factual learning and imperatives about the data. Therefore, each example is one conceivable acknowledgment (conceivable world) in the space of vulnerability, and the whole arrangement of tests uncovers the dispersion of the unverifiable information which need to demonstrate. The queries and inductions are then directed against this inference. MCDB [19], for instance, permits a client to characterize self-assertive variable age works that exemplify the database vulnerability. MCDB utilizes these capacities to pseudo-randomly produce acknowledged esteems for the indeterminate characteristics, and assesses questions over the acknowledged esteems. On contrasted with the factual social learning approaches, the majority of the above works are just centered around easier probabilistic models for query handling

3. Proposed System

A model to appraise connect presence in unverifiable informal communities in light of probabilistic thinking. The proposed framework is for the most part executed by MLN induction yet figures out how to give a harmony between MLN surmising precision and Bayesian derivation effectiveness. To start with, so as to keep up the framework of social diagrams, sort every one of the nodes by their degrees and build the relating k-backbone graph. At that point autonomous keeps running of the Random Walk Metropolis(RWM) testing is performed to investigate neighborhood social graphs. This technique applies MLN interface on the association of the considerable number of hubs in the k-backbone graph and all nodes found locally. What's more, with a specific end goal to help questionable probabilistic confirmation information, first characterize probabilistic social graphs by infusing vulnerability into social diagrams and formalizing the connection inquiries on them. At that point a model for anticipating probabilistic connections is proposed to answer such queries.[1] The Markov rationale systems used in taking care of incompletely rectify surmising rules which show up ordinarily in informal communities. This technique lessens the time and space costs by a few requests of size keeps up the inference exactness in a satisfactory level. It foresee a predetermined connection (e.g., fellowship) exists among the two connected individuals.

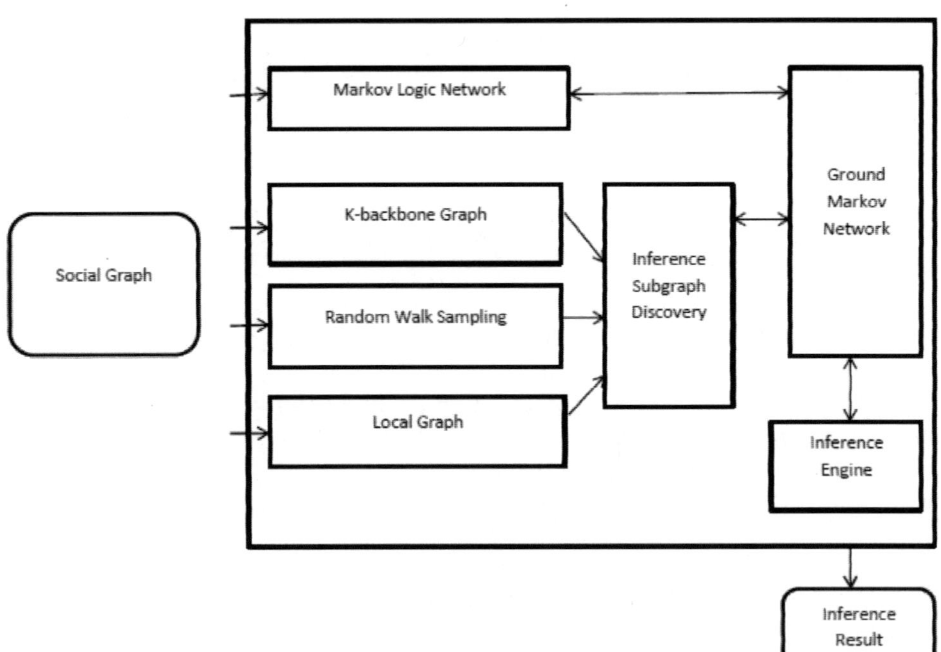

Figure 1. System Overview of Inference Method.

IOP Conf. Series: Materials Science and Engineering **396** (2018) 012021 doi:10.1088/1757-899X/396/1/012021

3.1. Probabilistic social graphs
The Probabilistic social graphs can be stored in a database. Database may have the following schema: <from, to, prob >[1]. Each pair of from and to contain the two IDs of the vertices associated to an edge. I prob denotes the weight of an edge, reflecting the probability of a friendship. There is a relation defined as a mutual friendship between people. For example, a record $< 1, 2, 1.0 >$ means that the person with ID 1 knows another person with ID 2 with a probability of 1.0.

3.2. Link Prediction
Given a probabilistic social graph G, the purpose of link prediction is to predict the probability that specific link exists between two nodes in G [1]. Query(A, B, knows) may be launched to investigate how well A knows B, i.e., from the probabilistic view, to estimate the likelihood that a potential friendship edge exists from A to B.

3.3. k-Backbone Graph
The weighted degree (WD) of a vertex u in a probabilistic social graph G, denoted as WDG(u), is defined to be the sum of the weights of all the edges incident to u.in other words, WDG(u) can be calculated as $WDG(u) = E(u)+W(e)$, where E(u) denotes the set of edges incident to u and W(e) means the associated weight of edge e in G.A k-backbone graph of a probabilistic social graph G is a subgraph of G which can be acquired by deleting from G all the vertices with the weighted degree less than k and all the associated edges.vi, add vi to the k-backbone graph if WDG(vi) is no less than k.

3.4. Random Walk
The random walk explore on a relatively small part of the graph by retrieving d-local graphs and then employing a MCMC-based sampler, the Random Walk Metropolis (RWM), in order to connect d-local graphs with the derived k-backbone graph.In Figure 2 vertex E is the k-backbone vertex.

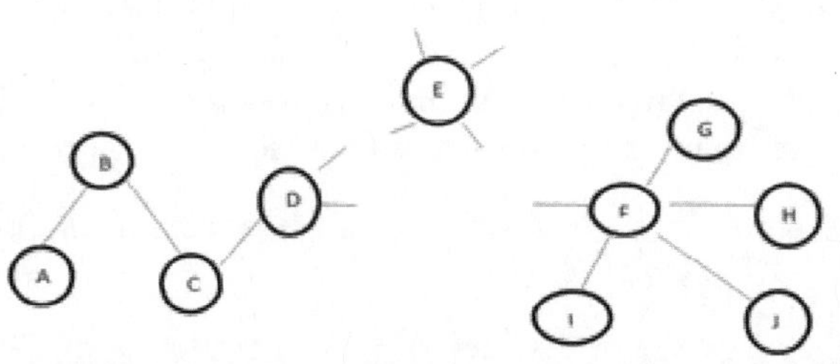

Figure 2. Random Walk on Social Graph

3.5. WalkSAT
WalkSAT calculation is utilized as a part of this model for nearby search. It adds irregular value to each variable in the formula.In the event that the task fulfills all standard procedures, at that point the calculation ends, restoring the task. Something else, a variable is flipped and the above is then rehashed until the point when all the standard procedures are fulfilled. MaxWalkSat is a change of WalkSAT intended to take care of the weighted satisfiability issue, in which each

IOP Conf. Series: Materials Science and Engineering **396** (2018) 012021 doi:10.1088/1757-899X/396/1/012021

standard procedures has related with a weight, amplifies the aggregate weight of the guidelines fulfilled by that task.

4. Inference Method

A basic inference task is finding the most probable state of the nodes given some evidence. This is known as MAP inference in the Markov system and MPE inference in the Bayesian network.This derivation strategies is utilized to discover reality task that maximize the whole of weights of fulfilled provisos. This should be possible utilizing any weighted satisfiability solver,thus this approach utilizes WalkSAT.

4.1. MAP Inference in Markov Network

Maximum a Posterior (MAP) is the issue of finding a most likely instantiation of an arrangement of factors in a Markov network, given (incomplete) prove about the complement of that set.
Direct Model:

$$P(X_1...N, w_1..N) = (\pi_{n=1}^{N} P(X_n|w_n))(\pi_{n=2}^{N} P(w_n|w_{n-1})) \tag{1}$$

MAP inference:
$\mathrm{w}_1..N = argmax_{w_1..N}[P(w_1...N|X_1..N)]$
$w_1..N = argmax_{w_1..N}[P(X_1...N, w_1..N)]$
$w_1..N = argmin_{w_1..N}[-log(P(X_1...N, w_1..N))]$

Substituting in:

$$w_1..N = argmin_{w_1..N}[-\sum_{n=1}^{N} -log(P(X_n, w_n)) - \sum_{n=2}^{N} -log(P(w_n, w_{n-1}))] \tag{2}$$

4.2. MPE Inference in Bayesian Network

Most Probable Explanation (MPE) is the issue of finding the in all likelihood arrangement of an arrangement of factors in a Bayesian network,given the complement of that set. Bayesian inference is a strategy for factual induction in which Bayes' theorem is utilized to refresh the likelihood for a theory as more proof or data ends up accessible Bayesian inference is firmly identified with subjective likelihood, frequently called Bayesian probability.Bayesian derivation infers the back likelihood as a result of two precursors, an earlier likelihood and a probability work got from a measurable model for the watched information. Bayesian derivation processes the back likelihood as indicated by Bayes' hypothesis:

$$P(X/E) = P(E/X).P(E)/P(X) \tag{3}$$

The evidence data E on values taken by some variables modify the probabilities of the rest of variable.

$$P(X) - - > P(X) = P(X|E) \tag{4}$$

Directed method:

$$XB =< G = A, B, C, D, E, P(A, B, C, D, E) > \tag{5}$$

Evidence : A = ai and B = bj

$$P(C = ck|A = ai, B = bj) = (\sum_{m,p} P(a_i, b_j, c_k, d_m, e_p)/(\sum_{k,m,p} P(a_i, b_j, c_k, d_m, e_p) \tag{6}$$

IOP Conf. Series: Materials Science and Engineering **396** (2018) 012021 doi:10.1088/1757-899X/396/1/012021

5. Experiment Validation

The table below shows the comparison between MAP inference method in MLN and MPE in Bayesian Network.

Table 1. Comparison between MAP and MPE

Inference Method	Computation	Algorithm	Efficiency	Accuracy
MAP	Difficult	Variable Elimination Algorithm	not efficient	Accurate.
MPE	Easier than MAP	Variable Elimination Algorithm	Efficient	not accurate.

6. Conclusion

This paper elaborate the inference procedure of the prediction system for each submitted ground rule.First derive the k-backbone Graph from social network graph. Then retrieve the d-local graphs for a link query. The k-backbone graph is used to connect with the d-local graphs.The probabilistic graph collects all the discovered nodes during the RWM sampling. As for each query, this method applies WalkSAT over its inference subgraph instead of on the original graph. The MAP and MPE inference method are used to predict the accuracy and efficiency of link prediction in social graph.Thus this proposed system predict a strong relationship such as friendship in social networks.In future implementation suggest how to calculate efficiency and accuracy of link prediction in social networks such as Facebook,Linked In etc. using MAP inference method in Markov model.

7. References

[1] *H. Chen, W. Ku, H. Wang, L. Tang, and M. Sun, LinkProbe: Probabilistic inference on large-scale social networks, in ICDE, 2013, pp. 290301.*

[2] *C.-Y. Lin, N. Cao, S. Liu, S. Papadimitriou, J. Sun, and X. Yan, Small-Blue: Social Network Analysis for Expertise Search and Collective Intelligence, , 2009, pp. 14831486.*

[3] *U. Kuter and J. Golbeck, Using probabilistic confidence models for trust inference in web-based social networks, ACM Trans. Internet Techn., vol. 10, no. 2, 2010.*

[4] *J. Tang, J. Sun, C. Wang, and Z. Yang, Social influence analysis in large-scale networks, 2009, pp. 807816.*

[5] *P. Oliveira and P. Gomes, Instance-based probabilistic reasoning in the semantic web, 2009, pp. 10671068.*

[6] *J. Zhu, Z. Nie, X. Liu, B. Zhang, and J.-R. Wen, Statsnowball: a statistical approach to extracting entity relationships, 2009,pp. 101110.*

[7] *J. W. Shavlik and S. Natarajan, Speeding Up Inference in Markov Logic Networks by Pre-processing to Reduce the Size of the Resulting Grounded Network, in IJCAI, 2009, pp. 19511956.*

[8] *S. M. Ross, Introduction to Probability Models, Ninth Edition. Academic Press, 2006.*

[9] *C. Andrieu, N. de Freitas, A. Doucet, and M. I. Jordan, An Introduction to MCMC for Machine Learning, Machine Learning, vol. 50, no. 1-2, pp. 543, 2003.*

[10] *R. Jampani, F. Xu, M. Wu, L. L. Perez, C. Jermaine, and P. J. Haas, MCDB: A Monte Carlo Approach to Managing Uncertain Data, 2008, pp. 687700.*

[11] *J. Leskovec, K. J. Lang, A. Dasgupta, and M. W. Mahoney, Community structure in large networks: Natural cluster sizes and the absence of large well-defined clusters, Internet Mathematics, vol. 6, no. 1, pp. 29123, 2009.*

[12] *J. Xie, J. Yang, Y. Chen, H. Wang, and P. S. Yu, A Sampling-Based Approach to Information Recovery, 2008, pp. 476485.*

[13] *H. Chen, W.-S. Ku, H. Wang, and M.-T. Sun, Leveraging Spatiotemporal Redundancy for RFID Data Cleansing, in SIGMOD Conference, 2010, pp. 5162.*

[14] *M. Yang, H. Wang, H. Chen, and W.-S. Ku, Querying uncertain data with aggregate constraints, in SIGMOD Conference, 2011, pp. 817828.*

[15] *J. Xie, J. Yang, Y. Chen, H. Wang, and P. S. Yu, A Sampling-Based Approach to Information Recovery, in ICDE, 2008, pp. 476485.*

[16] *P. Andritsos, A. Fuxman, and R. J. Miller, Clean Answers over Dirty Databases: A Probabilistic Approach, 2006, p. 30.*

[17] *N. Dalvi and D. Suciu, Efficient Query Evaluation on Probabilistic Databases, The VLDB Journal, vol. 16, no. 4, pp. 523544, 2007.*

[18] *D. Z. Wang, E. Michelakis, M. N. Garofalakis, and J. M. Hellerstein, Bayesstore: managing large, uncertain data repositories with probabilistic graphical models, vol. 1, no. 1, pp. 340351, 2008.*

[19] *R. Jampani, F. Xu, M. Wu, L. L. Perez, C. Jermaine, and P. J. Haas, MCDB: A Monte Carlo Approach to Managing Uncertain Data, in SIGMOD, 2008, pp. 687700.*

[20] *L. Mihalkova and M. Richardson, Speeding up inference in statistical relational learning by clustering similar query literals, 2009, pp.110122.*

[21] *D. Liben-Nowell and J. M. Kleinberg, The link prediction problem for social networks, 2003.*

[22] *V. Leroy, B. B. Cambazoglu, and F. Bonchi, Cold start link prediction, 2010, pp. 393402.*

[23] *L. Backstrom and J. Leskovec, Supervised random walks: predicting and recommending links in social networks, 2011, pp. 635 644.*

[24] *[L. Antova, C. Koch, and D. Olteanu, Query Language Support for Incomplete Information in the MayBMS System, in VLDB, 2007, pp. 14221425.*

[25] *R. Cheng, S. Singh, and S. Prabhakar, U-DBMS: A Database System for Managing Constantly-evolving Data, in VLDB, 2005, pp. 1271 1274.*

IOP Conf. Series: Materials Science and Engineering **396** (2018) 012022 doi:10.1088/1757-899X/396/1/012022

An Approach of RDD Optimization in Big Data Analytics

Athira V

PG Scholar,Department of Computer Science, St.Josephs College of Engineering and Technology, Palai, Kerala, India

E-mail: athira24v@gmail.com

Jeena Thomas

Assistant Professor,Department of Computer Science, St.Josephs College of Engineering and Technology, Palai, Kerala, India

E-mail: jeena.thomas@sjcetpalai.ac.in

Abstract. We live in the information age and new information is produced in each and every second. The information can be worlds data which is responsible for major technological changes that can bring new ways in decision making. Keeping this data for further analysis and computation is a difficult task. Several studies are done in this arena, making it effective for future figuring. Processing or analysing such huge amount of data is a challenging task. All the existing technologies contain certain bottlenecks and performance overheads, diverse challenges like scalability. Spark is the commonly used data analysis framework. Map Reduce is a computing paradigm and a popular model for distributed data analysis. This paper gives a survey about some enormous information technologies,how it will deal with huge data,and the difficulties in existing advances, and has additionally learned about a portion of the execution bottlenecks and preventive techniques, and the concentration at that point moves to the Resilient Distributed Dataset (RDD), and how it is optimized.

1. Introduction

The new media age has witnessed the growing of todays enterprises in an exponential rate day by day, along with the explosion of data and the databases, this has caused a big data problem faced by the industries due to its inability to manage or process this data within the time limit. Data is generated through several social networking sites, as a result of several transactions. The amount of data generated can be structured as well as unstructured [1]. Processing or analysing such huge amount of data is a challenging task. The blowing up of data has created a major challenge in the field of science and engineering. Data sets are fast growing and the conventional methods lacks solutions to manage easily. Existing solutions use files or some based on storing in databases still fails in handling and analysing data properly and to make it use in future. As data sets exceeds the capacity of the system, its analysis gets difficult and performance also gets limited. Data analysis is done by evaluating certain attributes and necessary data is extracted and transformed, types of data varying from simple to complex ones are extracted and perform multiple complex joins to these datasets. Explosion of data size makes it inefficient to store and process data, hence causing several challenges to preserve it for future computation. Challenges include limited scalability of I/O, scalability factor determines performance of application. Big data is large data that cannot be processed using traditional computing techniques. The volume of data Facebook and YouTube [19] handles will come under the category of big data. The data size varies from terabytes to petabytes of data as the data can be structured or unstructured. Big data is important as in this growing economy data is also growing in abundance due to the abundant uses of social networking sites, mobile and networking. More often there is a need to analyse this data to acquire the required information within a short time.

2. Related Works

Several researches are done aiming the improvement of performance of data analytics framework, most of them failed to understand properly the factors affecting the performance of the system. Kay Ouster hout [6] has developed a big data analytics framework for identifying the performance bottlenecks in a distributed computing framework, and have used it to analyse the performance of spark framework on two sql benchmarks and production workload, and he came to conclusion that cpu is often the bottleneck and by improving the network performance can improve the job completion time by 2 percent. In order to improve the performance hadoop and spark is widely used. Usually identifying performance bottlenecks is a problem due to parallelism, there will be multiple task performing.

Francisco Clemente-Castello and Bogdan Nicolae [7] discusses about enabling bigdata analytics in the hybrid cloud using iterative map reduce. There are different cloud computing models like private cloud, public cloud and a combination of private and public cloud called Hybrid cloud. Hybrid clouds can be made used in bigdata analytics and can hence improve the iterating level of map reduce applications. Using cloud can improve the environment like hiring some from off- premises also.

Bodgan Nicolae [9] introduced an approach that identifies and remove the data redundancy. This new approach minimizes the data that gets replicated, and compresses the data which is present above the required level using local storage instead of parallel file systems as it cause for a large amount of related data, and these local storage has limited capacity to hold data and also these local storages are prone to failures also, and if multiple requests arrive it will be difficult to handle in local storage, so to make high availability replication is usually done.

Nicolae [9] has also proposed redundancy elimination and replication as a co-optimized phase. In many of the high performance computing platforms [9] input and output will be the main performance bottlenecks. Yong chen and Xian-He sun [10] has proposed a collective input output strategy called layout aware collective input output (LACIO), which optimizes the input output performance and provides integration. Input output can cause performance degradation causing low latency, LACIO[10] improves the performance of the current systems, it does it by a certain file system calls, and hence improves the input output performance of parallel systems. Bigdata analytics plays an important role in many fields like science, medicine healthcare etc. [14]. As the data size has increased a lot there needed an efficient method for its computation and its analysis. In memory platforms always face challenges like data shuffling and it will diversely affect the performance and scalability of the system. Commonly used in memory platform is apache spark. There will be massive amount of data transfer that occur and some cases these platforms need to operate with scarce memory also. So to make shuffling easier efficient scheduling of the data transfer is required. There are many solutions exists that gives sub optimal performance and resource usage.

Bodgan Nicolae [11] gives a memory optimized data shuffling, this shuffle strategy dynamically adapts to the computation with minimum memory utilization and Aaron Davidson [8] discusses about optimizing shuffle performance in spark. Spark is a commonly used framework for performing in-memory computation. Shuffle performance in spark should be optimized so that it performs better than hadoop. There are different bottlenecks available which affect the performance, the existing solutions have identified some bottlenecks. The method has identified bottlenecks and has proposed some alternatives to mitigate the operating system overheads associated with these bottlenecks. It is shuffle file consolidation, the new simple solution has led to two times improvement in overall job completion time.

In map reduce there are map and reduce phases and shuffling occurs in between them. Shuffling is a part of reduce phase and it has a little role to deal with data. Shuffling can cause operating system overheads, and it causes input and output over the network. Several researches are done to avoid the operating system overhead, since shuffle phase is not connected to the semantics of the data, additional storage can significantly reduce the overhead due to the data transferred. Reducer will combine the results with the use of combiners.

Xiaoyi Lu and Dipti Shankar [11] talks about enhancing features on spark to improve the performance in shuffle phase. RDMA [11] based apache spark on high performance network. Apache spark is the most commonly used method for inmemory processing of real time data, in apache spark data can be loaded in memory and can query it repeatedly. It gives an abstraction of RDD, as it supports data lineage. Shuffle phase caused overhead and many research works are going on to mitigate it. Xiaoyi Lu [11] has proposed a data transfer strategy in the shuffle phase, which consumes only least amount of memory, it has given a design of RDMA based Apache spark and also has proposed block transfer service plugin which supports three shuffle scheme sort, hash and tungsten sort in spark. It is expected that in future can remove all the bottlenecks in spark and introduce methods to improve performance.

Author	Proposed Approach	Aim	Findings
1) Kay Ouster Hout	A big data analytics framework	To identify performance bottlenecks in distributed computing framework	• CPU is the main bottleneck • Improving network performance improves job execution time
2)Bodgan Nicolae	Approach done to minimize data that gets replicated	To identify and remove data redundancy	• Compressing data that exceed the storage level
3)Yong Chen	LACIO	Optimizing Input output performance and to provide integration	• Improves the performance of system • Include certain file system calls
4)Aaron Davidson	Proposed Shuffle file consolidation	To optimize shuffle performance	• Improved overall job execution time
5)Xlaoyi Lu	A data transfer strategy in shuffle phase	To remove all the bottlenecks.	• Consumes only less amount of memory • Support shuffle sort, hash and tungsten sort

Figure 1. Review of Literature

3. Big data Technologies

3.1. Hadoop

Hadoop [16] is a fault tolerant, cost effective, flexible and scalable computing solution. Most of the industries use hadoop to analyse their dataset. It is an open source software, and allows distributed processing of large datasets across clusters of computer. It is a simple programming model, it involves HDFS (Hadoop Distributed File System) [16], which is a distributed file system providing fault tolerance and it runs on a hardware. It store large datasets across multiple clusters of computer. It involves a master/slave architecture [18]. Its design is based on the Google File System [2]. It supports high- streaming read performance. It has a block structured file system, where files are divided into blocks, distributed and stored across different hadoop clusters. It has an open source implementation. There is a job tracker node also known as master node and a Task Tracker node called worker node. Client interact with job tracker, job tracker takes jobs from client and necessary operations according to the client request, decomposes the jobs into certain tasks. Task tracker contains slots and each slot will contain task. Task tracker receives task from job tracker and executes finally, and after completion it send notification back to the task tracker using a heartbeat message. If certain tasks are failed it executes the failed task again. Namenode manages and provides access to files by clients. HDFS replicates blocks to avoid missing of particular block. Data node sends feedback as heartbeat message to the namenode.

3.2. Map Reduce

Map reduce is a computing paradigm,[13] and a popular model for distributed data analysis, and it provides simplified data processing on large clusters. Its a popular model of its easiness to use, a programmer who does not have any experience in distributed computing systems can make use of this model. It provides functionality like fault tolerance [13], load balancing, hiding details of parallelization [2] etc. Programmers use map reduce to compute different programs relating to different data types. Problems written in Map Reduce are automatically parallized. Map Reduce provides scalable solution to process large amount of data. Map Reduce involves two parts a mapper part and a reducer part, mapper part will split up the data, and gives it as a particular key/value pair, this intermediate key/value pair is the input given to reducer, reducer combines this which forms the output. Result of a Map Reduce jobs are stored in a distributed file system.

3.3. Issues in Map Reduce

Hadoop/MapReduce is designed for batch processing of huge volumes of data but is not optimized for iterative computations, interactive mining and stream processing due to redundant data [24].The main problem is that both iterative and interactive applications need data sharing across multiple MapReduce steps. In MapReduce the data sharing between parallel operations is done by writing the intermediate results to distributed file system HDFS which adds overhead due to disk I/O, redundant and wasteful processing and data replication [25]. It also suffers from various Configuration and Automation issues which require numerous configuration parameters to set when deploying a Hadoop MapReduce cluster. These Programming Model issues makes it unsuitable for machine learning algorithms and graph processing which often require iterations or incremental computations [25].

4. The Need of RDD

(i) Issue with Small Files
Hadoop isn't suited for little information. (HDFS) Hadoop distributed file system does not have the capacity to productively bolster the irregular perusing of little documents as a result of its high limit plan.

(ii) Slow Processing Speed
MapReduce requires a great deal of time to play out these assignments in this manner expanding dormancy. Information is conveyed and handled over the bunch in MapReduce which builds the time and lessens preparing speed.

(iii) Support for Batch Processing only

Hadoop supports batch processing only, it does not process streamed data, and hence overall performance is slower. MapReduce framework of Hadoop does not leverage the memory of the Hadoop cluster to the maximum.

(iv) No Real-time Data Processing
Apache Hadoop is intended for group handling, that implies it take a gigantic measure of data as input, process it and create the outcome. Despite the fact that bunch preparing is extremely proficient for handling a high volume of information, yet relying upon the measure of the information being prepared and computational energy of the framework, a yield can be postponed altogether. Hadoop isn't reasonable for Real-time information preparing.

(v) No Caching
Hadoop is not efficient for caching. In Hadoop, MapReduce cannot cache the intermediate data in memory for a further requirement which diminishes the performance of Hadoop.

(vi) Lengthy Line of Code
Hadoop has 1,20,000 line of code, the quantity of lines creates the quantity of bugs and it will require greater investment to execute the program.

5. Spark RDD and basic operations in RDD

Resilient Distributed dataset (RDD) [3] is a memory abstraction, in memory computations can be performed using RDD, iterative algorithm and interactive data mining tools cannot handle the computing frameworks efficiently, RDD came motivated from it. Performance can be improved by keeping the data in memory, RDD provides fault tolerance. RDD provides coarse grained transformations rather the fine grained one. RDD is implemented in spark which is a commonly used computing platform. Map reduce has been used widely for data analytics where user can write programs in a distributed way in a fault tolerant manner. RDD allows data reuse. It allow results to remain the intermediate memory. Defining a programming interface is a challenge in RDD. RDDs apply same operation to many data item as it is based on coarse grained transformation. In RDD lost data can be acquired quickly, as it stores data in its lineage. RDD can be created by using data in a stable storage and using other RDD. Spark expose RDD via language interpreted API. Spark computes RDD lazily. Persist method [3] is use to make use of RDD for future operations. Spark keeps RDD in memory and it also distribute to disks in case of shorter memory.

In case of distributed computing datasets is split into smaller nodes and divided among to achieve speedup and to improve efficiency. Bodgan Nicolae [12] proposed a novel coded data delivery scheme in case of no excess storage, this new coded scheme exploits new coding opportunity called leftover combining to reduce communication overhead.Resilient distributed datasets is an important concept in spark. Spark RDD is fault tolerant and it can perform computations in a parallel manner. Consider RDD as a machine language, dataset and dataframe comes with RDD optimization, based on the distribution of the system. RDD will compute in case of two, four and eight distributed systems. An optimized RDD is developed from dataset and dataframe as a result of computer based optimization. Dataset and dataframe is based on dynamic language principle and can take python, scala, java and R.

Bigdata hadoop system is a distributed file based system. Data shuffling is a high I/O operation, it will have a input, map, reduce and an output, in between there is a shuffle phase. It is a store and forward pattern. While doing store and forward pattern there can be a lot of input output operation, to avoid this shuffling data has to go into a primary storage. In bigdata high input output happen as it has to go into a muliple file storage system and it causes a high store and forward mechanism.To do a I/O based lookup need to remove going to a primary storage. In a task to add six numbers from 1to 6 , consider all these numbers are in some files

IOP Conf. Series: Materials Science and Engineering **396** (2018) 012022 doi:10.1088/1757-899X/396/1/012022

read one value save it read another value save separetely as a two process, adding happens in between a shuffle phase, and in some other file storage. Every action or operation taken need a input output bound operation, all these operations takes place in the memory, it will be saved in systems primary memory. Do all these operations without looking secondary memory.

The task performed is a primary memory lookup opearation, its a DAG operation. A DAG structure will be computed for every operation. In spark this operation is done using RDD , it will have different computers for doing different operations and all thsese will be attached with primary memory, next figure 3 shows RDD distribution. RDD is immutable, RDD contains data in its lineage, hierarchy of data, will be saved in the environment. In RDD what ever data is present will remain and cannot edit it each and every time. There is no chance for the data to get changed. If x=abc, there there is no chance for x to become x= xyz. Each and everytime no need to go and look whether it is changed. It will be stored in memory, therefore dats shuffling is not required. Spark 6.0 is used for RDD optimization.

Table 1. Hadoop vs Spark

	Hadoop	Spark
`Data Storage`	The data is stored in the disc.	The data is stored in-memory.
`Computing`	Computing is based on the disc.	Computing relies on RAM.
`Fault Tolerence`	Fault tolerance is done through replication.	Fault tolerance is done through RDD.
`Design Aspect`	Batch processing only.	Interactive query processing
`Economic Aspect`	Less costly in comparison to spark.	More costly.

5.1. Optimizing RDD

From the distributed store try to query a distributed data set using RDD and spark SQL. No matter which language or which API is used, or whether it is written in java, scala, sql, or dataset and dataframe, first thing happens is constructing a logical plan that will tell the structure of computation. Instructions can be like take the data read it from a particular file, do a join, do a filter, do dataframe operations on static data etc. It is easy to understand as a batch.

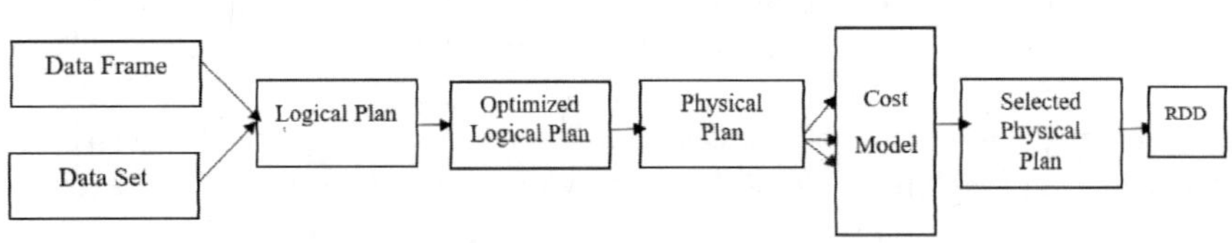

Figure 2. RDD Optimization

In the physical planning spark automatically runs queries in streaming fashion, it will be done continuously. Logical plan we will understand the structure of computation and in physical plan we will understand how it is going to perform after optimization. We use the catalystic optimizer and take the accurate plan and turn it into an incremental query plan. It is incremented by spark. There are four stages of plan,

(i) Parsed Logical Plan:It is the logical plan, applied by the spark optimize for query optimization.

(ii) Analyzed logical plan: It is the logical plan combined with catalog information, which is combined with name of datasets and types.
(iii) Optimized Logical Plan:It is applied to number of rows to simplify this query, we will do filter operation and then combine.
(iv) Physical Plan: Regression model is used for analyzing cost, cost model is similar to regression cost model.

In RDD there are lot of optimization plan, it can be different DAG operations and cost is applied for all. These DAG sets are created based on the datasets. Data will be applied and find which one will have least cost and that DAG will be selected and RDD will be created. Here RDD creation will be an optimization. Data is in table and it has attributes. If these attributes are unresolved there will be some rules to do that called catalystic rules. Developers will optimize this RDD based on these attributes, and then optimization will takes place using Catalystic Optimizer. Dataframe is data organized in tables or named columns. It is designed in such a way that to make data processing easier. Dataframes will allow developers to make changes in the data. Datadframe uses the caatlystic tree transformation framework. It will analyse the logical plan, and optimize it. For analysis it uses catalytic rules, and using that rules it resolve attributes that are not resolved. It will generate many physical plans, physical operators help for that, and it select a plan among them using the cost model.

6. Conclusion

Big data technologies takes a secondary file fetch for data shuffling,and faces many challenges in different cases like handling small files, batch processing, real time data processing and caching. To solve this difficulties spark takes an in-memory RDD based shuffling. All these comes under distributed computing events. Input output optimization leads to increase in data fetch and computation. In big data it takes a long time for data fetching, and results in time loss. So there is a need to minimize the input output computation. For that, every computation should be performed in primary memory. Encouraged by these results, we plan to broaden the scope of our work in future efforts.We analyzed the basic mechanism of hadoop and spark, and the various bottlenecks in the input-output mapping. For that, we considered the Resilient Distributed Dataset approach in both platforms. Based on the analysis, Spark found superior over the Hadoop. But there are various technologies like Apache Flink that are overtaking Spark. Like stream processing is much better using Flink then Spark as it is real time processing- 4G of Big Data.As a future analysis we can learn feature wise differences between Apache Spark vs Apache Flink to understand which is better and how.

7. References

[1] Tony Hey, Stewart Tansley, Kristin Tolle 2009 The Fourth paradigm data intensive scientific discovery,*Microosoft Reserch version 1.1*
[2] Jeffrey Dean and Sanjay Ghemawat 2004 Map Reduce: Simplified data processing on large clusters *Communications of the ACM, vol. 51, no. 1, pp. 107113.*
[3] Matei Zaharia, Mosharaf Chowdhury, Tathagata Das 2012 Resilient Distributed Datasets: A fault tolerant abstraction for in-memory cluster computing.*in NSDI12: The 9th USENIX Symposium on Networked Systems Design and Implementation, San Jose, USA, 2012, pp. 15 28.*
[4] Bodgan Nicolae 2013 Understanding vertical scalability of i/o virtualization of mapreduce workloads: challenges and opportunities,*in BigDataCloud 13: 2nd Workshop on Big Data Management in Clouds (held in conjunction with EuroPar13), Aachen*
[5] Zujie ken, weisong shi 2012 Workload characterization on a production hadoop clusters: A case study on Taobao,*978-1-4673-4532-3/12/IEEE*
[6] Kay ouster hout 2015 Making sense of performance in data analytics frameworks*in NSDI15: The 12th USENIX Conference on Networked Systems Design and Implementation, Oakland, USA, 2015, pp. 293307.*
[7] Francisco Clemente-Castello, Bogdan Nicolae 2015 Enabling bigdata analytics in the hybrid cloud using iterative mapreduce.*IBM Research*

[8] Aaron Davidson, 2013 Optimizing shuffle performance in spark *University of California, Berkeley - Department of Electrical Engineering and Computer Sciences, Tech. Rep.*
[9] Bodgan Nicolae 2015 leveraging naturally distributed data redundancy to reduce collective i/o replication overhead.*in IPDPS 15: 29th IEEE International Parallel and Distributed Processing Symposium, Hyderabad, India, 2015, pp. 10231032.*
[10] Yong chen, Xian- He Sun, 2015 LACIO:A new collective i/o strategy for parallel i/o systems.
[11] Xiaoyi Lu, Dipti Shankar, 2016 High performance design of Apache spark with RDMA and its benefits on various workloads.
[12] Bodgan Nicolae, 2015 Towards memory optimized data shuffling patterns for big data analyticsn *CCGrid16: 16th IEEE/ACM International Symposium on Cluster, Cloud and Grid Computing, Cartagena, Colombia, 2016, pp. 409 412, short Paper.*
[13] Yong chen, Xian- He Sun, 2015 On the worst-case communication overhead for distributed data shuffling.
[14] Bodgan Nicolae 2016 Leveraging adaptive i/o to optimize collective data shuffling patterns for big data analytics
[15] V Srinivas Jonnalagadda , P Srikanth, 2015, A Review Study of Apache Spark in Big Data Processing
[16] Aditya B. Patel, Manashvi Birla Addressing big data problem using hadoop and map reduce.
[17] HDFS Architecture Guide, available: *https://hadoop.apache.org/docs/r1.2.1/hdfs$_d$esign.pdf*
[18] Saeed Shahrivari, 2014 Beyond Batch Processing: Towards Real-Time and Streaming Big Data.
[19] Professional Hadoop Solutions,*https://www.wiley.com¿en-us/professional hadoop solutions-p-9781118611937*
[20] Vasiliki Kalavri, Vladimir Vlassov,2013 MapReduce: Limitations, Optimizations and Open Issues *12th IEEE International Conference on Trust, Security and Privacy in Computing and Communication, 2013, pp. 1031-1038, doi: 10.1109/TrustCom.126.*
[21] Mansi Shah1, Vatika Tayal2,2015 Fast Analysis of Sensor Data over MapReduce using Spark *International Journal of Advance Engineering and Research Development Volume 2,Issue 5*

IOP Conf. Series: Materials Science and Engineering **396** (2018) 012023 doi:10.1088/1757-899X/396/1/012023

A Study on Various Methods for Mining Energy Consumption pattern from Smart Home Big Data

Jemeema Maria John
P.G Scholar, Department of Computer Science and Engineering, St Joseph's College of Engineering and Technology,Palai, India
E-mail: jemeemajohn@gmail.com

Hima Anns Roy
Assistant Professor, Department of Computer Science and Engineering, St Joseph's College of Engineering and Technology,Palai,India
E-mail: himaannsroy@gmail.com

Abstract. Houses are the main source of electrical energy consumption. Getting information about energy consumption pattern from smart homes are of vital importance. It helps in analyzing the behaviour of inhabitants and that of the equipments. Smart meters in homes collect large volume of data each day. Data mining techniques can be used to get relevant information from the smart home big data. Data mining helps in easily predicting the consumption patterns of various equipments. Forecasting energy consumption pattern helps in better energy management plans. The techniques that are used for mining and predicting energy consumption patterns are mentioned in this paper.
keywords: Smart homes, Smart meter, big data, energy consumption, data mining.

1. Introduction

In todays world electrical energy is a major source of power generation [1]. The main consumers of electricity are houses. The major consumers of electricity are the various appliances that are used. Knowledge about the energy consumption pattern helps people in various ways. It helps in analyzing the consumption of electricity by appliances and devices. The prediction of energy consumption pattern or appliance usage pattern is also of greater importance. Energy consumption across the world is rapidly increasing due to various reasons. Higher energy consumption rates can occur due to various reasons. Some of them are increase in population, pressures for better living standards, large-scale industrialization in developing countries. Thus, for accurate investment planning of energy production generation and distribution a forecasting technique is essential.

It is very difficult to analyze millions of data records produced by smart meters. There are a lot of techniques available for analyzing smart meter big data. But these techniques are not very efficient. So a new technique called data mining was adopted. Data mining techniques can be used to extract relevant information from large amount of data called big data. The data mining techniques play a big role in obtaining energy consumption patterns from smart meter.

IOP Conf. Series: Materials Science and Engineering **396** (2018) 012023 doi:10.1088/1757-899X/396/1/012023

2. Data Mining

Data mining is the process of extracting relevant information from large collection of data. It can also be defined as analysing large data sets to find patterns to establish relationship among various factors. Data mining can be applied to extremely large collection of data sets. Generally data mining can be applied to any kind of data as long as the data are meaningful for a target application. The most basic forms of data for mining applications are database data, data warehouse data, and transactional data.

Data mining processes are very useful in collecting required information from big data. Big data represents huge volume of data. Big data sets are those that outgrow the simple kind of database and are not very easy to analyse, but with data mining techniques analysis on these big data sets are made easy. Thus datamining techniques are used for analyzing big data.

Some healthcare applications utilize data mining techniques to find out activity patterns of humans in smart homes. This information can be used to understand the life style of human and to identify any deviation from their daily routines. This is done by mainly analysing the appliance usage, appliance to appliance and appliance to appliance to time relationships [2].

3. Mining and Predicting Energy Consumption Patterns

3.1. DCMiner Algorithm

Based on the usage representation, Dynamic Correlation Miner(DCMiner) [3] algorithm is proposed to incrementally discover correlation patterns. DCMiner can incrementally mine the correlation patterns from evolving usage databases effectively and efficiently.

3.2. CTMiner Algorithm

Mining temporal patterns from time interval-based data is a difficult problem since the processing of complex relations among intervals may require generating and examining large amount of intermediate subsequence. A novel technique, called incision strategy and a new representation,known as coincidence representation are used to solve this problem.

An algorithm called CTMiner(Coincidence Temporal Miner) [4] is developed to discover frequent time-interval based patterns. The algorithm uses two pruning techniques to reduce the search space effectively. The pruning techniques used are called pre pruning and post pruning.

3.3. Incremental Mining

The key elements for understanding power consumption of a typical home are related to the activities the inhabitants are performing, the time at which appliances are used, and the inter-dependencies among appliances. This information can be obtained from context rich smart meters big data.

Furthermore, generation of energy consumption data from a smart meter is an ongoing continuous process and over period of time inter-appliance associations can change or new ones can establish. An incremental model for mining frequent power consumption pattern [5] is proposed. Here frequent patterns are extracted from data comprising of appliance usage tuples for 24 hours period, in a progressive manner.

3.4. Progressive Incremental Mining

Since houses are equipped with smart meters, appliance usage can be easily collected and stored in database. Analysing the database will help in studying the energy consumption behavioural pattern. Appliance-appliance and appliance-time relation are determined for getting more details.

Due to the continuous generation of energy consumption data, over a period of time, appliance associations can change. Therefore, they need to be captured regularly and continuously.

Progressive incremental data mining mechanism can be applied to energy consumption data to overcome these challenges [6].

Progressive incremental data mining mechanism can be done by four distinct phases: data preparation, frequent pattern mining and clustering, association rules generation, and visualization. Using the progressive mining techniques AoI (Appliances of Interest), the major energy consuming appliances for the house, which require equal attention from both consumers and utilities to attain energy efficiency.

3.5. Decision tree and time based multi label classifier
Buildings represent the first energy consumer. Home automation system is made up of household appliances that are linked via a communication network which allows interactions for control purposes. Load management helps inhabitants to adjust their power consumption. The energy management of residential houses requires the prediction of future usage of appliances.

This model tries to formalize such an approach using a time-series based multi-label classifier [7]. The classifier takes into account correlation between different appliances. Here, prediction results are shown for 1-hour.

The benefit of using only energy consumption reading for each appliance is many-fold. The objective of this model is to construct a system that is able to predict the appliance usage in housing which helps the system to organize energy production and consumption and to decide which appliance will be used at each hour.

3.6. Rule mining and DBSCAN clustering
Energy management in home is a key factor for creating a sustainable future for our society. Now a days a lot of tools are available for measuring energy usage in homes. One such model uses a technique called rule mining [8]. The model uses a rule mining algorithm to identify significant associations between energy usage and four key features. These key features are hour of the day, day of the week, use of other appliances in the home, and user-supplied annotations of activities such as working or cooking.

This model also utilizes the DBSCAN clustering method for grouping the various activities.DBSCAN is used to produce a unique profile for each device in the home. DBSCAN is a density-based algorithm that identifies clusters while excluding noise. DBSCAN is very helpful for as it does not require the number of clusters to be provided as input and it can be implemented very efficiently.

3.7. Clustering and Bootstrapping
Large quantity of information about, how customers use their energy is available through smart meters. Clustering the most important attributes of customers is a very common method for better understanding the residential energy behaviours that exist and has many applications [9].

Clustering methods can be applied to smart meter data. It helps distribution network operators(DNOs) to manage and plan the low voltage networks. This method proposes an in-depth analysis of customer smart meter data. Here four key time periods are used in which the data should be analysed. The time periods used are overnight, breakfast, daytime and evening.

The time periods not only helps to model peak time behaviours and their variation but also allows to reduce the number of attributes in the clustering implementation. It uses 10 distinct behaviour groups that describes the customers based on their demands.Finally, a bootstrapping technique is used to make the clustering reliable.

IOP Conf. Series: Materials Science and Engineering **396** (2018) 012023 doi:10.1088/1757-899X/396/1/012023

Table 1. Comparison Table

Techniques	Pros	Cons
DCMiner Algorithm	• Takes less memory space for execution.	• As the size of the database increases the processing time of DCMiner increases
CTMiner Algorithm	• Reduces search space • Consumes much smaller memory space.	• Requires additional computations.
Incremental mining	• Process only the incremental part,so execution time is faster.	• Costly in case of small amount of data.
Progressive incremental mining	• Execution time is faster.	• Costly in case of small amount of data.
Decision tree and time based multi label classifier	• It takes into account both conditional and unconditional label dependencies.	• Process of labeling to multi-label data is expensive and time consuming
Rule mining and DBSCAN clustering	• It does not require the number of clusters to be provided as input • can be implemented very efficiently	• It does not work well in case of data having high dimensionality
Clustering and bootstrapping	• Allows an experimenter to adjust results to reflect biases in their sample. • Helps to create more accurate customer profiles.	• The success depends on the data set it is used on. • Small differences in input can lead to substantial differences in output.

4. Conclusion

In this study, various methods that were used to analyse energy consumption pattern were discussed.It can be seen that a lot of techniques are used for obtaining the energy patterns. Data mining helps in extracting the energy patterns from smart home big data. It also helps in forecasting the consumption patterns of energy usage. As the world population is increasing day by day energy consumption is also increasing. So analyzing the energy consumption patterns helps us to understand the user behaviour and to optimize the power consumption accordingly. Using the monthly or daily data can help make accurate predictions. For example, by analyzing on more amount data, predictions can be done with less error rate. Since data mining plays huge role in healthcare sector, the forecasted results can be used in healthcare sector for predicting human behavioural anomalies.

.

5. References

[1] Noorollah Karimtabar ,Sadegh Pasban, Siavash Alipour, *Analysis and predicting electricity energy consumption using data mining techniques A case study I.R. Iran Mazandaran province* , 2nd International Conference on Pattern Recognition and Image Analysis (IPRIA 2015) March 11-12, 2015 .

[2] Abdulsalam Yassine; Shailendra Singh; Atif Alamri , *Mining Human Activity Patterns From Smart Home Big Data for Health Care Applications* , IEEE 2017.

[3] Andreas B. Y. Chiang , H.C. Hung , P. J. Chen, S. Y. Peng , Y. C. Chen, *Incrementally mining usage correlations among appliances in smart homes* , in Proc. 18th Int. Conf. Netw.-Based Inf. Syst. (NBiS), pp. 273279, 2015.

[4] J. Jiang, S. Lee , W. Peng, Y. Chen, *An Efficient Algorithm for Mining Time Interval-based Patterns in Large Databases* , Proceedings of 19th ACM International Conference on Information and Knowledge Management (CIKM10), pp. 4958, 2010.

[5] A. Yassine, S. Shirmohammadi, S. Singh, *Incremental mining of frequent power consumption patterns from smart meters big data* , in Proc.IEEEElectr. Power Energy Conf. (EPEC), pp. 16, Oct. 2016.

[6] A. Yassine, S. Singh, *Mining energy consumption behaviour patterns for households in smart grid* ,IEEE Trans. Emerg. Topics Comput. to be published...

[7] K. Basu, S. Bacha, V. Debusschere, *Appliance usage prediction using a time series based classification approach* , in Proc. IEEE 38th Annu.Conf. Ind. Electron. Soc. (IECON), pp. 12171222 ,May 2012.

[8] N. Banerjee, S. Rollins, *Using rule mining to understand appliance energy consumption patterns* , in Pervasive Computing and Communications (PerCom), 2014 IEEE International Conferen, pp. 2937, 3 2014.

[9] C. Singleton, P. Grindrod, S. Haben, *Analysis and clustering of residential customers energy behavioral demand using smart meter data* , IEEE Transactions on Smart Grid, vol. 7, no. 1, pp. 136144, 1 2016.

IOP Conf. Series: Materials Science and Engineering **396** (2018) 012024 doi:10.1088/1757-899X/396/1/012024

Pedagogue Performance Assessment (PPA) using Data mining Techniques

Ashily M Baby[1]

[1]Graduate Student,
Dept. of Computer Science and Engg., St. Joseph's College of Engineering and Technology, Palai, Kerala, India
E-mail: ashily77@gmail.com

Abstract. Educational Data Mining (EDM) is a developing domain that exploring pedagogical data by applying different machine learning techniques. It can be considered as a research field which provides intrinsic knowledge of teaching and learning process for effective education. The main objective of any educational institution is to provide quality education to their students. One way to achieve highest level of quality in higher education system is by discovering knowledge that predicts pedagogues performance. This paper presents an efficient system model for evaluation and prediction of pedagogue performance in higher institutions of learning using data mining technologies. To achieve the objectives of this work, three classification algorithms like decision tree algorithms(C4.5), support vector machines (SMO) and artificial neural networks (MLP) are chosen to build classifier models on a dataset. The classifier system was tested successfully using case study data from St. Joseph college of Engineering and Technology, Palai. The data consists of the feedbacks that got from the students. The aim of this work is to demonstrate the capability of EDM in illuminating the criteria of fruitful instructor performance as perceived by the students. The result shows the accuracy of classifier models that predicts the performance of pedagogues. Decision tree J48 have higher accuracy of 94.37% than SMO and MLP.

Keywords: Artificial neural networks, classification algorithms, decision trees, performance evaluation, support vector machines.

1. Introduction

One of the biggest problem of higher education organizations are the large amount of data and how to use them to improve quality of academic programs and services and the managerial decisions. A variety of "formal and informal" procedures based on "qualitative and quantitative" methods is utilized by advanced education institutions to tackle issues, which keep them far from accomplishing their quality goals. Techniques utilized in higher education for quality reasons are mainly based on predefined questions and charts to investigate the information. In expansion, these strategies do not have the capacity to uncover useful hidden data. Covered up information in substantial datasets is best analyzed with data mining methods. Data mining can be characterized as the automated process of removing useful knowledge and information that are obscure from large datasets. One of the common problems in higher education is the evaluation of pedagogues performances in a subject. The most widely applied tool to evaluate the pedagogues performance in a subject is through surveying students responses about the subject and its pedagogue through a questionnaire. Students are the only source of information about the learning environment, and they are the only ones who can rate the quality, the effectiveness, and the satisfaction of course content, method of instruction, textbook, and homework. Student evaluations are used primarily to improve course and teaching quality as well as a part of the evaluation process for staff assessment systems. The aim of this work is to demonstrate the capability of EDM in illuminating the criteria of fruitful instructor performance as perceived by the students. In this study, three classification techniques-decision tree algorithms, support vector machines (SMO) and artificial neural networks (MLP) are chosen to build classifier models on a dataset composed of the feedbacks of students to a subject evaluation questionnaire and the performances of these models are compared.

2. Related Works

Broad literature survey of the EDM research field are given by Romero and Ventura [1]. In this paper, they overviewed the use of data mining to traditional educational systems, surely understood learning content administration frameworks , specific web-based courses and versatile and keen web-based educational systems. Another broad survey was done by Baker and Yacef[2]. They investigated the past points of interest and current trends in the field of Educational Data Mining (EDM), thought about the methodological profile of research in the early years of EDM, and talked about trends and shifts in the examination led by this group. Another broad survey was done by Rajni and Malaya [3], concentrated on parts, research trends of EDM featuring its related tools, instructive outcomes and methods. They additionally featured the challenges in EDM. Another broad survey by Varun and Arupama [4] inspected the use of data mining procedures in higher instructive foundation to separate valuable data from big data sets and gave analytical tool to view and utilize this data for decision making processes by taking genuine illustrations. Chin-Chia Hsu and Tao Huang [5] directed an investigation on the utilization of data mining innovation to assess student's academic achievement through numerous channels of enrolment like joint enlistment enrolment, athletic enrolment and application enrolment. A comparative report was done by Osofisan and Olamiti [6] where they examined the academic background in relationship with the performance of students in a computer science programme in a Nigerian university. Their investigation demonstrated that the review got from senior secondary school examination (SSCE) in mathematics is the highest determinant of students' performance using the C4.5 learning algorithm in building the model of the student's performance.

3. Proposed Methodology

The proposed system aims at using the classification method of data mining for the prediction of pedagogues performance. The prediction model was developed using the Classification methods of the Data mining technique. The Neural Network data mining technique (the multilayer perceptron algorithm), Support Vector Machine (SMO), Decision Trees methods specifically the C4.5 algorithms were used and their performances were compared to each other.

3.1. Data Acquisition

The dataset used in this project was collected from St. Joseph College of Engineering and Technology, Palai. Student evaluation data has 13 variables including the name of faculty, the subject name, year, semester and questionnaire . Response values of these questions are of the form {1; 2; 3} where 1; 2; 3 represents the answers "Good", "Satisfactory", "Not satisfactory", for Q1 and 1; 2; 3 represents "Loud and clear", "Reasonably loud", "Too low" for Q2 and 1;2;3 represents "Very clear", "Somewhat clear", "Not clear" for Q3 and 1;2;3 represents "Execellent", "Reasonably good", "Occasional disorder in class" for Q4 and 1;2;3 represents "Too fast", "Good", "Too slow" for Q5 and 1;2;3 represents "Yes", "To some extend", "No" for Q6 and 1;2;3 represents "Very easy to understand" , "Somewhat easy to understand", "Difficult to understand" for Q7 and 1;2;3 represents "As a role model", "Good", "Satisfactory" for Q8. The last variable is dichotomous variable measured on a nominal scale in the form of {1; 2} where 1 stands for "Good" and 2 for "Poor". Details of the variables used in this study are shown in Table I. Variable Q9 is the class variable that is to be predicted. The value "Good" for this variable is taken as positive class label whereas "Poor" is assumed to be negative.

Table 1. Performance Summary of Models

Variable	Description	Possible Values
Q1	Knowledge of the pedagogue	1,2,3
Q2	Pedagogue's voice	1,2,3
Q3	Writing and sketeches	1,2,3
Q4	Control and command over class	1,2,3,
Q5	Speed of presentation	1,2,3,
Q6	Pedagogue encourages questioning	1,2,3
Q7	Thw way the subject taught by the pedagogue	1,2,3
Q8	Rating of the pedagogue	1,2,3
Q9	Overall performance	1,2

3.2. System Design

The tasks involve machine learning and classification algorithms; hence in the research work a three-layered classifier system in figure 1 was designed to achieve the objective of the work. Layer 1 consists of an Artificial Neural Networks (MLP), layer 2 is made up of Decision Trees classifiers and layer 3 made up of Support Vector Machine (SMO). These classifiers have been selected because of their performances in various domains. They have both been successfully applied to a variety of real-world classification tasks in industry, business, science and education with good performances. The Neural Networks is known for its predictive accuracy, ability and aptitude to learn and remember. Decision Tree classifiers are considered "white box" classification model as they can provide explanation for their models and can be used directly for decision making. These abilities and aptitude are best suited and of good requirement for any effective and efficient intelligent system.

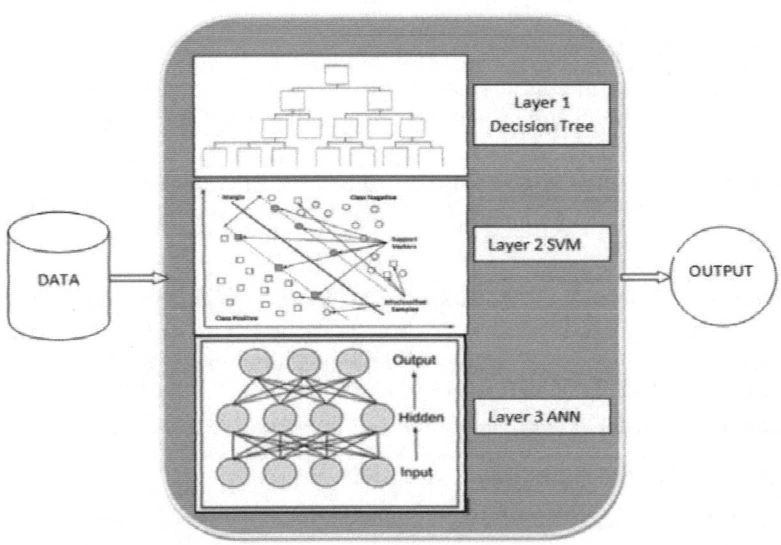

Figure 1. Three Layered Classifier System

3.3. Analysis Proposed Model

The goal of the proposed system it to aid higher institution management in determining teachers' performance and recommend necessary actions to be taken on individual teacher based on the prediction from the intelligent evaluation system. The proposed system framework includes three components: The first and the second components take care of data acquisition and storage, responsible for storing teachers' data, gathered from different data sources proposed in a data warehouse. The third component is model building, responsible for obtaining knowledge about the teachers, through appropriates classification models. Different classification algorithms are proposed in search for the best model with high predictive accuracy.

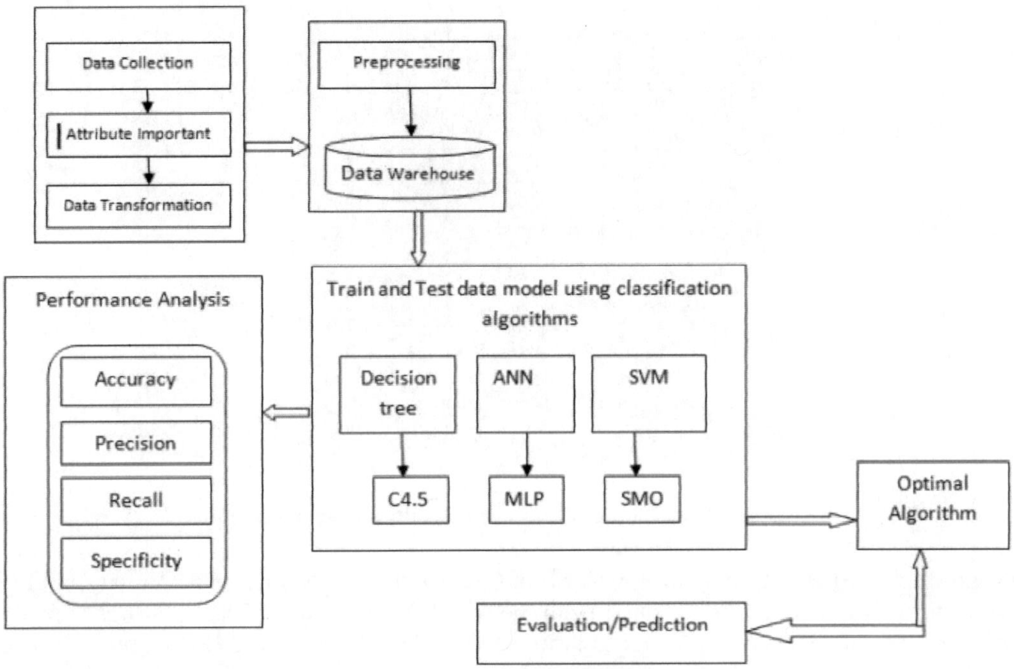

Figure 2. Architecture of Proposed System

4. Results and Discussion

In this study, three classification models are generated: decision tree algorithms (C4.5), Support Vector Machine(SMO) and Artificial Neural Networks(MLP). The performances of these models are evaluated on the test data in terms of accuracy, precision, recall, and specificity. Accuracy values, which assess the effectiveness of the models, are all at least approximately 90%, C4.5 have 94.37%, SMO have 90.85% and MLP have 92.96%. C4.5 classifier is the best in performance according to accuracy followed by MLP, and SVM is the worst. Precision, which assesses the predictive power, again indicated C4.5 as the best classifier; however, MLP and SVM also show high predictive power. According to MCC and F-measure C4.5 is best. As a result, from the SVM classifier, and from the artificial neural network classifier, the C4.5 is comparatively better. Ultimately, C4.5 can be considered as the outstanding classifier among all according to the given performance measures. The weighted averages of the models were compared using different performance measures like True Positive Rate, False positive rate, Precision, Recall, MCC, F-Measure, ROC area, PRC area.

Table 2. Performance Summery of Models

Algorithm	TP Rate	FP Rate	Precision	Recall	MCC	F-Measure	ROC Area	PRC Area
J48(C4.5)	1.000	0.667	0.986	1.000	0.993	0.573	0.667	0.986
SVM(SMO)	1.000	1.000	0.980	1.000	0.990	0.000	0.500	0.980
ANN(MLP)	1.000	1.000	0.980	1.000	0.990	0.000	0.900	0.998

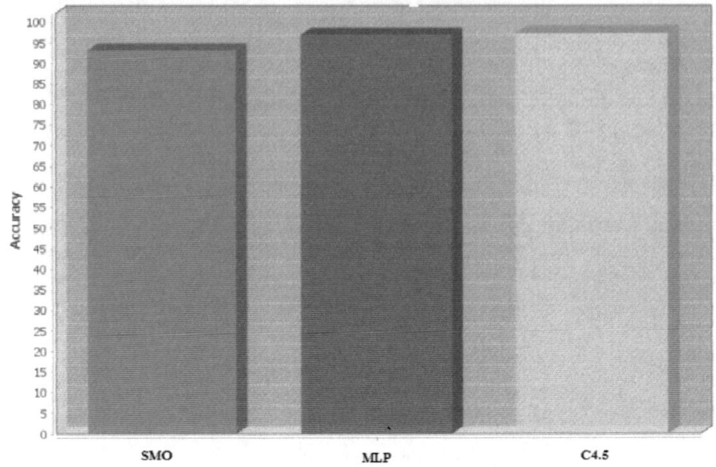

Figure 3. Accuracy Graph of Classification Algorithms

C4.5 algorithm predicts better than the SVM and MLP algorithms since its accuracy is the highest compared to others. The results obtained from the analysis demonstrated a slight higher performance of model built with decision tree (C4.5 algorithm) over neural network (MLP algorithm). Both C4.5 and MLP algorithms results show great superiority over SVM algorithm in terms of performance. Also the rules generated makes C4.5 decision tree algorithm clearer and understandable. The figure 3 shows the accuracy graph of classification algorithms, the C4.5 and MLP have much accuracy than the SMO.

5. Conclusion

Data mining techniques are applied in higher education more and more to give insights to educational and administrative problems in order to increase the managerial effectiveness. Based on student's perception instructor performances are found and these can help instructors to improve their performances. The aim of this project is to show the potential of EDM in enlightening the criteria or measures of effective instructor performance as perceived by the students. The Decision tree J48(C4.5) is the best algorithm suitable for predicting teachers' performance in relation to other algorithms in this work. Our results also show the performance of the teachers in this study. The outcome of this project demonstrated that data mining techniques can be applied in the prediction of teachers' performance. This project can be improved upon by using other classification algorithms and other data mining techniques such as, Naïve Bayes classifier, genetic algorithm as well as data from other universities. The data set can also be expanded with more distinctive attributes to get improve upon the classification accuracy and make the work more robust.

References

[1]C. Romero, S. Ventura, 2007 "Educational data mining: A Survey from 1995 to 2005", Expert Systems with Applications, (33), pp. 135-146.

[2]Baker, R. S. J. D., and K. Yacef. 2009 "The State of Educational Data Mining in 2009: A Review and Future vision", Journal of Educational Data Mining 1 (1): 3–17.

[3]Rajni J. and Malaya D.B , 2013 "A Survey on Educational Data Mining and Research Trends", International Journal of Database Management Systems (IJDMS).

[4]VarunKumar and Anupamachada, 2011 "An Empirical Study of the Applications of Data Mining Techniques in Higher Education", International Journal of Advanced Computer Science and Applications, Vol. 2, No.3.

[5]Chin Chia Hsu and Tao Huang, 2006 "The use of Data Mining Technology to Evaluate Student's Academic Achievement via multiple Channels of Enrolment", An empirical analysis of St. John's University of Technology.

[6]Osofisan A.O. and Olamiti A.O, 2009 "Academic Background of Students and Performance Computer Science Programme", Nigerian University.

[7]Asanbe M.O, Osofisan A.O, William W.F, 2016 "Teachers' Performance Evaluation in Higher Educational Institution using Data Mining Technique", International Journal of Applied Information Systems (IJAIS) – ISSN : 2249-0868.

[8]Ankita A Nichat, Dr.Anjali B Raut. 2017 "Predicting and Analysis of Student Performance Using Decision Tree Technique", International Journal of Innovative Research in Computer and Communication Engineering, Vol. 5, Issue 4.

[9]Ahmed Mohamed Ahmed, Ahmet Rizaner, Ali Hakan Ulusoy. 2016 "Using data mining to predict instructor performance", 12th International Conference on Application of Fuzzy Systems and Soft Computing, ICAFS.

[10]Aranuwa Felix Ola, Prof. Sellapan Pallaniappan. 2013 "A data mining model for evaluation of instructors' performance in higher institutions of learning using machine learning algorithms", International Journal of Conceptions on Computing and Information Technology
Vol. 1, Issue 2, ISSN: 2345 – 9808.

[11]Randa Kh. Hemaid, Alaa M. El-Halees. 2015 "Improving Teacher Performance using DataMining", International Journal of Advanced Research in Computer and Communication Engineering, Vol. 4, Issue 2.

[12]Mardikyan S, and Badur B. 2011 "Analyzing teaching Performance of Instructors Using Data Mining techniques", Informatics in Education.

[13] Musthafa Agaoglu. 2016 "Predicting Instructor Performance Using Data Mining Techniques In Higher Education", IEEE Access, Volume 4.

[14]A. M. Abaidullah, N. Ahmed, E. 2015 "Identifying hidden patterns in students' feedback through cluster analysis", Int. J. Comput. Theory Eng.,vol. 7, no. 1, pp. 1620.

[15] Randa Kh. Hemaid, Alaa M. El-Halees. 2015 "Improving Teacher Performance using Data Mining", International Journal of Advanced Research in Computer and Communication Engineering,Vol. 4, Issue 2.

[16] A. Pena-Ayala. 2014 "Review: Educational data mining: A survey and a data mining-based analysis of recent works", Expert Syst. Appl., vol. 41.

[17] N. Delavari, S. Phon-Amnuaisuk, and M. R. Beikzadeh. 2007 "Data mining application in higher learning institutions", Inform. Edu.-Int. J., vol. 7

[18] M. Goyal and R. Vohra. 2012 "Applications of data mining in higher education", Int. J. Comput. Sci. Issue, vol. 9.

IOP Conf. Series: Materials Science and Engineering **396** (2018) 012025 doi:10.1088/1757-899X/396/1/012025

Review on Energy Harvesting and Data Collection

Thasni Noushad [1] and Kishore Sebastian [2]

1 P G Scholar, Department of Computer Science and Engineering, St.Joseph's College of Engineering and Technology, Palai, Kerala, India.
2 Assistant Professor,Department of Computer Science and Engineering, St.Joseph's College of Engineering and Technology, Palai, Kerala, India.

E-mail: thasnin95@gmail.com

Abstract. Energy harvesting is important in the case of recharging or repay of battery. Energy harvesting means that energy is derivered from external sources like solar energy, thermal energy, kinetic energy and wind energy. Two or more sensors are attached to a body forms Body Sensor Networks (BSNs). BSN s are mainly focused in healthcare domain. it allows patients, physiciants, caregivers, scientists, etc. It useful for continuous monitoring of patients (Diagnosis patients, treatment of chronic obstructive pulmonary disease patients and other health conditions). Doctors made interaction between the data collected body sensors and personal digital assistant (PDA). Disadvantage of existing system is batteries cannot be easily replaced, met must be near a transmitter, expensive, etc. Energy Harvesting and Data Collection (EHDC) focuses two types of energy harvesting techniques such that solar and thermal energy harvesting and two types of data collections are used :indoor and outdoor data collection. Harvested energy can be predicted by using Kalman Filter. EHDC provide mobile healthcare services and does not need recharging .

1. Introduction

Body Sensor Networks (BSNs) are cyber-bodily sensor structures which can be able to amassing, processing, and speaking physiological statistics from the body and environmental statistics from the region surrounding it.BSN overcome the weakness of traditional remote hospital system.Recent technology provide full continuous monitoring of patients beyond the clinic and also overcome the drawback of limitted battery life time or must be replace or recharging battery [26]. power harvesting is vast answer for the limitted battery life time of body sensor community Three stages of healthcare monitoring system are sensors, data hub and hospital system. Although there are many applications for BSNs such as tracking, scientists, remote patient monitoring in healthcare system [27].Researchers have necessity methods of separate magnanimous supervise such as self-restrained unmixed-relate and abode hardiness technologies to vestige symptoms and become recommendations to patients or to occur when inevitable. sometimes, patient self-report tends to be imprecise because it can be difcult for patients to accurately represent their own symptoms. BSN network provide an alert message to the hospital system ,if due any changes in vital parameter.

Automatic recognition of daily living activities can be primarily classified in two -wearable and non-wearable system. This paper focused on two types of energy harvesting technique-solar and thermoelectric energy harvesting. Data collection are done both indoor and outdoor condition [27].

2. Body Area Sensor Network (BASN)

A body sensor network (BSN),is also Known as wireless body area network (WBAN) or a body area network (BAN) [1]. BAN devices may be fixed in the body, or may be attend the device which human can move in distinct places, in cloths pocket, by hand. That is it can fix anywhere in the body. All body sensors unit (BSU) are controlled by a single central control unit (CCU). Initial application of BAN are originally in the hospital system for the continuous monitoring of patients suffering from chronic obstructive pulmonary diseases (COPD) such as arthritis, asthma, cancer, diabetes and some viral diseases such as hepatitis C and HIV/AIDS [2]. Body sensors are placed on a patient body, it can alert the hospital system, even before they have a heart attack or any emergency situation, through measuring changes in their vital parameters. Other applications of BAN are include Tracking, sports, Scientists (measuring body variations) military, or security. The six operational characteristics in BASN- sensing, signal processing, communication, caching, feedback control and energy harvesting [1].

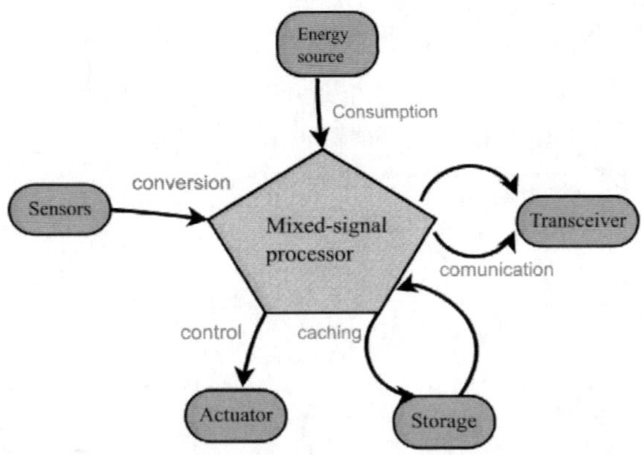

Figure 1. Figure 1. BASN node architecture

As figure 1 shows BASN node architecture. The main components are sensors, energy source, transceiver, storage, actuator, and sensors. Sensor is a device used to detect events or detect changes in to environment and send information from one to another device. Sensors are classified in to three categories- physiological sensors, biokinetic sensors, ambient sensors. Physiological sensors capture signal from sensors are placed on the patient body. Different physiological signal processing (PSP) have been developed such as electrocardiography (ECG), electroencephalography (EEG) and electromyography (EMG). Biokinetic sensors measurement acceleration and sudden rank of rotation trace from human motion. Ambient sensors metric environmental phenomena, such as dampness, knowledge, correct squeezing horizontal, and mixture. Aggregator is used to combining data from multiple sensors on the body and act as bridge between the nodes higher level infrastructure and provide user interface. long term information series regularly needs no actual-time aggregation, on-node storage is an inexpensive solution for archiving statistics, thereby growing battery lifestyles.Disadvantage of lithium based battery is high energy density, limited capacity and recharge/replace battery. To overcome these use super capacitor or carbon nanotube. It leads to improve the battery capacity. Wireless transceiver can reduce power consumption by data rate extracting only the essential information and limited transmission range can reduce power consumption. Automatic recognition of daily living activities can be broadly classified in to two- wearable and non-wearable system.

IOP Conf. Series: Materials Science and Engineering **396** (2018) 012025 doi:10.1088/1757-899X/396/1/012025

3. Wearable System

Wearable system classified in to two categories- wired wearable sensor network and wireless sensor network. Its design depends on low cost, intelligent, tiny and lightweight sensors that can be placed on the patient body. Wearable sensors are proposed for automatic recognition of daily living activities[29].

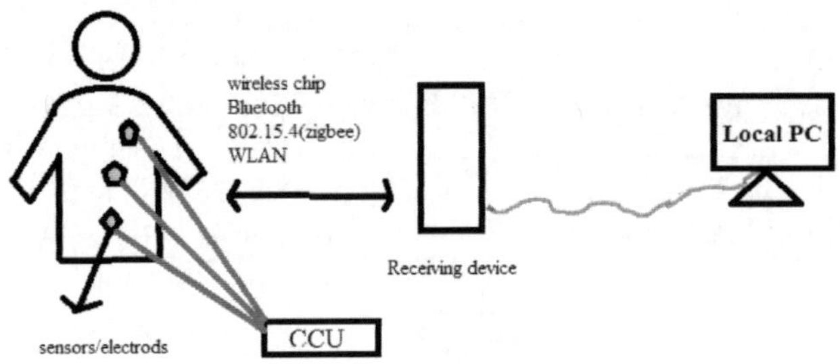

Figure 2. Wearable wired BASN

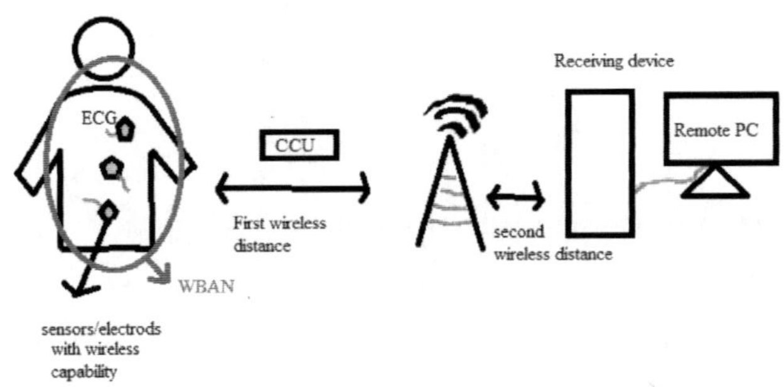

Figure 3. Wearable wireless BASN

Figure 2 and 3 shows primarily three parts body area network, data aggregation or communication part and hospital system. Figure 2 defines two or more sensors are attached to the body and that are interconnected to a common control unit. CCU which control overall sensors. If the connection between the BAN and CCU is a wired connection then the sensor signal is transmitted only the limited distance, that is signal is send to the local PC [28]. Figure 3 shows the wearable wireless connection of BSN. The WBSN are connected to the CCU through without any wired connection (ie, wireless connection). Two transmissions are possible , first the transmission between WBSN and CCU through ZigBee or Bluetooth and the next transmission from CCU to hospital system. The example of wearable wired system is take blood pressure measurements. Blood pressure sensors (Physiological sensors) are used and the corresponding result is obtained in local PC. Commonly used wearable wireless system is wrist band. Human beings are warm blooded , they can be used as a heat source attached to the skin. Small body temperature sensors (physiological sensors) is attached to the wrist band and monitor the temperature of a body [4].

4. Non-Wearable System

Non-wearable sensors are used for continuously monitor of patients activity of daily living (ADL) [29]. If any abnormal situation (eg: Heart Attack) occurred an alert message is send to receiver station (eg: hospital system) based on the changes in vital parameters. Non- wearable sensors include self- report instruments and devices installed in the home (eg: cameras, thermal sensors, infrared sensors). These devices like cameras or sensors are installed in to the home or office and automatically recognize the daily living activities of a human being like walking, standing, sitting etc. Non-wearable devices are under the category of ambient sensors. Ambient sensors are used to measure the environmental parameters (eg: temperature, humidity etc.)[1]. Ambient sensors are fixed sensors which are communicate through fiber optic support around 20-30m.

5. Design Challenges

while designing self-powered sensor system sources which includes sun, thermoelectric , piezoelectric, RF, wind ,etc. have been explored. regarding self -powered BSNs availability is a main factor, however also wearability has excessive impact when dening the source for use based totally on these considerations for self-powered BSNs sun, thermoelectric and piezoelectric strength harvesting are commonly selected as the power supply alternatives. Gaps from overview is energy harvesting techniques focused on out of doors conditions only most commonly used lithium based battery has limited capability and want recharge or replace frequently.To overcome this solution use supercapacitors and carbon nanotube [1].

6. Energy Harvesting and System Wearability Optimization

Two types of energy harvesting techniques-solar energy and thermoelectric enertgy harvesting. Existing system solar energy is harvested from outdoor condition only.This paper focused on solar energy is harvestedv from both indoor and outdoor condition. In the case of solar cell , it is clear that the ambiant illumination level defines the maximum amount of harvested energy, but it also has been stuidied how the angle of incident of light impacts the efficiency [27]. Sensors are attach to the patient body in some manner, then both illumination and angle of incident changes when a person move around during a typical day. The AM-1417CA solar cell and NOA 1212 sensor are attach on the body.solar cells are directly captured hea energy when the molecules are collide,chemical reactions occured. Chemical energy convert in to electrical energy.Similarly for the thermoelectric generator(TEG) temperature difference across the device determine the avilable power. TEG is also known as seedback temperature is a solid state device that convert heat energy in to electrical energy. Sp5424-04Ac thermoelectric generator is used and a heat sink was attached to the cold side of TEG. A MAX6605 body temperature sensor is placed inner side of TEG.

7. Energy prediction model

Harvested energy can be predicted by using kalman filter. Prediction usually mean that amount of energy harvest from sources. Kalman filter is used to estimate the obsevation and to predict harvested energy.Kalman filter depending parameters are two types Environmental and behavioral parameter and harvested energy. Environmental parameter consist of light intensity,temperature difference and activity level. Harvested energy parameter depends on solar energy and thermal energy.

8. Graphical rule editor

Emergency services provide by using graphical rule editor. Figure 4 shows two measurements. If the current measurement value are compare with predefinely set value. For example Figure 3 shows current pulse reading is greater than predefinly set value of pulse (120) and if the current

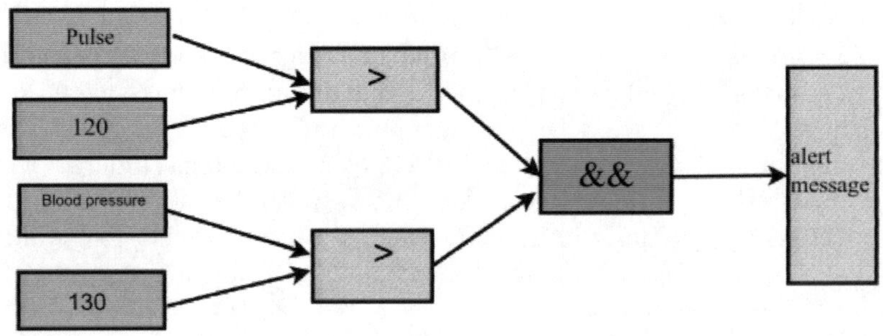

Figure 4. Graphical rule editors

blood pressure (BP) reading greater than predefinly set value of BP (130). If exceed the value then corresponding alert or message is send to the emergency service.

9. Conclusion

To improve the quality of life for the elderly persons with chronic diseases ,propose a small cell based sensor network for home monitoring. Two types of energy harvesting techniques are used solar and thermoelectric. Data collections are done in both indoor and outdoor conditions. The EHDC platform is designed to characterize the relationship between environmental and human behavioral conditions and real-world energy harvesting dynamics[27]. Human beings are warm-blooded,therefor TEGs are attached to the skin and can be used as a heat source. With this consideration for people to wear these devices, it is common to attach TEGs on the wrist or arm.The feasibility of batteryless operation of wireless sensor nodes using solar energy harvesting at locations where the amount of solar radiation is severely limited and seasonal variations are large. Kalman Filter (KF) based model to predict the harvested energy based on environ- mental and human behavioral parameters. KF is an estimation model that combines several previous observations to produce an estimate for the desired unknown as it is presented in [6]. The proposed model was adapted from this work and consists of state process and measurements by sensors; the state process is dened by the environmental and human behavioral parameters since they represent the physical principle of self-powered energy harvesting while the measurement values are acquired from sensors.

10. References

[1] A. Hanson, H. C. Powell, A. T. Barth, K. Ringgenberg, B. H. Calhoun, 2009 Body area sensor networks: Challenges and opportunities *IEEE Comput.,42, no. 1, pp. 5865.*

[2] T. Falck, J. Espina, J. Ebert and D. Dietterle, 2006 BASUMA-The Sixth Sense for Chron- ically Ill Patients *Int. W. Wearable and Implantable Body Sensor Networks* 124-34

[3] S. Bader and B. Oelmann 2010 Enabling battery-less wireless sensor operation using so- lar energy harvesting at locations with limited solar radiation*Proc. - 4th Int. Conf. Sens. Technol. Appl. SENSORCOMM* **41**

[4] Leonov, Vladimir 2013 Thermoelectric Energy Harvesting of Human Body Heat for Wearable Sensors ,*IEEE Sensors Journal)* 2284-2291

[5] J. R. Piorno, C. Bergonzini, D. Atienza, and T. S. Rosing 2009 Prediction and manage- ment in energy harvested wireless sensor nodes 1st Int. Conf. Wirel. Commun. Veh. Technol. Inf. Theory Aerosp. Electron. Syst. Technol. Wirel. VITAE

[6] R. E. Kalman and R. S. Bucy 1961 New Results in Linear Filtering and Prediction The- ory*Journal of Basic Engineering*

[7] v. A. Maciuca, M. Strutu, D. Popescu and G. Stamatescu 2013 Cell-based Sensor Network for Complex Monitoring at Home of Patients with Chronic Diseases *Int. S. Electri- cal and Electrnics Engineering*

[8] M. Gors, M. Albert, K. Schwedhelm, C. Herrmann and K. Schilling 2016 Design of an Advanced Telemedicine System for Remote Supervision " *IEEE J. Systems* 1089-1097

[9] A.Q. Javaid, N. F. Fesmire, M. A. Weitnauer and O. T. Inan 2015 Towards Robust Estima- tion od Systolic Time Intervals Using Head-to-Foot and Dorso-Ventral Components of Sternal Acceleration Signals *12th Int. Conference on Wearable and Implantable Body Sensor Networks (BSN)*
[10] V. Misra, A. Bozkurt, B. Calhoun, T. Jackson, J. Jur, J. Lach, B. Lee, J. Muth, O. Oralkan, M. Ozturk, S. Trolier-McKinstry, D. Vashaee, D. Wentzloff, and Y. Zhu 2015 Flexible Technologies for Self-Powered Wearable Health and Environmental Sensing *Proceedings of the IEEE* 665-681.
[11] Y. K. Tan and S. K. Panda 2011 Energy Harvesting from Hybrid Indoor Ambient Light and Thermal Energy Sources for Enhanced Performance of Wireless Sensor Nodes *Ind. Electron. IEEE Trans*44244435
[12] Seshan.C 2010 Cell efciency dependence on solar incidence angle *35th IEEE Photo- voltaic Specialists Conference*
[13] Fan, D., Ruiz, L. L., Gong, J., Lach 2016 Proling, modeling, and predicting energy harvesting for self-powered body sensor platforms *13th Int. Conference on Wear- able and Implantable Body Sensor Networks (BSN)* 402-407
[14] Leonov, Vladimir 2011 Energy harvesting for self-powered wearable devices *In Wear- able Monitoring Systems,* 2011.
[15] C. Viehweger, M. Baldauf, T. Keutel, and O. Kanoun 2012 Energy Prole Analysis by Simulation for the Design of Energy Harvesting Systems *9th International Confer- ence on Systems, Signals and Devices (SSD)* 2012.
[16] A. Kansal, J. Hsu, S. Zahedi, and M. B. Srivastava 2007 management in energy harvesting sensor networks *ACM Trans. Embed. Comput. Syst*
[17] M. Hassan and A. Bermak 2012 Solar harvested energy prediction algorithm for wireless sensors *4th Asia Symp. Qual. Electron. Des. ASQE*
[18] J. Lu, S. Liu, Q. Wu, and Q. Qiu 2010 Accurate modeling and prediction of energy avail- ability in energy harvesting real-time embedded systems *Green Comput. Conf.*
[19] P. Lee, Z. A. Eu 2011 Empirical modeling of a solar-powered energy harvesting wireless sensor node for time-slottedoperation*IEEE Wirel. Commun. Netw. Conf.*
[20] V. Leonov, T. Torfs, P. Fiorini, and C. Van Hoof 2007 Thermoelectric converters of hu- man warmth for self-powered wireless sensor nodes *IEEE Sens. J*
[21] D. C. Hoang, Y. K. Tan, H. B. Chng, and S. K. Panda, 2009 Thermal energy harvesting from human warmth for wireless body area network in medical healthcare system. *Proc. Int. Conf. Power Electron*
[22] Y. K. Tan and S. K. Panda, 2011 Energy Harvesting from Hybrid Indoor Ambient Light and Thermal Energy Sources for Enhanced Performance of Wireless Sensor Nodes *Ind. Electron. IEEE Trans*
[23] C. Petrioli and D. Spenza, 2012 Pro-Energy: a novel energy prediction model for solar and wind energy harvesting Wireless Sensor Networks.
[24] Roy, A., A. Klinefelter, F. Yahya, X. Chen, P. Gonzalez, C. J. Lukas, D. Akella, J. Boley, K. Craig, M. Faisal, et al.,2015 A 6.45LINK http://rlpvlsi.ece.virginia.edu/node/370 hor/319 nd Prediction Theory, wind energy harvesting Wireless Sensor Networks,*IEEE Transactions on Biomedical Circuits and Systems* 862-874.
[25] Kamakshi, A. D., A. Shrivastava, and B. H. Calhoun, 2016 CMOS Temperature Sensor for Ultra-Low-Power IoT Applications,*J. Low Power Electron. Appl. (JLPEA).*
[26] J. Dieffenderfer, H. Goodell, S. Mills, M. Mcknight, S. Yao, F. Lin, E. Beppler, B. Bent, B. Lee, V. Misra, Y. Zhu, O. Oralkan,2016 Low power wearable system for continuous monitoring of environment and health chronic respiratory diseasesv *IEEE Journal*
[27] Dawei Fan ,Luis Lopez Ruiz, 2017 EHDC An energy harvesting modeling and profiling platform for body sensor networks *IEEE j. biomedical*
[28] S.Majumder,T. Mondal, M.J Deen, 2017 wearable sensors for remote health monitoring*journals sensors*
[29] N.Hedge 2017 Automatic recognition of activities daily living utilizing insole based and wrist worn wearablr sensors *IEEE Journal*

IOP Conf. Series: Materials Science and Engineering **396** (2018) 012026 doi:10.1088/1757-899X/396/1/012026

Novel Web Service Based Fingerprint Identification Using Steganography and Xml Mining

Praseetha V M[1], Ayush Dattagupta[1], Suma R[2], S. Vadivel[1]

[1] Department of Computer Science, BITS Pilani,
International Academic City, Dubai, UAE.
[2] Department of Computer Science, SJCET, Choondacherry,
Pala, Kottayam

praseethasunil@gmail.com

Abstract. Fingerprint recognition has been considered as the most popular and reliable system for web based person authentication. Fingerprints have unique patterns that can be used to distinguish one fingerprint from another. The fingerprint pattern can be used to enhance the security of existing authentication system by adding a layer of biometric security. In this paper we propose a novel algorithm that is able to distinguish the pattern of one fingerprint from the pattern of another fingerprint using XML mining. But the valuable fingerprint data of a user is at risk as it can be hacked during web based authentication and the data can be easily compromised. Therefore we further enhance the security by extracting the fingerprint features into an xml file and embed the same into an image file using steganography. The encoded file can be used for safe transmission of fingerprints across the web for bio metric authentication. Our proposed system uses a novel method in which xml mining is applied at the server side to make the fingerprint identification faster.

1. Introduction

Automatic personal identification is very important as our day to day life is getting digitized. Two categories of traditional personal identification are there: they are token based and knowledge based [1]. In the token based approach a person is identified based on what he/she has (physical key, ID card, passport, Badges, etc.) and in knowledge based approach a person is identified based on what he/she knows (ID number, password, PINs, etc.). However both these approaches have certain limitations. In the token based approach, the token can be easily stolen, lost, shared or can be duplicated. In knowledge based approach, the knowledge can be guessed, shared or forgotten. Biometric authentication/ identification systems overcome the above mentioned limitations and are widely accepted. Biometric based approaches are considered as the most promising option for identifying individuals. Among the various approaches fingerprint recognition is one of the oldest and most popular technique for recognizing people [2].

Fingerprint recognition is used in many areas like access control [3], law enforcement, forensic science to aid criminal investigation [4], biometric smart cards etc. The popularity is mainly because of the uniqueness of fingerprint images, the availability of inexpensive fingerprint readers and the fact that criminals often leave their fingerprints at crime scenes. A fingerprint is a unique pattern of ridges and valleys on a finger. There are different stages in a fingerprint recognition system as shown in figure 1. They are mainly fingerprint acquisition, feature extraction, and matching. The traditional method for acquiring fingerprint was by using ink and paper. But now, many modern techniques are there to get an image of the fingerprint. They include optical, capacitive, ultrasound and thermal fingerprint readers.

Fingerprint matching is very challenging since two images acquired at different time exactly under same conditions need not be exactly the same. The difficulty in matching is due to several reasons [5]. The translation and rotation of the fingerprint images, application of a poor feature extraction algorithm, displaced, false and missing minutiae and the non linear deformations of the images are some of the reasons. To overcome these variations a powerful matching algorithm is needed. So the matching algorithm should be invariant to translation and rotation of the images. It should return the correct result while comparing fingerprints from the same finger even the feature extractor has missed some features or even when the fingerprint images are affected by non linear distortions.

The fingerprint matching algorithm can generally be classified into three different classes.

- Correlation based matching algorithms [6] [7]which use superimpose two fingerprint images to find the correlation among pixels for different displacement and rotation.
- Minutiae based matching algorithms [8] [9]which use the extracted minutiae of two fingerprints to find the matching pairs of minutiae.
- Non minutiae based matching algorithms [10] [11] which use the orientation, shape or frequency of ridges to perform matching between two fingerprints.

Among the three classes minutiae based algorithms are the most common and minutiae based matching is considered as a point pattern matching problem.

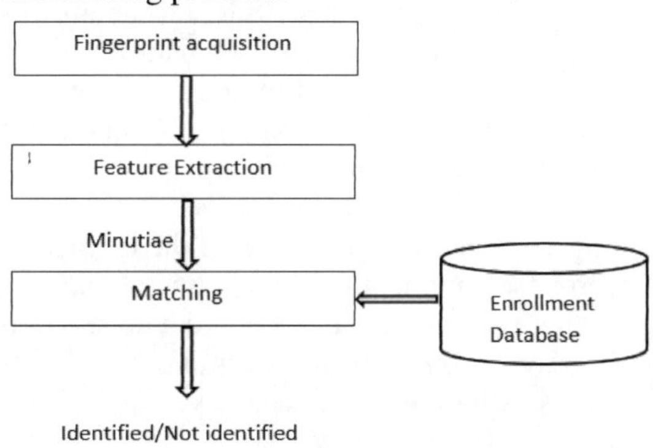

Figure 1. Different stages in fingerprint recognition

Our goal is to develop a soap enabled web service to perform authentication of a person using fingerprint. Towards this at the client end the finger print features are extracted, converted the features as an XML file and embedded the XML file into an image as part of improved security using steganography. The resultant image is sent in the form of a SOAP request to a web service. The hosted web service will extract the XML finger print data from the image and do the matching using XML data mining. In this paper we propose a secure, fast and accurate model for fingerprint verification.

The organization of this paper is as follows. Section 2 describes about the related works in this area, Section 3 explains the proposed model, Section 4 is about the extraction of features from a fingerprint image and creating the XML file using the minutiae triplets. Section 5 explains the technique of steganography to hide the fingerprint features. Section 6 explains the process of XML mining and fingerprint matching. In section 7 the results are discussed and section 8 is the conclusion and future scope of this work.

2. Related works

2.1. Transmission of Biometric Data

Lot of risk factors are there when biometric data is transmitted over communication channels. In the papers [12] [13] [14], the authors have mentioned about data hiding techniques using which biometric data are made hidden within another media. In [12], the authors use watermarking to hide the minutiae

IOP Conf. Series: Materials Science and Engineering **396** (2018) 012026 doi:10.1088/1757-899X/396/1/012026

of fingerprint in a face or synthetic fingerprint. A random cover image is used to hide the biometric data in [15]. Before hiding the authors encrypt the biometric data using a key and then by using error correcting code the encrypted data is encoded. Finally the encoded biometric data is embedded bit by bit using the sign of discrete cosine transform. In [16] the authors describe about multimodal biometric system employing both fingerprint and face information using watermarking techniques. They tested two different scenarios where in the first scenario they used a fingerprint image as a cover work and hide facial features into it. In the second scenario they hide fingerprint features into a facial image. By comparing the two scenarios the authors have found that hiding fingerprint features in a facial image is better than hiding facial features in a fingerprint image in terms of the verification accuracy of the watermarked image.

2.2. Fingerprint Recognition

Fingerprint recognition can be considered as a complex pattern recognition problem. The design of accurate feature extraction and matching algorithms is a very complicated and challenging task. Researchers have proposed many methods for fingerprint identification.

A fingerprint recognition system which uses Fast Fourier Transform and Gabor filters is explained in [17]. In this method two fundamental types of minutiae, endpoints and bifurcations, are extracted and these extracted features are used to perform fingerprint recognition. A new method for automatically identifying rare features in fingerprints based on combining level 1 features and minutia-based triangular descriptors is described in [18]. A fingerprint feature that is rare has higher discriminatory power when it is identified in a print (latent or otherwise), and multiple rare features in a single print can increase discriminatory power dramatically. The authors observed that some rare 9-point feature was found within every fingerprint that was tested and some of these rare 9-point features occurred only once in 1000 fingerprints. A robust damaged fingerprint recognition algorithm, with simpler recognition process Convolution Neural Network (CNN) of deep learning is explained in [19].

A method based on local minutia topology which generated by local minutia around the core point is proposed in [20]. The topological relationship is built on the extracted minutiae and the core point. The authors found that the proposed model has high computational efficiency and a significant improvement on robustness. Wang Yuan et al. [21] proposed a novel approach for minutiae filtering in fingerprint images. They use directionally selective steerable wedge filters to discriminate between minutiae and non-minutiae regions with reasonable accuracy.

3. Proposed Model

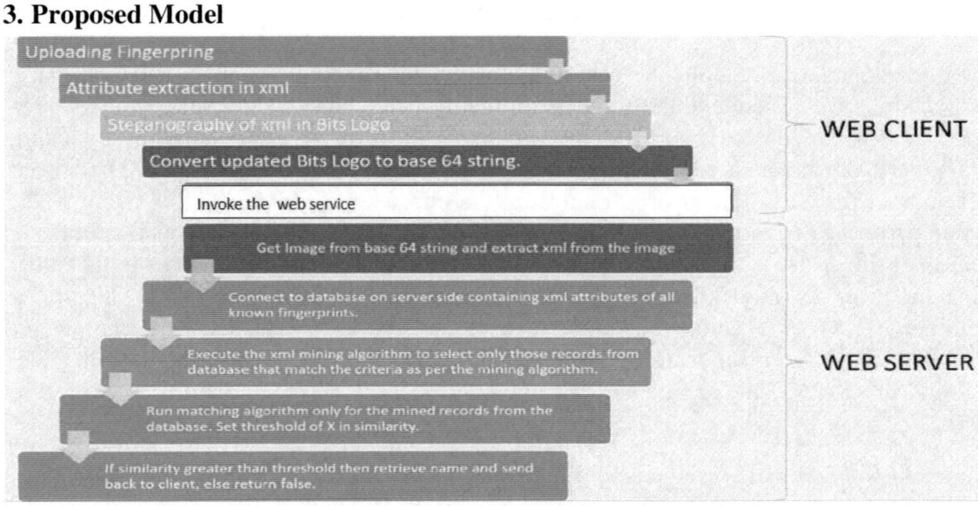

Figure 2. Proposed Web service model

Service Oriented Architecture (SOA) is a platform independent design which give emphasis to reusability and interoperability. Regardless of the platform, the services provide a common way of interaction [22]. Web services provide services which are easily accessible [23] over a network and they are considered as common means to exchange data and information over the network. The client program will extract the minutiae of the fingerprint image to be identified and these minutiae are then embedded in a host image as an xml file. Thus the features are protected using steganography and then the host image with embedded fingerprint features are send to the web server by invoking a SOAP enabled web service where the identification is taking place.

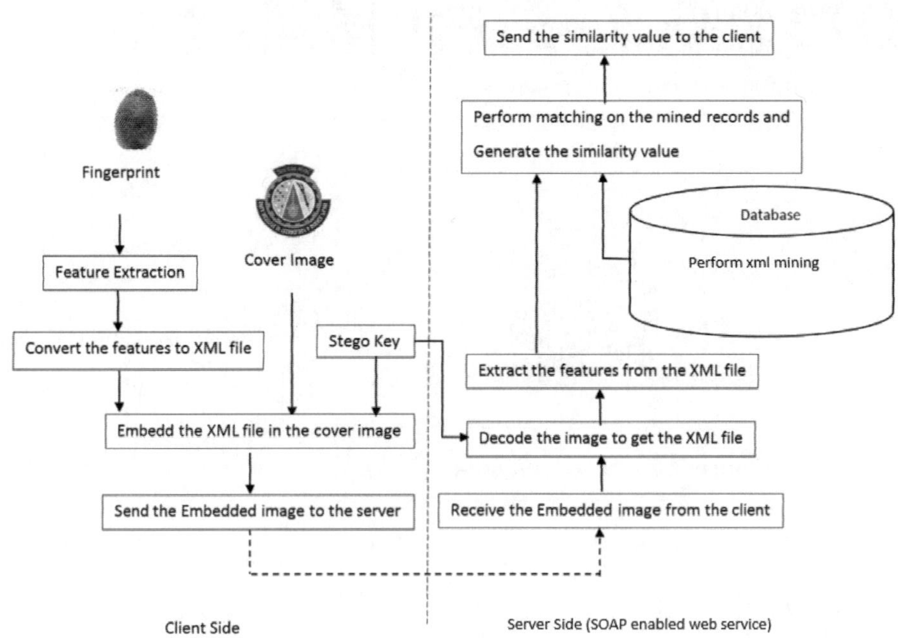

Figure 3. Client and server side operations of the proposed model

The client program will extract the minutiae of the fingerprint image to be identified and these minutiae are then embedded in a host image as an xml file. Thus the features are protected using steganography and then the host image with embedded fingerprint features are send to the web server by invoking a SOAP enabled web service where the identification is taking place. On the server an xml mining algorithm is executed to filter out the fingerprints from the database which contain the xml files corresponding to each fingerprint and to make the identification process easier. We have used a faster matching algorithm based on minutiae triplets known as the M3gl algorithm [24]. The result of the identification is then send back to the client.

4. Feature Extraction and XML file Generation

The three main categories of techniques used for fingerprint recognition are the minutiae-based (feature-based) matching, the pattern-based (or image-based) matching, and correlation-based matching [25]. Among the three, minutiae based technique is widely used as it is computationally inexpensive. The minutiae-based technique utilizes the rich local features of a fingerprint like ridge ending, bifurcation, crossover, island, lake etc. Figure 4. shows some of the important features that we need to extract from a fingerprint image in order to make an accurate recognition.

Here, we are considering the features like ridge ending, bifurcation and short ridge or independent ridge. The extracted fingerprint features are stored as (x,y,θ) where x and y give the coordinate of a particular feature and θ gives the angle of orientation of the feature. The extracted minutia data is converted to xml using xml serialization and some modifications to the framework classes to support xml formats. The modification have been made to the original framework itself and the changes can be observed in the Mtriplet extractor class of the framework.

IOP Conf. Series: Materials Science and Engineering **396** (2018) 012026 doi:10.1088/1757-899X/396/1/012026

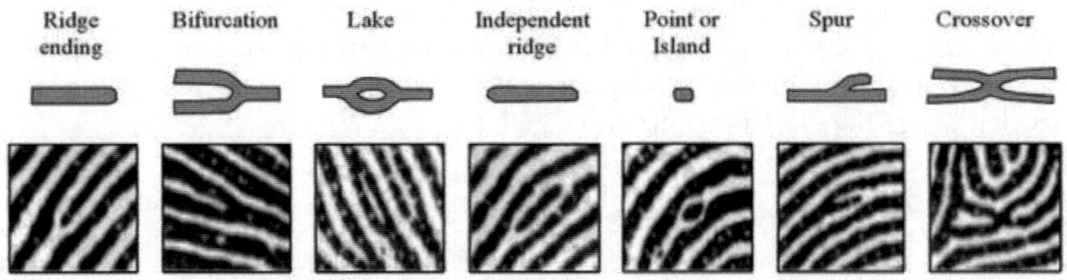

Figure 4. Fingerprint features

Figure 5. Extracted minutiae


```
<?xml version="1.0"?>
- <ArrayOfMinutia xmlns:xsd="http://www.w3.org/2001/XMLSchema" xmlns:xsi="http://www.w3.org/2001/XMLSchema-instance">
  - <Minutia>
      <X>342</X>
      <Y>74</Y>
      <Angle>3.5220990180969238</Angle>
      <MinutiaType>End</MinutiaType>
      <Flag>false</Flag>
    </Minutia>
  - <Minutia>
      <X>280</X>
      <Y>117</Y>
      <Angle>3.1415927410125732</Angle>
      <MinutiaType>End</MinutiaType>
      <Flag>false</Flag>
    </Minutia>
  - <Minutia>
      <X>311</X>
      <Y>159</Y>
      <Angle>0.46364760398864746</Angle>
      <MinutiaType>End</MinutiaType>
      <Flag>false</Flag>
    </Minutia>
  - <Minutia>
      <X>248</X>
      <Y>163</Y>
      <Angle>2.8564009666442871</Angle>
      <MinutiaType>Bifurcation</MinutiaType>
      <Flag>false</Flag>
```


Figure 6. XML file created with the extracted minutiae of finger print

5. Steganography for hiding minutiae

Lot of security risks are there while transmitting the sensitive fingerprint minutiae over the internet. The following figure shows the various security risks [26] that can occur from fingerprint image sensing to fingerprint matching.

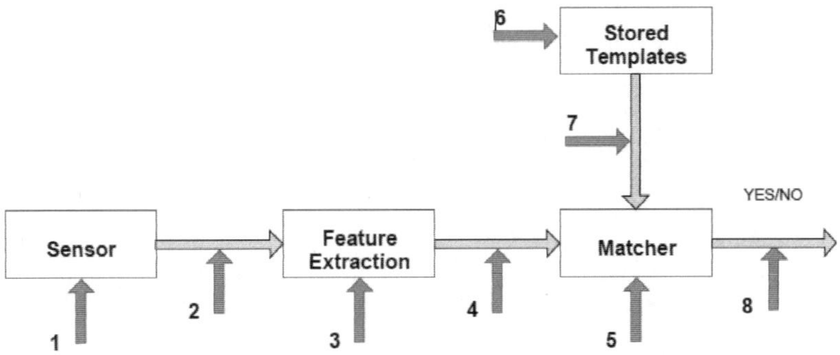

Figure 7. Security risks at various stages of fingerprint verification

1. Fake Biometrics at sensor: A reproduction of biometric feature is present such as face mask, a fake finger or a fake signature.
2. Resubmitting digitalized feature: In this type of attack a previously stored feature is resubmitted such as presenting an earlier recorded copy of fingerprint.
3. Overriding the biometric extraction program: A Trojan horse is used, the Trojan produces preselected features by the intruder.
4. Tampering with the extracted features: Changing the extracted features while it is being sent to the matcher.
5. Corrupting the matcher: Using Trojan horse, matcher is corrupted i.e. it selects the data which is preselected by the intruder.
6. Tampering with the database: As the database is spread over a number of server an intruder tries to modify one or more template, which results in authorization of a fraud data.
7. Attacking the data sent to matcher from the database: The stored data is communicated to matcher which can be intercepted and modified.
8. Overriding the final decision: Overriding the final yes/no result.

Our application primary focuses on enhancing security at point 4 of attack. In most verification systems the first three steps occur on the client side device which includes getting the fingerprint and extracting the features. Then these features are sent via the internet to a web service or a server that does the matching making this point of transmission of data from client to server most vulnerable. Therefore our program enhances the security at this level. This is done by the following methods:

- The extracted features are converted to xml format. This makes the data less readable and also creates the possibility of performing xml encryption of the features.
- The extracted xml is further secured by using a process of steganography.
- The steganography images are converted to base64 string and there can be an encryption at this level too.

Steganography is the method of hiding important data in an unrelated piece of information [27]. The security of biometric data can further be ensured by using steganography. Three basic types of steganography techniques are Pure Steganography, Secret key Steganography and Public key Steganography. Among the three different techniques pure steganography is the one used in most of the applications since it avoids the need to share the stego key between the communication parties.

The pure Steganography can be defined as the quadruple (C, M, D, and E) [28] where:
C: the set of possible covers,
M: the set of secret massage with $|C| \geq |M|$,
E: $C \times M \rightarrow C$ the embedding function,
D: $C \rightarrow M$ of the extraction function with the property that $D(E(c,m))=m$ for all $m \in M$ and $c \in C$

IOP Conf. Series: Materials Science and Engineering **396** (2018) 012026 doi:10.1088/1757-899X/396/1/012026

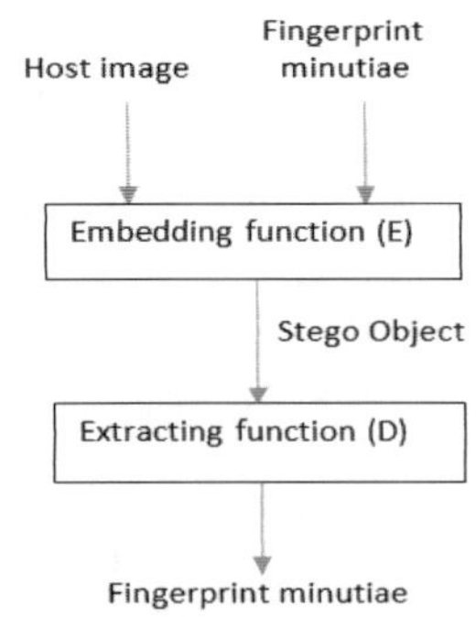

Figure 8. Steganography applied for fingerprint minutiae

We have used pure steganography to hide the fingerprint minutiae. The technique used to implement steganography here is LSB replacement logic. We hide the fingerprint minutiae in the image by hiding a bit of the minutiae in every least significant bit of every byte. We can hide three bits per pixel as there are 3 bytes for every pixel in a color image. The reverse process is done at the receiving end. The LSB is removed from every byte of the image until an ending byte is found and appended to the text file to get the original fingerprint minutiae.

```
For each pixel in the image
  If the whole XML file is not embedded
    Clear the LSB of R, G, B values
    Get the character to be embedded and convert it to integer
    Hide the 8 bits of the character in the
    LSBs of the RGB values of consecutive pixels
Add 8 consecutive zeros to indicate the end of text
```

Figure 9. Algorithm for pure steganography

The encoded images are converted to Base 64 strings. The base 64 strings are sent to a soap web service that computes the similarity between 0 and 1, where 0 means completely different and 1 means completely same.

6. Xml Mining and Fingerprint Matching

On the server side the database contains the XML files corresponding to the extracted features of each fingerprint. These XML files are grouped into three different classes based on the number of bifurcations and number of ridge endings. There are finite group of Minuta based finger print features which can be formed into definite number of groups based on the presence or absence of some features.

XML mining is a form of semi structured data mining which uses trees to identify schema and mine based on different techniques. The current application uses structure based/ schema based xml mining techniques for static xml documents, as the pattern of the fingerprint xml is fixed. Our algorithm compares the schema of the incoming xml file with the schema of the xml files in the database. This comparison is done with the tree structure of the xml document. Then the files whose schema matches with the schema of the incoming file are selected. Then we apply a matching algorithm on these mined files to get the exact match of the fingerprint.

The matching technique uses the M3gl algorithm proposed in [24]. The algorithm is based on minutiae triplet similarity measure. m_1, m_2 and m_3 are three minutiae arranged clockwise. d_i is the Euclidean distance between the minutiae other than m_i. α_i is the angle required to rotate the direction of a minutiae to coincide with the vectors associated with the minutiae and β_i is the angle required to rotate the direction of the minutiae to coincide with the other minutiae.

IOP Conf. Series: Materials Science and Engineering **396** (2018) 012026 doi:10.1088/1757-899X/396/1/012026

ach figure should have a brief caption describing it and, if necessary, a key to interpret the various lines and symbols on the figure.

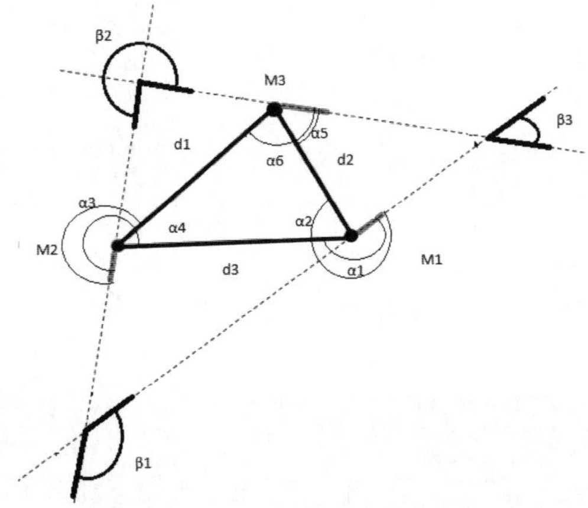

Figure 10. Components of m-triplets feature representation

For finding the m-triplets for a given minutiae set M, for each minutiae m in M its nearest c minutiae are found out. By including m and two of its nearest minutiae all possible m-triplets are computed discarding duplicates. After computing the m-triplets local minutiae matching is done to find similar m-triplets and then to find the local matching minutiae pairs of the query and template fingerprints. Then global minutiae matching is done in which every minutiae pair is considered as a reference pair for fingerprint rotation. A query minutiae transformation for each reference pair is done in the global minutiae matching. Finally the similarity score is computed as $n^2/|X||Y|$ where n is the number of matching minutiae pairs and X and Y are the template and query fingerprint minutiae.

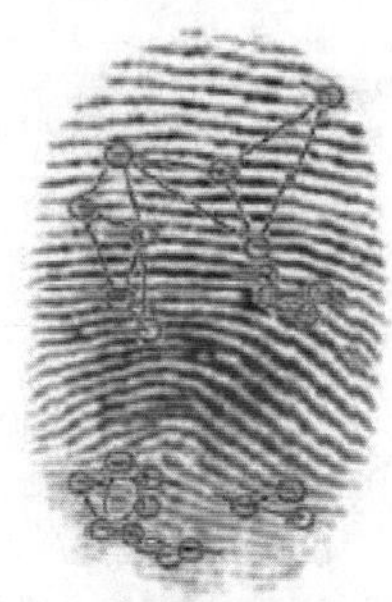

Figure 11. Extracted triplets

7. Results

We have successfully implemented a model by which we can transmit and verify fingerprints in a secure, faster and accurate way.

```
POST /Service.asmx HTTP/1.1
Host: localhost
Content-Type: application/soap+xml; charset=utf-8
Content-Length: length

<?xml version="1.0" encoding="utf-8"?>
<soap12:Envelope xmlns:xsi="http://www.w3.org/2001/XMLSchema-instance" xmlns:xsd="http://www.w3.org/2001/XMLSchema" xmlns:soap12="http://www.w3.org/2003/05/soap-envelope">
  <soap12:Body>
    <matchFingerprints xmlns="http://fingerprintmatcher.com/">
      <bit1>string</bit1>
      <bit2>string</bit2>
    </matchFingerprints>
  </soap12:Body>
</soap12:Envelope>
```

Figure 12. Sample SOAP request

Using the extracted minutiae, we created an XML file and this XML file is embedded in an image for transferring the minutiae securely. The SOAP web methods and the HTTP request and response are shown in the below figures. The server calculates the similarity value by matching with a template fingerprint and the similarity value is returned to the client.

```
HTTP/1.1 200 OK
Content-Type: application/soap+xml; charset=utf-8
Content-Length: length

<?xml version="1.0" encoding="utf-8"?>
<soap12:Envelope xmlns:xsi="http://www.w3.org/2001/XMLSchema-instance" xmlns:xsd="http://www.w3.org/2001/XMLSchema" xmlns:soap12="http://www.w3.org/2003/05/soap-envelope">
  <soap12:Body>
    <matchFingerprintsResponse xmlns="http://fingerprintmatcher.com/">
      <matchFingerprintsResult>double</matchFingerprintsResult>
    </matchFingerprintsResponse>
  </soap12:Body>
</soap12:Envelope>
```

Figure 13. Sample SOAP response

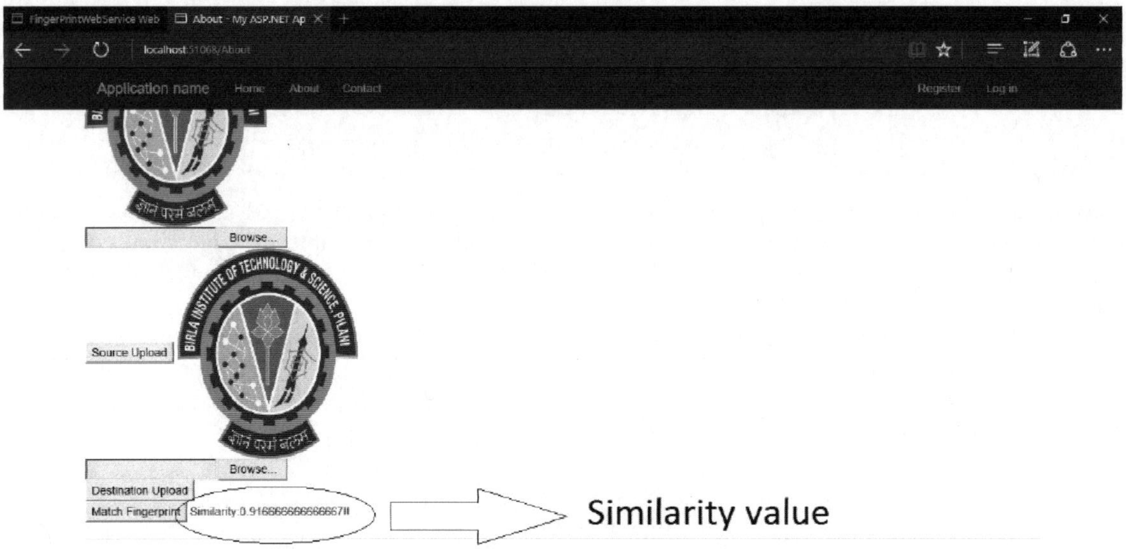

Figure 14. Final similarity of two different fingerprints of the same person

Our experiment compared the performance of fingerprint identification without xml mining and with xml mining. It is observed that the time required for the fingerprint identification can be reduced considerably if xml based mining is used.

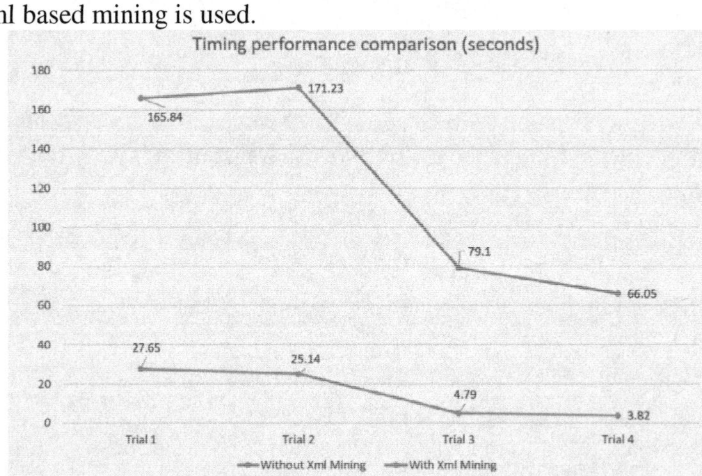

Figure 15. Performance comparison

IOP Conf. Series: Materials Science and Engineering **396** (2018) 012026 doi:10.1088/1757-899X/396/1/012026

8. Conclusion and Future Scope

The implementation of the proposed mechanism will ensure end to end security between a client side web application and a server based RPC web service. Converting the features in xml format is a very good technique as it gives programmers the independence to use it in multiple ways and makes transmission very easy and lightweight as xml format for data transmission is used in web and cloud based applications extensively. It is a very fast and very secure method for comparing two fingerprints on an online platform and is easy to use and understand.

The data can be encrypted twice before it is send to the service for computation. Double encryption means double protection. Xml encryption can be performed at the extraction level and the extracted xml can be encrypted using a Hash Key technique. The base 64 strings can also be encrypted using AES or RSA or other such encryption algorithms adding another layer of security to the data just before it is passed to the web service. The framework can be extended so that instead of comparing two fingerprints the fingerprint can be uploaded and can be compared with a list of fingerprints already stored in the database in base 64 encrypted format. The Graphical user interface of the web application can be improved. The security implementation may be extended to other biometric methods of verification like iris scanning, palm recognition etc.

9. References

[1] Jayaraman, in *Digital Image Processing*, Tata Mc Graw Hill, pp. 434-435.

[2] A. Babich, "Biometric Authentication. Types of biometric identifiers," Haaga-Helia University of Applied Sciences, 2012.

[3] Priyanka, "Fingerprint recognition techniques and its applications," in *International Conference on Advances in Engineering and Technology Research (ICAETR)*, Unnao, 2014.

[4] M. Saini and A. K. Kapoor, "Biometrics in Forensic Identification: Applications and Challenges," *Journal of Forensic Medicine,* 2016.

[5] W. Sheng, G. Howells, M. Fairhurst and F. Deravi, "A Mementic Fingerprint Matching Algorithm," *IEEE Transactions of Information Forensics and Security,* vol. 2, no. 3, pp. 402-412, 2007.

[6] T. Hatano, T. Adachi, S. Shigematsu, H. Morimura, S. Onishi , Y. Okazaki and H. Kyuragi, "A fingerprint verification algorithm using the differential matching rate," in *Proceedings of the 16th International Conference on Pattern Recognition*, Quebec City, Quebec, Canada, 2002.

[7] A. LindosoLuis, L. Entrena, J. Liu-Jimenez and E. San Millan, "Correlation-Based Fingerprint Matching with Orientation Field Alignment," in *International Conference on Biometrics*, 2007.

[8] X. Jiang and W.-Y. Yau, "Fingerprint minutiae matching based on the local and global structures," in *Proceedings of 15th International Conference on Pattern Recognition*, Barcelona, Spain, 2000.

[9] X. Luo, J. Tian and Y. Wu, "A minutiae matching algorithm in fingerprint verification," in *Proceedings of 15th International Conference on Pattern Recognition*, Barcelona, Spain, 2000.

[10] J. ChengYang and D. Sun Park, "A fingerprint verification algorithm using tessellated invariant moment features," *Neurocomputing,* vol. 71, no. 10-12, pp. 1939-1946, 2008.

[11] L. Nanni and A. Lumini, "Descriptors for image-based fingerprint matchers," *Expert Systems with Applications,* vol. 36, no. 10, pp. 12414-12422, 2009.

[12] A. K. Jain and U. Uludag, "Hiding Fingerprint Minutiae in Images," in *Third Workshop on Automatic Identification Advanced Technologies (AutoID)*, Tarrytown, New York. USA, 2002.

[13] A. K. Jain and U. Uludag , "Hiding a Face in a Fingerprint Image," in *16th International Conference on Pattern Recognition*, Canada, 2002.

[14] A. Jain and U. Uludag, "Hiding biometric data," *IEEE Transactions on Pattern Analysis and Machine Intelligence,* vol. 25, no. 11, pp. 1494 - 1498, 2003.

[15] N. Agrawal and M. Savvides, "Biometric data hiding: A 3 factor authentication approach to verify identity with a single image using steganography, encryption and matching," in *IEEE Computer Society Conference on Computer Vision and Pattern Recognition*, Miami Beach Florida, 2009.

[16] Y. Chung, D. Moon, K. Moon and S. Pan, "Hiding Biometric Data for Secure Transmission," in *International Conference on Knowledge-Based and Intelligent Information and Engineering Systems*, 2005.

[17] G. Aguilar, G. Sanchez, K. Toscano, M. Salinas, M. Nakano and H. Perez, "Fingerprint Recognition," in *Second International Conference on Internet Monitoring and Protection ICIMP 2007*, San Jose, CA, USA , 2007.

[18] I. Munagani , M. S. Hsiao and L. Abbott, "On the Uniqueness of Fingerprints via Mining of Statistically Rare Features," in *IEEE International Symposium on Technologies for Homeland Security (HST)*, Waltham, MA, USA, 2015.

[19] W. Yani, W. Zhendong, Z. Jianwu and C. Hongli, "A Robust Damaged Fingerprint Identification Algorithm Based on Deep Learning," in *Advanced Information Management, Communicates, Electronic and Automation Control Conference (IMCEC), IEEE*, Xi'an, China , 2016.

[20] S. Bei , L. Wusheng , D. Liebo and L. Qin, "A fingerprint identification algorithm based on local minutiae topological property," in *IEEE International Conference on Data Science in Cyberspace (DSC)*, Changsha, China , 2016.

[21] W. Yuan , Y. Lixiu and Z. Fuqiang, "A Real Time Fingerprint Recognition System Based On Novel Fingerprint Matching Strategy," in *8th International Conference on Electronic Measurement and Instruments, ICEMI '07.* , Xi'an, China, 2007.

[22] S. Kumari and S. K. Rath, "Performance comparison of SOAP and REST based Web Services for Enterprise Application Integration," in *International Conference onAdvances in Computing, Communications and Informatics (ICACCI)*, Kochi, 2015.

[23] S. G. C. Isaac and V. U. Devi, "Efficient Querying and SOAP Based Streaming of Multimedia Content Using WEB Services," in *International Conference on Intelligent Computing Applications (ICICA)*, Coimbatore, 2014.

[24] M. A. Medina-Perez, M. Garcia-Borroto, A. E. Gutierrez-Rodriguez and L. Altamirano-Robles, "Improving Fingerprint Verification Using Minutiae Triplets," *Sensors,* pp. 3418-3437, 2012.

[25] J. F. Lim and R. K. Y. Chin, "Enhancing Fingerprint Recognition Using Minutiae-Based and Image-Based Matching Techniques," in *1st International Conference on Artificial Intelligence, Modelling and Simulation (AIMS)*, Kota Kinabalu, 2013.

[26] TanishaAggarwal and C. K. Verma, "Fake Fingerprint Detection Methods," *IJITKM,* no. Special Issue, pp. 61-69, 2014.

[27] C. Whitelam, N. Osia and T. Bourlai, "Securing Multimodal Biometric Data through Watermarking and Steganography," in *IEEE International Conference on Technologies for Homeland Security (HST)*, Waltham, MA, 2013.

[28] Z. K. AL-Ani, H. O.Alanazi , A. Zaidan and B. Zaidan, "Overview: Main Fundamentals for Steganography," *Journal Of Computing,* vol. 2, no. 3, pp. 158-165, 2010.

Security mechanisms and Vulnerabilities in LPWAN

Smilty Chacko.

P.G Scholar , Department of Computer Science and Engineering,
St. Joseph's College of Engineering and Technology, Palai, Kerala.

Email: smiltychacko1993@gmail.com

Mr. Deepu Job.
Assistant Professor, Department of Computer Science and Engineering,
St. Joseph's College of Engineering and Technology, Palai, Kerala.

Email: deepu.job@sjcetpalai.ac.in

Abstract. The count of Internet of things (IoT) devices getting connected to internet are exponentially growing and is proposed to be 75 billion by 2025. IoT devices uses Low Power Wide Area Network (LPWAN) that are precisely build to operate at a lower cost with greater power efficiency and long range connectivity than traditional wireless networks. LPWAN are inherently prone to security threats since they are also wireless communication network. Secure communication is essential for LPWAN and most LPWAN technologies uses simple cryptography methods. Despite of this security mechanisms, LPWAN is vulnerable to wide range of attacks. In this paper Network architecture and Security mechanisms of LPWAN technologies (LoRa, Sigfox, NB-IoT and DASH7) are discussed and are compered in terms of IoT factors. As a case study, vulnerabilities of LoRa is also being analyzed.
Keywords: LPWAN, Wireless network, LoRa, IoT.

1. Introduction

Internet of things (IoT) is defined as a network of devices or physical objects with electronics, sensors, software and network connectivity that enables them to exchange information and communicate with each other. IoT applications can be used for smart homes, smart metering, factory monitoring, agriculture, smart buildings etc. In IoT wireless technologies such as Bluetooth, WiFi, ZigBee etc. are used to meet the communication requirements, but they can offer only a short range connectivity. In order to overcome the limitations of short range protocols Low Power Wide Area Networks (LPWAN) are introduced, which offer a long range connectivity in the order of kilometers. LPWAN [1] is getting wide acceptance in industrial and research communities due to its low power long range characteristics. LPWAN [2] can offer long range connectivity up to 2 to 50 km depending upon rural and urban areas. Long battery life is one of the key feature of LPWAN technologies. The data rate offered by LPWAN is from 100 bps to 250 kbps and data rate is selected based on range requirements. LPWAN is suitable for IoT applications since it need to send only small amount of data to a long range. There are many LPWAN technologies and among them LoRa, Sigfox, NB-IoT and DASH7 are getting wide acceptance now a day which have many technical differences.

The paper is organized as follows: the first section discusses various LPWAN network architecture. In second section, since security is an essential requirement of wireless communication technologies, how security is assured by these technologies is discussed and a comparison is made based on the IoT factors such as Quality of Service, Battery life & latency, Network coverage & range, Deployment model, Cost etc. In the third section vulnerabilities in LoRa is discussed as a case study.

2. Technical differences: LORA, SIGFOX, NB-IOT and DASH7

IoT applications require long battery life in order to avoid costly replacements. Many IoT use cases are used within buildings and current technologies have limited coverage and are too costly too. In order to fill this gap LPWAN technologies are introduced. In this section we discuss about technical features of LPWAN technologies such as LORA, SIGFOX, NB-IOT and DASH7.

2.1. LoRa

LoRa is a radio modulation technology by Semtech Corporation. LoRa [3] provides connectivity about 15 to 20 km using chirp spread spectrum technique in which entire band width is used to transmit single signal. LoRa [8] uses 868 MHz to 900 MHz ISM bands for its operation and data rate is 0.3 kbps. Since LoRa have a long battery life, cost of replacement of devices can be reduced and its deployment is not so complex. LoRa use symmetric key cryptography to ensure security to its devices. Chirp spread spectrum used by LoRa can reduce the signal degradation, noise etc. while transmitting signal. LoRa devices support bidirectional communication and it communicates to LoRa gateways. This gateways are connected to the network server and it is connected to the applications. LoRa can use different topologies like star, mesh etc.

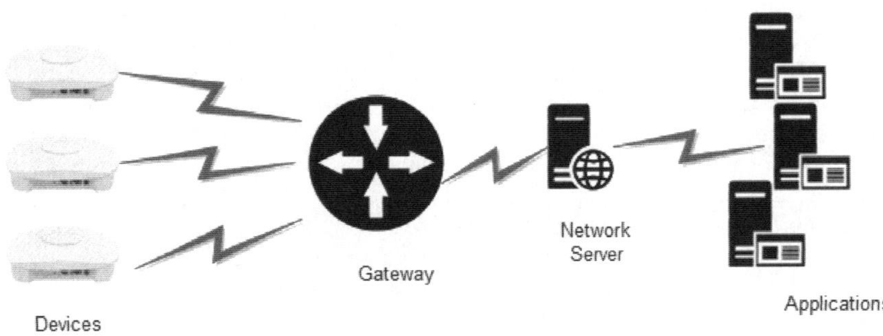

Figure 1. LoRa Network Architecture

2.2. Sigfox

Sigfox is a LPWAN technology which provides low power low data and low cost communication devices. It operates in global network and is used for IoT communications. Even though there ara plenty of wireless technologies Sigfox is widely accepted due its cheaper connection and an extended battery life. Sigfoix [2] enables new IoT applications and it can provide backup connectivity for higher bandwidth devices. Communication using Sigfox can be performed when it detect an event or measure something, it will power on the communication module and send the message. The message is then picked up by the network and the data is received on your server. Sigfox is designed to maximise the energy efficiency. Sigfox consumes very low power when it transmit a data and no maintenance is required. Depending on the topology and environment Sigfox can provide long range connectivity about a few kilometres to tens of kilometres. Sigfox uses unlicensed ISM bands and it uses frequencies, 868 MHz in Europe and 915 MHz in US. Sigfox payload size is 12 bytes anda data rate is 100 bps.

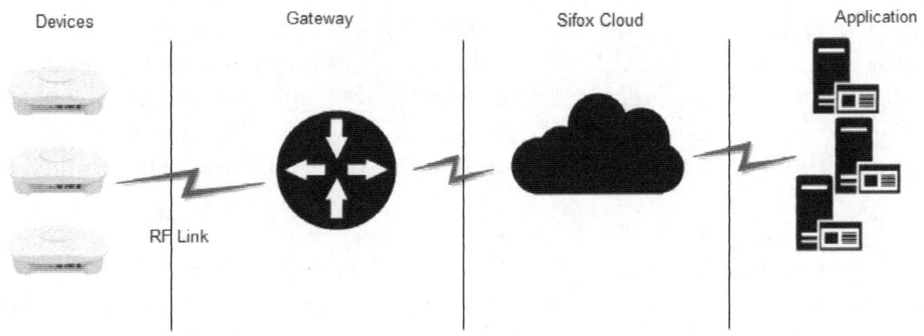

Figure 2. Sigfox Network Architecture

2.3. NB-IoT

Narrow band IoT (NB-IoT) [7] is a technology that offer long battery life and long range connectivity. NB-IoT use cellular telecommunication band to connect wide range of devices. NB-IoT [8] is standardised by third generation partnership project (3GPP). The frequency bandwidth of NB-IoT is 700 MHz, 800 MHz and 900 MHz and NB-IoT is based on LTE (long term evolution) protocol. The functionality of LTE is reduced by NB-IoT to its minimum and it enhances the functionalities as required by IoT applications. NB-IoT improves the power consumption rate of devices, spectrum efficiency and system capacity. NB-IoT can offer a very low power consumption and a battery life more than ten years. The capacity of NB-IoT is about 5000 connections per cell and it support 40 device per household. The data rate offered by NB-IoT is 200 kbps and the transmitting power is +23 dB. Device cost of NB-IoT is less than dollar 5 per module.

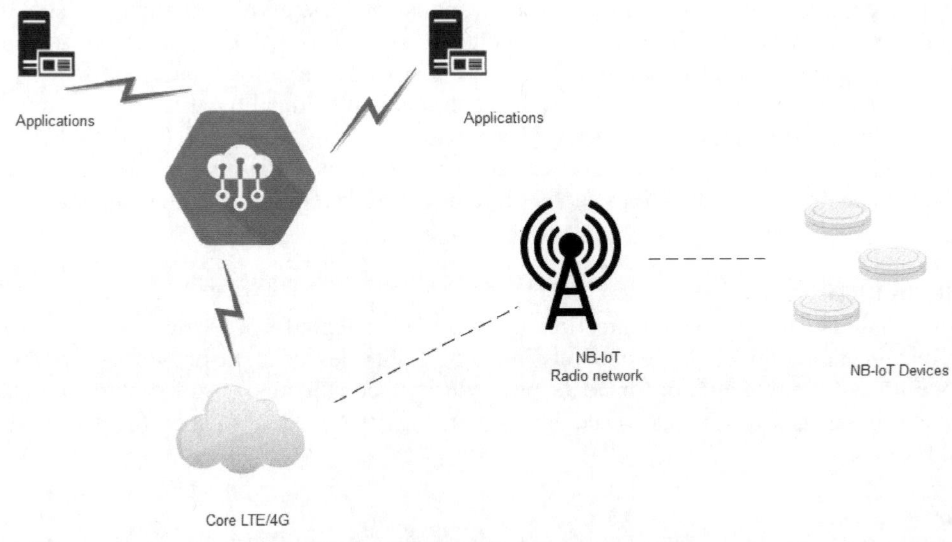

Figure 3. NB-IoT Network Architecture

NB-IoT supports three frequency spectrum that is standalone, In-Band and Guard Band. A new bandwidth is used by standalone deployment type and a bandwidth reserved in guard band deployment type. NB-IoT can be used for personal use, public use, IoT applications and industrial use and it is a secure and reliable technology.

IOP Conf. Series: Materials Science and Engineering **396** (2018) 012027 doi:10.1088/1757-899X/396/1/012027

2.4. DASH7

DASH7 Alliance protocol is an Actuator network protocol and it is an open source wireless sensor network. DASH7 operates in 433 MHz, 868 MHz and 915 MHz unlicensed ISM bands. DASH7 can provide a range up to 2km. For security AES 128 bit shared key encryption is used. Data rate offered by DASH7 is 28 kbps and it uses wakeup signal to achieve low power, low latency and elegant architecture. Wakeup mechanism needs a little overhead and latency of only 2.5 to 5 seconds. DASH7 can offer months or years of battery life and it can use calculator-style solar cell, energy harvesting and flat-flex batteries.

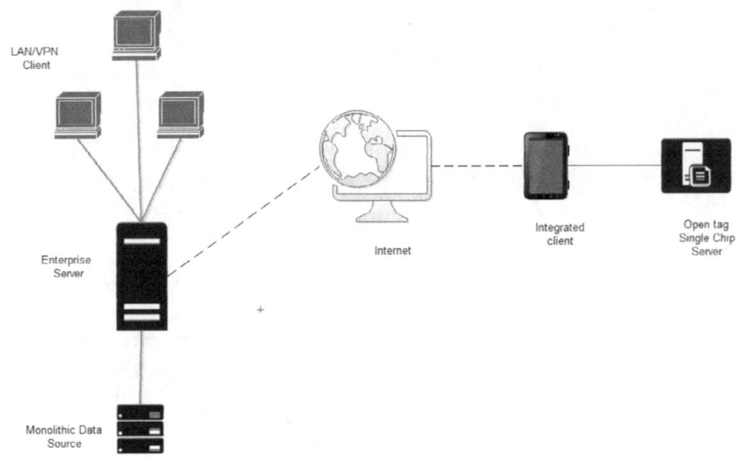

Figure 4. DASH7 Network Architecture

DASH7 [10] uses BLAST networking technology. BLAST means Bursty Light data Asynchronous and Transitive. In bursty, data transfer is abrupt and it does not contain contents like audio and video. In light, multiple consecutive packet transmission is generally avoided and the packet size is limited to 256 bytes. DASH7 communicates through command- response and therefore periodic hand shaking is not required. The whole network of devices is considered as transitional and there is no need routing information or management information. DASH7 use star or tree topology and there will be only one hop between the endpoint and gateway. The packets are acknowledged by at least one gateway and it provide sensor to cloud communication. All OSI layers are covered by DASH7. DASH7 [11] file system contains collection of structured data elements with their properties such as permission and storage class. DASH7 have wide range of applications such as bird tracking, green house monitoring applications etc.

3. Security in LPWAN

Security is a challenge for IoT communication. IoT broadens the platform of attack due to its long range connectivity and comparatively large transmission time. Every device in the network is a potential point of vulnerability and every technology have its own security mechanism to ensure secure communication. In the next section, the mechanism used by different LPWAN technologies to ensure security is discussed [12].

3.1. Security in LoRa

Security in LoRa [5] is handled using AES-128 encryption method. It provides multiple layer of encryption in LoRaWAN. In LoRa, network and application keys are used to ensure security to the packets in the network. Using 128 bit AES application key (Appkey) two session keys, Network session key and Application session key are generated. A message integrity code [6] is generated and verified using Network session key which is shared by end devices and network server. This message integrity code is used to assure integrity of the message. It can be used to create a unique signature for each device. Application session key is used for encryption and decryption of data. An XOR operation is used to encrypt each message and it also uses a key stream generated by network session key, application session key and uplink and downlink message counter to generate encrypted payload. But the length of the message before and after encryption is same in LoRa and it provides an opportunity to the malicious entity to recreate the key stream from encrypted message. Despite of the security mechanism LoRa devices are susceptible to jamming attack, compromising device and network keys, replay attack and wormhole attack.

3.2. Security in Sigfox

In Sigfox security challenge is addressed through a systematic process. Sigfox is one of the most secure LPWAN technology and it is unique in design. Sigfox devices predominantly operate in offline with a built-in behaviour. When Sigfox needs to transmit or receive data from the internet, the Sigfox device will broadcast a radio message. This broadcasted message is received by base stations and the message is then transmitted to Sigfox core network, which is then delivered to corresponding IoT applications. This Sigfox network architecture provides an air gap and it is not possible to access an end point through internet maliciously.

In Sigfox, [9] each device is given by a unique symmetrical authentication key during the manufacturing time itself. If one of the device is compromised, it can only make a limited impact on the network. A cryptographic token is computed using this authentication key for each sending or receiving message. This token is used for authentication of sender and integrity of message. In order to detect the replay attempts sigfox core network uses a sequence counter contained in each Sigfox message. Since Sigfox deice are not IP addressable there is no over the air activation (OTA) mechanism and it reduces the possibility of attacks. In Sigfox a user can choose whether or not encrypt the message using the encryption solutions provided by Sigfox. The users are also allowed to use their own encryption methods if necessary.

3.3. Security in NB-IoT

NB-IoT inherits LTE's authentication and encryption. NB-IoT [8] consists of three layers, perceptron layer, transmission layer and application layer. Security requirement of NB-IoT aims this three layer architecture. The bottom layer is the perceptron layer and it is susceptible to both active and passive attacks. In passive attack the attacker will simply monitor the network traffic and active attack will affect the integrity of message, falsification of data enc. To prevent this cryptographic algorithms can be used to encrypt the data, integrity authentication and verification. In perceptron layer, each node can directly communicates with base station and this routing security problems during networking can be avoided.

In transmission layer NB-IoT change the network deployment which is complicated and the gateway collects information which is then feeds back to base station. By this, many problems like multi-network networking, high capacity battery, and high cost etc. despite of security threats are there such as access to high capacity NB-IoT terminals, open network environment etc. Application layer in NB-IoT store massive amount of data efficiently. Therefore the main security requirements are identification and processing of this massive heterogeneous data, integrity and authentication of data and access control of data.

3.4. Security in DASH7

To implement an open source code (OpTag) is used and it was released to debug the source code and performance test. To ensure confidentiality of data link layer OpTag beta two includes software based advanced encryption standard (AES) but network layer security is not implemented [11]. To ensure security of data link layer and network layer DASH7 [10] AES-counter with CBC-MAC is used. CCM can be operated by using AES 128 bit block cipher and the message is encrypted using the counter mode. CBC-MAC mode is used to compute the authentication field. In DASH& the sender will generate a packet and the packet is then encrypted using counter mode and it is then authenticated with CBC-MAC. After encryption and authentication, the packet is transmitted through air. The receiver receives the

IOP Conf. Series: Materials Science and Engineering **396** (2018) 012027 doi:10.1088/1757-899X/396/1/012027

packet and it will then decrypt the packet and verify it. If it is a legitimate packet the process will be continued.

Table 1. Comparison of LPWAN Technologies

	LoRa	NB-IoT	Sigfox	DASH7
Bandwidth	125 KHz	180 KHz	100 Hz	25 KHz/200KHz
Frequency	Below 1 GHz	Below or above 1GHz	Below 1 GHz	Below 1 GHz
Downlink peak data rate	50 Kbps	250 Kbps	600 bps	55.55 Kbps
Uplink peak data rate	50 Kbps	250 Kbps	100 bps	9.6 Kbps
Module cost	Low	Low	Very low	Low
Data confidentiality	Yes	Yes	No	Yes
Authentication and encryption	Yes(AES 128)	Yes (LTE Encryption)	No	Yes
Bidirectional	Yes/Half-duplex	Yes/Half-duplex	Limited/Half-duplex	Yes/Half-duplex
Standardization	LoRa Alliance	3GPP	Sigfox company	DASH7 Alliance
Range	15-20 km	22 km	30-50 km	2 km
Battery life	10 years	10 years	10 years	10 years

4. Security Vulnerabilities of LPWAN by example of LoRaWAN

LoRaWAN [3] provides end-to-end security using application and network keys. However, an attacker with physical access may compromise the LoRa end-devices. If an attacker gains physical access to a device, he/she may extract the keys. Usually, end-devices contain a LoRa radio module and a host microcontroller unit (MCU). The radio module communicates with the host microcontroller via UART or SPI interface. Commands and data exchanges between the host and the radio module can be intercepted using external hardware. For instance, if UART interface is used between two ICs, a basic FTDI interface can be used to extract all the key exchanges. Contemporary radio modules on the market does not provide built-in encryption support to secure the interactions between the host microcontroller and radio module. In such cases, there is no way to understand that the commands sent to the radio module were issued by the host MCU or a malicious entity. Also a malicious entity could intercept all the data exchanges between the host MCU and the radio module and use this intercepted information to create a mock device with the same credentials or manipulate the data payload. Therefore, application developers must not perform sensitive operations such as setting security keys for each data transmissions as it may expose critical information to malicious entities.

4.1 Compromising devices and Network keys

LoRa

To prove the feasibility of this attack, the Xignal mousetrap was used as a target device. The hardware unit was tampered to expose the UART serial lines between the MCU and the LoRa radio module. A regular FTDI chip was connected to the serial line to read intercept all the transactions between them. Whenever the mousetrap was reset, the host MCU issues commands to configure the network keys of the radio module. Using these keys and custom LoRa device, we impersonated a LoRa mouse trap and sent data as if it were coming from the mouse trap.

4.2 Jamming attack
Jamming attack [5] is one of the main problems for IoT. Malicious entities can transmit a powerful radio signal in the same strength of application devices and interrupt the radio transmissions. It is possible to jam LoRa devices using commercial-off-the-self LoRa hardware. Concurrent LoRa transmissions at same frequency and spreading factor can interfere with each other. This vulnerability in LoRa physical layer allows malicious entities to use commercial-off-the-shelf (COTS) LoRa devices to jam LoRa networks. It is hard to prevent jamming attack because there are workarounds to overcome this issue. Firstly, the jamming of the whole network or frequency can be easily detected since all the devices that communicate in that frequency would suddenly start to drop out from the network. By detecting such behaviour, network administrators can take appropriate action to prevent the impact of jamming.

4.3 Replay Attack
A replay attack [13] is an attack on security protocol, re-sending or repeating the valid data transmission by the malicious entity. The main purpose of this attack is fooling the device or module by using handshake messages or old data from the network. In order to perform the attack in wireless networks, the entity should know the communication frequencies and channels to sniff data from transmission between devices. In LoRaWAN, it is not possible to decrypt transmissions between end-devices and gateways without AppSKey, since the entire payload of the LoRaWAN message are encrypted by it. Additionally, since tampering with the data will make the MIC check fail, it is not possible to do it without NwkSKey. Although the malicious entity can resend the message consecutively, using frame counters which are defined in LoRaWAN specifications these messages or attacks can be detected and discarded. Once the end-device is activated, these counters are both set to 0 and each message coming from the gateway or the device increments counters. If a message is received with a lower frame counter than the last message, it is ignored. However, the LoRaWAN specification handling off frame counters is specifically left to the application and developer. Therefore, networks which do not track these frame counters could be vulnerable to replay attacks.

4.4 Wormhole Attacks
End-devices in LoRaWAN network can be jammed by using off-the-shelf hardware. Together with replay attack, a wormhole attack can be performed against LoRaWAN network. In this type of attack, one malicious device captures the packets from one device and transmits them to other distant located device to replay the captured packet. This can easily be launched by malicious entity without prior knowledge of the network or cryptographic mechanism.

In LoRaWAN network, a wormhole attack [14] could be performed by using two type of device that are sniffer and jammer. The sniffer captures packets and, signals to the jammer to notify that it captured the packet. The captured packet never reaches to the gateway and, validation of captured message stays valid. The captured message can be replayed any time. Gateway and Network server forwards to the packet to the application layer. Therefore, the important alarm messages can be jammed and regular messages which are captured before and never reached to the gateway, could be sent to the gateway as if there is no alarm. Since there is no time-related information in LoRaWAN messages, it is hard to detect this attack in LoRaWAN networks. A general mechanism, called *packet leashes* is used for detecting and thus defending against wormhole attacks.

5. Conclusion
This paper summarizes the network architecture, security mechanisms and vulnerabilities of LPWAN (LoRa, Sigfox, NB-IoT and DASH7). Each technology has its own advantages and limitations in term of IoT factors. NB-IoT and Sigfox offer long range connectivity and low cost devices. Most of the technologies offer long battery life, reliable communication. Among them NB-IoT will give high-value in IoT markets. NB-IoT offers low latency and high quality of services. NB-IoT use LTE encryption while other technologies use AES encryption methods. The actual battery life, security and performance of NB-IoT is currently an open question. From the case study of LoRa and its security vulnerabilities it

is clear that LoRa devices are prone to various security attacks despite of the security mechanism offered by it.

6. Research challenges and future scope

LPWAN still have problems that are to be resolved, such as secure key management and authentication. Most of the LPWAN technologies uses Advanced Encryption Standard for security which are not designed to run in low power devices. AES can be replaced by light weight block cyphers specially designed and build to run in low resource devices. Future research have to be made to make LPWAN a promising network for the 5th generation (5G) wireless mobile communication that would lead to a global LPWAN solution for IoT applications.

7. References

[1] Lukas Krupka; Lukas Vojtech; Marek Neruda *The issue of LPWAN technology coexistence in IoT environment*, 2016 17th International Conference on Mechatronics - Mechatronika (ME) on 2016.

[2] Usman Raza, Parag Kulkarni, and Mahesh Sooriyabandara *Low Power Wide Area Networks: An Overview*, IEEE Communications Society on, 16 January 2017.

[3] Emekcan Aras, Gowri Sanker Ramachandran, Piers Lawrence, Danny Hunghes *Exploring the Security Vulnerabilities of LoRa*, 3rd IEEE International Conference on Cybernetics on 2017.

[4] Rashmi Sharan Sinha, Yiqiao Wei, Seung-Hoon Hwang *A survey on LPWA technology: LoRa and NB-IoT*, ICT Express, March 2017.

[5] Sigfox Technical Overview, available at https:// www.disk91.com/wpcontent/uploads/2017/05/4967675830228422064.pdf, May 2017

[6] Kais Mekkia, Eddy Bajica, Frederic Chaxela, Fernand Meyerb *A comparative study of LPWAN technologies for large-scale IoT deployment*, ICT Express, 4 January 2018.

[7] Hwajeong Seo and Howon Kim, Member, KIICE *Network and Data Link Layer Security for DASH7*, journal of information and communication convergence engineering, Sep. 2012.

[8] Glenn Ergeerts, Maciej Nikodem, Dragan Subotic, Tomasz Surmacz, Bartosz Wojciechowski, Paul De Meulenaere, Maarten Weyn *DASH7 Alliance Protocol in Monitoring Applications*,10th International Conference on P2P, Parallel, Grid, Cloud and Internet Computing, 2015.

[9] *LPWA Technology Security Comparison*, A White Paper from Franklin Heath Ltd, 02 May 2017. Available: http://www.huawei.com/en/events/mwc/2016/summit/global-nb-iot/nb-iot-enabling-new-business-opportunities

[10] Sunakshi Jaitly; Harshit Malhotra; Bharat Bhushan *Security vulnerabilities and countermeasures against jamming attacks in Wireless Sensor N etworks: A survey*, International Conference on Computer, Communications and Electronics, 2017.

[11] J. P. Tomás, *Operators in Korea, Netherlands deploy LoRa networks for IoT* 2016, Available:http://www.rcrwireless.com/20160704/carriers/operatorskoreanetherlandsdeploy-lora networks-iot-tag23.

[12] Yonghua Song, Jin Lin, Ming Tang, Shufeng Dong *An Internet of Energy Things Based on Wireless LPWAN*, Engineering, August 2017.

[13] *Narrowband-IoT: pushing the boundaries of IoT*. Available: vodafone.com/iot/nb-iot.

[14] Alex Goldschmidt *Replay Attack Vulnerabilities and Mitigation Strategies*, www.cs.tufts.edu/comp/116/archive/fall2015/agoldschmidt.pdf on 2015.

[15] Y.-C. Hu, A. Perrig, and D. Johnson, *Packet leashes: a defense against wormhole attacks in wireless networks*, in 22nd Annual Joint Conference of the IEEE Computer and Communications on 2003.

Survey on Encryption Approaches for Secure Face Biometrics

Kavya R
St.Josephs College Of Engineering And Technology,Palai
E-mail: kavyar132@gmail.com

Angitha George
St.Josephs College Of Engineering And Technology,Palai
E-mail: angitha.george@sjcetpalai.ac.in

Abstract. Biometric system is twisting up logically basic since they give strong moreover, compelling techniques for character affirmations. The face recognition is the most trademark methods for biometric conspicuous verification and give versatile biometric affirmation and is a development prepared for serving an extensive variety of security application. This paper deals with the examination of different encryption approaches like AES, DES, and RSA that can be used for go up against face recognition security suggests and moreover analyze DNA cryptography. With the examination of these systems, AES gives sublime security while considering face affirmation.
Keywords: Biometric, Face Recognition, AES, DES, RSA, DNA cryptography.

1. Introduction

The multiplication of security issues identified with the validness of the people, there is a solid requirement for confirmation and validation frameworks. Today human confirmation can be classified as What you know, e.g secret key, individual check number; What you have, for example, a token, smart card and so forth and What you are, biometrics.

The initial two classifications are less secure techniques when contrasted with the biometrics.Brute compel assaults, Shoulder surfing, and phishing assaults are influenced by the passwords and individual check numbers. The initial two verification classes are shared among the collaborators and gatherings, which influence the security levels of the framework.

Biometric verification outperforms What you know and what you have. Biometric depends on the inherent parts of the human. Exact distinguishing proof, responsibility, simple and safe for utilize and remarkable individual ID are the principle advantages of biometric validation and assuming control conventional passwords and Id card confirmation. In light of the special highlights and about difficult to produce, individual to be distinguished to be physically present for the purpose of the confirmation. Physically, the individual to be distinguished at the present purpose of the check procedure. This trademark eases the confinement of utilizing a secret word or token that can't separate between an honest to goodness client and a gatecrasher. Also, biometric highlights are an effective weapon against the dismissal of a proposition or thought.

IOP Conf. Series: Materials Science and Engineering **396** (2018) 012028 doi:10.1088/1757-899X/396/1/012028

Biometric security giving access to an individual in light of check of one's physical attributes.Biometric data is can't transferable and share-able.The possibility of assault is less when contrasted with different frameworks.

These days the installment approaches by means of the web and the system have been developing at an enraged pace. Along these lines, the assortment of the electronic installment frameworks comprises of various kinds to accomplish a solid level of the security. Be that as it may, in parallel, the assaults techniques and methodologies are as cutting edge as the security arrangements. In this thorough overview, we attempted to specify distinctive encryption approaches for securing face biometrics. Encryption calculations utilized for securing information by encryption and unscrambling process.

2. Background

This area gives an outline of the diverse encryption calculations for the face acknowledgment biometric framework. An encryption calculation is a segment of information transport security and the way of changing over plain content to figure content, and after that back to the plain content for securing electronic information. We are adopting three encryption strategies those are DES, RSA, and AES.

- Data Encryption Standard (DES): DES (Data Encryption Standard) algorithm used as a standard practice for securing unclassified and sensitive data.
- Advanced Encryption Standard (AES): The benefits of AES is the high speed.
- Rivest-Shamir-Adleman (RSA): The Public key algorithm for encrypting the data.
- DNA Cryptography: New emerging domain form on DNA computing.

2.1. Face Recognition

Face Recognition is the most normal methods for biometric distinguishing proof and give adaptable biometric confirmation facial acknowledgment is an intense innovation fit for serving a wide range of security application. Facial acknowledgment can substitute the watchword for the gadget and the client account get.

Facial acknowledgment frameworks are computerized frameworks that confirm the people by contrasting the facial qualities, for example, remove between eyes, mouth, nose, and ear. These qualities are broke down and contrasted all together with distinguishing or confirm the people.

Biometric face system follows the following stage of operations:

- Capture the biometric sample of the person.
- Extract set of relevant features from the captured sample.
- Compare the extracted feature set against the template set in the database.

Various calculations, these are identified with the face acknowledgment. Face acknowledgment calculation can be partitioned as two: geometric component based and appearance-based. Appearance-based techniques incorporate Eigenfaces [1],principal segment examination (PCA) and direct discriminant investigation (LDA), Independent Component Analysis (ICA), Kernel Principal Component Analysis (KPCA) [9], Kernel Fisher Discriminant Analysis (KFDA) [3], General Discriminant Analysis (GDA), and Support Vector Machine (SVM) [10]. The disadvantage of appearance-based techniques is that the face acknowledgment under a specific lighting can be performed dependably when the face has been already observed under comparable conditions. In appearance-based strategies, the caught highlights are the worldwide highlights, so these are hard to deal with.

The principal point of the geometric component based technique is to locate an arrangement of agent highlights from the image.Points, surfaces, and bends are joined together to shape the geometric features.These highlights of the pictures are seen by the element recognition strategies.

Geometric component based strategies are powerful in perspectives, however, include extraction process are touchy. The geometry highlight based techniques incorporate an itemized examination of the components or structure of nearby facial highlights. The geometry highlight based strategies incorporate Active Shape Mode and Local Feature Analysis (LFA). Acknowledgment of appearances from still pictures or 2D pictures is a troublesome issue, with the clarification, particular position and demeanor changes in the pictures make factual contrasts and face character winds up shadowed by these components.

2.2. DES (Data Encryption Standard)

In 1975 at IBM lab, Horst Feistel created DES. In 1978, the National Bureau of Standards, now called NIST has endorsed the DES. The DES was systematized by the American National Standard Institute(ANSI) as DEA (Data Encryption Algorithm).

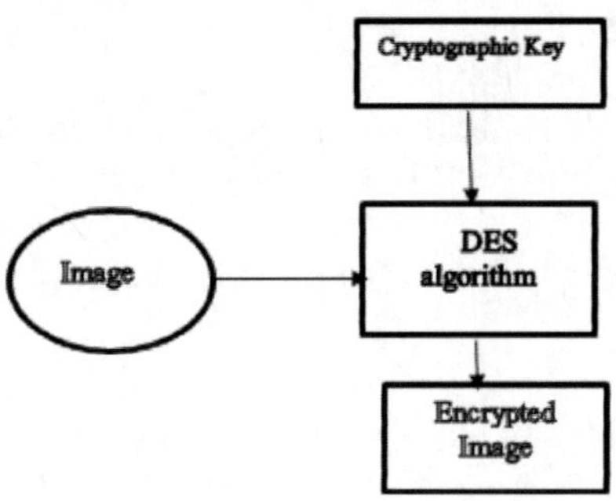

Figure 1. DES ENCRYPTION utilizing a picture.

The most generally utilized symmetric square figure depends on the DES. For DES, information is scrambled in 64-bit pieces utilizing 56-bit keys and is a predecessor. Each square is scrambled in disengagement, which is a security helplessness. The key length of DES is expanded by the Triple DES. In triple DES, three DES task is performed on each square. By ordering encryption with key 0, a decoding with key 1 and an encryption with key 2. These keys might be interconnected.

DES displays exemplary Feistel structure, ensuing occasions, and open investigation appear in actuality configuration was suitable. DES has turned out to be broadly utilized, particularly in financial requests yet now it is supplanted by AES. For the DES encryption process, two contributions to the encryption work: the plain which is of 64-bits long and key size is 56-bits long. The little changes in the plaintext or key roll out critical improvements in the ciphertext. The change in plaintext or one piece of key deliver changes in numerous bits of the ciphertext(Avalanche impact). DES uncover solid torrential slide impacts.

2.3. Advanced Encryption Standard (AES)

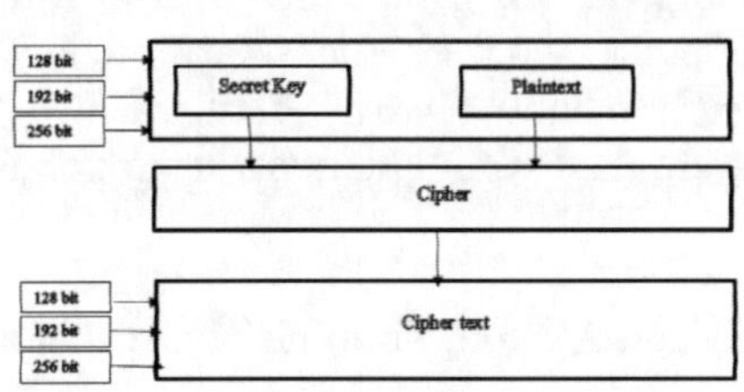

Figure 2. AES Design.

AES is piece figure proposed to trade DES for business applications. It utilizes 128 piece square size and a key size of 128,129 or 256 bits. AES is to a great degree effective in a 128-bit

IOP Conf. Series: Materials Science and Engineering **396** (2018) 012028 doi:10.1088/1757-899X/396/1/012028

frame; AES utilize keys of 192 and 256 piece for encryption reason. AES considered impermeable to all assaults, except for beast compel, which endeavor to decode utilizing all conceivable blend in 128, 192, or 256-bit figure. AES is a symmetric figure actualized in both equipment and programming to secure computerized information.

AES does not utilize Feistel structure, rather each full round comprise of four separate functions: byte substitution, permutation, arithmetic activities and XOR with a key.In AES, all tasks are performed on 8-bit bytes.Ciphertext takes a plaintext piece size of 128 bits or 16 bytes. The key length can be 1624 or 32bytes(128,192, 256 bits) and the calculation alludes as AES-128, AES-192, and AES-256 relying upon the key length.

2.4. RSA

In 1978, Ron Rivest, Adi Shamir, and Leonard Adleman composed the RSA. Key generation, encryption, and decoding are the real strides for the RSA.RSA conspire is a square figure, the plaintext and ciphertext are the whole numbers in the vicinity of 0 and n-1 for some n and run of the mill estimate for n are 1024-bits or 309 decimal digits.

In advanced mark and encryption plans, RSA most generally utilized. Its security depends on the diculty of figuring substantial numbers. An advanced mark plot in light of RSA keeps running as takes after: assume p and q are two huge primes.

Compute $n = p \times q$

and choose e and d such that

$e \times d \bmod (p - 1)(q - 1) \equiv 1$.

Each user has a key pair, which includes a private key and a public key. Suppose d denotes the users private key, and (e, n) denotes the users public key. When Alice wants to sign a message M and sends it and its digital signature to Bob, Bob can verify whether the signed message was really signed by Alice as follows:

Sign:

- Alice computes $s = M^d \bmod n$, where d is Alices private key.
- The value s is Alices signature on M. Then Alice sends Bob (M, s).

Verify:

When Bob receives these messages from Alice, he can verify whether s is Alices signature on M by checking $s^e = M \bmod n$, where e is Alices public key.

2.5. DNA Cryptography

In 2005, Kazuo Tanaka proposed the DNA cryptographic approach in light of open key.DNA Cryptography can be characterized as a concealing information regarding DNA Sequence. Cryptographic strategy in which each letter of the letters in order is changed over into an alternate blend of the four bases that make up the human deoxyribonucleic corrosive (DNA). It comprises of regular cryptographic methodologies key age, encryption and unscrambling process.

The DNA strand made up of agglomeration of nucleotides, which comprise of four nitrogen bases: Adenine, Cytosine, Thymine, Guanine. The letter A, C, T, and G are nucleotide bases. Adenine (A) make combine with Thymine (T) and Cytosine(C) with Guanine(G).These bases make the match with each other to frame base sets, to make twofold helix structure of DNA.DNA cryptography gives Unbreakable security.

DNA cryptography for the biometric confront framework come behind the resulting stages :

- Image is used under the form of ASCII code..
- Numbers are grouped into blocks and encrypted.
- Encoded message is changed into binary format.

- Digits are grouped into two and substituted as A- 00, T-01, G-10, C-11.
- Fit primers on either the side of the code.

3. Comparative Study on algorithms

Table 1. Comparative study of AES,DES and RSA.

Parameters	DES	AES	RSA
Developed	1977	2000	1978
Keysize	56 bit	128,192,256 bits	1023 bits
Blocksize	64 bits	128bits	Mim 512 bits
Cipher/ Decipher key	Same	Same	Different
Algorithm	Symmetric	Symmetric	Asymmetric
Encryption	Moderate	Faster	Slower
Decryption	Moderate	Faster	Slower
Security	Slightly Secure	Excellent	Insecure

4. Conclusion

The understood overview for the most part centered with considering around various encryption approaches utilized for confront acknowledgment. Encryption calculation assumes an imperative part of security. In this review, we break down various encryption approaches like AES, DES and RSA calculations and discovered AES indicates critical security while thinking about DES and RSA. DNA cryptography is a rising innovation give unbreakable security. The near table demonstrates the confirmation of the investigation.

References

[1] Michael Goh and David Ngo, "PalmHashing: a novel approach for dual-factor authentication", Pattern Analysis and Applications, Vol. 7, Issue 3, pp 255268, 2004.
[2] Abate. F, Nappi M, Riccio D, and Sabatino G, Face recognition: A survey, Pattern Recognition Letters, Vol. 28, pp. 1885-1906, 2010.
[3] M. Kumar and E. G. Dharma, A comparative analysis of symmetric key encryption algorithm, IJARCET, vol. 3, no. 2, 2014.
[4] Achermann B., Jiang X., and Bunke H, Face recognition using range images, In International Conference on Virtual Systems and MultiMedia, pp. 129136, 1997.
[5] Nyo Htweand Nu War, "Human Identification Based Biometric Gait", International Journal of Advanced Research in Computer Engineering Technology, Vol. 2, no 5, 2013.
[6] E. Biham and A. Shamir, "A differential cryptoanalysis of data encryption standard", Springer-Verlag, 1993.
[7] S. Basin, International data encryption algorithm (idea) a typical illustration, Journal of global research in computer science (JGRCS), Vol. 2, no 7, 2011.
[8] A. Kakkar and M. L Singh, Comparison of Various Encryption Algorithms and Techniques for Secured Data Communication in Multimode Network, Published in International Journal of Engg and Technology(IJET), Vol. 2, no. 1, 2012.
[9] Andrew Teoh, BengJina David Ngo, and ChekLinga Alwyn Gohb, "Biohashing: two-factor authentication featuring fingerprint data and tokenized random number", Pattern Recognition, Vol. 37, no 11, 2004.
[10] L. Singh and R. K. Bharti, "Comparative performance analysis of cryptographic algorithms", International Journal of advanced research in computer science and software engineering (IJARCSSE), Vol. 3, no. 11, 2013.
[11] W. Zhao, R. Chellappa, P. J. Phillips and A. Rosenfeld, Face Recognition: A Literature Survey, ACM Computing Surveys, Vol. 35, no. 4, pp. 399458, 2003.
[12] Joseph P. Campbell, Face Recognition: A Tutorial, Proceedings of the IEEE, Vol. 85, no. 9, September 1997.

IOP Conf. Series: Materials Science and Engineering **396** (2018) 012028 doi:10.1088/1757-899X/396/1/012028

[13] Anna Labb and Annie Prez, "AES Implementation on FPGA: Time - Flexibility Tradeoff", 12th International Conference on Field-Programmable Logic and Applications, pp 836-844, 2002.
[14] D. Coppersmith, "The data encryption standard (DES) and its strength against attacks", IBM Journal Research Develop., vol. 38, no. 3, pp. 243 -250, 1994.
[15] R. Davis, The data encryption standard in perspective, Communications Society Magazine, IEEE), pp. 59, 2003.
[16] A.Sindhuja, B.Sri Nandhini, S.Bharathi, and I.Kalaimathy, "A Hybrid Authentication System Using Biometrics and Text Compression", International Journal of Scientific and Research Publications, Vol. 3, no. 3, 2013.
[17] Dr. Prerna Mahajan and Abhishek Sachdeva, "A Study of Encryption Algorithms AES, DES and RSA for Security", Global Journal of Computer Science and Technology Network, Web Security, Vol. 1, no. 15, 2013.

IOP Conf. Series: Materials Science and Engineering **396** (2018) 012029 doi:10.1088/1757-899X/396/1/012029

A Review on Cybersecurity Threats and Statistical Models

Feba Babu and Kishore Sebastian
1) P,G Scholar, St Joseph College of Engineering and Technology, India
E-mail: febabinu04@gmail.com

Kishore Sebastian
2)Asst.Proessor, St.Joseph College of Engineering and Technlogy,India
E-mail: kishorekinattukara@gmail.com

Abstract. Data generated in past two years is more than previous history tutorials with the help of evolution of technology.Big data is a term using for abundant structured and unstructured data,it becomes difficult to process,store,analyze and visualize using on hand database system tools.This big data systems are facing attack targets.Some statistical models are helps to find out problems of the cybersecurity.In this paper studied on Big data,cybersecurity,categories of cybersecurity threats,main challenges of cybersecurity,and study on statistical models for vulnerability prediction.

1. Introduction

Big data systems are very essential part of this modern organisations,because now a days we are living in digital world so every 60 seconds millions of data is getting generate.This abundant data just beyond the technology's.As little as 5 years ago almost people only thinking how to store tens to hundred of gigabytes data in our personal computer? but today people thinking how to store tens to hundred of terabytes data?

IBM survey provided that every day 2.5 quintillion bytes of knowledge square measure created most that 90%of the info within the world nowadays has been created within the last 2 years[1]. Intels Infographic reveals each sixty seconds, 639,800GB of worldwide information is transferred,One minute of net time, 204 million e-mails sent. on-line denizens read twenty million photos on Flickr. Twitter processes 100,000 new tweets and 320 new Twitter accounts are created[2].

At the same time of data collection need to improve cyber security.This paper mainly focus on cyber security threats and statistical models.The survey Paper starts with Big data,Cyber security,Threats,Challenges and Statistical models for vulnerability prediction.

2. CONCEPT OF BIGDATA

Bigdata is collection of structured,semi structured and unstructured data.It's so large and complex to manage with traditional systems.

Bigdata have three diamensions:Data Volume,Data Velocity and Data Variety. Volume is that the quantity of information in terabytes and petabytes, variety is that kinds of information that's audio, video,image,doccuments etc. and velocity is that the speed of information process therefore time is extremely vital in here.

3. CYBERSECURITY

Investopedia defines[3],Cybersecurity refers to the measures taken to keep electronic information private and safe from damage or theft. It is conjointly accustomed make certain these devices and information aren't abused. Cybersecurity applies to both software and hardware, as well as information on the Internet, and can be used to protect everything from personal information to complex government systems.The major cyber-security issues for the most part focused on software vulnerabilities, organisations must maintain effective vulnerability management programs including remediation, identification,reporting and assessment.

Cyber security is that the main a part of technologies practices and processes designed to to shield networks ,computers ,programs from attack ,damage or unauthorized access .In computing context ,The word security implies cyber security.All the sectors like financial institutions,business,Governments,corporations etc collect ,process and store confidential information on computers and transmit that data across networks or other computers .So growing volume of cyber attacks.

The main elements of cybersecurity:Application security, Information security, Network security.

4. CYBERSECURITY CHALLENGES

To protect information and infrastructure in cyberspace, decrease vulnerabilities,and minimize injury from cyber incidents through a mixture of institutional structures,build capabilities to stop and reply to cyber threats.Some challenges are:

- a commentary from CSI [4],Create mechanisms for cover of non-public knowledge in third party domain specifically,cloud suppliers, social networks, outsources throughout varied phases of its life cycle i.e., transmission, process or storage.
- Create mechanisms which ensure trust in dynamic environment where identities are protected, anchors of trust exist and those interacting are trustworthy. This is in a transparent domain.
- Create new flexible access control technologies which are ethical, less dependent on dynamic identities, using more reliable way of management in a distributed world.
- cyber security is that the quickly and perpetually evolving nature of security risks.
- Budget is very difficult for security professionals to acquire the budget needed for a proper cyber security program.
- Proactive approach checking available information and apply predictive and behavioral analytic tools to find out threat, detect the actual threat, gather intelligence regarding the attack, and execute an enterprise wide response before the threat becomes significant.

5. CYBERSECURITY THREATS

According to[5],many types of threats are on the market and may be categories:

5.1. Trojan horse

This computer virus threats will do something,record your passwords by work keystrokes and record each move and hijacking digital camera to look at

5.2. Malicious spyware
Recorded information is sporadically sent back to the originating cybercriminal over the net.

5.3. Computer worm
Computer worm it's a code program which will copy itself and sent from one pc to a different, with none human interaction. It will replicate with nice volume and with nice speed.

5.4. Botnet
Botnet it's a bunch of computers or systems connected to the net and it compromised by a hacker employing a computer program for obtaining informations from computers.

5.5. Spam
Spam Spam is nothing however unwanted messages from your email box.This Unwanted unsolicited mail contain links that once clicked on may visit a web site that installs malicious code onto pc.

5.6. Phishing
It may be capture all variety of sensitive data like mastercard details,usernames and passwords by impersonating as a trustworthy entity.

5.7. SQL Injection Attack
SQL is structured programming language it can communicate with databases,this injection attacks mainly targets on customers private information from the web site, like mastercard numbers, usernames and passwords.

5.8. Cross-Site Scripting (XSS)
XSS is incredibly the same as SQL Injection Attack,and it's additionally injecting malicious code into a web site.

5.9. Buffer Overflows
This threat mainly doing attacker to control its internal variables.Modern programming languages like C, Java and Perl these languages will be reduce the chances of coding errors and creating buffer overflow vulnerabilities.

Other cybersecurity vulnerabilities are Memory corruption, Gain privileges,Exploits etc.

6. STATISTICAL MODELS FOR VULNERABILITY PREDICTION
The main ojective of this section is to investigate vulnerability prediction with the help of statistical models.Here going to describe 3 models:Copula,GARCH,ARIMA.

6.1. COPULA
Copula is very effective approach for vulnerability prediction,it is a joint distribution perform of random vectors with standard uniform marginal distributions.

It have a vital role in trendy statistics and estimator risk analysis to decompose n dimensional distribution function F into two elements specifically the marginal distributionF_i and also the copula C, capturing the dependence relationship of the marginals.

The copula is formally defined as:

Let X = $(X_1, ..., X_n)$ be a random vector with joint distribution function F and marginal distribution functions $F_i, X_i \sim$ $F_i, 1 \leq$ i $\leq$ n .

A distribution function C with uniform marginals on [0, 1]] is named a copula of X if,

$$\mathrm{F} = \mathrm{C}(F_1, ..., F_n)$$

Then Copula for continuous marginals.Here consider C is the joint distribution function of $(F_1(X_1), ..., F_n(X_n))$ and $\mathrm{F}_i(X_i)\sim$ U(0; 1)

Copula C can be

$$\mathrm{C}(u_1, ..., u_n) = \mathrm{F}_X(F_1^-1(u_1), ..., F_n^-1(u_n)))$$

In Copula model,after determine the proper marginal forms then modeling their joint distribution,in this section considering bi-variate case on overflow vulnerability,then find out observed bi-variate dependence structure using a medium and positive relationship between these two disclosures,through estimated marginals.Then simulate bi-variate dependence structure these two disclosures.According to [7] simulate 200, 000 pairs of overflow disclousres.So this study surely said copula model effectively preserve their long-term and persistent dependece relationship.

6.2. ARIMA

ARIMA stands Autoregressive Integrated Moving Average,this approach provide to time series forecasting.ARIMA models aims to describe the autocorrelations in the data.

Procedure for ARIMA Model[6] :

- plot the data,identify all the patterns usal and unusal.
- If required, transform the information to stabilize the variance.
- If remodeled information are non-stationary: then calculate the variations of the information till the information are stationary.
- Calculate the ACF/PACF.
- Try the AICc to find out a better model..
- Check the residuals,if it fail try a modified model.
- calculate forecasts.

In ARIMA Model(Fig:1),after the plot standardised residuals to visually examine if residuals forms a accurate approximation to a white noise process or not.But ARIMA model[7] provide better fit for buffer overflow disclosure data.ARIMA models can be capture the mean behaviour of each time series based on the in-sample data.ARIMA model is also better for predict Buffer Overflow.

6.3. GARCH

The GARCH (generalized autoregressive conditional heteroskedasticity)This method is usually most popular by financial modeling professionals as a result of it provides a more real-world context for predict the costs and rates of financial instruments.It is the extension of the ARCH method to the GARCH method.GARCH technique is extremely helpful because of it can be use in two things :predict and simulate with GARCH model.

Procedure for GARCH Model:

- Choose best fitting AR model.
- calculate and plot the autocorrelations.
- calculate forecast.

In GARCH model,it provide suitable model for capturing conditional variances in all the empirical data except for the buffer overflow vulnerability,model provides a better fit.

IOP Conf. Series: Materials Science and Engineering **396** (2018) 012029 doi:10.1088/1757-899X/396/1/012029

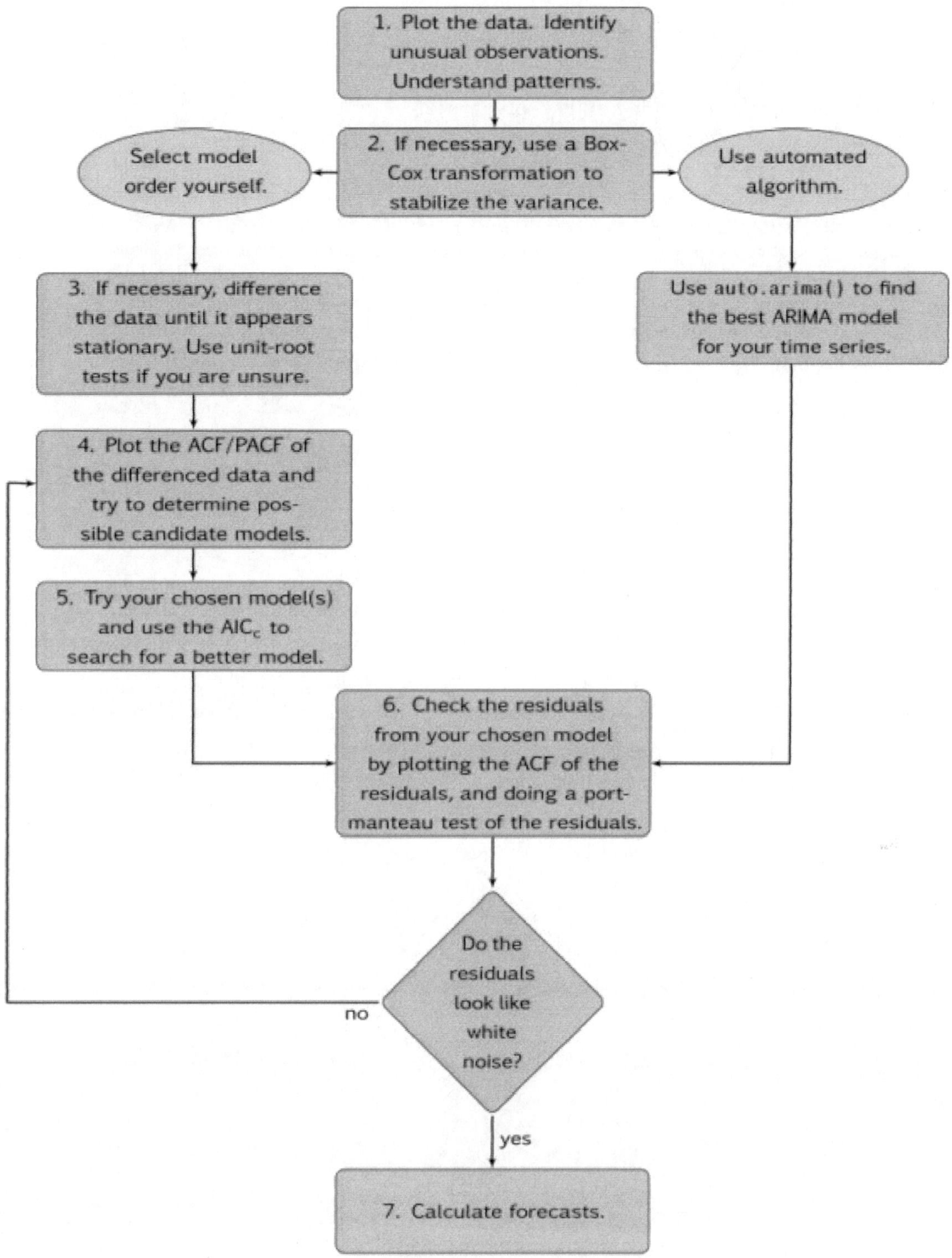

Fig:1-process summary-ARIMA Model

COPULA,ARIMA,GARCH these models will be use for predict on top of cybersecurity threats victimization these procedure.

IOP Conf. Series: Materials Science and Engineering **396** (2018) 012029 doi:10.1088/1757-899X/396/1/012029

7. CONCLUSION

The Bigdata diamensions are volume, velocity and variety can be used to solve cybersecurity issues efficiently.This paper studied on Bigdata,cybersecurity,challenges of cybersecurity,threats and statistical models for vulnerability predication.A testing genuine digital security issue in the space of abusing helplessness exposure incline with complex multivariate time arrangement information.In this article,three part analysis technique to detect buffer overflow vulnerability and thorough measurable system towards extending this study about the exposure progression and their interesting reliance structures.COPULA,ARIMA,GARCH viably contemplated on multivariate time arrangement information time-invariant reliance connections in the information. This article demonstrated that a copula based displaying approach can successfully recognize and save idle reliance structure in multivariate time arrangement information.The fitted copula display effectively caught the coveted inert structure.

In the future,a superior comprehension about how unique powerlessness furthermore, abuse exposures communicate with each other through rich recorded information in reality digital security situation.

8. REFERENCES

(i) Sachchidanand Singh and Nirmala Singh Big Data Analytics,2012
(ii) https://www.cnet.com/news/intel-reveals-what-happens-in-a-single-internet-minute/
(iii) https://www.investopedia.com/terms/c/cybersecurity.asp
(iv) N.J.Rao, Cybersecurity: Issues and Challenges,CSI Communications, May 2015.
(v) https://blogs.cisco.com/smallbusiness/the-10-most-common-security-threats-explained.
(vi) https://www.otexts.org/fpp/8/7
(vii) MingJian Tang, Mamoun Alazab,and Yuxiu Luo Big Data for Cybersecurity: Vulnerability Disclosure Trends and Dependencies,2016

IOP Conf. Series: Materials Science and Engineering **396** (2018) 012030 doi:10.1088/1757-899X/396/1/012030

A Review on BlockChain Security

Remya Stephen
PG Scholar, St.Joseph College of Engineering and Technology, INDIA
E-mail: remyastephen93@gmail.com

Aneena Alex
Asst.Professor, St.Joseph College of Engineering and Technology,INDIA
E-mail: aneenaalex@gmail.com

Abstract. Blockchain is a decentralized technology. It has extensive power to solve business problems. Cryptography secures the records in a blockchain transaction and each transaction is tied to previous transactions or records. Blockchain transactions are validated by algorithms on the nodes . A single entity cannot create a transaction. Finally, blockchains provide transparency, giving each participant the ability to monitor the transactions at any time. Smart contract make secure transaction which helps to avoid third party disruption .Ethereum is a decentralized platform that runs smart contracts. This enables developers to create markets move funds in accordance with instructions given long in the past. The main features of blockchain are Decentralization, Immutability Faster dealings, Transaction and validation happens in seconds etc..

1. Introduction

Blockchain technology has huge potential with a variety of applications and provides wide opportunities for various infrastructure. The technology encourage resource management and makes communication both secure and efcient. Trust is increased when conducting nancial transactions among parties using Blockchain, as it reduces the chances of swindle and automatically produces a record of activities. Creating an automated background check of any member of the system. Due to its decentralized properties, Blockchain creates reliability and reduces the risk faced when looking to enter a business agreement with an unfamiliar party.

Today all people were using advanced technology for communication through internet. Voice call, video call, messages, pictures, are travel directly from sender to receiver over the internet. For this transaction, must maintain a trusted third party between these sender and receiver. When it comes in the case of money transaction, people have to trust a third party for complete this, in traditional system. But in the case of blockchain it will give a perfect security in transaction. A block should record every transaction, it will act like a record book. Once completed a transaction a block goes into the blockchain as a permanent database. If a block is completed a new block is added with this or a new block is generated. Every block carry a hash of the previous block.

IOP Conf. Series: Materials Science and Engineering **396** (2018) 012030 doi:10.1088/1757-899X/396/1/012030

2. KEY ATTRIBUTES OF BLOCKCHAIN TECHNOLOGY

2.1. Decentralization

Decentralization is completely different from centralization. It provide more security and exibility than the centralized application. Quick decision making is required and therefore many organizations opted for decentralization[4]. In the centralized environment everything is done in the same location. Where decentralized environment works in different location. It has the ability to provide both efciency and innovation. Efciency deals with saving cost and time ,should provide a better result. Innovation provide new idea. That should be a new benets.

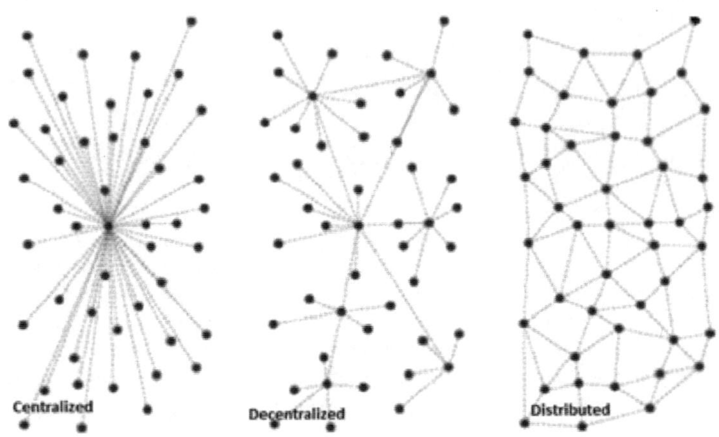

Figure 1. Network Comparison

2.2. Trust

In Blockchain technology each block contain information about the previous block. It will provide an authentication mechanism during the transaction. There is no third party communication. Instead of this that will use a public ledger .All transaction should be automatically recorded in this ledger

2.3. Transparent

In blockchain stored data are transparent and also immutable. That is why blockchain are trusted.

3. PUBLIC AND PRIVATE BLOCKCHAIN

Blockchain is mainly divided in to three types. Public,Private and consortium blockchain.This paper focusing on the comparison between public and private blockchain based on some recent property. Public and private blockchain should have many similarities as well as differences in it's functionality[9].

Property	Public Blockchain	Private Blockchain
Read Permission	Public	Could be public or restricted
Immutability	Approximately beyond to tamper	Could be tampered
Efficiency	Low	High
Centralized	Yes	No

Table 1. Comparison table

4. CHALLENGES IN BLOCKCHAIN

Blockchain has faced some challenges in industry especially during bitcoin transaction. Those problems inherent to functions of the Bitcoin blockchain are compiled below.Summaries some typical challenges below[10].

4.1. Scalability

The amount of transactions are increasing day by day. Most of the companies were suggesting blockchain for their transaction process. All transaction have to be stored and it will be validated. The capacity of the block will very small. Some transaction must be delayed due to miner prefer high transaction fee to those transaction. So the large block size will lead to reduce the propagation speed. There fore scalability problem is quite though.

There some methods to avoid the scalability problem in blockchain

- Storage Optimization of blockchain
- Redesigning blockchain

4.2. Privacy Leakage

Users believed that blockchain gives better privacy when handling sensitive data. In blockchain users could only generate address instead of their identity. In 2013Meiklejohn and in 2016 Kosba shows that blockchain cannot guarantee the transnational privacy. Recent study shows that bitcoin transactions are linked together to an account address to reveal the identity of user. The problems were leak out the users identity

Elliptic Curve Diffie- Hellman- Merkle(ECDHM) can be used to overcome this problem. It will deals with public and private key. This will exchange shared secrets between two people.It will helps to secured message transaction0n over the internet. A secured platform like smart contract and ethereum is also developed to keep secured transaction

4.3. MITM Attack

MITM [11] means Man in The Middle Attack. This is known as third party interaction. Here a user came in the middle, who may have a forged public key. By using this key, he can easily decrypt the sensitive data. In blockchain public key is distributed across the participating nodes. Each block should be connected with a link to previous and following blocks. Because of this, the public key is immutable it should not attack by any forged keys.

4.4. DDoS Attack

Distributed Denial of Service attack (DDoS) is an attack [11] which is target to attack one particular system Such as c computer, website, server or other network resources. Therefore, the incoming messages or connection to the target system may slow down, or even crash or shut down. Especially in blockchain DDoS attack create some significant business risk. Practically it is impossible to prevent this type of attack.

Flow analytic device is one of the common solution for DDoS attack. This device will be watch and react against the attackers. This will be telling what action to take next. This device would help to clear the traffic. Because it will redirect the bad traffic.

5. SECURITY FEATURES OF BLOCKCHAIN

- Use ledger. Ledger should record each and every transaction in a blockchain. This ledger is immutable. Existing data cannot be edited or deleted. In blockchain technology these ledger is decentralized application. So, no one can access the transaction or even any sensitive data from this ledger [12]. People can only read the information from a ledger.
- Another type of security feature is the chain of block. In blockchain each block should contain a hash value. These blocks are connected by its previous hash. Suppose an attacker came to correct the data, then its hash will be changed. It will affect the overall chain. So, it will increase the protection of sensitive data or information.
- Blockchain technology is a decentralized application. Mainly it will support peer to peer communication. So, in a network node is considered as computers. These thousands of nodes should have the copy of distributed ledger. This should be authenticating the transaction. If any of the node does not agree a transaction, then it cannot be proceeding. So, it will be cancelled[12]. This will protect from a fraud transaction.

6. APPLICATIONS OF BLOCKCHAIN TECHNOLOGY

6.1. Blockchain for health care industry

In today's life patient doesnt like to reveal there treatment details to outsider. In this case patient can use this technology to keep secure all information from others. This blockchain can be used as a website, or mobile app. Each and every user in a blockchain has two keys. Public key and private key. By using this only who can make a transaction. For example there two person Alice and Bob. Alice want to send some secure data to bob. So that Alice sign a digital signature by using here private key. That means private key is always act like as a password. Then shew will hash the data by using here public key and generate an address. Then bob validate the digital signature. If it is validated they will make a transaction. So by using these type of security methods, potions information can be protected from others.

6.2. Electronic medical records.

Patients can handle electronic medical records by using the blockchain technology. Most of the health cares institution should not allow patient to access their medical data. Patients are becoming disappointed about the privacy of their medical records. This all can be avoided by blockchain.

In handling electronic medical records, blockchain should deal with different frame work for managing the authentication, confidentiality, and accountability. It is mainly used when handling the sensitive data. Online electronic records in blockchain will operated as decentralized application. In centralized environment all application should be done at one location. But in decentralized environment application should be done in different location.

Electronic medical records should affect some challenges and limitations. This system will face some important challenges during the implementation of personally controlled system. That is this personally controlled records would replace provider or hospital records. Some segment of the personally controlled records would be downloaded in to the institutional record to tribute the existing data

This challenges can be avoided by blockchain. Because blockchain leads a key exchange based transaction between two parties. Their personal identity does not revel to any others. Because they are only providing their key identity. Each and every user in a blockchain should have one public key and one private key.

6.3. Blockchain to protect personal data

Today there is recent increase in reported incident of security problem in users personal data. Because of this there is a third party control over the data, who will collect all personal information. Blockchain can eliminate this third party and can transfer directly between two parties.

The amount of data recently increasing in our world. Facebook, is the largest online social-network, collected 300 petabytes of personal data. Personal data or sensitive data should not be secure in the hand of third parties. They are tried to attack and misuse. Blockchain helps users that not required to trust any third party. Blockchain recognizes the users as the owners of their personal data. Blockchain should have its own rules and regulation. It is known as smart contract. Before starting a transaction the gateway keeper should create some rules and will written as a contract. It will make a peer to peer communication. Bitcoin has demonstrated in financial space that is trusted and computing is possible in decentralized network. Blockchain is mainly proposed to handle the bitcoin, it is a digital currency.

6.4. Bitcoin

In 2008 bitcoin is invented as a new way to send value over the internet. Vitalik Buterin,is a programmer, introduced bitcoin in 2011. [2]

Bitcoin is a digital currency, created and held electronically. It is operated as a decentralized application. That directly control the transfer of digital currency. Value of bitcoin increasing in recent year. Bitcoin sets out to solve the distributed tracking and validation of transactions is one ot its main problem. It will keep the full history of transaction. Blockchain is mainly developed to transact this digital currency. If the user wants the recent history, then who can filtering it. Before making a transaction the rules and regulations of this will be written as a contract form known as smart contract. Transaction is only possible between two persons, before making it sender side should enter a digital sign. The transaction is validated by this digital sign. If this sign is validated transaction is proceed, that means botcoin should transacted.

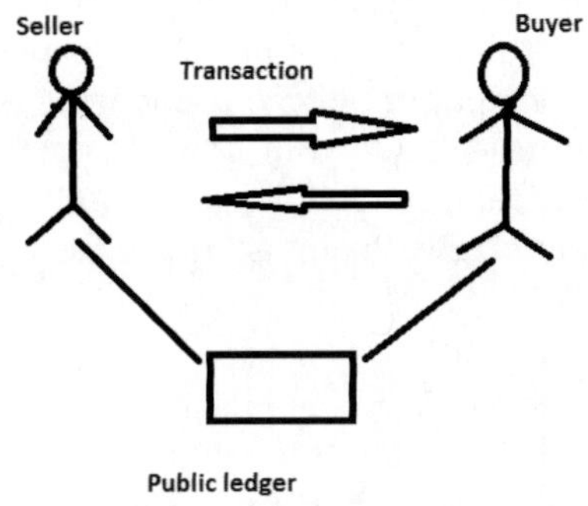

Figure 2. Ledger

IOP Conf. Series: Materials Science and Engineering **396** (2018) 012030 doi:10.1088/1757-899X/396/1/012030

6.4.1. SMART CONTRACT: Smart contract is also known as a crypto contract. Smart contract were first proposed by Nick Szabo in 1994. It is a computer program, it directly control the transfer of digital currency. These contracts are stored on blockchain technology. Smart contract is a decentralized system. It is existed in two parties[6].

There is no need to pay a middleman. It will saves time and conflict. Ledger is used in smart contract. It is a decentralized application.

6.4.2. LEDGER: Ledger is a decentralized application assigned to each user in a blockchain. After completing a transaction it will be recorded in the ledger automatically. For example there is two person A and B. The person A must give 100 rupees to person B. There is some other person is this blockchain. They also have a separate ledger. This data will be automatically updated in everyones ledger. Person A said that who should only give 10 rupees to B. Then there will be a voting mechanism. This voting mechanism can prove that the statement by person A is not validated. Therefore it will be rejected.

This Public ledger is best suited for all crypto currency transaction. There is no central administrator or centralized data storage.

6.4.3. Ethereum: Ethereum is a decentralized platform that runs the blockchain. That enables developers to build and deploy decentralized applications. Ethereum is similar to Bitcoin. Because it is a distributed public blockchain network. But there are some technical difference between these two. For each ethereum application the network needs to keep the current state information. Including all users balance ,smart contract code and where its all stored.[3]

Ether is a crypto currency, its blockchain platform is developed by Ethereum

7. OTHER APPLICATIONS

Blockchain have lot of applications in industry.Some of them are given below(fig:3). Today banking professionals were tried to adapt blockchain to make secure transaction. Finance is one of the most evident application of blockchain.

8. ANALYSIS

This paper tells about security of blockchain technology. From this review paper we can analyses that blockchain technology has facing some security issues. These security issues should affect the transactions also. There is a chance for different types of attacks in this technology, and also give some solutions to these issues. There are mainly three types of blockchain networks. Public, private and consortium. This review paper only concentrating on both public and private blockchain. Gives a short study between these by a comparison table.

Blockchain technology is an increased trending technology. Many applications are developed based on this technology. This review paper gives some solution for recent issues of blockchain.

9. CONCLUSION

Blockchain is an amazing topic in recent year, it will support different applications. Blockchain will give Better security during transaction of any value. This technology is mainly proposed to handling bitcoin transaction.Smart contract, Ethereum and distributed ledger are some applications of blockchain, This will also give more security.

Best suited and mostly used application of blockchain bitcoin. Blockchain gives faster and cheaper transaction than any other application. It will provide a better security especially to sensitive data. Blockchain applications often see additional benefits in its transparency and immutability.

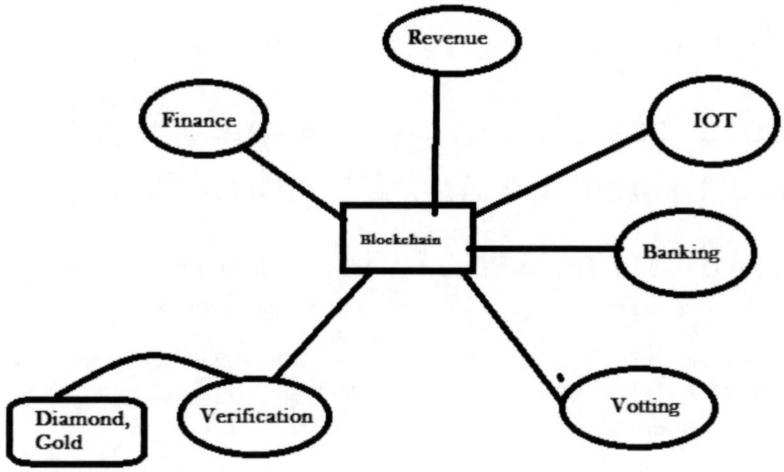

Figure 3. Applications

10. REFERENCES

[1] http://www.blockchain4innovation.it/wp-content/uploads/sites/4/2017/05/Blockchain-
[2] https://www.coindesk.com/information/who-created-ethereum
[3] https://www.coindesk.com/information/how-ethereum-works
[4] A Survey of blockchain security issue and challenges(Iuon-Chang Lin1,2 and Tzu-Chun Liao2)[jan-12- 2017]
[5] Public standares and patients controll:how to keep electronic medical records accessible but private(Kenneth D Mandl,Peter Szolovits,Issac S Kohane)[3 february 2001]]
[6] https://blockgeeks.com/guides/smart-contracts/
[7] A. Azaria, A. Ekblaw, T. Vieira, and A. Lippman, Medrec: Using blockchain for medical data access and permission management, in 2016 2nd International Conference on Open and Big Data (OBD), Aug 2016, pp. 2530.
[8] G. Zyskind, O. Nathan, and A. . Pentland, Decentralizing privacy:Using blockchain to protect personal data, in Security and Privacy Workshops (SPW), 2015 IEEE, May 2015
[9] https://www.researchgate.net/publication/319058582 Blockchain Challenges and Opportunities A Survey
[10] http://www.meti.go.jp/english/press/2016/pdf/0531 01f.pdf
[11] https://www.dotmagazine.online/issues/innovation-in-digital-commerce/what-can-blockchain-do/security-and-privacy-in-blockchain-environments
[12] https://www.business2community.com/tech-gadgets/issues-blockchain-security-02003488

A Study of Next Generation Sequencing Data, Workflow, Application and Platform Comparison

AswathyThankachan[1] and Mr Bino Thomas[2]

1 P G Scholar, Department of Computer Science and Engineering, St.Joseph's College of Engineering and Technology, Palai, Kerala, India.
2 Assistant Professor,Department of Computer Science and Engineering, St.Joseph's College of Engineering and Technology, Palai, Kerala, India.

E-mail: aswathy911@gmail.com, binothomas@sjcetpalai.ac.in

Abstract.
DNA sequencing determines the precise order of nucleotides within a DNA molecule.Next generation Sequencing (NGS) is an efficient parallel high throughput DNA Sequencing Technology which revolutionizing the genomic research. Earlier methods that are quite expensive give rise to different sequence comparison techniques. NGS used for faster detection of variants in human genome and give rise to accelerated response to disease detection like cancer, hepatitis etc. In this paper work flow of data analysis part of NGS are clearly discussed. The different NGS platforms, its applications and different sequencing comparison techniques are also mentioned here.

Keywords:Sanger sequencing, Maxam-Gilbert Sequencing, Needleman-Wunsch algorithm

1. Introduction

DNA Sequencing is a fundamental terminology in Bioinformatics. Genes are the fundamental element in a cell that makes up the organism .The process of determining the precise order of nucleotides Adenine(A) ,Guanine(G), Cytosine(C) ,Thymine(T) within a DNA molecule is referred to as DNA sequencing. The order of the bases along DNA contains the complete set of instructions that make up the genetic inheritance. The key principle of DNA sequence generation method is to evaluate the sequencing with very high accuracy and reliability.

Next generation sequencing (NGS) Technology which is a high through put parallel sequencing technology which accelerated the genomic research.NGS deals with enormous amount of data as the different combination of nucleotides which is then converted into a string problem in computer science.[12] Applications of NGS includes personalized medicine, Genetic Disorder, clinical diagnosis of different viral disease like cancer, hepatitis etc.It provide greater opportunities for fast and low cost sequencing in the modern times. Viral transmission like hepatitis, cancer, HIV can be efficiently detected by using NGS Technologies. International research project like The Human Genome Project (HGP) is a remarkable achievement of NGS. It is a program that aims to provide complete mapping of all set of genomes and will act as a world resource of detailed information about the function, structure and organization of the complete set of human genomes. After HGP the time, cost of comparison started decreasing

IOP Conf. Series: Materials Science and Engineering **396** (2018) 012031 doi:10.1088/1757-899X/396/1/012031

while amount of scientific knowledge has continued growing exponentially. So NGS has its own significance to understand the code of life.

2. Old Sequencing Methods

Human genome consist of 3 billion base which encapsulated 23 pairs of chromosomes inside the nucleus of a cell.DNA is having a double helix structure get attached with each other based on the complimentary pairs.(A-T) and (C-G).Sequencing of the Single DNA strand can be done using different methods.

2.1. Sanger sequencing

In 1977 Federic Sanger [1] did First DNA Sequencing called Sanger sequencing or chain termination method.It uses enzymatic synthesis to fragments DNA strands. DNA template, primer, polymerise and florescent nucleotides are the commonly used enzymatic synthesis. Strands are subjected to electrophoresis and visualized by UV light. The output obtained is termed as reads ie the Sequence of DNA strand is obtained as the combination of A C T G. The method is expensive time consuming more over it provides high quality reads. The use of dideoxynucleotide triphosphates (ddNTPs) as DNA chain terminators is the key point. Non-specific binding of the primer to the DNA results in deformalities in reads-out of the sequence.

2.2. Maxam-Gilbert

Maxam-Gilbert Chemical Sequence Method is more effective yet still limiting method used chemical sequencing. Strands are subjected to gel electrophoresis and precious order is determined. The method is best suited for heterogeneous DNA sequences .But it employs harmful radioactive chemicals and was difficult to scale up for high-throughput, robust sequencing. Sangers Sequencing is most preferred Compared with Maxam-Gilberth method because of its higher efficiency and through put.

2.3. Dye-Terminator Sequencing

Other methods include Dye-terminator sequencing utilizes labelling of the chain terminator ddNTPs, which permits sequencing in a single reaction. The variations in dye effects results in the unequal peak heights and shapes in the electronic DNA sequence trace chromatogram due to the incorporation of the dye-labelled chain terminators into the DNA fragment. The common challenges of DNA sequencing include poor quality in the first 15-40 bases of the sequence and can sequence upto 700- 900 bases.

3. Next Generation Sequencing Data

Next generation Sequencing (NGS) is an efficient parallel high throughput DNA Sequencing Technology. High Throuhput indicates that large scale replication of genomes is possible for further evaluation. NGS replaces enzymes synthesis by machines and makes the processing much faster. Fig 1 shows NGS pipeline

3.1. NGS Library Preparation

Sample DNA fragments are taken to library preparation. It is the process of preparation of targeted DNA into a form compatible with sequencing data. Library preparation consists of steps like DNA fragmentation, adaptor sequencing, size selection and final library quantification and QC. After this fragments are taken to DNA sequencing machines.

IOP Conf. Series: Materials Science and Engineering **396** (2018) 012031 doi:10.1088/1757-899X/396/1/012031

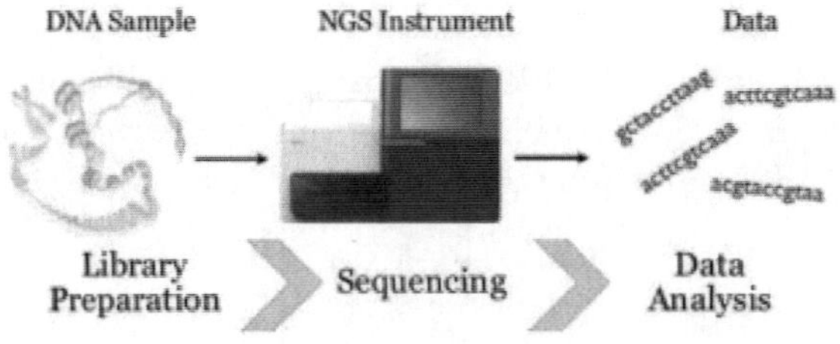

Figure 1. NGS pipeline

3.2. NGS Platforms

NGS systems are typically represented by SOLiD/Ion Torrent PGM from Life Sciences, Genome Analyzer/HiSeq 2000/MiSeq from Illumina, and GS FLX Titanium/GS Junior from Roche. Depending upon the application, quality and number of reads and run time each sequencer is selected. Table 1.1 shows some platform compared based on read length, its prons and cons.

Table 1.1 :Platform Comparision

Platforms	Read Length	Advantage	Disadvantage
454 GS Roche	700-1000bp	Long read	Expensive
HiSeqIllumina	2000- 2*150 bp	Low cost	Short read, long structural variants
Life Technology Solid 4	35-50 bp	High accuracy	Expensive, long run time

3.2.1. Advanced NGS platforms Pacific Biosciences (Pac Bio) Single Molecule Real Time (SMRT) sequencing, the Illumina Tru-seq Synthetic Long-Read technology, and the Oxford Nanopore Technologies sequencing platform are the advanced version of NGS. This provide direct sequencing of single DNA molecule having read lengths and run time which simplifies resequencing of genomes, to achieve a higher level of accuracy. Nanopore released a device called MinION which is a palm sized sequencing device that helps the real time analysis of single molecules

3.3. Application

There are different applications for NGS that includes personalized medicine, Genetic Disorder, clinical diagnosis of different viral disease like cancer, hepatitis etc. Applications of NGS such as whole genome sequencing, target sequencing, gene expression profiling, chromatin immune precipitation sequencing, and small RNA sequencing that are used to accelerate biological and biomedical research [2]. The Human Genome Project (HGP) is an international research program that aims is to provide complete mapping of all set of genomes and will act as a world resource of detailed information about the function, structure and organization of the complete set of human genomes.

3.4. Work flow and data Analysis

The Graphical Pipeline for Computational Genomics (GPCG) [4] represents the workflow for basic sequence alignment and data quality control, analysis of single nucleotide polymorphism. Analytical steps starts with the quality assessment of raw data, read alignment, variant identification calling, annotation and visualization of result in a phylogenetic tree.

The first analysis step after completing the sequencing run is to evaluate the quality of raw reads and to remove, trim or correct reads that do not meet the defined standards. The NGSQC Toolkit and PRINSEQ are able to handle FASTQ files that are capable of filtering and trimming reads. Those who meet certain quality standard are aligned to an existing reference genome. The University Of Santa Cruz (UCSC), and the Genome Reference Consortium (GRC) are two main sources for the human reference genome assembly which provide alignment programs to refer the human genome for MAQ, BWA, SOAP, Bowtie/Bowtie2 are examples of alignment programs.

If the sequence doesnt have a reference genome then it is called Denovo assembly. It creates a transcriptome without the aid of a reference genome and sequenced separately then added to database for further observations. It identifies structural variants and complex rearrangements, such as deletions, inversions, or translocations and generally produces fragmented assemblies Ex: DNASTAR .In the variant identification, SAM File is obtained and compressed into BAM file, which is a universal file format for mapped sequenced reads that contains the sequence score of each reads

Predicting the functional impact of Data is done by using Computer-aided annotation that enables research groups to filter and prioritize potential disease-causing mutations for further analysis. The annotation tools and strategies that have been developed to retrieve information and test hypotheses about the functional role of variants present in the human genome. NGS visualization tools should support users by displaying aligned reads, mapping quality and identified mutations combined with annotations from various public resources.

3.5. Database

Tools have been developed for the effective and efficient mining of sequence information. These include Locus-Link, which provides a gene-centric view of sequence-based information. The National Centre for Biotechnology Information Map Viewer, the University of California Santa Cruz Genome Browser, and the European Bioinformatics Institutes Ensemble system are some of major genome browsers which consists of DNA,RNA,Protein,Disease,Expression databases[5][6].

3.6. Comparison Techniques

Sequence comparisons lie at the heart of all bioinformatics. Edit distance is a way of quantifying how similar two strings were. It is the minimum number of operations required to transform one string into the other. Commonly used sequence comparison techniques were:

Multiple Sequence Alignment (MSA)[8]: used for homologous sequence comparisons of three or more biological sequences, generally protein,DNA,or RNA.MSA programs use heuristic methods because identifying the optimal alignment between more than a few sequences of moderate length is computationally expensive. Several high speed multiple sequence alignment program includes MAFFT (Multiple Alignment Using Fast Fourier Transform) can align up to 500 sequences [9].

Needleman-Wunsch algorithm [10] is a smart way to reduce the massive number of possibilities of sequence that are needed to be considered. It is developed by IBM for dynamic programming that runs in O(mn) time. The basic idea is to build up the best alignment by using optimal alignments of smaller sub sequences. It makes use of the concept of longest common substring.

The P-Bigram method [8] is method used to find the similarity problem in DNA sequence by integrating the dynamic programming concept. Editing operations include deletion, insertion, substitution two characters. This method provided an efficient algorithm that locates all minimum operation in a string. The proposed method was tested on 5 DNA matching datasets with representative two characters based string measures which runs in O(n*n) time.

De Bruijn graph also [11] called Kmer graph, is another efficient way to represent a DNA sequence in terms of its k-mer components. It is based on assembly approach and was proposed to handle the assembly of repetitive regions better. Sequence is constructed from an NGS library, and then the genome is derived from the de Bruijn graph. The reads are k-mers and each one corresponds to one edge in the graph. The vertices are (k-1) mers that appear in some read, and edges defined by overlap of k-2 nucleotides. Small values of k produce small graphs and thus eliminate repetitive sequences and thus remove the computational complexity

4. Conclusion

The use of Next generation sequencing Data has its own relevance in the field of Bioinformatics to understand the code of life. The high through-put sequence comparison helps to analysis the sequence similarities of human genome. Different platforms are selected based on the application with which we are dealing with. The work flow and data analysis of sequences are gone through different stages in order to determine the precise sequence. Sequencing similarity can be compared by techniques speed ups the faster detection of abnormal behavior in human genome. The use of NGS include efficient detection and faster response rate of viral transmission like Hepatitics, tumor etc. A web system called the Global Hepatitis Outbreak and Surveillance technology (GHOST)[3] developed for advanced molecular detection (AMC) for Hepatitis C virus transmission is a remarkable application of NGS. GHOST aims to improve the accuracy and response time of transmission detection by integrating epidemiological evidence. The tool is available to public health laboratories to identify outbreaks by simply uploading viral sequences.

References

[1] F. Sanger, S. Nicklen, and A. R. Coulson 1977 DNA sequencing with chain-termination inhibitors *74(12):5463-7.*

[2] Chien-Yueh Lee, Yu-Chiao Chiu, Liang-Bo Wang, Yu-LunKuo 2013 feb Common applications of next-generation sequencing technologies in genomic research *Translational Cancer Research Vol 2, No 1*

[3] Inna Rytsareva , David S. Campo, Yueli Zheng ,Sharma V. Thankachan, 2015 Efficient detection of viral transmissions with Next-Generation Sequencing data *Fifth IEEE International Conference on Computational Advances in Bio and Medical Sciences (ICCABS 2015))*

[4] Federica Torri Ivo D. Dinov Alen Zamanyan Sam Hobel Alex Genco Petros Petrosyan, 2012 Sep. Next Generation Sequence Analysis and Computational Genomics Using Graphical Pipeline Workflows *Genes (Basel) 2012 Sep doi: Genes (Basel) 3(3): 545575.*

[5] Andreas D Baxevanis, 2003, Using Genomic Databases for Sequence-Based Biological Discovery *BMC informatics, Mol Med 2003 Sep*

[6] Dong Zou, Lina Ma, Jun Yu,ZhangZhang , 2015 Feb "Biological Databases for Human Research", Proteomics Bioinformatics 2015 Feb; 13(1): 5563.

[7] Veronique G. LeBlanc, and Marco A. Marra). 2015 Sep,Next-Generation Sequencing Approaches in Cancer,*BMC Bioinformatics Cancers (Basel). 2015 Sep; 7(3): 19251958.*

[8] Patsaraporn Somboonsak, Mud-ArmeenMunlin , 2011, AA New Edit Distance Method for Finding Similarity in Dna Sequence, *World Academy of Science, Engineering and Technology International Journal of Biological, Biomolecular, Agricultural, Food and Biotechnological Engineering Vol:5, No:10*

[9] JurateDaugelaite Aisling O, Driscoll and Roy D. Sleator,2013 An Overview of Multiple Sequence Alignments and Cloud Computing in Bioinformatics, *BMC Genomics Volume 2013 , Article ID 615630, 14 pages*

[10] http://vlab.amrita.edu/?sub=3brch=274sim=1431cnt=1

[11] http://www.homolog.us/Tutorials/index.php?p=1.4s=1

[12] https://en.wikipedia.org/wiki/DNA$_s$*equencing*

IOP Conf. Series: Materials Science and Engineering **396** (2018) 012032 doi:10.1088/1757-899X/396/1/012032

A Study on Packet Loss Reduction methods and Node Registration methods in AODV for MANET

Achu Anna Antony [1], Bino Thomas [2]
[1]St. Joseph's College of Engineering and Technology, Palai
[2] Assistant Professor,St. Joseph's College of Engineering and Technology, Palai

E-mail: [1]achuannaantony2@gmail.com, [2]binothokar@gmail.com

Abstract. Mobile Ad-hoc NETwork (MANET) is a restless self-forming, infrastructure-less network of mobile nodes in a wireless connection. As there is a high hike in the use of mobile devices and wireless networks over past years, MANET has become one of the vital networks used for communication. A routing protocol is used for distributing information that allows selecting routes between two nodes in a network. Ad- hoc on- demand distance vector (AODV) protocol is widely used protocol for routing in mobile ad- hoc network. Packet loss is one of the significant problems that happen in the mobile ad- hoc networks while routing. A packet consists of the unit of data which is routed between source and destination in a network. Packet loss happens when one or more packets across devices in a network drop before reaching the destination node. A node density method is proposed in this study to alleviate the packet loss problem to an extent. Due to network infrastructure of MANET dynamically changes, mobile ad- hoc network is very vulnerable to attacks. As for the security purpose, we also propose a bloom filter method, which can be used to register the mobile nodes that are participating in a network in- order to restrict attacker nodes or foreign nodes to participate in the packet transmission.

Keywords- Bloom-filter, mobile ad-hoc network, routing protocol, neighbor node.

1. Introduction

Usage of mobile nodes these days has increased and the communication enhancement in its network becomes crucial. Ad hoc networks are generally used by military, rescue teams, personal electronic device networking, maritime communications etc. These users cannot rely on the centralized network [10]. The main factors affected in ad-hoc networks are routing and the characteristic of wireless communication. In ad hoc, a node can communicate only with nodes in its area and to communicate with other nodes uses a routing algorithm. [10]. Mobile Ad-hoc network (MANET) is a collection of mobile nodes that constitute a network with no central admin [1]. A MANET can change location and is a kind of ad-hoc network. MANET has its property that it can configure itself. The advantage of a decentralized network is that they are more robust than centralized networks due to its multi-hop pattern.

Packet loss in transmission is one of the major limitations in the mobile ad-hoc network. As one node moves away from the network, the connection gets lost and the packet drop may happen and also because of congestion packet loss happens. Congestion happens when many

IOP Conf. Series: Materials Science and Engineering **396** (2018) 012032 doi:10.1088/1757-899X/396/1/012032

demand request gathers and when there is a shared medium [11]. To send packets to the Internet, a MANET node acquires information about an Internet Gateway and establishes appropriate routes to this gateway [2]. There are generally three kinds of protocols used like proactive, reactive and hybrid and is shown in fig 1.

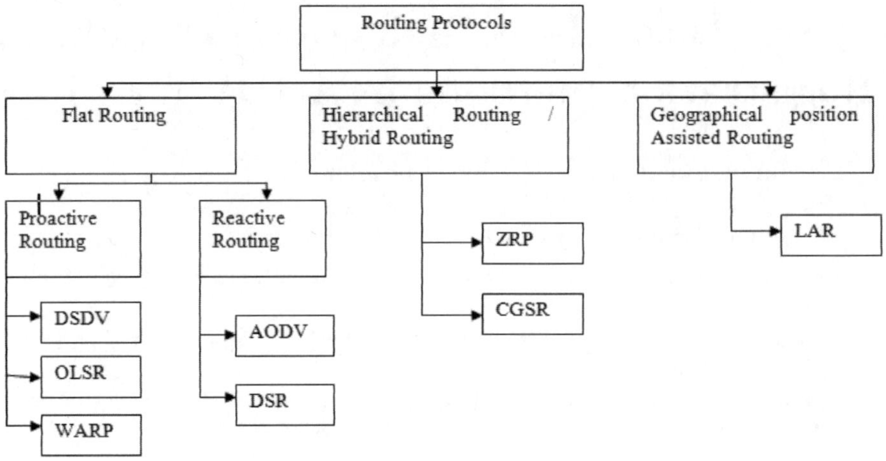

Figure 1. Ad- hoc routing protocols.

Considering AODV and DSDV routing protocols, AODV is better [11]. AODV routing protocol is a distance vector routing protocol and uses destination sequence numbers to determine the freshness of routes. AODV mostly maintain only active routes. Due to openness, dynamic, infrastructure-less nature, MANET is vulnerable to various attacks. One of these possible attacks is a Black Hole Attack.

A proposed node density method in this survey can be added along with the AODV protocol to send packets from node to destination with less packet loss. This survey paper adopts bloom filter technique, to reduce the attacks from foreign nodes by making member nodes to register into the bloom filter.

2. Background and Related Works

2.1. Type of Attack and Node Registration Method

2.1.1. Attack in AODV .

Black Hole Attack In AODV Protocol [13]: AODV Routing Protocol is used to find a path from source to the destination in an ad hoc network. All mobile nodes are needed to use routing control messages to find the path to the destination. There are three types of routing control messages in AODV protocol and they are Route Requests (RREQs), Route Reply (RREP), Route Error (RERR) which are used to find a path to the destination [14]. The AODV uses a destination sequence number for each route entry. The destination sequence number is generated by the destination when a connection is requested to it. The principle of this protocol is higher the destination sequence number; fresher is the route [14]. When the source node S wants to communicate with destination D as shown in the figure2, it broadcasts RREQs messages to the neighbor nodes. These neighbors check their routing table whether there is a path to the destination or not. If not then they also forward RREQs of the source node until the message is received by the intermediate node having a path to the destination or to destination node itself. If the node having a path to destination receives RREQ, it sends route reply message to the source node. In this way, source node selects the shortest path to the destination node with the greater sequence number of route reply message. If any link break occurs then RERR message send to the source node. In Black Hole Attack in MANET, assume malicious node M as Source S broadcasts its RREQs to connect with destination D, all intermediate node check their routing table whether there is a path to the destination D. Here Node M as malicious does not check its routing table and send false route reply packet to the source S with greater forged sequence number [15] than expected that it is having a path to destination D. As malicious node M does not check its route table, this reply reaches the source node faster than the normal nodes [16] source node select this path and send all of its data to the node M. The node M receives this data packets and deprive of the destination node. As this packets never reach the destination node D the attack is called as a black hole attack [17]. .

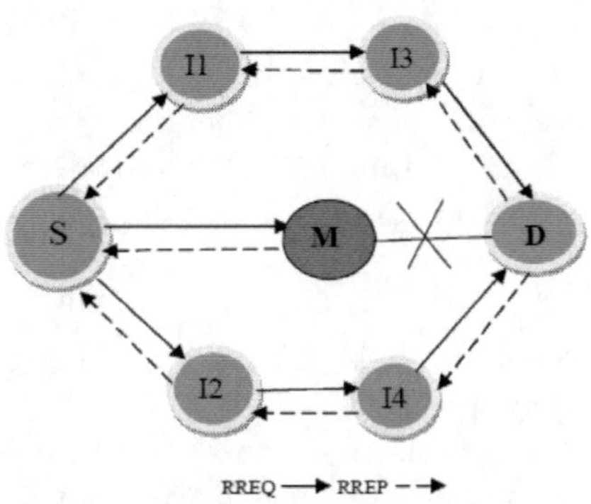

Figure 2. Black hole attack.

Spoofing Attack [12]: In spoofing attack, the attacker assumes the identity of another node in the network; hence it receives the messages that are meant for that node.

Man- in- the- middle attack: A malicious node sites between the source and destination and then sniffs any information being sent between two nodes.

Denial of service [12]: Attacker tries to block the specific network service or network operation.

Jamming Attack [12]: In this attack, the malicious node determines the frequency at which destination node is receiving signal. And then it transmits signal on that the same frequency.

2.1.2. Node Registration .
For node registration security, the survey is done with three methods which are proposed in [8], [12] and [4]. Node registration method described in [8] is, admin of the network tries to configure the nodes to give access. When other nodes request access, a token is generated and this will be valid only for some constant time. The token is then generated and is distributed to the requested node. As we know MANET does not have any central admin that is fixed, this approach will not be much feasible.

MANETs security attacks and proposed nodes registration based technique explained in [12] by using cryptography functions. The registered nodes will only have the provision to assign tokens to validated nodes and are confirmed by the registered node from the node id and home address in a request packet. Token exchange can lead to delay in packet transmission. Steps in this method proposed is as follows:

(i) Node is registered(RN) and generates registration token
(ii) Request RN for nodes communication
(iii) Checks node id and address.
(iv) After confirmation, token is generated
(v) After tokens are received, communication is initiated.

The bloom filter algorithm for node registration purpose explained in [4], only need is to register the node while joining the network and the foreign nodes will not be allowed to register into the bloom filter. When you insert a new node in a simple array or list, the index, where this data would be inserted, is not determined from the value to be inserted. That means there is no direct relationship between the key(index) and the value(node). As a result, if you need to search for a value in the array you have to search in all of the indexes. Now, in hash tables, you determine the key or index by hashing the value. Then you put this value in that index in the list. That means the key is determined from the value and every time you need to check if the value exists in the list you just hash the value and search on that key.[19]

2.2. Protocols

There are different routing protocols in MANET and is categorized into flat routing, hybrid routing and geographic routing shown in fig 1.
In the flat routing protocol, Routing table represented each network Identity individually.Flat routing protocols are two types, Proactive Routing (Table Driven) Protocol and Reactive Routing (On demand) Protocols. Proactive Routing is divided among DSDV, WAR, and OLSR [18]. Hybrid Routing- As the size of the wireless network increases, flat routing protocols will produce much more overhead for the MANET so hierarchical Routing can be used instead.Protocols are ZRP and CGSR [18]. The geographical position of a moving node is used in Geographical position Assisted Routing algorithms. The study mainly focuses on reactive routing protocols. Rather than AODV protocol, there is DSR protocol that comes under reactive protocols topology. In table 1 in section 3, it is clearly shown why AODV is efficient than DSR protocol.

Packet Loss Comparison for AODV and DSDV: The comparison of different packet losses for AODV and DSDV is as follows [11].

- Total packet loss: The total packet loss for DSDV will be 10percent to 20percent higher than that of AODV regardless of pause time or a number of connections.
- Congestion-related packet loss: DSDV loses more packets compared to AODV due to congestion. Fig 3 from [11] shows packet loss variation graph between AODV and DSDV.

So we can choose AODV over DSDV

2.3. Packet loss reduction method

In [9] when source node receives Warning Message from any node in the route then it will stop sending packets on that route. Here extra parameters are needed to be defined rather than finding the route. But in the proposed study, a method to find the route is the only factor needed to be considered. The steps for this method[9] is as follows

IOP Conf. Series: Materials Science and Engineering **396** (2018) 012032 doi:10.1088/1757-899X/396/1/012032

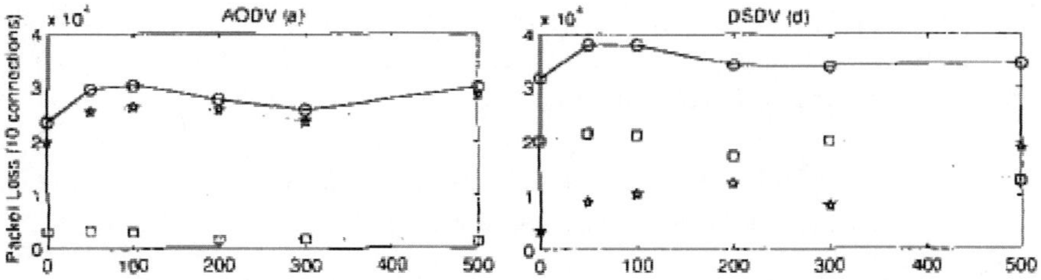

Figure 3. Packet Loss Comparison Graph.

(i) Transmit packets from source to destination.
(ii) For every node on the ongoing route do If the energy level is critical, then Node N will generate a Warning Message to the source. Here Node N is any node on the ongoing route.
(iii) If a Warning Message from any routing host is received, THEN
(iv) Source will not send a single packet in this current route.
(v) It will discard the current route from its cache.
(vi) If any messages need to be transmitted to the destination then the source will use another route from its cache. Else Source will forward packets on the current route.

3. AODV with Node Density Method and Node Registration method

3.1. Packet Loss Factors in Ad- hoc Network

Packet loss can occur in ad hoc networks where compromised nodes are not present. This packet loss happens mainly because of the following factors [7]:

3.1.1. Congestion in Network In mobile ad hoc networks, congestion is the main factor for packet loss. As the traffic increases packets may not reach the destination and packet loss happens.

3.1.2. Path Change and Noise In mobile ad-hoc networking, the path condition cannot be made unseen since it changes its path frequently. Presence of noise and fading of the transmitted signals are among the channel conditions that can lead to packet loss or bit errors in the transmitted signal. Because of these factors, packets can get dropped.

3.1.3. Energy Constraints Nodes in mobile ad hoc networks have limited energy resource. As the power of nodes decreases, the low energy nodes can get disconnected which may lead to packet loss.

3.2. Route Discovery by Neighbour Node Density Method

The mobile ad-hoc network is dynamic in nature the routing protocol will be preferred on the basis of administrative distance value allotted to each path in the network [3]. As an enhancement to AODV protocol, the node density feature can be embedded. When a node is about to get disconnected, the packets must be transferred to the neighbor node which has a maximum number of neighbors.

IOP Conf. Series: Materials Science and Engineering **396** (2018) 012032 doi:10.1088/1757-899X/396/1/012032

3.3. MANET Protocols

About MANET protocols, in section 2 it is explained in detail. MANET has mobile nodes and wireless links. As in MANET, the nodes move frequently which tends to change its topology frequently. These nodes have power constraints and are pervasive devices.

In MANET, dynamic routing is considered because, in dynamic routing, frequent topological changes are possible and also provides potential network partitions. There are proactive and reactive protocols based on topological routing. Proactive protocols like DSDV have high overhead independent of data traffic. But in reactive, on-demand routing enables overhead to scale with data traffic. Some reactive protocols are Dynamic Source Routing (DSR) and Ad-hoc On-demand Distance Vector (AODV). AODV can be used over DSR. Table 1 describes the difference between AODV and DSR.

Table 1. AODV and DSR difference.

Properties	AODV	DSR
Path Information	Limited	Detailed
Route Discovery times	Many	Few
Traffic when RREQ	Low	High
Route path life time	New	Unknown
Delete path when RERR	Delete all	Delete some nodes

The RREQ is the Route Request message and RERR is the Route Error message. By comparing both protocols, AODV is efficient that DSR.

The Ad hoc On-Demand Distance Vector (AODV) routing protocol offers fast conversion to dynamic link conditions, low processing and memory overhead, low network utilization, and determines unicast routes to destinations within the ad hoc network [6]. The routing table of AODV protocol is shown in table 2.

Table 2. AODV routing table.

Routing Table
destination
sequence number
$hop_{c}ount$
$next_{h}op$
$expiration_{t}imeout$

So in this study, it is evident that AODV is better considering flat type routing protocols.

3.4. Bloom Filter

Bloom filter is an efficient method for registering nodes that help to know if any attacker node is trying to be the part of the network. The bloom filter is used for various network security areas [4]. In [4] a proposed method, an efficient identity-based key management (IBKM) scheme, which exploits the bloom filter to authenticate the communication sensor node with storage efficiency. Example of a bloom filter is shown in figure 1. The basic bloom filter supports two operations[19]: test and add. The test is used to check whether a given element is in the set or not. Add simply adds an element to the set. Add function can be used to add member nodes to the network. Test function can be used to check whether there are any attacker nodes present.

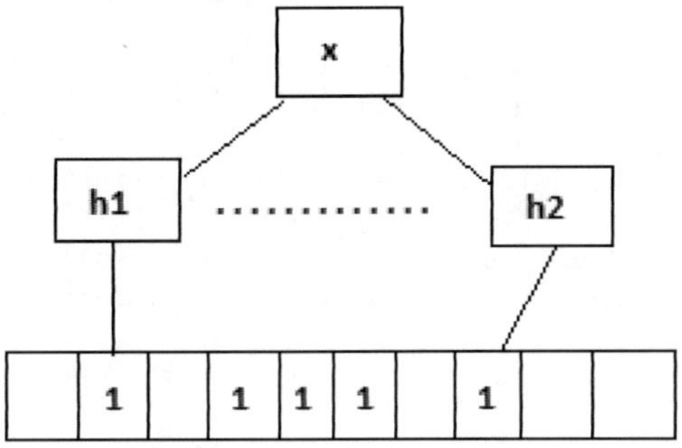

Figure 4. Example of Bloom Filter.

4. Conclusion

In this paper done a detailed study on packet loss reduction methods and node registration methods in AODV for MANET. Mobile ad-hoc network is a self-configuring, dynamic network of the mobile node in a wireless connection. AODV routing protocol is a distance vector routing protocol and uses destination sequence numbers to determine the freshness of routes. Bloom filter is the method adopted in this study and it will register all the member nodes in a network which transfer common packets and if any attacker node tries to join the network then the hash of group nodes will be different from the attacker node. Thus the attacker node will not be able to join the network. For routing, AODV protocol can be used which is already widely used a routing protocol. In order to reduce the packet loss in a network due to disconnectivity of nodes, the node density feature can be considered. In future, more on security and packet loss reduction methods will be concentrated.

5. References

[1] Aqeel Taha, Raed Alsaqour, Mueen Uddin, Maha Abdelhaq, and Tanzila Saba, May 24, 2017 *Energy Efficient Multipath Routing Protocol for Mobile Ad-Hoc Network Using the Fitness Function* (IEEE access)

[2] Jain S Charles E. Perkin, Jari T. Malinen, Ryuji Wakikawa, Anders Nilsson 2002 *Internet connectivity for mobile ad hoc networks* (Wirel. Commun. Mob. Comput. 2:465482)

[3] V. Karthikeyan, A.Vinod P. Jeyakumar Y*An Energy Efficient Neighbour Node Discovery Method for Wireless Sensor Networks* (India)

[4] Achu Anna Antony, Ashily M Baby, Kavya Rajeev, Bino Thomas 2017 *Cooperative game theoretical approach for efficient node discovery in mesh network with bloom filter* (International Journal of Advances in Electronics and Computer Science, ISSN: 2393-2835)

[5] Zhongyuan Qin , Xinshuai Zhang , Kerong Feng, Qunfang Zhang and Jie Huang *An Efficient Identity-Based Key Management Scheme for Wireless Sensor Networks Using the Bloom Filter* (ISSN 1424-8220)
[6] C. Perkins, E. Belding-Royer, S.Das, *Ad hoc On-Demand Distance Vector (AODV) Routing* (Network Working Group)
[7] Kennedy Edemacu, Martin Euku and Richard Ssekibuule 2014 *Packet drop attack detection techniques in wireless ad hoc networks: a review*)Vol.6, No.5(International Journal of Network Security Its Applications (IJNSA))
[8] Manoj V. Moril, G.B. Jethava 2013 *Node registration in MANET* (International International Journal of Emerging Trends Technology in Computer Science (IJETTCS)) Volume 2, Issue 1, January February 2013
[9] Rezvi Shahariar and Abu Naser *Reducing Packet Losses in Mobile Ad Hoc Network Using the Warning Message Generated from a Routing Node* (Dhaka Univ. J. Sci. 62(2): 141-145, 2014 (July)) April 2014
[10] Martinus Dipobagio *An Overview on Ad Hoc Networks* (Institute of Computer Science (ICS))
[11] Yi Lu, Yuhui Zhong, Bharat Bhargava 2003 *Packet Loss in Mobile Ada Hoc Networks* (Purdue University,Purdue e-Pubs)
[12] Rashid Jalal Qureshi,Khalid Haseeb,Muhammad Arshad , Huma Javed,Haleem Farman Sep 2012 *A Novel Technique Based on Node Registration in MANETs* (IJCSI International Journal of Computer Science Issues, Vol. 9, Issue 5, No 3)
[13] Nitesh Funde, P. R. Pardhi May 2014 *Analysis of Possible Attack on AODV Protocol in MANET* (International Journal of Engineering Trends and Technology (IJETT) Volume 11 Number 6)
[14] C. Perkins. July 2003 *(RFC) request for Comments-3561* (Category: Experimental, Network, Working Group)
[15] Dokurer,Seimih Septeber 2006*(RFC) Simulation of Black hole Attackin wireless ad-hocNetworks* (Masters Thesis Atihm University)
[16] Hsun Tseng Li-Der Chou and Han-Chieh Chao *A survey of black hole attacks in wireless mobile ad hoc networks Fan*
[17] Ebrahim Mohamad, Louis Dargin *Routing Protocols Security* (In:Ad Hoc Networks. A Thesis at Oakland University School)
[18] Neeraj Verma,Sarita Soni 2017 *A Review of Different Routing Protocols in MANET* (International Journal of Advanced Research in Computer Science)ISSN No. 0976-5697, vol 8
[19] Ahmed shamim hassan 2017*Probabilistic Data structures: Bloom* ([Online]: https://hackernoon.com/probabilistic-data-structures-bloom-filter-5374112a7832)

A survey on different privacy-preserving authentication schemes in VANET

Deepu Mathew
St. Joseph's College of Engineering and Technology, Palai
E-mail: vrossy.mathew608@gmail.com

Hima Anns Roy
St. Joseph's College of Engineering and Technology, Palai
E-mail: himaannsroy@gmail.com

Abstract. A vehicular ad hoc network(VANET) system used in modern transportation system that improves the road safety and other efficiency parameters. In this technology it uses cars as node in a network to develop a mobile network. In vanet, the distance between cars around 100 to 300 meters and only those ones are allowed to connect together. The broadcast of messages by vehicles are in an open-access environment. It makes several critical security issues. A mishandling of this information may cause several critical issues like traffic accident and other traffic problems. So authentication of vehicles is necessary to improve the safety in vanet. During authentication a vehicles confidentiality-related data, such as identity of user and information about locations must be kept private. There are different types of authentication schemes are available and preserving the privacy related data of vehicles are different in different authentication schemes. This survey aims to find the better schemes for privacy preserved authentication methods in the VANET system.
Keywords:CRL(Certification Revocation List),RSU(Road Side Unit),TA(Trusted Authority).

1. Introduction

Related to road security Vehicular ad hoc network (VANET) is one of the popular concept applications in many countries. It is an important part of the modern transportation systems. The aim of Vehicular Ad Hoc Network (VANET) is that to provide the road safety and comfort by decreasing the congestion in roads. In a VANET, each vehicle is use an on-board unit (OBU) and road-side units (RSU) installed at road side. A trusted authority (TA) is installed in background and other application servers are also running in back ground for supporting the proper working of the system. Communication between OBUs and RSUs is done by using a dedicated short range communications protocol using wireless channel. RSUs, TA, and other application servers are communicate with each other using a fixed secure network, Internet.

VANET offers secure driving experience and road safety, and other traffic related services. The basic use of a VANET is to grant vehicles to broadcast secure messages , message contains information like vehicular speed, traffic accident information to other nearby vehicles. This method of communication is called as vehicle2vehicle or V2V communications and communication between vehicle and RSUs is known as vehicle2infrastructure or V2I communications.

The broadcasting of messages in an open-access network may cause security and privacy critical and other issues in the VANET. A mishandling of this message may cause to a traffic accident and loss of human lives, therefore vehicle authentication is an important part in this system. Therefore the main aim of designer is to make VANET more and more secure. During authentication time a vehicles privacy-related data, such as personal information and information about the location must be kept private. There are different types of authentication schemes are available and preserving the privacy related data of vehicles are different in different authentication schemes.

2. Related Works

In the last few years, many privacy preserving authentication methods have been proposed such as pseudonym based methods , group signature-based schemes, ID-based schemes, symmetric cryptography-based methods.

2.1. Pseudonym based schemes

Message authenticity is compulsory to defend VANETs from outsiders, as well as bad insiders. But since safety messages will not contain any responsive information, condentiality is not mandatory .As a result, the transferring of safety messages in a VANET requires authentication but not encryption. Digital signatures are a good methods in the VANET setting, because safety related messages are normally standalone, and should be transform to receivers as fast as possible. A work by Raya and Hubaux [8], in which they described the safety and privacy requirements for VANET and suggest one of the initial pseudonym-based schemes, many other authors have followed their work and proposed a lot of pseudonym based and group signature-based schemes. The pseudonyms-based schemes are implemented by using Public Key Infrastructure (PKI). PKI based certicates are attached with the message that are signed with the equivalent private keys. Each certicate contains a pseudo identity. The relation between the pseudo identity and the certicate is known to the issuing authority, called certication authority (CA).

Raya et al. [2] introduce a hardware security module (HSM) or a Temper Proof Device (TPD) that is used to protected the cryptographic stuff stored in a vehicles OBU. This scheme suffer from the storage and communication overhead of pseudonymous certificates. Usage of Certification Revocation List(CRL) is a major drawback of this method. When revoking a vehicle, all the certicates issued to that vehicle needed to be included in the CRL whose size grows exponentially. Therefore, there is an extra overhead included in the distribution, storage and checking of CRL.

Zhang et al. [3] recently proposed a better scheme that requires a realistic TPD in its place of an ideal TPD. This method also provides conditional anonymity. Sun et al. [4] try to reduce the CRL by introducing a hash chains and use a proxy re-signature scheme in order to advance the time needed for update the CRL.

Lu et al. [5] propose a conditional privacy preserving approach. A vehicle wants to get short-time pseudonym keys from RSU and therefore, this method requires pervasive operation of RSUs. A major drawback of this system is that the trusted authority needs to frequently distribute the RSUs with updated CRL.

Rajput et al. [6] proposed a hierarchical pseudonymous-related method that requires vehicle to get primary pseudonym from the CA and secondary pseudonyms from the RSU. However, their method assumes the pervasive implementation of RSUs.

The approaches presented in [7] and [1] use identity based cryptography [9] where the public key is a recognizable identity and the equivalent secret key is generated by a Trusted Authority (TA). In order to provide privacy, the identifiable identity is concealed with the assist of a pseudonym and therefore, these schemes suffer from the pseudonym management overhead.

Zhang et al.[8] introduce a method that describes an identity based verification scheme that provide pseudo identity based certificates and the secret keys with the help of a TPD.

2.2. *Group signature based scheme*

In group signature method schemes [10],[11],[12] a vehicles real identity is concealed among a group of vehicles. The vehicles in a group sign messages with individual private keys and the receiver veries the message with the groups public key. In case of a harmful message, the group manager is clever to trace the malicious member and later revokes it . Calandriello et al. [11] proposed a method that integrates both the group signature-based and pseudonymous-based schemes. However, the scheme is computationally costly as it needs to check a message against the blocked vehicles. Xiong et al. [14] use revocable ring signatures invented by Liu et al. [13]. The scheme describes conditional anonymity but suffers from the sharing of revocation information to all the vehicles. The important works in the area of privacy preserved authentication schemes in VANET are listed in the table1 along with the key results of that works.

Table 1. Title and Key Results in the area of authentication schemes in VANET.

Sno	Title	KeyResults
1	Security of Vehicular Ad Hoc Networks	Earliest pseudonym-based scheme
2	Secure vehicular communications	Hardware Security, Temper Proof
3	Distributed privacy-preserving authentication in VANET	Realistic TPD
4	Efficient pseudonyms Validation For vehicular communication	Hash chains, proxy re-signature
5	ECPP-secure vehicular communication	Short-time pseudonym keys
6	Hierarchical privacy preserving protocol-VANET	Primary and secondary pseudonyms
7	Scalable robust authentication protocol- vehicular comm.	RSUs to act as group managers

3. Comparison of different methods

Pseudonym scheme and group signature methods are the two main schemes in the area of privacy preserved authentication in VANET. The work done by Raya and Hubaux [4] is the first work proposes the pseudonym based scheme. This work use Public key infrastructure scheme.CA. Raya et al. [2] come up with more improvements by introducing a hardware security module (HSM) or a Temper Proof Device (TPD) that is used to protect the cryptographic content stored in a vehicles OBU. This work is more efficient than their previous works. Usage of HSM and TPD increases the efficiency of the system. But the system is suffered in the case that there is more overhead involved in the distribution and storage of pseudonym certificates. Certificate Revocation List usage is another major drawback of this system.

Zhang et al[3] introduced realistic TPD in place of an ideal TPD which overcome the disadvantages of work done by CA. Raya et al. [2]. The usage of realistic TPD improves the overall working of the system. The scheme also offers conditional anonymity. The time required to update the CRL is one of the main problem faced by the works done in this area. The work by Sun et al. [4] reduces the CRL by proposing hash chains and uses a proxy re-signature scheme in order to get better the time required to renew the CRL. The work by Lu et al. [5] , the concept of short-time pseudonym keys requirement have some disadvantages that is the trustworthy authority needs to regularly issue the RSUs with updated CRL.Hierarchical pseudonymous-based approach that requires vehicle to get principal pseudonym from the CA and secondary pseudonyms from the RSU. But this system a pervasive use of RSUs is needed.

In the case group signature based methods in all works the vehicles real identity is concealed among a group of vehicles. The approach in [12] uses the RSUs to work as group managers in order to supervise and maintain the vehicles. In this work the main concept is that a group manager who have the central control to manage and maintain vehicles. The pervasive RSU deployment is need in this case but it can have a depressing impact on the overall performance. The scheme in the work by Liu et al. [13] provides conditional anonymity but suffers from the delivery of revocation list information to all the vehicles.

4. Results and Conclusion

The survey reveals that the pseudonym based schemes mainly suffer from the storage, communicational, computational overhead because of the usage of CRL. The number of vehicles revoked is more, CRL is increment exponentially. The group signature-based schemes suffer the overhead related to group management. Another disadvantage is the trust related problems with the vehicle acting as a group manager. The group signature based methods also have problems like computational overhead in terms of pairing-based calculations.

So the proper usage of a hybrid method that combines both two methods may leads to a better system which overcome the issues in each individual method.

5. References

[1] L. Zhang, C. Hu, Q. Wu, J. Domingo-Ferrer, and B. Qin, Privacypreserving vehicular communication authentication with hierarchical aggregation and fast response, IEEE Trans. Comput., vol. 65, no. 8, pp. 25622574, Aug. 2016.
[2] M. Raya, P. Papadimitratos, and J.-P. Hubaux, Securing vehicular communications,IEEEWirelessCommun.,vol.13,no.5,pp.815,Oct.2006
[3] L. Zhang, Q. Wu, J. Domingo-Ferrer, B. Qin, and C. Hu, Distributed aggregate privacy-preserving authentication in VANETs, IEEE Trans. Intell. Transp. Syst., vol. 18, no. 3, pp. 516526, Mar. 2017.
[4] Y. Sun, R. Lu, X. Lin, X. Shen, and J. Su, An efcient pseudonymous authentication scheme with strong privacy preservation for vehicula rcommunications, IEEE Trans. Veh. Technol., vol. 59, no. 7, pp. 35893603, Sep. 2010.
[5] R.Lu, X.Lin, H.Zhu, P.-H.Ho, and X.Shen,ECPP:Efcient conditional privacy preservation protocol for secure vehicular communications, in Proc. IEEE INFOCOM, Apr. 2008, pp. 1229123
[6] U. Rajput, F. Abbas, and H. Oh, A hierarchical privacy preserving pseudonymous authentication protocol for VANET, IEEE Access, vol. 4, pp. 77707784, 2016.
[7] J. Li, H. Lu, and M. Guizani, ACPN: A novel authentication framework with conditional privacy-preservation and non-repudiation for VANETs,IEEETrans.Parallel Distrib.Syst.,vol.26,no.4,pp.938948, Apr. 2015.
[8] M.Rayaand , J.P.Hubaux,Securingvehicularadhocnetworks,J.Comput. Secur., vol. 15, no. 1, pp. 3968, 2007.
[9] A. Shamir, Identity-based crypto systems and signature schemes, in Advances in Cryptology. Berlin, Germany: Springer, 1984, pp. 4753.
[10] Lin, X. Sun, P.-H. Ho, and X. Shen, GSIS: A secure and privacypreserving protocol for vehicular communications, IEEE Trans. Veh. Technol., vol. 56, no. 6, pp. 34423456, Nov.2007.

IOP Conf. Series: Materials Science and Engineering **396** (2018) 012033 doi:10.1088/1757-899X/396/1/012033

[11] G. Calandriello, P. Papadimitratos, J. P. Hubaux, and A. Lioy, Efcient and robust pseudonymous authentication in VANET, in Proc. 4th ACM Int. Workshop Veh. Ad Hoc Netw., 2007, pp. 1928.
[12] L. Zhang, Q. Wu, A. Solanas, and J. Domingo-Ferrer, A scalable robust authentication protocol for secure vehicular communications, IEEE Trans. Veh. Technol., vol. 59, no. 4, pp. 16061617, May 2010.
[13] D. Y. W. Liu, J. K. Liu, Y. Mu, W. Susilo, and D. S. Wong, Revocable ring signature, J. Comput. Sci. Technol., vol. 22, no. 6, pp. 785794, Nov. 2007.
[14] H. Xiong, K. Beznosov, Z. Qin, and M. Ripeanu, Efcient and spontaneous privacy-preserving protocol for secure vehicular communication, in Proc. IEEE Int. Conf. Commun., May 2010, pp. 16.

IOP Conf. Series: Materials Science and Engineering **396** (2018) 012034 doi:10.1088/1757-899X/396/1/012034

Study On Early Detection Of Autism Using Genetic And Kinematic Biomarker

Deepthi V[1] and Vineetha S[2]
[1] Student of Computer Science &Engineering , RIT ,Kottayam.
[2] Faculty of Computer Science &Engineering , RIT ,Kottayam.

E-mail: deepthithulasy@gmail.com

Abstract. Over these years prevalence of autism has shown a drastic increase. A great deal of progress has been made in recognizing autism but still a lot to do. The impact of autism on the individual differ due to difference in the degree of severity and the wide range of symptoms. Early detection can in turn ameliorate overall development of child and help to shorten the diagnostic odyssey that many families experience. One of the great things about machine learning is the capability of computers to work on the actual data to find the complex relationships through different algorithms which in turn result in accurate predictions. Thus by examining the various risk factors that contribute to autism, machine learning approach seems to be beneficial for identifying the potential markers for autism. This study covers machine learning approaches that are used for identification of autism which facilitate early detection and better diagnosis of autistic individuals using different biomarkers.

1. Introduction

Autism seems to be a neurodevelopmental disorder that become visible during the early childhood. It is a wide-spectrum disorder, which indicates that its degree of severity and symptoms may vary from one individual to another. With the numbers increasing annually, the Centre for Disease Control found out that one in 68 has been diagnosed with an autism spectrum disorder.

Real cause of autism cannot be confined to a single known symptom. Even though the symptoms vary but all reveal the deficit in social communication, motor- impairments and restricted behaviour or interest [1]. Presently there is no medical test for autism diagnosis. Doctors diagnose ASD by looking at a child's etiquette and progress. Young children with ASD can reliably diagnosed by age two. Assessment of adults and children for ASD is based on the behavioural variations in communication, social interactions as reported by parents or teachers.

Detection of autism in adults is a cumbersome procedure because in adults, many symptoms can blend with some of other mental health, motor impairment disorders thus misinterpretation of actual diseases can in turn result in a dreadful life without proper diagnosis and effective treatment mechanisms .However, providing effective and early diagnosis of autism could be useful for overcoming the difficulties that would in turn persist throughout life of an autistic person and to identify their strengths for leading normal life.

Researchers found that different arguments ponder regarding the cause of autism but a unique consensus which suggest that early intervention can in turn result in better outcomes. Instead of focusing on the scores from the behavioural exams and questionnaires for diagnosis of autism, researcher's interest lies on using different machine learning approaches which distinguishes autistic individuals from others.

Machine learning is a powerful computational tool that support different application domains by learning complex relationships or patterns from large datasets to make accurate conclusions [10]. Better disease assessment can be made with predictive analysis of health data and offering more appropriate treatment mechanisms which is a hot research area nowadays. Supervised learning is an important part of machine learning [11] which employs a rule based approach by examining empirical datasets for building accurate predictive models for example patients chance of having a disease based on risk factors like genetic information. Movement reflects the way people interact with and sense their environment. Thus by focusing on unique ways in which the autistic brain processes different movements, could be helpful for diagnosis. It was found that untreated motor impairments can strongly affect the overall development of autistic individuals [3].

Only few studies put in machine learning to identify autistic individuals using kinematic data. A proof of concept study called reach to drop task [2] was conducted based on the upper limb movement analysis .It was found that a goal orientated movement can be pivotal in distinguishing autistic individuals. Many studies revealed strong evidence of genetic basis of autism which showed how different gene mutations contribute to the risk of autism. Various twin studies [4, 5] analogized the concordance rate which was found to be higher in monozygotic twins. Such studies provide a clear understanding regarding influence of genetic factors as an indicator for autism detection. Better treatment can be imparted by understanding the changes in the biological pathway and the abnormal proteins produced.

This paper presents a survey on machine learning approach that uses motion and genetic factors as biomarkers for autism identification. Thus early intervention of autism using machine learning technique opens up a new way for autistic individuals to develop a potential for leading a better life by improving their behavioural and emotional skills.

2. Kinematic Indicators

In the past decades, more consistent and troubling, features of autism, such as social problems and difficulties with communication were explored for diagnosis of autism. From 1980s, standardized tests for motor skills were began to be examined. The motion impairments can vary from one individual to other with different degree of severity. Most people with autism were found to lack effective sensory motor responses such as movement coordination towards achieving a certain goal, unstable balance and unusual gait patterns [6, 7].

It was perceived that children with adaptive behavioural skills are those with good motor skills. Thus the motor impairments are highly correlated with practical difficulties in habitual life skills like writing, walking etc. [8, 9].Untreated motor problems can aggravate the impact of psychological and behavioural problems of autistic individuals. Motor abnormalities are found to be an important endophenotype for autism thus contributing to remarkable exploration in appropriate treatment mechanisms [12].

Proper characterization of kinematic impairments that gathers both spatial and temporal nature of movements could result in understanding the aetiology of motor impairments and contribute to diagnostic classification of ASD [13]. Here kinematic parameters of imitation of hand movement were taken for extracting the most desired features using machine learning which may classify autistic individuals and prove validity of motor bio signatures for preliminary diagnosis. Their result included two imitation condition and nine motion features extraction with a classification accuracy of 93%.

Movement ABC [14] was one among the frequently used evaluation method for many previous work in which four types of item assessment was done such as ball skill, static and dynamic balance etc. This method serves as a gauge in the recognition of motor impairments in children but motion patterns are poorly designated without considering the spatial temporal nature of movements. Poor motor coordination [9] can often limit the self-perception of an individual in their capabilities and handling of their environment and situation. However this lack of perceived competence can lead to anxiety and low self-worth.

Facial expressions [16] of children were recorded by direct motion capture and for these expressions pairwise causal dependencies between facial regions were studied. The complexity inspection of facial expressions were carried out using multiple scale entropy method. Six facial conditions were tested and found that highest complexity is observed in check regions for autistic children than in eye region for emotions like anger, joy, sadness and surprise. Dynamic time wrapping method and local complexity based similarity analysis showed that the major facial expression differences come from eye region in autistic individuals.

A number of studies applied the machine learning techniques for differentiating individuals with autism from others. Multivariate pattern analysis (MVPA)[19] conducted using two different fMRI experiments with social stimuli where support vector machines and recursive feature elimination method try to find out the neural markers for autism diagnosis Machine learning algorithms were employed for extracting relevant information from Autism Diagnostic Observation Schedule-Generic (ADOS) [20, 21]. Here behavioural biomarkers were explored for autism detection. Stahl et al. [15] examined the infant event related potential and discussed the disadvantages of the commonly used methods such as mean group comparison .Discriminant function analysis, regularized function analysis and support vector machines were employed that avoid the multiple testing problem and errors of group mean comparisons. These classification measures were capable of increasing the distinct power of event related potential measurements. Here the purpose of machine learning methods to discriminate infant groups with high or low risk for detection of autism were discussed.

In Wilson et al. [17] cognitive skills were evaluated for discriminating autistic individuals deviating from behavioural diagnosis. Cognitive abilities are those voluntary responses of a person such as perception, motor skills, attention which contribute to a bank of knowledge based on combining the extracted information. Different range of cognitive skills were sampled. Tests valuating the symptom and intellectual capabilities along with neuropsychological tests were carried out. Support vector machines were used for classification using these twelve cognitive scores .Five cognitive task offers a significant difference and the important thing was that motor task fall under this category. However the cognitive classification models along with exploration of different empirical datasets were identified to be vital for autism diagnosis.

With the help of data from Autism Diagnostic Interview-Revised (ADI-R)[18] and Social Responsiveness Scale (SRS) an attempt was made to design more effective algorithms using machine learning with support vector machine was proposed to classify masses of verbal individuals with and without ASD .Greedy forward feature selection method provides information about most critical codes utilised for evolving a screener algorithm and provides an insight for improving the screening and diagnosis tools for ASD.

Artificial neural networks are adaptive systems influenced by the information processing mechanism of human brain. Neural networks are well-adapted to identifying non-linear patterns, in which there isn't a direct, one-to-one relationship between the input and the output and could be employed in many application domains. Hence it can form a predictive model with high noise tolerance with high performance. Grossi et al. [22] assessed the risk factors during pregnancy in autism development. The frequency of potential risk factors in autistic and control groups were compared. This work demonstrate the deployment of artificial neural networks predictive capacity that differentiate different risk factors with an accuracy of 80% and could be used as diagnosis tool.

Artificial neural networks usage for analysing the effect of treatment in autism spectrum disorder was proposed in the work of Narzisi et al. [23]. The Auto Contractive Map (Auto-CM) a kind of ANN with a three-layer architecture. It include an Input layer, with inputs captured from the environment, a Hidden layer in which modulation is carried out, and an Output layer, which make use of a feeds back mechanism based on the previously received and processed stimuli, here the association among variables were created as semantic connectivity map. ADOS computed severity score (ADOS-CSS) which distinguished children who actively respond to treatments versus non responders. ADOS-CSS indicate the extremity of autism symptoms.

Table 2.1: Summarizing the impact of motion parameters for autism identification.

Selection Criteria	Importance
Kinematic factors	Highly Quantifiable and measurable
	Motor-abnormalities act as endophenotype for autism, exploring better treatment mechanisms.
	Better characterization of motor functions were critical for improving the behavioural and functional characteristics.
	Provides a method for examining the variations in neural circuits leading to autism
	Predicting the severity by analysing the repetitive behaviour of autistic individuals.

3. Genetic Indicators

The gene susceptibility to autism has shown a remarkable development with the ongoing researches. Several genomic studies showed that many genes and chromosomal locations were associated to ASD. Advanced researches with new methods identified some significant mutations and the chromosomal instability which were analysed as an important risk factors for ASD. Here the main focus is to find whether the genetic factors provide the basis for autism biomarkers in machine learning scenarios.

One way of knowledge about the genetic influence in autism is through twin studies [4,5]. Researchers compared twin pairs both identical and fraternal pairs to see if the pattern of a disease or trait's inheritance differs between them. Twin studies illustrated the phenotypic variation due to genetic factors to be about 90%.

Family studies inspected the siblings of individuals known to have autism. Family history data analysis [24, 25] showed that the some of the problems experienced by the relatives have a close resemblance with behavioural characteristic commonly found in autistic persons. Reduced intrafamily variance provides an evidence related to hereditary influence in autism. Multiple predisposing factors were involved in autism which resulted in the variation of severity of disease and the phenotype [32]. The large degree of convergence from the causal factors obstructs the identification of predisposing factors and their development pathways.

The variation in the gene expressions were evaluated from blood samples collected which were used to differentiate between ASD individuals and unaffected controls. Glatt et.al [29] proposed a transcriptome-wide biomarker classifiers with high accuracy from the blood samples of different children having or not having autism. Gene expression profiling of mRNA samples collected from blood were examined by microarray to indicate that the abnormalities observed in these expression profiles mark the probability of autism occurrence. Another work focused on a class prediction algorithm [30] applied to find genetic mutations in lymphoblastoid cell lines (LCL). A rule based machine learning methods captured the relevant mutations that may result in autism by identifying the significant variation in genetic expression. ADI-R questionnaire scores were obtained by microarray analysis. Prediction was based on observed gene mutations with an accuracy of 94%.This work suggests the genes may be useful biomarkers for diagnosis of subtypes of idiopathic autism.

Kong et.al [26] conducted a largest transcriptome study that distinguished autistic individuals from the age/sex-matched controls. Repeated leave-group out cross-validation (LGOCV) strategy was used to build the predictive model. It was found that a favourable comparison occurs with DNA-based tests currently proposed for ASD diagnosis and ASD55 predictor's classification ability. Age of the individuals can significantly influence gene expression but ASD55 genes were not significantly correlated with age except two. Fisher's exact test was used to find the biological pathway of differentially expressed genes. The outcome indicated feasibility of blood based transcriptomic approach for ASD detection and hence it serves as a diagnosis tool.

In toddlers early identification of individuals at risk for autism spectrum disorder (ASD) was difficulty due to the impact of overlapping of symptoms with other disorders along with genetic diversity . Pramparo et.al [27] analysed illumina microarrays which included the leucocytes collected from different sets of children below four years. Differentially coexpressed genes were identified and the symptomatic categorization of children with ASD were obtained with 83% accuracy. Logical regression and curve analysis evaluated the classifiers performance. It showed the possibility of using blood based clinical test in pediatric diagnostic settings and the use of genetic biomarkers in machine learning.

Genetic cause [31] including rare de novo and inherited mutations or chromosomal rearrangements were investigated by conducting whole-exome sequencing (WES) that is a genomic technique for sequencing all of the protein-coding genes in a genome and a transcriptomic approach . This strategy identified the rarely inherited mutations that result in greater probability autism emergence and monogenetic forms of ASD. Dong Hoon [28] employed the transcriptomic approach which identifies the gene expressions suitable for diagnostic prediction of young adults with ASD. Both supervised and unsupervised machine learning methods were taken in to consideration. Supervised learning used three different machine earning algorithms, such as the support vector machine (SVM) K-nearest neighbours (KNN) and linear discriminant analysis (LDA) and other by clustering. Thus the results of this and four previous studies suggest that gene expression profiles in the peripheral blood samples serves as biomarker to predict the ASD risk in all type of age groups .

Voineagu et.al [33] distinguished the mRNA molecules expressed from the genes in autistic and normal brain by gene co-expression network analysis and revealed molecular abnormalities in autistic individuals. The experiments were carried out on post-mortem brain tissue samples from 19 autism cases and 17 controls and found that differentially expressed genes of autistic individuals concentrated more in cerebral cortex. Gene expression changes related to many diseases and hereditary characteristics were modelled using co-expression network.

Chromosomal micro array [34] is more powerful detecting capability than G-banded karyotype and fragile X DNA testing in patients with ASD. The benefits [35] of array comparative genomic hybridization which is a strong disease gene identification mechanism not only pave way for effective diagnosis but also helpful in the determining ASD genes with chromosomal imbalances. Hence this array offers an important aspect for genetic analysis. An attempt to identify the structural variations in ASD using both high-resolution microarray technologies and karyotyping was suggested and potential contribution of gene variation in ASD were identified [36].This also supported the use of microarray analysis for clinical characterization in the detection of ASD.

Reviews on clinical and research genetics [37] gathered all genes and recurrent genomic imbalances which can become a cause of ASD. This work found that autism is a behavioural demonstration several genes and genomic disorders from the microarray analysis. Whole exome and genome sequencing act as genetic signatures for ASD identification.

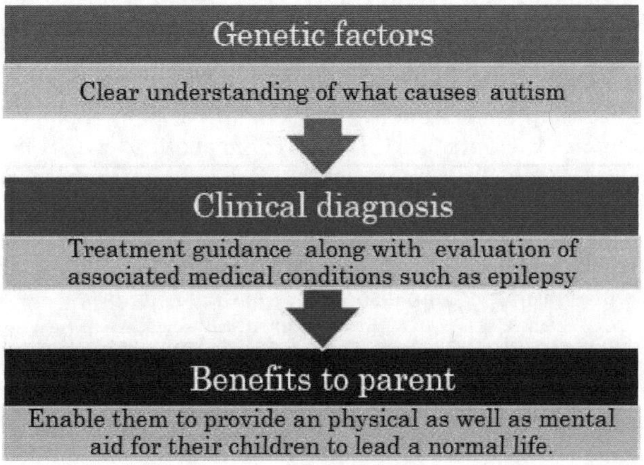

Figure 3.1: Importance of genetic factors for autism identification.

The advancement in autistic detection for improved treatment and diagnosis can be made using the genetic biomarkers. Thus the machine learning methods using genetic indicators result in early detection of autism which reduces the stressful diagnostic odyssey.

4. Summary

However it was shown that both genetic and motion parameters are relevant in their own aspect for autism detection .Thus the benefits offered by these two biomarkers can be combined to facilitate better diagnosis and effective treatment mechanisms through a machine learning based approach. As the neurology of autism on a genetic level may be complicated due to the heterogeneity of motor deficits ,the kinematic factors which involve the hand movements is helpful and a blood based gene expression analysis would be helpful to find the mutated genes that offer the risk of autism. This combination pave way in early intervention of autism than those studies whichinvolve these biomarkers independently.

IOP Conf. Series: Materials Science and Engineering **396** (2018) 012034 doi:10.1088/1757-899X/396/1/012034

5. Conclusion

In this paper a brief review of different machine learning approaches employed in identifying autistic individuals are presented .Early intervention of autism reduces the diagnosis stresses faced by many families and appropriate treatment mechanisms as early as possible which in turn helps the autistic individuals to lead a normal life in the society. Motor impairments and the genetic factors have strong influence in autistic individual .Hence they found to be effective as a diagnosis biomarkers for autism detection using machine learning algorithms with better accuracy and computation speed. Heterogeneity of symptoms along with the lack of knowledge about gene susceptibility paved way for researchers to explore new domains for autism identification. Thus machine learning prediction by analyzing behavioural and hereditary features provide an aid in developing a better diagnostic and screening tool for ASD.

6. References

[1] American Psychiatric Association, Diagnostic and statistical manual of mental disorders (5th Ed.),Arlington, 2013, *American Psychiatric Publishing.*

[2] Crippa A, Salvatore C, Perego P, Forti S, Nobile M et.al, 2015, Use of machine learning to identify children with Autism and their motor abnormalities, *Journal of autism and developmental disorders.*

[3] Fournier KA, Hass CJ, Naik SK, Lodha N, Cauraugh JH, Motor coordination in Autism Spectrum Disorders: A synthesis and meta-analysis, *Journal of autism and developmental disorders.*

[4] Steffenburg S, Gillberg C, Hellgren L, Andersson L, Gillberg IC, et al.,1989,A twin study of Autism in Denmark, Finland, Iceland, Norway and Sweden, *Child Psychol Psychiatry* **30**: 405–416.

[5] Bailey A, Le Couteur A, Gottesman I, Bolton P, Simonoff E, et al. 1995, Autism as a strongly genetic disorder: evidence from a British twin study, *J Psychol Med* **25***:* 63–77.C.

[6] Gowen E, Hamilton A, 2013, Motor abilities in autism: A review using a computational context, *Journal of autism and developmental disorders*.

[7] Bhat AN, Landa RJ, Galloway JCC, 2011, Current perspectives on motor functioning in infants, children, and adults with autism spectrum disorders , *Physical Therapy.*

[8] Jasmin E, Couture M, McKinley P, Reid G et.al, 2009, Sensory-motor and daily living skills of preschool children with autism spectrum disorders, *Journal of autism and developmental disorders.*

[9] MacDonald M, Lord C, Ulrich D, 2013, The relationship of motor skills and adaptive behavior skills in young children with autism spectrum disorders, *Research in autism spectrum disorders*.

[10] Mitchell TM, 1997, Machine learning(McGraw-Hill Boston).

[11] Kotsiantis SB, Zaharakis I, Pintelas P, 2007, Supervised machine learning: A review of classification techniques.

[12] Esposito G, Pasca SP, 2013, Motor abnormalities as a putative endophenotype for Autism Spectrum Disorders,*Front Integr Neurosci*.

[13] Baihua Li1, Arjun Sharma, James Meng,et.al, 2016, Applying machine learning to identify autistic adults using imitation: An exploratory study.

[14] Henderson SE, Sugden DA, Barnett AL, Smits-Engelsman C,1992, Movement assessment battery for children,*Psychological Corporation London*.

[15] Stahl D, Pickles A, Elsabbagh M, Johnson MH, Team B et al,2012,Novel machine learning methods for ERP analysis: a validation from research on infants at risk for autism,*Developmental neuropsychology.*

[16] Guha T, Yang Z, Ramakrishna A, Grossman RB, Hedley D, Lee S, et al, On quantifying facial expression-related atypicality of children with autism spectrum disorder. *IEEE Int Conf Acoustics, Speech and Signal Processing (ICASSP).*

[17] Wilson CE, HappeÂ F, Wheelwright SJ, Ecker C, Lombardo MV, Johnston P, et al,2014, The Neuropsychology of Male Adults With High-Functioning Autism or Asperger Syndrome,*Autism Research.*

[18] Bone D, Bishop S, Black MP, Goodwin MS, Lord C, Narayanan SS,2016,Use of machine learning to improve autism screening and diagnostic instruments: effectiveness, efficiency, and multi-instrument fusion, *Journal of Child Psychology and Psychiatry*.
[19] Chanel G, Pichon S, Conty L, Berthoz S, Chevallier C, Grèzes J,2016, Classification of autistic individuals and controls using cross-task characterization of fMRI activity, *NeuroImage: Clinical.*
[20] Kosmicki J, Sochat V, Duda M, Wall D,2015,Searching for a minimal set of behaviors for autism detection through feature selection-based machine learning, *Translational psychiatry*.
[21] Wall D, Kosmicki J, Deluca T, Harstad E, Fusaro V,2012, Use of machine learning to shorten observation based screening and diagnosis of autism,*Translational psychiatry*.
[22] Grossi E, Veggo F, Narzisi A, Compare A, Muratori F,2016, Pregnancy risk factors in autism: a pilot study with artificial neural networks.,*Pediatric research*.
[23] Narzisi A, Muratori F, Buscema M, Calderoni S, Grossi E,2015, Outcome predictors in autism spectrum disorders preschoolers undergoing treatment as usual: insights from an observational study using artificial neural networks,*Neuropsychiatric disease and treatment.*
[24] Bolton P, Macdonald H, Pickles A, Rios P, Goode S, et al,1994, A case-control family history study of autism, *J Child Psychol Psychiatry* **35:** 877–900.
[25] Bailey A, Palferman S, Heavey L, Le Couteur A ,1998, Autism: the phenotype in relatives, *J Autism Dev Disord* **28**: 369–392.
[26] Kong SW, Collins CD, Shimizu-Motohashi Y, Holm IA, Campbell MG, Lee IH, et al,2012, Characteristics and predictive value of blood transcriptome signature in males with autism spectrum disorders,*PLoS One*
[27] Pramparo T, Pierce K, Lombardo MV, Carter Barnes C,Marinero S, Ahrens-Barbeau C, etal,2015, Prediction of autism by translation and immune/inflammation coexpressed genes in toddlers from pediatric community practices.,*JAMA Psychiatry* .
[28] Dong Hoon Oh1, Bin Kim, Seok Hyeon et.al,2017, Predicting Autism Spectrum Disorder Using Blood-based Gene Expression Signatures and Machine Learning.
[29] Glatt SJ, Tsuang MT, Winn M, Chandler SD, Collins M,Lopez L, et al,2012, Blood-based gene expression signatures of infants and toddlers with autism. *J Am Acad Child Adolesc Psychiatry.*
[30] Hu VW, Lai Y,2013, Developing a predictive gene classifier for autism spectrum disorders based upon differential gene expression profiles of phenotypic subgroups. *N Am J Med Sci* (Boston) .
[31] Codina-Solà M, Rodríguez-Santiago B, Homs A, Santoyo J, Rigau M, Aznar-Laín G, et al,2015, Integrated analysis of wholeexome sequencing and transcriptome profiling in males with autism spectrum disorders, *Mol Autism*
[32] Belmonte MK, Cook EH, Jr., Anderson GM, Rubenstein JL, Greenough WT, et al,2004,Autism as a disorder of neural information processing: directions for research and targets for therapy.
[33] Voineagu I, Wang X, Johnston P, Lowe JK, Tian Y, et al,2011, Transcriptomic analysis of autistic brain reveals convergent molecular pathology.
[34] Shen Y, Dies KA, Holm IA, Bridgemohan C, Sobeih MM, et al,2010, Clinical genetic testing for patients with autism spectrum disorders.
[35] Jacquemont ML, Sanlaville D, et al,2006, Array-based comparative genomic hybridisation identifies high frequencyof cryptic chromosomal rearrangements in patients with syndromic autismspectrum disorders.
[36] Marshall CR, Noor A, Vincent JB, Lionel AC, Feuk L, et al,2008, Structural variation of chromosomes in autism spectrum disorder.
[37] Betancur C,2010, Etiological heterogeneity in autism spectrum disorders: more than 100 genetic and genomic disorders and still counting.

IOP Conf. Series: Materials Science and Engineering **396** (2018) 012035 doi:10.1088/1757-899X/396/1/012035

Synthesis, structure and characterization of Co doped CeO_2 nanoparticles

Saurabh Tiwari[a], Nivedha Balasubramanian[b], Sajal Biring[c], Somaditya Sen[a,*]

[a]Metallurgical Engineering and Material Sciences, Indian Institute of Technology Indore, India

[b]Department of Physics, PSG College of Arts and Science, Coimbatore, India

[c]Electronic Engg., Ming Chi University of Technology, New Taipei City, Taiwan

E-mail: sens@iiti.ac.in

Abstract. Nanocrystalline Co doped CeO_2 particles were synthesized using simple sol-gel method. The structural, vibrational and photoluminescence properties of synthesized nanoparticles were investigated using X-ray power diffraction (XRD), UV-visible spectroscopy, Raman spectroscopy and Photolumiscence (PL) spectroscopy. XRD showed that the lattice shrink and crystallite size decreases with doping which leads to lattice distortion and generation of strain. The bandgap is red shifted due to formation of defects states between bandgap as confirmed by urbach energy calculation. The PL intensity of Co doped CeO_2 reduces which is due to the non radiative emission centres. Raman analysis showed characteristic peak at 460.8 cm^{-1} of Ce-O vibrational unit. A weak peak emerges with doping at ~ 540-640 cm^{-1} hints generation of oxygen vacancies related defects which also acts as emission quenching center.

1. Introduction

Cerium is one of the most abundant rare earth elements (66.5 ppm in earth crust) and its oxide (CeO_2) has high dielectric constant and bandgap. Nanocrystalline CeO_2 generally shows different physical and chemical properties than its bulk form and has superior physical and chemical properties which make it useful for various applications. It is used as electrolyte in solid oxide fuel cells (SOFC), as catalysis to controlling emission from vehicles, solar cells, as gas sensor, as phosphors, as polishing material etc. [1–6]. For tailoring the properties of material doping is one of the efficient tools. It is well observed that transition element doping in CeO_2 enhance its properties and increase its functionality. Among transition element doping Co doping has attracted much interest. As Ranjith *et al* reported enhancement in ferromagnetism at room temperature with Co doping in CeO_2 [7], Saranya *et al* reported increase in photocatalytic behavior of CeO_2 with Co doping using UV and visible light [8], Abbas *et al* found differential cytotoxicity of Co doped CeO_2 nanoparticles i.e does not affect healthy cell but inhibit the viability of Neuroblastoma cancer cell [9].

Different methods are used for preparing CeO_2 nanoparticles as wet chemical method, solution evaporation and decomposition method, solgel method, etc. We have used simple solgel method for synthesis of all samples since it is cost effective, low temperature and effective method for getting better homogeneity. The synthesized $Ce_{1-x}Co_xO_2$ nanoparticles were characterized for structural, vibrational and photoluminescence properties using XRD, UV-visible absorption, Raman and Photoluminescence spectroscopy. XRD shows generation of strain with doping which may be due to generation of defects and lattice distortion. Co doping affects bandgap and PL emission, the effect of oxygen vacancies discussed.

2. Experimental

Solgel method was used for synthesizing $Ce_{1-x}Co_xO_2$ samples. Cerium nitrate ($Ce(NO_3)_3.6H_2O$) and Cobalt nitrate ($Co(NO_3)_2.6H_2O$) of analytical grade were obtained from Alfa Aesar. These precursors were dissolved in double distilled de-ionized water and stirred in magnetic stirrer for 1 h. After that these two solutions were mixed and stirred for 2 h for homogeneous mixing. A solution of citric acid and glycerol was used for forming chelating chain and burning of the gel. This solution was mixed with the precursor's solution. The resulted solution was stirred and heated for 2 h at 80^{0}C which finally gives a black powder which was calcined at 450°C for 6 h to get rid of carbon and nitrogen compounds. $Ce_{1-x}Co_xO_2$ samples were obtained for x=0 and x=0.05, which were further characterized.

X-ray Diffractometer (Bruker D2 Phaser) was used for investigating structural properties. UV-vis spectrophotometer (Shimadzu (UV-2600)) used for analyzing optical bandgap energy. Photoluminescence study was done using Perkin Elmer fluorescence spectrometer LS55, excitation wavelength 340nm. Vibrational properties were assessed using High Resolution Raman Micro Spectrometer, (HORIBA Scientific, 632.8nm).

3. Result and discussion

Fig. 1(a) shows the XRD spectra for synthesized $Ce_{1-x}Co_xO_2$ samples. All Peaks matches with prescribed CIF file-4343161 (Crystallography open database) confirming single phase cubic fluorite structure formation of CeO_2. No peaks correspond to oxide of Co or other impurities are present.

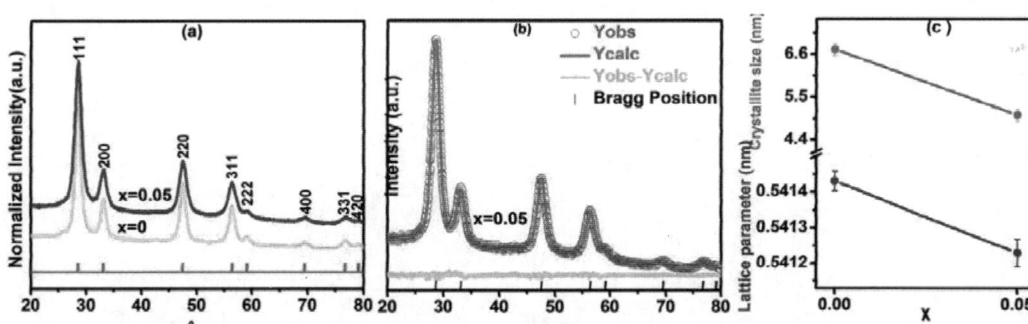

Fig.1. (a) XRD spectra of undoped and Co doped nanoparticles; (b) Rietvild refinement of Co doped CeO_2; (c) Lattice parameter and crystallite size variation with doping.

To observe the effect of Co doping in CeO_2, Reitveld refinement [fig. 1 (b)] was done using Fullprof suite software. Co doping cause lattice shrinkage as lattice parameter decreases with doping [fig. 1 (c)] this may be due to smaller ionic radii of $Co^{2+/3+}$ (0.75-0.9Å) than that of Ce^{4+} (0.97Å).

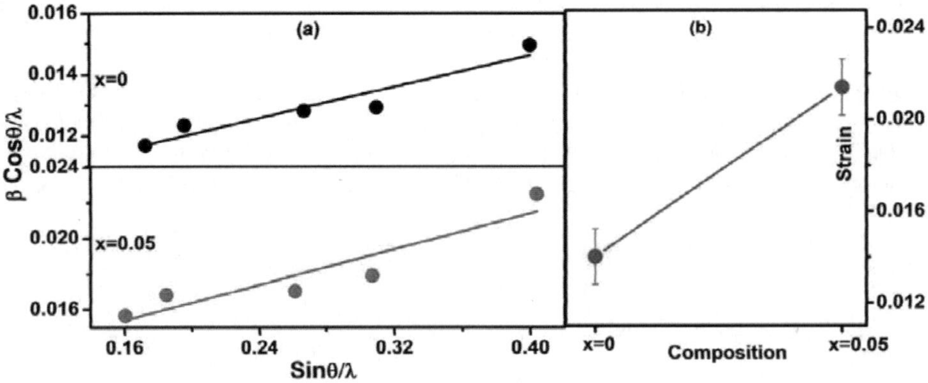

Fig. 2. (a) Williamson-Hall plot; (b) Strain variation with doping

This shrinkage should lead to lattice distortion and strain. Strain is calculated using Williamson-Hall (W-H) equation [10]; $\beta Cos\theta/\lambda = 1/D + \eta Sin\theta/\lambda$, where β is the full width half maxima (FWHM), θ is the angle of diffraction, λ is the wavelength of X-rays and D is crystallite size. A plot between $\beta Cos\theta/\lambda$ and $Sin\theta/\lambda$ plotted, slope η gives strain [Fig. 2 (a)]. It is found that strain is increasing with doping as expected [Fig. 2 (b)]. Crystallite size is calculated using formulae; $D = k\lambda/\beta cos\theta$ (k=0.9), found decreasing with doping this may be due to strain which restrict crystallite growth.

Optical properties of CeO_2 are of great interest in tuning its catalytic, biomedical and other activities. UV-visible spectroscopy depicts the absorption spectra of $Ce_{1-x}Co_xO_2$ (x=0 and x=0.05) nanoparticles samples [Fig. 3 (a)]. Kubelka-Munk function [11], $F(R) = (1-R)^2/2R$ used for calculating the bandgap as previously reported [12]. Co doping cause reduction of bandgap from 3.05 (x=0) to 2.52 eV (x=0.05) [Fig. 3 (b)]. This lowering of bandgap may be due to formation of some impurities band due to defects between valance band (VB) and conduction band (CB) [13]. Urbach energy (E_U): $\alpha = \alpha_0 * exp(E/Eu)$ accounts for defects in lattice, where α is coefficient of absorption, E is energy in eV.

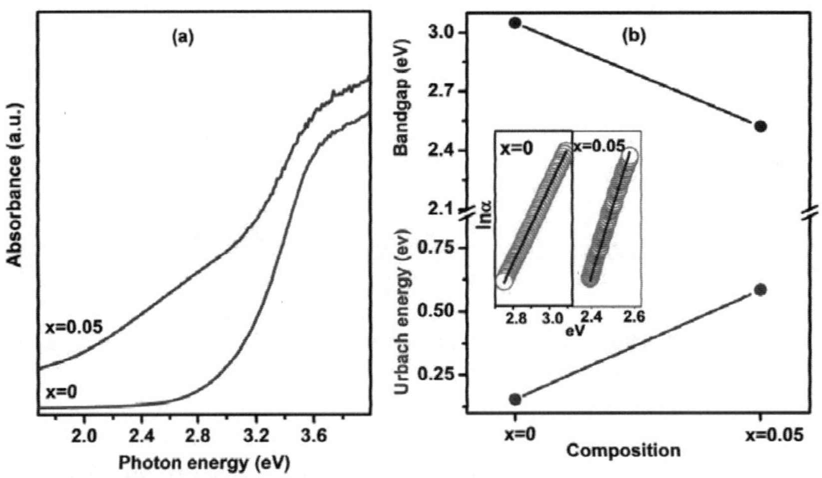

Fig. 3. (a) UV-visible absoption spectra; (b) Change in bandgap and urbach energy with doping (-inset shows urbach fitting)

The inverse of the slope of the plot between lnα and E give urbach energy. Urbach energy is increases with doping which confirm that defect states are increasing with doping [Fig. 3 (b)]. The effective gap between VB and CB reduces due to the deformation of electronic state.

Raman spectroscopy is very effective tool for obtaining structural information of fluorite structure because it is very sensitive to oxygen lattice vibration. [Figure 4 (a)] shows the Raman spectra of undoped and Co doped samples. A peak at 460.8 cm^{-1} is a characteristic F_{2g} symmetrical stretching mode of Ce-O vibrational unit [14]. This peak becomes broader and shift to lower wave number this may be due to generation of strain and reducing of Ce-O vibrational unit [Fig. 4 (b)]. A weak broad peak ~ 540-640 is due to generation of oxygen vacancies related defects Figure 4 (a)] [15]. Raman analysis supports urbach energy calculation regarding generation of defects which increases with doping.

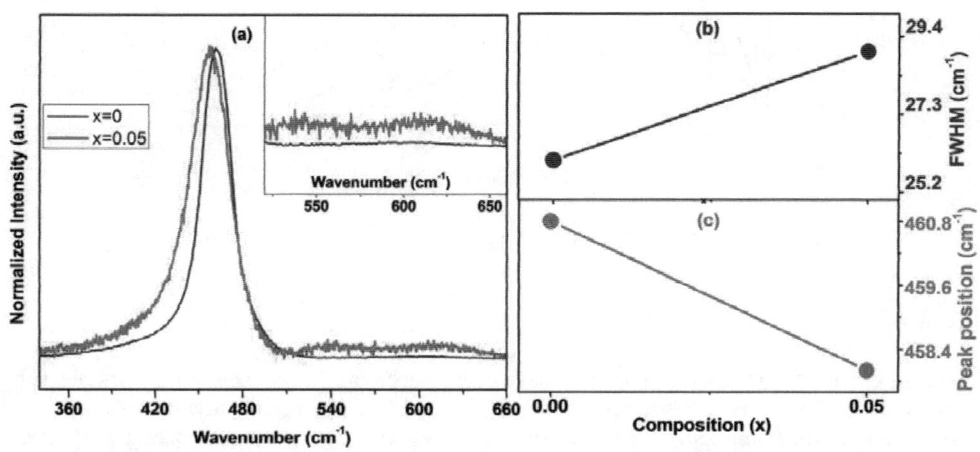

Fig. 4. (a) Raman spectra of $Ce_{1-x}Co_xO_2$ samples; (b, c) Variation in FWHM and peak position with doping.

The photoluminescence (PL) is an effective tool for analyzing defects and recombination process. Fig.5 shows the PL spectra of CeO_2, the peaks lying in between 395 and 529 are due to oxygen vacancies, surface defects and switching of charge between conduction band Ce^{4+} (4f) and valance band O^{2-} (2p) [16]. The emission bands are at ~424 nm correspond to surface defects, the band at ~457 nm and a feeble band at ~485 nm are considered as transition from ionized oxygen vacancies to the valance band and a band at ~528 nm reported as band due to defects as oxygen vacancies [17]. The PL emission intensity decreasing with doping, this infers the formation of defects states between valance and conduction band as confirmed by Raman and urbach calculation. The defects presents on surface and in bulk have different emissions one has non-radiative and other is having radiative emissions which affect the PL emission which depend on the nature of the oxygen vacancies. With Co doping in CeO_2, on the surface and on grain boundaries there is increase in oxygen vacancies which acts as trap centers leads to non-radiative recombination center and repress the emission intensity. Co doping causes high oxygen vacancies in CeO_2 matrix due to which spatial overlapping of electron and hole wave function obstructed which in turn prolong the recombination of charge carriers. Hence with Co doping surface defects increasing which cause non radiative emission and reduces the PL emission.

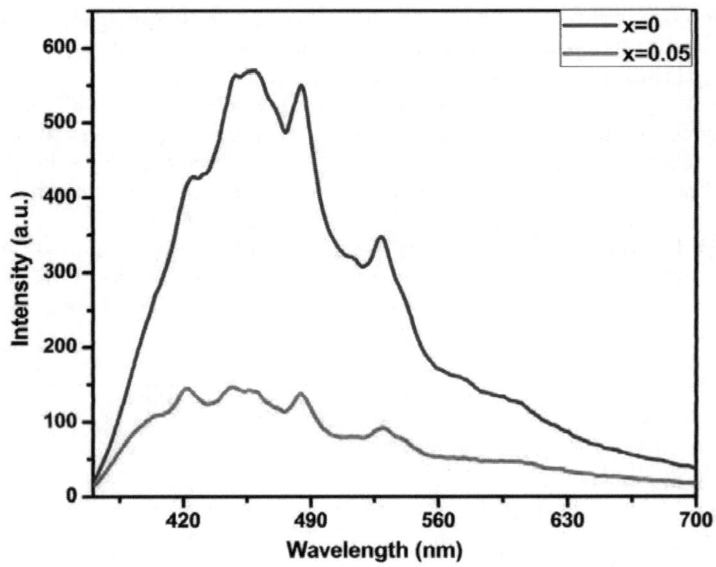

Fig. 5. Photoluminscence (PL) spectra of Ce1-xCoxO2 samples.

4. Conclusion

$Ce_{1-x}Co_xO_2$ (x=0 and x=0.05) nanoparticles have been successfully synthesized by simple sol-gel method. XRD and Raman analysis confirms the formation of cubic fluorite structure of undoped and Co doped CeO_2. Lattice shrinks and strain increases with doping. The estimated bandgap of undoped CeO_2 is 3.05 eV which decreases to 2.52 eV with Co doping. Urbach energy shows defects states are increasing which causes reduction of bandgap. A weak and broad peak ~540-640 pointing out that oxygen vacancies related defects are increasing with doping. The PL spectrum of CeO_2 has emission band from ~395-529 nm. The PL emission decreases with Co doping due to formation of oxygen vacancies related defects which act as non radiative emission center.

Acknowledgements

The authors would like to thank IIT Indore for the funding, Dr. Vipul Kumar Singh (IIT Indore) for Photoluminescence spectroscopy, Dr. V. K. Jain (Amity University, Noida, India) for UV-Vis measurement facility, Dr. Manoj kumar (IISER, Bhopal, India) for Raman spectroscopy.

Reference

[1] Murray E P, Tsai T and Barnett S A 1999 A direct-methane fuel cell with a ceria-based anode *Nature* **400** 649–51

[2] Fu Q, Saltsburg H and Flytzani-Stephanopoulos M 2003 Active Nonmetallic Au and Pt Species on Ceria-Based Water-Gas Shift Catalysts *Science* **301** 935–8

[3] Corma A, Atienzar P, García H and Chane-Ching J-Y 2004 Hierarchically mesostructured doped CeO2 with potential for solar-cell use *Nat. Mater.* **3** 394–7

[4] Izu N, Shin W, Murayama N and Kanzaki S 2002 Resistive oxygen gas sensors based on CeO2 fine powder prepared using mist pyrolysis *Sens. Actuators B Chem.* **87** 95–8

[5] Morshed A H, Moussa M E, Bedair S M, Leonard R, Liu S X and El-Masry N 1997 Violet/blue emission from epitaxial cerium oxide films on silicon substrates *Appl. Phys. Lett.* **70** 1647–9

[6] Lee S-H, Lu Z, Babu S V and Matijević E 2002 Chemical mechanical polishing of thermal oxide films using silica particles coated with ceria *J. Mater. Res.* **17** 2744–9

[7] Ranjith K S, Saravanan P, Chen S-H, Dong C-L, Chen C L, Chen S-Y, Asokan K and Rajendra Kumar R T 2014 Enhanced Room-Temperature Ferromagnetism on Co-Doped CeO2 Nanoparticles: Mechanism and Electronic and Optical Properties *J. Phys. Chem. C* **118** 27039–47

[8] Saranya J, Ranjith K S, Saravanan P, Mangalaraj D and Rajendra Kumar R T 2014 Cobalt-doped cerium oxide nanoparticles: Enhanced photocatalytic activity under UV and visible light irradiation *Mater. Sci. Semicond. Process.* **26** 218–24

[9] Abbas F, Iqbal J, Jan T, Naqvi M S H, Gul A, Abbasi R, Mahmood A, Ahmad I and Ismail M 2015 Differential cytotoxicity of ferromagnetic Co doped CeO2 nanoparticles against human neuroblastoma cancer cells *J. Alloys Compd.* **648** 1060–6

[10] Kuriakose S, Satpati B and Mohapatra S 2014 Enhanced photocatalytic activity of Co doped ZnO nanodisks and nanorods prepared by a facile wet chemical method *Phys. Chem. Chem. Phys.* **16** 12741–9

[11] Singh P, Choudhuri I, Rai H M, Mishra V, Kumar R, Pathak B, Sagdeo A and Sagdeo P R 2016 Fe doped LaGaO3: good white light emitters *RSC Adv.* **6** 100230–8

[12] Tiwari S, Bajpai G, Srivastava T, Viswakarma S, Shirage P, Sen S and Biring S 2017 Effect of strain due to Ni substitution in CeO2 nanoparticles on optical and mechanical properties *Scr. Mater.* **129** 84–7

[13] Veis M, Kucera M, Zahradnik M, Antos R, Mistrik J, Bi L, Kim H-S, Dionne G F and Ross C A 20140507 Optical and magneto-optical properties of Co-doped CeO2−δ films in the 0.5 to 4 eV range *J. Appl. Phys.*

[14] Weber W H, Hass K C and McBride J R 1993 Raman study of ${\mathrm{CeO}}_{2}$: Second-order scattering, lattice dynamics, and particle-size effects *Phys. Rev. B* **48** 178–85

[15] Bernardi M I B, Mesquita A, Béron F, Pirota K R, Zevallos A O de, Doriguetto A C and Carvalho H B de 2015 The role of oxygen vacancies and their location in the magnetic properties of Ce1−xCuxO2−δ nanorods *Phys. Chem. Chem. Phys.* **17** 3072–80

[16] Maensiri S, Masingboon C, Laokul P, Jareonboon W, Promarak V, Anderson P L and Seraphin S 2007 Egg White Synthesis and Photoluminescence of Platelike Clusters of CeO2 Nanoparticles *Cryst. Growth Des.* **7** 950–5

[17] Wang G, Mu Q, Chen T and Wang Y 2010 Synthesis, characterization and photoluminescence of CeO2 nanoparticles by a facile method at room temperature *J. Alloys Compd.* **493** 202–7

IOP Conf. Series: Materials Science and Engineering **396** (2018) 012036 doi:10.1088/1757-899X/396/1/012036

Multirate and Filterbank Approaches in Digital Hearing Aid Design: A Review

Tomson Devis
Research Scholar, Dept. of Electronics and Communication Engineering,
Rajiv Gandhi Institute of Technology, Kottayam, Kerala, India - 686501
tomsondevis@gmail.com

Manju Manuel
Associate Professor, Dept. of Electronics and Communication Engineering,
Rajiv Gandhi Institute of Technology, Kottayam, Kerala, India - 686501
manju.manuel@gmail.com

Abstract. Hearing is considered as the most important senses of humans because it connects us to the world always. And the communication between humans are mostly by speaking and listening. So hearing loss can have a significant affect on our life, from our work to our connections and enthusiastic prosperity. A hearing aid is an electro-acoustic gadget, which is intended to increase sound, with the point of making speech more comprehensible. The fundamental task of the hearing aid is to selectively amplify sound signals to such an extent that the handled sound matches ones audiogram. The emerging demand for personalized hearing assistants requires the filter bank to be capable of decomposing the sound waves in accordance with the characteristic of the patients hearing loss. In this review paper, the application of multirate frameworks and filter banks for different procedures in personalized hearing assistants are investigated.

1. Introduction

Hearing aids are gadgets that mostly defeat sound-related shortfalls and are regularly utilized to make up for hearing-loss in hearing-impeded individuals. The fundamental target of a hearing aid is to fit the dynamic range of speech and regular sounds into the limited dynamic range of the hindered ear. With a specific end goal to accomplish a superior comprehension of this gadget and its capacity, we have to clarify how the sound is perceived by the human auditory framework and what are the issues experienced by hearing-disabled individuals [1].

The most well-known sort of hearing loss is sensorineural hearing loss (SNHL) in which the underlying cause lies in the vestibulocochlear nerve, the inner ear and the sound handling centers of the brain. Generally an issue with the cochlea, in the internal ear, in which the cilia don't distinguish clues of specific frequencies as they should. Most hearing impedances fall into this class, so this is the issue that most hearing aids try to address. Sensorineural disability reduces the capacity of a man to recognize and detect one frequency within the presence different frequencies. Thus, a man with hearing misfortune has diminished capacity to hear a sound that quickly takes after, or is quickly trailed by an alternate sound. This diminished frequency makes it more probable for a hearing-impeded individual that noise will mask speech [2].

These days, multirate frameworks are applied in relatively every field of human interventions. Different applications of multirate frameworks are in speech processing for hearing aids, speaker and speech recognition, spectrum sensing in cognitive radios, signal compression for broadcasting, transmultiplexer design etc [3].

Whatever remains of this paper is sorted out as follows. As a matter of first importance, the basics of multirate frameworks and filter banks are introduced in section 2. In section 3, an overview of hearing aid systems are made. Subtle elements of auditory compensation techniques are talked about in section 4. In section 5, feedback or echo cancelation is talked about. The procedures relating to noise reduction is said in section 6. In section 7, speech enhancement for portable hearing assistants are clarified. At long last, conclusions are made in Section 8 and acknowledgements are specified in section 9.

2. Multirate Systems and Filterbanks

In digital signal processing, Multirate Systems utilizes different sampling rates at various parts of the framework. Multirate Systems primarily comprises of various operations like decimation, interpolation and resampling [4].

Decimation or down-sampling is the way toward lessening the sampling rate of a signal. The decimation factor is the ratio of the input sampling rate to the output sampling rate. It is normally symbolized by a positive integer M. It is performed by taking each Mth sample and disposing of every single other samples. This causes M fold expansion of the spectrum in frequency domain. An anti-aliasing filter with a cutoff of π/M is utilized to band-limit the input signal $x(n)$. The decimation procedure in time domain and frequency domain are expressed in (1) and (2) respectively, where $x(n)$ is the input signal and $y(n)$ is the output signal.

$$y(n) = x(Mn) \tag{1}$$

$$Y(z) = \frac{1}{M}\sum_{k=0}^{M-1} X(z^{\frac{1}{M}}W_M^k) \tag{2}$$

Interpolation or up-sampling is the way toward expanding the sampling rate of a signal. The interpolation factor is the ratio of the input sampling rate to the output sampling rate. It is generally symbolized by a positive number L. It is performed by including $L-1$ zeros between every sample. This produces $L-1$ images of the spectrum in frequency domain. An anti-imaging filter with a cutoff of π/L is utilized to expel these repetitive images. The interpolation procedure in time domain and frequency domain are given in (3) and (4) respectively, where $x(n)$ is the input signal and $x_u(n)$ is the output signal.

$$x_u(n) = \begin{cases} x(n/L), & n = 0, \pm L, \pm 2L, ... \\ 0, & otherwise \end{cases} \tag{3}$$

$$X_u(z) = X(z^L) \tag{4}$$

Resampling combines interpolation and decimation to change the sampling rate by a rational factor [5]. The resampling factor is typically spoken to by L/M. Interpolation is always preceeded by decimation. The anti- aliasing and anti-imaging filters are combined to a single filter, which has a cut-off of $min(\pi/L, \pi/M)$.

If a sequence $x(n)$ is band-limited, then it is possible to decimate it by use of appropriate multirate techniques. The desire to reduce the sampling rate whenever possible is completely understandable since it usually reduces the processing as well as the storage requirements. However if $x(n)$ is not band-limited but has most of its energy in the low frequency region, some

IOP Conf. Series: Materials Science and Engineering **396** (2018) 012036 doi:10.1088/1757-899X/396/1/012036

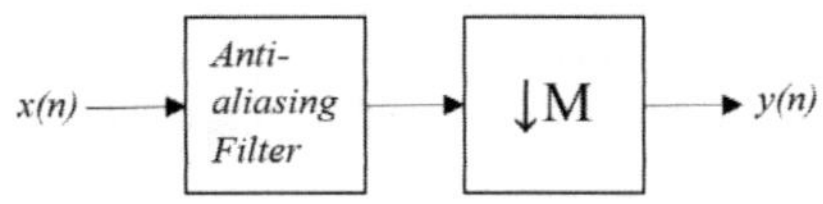

Figure 1. Block diagram of a decimator.

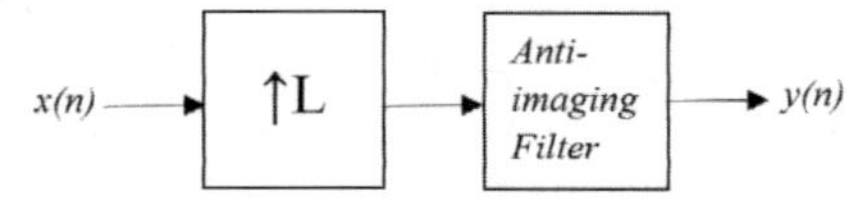

Figure 2. Block diagram of an interpolator.

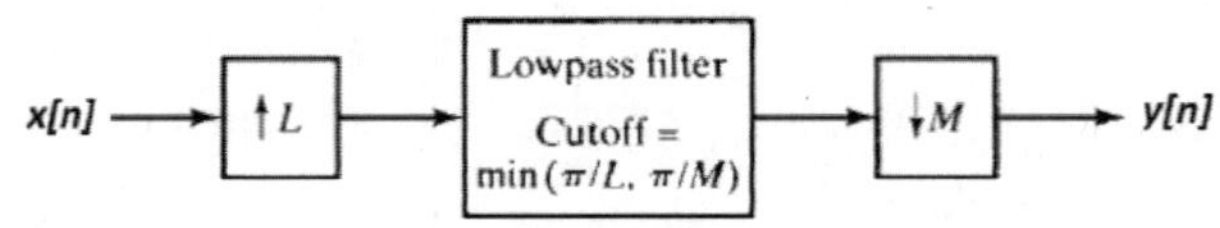

Figure 3. Block diagram of a resampler.

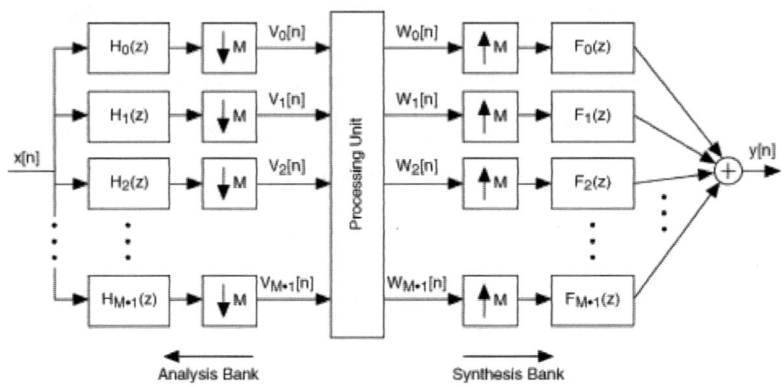

Figure 4. Block diagram of an M-band analysis-synthesis filterbank.

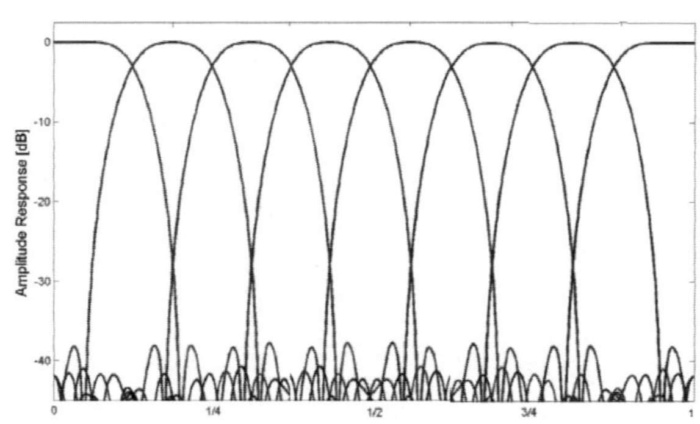

Figure 5. Frequency Response of an eight channel uniform filterbank.

kind of data rate reduction is feasible even though it cannot be decimated without aliasing. This is indeed achieved by a technique called sub-band decomposition in which the average number of bits per sample is reduced while the average number of samples per unit time is unchanged. Sub-band Decomposition is implemented using filterbanks [6].

A digital filter bank is an accumulation of digital filters having a common input or a common output as shown in Fig. 4. The left framework is called an analysis bank and its filters $H_k(z)$

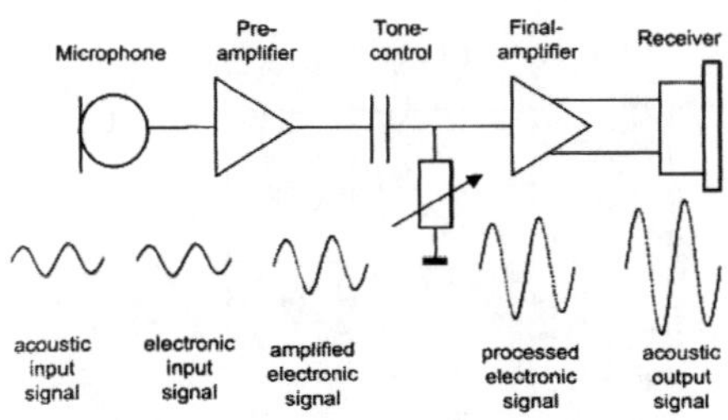

Figure 6. Block diagram of an analog hearing aid.

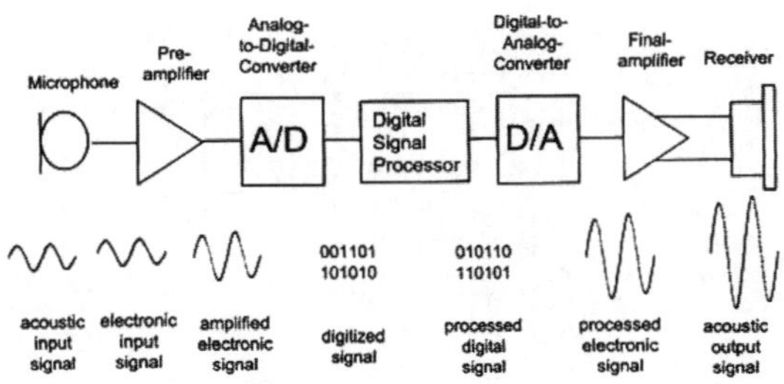

Figure 7. Block diagram of a digital hearing aid.

are the analysis filters. An input signal is splitted into M signals $v_k(n)$ typically called subband signals. We can process these individual bands independently and proficiently. The framework on the right part of the figure is a synthesis bank and $F_k(z)$ are the synthesis filters which joins the M-subband signals into a single signal $\hat{x}(n)$. Frequency responses can be marginally overlapping, non overlapping or very much overlapping, depending on the application. The magnitude response of a uniform filterbank with $M = 8$ is shown in Fig. 5.

3. Hearing Aid System Overview

Analog hearing aids give straightforward enhancement of the sound grabbed by the microphone and reproduced louder by the receiver or loud speaker. The amplification can be shaped marginally to consider distinctive hearing losses in spite of the fact that there is not always much flexibility. Many types of analog hearing aids can be programed by the user. They have a microchip which enables the device to have settings customized for various listening situations, for example, in a quiet place like in a library, or in a noisy place like in a hotel, or in an expansive region like a soccer field. The analog programmable hearing aids can store different programs for the different situations [7]. The client can easily select these modes when his listerning condition changes. An overview of analog hearing aid is shown in Fig 6.

Digital hearing aids take the sound waves from the microphone and change them to advanced digital binary code by the utilization of digital signal processing (DSP) strategies. This computerized data would then be able to be controlled in numerous sophisticated ways by the micro chip inside the amplifier, before being transformed back into ordinary analog signals and conveyed to the speakers. An overview of digital hearing aid is shown in Fig 7. Digital hearing aids, usually offer more highlights and adaptability, and are commonly client configurable. They have various memories to store more condition specific settings than analog hearing aids. They can also utilize modern algorithms to recognize and decrease background noise, to expel feedback and whistling, or to specifically distinguish the sound of voices and track them utilizing directional microphones . The distinctive aspects relating to digital hearing aid systems incorporate auditory compensation, feedback cancellation, noise reduction, speech enhancement etc [8].

4. Auditory Compensation

An audiogram is a diagram of the hearing test, that demonstrates the mildest sound a person can hear at various pitches or frequencies [9]. In the diagram, the Y axis shows the intensity or response of the ear estimated in decibels (dB) relating to the frequency in hertz (Hz) set apart on the X axis. An 'O' often is utilized to represent the responses for the right ear and a 'X' is utilized to represent the responses for the left ear. Curves showed in decibels generally shows the individual hearing limit of a man compared with the typical normal hearing, which is at 0

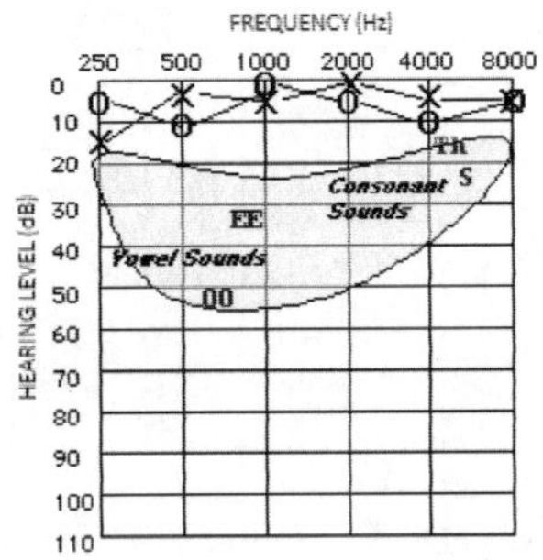

Figure 8. Audiogram of normal hearing.

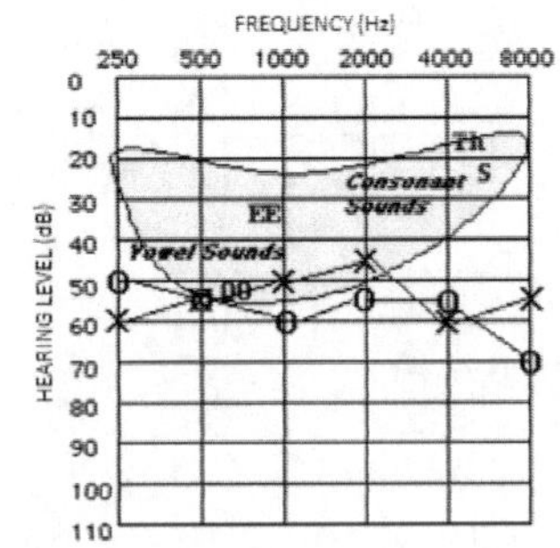

Figure 9. Audiogram of moderate hearing loss.

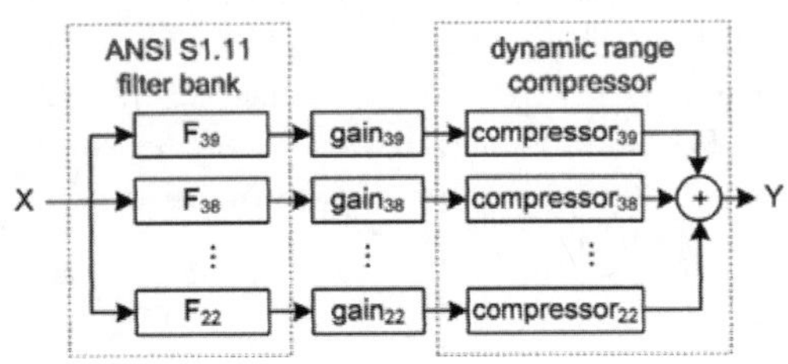

Figure 10. ANSI S1.11 filterbank parallel structure.

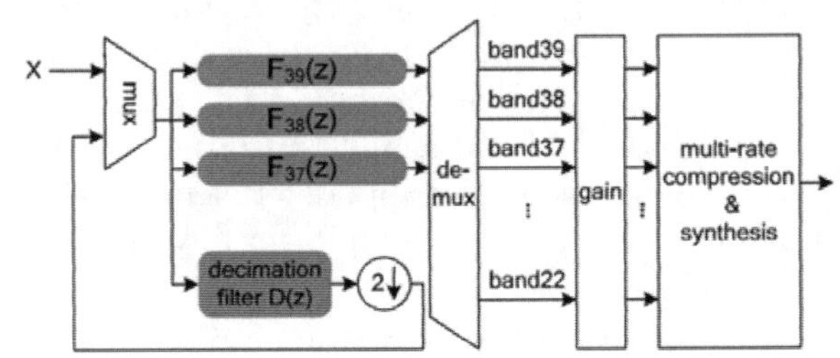

Figure 11. ANSI S1.11 filterbank multirate structure.

dB. Because of individual contrasts, all limits up to 20 dB are considered as would be normal. 20-40 dB is mild hearing loss, 40-70 dB is moderate hearing loss, 70-90 dB is severe hearing loss and above 90 dB is considered as profound hearing loss. Fig 8 shows the audiogram of a person with normal hearing. Fig 9 shows the audiogram of a person with moderate hearing loss in all frequencies. Auditory Compensation or audiogram matching is the process of adjusting the gains of individual bands of the filterbanks based on the patients hearing profile [10].

ANSI S1.11 provides the performance specification for the octave and fractional-octave-band filter bank. The ANSI S1.11 standard characterizes 1/3-octave bands covering the frequency range of 0-20 kHz. Every 1/3-octave band is indicated by its mid-band or central frequency f_m and the bandwidth Δf. The mid-band frequency of the nth band, denoted by $f_m(n)$, is defined in (5).

$$f_m(n) = 2^{\left(\frac{n-30}{3}\right)} \times f_r \tag{5}$$

where, f_r the reference frequency, is set to 1 kHz.

In the ANSI S1.11 standard, three classes of filters are portrayed [11], i.e., class-0, class-1, and class-2. For each class of filters, a few parameters with respect to performance necessities are concerned, which incorporate the relative attenuation, the linear operating range, environmental sensitivities (e.g., humidity and temperature), the maximum output signal, the terminating impedance, etc. The specifications in either class-0 or class-1 filters are more stricter than that of class-2 filters. However, the ANSI S1.11 class-2 1/3-octave band has the comparable stopband attenuation requirement compared with other filter banks for hearing aids. Then again, the sampling rate is set to 24 kHz in order to provide the great sound quality [12].

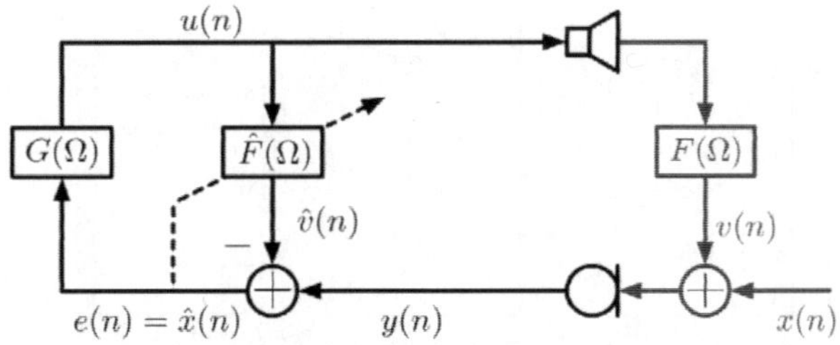

Figure 12. Adaptive feedback cancellation system.

5. Feedback Cancellation

Whenever a sound signal is recorded by a microphone, processed and then instantly played back by a loudspeaker nearby the microphone, it is unavoidable that the loudspeaker signal is fed back to the microphone. The depicted phenomenon is called acoustic feedback and it appears in public address systems and in hearing aids. Acoustic feedback creates a closed signal loop, which, under specific conditions, causes a phenomenon called howling effect. On account of listening devices, the howling effect is extremely irritating for the hearing aid client and his environment [13].

Numerous answers for this issue have been proposed, including phase modulation, gain reduction, filtering in spatial domain and modeling the acoustic environment. The most common approach to deal with acoustic feedback reduction is the adaptive feedback cancellation (AFC) algorithm [14]. The acoustic feedback path is demonstrated by an adaptive filter and subtracted from the input microphone signal. Thus, the AFC algorithm can cancel the acoustic feedback totally when the adaptive filter converges to the perfect solution, i.e., the acoustic feedback path. Fig. 12 displays the block diagram of a typical AFC system.

The idea of the adaptive feedback cancellation (AFC) approach is to assess the feedback path by an adaptive filter and then to subtract the assessed feedback from the recorded signal before it is amplified and played back. Because of the adaptive filter, the technique can adjust quick changes of the feedback path, which is often the case in hearing aid situations. To calculate the adaptive filter coefficients the normalized least mean squared (NLMS) algorithm is used which minimizes the power of the error between the microphone signal and the assessed feedback signal, the error is denoted as $e(n)$.

A idea to conquer this issue is to separate the sound signals into various frequency bands by utilizing a filterbank and adapt each band independently [15]. Since we need to adjust in each band independently, we apply the normalized least mean square (NLMS) algorithm based calculation in each band. The advantage of this method is that we can adapt only in those bands where the feedback is strong. This prompts less adjustment where the correlation occurs. In this way, the bias is decreased, the sound signal is distorted to a lesser extent and the possible gain is increased. The remaining task is to automatically control the adjustment in each band, so that the echo is reduced to a feasible extent, the howling is prevented and the sound quality is preserved for all different scenarios and sound signals [16].

6. Noise Reduction

The most clear impact of noise for individuals with hearing loss, is its effect on their ability to converse. It impacts their capacity to convey each time they endeavor to have a discussion with someone in a noisy place. Every day we are gone up against with wanted and unwanted noise. Wanted noise like music or unwanted noise like a noisy group or background traffic noise. The cause of the noise doesn't make a difference, regardless of whether it is viewed as attractive or bothersome, it can still tamper with a man's capacity to get it [17]. Individuals with hearing

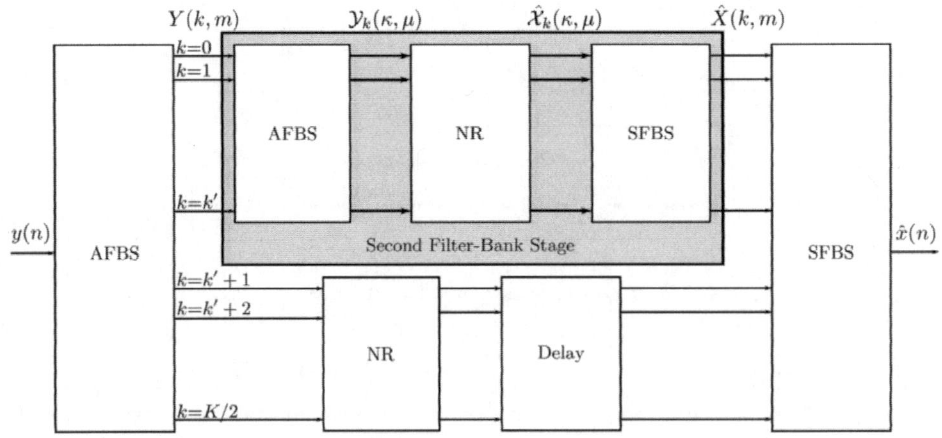

Figure 13. Overview of two-stage filterbank system for noise reduction.

aids need to have the capacity to convey in all circumstances, much the same as individuals without hearing loss. There are various hearing aids includes that work in various different approaches to enhance speech perception in the presence of background noise for a hearing aid client. Directional microphones are one and noise reduction is the other.

A two stage filterbank system is proposed by A. Schasse et. al. [18] for reducing the single channel noise in hearing aids. Fig. 13 demonstrates the above two stage filterbank system. The DFT coefficients processed in the Filterbank System (FBS) can be deciphered as subband signals with time index m and subband index k. In this way, we can easily realize a second stage FBS to perform a further frequency analysis and synthesis of each subband signal. This prompts a short-time frequency-domain representation of the kth subband signal $Y(k, m)$ characterized as $Y_k(\kappa, \mu)$. The indexes μ and κ demonstrate the time allotment and frequency bin within the second filter-bank stage. While applying no decimation in the second stage, the time allotment indexes of the first and second FBS are the same.

The time-domain signal $y(n)$ is decayed into k frequency bands by the primary analysis filter-bank system (AFBS). The secondary stage of filterbank is only applied in lower frequency channels up to the subband index k', while the rest of the bands are postponed w.r.t. the implementation of the second stage. At long last, the time domain signal $\hat{x}(n)$ is the output of the last synthesis filterbank system (SFBS). A noise reduction algorithm, for example, the Wiener filter can be actualized in the second stage. All smoothing parameters, must be adjusted to the new channel bank settings to accomplish a similar measure of reduction in variance. The challenge of this two-stage design, is to accomplish a low overall signal delay [18].

7. Speech Enhancement

In ordinary circumstances, an individual can experience a variety of acoustic situations. For a person with a hearing aid following speech in various kinds of background noise can regularly face a challenge. Thus, assessing the signal-to-noise ratio (SNR) is a key factor to look at the hearing-aid design. Speech enhancement strategies can extract the speech signal in noisy environment. It is accounted that each improvement of $1dB$ in noisy speech can bring about a 10% promotion in clarity of speech [19]. So, when the speech enhancement technique is utilized for a pre-treatment of noisy speech signal preceding to speech recognition, it will enhance the comfort of hearing aid in noisy environment and improve the comprehensibility of speech. Consequently, speech enhancement strategy for enhancing performance of hearing aid systems is desirable.

Speech enhancement technology can separate the speech signal in noisy condition and improve the recognition rate. The hearing aid system using Kalman filter based speech enhancement can increase the rate of speech recognition and the hearing comfort of hearing disabled people in a noisy situation, compared with the hearing aid system using Wiener filter based speech enhancement [20].

8. Conclusion

Various applications of Multirate signal processing techniques and filterbanks in the field of hearing aids are reviewed in this paper. Nowadays, Multirate filterbanks are very commonly used on various parts of the hearing aids. These filterbanks are made reconfigurable to adjust the hearing aid suitable for different patients with individual hearing profiles. The different aspects pertaining to hearing aid systems include auditory compensation, feedback cancellation, noise reduction, speech enhancement etc are mentioned in this review paper.

References

[1] A. R. Moller, *Hearing: Anatomy, Physiology and Disorders of the Auditory System*, 2nd ed. New York, USA: Academic, Sep. 11, 2006.
[2] H. Dillon, *Hearing Aids*, New York: Thieme Medical Publisher, 2001.
[3] J. M. Kates, *Digital Hearing Aids*, Plural Pub., 2008.
[4] S. K. Mitra, *Digital Signal Processing, A Computer-Based Approach*, 2nd ed. Reading, PA: McGraw-Hill Irwin, 2001.
[5] P. P. Vaidyanathan, *Multirate Systems and Filter Banks*, New Jersey: Prentice Hall, 1993.
[6] R. Crochiere and L. Rabiner, *Multirate digital signal processing*, Englewood Cliffs, NJ, USA: Prentice-Hall, 1983.
[7] Christopher Schweitzer, "Development of Digital Hearing Aids", *Trends in Amplification*, Vol. 2, No. 2, 1997.
[8] A. M. Engebretson, "Benefits of digital hearing aids", *IEEE Eng. Med. Biol. Mag.*, Vol. 13, No. 2, pp. 238-248, Apr. 1994.
[9] Yu-Ting Kuo, Tay-Jyi Lin, Yueh-Tai Li, Wei-Han Chang, Chih-Wei Liu and Shuenn-Tsong Young "Design of ANSI S1.11 Filter Bank for Digital Hearing Aids", *14th IEEE International Conference on Electronics, Circuits and Systems*, pp. 242-245, 2007.
[10] Yu-Ting Kuo, Tay-Jyi Lin, Yueh-Tai Li, Chou-Kun Lin and Chih-Wei Liu, "Low-power ANSI S1.11 filter bank for digital hearing aids", *International Conference on Signals and Electronic Systems*, pp. 347-350, 2008.
[11] Yu-Ting Kuo, Tay-Jyi Lin, Yueh-Tai Li, Chou-Kun Lin and Chih-Wei Liu, "Ultra low-power ANSI S1.11 filter bank for digital hearing aids" *Asia and South Pacific Design Automation Conference*, pp. 115-116, 2009.
[12] Yu-Ting Kuo, Tay-Jyi Lin, Yueh-Tai Li, and Chih-Wei Liu "Design and Implementation of Low-Power ANSI S1.11 Filter Bank for Digital Hearing Aids", *IEEE Transactions on Circuits and Systems*, Vol. 57, No. 7, July 2010.
[13] S. Haykin, *Adaptive Filter Theory*, Upper Saddle River, NJ, USA: Prentice-Hall, 2002, Vol. 2.
[14] T. van Waterschoot and M. Moonen, Adaptive feedback cancellation for audio applications, *Signal Process.*, Vol. 89, No. 11, pp. 2185-2201, 2009.
[15] Falco Strasser and Henning Puder, Subband feedback cancellation with variable step sizes for music signals in hearing aids, *in Proc. IEEE Int. Conf. Acoust., Speech, Signal Process. (ICASSP)*, May 2014, pp. 8207-8211.
[16] Falco Strasser and Henning Puder "Adaptive Feedback Cancellation for Realistic Hearing Aid Applications", *IEEE/ACM Transactions on Audio, Speech and Language Processing*, Vol. 23, No. 12, Dec 2015.
[17] A. Schasse, R. Martin, W. Sorgel, T. Pilgrim, and H. Puder, Efficient implementation of single-channel noise reduction for hearing aids using a cascaded filterbank, *in Proc. ITG Symp. Speech Commun.*, Braunschweig, Germany, Sep. 2012, pp. 287-290.
[18] Alexander Schasse, Timo Gerkmann, Rainer Martin, Wolfgang Sorgel, Thomas Pilgrim, and Henning Puder "Two-Stage Filter-Bank System for Improved Single-Channel Noise Reduction in Hearing Aids", *IEEE/ACM Transactions on Audio, Speech and Language Processing*, Vol. 23, No. 2, Feb 2015.
[19] P. Vary, An adaptive filter bank equalizer for speech enhancement, *Signal Process.*, Vol. 86, No. 6, pp. 1206-1214, Jun. 2006.
[20] Zheng Gong and Youshen Xia, "Two Speech Enhancement-Based Hearing Aid Systems and Comparative Study" *5th International Conference on Information Science and Technology (ICIST)*, 2015.

IOP Conf. Series: Materials Science and Engineering **396** (2018) 012037 doi:10.1088/1757-899X/396/1/012037

Compressive Sensing Recovery Algorithms and Applications- A Survey

Sreeharitha S

Dept. of Electronics and Communication Engineering
St. Joseph's College of Engineering And Technology, Palai, Kerala

E-mail: sreeharithaks1993@gmail.com

Sabarinath G

Dept. of Electronics and Communication Engineering
St. Joseph's College of Engineering And Technology Palai, Kerala

E-mail: sabari@sjcetpalai.ac.in

Babita R Jose

Division of Electronics
CUSAT

Abstract. Compressive sensing is an efficient method of acquiring signals or images with minimum number of samples, assuming that the signal is sparse in a certain transform domain. Conventional technique for signal acquisition follows the Shannon's sampling theorem, which requires signals to be sampled at a rate atleast twice the maximum frequency (i.e $f_s \geq 2f_m$). As compared with this traditional acquisition technique, compressive sensing technique captures wide range of signals at a rate significantly lower than Nyquist rate without losing the imperative information, so this technique can be widely used in MRI. Suitable reconstruction algorithms are needed for recovering the original signals from compressed sampled signal. This paper introduces a survey of the various reconstruction algorithms which might enable the use of this technology for wide spread hardware combatiable implementation in the near future.

1. Introduction

The basic concept behind compressive sensing (CS) is to directly collect the signal in compressed form Figure 1. According to the Shannon/Nyquist sampling theorem, analogue signal can be recovered perfectly, if it can be sampled at a rate at least twice the maximum frequency component present in the signal (Nyquist 1928; Shannon 1949). CS builds on the work of Candes, Romberg, and Tao [1] and Donoho [2], were possibly the first set of researchers who had mathematically shown that a sparse signal can be recovered from a few set of measurements. Since there were only a few set of measurements, these researchers termed it as compressive sensing. It offers the possibility to capture less data then is commonly done, but still to be able to reconstruct the entire information. This can also appear as a consequence of omitting samples that are exposed to different kinds of noise or losing some parts of the signal during the transmission. These omitting samples recovered by using suitable CS reconstruction algorithms.

IOP Conf. Series: Materials Science and Engineering **396** (2018) 012037 doi:10.1088/1757-899X/396/1/012037

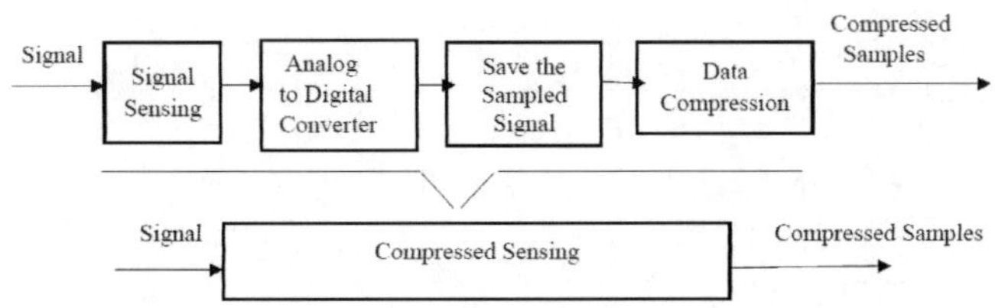

Figure 1. Traditional Vs Compressed Sensing

2. Compressive Sensing Essentials

Nearly all the digital camera records every pixel in an image, which is immediately discarded for reducing the storage space for saving an image. A natural question asks why we need to acquire this abundance of data, just to throw most of it away immediately. This notation sparked the theory of compressive sensing. As an alternative to the traditional sampling theory, compressive sensing approach provides grate quality to the signal without increasing the amount of data required for reconstruction. Thus CS specifically important in applications related to medical signal processing. The important feature of Compressed Sensing is to avoid compression after acquisition and to directly acquire data in the compressed form. To make this possible we should know about two terms, Sparsity and Incoherence.

2.1. Sparsity

Sparsity is an important term used in compressed sensing. In particular most of the signals are sparse i.e most values are zero and only a few contain essential data when represented in certain domain. Figure. 2, shows sparsity in time domain. Figure 2 (a) the actual signal that has 256 samples with only 5 non zeros, which is very sparse. Figure 2 (b) is the number of measurements to be taken, here we take only 64 random measurements. Figure 2 (c) is the recovery result by taking only 64 random measurements, which specify that the actual signal can be recovered from few set of measurements. In the frequency domain, it is completely shown by two peaks. In, Figure 2, the time domain signals was already in its sparse domain, hence representation basis ψ was simply the identity matrix I_n. However in Figure 3, ψ transforms a time domain signal in to the frequency domain, ie the DFT matrix Figure 3 (b). Figure 3 (c) shows the number of measurements to be taken, here we take only 64 random measurements. The original time domain signal had 256 values, by taking only 64 random measurements from it we can recover all 256 original values Figure 3 (c). This differ from the example sparse in time domain, is that the signal of interest was not in the time domain, but in the frequency domain.

2.2. Incoherence

Compressed sensing is based on a choice of a sensing basis ϕ relative to representation basis ψ, which use an incoherence property μ for measure the correlation between vectors from each basis. The minimum correlation between any two elements of two different matrices is called coherence, it is denoted by $\mu(\phi,\psi)$. We can reconstruct signals with fewer measurements if the value of incoherence is to small [3].

3. Reconstruction Algorithms

The important feature of compressive sensing is that it requires an efficient reconstruction algorithms. The reconstruction of compressed sampled signal involves solution of an

IOP Conf. Series: Materials Science and Engineering **396** (2018) 012037 doi:10.1088/1757-899X/396/1/012037

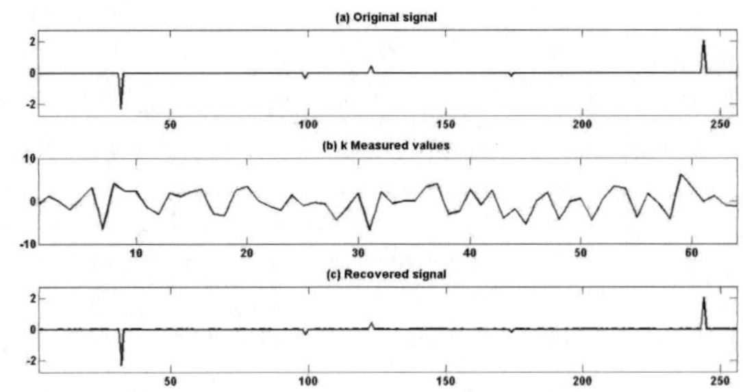

Figure 2. Sparsity in time domain

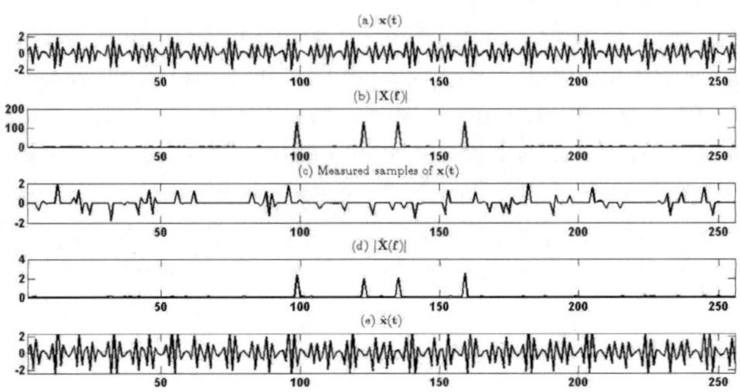

Figure 3. Sparsity in frequency domain

underdetermined system of linear equations and therefore has infinitely many solutions. The basic sensing equation is given by

$$y = Ax \tag{1}$$

where, y is the measured values, A is the sensing matrix, x is the original Signal. Reconstruction requires the solution of equation for x. Direct solution of the equation never yields a unique solution even with the prior knowledge of the sparseness of the solution. The inverse equation is given by

$$\hat{x} = (A^T A)^{-1} A^T \tag{2}$$

The signal reconstruction process is essentially to choose the best estimate of the original signal from all the possible solutions obtained from the above inverse equation. This may be achieved by the convex optimization algorithm. Various reconstruction algorithms used for reconstruction of compressed sampled signal may be classified as shown in Figure 4.

3.1. Convex Relaxation

Convex Relaxation is the class of algorithms which can solve reconstruction issues through linear programming. Basis Pursuit, Basis Pursuit De-Noising, Least Absolute Shrinkage and Selection Operator(LASSO) are some examples of such algorithms.

Basis pursuit is the mathematical optimization of a problem in the form of

$$min||x||_1 \qquad subject\ to\ \ y = Ax \tag{3}$$

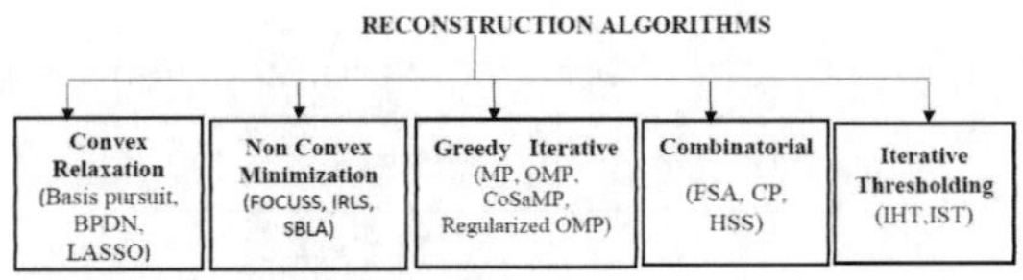

Figure 4. Reconstruction algorithms

These types are mainly used in cases where there is an underdetermined signal of linear equation of $y = Ax$. This algorithm is more complex and time consuming, so it cannot be applicable in time critical reconstruction applications. Figure. 5, represents reconstruction of a input signal of length 256 by using basis pursuit algorithm. In this experiment we take 64 samples, for reconstructing signal of length 256 with sparsity 8. From the experiment we can conclude that when sparsity level increases accuracy decreases and will get approximately accurate signal. The signal reconstruction quality is analysed via the objective test method using some methods including Signal-to-Noise Ratio (SNR), Relative Error (RE)etc..as shown in Table 1 and Table 2. From Table 2 , M>=K*log(N/K) we can see that greater the quality of signal other wise signal quality will be decreased.

In 1996, Tibshirani [4] introduced the method Least Absolute Shirnkage And Selection Operator(LAASO). It is a commonly used technique in sparse modeling, originally developed in the field of statistics. As in the case for compressive sensing, LASSO also utilizes $l1$ norm, but the solution x is obtained as

$$x_{LASSO} = argmin||y - Ax||^2{}_2 + \mu||X||_1 \tag{4}$$

Where, $\mu > 0$ is a fixed regularization parameter. The first term in the argument of the right-hand side describes goodness of fit, which is commonly used in least-squares fitting, and the second term is the penalty term based on the $l1$-norm. The regularization parameter μ adjusts degree of sparsity by changing the weight of the $l1-$norm penalty.

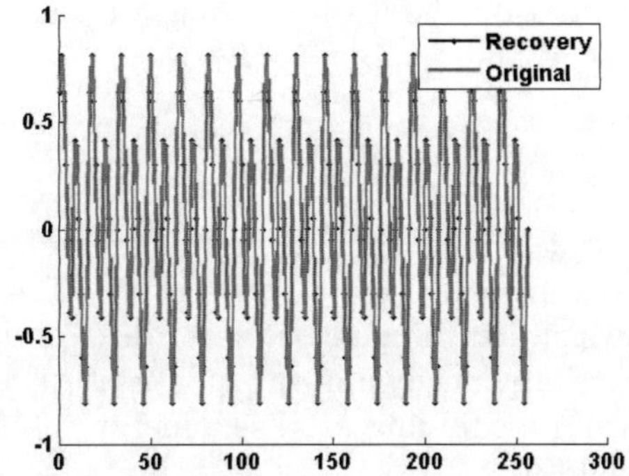

Figure 5. Reconstructed output data using BP

For instance, a large μ prefers a solution with very few non-zero components, yet $\mu = 0$ provide no sparsity. Optimization of μ in LASSO is a major problem. Information criteria or cross

IOP Conf. Series: Materials Science and Engineering **396** (2018) 012037 doi:10.1088/1757-899X/396/1/012037

Table 1. Sparsity level Vs Error rate for reconstruction using Compressive Sensing

Sr. No	Sparsity	Error
1	10	6.2780e-14
2	20	8.1502e-14
3	30	8.7132e-14
4	40	3.3142
5	50	21.7598

Table 2. Number of measurements Vs SNR for reconstruction using Compressive Sensing N=256, Sparsity(K)=10

Sr. No	Measurement Number(M)	SNR
1	512	6.0206
2	64	6.0206
3	20	5.6047
4	15	-2.9872
5	10	-0.4331

validation methods can be used for this optimization. Another method is to adjust μ, based on the estimate of the sparsity, which might be obtained from some a priori information or theoretical predictions. Optimized value of μ is dependent of characteristics of problems to be solved (i.e., the shapes of observation matrix A and also image x). LASSO yields a sparse solution of which the parameter vector μ has nearly few nonzero coefficients. In contrast, the l_2 regularised least squares generally has all coefficients nonzero. Because it actively reduces the number of variables, the lasso is useful in some situation. The basic theory behind this algorithm is to shrinks some coefficients and hence tries to retain the good feature of both subset selection and ridge regression. Amusing work on the lasso is being carried out in many fields, including statistics, engineering, mathematics and computer science.

3.2. Non Convex Minimization Algorithm

Replacing l^1 norm in equation (1) with the l^p quasi-norm ($0 < \mathrm{p} < 1$), for recover x from fewer linear measurements. ie, $||x||_p$ subject to y=Ax. In practical case convex optimization problem can be solvable using Non convex minimization algorithms [5]. Focal Underdetermined System Solution (FOCUSS), Sparse Baysian Learning algorithms are some examples of such algorithms. In medical imaging tomography Non-Convex optimization used. Focal underdetermined system solution(FOCUSS) is an algorithm for finding localised energy solution from limited data. The inverse problem with sparse solutions.

$$y = Ax + \nu \tag{5}$$

Where y is the observed data, A $\in R^{M\times N}$ is the generating matrix, x is the sparse solution and ν is the additive noise. Sparse vector selection and a dictionary learning step are the main steps involved in this algorithm. FOCUSS also used to solve for sparse solution of linear inverse problems when A is known and perform the vector selection step of the algorithm. The dictionary learning step is performed by gradient descent. Sparse Bayesian Learning (SBL) is an efficient technique to perform sparse processing, which is build by a Bayesian framework,

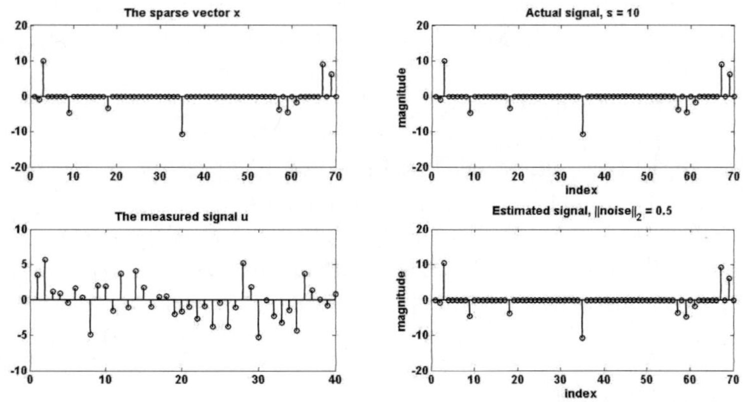

Figure 6. Reconstruction of signal using OMP Algorithm

approximately solves a non-convex optimization problem using fixed point updates. Sparse Baysian Learning algorithms was introduced for regression and classification problems in the context of machine learning it has been used since for signal processing with various modifications and extensions. Since SBL does not impose explicitly any sparsity constraints but determines sparsity automatically.

3.3. Greedy Iterative Algorithm
The greedy algorithms are based on finding the elements of the transform matrix called dictionary that best matches the signal through iterations. These algorithms are less computationally complex and therefore much faster compared to the l_1-norm based optimization techniques, but are also less precise. Matching pursuit(MP), Orthogonal Matching Pursuit(OMP) and Compressive Sampling Matching Pursuit(CoSaMP) are the commonly used greedy iterative algorithms[9]. The matching pursuit (MP) is a sort of exhaustive search that attempts to select in a greedy fashion the subset columns (the support-set that identified the locations of the non-zero elements in a sparse signal-vector) of the dictionary that match the measurements the most. The columns with the biggest inner product with the signal are selected first. The method then optimises the selection by least squares in every iteration, subtracting from the signal an approximation of the selected columns, until the signal is well decomposed (the residual is less than a defined threshold).

MP was extended to orthogonal MP (OMP) of higher computational cost and more accurate results. OMP calculates in every iteration the projection of the signal on all dictionary columns selected, and updates the coefficients. OMP works by finding a basis vector in ϕ that maximizes the correlation with the residual (initialized to y), and then recomputing the residual and coefficients by projecting the residual on all atoms in the dictionary using existing coefficients. OMP has the major advantage that, each column vector of the sensing matrix picked up once wont be picked up in the next iteration. This is achieved by maintaining an atom index. which is updated on each iteration. As, a result OMP produces an estimate of the sparse signal in m iterations. The case $Ax = b$ is considered. Both A and b are created with random values. Using OMP, a sparse x is found. Figure.6. shows the result.

Regularised OMP is an iterative algorithm which is also based on OMP with some differences. The algorithm selects many vectors at each iteration not like OMP, which at each step only selects one vector. CoSaMP is the latest algorithm developed based on OMP. The performance of CoSaMP is not accurate as OMP algorithms, but using CoSaMP recovery is extremely fast.

Without the noise, the recovery resuts by OMP are much more accurate than the results by CoSaMP. The advantage with CoSaMP is that it works well when the samples are disrupted with noise.

3.4. Combinatorial / Sublinear Algorithms

By this method, acquire highly structured samples of the signal and reconstruction by group testing [8]. This class include Fourier sampling uses random(but structured) time samples to recover signals that are compressible with respect to the discrete Fourier basis. Fourier Sampling Algorithm, Chaining Pursuit Heavy Hitters on Steroids (HHS) are the commonly used combinatorial algorithms. These are fast and efficient as compared to convex relaxation or greedy algorithms.

3.5. Iterative Thresholding Algorithms

For this class of algorithms, correct measurements are recovered by soft or hard thresholding. Iterative Hard Thresholding is a class of low computational algorithms, which has been proposed for reconstruction of sparse signals , starting from x^0=0, IHT iteratively finds the sparset solution of y=Ax using the following iteration,

$$x^n + 1 = H_k(x^n + \phi^T(y - \phi x^n)) \tag{6}$$

Where H_k is a nonlinear operator that sets all but the k largest (in magnitude) coefficients of its input vector to Zero. It can be shown that if $||\phi||_2$ <1 the guaranteed to coverage to the local minimum of the cost function $||y{-}\phi x||_2$ under the constraint that x is k sparse.

Variations of the IHT includes Message Passing[11], Expander Matching Pursuit, Sparse Matching Pursuits. Sequential Sparse Matching Pursuits, and Belief Propagation. Variations of the IST include fast iterative soft thresholding (FISTA) [10].

4. APPLICATIONS

4.1. Medical Imaging

CS reduces the sampling rate without losing the useful information, so this technique can be used in MRI. MRI uses magnetic fields to excite hydrogen atoms in the body, which result in measurements of the Fourier transform of the image. The number of measurements to be captured is proportional to the scan duration. Longer scan are unpleasant for patients and also more prone to motion artifacts and impossible for certain parts of the body (example. those affected by breathing). However, the introduction of CS based techniques provide great quality to the image and reduce the scan duration through reduction in the number of collected measurements[11].

4.2. Radar

The main function of Radar system is to target detection and range determination, also it's function extends to air and traffic control, air defence systems, antimissile systems etc. In radar signal processing in order to determine the target location, large bandwidth signals needs to be launched, so it requires a very high signal sampling rate and system calculations. Radar echo signal is highly sparse, so compressed sensing theory can be used to reduce the sampling rate and in the lower sampling rate, target parameter can be estimated from radar echo. Compressed sensing enables the achievement of better target resolution with minimum number of samples.

4.3. CS in Cameras

The single- pixel camera developed at Rice University is a one of the most important application of compressed sensing [12]. Single pixel camera mainly consists of a Digital Micro mirror Device

IOP Conf. Series: Materials Science and Engineering **396** (2018) 012037 doi:10.1088/1757-899X/396/1/012037

(DMD), two lenses, a photon detector to sense data and an electronic switching device to flip the mirrors at high speed. Each mirror in Digital Micromirror Device (DMD) array performs one of these two tasks: either reflect light towards the sensor or reflect light away from it. Therefore, light received at sensor (photodiode) end is weighted average of many different pixels, whose combination gives a single pixel. A stranded digital camera uses a massive CCD arrays to collect millions of pixels, which are typically compressed to reduce their cost for storage and transmission. For example, a 20 mega pixel camera measure 20,000,000 pixels using a CCD or CMOS light sensors. On the other hand, the single pixel camera, uses one light sensor to measure the entire image, compress the image data before the pixels are recorded. As a result it's able to capture an image with minimum number of samples.

5. Conclusions

A review of the compressed sensing is discussed, which indicates the proper selection of sparse representation of the signals, selection of measurement matrix for acquisition, algorithms for reconstruction and also discussed compressive sensing applications. Here we can see that, algorithms used for the signal reconstruction focus to find the solution of an underdetermined system of linear equations using sparseness constraints. Generally most of the natural signals are sparse or sparse under some domain. If we can find the location of those non-zero entries, we can reconstruct the signal uniquely. The traditional process of signal compression is quite costly, because this acquisition system acquire all samples of the original signal and a significant portion of which is immediately discarded. The new idea of signal compression combines signal acquisition and compression as one step, which reduces the overall cost significantly. As compared with the traditional Nyquist-Shannon sampling theory, CS afford grate quality to the signals without increasing the quantity of required data , which means that the original signal can be reconstructed with minimum number of samples ($f_s < 2f_m$). This paper broadly speaking the survey of various algorithms for reconstructing compressed sampled signals and compares their complexity. In future work, test various algorithms to reconstruct compressed sensed images can be done.

References

[1] E. J. Candes, J. Romberg, T. Tao, Feb. 2006, *Robust uncertainty principles: exact signal reconstruction from highly incomplete frequency information*, IEEE Transactions on Information Theory, vol. 52, no. 2, pp. 489-509.

[2] D. Donoho, 2006, *Compressed sensing* IEEE Transactions on IT, vol. 52, no.4, , pp. 1289 - 1306.

[3] E. Candes and J. Romberg, 2007, *Sparsity and incoherence in compressive sampling*, Inverse Problems, vol. 23, p. 969.

[4] R. Tibshirani, 1996, *Regression shrinkage and selection via the Lasso* vol.11, pp. 125-256

[5] R. Chartrand, 2007, *Exact reconstruction of sparse signals via nonconvex minimization* IEEE Signal Process. vol. 14, pp. 707-710.

[6] A. Maleki, 2009, *Coherence analysis of iterative thresholding algorithms,* in Communication, Control, and Computing, pp. 236-243, IEEE, 2009.

[7] Donoho D. L., Maleki, A., 2009, *Message Passing Algorithms for Compressed Sensing.*

[8] S. Muthukrishnan, March. 2006, *Combinatorial algorithms for Compressed Sensing.* In Proc. 40th Ann. Conf. Information Sciences and Systems, Princeton.

[9] S. Budhiraja, 2015, *A Survey of Compressive Sensing Based Greedy Pursuit Reconstruction Algorithms*, vol. 10, no. 10, pp. 1-10.

[10] A. Fiandrotti, S. M. Fosson, 2013, *Parallel Iterative Soft Thresholding algorithm for sparse image recovery,* pp. 325-328.

[11] U. Gamper, P. Boesiger, 2008, *Compressed sensing in dynamic MRI,* vol. 59, no. 2, pp. 365-373.

[12] M. Duarte, M. Davenport, and R. Baraniuk, 2008, *Single-pixel imaging via compressive sampling,* IEEE Signal Processing Magazine, vol. 2, no. 25, pp. 83-91.

IOP Conf. Series: Materials Science and Engineering **396** (2018) 012038 doi:10.1088/1757-899X/396/1/012038

A Review on White Blood Cells Segmentation

NIMMY THOMAS
M.Tech Student, Electronics and Communication Engineering
St. Joseph's College of Engineering And Technology, Palai, Kerala
E-mail: nimmyt007@gmail.com

SREEJITH V.
Assistant Professor Dept. ECE
St. Joseph's College of Engineering And Technology Palai, Kerala
E-mail: sreejithvasu@gmail.com

Abstract. Currently blood smear's evaluation is important in the diagnosis of different diseases. Usually,the hematologists are curious on diagnosing white blood cells (WBCs) only. White Blood Cell (WBCs) detection is a vital evaluation method. In this paper,a review on some of the common segmentation methods are discussed. It includes K- mean segmentation,Otsu thresholding and Color based segmentation. This review also helps to get more details about the different types of segmentation and classification process.

1. Introduction

Blood is a life-giving fluid that transports oxygen and gives energy to the cells and carries away carbon dioxide and other waste products. Blood circulating in our body consist of 55 percent plasma, 40 percent Red Blood Cell(RBC), 4 percent of Platlets and 1 percent White Blood Cells(WBC).WBCs are important for healthy immune system.Neutrophils, eosinophils (acidophilus), basophils, lymphocytes, and monocytes are the five types of WBCs. All these cells have nuclei, which differentiate them from the other blood cells. The normal white cell count is between $4\times 10^9/L$ and $11\times 10^9/L$. Abnormalities in blood cells are determined by blood smear test. Irregularity in blood cells result in variations in the number of WBCs. The two types of blood count needed for the diagnosis of blood are Complete Blood count (CBC) and the Differential Blood count(DBC). The most commonly performed hematologic test is Complete Blood count (CBC) and the Differential Blood count(DBC).

Detection of the WBCs efficiently diagnosis various diseases. Basically it consist of four steps: preprocessing stage, image segmentation of image, feature extraction and classification as shown in figure 2. The precision of the extracted feature and classification depends on segmentation of WBCs.

2. A review on different Segmentation Techinques

Figure 2 shows the simple block diagram of a WBCs detection system. It includes various functional modules. The input image of blood smear is fed to the system. It consist of four

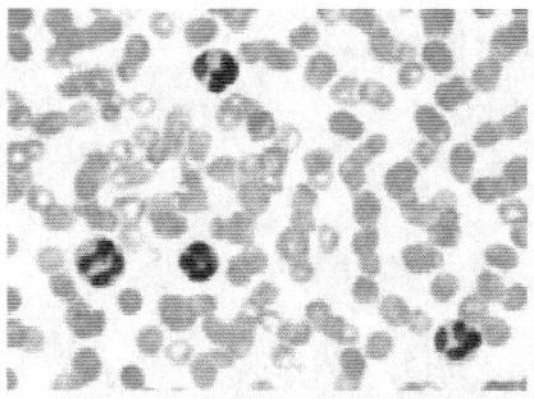

Figure 1. Blood Smear Microscopic Image

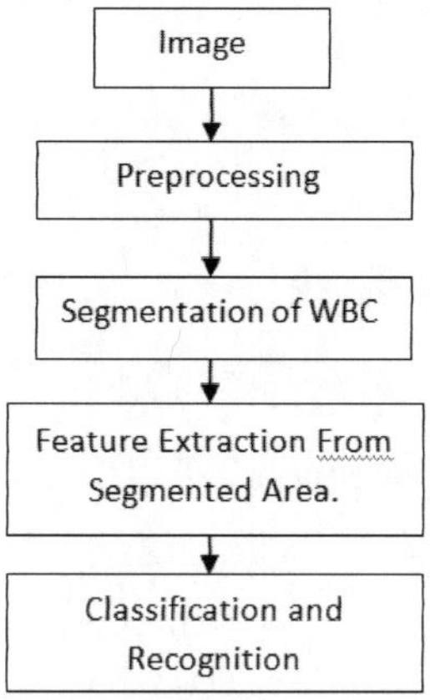

Figure 2. Common Block Diagram for WBC detection

stages: Preprocessing stage, image segmentation, feature extraction and classification. The precision of detection depends on different segmentation process.

2.1. Preprocessing Stage
The pre-processing is a significant stage in image processing in order to improve the image data that avoid unwanted distortions and highlight important blood smear image feature for next processing. Operations like image enhancement,image contrasting, image sharpening and noise filtering gray level conversion are performed in images for better results. In image enchancement process operations like Median and Max/Min filtering, image subtraction, Histogram Equalization, Image smoothing,Neighbourhood Averaging,Image sharpening. The visual appearance and contrast of an image can be improved by histogram equalization. The aim of image smoothing is to diminish the effects of camera noise within an image, spurious pixel and missing pixel values. Image sharpening is to overcome blurring produced by blood smear image and to enhance fine details in the image.

2.2. Segmentation Of WBCs.
Segmentation means dividing the image into different part for various applications. Because of The complex nature of the blood cells and uncertainty in the microscopic blood smear image, WBCs segmentation hard and challenging. Image segmentation methods inlcude threshold-based, edge-based, region-based or clustering methods are used.

2.2.1. K-means clustering is one of the easiest unsupervised learning algorithms based on distance. This algorithm takes distance as calculating parameter of dissimilarity. It divides

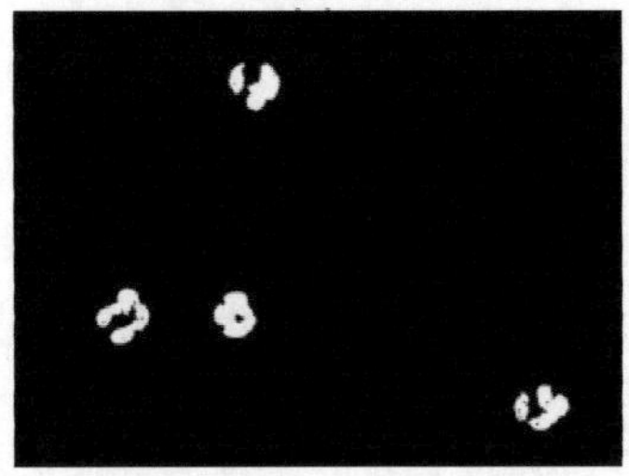

Figure 3. Segmented WBCs using K Mean.

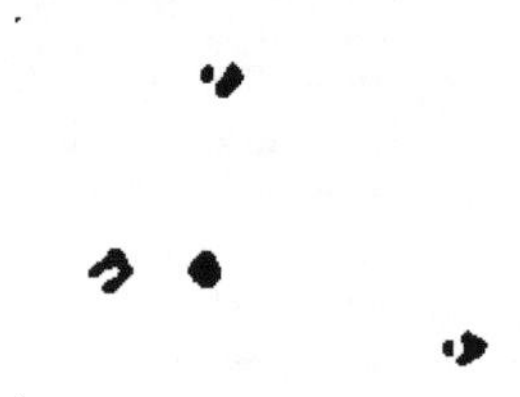

Figure 4. Segmented WBC using Otsu

p objects into k clusters in which each object is a part of the cluster with the closest mean. In WBCs detection, blood smear image clustering is done based on the color information of RBCs, WBCs and plasma. This method gives absolutely k different clusters of strong possible distinction. The aim of K-Means clustering is to minimize total the squared euclidean distance:

$$\sum_{i=1}^{m}\sum_{j=1}^{n}||x_i - c_j||^2$$

c is the number of cluster centers.

c_i is the number of data value in *ith* cluster.

2.2.2. Otsu Thesholding Otsu method is commonly used as a result of its easy calculation. Otsu segmentation is threshold based segmentation. It finds a measure of spread for the pixel levels that either fall in object or background via iterating all the feasible threshold values. The aim is to keep clusters as tight as possible by finding a threshold value that minimize variance within the class. This algorithm gives good result in WBCs segmentation since the object area is small compare to background area.

2.3. Feature Extraction

Choosing the best diagnostic feature is almost important task in WBCs detection.This is a process of redefining a large set of similar data into a set of features (or feature vector).For detecting WBS cells several features have been observed viz,area, perimeter and circularity, convex areas, solidity, major axis length, orientation, filled area, eccentricity because structure of the nucleus is relevant feature for differentiation of cells. All the extracted features are choosen from the binary converted image of the nucleus with pixels of non-zero value. The area of blood cell within the image region is calculated from the total number of non-zero pixels.

Perimeter is determined by finding distance between adjacent boundary pixels. Circularity changes with surface irregularities and is defined as,

Circularity = 4* Pi* Area/ Perimeter

2.4. Classification

Classification of WBCs depend on characteristics of its cytoplasm and nucleus. The feature extracted will effect the classifer performance. Classification methods used include K nearest neighbor (kNN),Learning Vector Quantization, MultiLayer Perceptron, and Support Vector Machine.kNN classifier is used, where the pixel is classified according to the majority vote of the k closest training data. The kNN classifier does not make underlying conclusion about the statistical structure of the data because it is a non parametric technique.

3. Conclusions

A review of various WBCs detection methods have been discussed in this paper.Most of the methods are image enhancement and noise filtering. Feature extraction has an important role in the process because it effect the result. Segmentation and classification depend on extracted features. Identifying each type of white blood cell has a challenging part because of its irregularity in shapes. So this review helps to get more details about the different types of segmentation and classification process.

Acknowledgments

I would like to show my sincere regards to my supervisor, Prof. Sreejith V. , whose guidance made this work possible. His knowledge and encouragement has invaluable help in producing this work.

References

[1] Piuri V, Scotti F., Feb. 2004, *Blood leukocytes classification by microscopic images.* information, IEEE International Conference on Intelligence for Measurement Applications.

[2] Nisha Ramesh, Bryan Dangott, 2012 *Isolation and two-step classification of white blood cells in blood smears.* IEEE Transactions on IT, vol. 52, no.4, , pp. 1289 - 1306.

[3] D T Pham ,*Selection of K in K-means clustering*,Manufacturing Engineering Centre, vol. 14, no. 5-6, pp 629-654, , doi:10.1007/s00041-008-9035-z

[4] Congcong Zhang, 2013 *White Blood Cell Segmentation using Color Space Based K-Means* Sensors 2014, 14, 16128-16147; doi:10.3390/s140916128

[5] Al-Daoud, M. B., 2007 *Fast K-means clustering algorithms*, University of Leeds, vol. 23, p. 969,

[6] Putzu, L.; Di Ruberto, C., 1996 *White blood cells identification from microscopic blood images.*Biomedical Engineering, Guangzhou, China, 1 November 2013

[7] Saraswat, Arya, 2007 *Automated microscopic image analysis for WBCs identification: A survey* Lett., vol. 14, pp. 707710, .

[8] Huang, D.C.; Hung, *Leukocyte nucleus segmentation and recognition in color blood-smear images.*IEEE International Instrumentation and Measurement Technology Conference (I2MTC),13 May 2012; pp. 171176.

IOP Conf. Series: Materials Science and Engineering **396** (2018) 012039 doi:10.1088/1757-899X/396/1/012039

A Simple and Robust Strategy for Background Removal from Brain MR Images

Rameesa Mol J, **Rashmi A G** , **Jiss Mohan K** , **Alsaheer S.S** , **Mable A** and **Justin J**
School of Electronics, St Joseph's college of engineering and technology, Pala, Kerala

E-mail: rashmi.ann21@gmail.com

Abstract. This paper proposes a simple and robust method for elimination of noisy background region from the brain Magnetic Resonance Images. Background beneath the morphological structures presents strong pixel intensity inhomogeneity. This pixel intensity inhomogeneity may get preserved during non-linear edge sensitive smoothening and further may get enhanced during contrast enhancement. This inhomogeneous and noisy background may further hamper the accuracy of skull stripping and segmentation of morphological regions of interest. The skull stripping becomes easier if background could be eliminated prior to it. Background elimination is accomplished by multiplying the raw MR image with a multiplication mask synthesized by gradient based thresholding and a series of morphological operations. Qualitative evaluation of the results confirms the efficiency of the proposed method.

1. Introduction

In Magnetic Resonance (MR) images the background region, behind the morphological regions of interest may be noisy and usually exhibit very strong intensity inhomogeneity. In some areas of the brain MRI, these inhomogeneity forms edges equally strong as the edges in morphological region of interest as visible in fig. 1. Even though the background region gets eliminated during skull stripping, the presence of this noisy background makes the skull stripping more challenging and even questions the success of gradient or edge based approaches. Moreover, the background edges may get preserved during edge preserving noise restoration like bilateral, anisotropic diffusion filtering and non-local means filtering. Further, this noisy background will also be enhanced during contrast enhancement. This may hamper the segmentation accuracy of region of Interest (ROI). Even though the problem of noisy background looks primitive, the literature addressing this issue is limited. Kalicka *et. al* [1] presented a method of removing the noisy background from the sequence of magnetic resonance imaging (MRI) scans, using a mask which was obtained by applying morphological operations of dilation, erosion, opening and closing to the binary image generated from MRI by intensity threshold. The results of removing the background from scans with the use of different shapes and sizes of structuring elements are presented. Mohsin *et. al* [2] is an attempt to improve the efficiency of stripping skull in MRI, using systematic application of "Erosion" with AOI (Area of Interest) approach after the detection of false-background. Before applying "Erosion", false back ground is detected. The skull boundary was identified through Dilation and then used scan. line algorithm to fill the false background area. Consequently "Erosion" algorithm erodes the AOI, resulting in the stripping of skull.

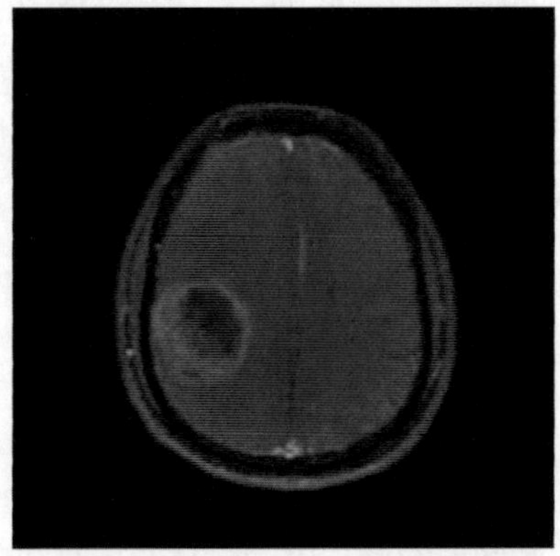

Fig. 1: Axial plane T1 contrast brain MRI

(Image Courtesy: Hind Labs, Kotayam Medical College Kerala)

This paper describes an efficient strategy for removing noisy background from brain MRI. In the forthcoming discussions, mathematical basis of gradient based thresholding as well as gradient estimation are incorporated, followed by the particulars of morphological operations and the corresponding structuring elements, used to generate the multiplication mask used in background removal. The outcome of each stage of morphological operation is also demonstrated.

2. Materials and Methods

Axial Plane T1 contrast enhanced (Series: AX T1 SE FS+C, Spin Echo Sequence (SE)) MR images (courtesy: Hind Labs, Govt. Medical College Kottayam, Kerala) were selected for the experimental evaluation of the proposed background method. The specification of MR equipment is; Manufacturer: GE Medical Systems, Model Name: Signa HDxt, Acquisition Type: 2D and 1.5T field strength. Experimental evaluation of background elimination was performed on Matlab®, (Version: 7.12.0.635 (R2011a)) Image Processing Tool Box.

Background elimination was accomplished by element-wise multiplying the original MR image with a multiplication mask, which was constructed by a gradient based thresholding followed by a series of morphological operations on the binary edge image. These morphological operations include dilation, hole filling, border clearing and erosion. Gradient was computed with Sobel mask. The principle of gradient based threshold is [3],

$$g_x = \frac{\partial f(x,y)}{\partial x} \quad (1)$$

$$g_y = \frac{\partial f(x,y)}{\partial y} \quad (2)$$

$$G(x,y) = \sqrt{g_x^2 + g_y^2} \quad (3)$$

$$\hat{f}(x,y) = \begin{cases} 1, & \text{if } G(x,y) \geq G_T \\ 0, & \text{else} \end{cases} \quad (4)$$

$\hat{f}(x,y)$ is the edge image generated from gradient based threshold. G_T is the adaptive gradient threshold and G(x,y) is the gradient magnitude. Gradient is estimated from gradients along x and y directions, say, g_x and g_y respectively. For dilation of the traced edges, horizontal and vertical

IOP Conf. Series: Materials Science and Engineering **396** (2018) 012039 doi:10.1088/1757-899X/396/1/012039

structuring elements or strel objects with three neighbours or length 3 were used. Erosion was performed with a diamond strel object having five neighbours or with a length one. The Sobel discrete derivative operator along both x and y directions and structuring elements used for erosion and dilation are furnished in fig. 2-fig. 5, respectively

-1	-2	-1
0	0	0
+1	+2	+1

Fig.2 : Sobel derivative operator along x-direction

-1	0	+1
-2	0	+2
-1	0	+1

Fig. 3: Sobel derivative operator along y-direction

	1	
1	1	1
	1	

Fig.4 : Diamond structuring element for erosion

1
1
1

1 1 1

Fig.5 : Horizontal and vertical flat structuring elements for dilation

3. Results

Fig. 6 shows the output images at intermediate stages of generation of multiplication mask, raw MR image, carrying noisy background and the same image after background removal. It is obvious from the gradient image that the edges present in the noisy background are equally strong as that in morphological structures. A reduced gradient threshold allows the detection of more edge content. The dilation interconnects and links the edges in the binary image, from gradient based thresholding. Completely connected edges are a mandatory requisite for the hole-filling to be effective. Border clearing eliminates the boundary '1's in the multiplication mask and '1's in its four neighborhoods and smoothen the irregularity at the outer boundary of binary multiplication mask. The erosion excludes the edges present in the noisy background.

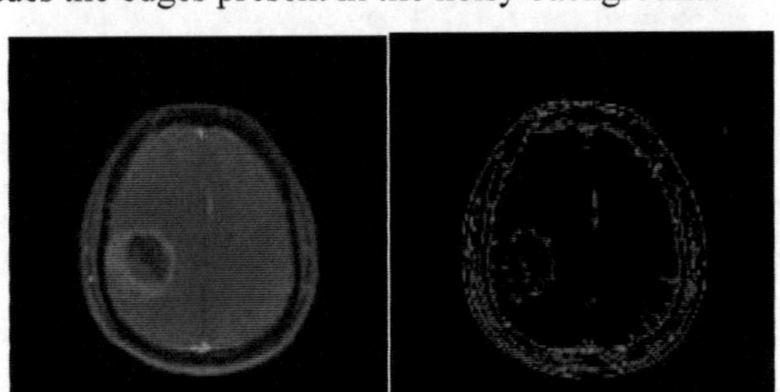

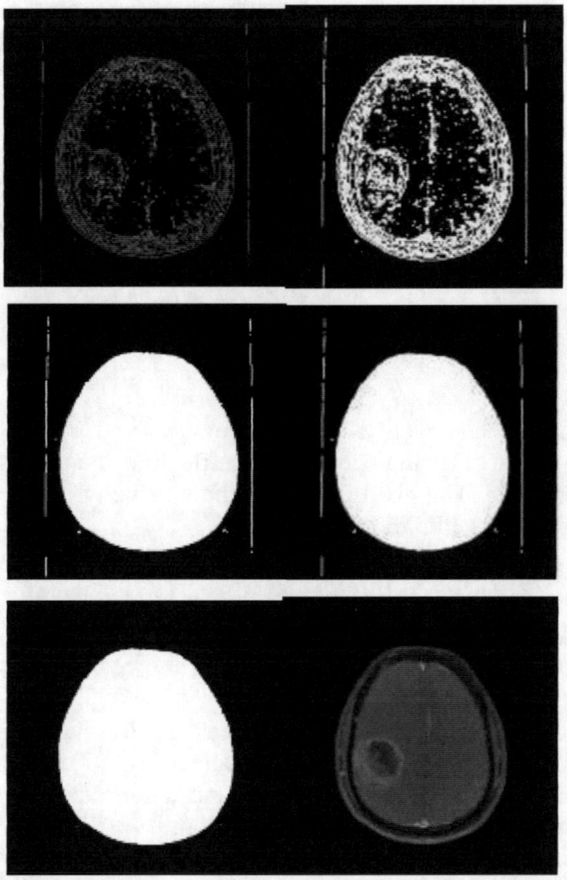

Fig. 6: (a) Raw MR image (b) Binary gradient mask with higher threshold (c) Binary gradient mask with lower threshold (d) Binary gradient mask after dilation (e) Binary gradient mask after hole filling (f) Binary gradient mask after border clearing (g) Multiplication mask after erosion (h) Image after background elimination

(Image Courtesy: Hind Labs, Kotayam Medical College Kerala)

Fig. 7 confirms the efficacy of the proposed background elimination strategy on multiple test MR images. The strategy performs well on the entire test images equally.

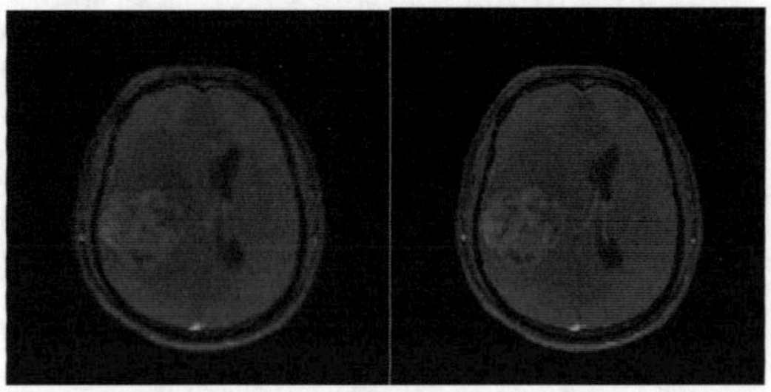

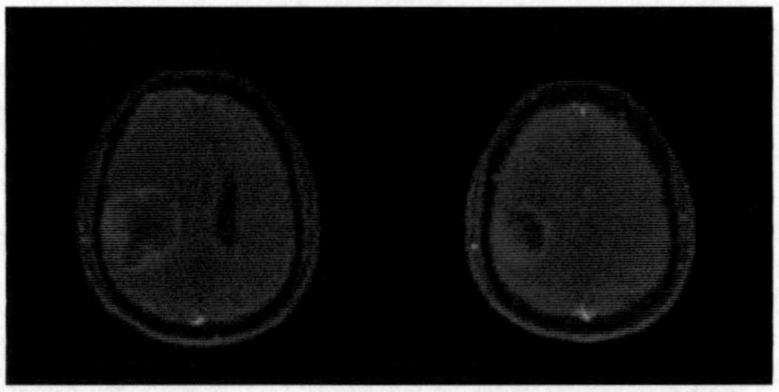

Fig. 7: Raw MR images and image after background elimination.

4. Conclusion

The proposed strategy has been proven to be a viable solution for removing the noisy background from the MR images. The skull stripping becomes more flexible and allows wide choice of methods as noisy background is eliminated. The pixel intensity inhomogeneity in the noisy background does not get preserved or enhanced as it is eliminated prior to it.

[1] Kalicka, R.; Pietrenko-Dabrowska, A.; Lipinski, S., "Efficiency of new method of removing the noisy background from the sequence of MRI scans depending on structuring elements used to morphological processing," Information Technology, 2008. IT 2008. 1st International Conference on, vol., no., pp.1,4, 18-21 May 2008 doi: 10.1109/INFTECH.2008.4621682

[2] Sajjad Mohsin, Sadaf Sajjad, Zeeshan Malik, and Abdul Hanan Abdullah, “Efficient Way of Skull Stripping in MRI to Detect Brain Tumor by Applying Morphological Operations, after Detection of False Background”, International Journal of Information and Education Technology, Vol. 2, No. 4, August 2012.

[3] R. Gonzalez and R. Woods Digital Image Processing, Addison-Wesley Publishing Company, 1992.

[4] Acharya and Ray, Image Processing: Principles and Applications, Wiley-Interscience 2005 ISBN 0-471-71998-6.

[5] HanizaYazida and HamzahArofb, Gradient based adaptive thresholding, Journal of Visual Communication and Image Representation, Volume 24, Issue 7, October 2013, Pages 926–936.

[6] Nobuyuki Otsu, A Threshold Selection Method from Gray-Level Histograms, IEEE Transactions on Systems, Man and Cybernetics 9(1):62--66 (1979).

[7] B. Jähne, H. Scharr, and S. Körkel. Principles of filter design. In Handbook of Computer Vision and Applications.Academic Press, 1999.

IOP Conf. Series: Materials Science and Engineering **396** (2018) 012040 doi:10.1088/1757-899X/396/1/012040

A Review on Visual Recognition of RGB Images and Videos by Learning from RGB-D Data

ALPHONSA JOSE
M.Tech Student, Electronics and Communication Engineering
St. Josephs College of Engineering And Technology, Palai, Kerala
E-mail: alphonsajose80@mail.com

SREEJITH V
Assistant Professor Dept. ECE
St. Josephs College of Engineering And Technology Palai, Kerala
E-mail: sreejith.vasu@sjcetpalai.ac.in

Abstract. Domain adaptation provides the possibility of research that allows changes in data distribution across training and testing datasets. Recognizing the RGB images by learning RGB-D data contains the additional depth information. The unsupervised domain adaptation (UDA) take advantage of the additional depth features. UDA deals with domain mismatch between the source and the target.The various adaptation techniques deals with the source and target domain. The domain mismatch is minimized by describing a projection matrix that is optimized by reducing the Maximum Mean Discrepancy (MMD) and aligning the source and target domains. To optimize the depth information the correlation between different types of features are to be maximized. Inorder to simultaneously cope with the domain mismatch issues, a unified framework called domain adaptation from multi-view to single-view (DAM2S) is learned. The effectiveness of the proposed methods for recognizing RGB images and videos by learning from RGB-D data is demonstrated by comprehensive experiments for object recognition, cross dataset and cross-view action recognition.

1. Introduction

Visual recognition gives the ability to see, hear or become aware of something through the senses and uses deep learning algorithms to analyse images that can give insights into ones visual content. One can organize images libraries, understand an individual image, recognize food, detect faces, and create custom classifiers for specific results that are tailored for the needs. Visual recognition quickly and accurately tag, classify and search visual content using machine learning. Visual recognition understands the contents of images, analyse images for scenes, objects, faces, colors, food, and other subjects that gives insights to the visual content. Humans have the remarkable ability to organize their perceptual input instead of a collection of values associated with individual photoreceptors, to perceive a number of visual groups, usually associated with objects or well-defined paths of objects. Algorithms and representations is used to understand the methods to recognize objects, scenes, place and peoples.There is an increasing interest in developing new technologies using depth images and videos for various visual recognition tasks. It is obtained by capturing depth information with the RGB-D equipments namely Kinect sensors. The depth information cannot be applied to most ordinary visual recognition applications, in which images and videos are captured by conventional RGB cameras. Several labelled RGB-D datasets [1] are released for visual recognition tasks.

2. Domain Adaptation Methods

The domain adaptation is related to machine learning methods. The distribution of test data is different from that of training data [2]. The related works includes several adaptation methods. This includes the (a). Unsupervised Domain Adaptation (UDA) methods (b). Heterogeneous Domain Adaptation (HDA) methods (c). Multi-view Domain Adaptation method (e). Multi Domain Adaptation from Heterogeneous Sources (MDA-HS).

2.1. Unsupervised Domain adaptation Method

It contains no labelled data in the target domain [3].Domain adaptation in deep architectures is adapted that can be trained on large amount of labeled data from the source domain and large amount of unlabeled data from the target domain.

2.2. Heterogeneous Domain Adaptation Method

The method does not require any labelled target domain samples [4]. In contrast this have both visual and depth features in the source domain, while the depth features are not available at the testing stage.

2.3. Multi-view Adaptation Methods

The samples in the source and the target domains have multiple type of features[5].

2.4. Multi-view Domain Adaptation from Heterogeneous Methods

The samples from the target domain have all types of features from all source domains. In contrast, it have single view features in the target domain [6].

3. Domain Adaptation With Additional Features

The approach to domain adaptation that is appropriate exactly in the case when one has enough target data to do slightly better than just using only source data is described.

In the framework, the visual features and depth features from the RGB images/videos and depth images/videos is extracted. To handle the distribution mismatch, we learn a common subspace that is parameterized by a matrix P for the visual features from two domains.

3.1. Learning Projection For Domain Adaptation

Two strategies for learning the projection matrix P is proposed. The strategies include reducing the Maximum Mean Discrepancy (MMD) and aligning two subspaces. The first strategy is to minimize the Maximum Mean Discrepancy (MMD) criterion, which is widely used to measure the distribution difference between the data sampled from two datasets. The second strategy seeks a domain invariant feature space by learning a mapping function which aligns the source subspace with the target one.

3.2. Incorporating Depth Infomation

Inorder to effectively incorporate the depth features to learn more robust classifiers it includes two different strategies, one is maximizing the feature correlation where Canonical Correlation Analysis (CCA) is one pioneering work in two-view learning, which aims to learn two projection matrices to map the training samples with different features into a common subspace, such that the feature correlation can be maximized. Kernel Canonical Correlation Analysis (KCCA) is an extension of CCA by applying the kernel trick.The two classifiers for visual and depth features based on the projected visual and depth samples in the common subspace is learned. The objective function is formulated as the DAM2S-A algorithm. The second is enforcing the classifier consistency done by using the MMD based regularizer with the objective function formulated as the DAM2S-B algorithm and the Kernel SA based regularizer with the objective function formulated as the DAM2S-C algorithm.

3.3. Summary of DAM2S Approaches

The domain adaptation with additional features proposed two regularizers to reduce the domain distribution mismatch and two forms of regulrizers for incorporating the depth features, which leads to three DAM2S algorithms, DAM2S-A, DAM2S-B and DAM2S-C, the MMD based regularizer in DAM2S-A and DAM2S-B , and the kernel SA based regularizer in DAM2S-C. For incorporating the depth information, we utilize the feature correlation maximization based regularizer in DAM2S-A, and the classifier consistency based regularizer in DAM2S-B and DAM2S-C. the parameter matrix P or parameter matrices (P, Q) can be written in a single matrix in the dual form.

4. Evaluation of Effectiveness

The effectiveness of our proposed three algorithms for different visual recognition tasks, including object recognition, cross-dataset human action recognition, and cross-view

IOP Conf. Series: Materials Science and Engineering **396** (2018) 012040 doi:10.1088/1757-899X/396/1/012040

human action recognition is evaluated.

4.1. Object Recognition

The proposed three DAM2S algorithms is evaluated for object recognition by using the RGB-D Object dataset as the source domain and the Caltech-256 dataset [7] as the target domain. The RGB-D Object dataset contains the color and depth images of different objects from 51 categories.

4.2. Cross-Dataset Human Action Recognition

For cross-dataset action recognition, the Hollywood 3D Dataset are used as the source domain, and the Hollywood2 dataset as the target domain. The Hollywood2 dataset is a widely used benchmark dataset for human action recognition, which contains 1, 707 (823 in the training set and 884 in the test set) RGB videos from 12 human actions cropped from the Hollywood movies. Similarly, the Hollywood 3D dataset contains 650 RGB-D video clips from 14 human actions cropped from the Hollywood 3D movies.

4.3. Cross-View Human Action Recognition

A recently released multi-view RGB-D action dataset ACT42 is used for the experiments, which contains 2, 648 RGB-D videos from 14 human actions captured by Kinect from four different view points. To evaluate the proposed algorithms, we use the RGB-D videos from the first two views as the source domain, and use only the RGB videos from the remaining two views as the target domain, which leads to 1, 324 RGB-D videos in the source domain, and 1, 324 RGB videos in the target domain.

5. Conclusions

A new framework is discussed for recognizing the RGB images and videos by learning from a set of labelled RGB-D data. A review of different adaptation techniques provides the advantage of understanding the domain mismatch between the source and the target. The framework for domain adaptation is formulated as a new Unsupervised Domain Adaptation (UDA) problem, in which both visual and depth features are considered in the source domain and the visual features are considered in the target domain. Three DAM2S algorithms are described by taking into consideration the various datasets. The algorithms take into account the depth features in the source domain and will reduce the domain distribution mismatch between the source and target domains. Inorder to evaluate the effectiveness of the three DAM2S algorithms, comprehensive experiments for object detection, cross-dataset human action recognition and cross-view human action recognition have been demonstrated. The comparison between the three algorithms are discussed. The algorithm play a crucial role in computer vision.It is actually a measurement that computes the properties of 3D world by identifying the visual data and information. The algorithms are used to understand the methods to recognize objects, scenes, place and peoples is the method of visual recognition.

References

[1] Wen Li, Lin Chen, Dong Xu, Senior Member, IEEE, and Luc Van Gool Visual Recognition in RGB Images and Videos by Learning from RGB-D Data,IEEE Trans. on Pattern Analysis and Machine Intelligence , TPAMI.2017.

[2] K. Lai, L. Bo, X. Ren, and D. Fox,A large-scale hierarchical multiview RGB-D object dataset, ICRA, 2011.

[3] L. Duan, I. W. Tsang, D. Xu, and T. Chua, Domain adaptation from multiple sources: A domain-dependent regularization approach, ,T-NNLS, vol. 23, no. 3, pp. 504518, 2012.

[4] Raghuraman Gopalan, Ruonan Li, and Rama Chellappa Center for Automation Research, University of Maryland, College Park, MD 20742 USA , Domain Adaptation For Object Recognition.

[5] B. Kulis, K. Saenko, and T. Darrell, What you saw is not what you get: Domain adaptation using asymmetric kernel transforms,CVPR,2011.

[6] D. Zhang, J. He, Y. Liu, L. Si, and R. D. Lawrence, Multi-view transfer learning with a large margin approach,KDD, 2011.

[7] G. Griffin, A. Holub, and P. Perona, Caltech-256 object category datasetCalifornia Institute of Technology, Tech. Rep., 2007.

IOP Conf. Series: Materials Science and Engineering **396** (2018) 012041 doi:10.1088/1757-899X/396/1/012041

Comparative Analysis of Full Adder Circuits

ANEELA ACHU MATHEW
M.Tech Scholar, Electronics and Communication Engineering
St. Joseph's College of Engineering And Technology, Palai, Kerala
E-mail: aneelamathew94@gmail.com

SREESH P R
Assistant Professor, Dept. of Electronics and Communication Engineering
St. Joseph's College of Engineering And Technology Palai, Kerala
E-mail: sreeshpr@sjcetpalai.ac.in

Abstract. In this paper, comparison of various full adder circuits is analysed. Full Adder circuits are extensively used in digital design. Here different types of full adder such as Conventional CMOS, based on Ex-OR/Ex-NOR, Pass Transistor Logic(PTL) and Gate Diffusion Input(GDI)technique are done. From this, GDI technique takes less number of transistors and therefore consumes less power. In GDI- 14T, 12T which is based on MUX and a proposed 10T is implemented. The power consumption is less for 12T whereas better output is obtained for 10T. Simulation results are obtained in 180nm technology using Mentor Graphics Tool.

1. Introduction

Now a day, a tremendous increase in electronic industry is seen. The use of portable devices are also increasing. The evolution of nanotechnology has become a boon to the electronic manufacturing. As the channel length reduces, maximum number of transistors that can be embedded into a single chip increases. This gives rise to improved functionality. The current demands are high speed, low power, high throughput. So in VLSI, the circuits need good output with less power.

Full Adders are widely used in different combinational circuits, Application Specific Integrated Circuits(ASICs). It is an essential component for the design and development of all type of processors like Digital Signal Processors(DSPs), Microprocessors etc. It is the fundamental gate in many arithmetic circuits, such as adders and multipliers. Thus, upgrading the performance of the full adder block leads to enhancement of overall system performance. In order to attain low power for large circuits, the basic component itself needs to be minimized in terms of power. Different low power techniques are available. The most widely used methods include PTL, which requires less number of transistors as compared with conventional CMOS circuits. The problem with this is improper output voltage swings due to difficulty of PMOS to turnoff. This even increases static power dissipation[5]. Other existing low power technique includes Hitachi CPL[4] and DPL[6]. Older low power technique includes Dynamic logic, Domino logic, NORA logic, Differential Cascode Voltage Switch(DCVS), MOS Current Mode Logic(MCML) and Clocked CMOS(C2CMOS[8][9]. The latest low power circuits include GDI technique which was introduced by A. Morgenshtein, A.Fish and Israel Wagner in 2002[10]. This technique includes a unique cell library which can perform nearly eight different functions. Therefore, large number of functions can be realized easily. Also as number of transistors required is less and thus the power consumption is less.

In this paper different full adder circuits are compared. Section 2 includes a basic full adder circuit. Section 3 describes of various existing low power techniques for full adder including Conventional CMOS, Transmission Gates(TG) based on Ex-OR/Ex-NOR, NMOS based PTL. Section 4 consists of basic GDI cell, Full adder analysis with GDI 14T, 12T based MUX and includes a proposed 10T based on GDI. Section 5 consists of Simulation Results and Comparisons based on number of transistors used, power consumption and output swing. Section 6 concludes the paper.

2. Basic Full Adder

A basic full adder circuit is used to add three one-bit binary numbers A, B and C. It produces two one-bit outputs Sum and Carry out. The full Adder can be realised with a Boolean function such as:

$$Sum = A \oplus B \oplus C \quad and \quad Carry \;= AB + BC + AC \tag{1}$$

For an n-bit full adder carry in(C) is the carry out(carry) of the previous bit addition operation. Similarly, the carry out is fed to another full adder as the carry in. Here the Sum operation is obtained by performing Ex-OR operation between three inputs, whereas the Carry output is obtained from AND and OR operations. It is observed in the equation (1).

3. Various Existing Low Power Techniques For Full Adder Circuits

3.1. Conventional CMOS Circuit

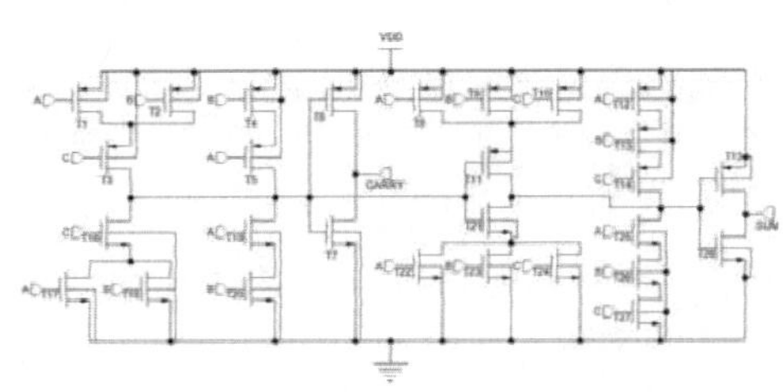

Figure 1. Conventional CMOS of 28T Full Adder

In a basic CMOS circuit consists of PMOS and NMOS transistors arranged complementary. PMOS transistors are present at pull up and NMOS transistors are complementary arranged at pull down. The number of transistors are large since if n PMOS is present then n NMOS are also present. The proper output swing is obtained. Since number of transistors are large, therefore area get increased and power consumption is also large. The Fig.1 shows the circuit representation of a conventional CMOS based full adder circuit. In this conventional circuit, in order to represent one-bit full adder there requires 14T PMOS and 14T NMOS. The pull up network consists of PMOS only and pull down with NMOS only. This wide area circuit consumes high power.

3.2. Based on Ex-OR/Ex-NOR and Transmission gate

The Ex-OR/Ex-NOR gate is the basic building block of adder, comparator, and parity generator/checker and encryption processor. This method provides better swing and less

IOP Conf. Series: Materials Science and Engineering **396** (2018) 012041 doi:10.1088/1757-899X/396/1/012041

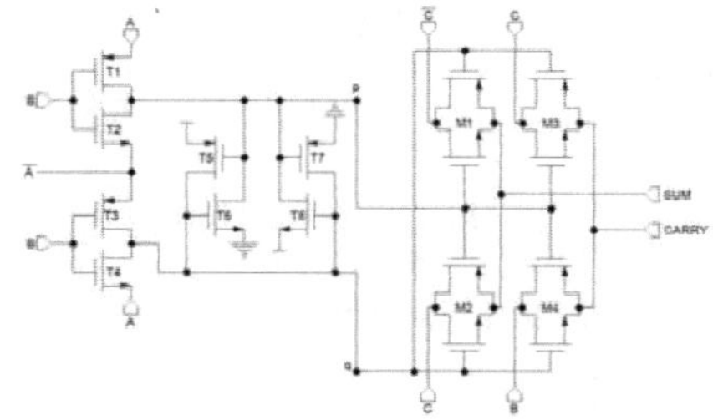

Figure 2. Ex-OR/Ex-NOR based Full Adder

propagation delay. A full adder is constructed with the Ex-OR/Ex-NOR circuit and four transmission gates[7]. The circuit is as shown in Fig.2. Proper output swing is obtained by connecting M1, M2, M3 and M4 transmission gates. The Sum operation is obtained by connecting M1 and M2 transistors whereas connecting M3 and M4 transistors to Ex-OR/Ex-NOR outputs results in Carry out. The constructed full adder logic is represented in the following equations.

$$Sum = (A \oplus B)\overline{C} + \overline{(A \oplus B)}C \quad and \quad Carry \ = C(A \oplus B) + B(A \oplus B) \qquad (2)$$

The point p gives Ex-or output and q gives Ex-NOR output in Fig.2. For sum operation, consider Ex-OR output is logic 1 then Ex-NOR output is logic 0, so M1 transmission gate turns ON and M2 transmission gate turns OFF. Therefore, the sum output of the circuit is the inverted input of C. Now Ex-OR output is logic 0 then Ex-NOR output is logic 1, M1 transmission gate turns OFF and M2 transmission gate turns ON. Therefore, the sum output of the circuit is same as the input C. The circuit operation of the carry output is as follows. When Ex-OR output is logic 1 and Ex-NOR output is logic 0, M3 transmission gate turns ON and M4 transmission gate turns OFF. Therefore the carry output of the circuit is same as that of C. When Ex-OR output is logic 0 and Ex-NOR output is logic 1, M4 transmission gate turns ON and M3 transmission gate turns OFF. Therefore, the carry output of the circuit is same as input B[7]. This can be explained by the above equation (2). Here there are 20T are needed overall so the total power consumption is large. But as compared to conventional CMOS 28T the power consumption is less.

3.3. NMOS based PTL Full Adder

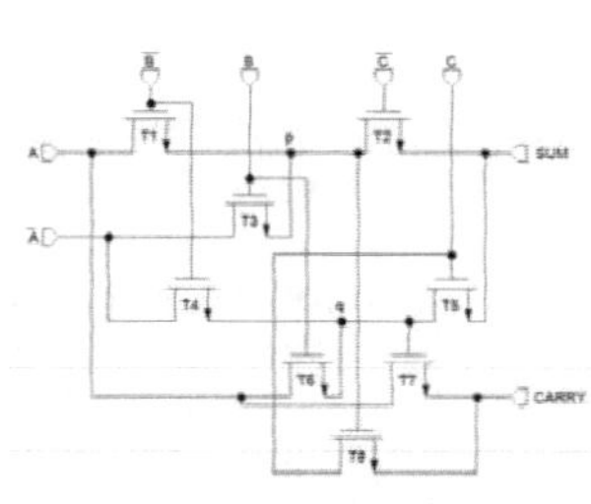

Figure 3. NMOS based PTL Full Adder

The Fig.3 shows a PTL logic which is analysed completely with NMOS circuits[2]. In total there are 12T required. The circuit requires 8T and to implement$\overline{B}$ and $\overline{C}$, four more transistors are needed. Since inverters are present, static power dissipation increases. The T1 transistor generates output A$\overline{B}$ while T3 generates $\overline{A}$B. The two outputs together generate A⊕B at point p. Outputs of T2 and T4 are $\overline{AB}$ and AB respectively. They produce A⊙B after meeting at point q. The logic at the outputs of T5 and T6 are (A⊕B)$\overline{C}$ and (A⊙B)C.The output logics of T7 and T8, (A⊙B)A and (A⊕B)C, together generate Carry output.

$$\begin{aligned} Sum &= (A \oplus B)\overline{C} + \overline{(A \oplus B)}C &&= A \oplus B \oplus C \\ Carry &= (A \odot B)A + (A \oplus B)C &&= AB + (A \oplus B)C \end{aligned} \tag{3}$$

Usually PTL is more advantageous than CMOS. Low node capacitance leads to high speed and less number of transistors leads to low power. The basic drawback includes at low voltages the operation become slower due to threshold drop across the pass transistors.

4. A Basic GDI Cell

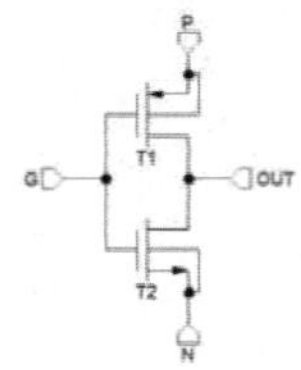

Figure 4. Basic GDI cell

Many low power techniques exist. The latest is the Gate Diffusion Input(GDI) technique[10]. GDI is a new low power technique used in digital circuits. More than one function can be implemented by using this method. A basic GDI cell resembles that of a basic inverter, it has three inputs G (common gate input of NMOS(T2)and PMOS(T1)) P (input to source/drain of PMOS(T1)) N (input to source/drain of NMOS(T2)). The number of transistors required are very less, so the required area and total power needed are also less. GDI plays a vital role in designing complex circuits with multiple functionality. Fig.4 shows a basic GDI cell. A GDI cell can be used to implement eight functions and they are shown in Table.1. Therefore, the speciality of this technique is that it has a unique cell library. The cell working is completely depends on the proper inputs that are applied to the three inputs. So it is easy to implement any complex circuitry. The power consumption is also very less as compared to other existing full adder circuits.

Based on GDI different ways of full adder can be implemented[11]. Out of this two of them are as follows:

4.1. Simple GDI Full Adder

This Full Adder is implemented with 14T. The Fig.5 shows block diagram representation of the full adder. Each block in the diagram represent a GDI cell. Each GDI cell consists of two transistors, so in total 14T is present. The number of transistors used is very small and i.e., required area and power consumption has reduced. The analysis is such that:

$$Sum \ = A \oplus B \oplus C \quad and \quad Carry \ = AB + BC + AC \ = AB + (A \oplus B)C \tag{4}$$

To implement Sum in GDI according to the above equation requires four cells i.e. two of them for inverters $\overline{B}$ and $\overline{C}$, and two of them for Ex-OR operations. Similarly, to implement carry three more cells are needed i.e. for two AND operation and one OR operation. So in total seven cells and 14T are needed. The drawback of this circuit is such that output swing is not obtained.

IOP Conf. Series: Materials Science and Engineering **396** (2018) 012041 doi:10.1088/1757-899X/396/1/012041

Table 1. Various Functions of a GDI Cell

G	P	N	Out	Function
A	1	0	$\overline{A}$	Inverter
A	B	0	$\overline{A}$B	Function1
A	1	B	$\overline{A}$+B	Funcion2
A	B	1	A+B	OR
A	0	B	AB	AND
A	B	C	$\overline{A}$B+AC	MUX
A	B	$\overline{B}$	$\overline{A}$B+A$\overline{B}$	XOR
A	$\overline{B}$	B	$\overline{AB}$+AB	XNOR

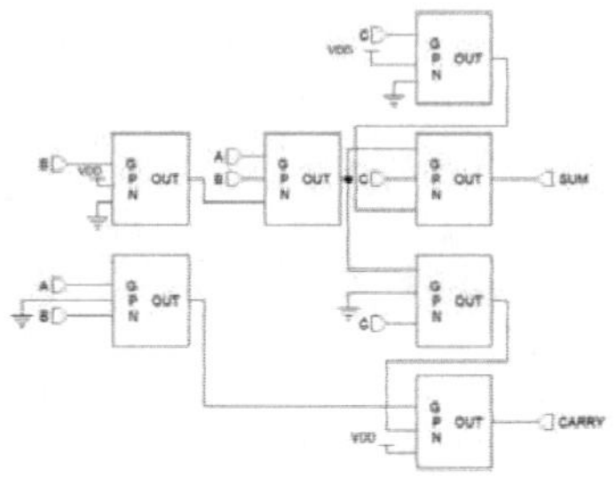

Figure 5. Block diagram of 14T GDI

4.2. GDI MUX Based Full Adder

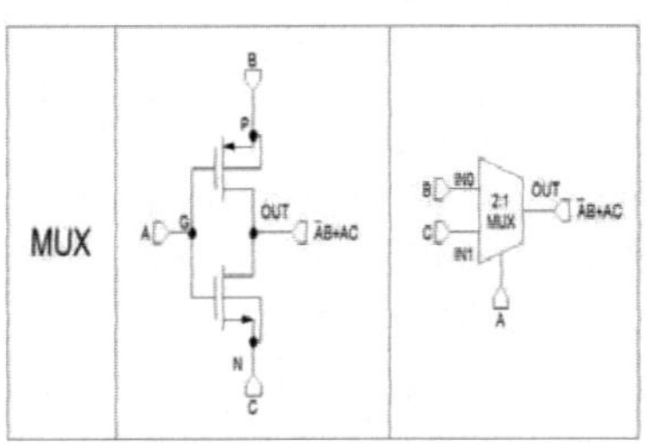

Figure 6. GDI MUX

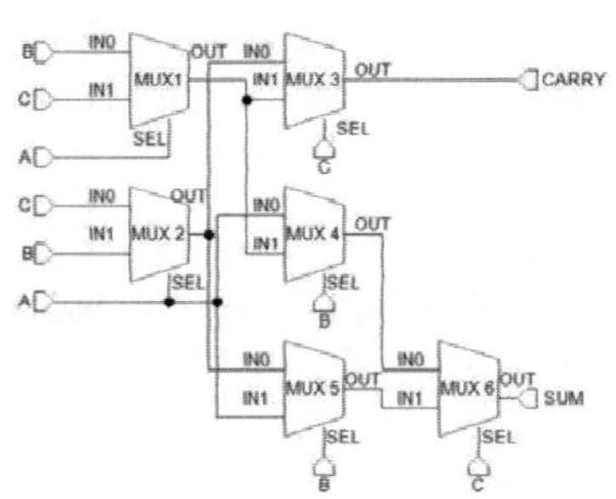

Figure 7. Block diagram representation of 12T GDI MUX Adder

This requires less number of cells as compared to the simple GDI cell. The full adder is analysed completely on multiplexers. The Sum and Carry is minimized in terms of MUX. Therefore, only single operation of GDI 2:1 MUX is basically done. The Fig.6 shows conversion of MUX into a GDI cell[1].

The operation is such that when the input A is high, NMOS is ON and C is obtained at the output otherwise B. The equation can be represented as:

$$Output = \overline{A}B + AC \tag{5}$$

IOP Conf. Series: Materials Science and Engineering **396** (2018) 012041 doi:10.1088/1757-899X/396/1/012041

Equation (5) is the simple equation for the Multiplexer. The Sum and Carry are analysed based on Equation(6) as follows.

$$
\begin{aligned}
MUX1 &= \bar{A}B + AC \quad and \quad MUX2 = \bar{A}C + AB \\
MUX3 &= [(\bar{A}C + AB)\bar{C} + (\bar{A}B + AC)C] = AB\bar{C} + \bar{A}BC + AC = AB\bar{C} + \bar{A}BC + A(B+\bar{B})C \\
&= AB\bar{C} + \bar{A}BC + ABC + A\bar{B}C = AB\bar{C} + ABC + \bar{A}BC + ABC + A\bar{B}C + ABC \\
&= AB(C+\bar{C}) + BC(A+\bar{A}) + AC(B+\bar{B}) = AB + BC + AC = Carry \\
MUX4 &= A\bar{B} + (\bar{A}B + AC)B \quad and \quad MUX5 = (\bar{A}C + AB)\bar{B} + AB \\
MUX6 &= [A\bar{B} + (\bar{A}B + AC)B]\bar{C} + [(\bar{A}C + AB)\bar{B} + AB]C = A\bar{B}\,\bar{C} + \bar{A}B\bar{C} + \bar{A}\,\bar{B}C + ABC \\
&= A \oplus B \oplus C = Sum
\end{aligned} \tag{6}
$$

During the transition of output from one logic level to other both the NMOS and PMOS transistors become active and provides a short circuit path directly between supply to ground which increases the power consumption of the circuit. As the 12-T full adder is made of GDI based MUX, it does not provide direct connections between supply and ground, so the probability of a getting short circuit current during switching can be considerably reduced; i.e. the power consumption due to short circuit current is considerably small. Again, in the proposed 12T full adder, all the select line of the MUX i.e. the G nodes of the GDI cells are directly connected with the input signals, results a much faster transition (less delay) in its output signals. As a result, the power consumption of the final pad out stage is low and it can provide faster Sum and Carry outputs[1]. The Fig.7 represent block diagram representation of the GDI 12T based full adder.

4.3. Proposed 10T GDI Full Adder

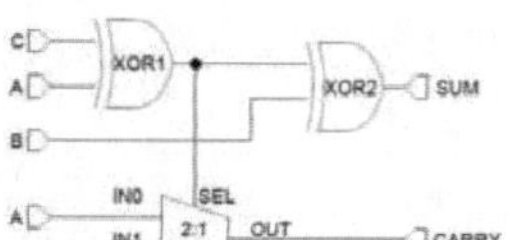

Figure 8. Gate level representation of full adder

The proposed full adder includes basically two functions, two Ex-OR operation and one MUX operation[3]. In GDI, to perform Ex-OR four transistors are required and for MUX two transistors are needed. The Fig.8 shows the gate level representation of the proposed 1- bit full adder. As the total transistors required is less, the power consumption has also decreased. The area needed is less.This Sum is directly represented in the Fig.8.

LOGICAL ANALYSIS

The circuit consists of two Ex-OR operations and a MUX[3]. The Sum and Carry are given by:

$$
\begin{aligned}
Sum &= A \oplus C \oplus B \\
Carry &= \overline{(A \oplus C)}A + (A \oplus C)B = \overline{(\bar{A}C + A\bar{C})}A + (\bar{A}C + A\bar{C})B = (\bar{A}\,\bar{C} + AC)A + (\bar{A}C + A\bar{C})B \\
&= AC + \bar{A}BC + AB\bar{C} = AC(B+\bar{B}) + \bar{A}BC + AB\bar{C} = ABC + A\bar{B}C + \bar{A}BC + AB\bar{C} \\
&= ABC + A\bar{B}C + ABC + \bar{A}BC + ABC + AB\bar{C} = AC(B+\bar{B}) + BC(A+\bar{A}) + AB(C+\bar{C}) \\
&= AC + BC + AB
\end{aligned} \tag{7}
$$

IOP Conf. Series: Materials Science and Engineering **396** (2018) 012041 doi:10.1088/1757-899X/396/1/012041

From the equation (7) sum is obtained by performing A $\oplus$ C initially and then Ex-ORing with B. For Carry operation a single MUX is used such that A $\oplus$ C acts as selection line(SL). Depending on the SL, if SL is 0 then A is taken at the output, else B is taken. Also in order to perform Ex-OR operation twice, C and B needed to be inverted. Therefore, in total, 10T is required.

5. Simulation and Results

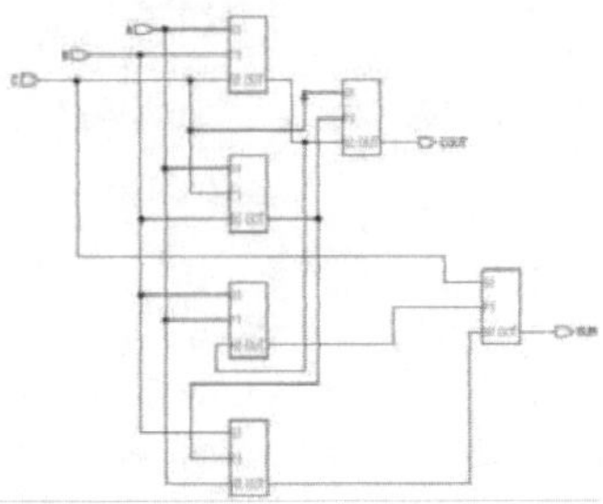

Figure 9. Full adder with 12T GDI MUX

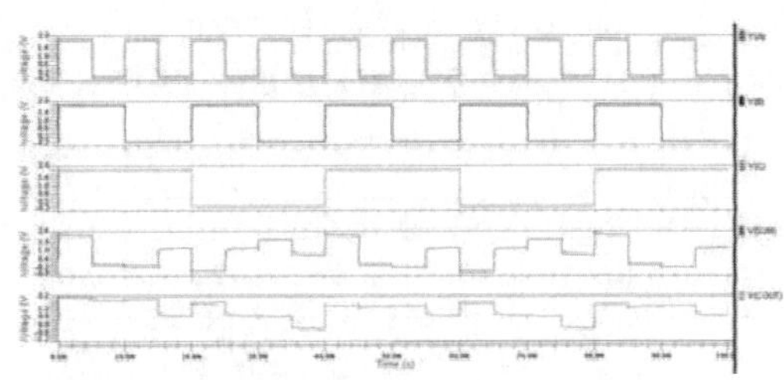

Figure 10. Waveform of 12T GDI Full Adder

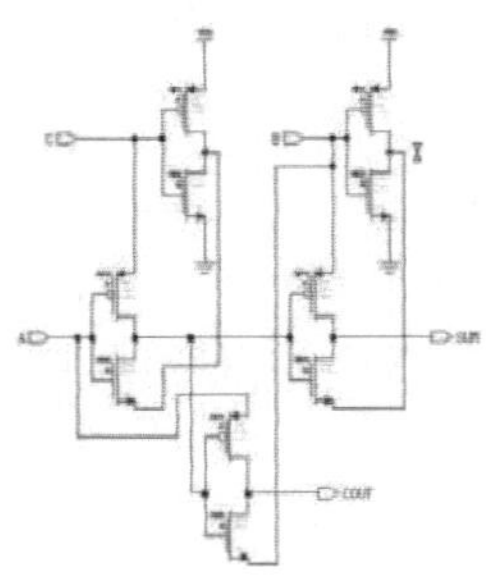

Figure 11. Schematic of 10T Full Adder

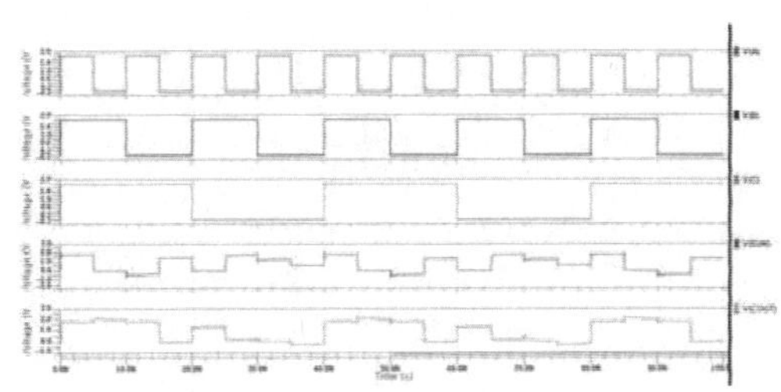

Figure 12. Waveform of 10T Full Adder

The analysis is done using Mentor Graphics Pyxis Version 10.1 Tool. It is done in 180nm technology. Therefore L is taken as 0.18u and width for PMOS is 0.9u, NMOS is 0.36u. Thus the W/L ratio is 2.5. The Fig.9 shows the schematic of 12T GDI MUX based full adder. The Fig.10 shows the waveform of the 12T Adder. It was observed that output swing is not obtained in this case.

The Fig.11 shows the schematic of Full Adder with 10T. Each PMOS and NMOS forms as a GDI cell. The Fig.12 shows the waveform. The proposed full adder circuit produces better swing as compared to 12T.

The 10T Full Adder provides low power consumption as compared with simple GDI full adder whereas 12T produces the minimum power consumption as compared to all the performed full adders. The Table 2 shows the comparison of different types of full adder including basic CMOS circuits and also GDI circuits. Also the table shows output swing for these circuits. Here, for the earlier CMOS circuits produce better swing but corresponding power consumption of such

IOP Conf. Series: Materials Science and Engineering **396** (2018) 012041 doi:10.1088/1757-899X/396/1/012041

Table 2. Comparison of Different Full Adder Circuits

Sr.No	Full Adder	No.of Transistors	Power Consumption(in pW)	Output Swing
1	Conventional CMOS	28	370.61	Yes
2	Ex-OR/Ex-NOR and TG	20	334.01	Yes
3	Basic GDI	14	147.44	No
4	NMOS based PTL	12	280.53	No
5	GDI MUX	12	34.63	No
6	Proposed	10	38.31	Better

circuit is large. When the number of transistors decreases the power consumption get decreased. As compared with 12T full adder, the power consumption of 10T is high. It is due to a single operation is performed in 12T and for 10T inverter, Ex-OR and MUX operation is done. But better output is obtained for 10T.

6. Conclusion

The current trends in VLSI technology includes the reduction of channel length. As the channel length decreases the source and drain of a transistor becomes closer. This causes many short channel problems and thereby increasing power consumptions especially static power. The latest research is related to minimizing power by reducing number of transistors. The low power techniques include GDI, which is a two transistor cell. Using this technique multiple functions can be implemented. As the number of transistors is less the power consumption is low. The proposed full adder uses 10T and attains low power with better swing output. Thus this proposed full adder can be widely used in DSPs, ASICs and multipliers.

References

[1] Biswarup Mukherjee, Aniruddha Ghosal, Design and Study of a Low Power High Speed Full Adder Using GDI Multiplexer,*IEEE 2nd International Conference on Recent Trends in Information Systems (ReTIS)*, 2015.

[2] Debanjana Datta, Debarshi Datta, A Novel Power Efficient N-MOS Based 1-Bit Full Adder, *IEEE*, 2016.

[3] Biswarup Mukherjee, Biplab Roy, Arindam Biswas, Aniruddha Ghosal, Design of a Low Power 4x4 Multiplier Based on Five Transistor (5-T) Half Adder, Eight Transistor (8-T) Full Adder and Two Transistor (2-T) AND Gate, *IEEE*, 2015.

[4] Yano, K, et al, "A 3.8 ns CMOS 16*16b multiplier using complementary pass transistor logic", *IEEE J. Solid State Circuits, Vol 25, p388-395*, April 1990.

[5] Jaume Segura, Charles F. Hawkins,"CMOS electronics: how it works, how it fails, Wiley" *IEEE*,2004, page 132.

[6] Makoto Suzuki, et al, "A 1.5 ns 32 b CMOS ALU in double pass transistor logic",*ISSCC Dig. Tech. Papers*, pp 90-91, February 1993.

[7] Sarada Musala, Rajasekhara Reddy, Implementation of a Full Adder Circuit with New Full Swing Ex-OR/Ex-NOR Gate *IEEE Asia Pacific Conference on Postgraduate Research in Microelectronics and Electronics (PrimeAsia)*,2013.

[8] Mohamed W. Allam, New Methodologies for Low-Power HighPerformance Digital VLSI Design, *PhD. Thesis, University of Waterloo, Ontario, Canada*, 2000.

[9] A.Bazzazi and B. Eskafi, "Design and Implementation of Full Adder Cell with the GDI Technique Based on 0.18m CMOS Technology", *International MultiConference of Engineers and Computer Scientists (IMES) Vol II*, March 17 - 19, 2010, Hong Kong.

[10] Arkadiy Morgenshtein, Alexander Fish, and Israel A. Wagner, "GateDiffusion Input (GDI): A Power-Efficient Method for Digital Combinatorial Circuits", *IEEE Transaction on VLSI Systems*, Vol. 10.

[11] Sudeshna Sarkar, Monika Jain, Arpita Saha, Amit Rathi, Gate Diffusion Input: A technique for fast digital circuits (implemented on 180 nm technology), *IOSR Journal of VLSI and Signal Processing (IOSR-JVSP)* Volume 4, Issue 2, Ver. IV (Mar-Apr. 2014), PP 49-53.

IOP Conf. Series: Materials Science and Engineering **396** (2018) 012042 doi:10.1088/1757-899X/396/1/012042

Low Power Wallace Tree Encoder For Flash ADC

AJANYA M P

M.Tech Student, Electronics and Communications Engineering
St. Josephs College of Engineering and Technology, Palai, Kerala
E-mail: ajanyamp26@gmail.com

GEORGE TOM VARGHESE

Associate Professor, Dept. of ECE
St. Josephs College of Engineering and Technology, Palai, Kerala
E-mail: georgetomvarghese@gmail.com

Abstract – Flash ADC is the fastest analog to digital converter. It is also known as parallel ADC. It consists of a resistor ladder, a comparator and an encoder circuit. The encoder circuit converts the thermometer code which is the output of comparator into binary code. The efficiency of encoder is very important. It should be capable of reducing the bubble errors and should dissipate low power. Wallace tree encoder is well efficient in reducing all the bubble errors. This paper proposes a low power Wallace tree encoder for Flash ADC. It is having a power dissipation of 939.43pW and delay of 5.38nS. The proposed encoder is implemented in 180nm CMOS technology with 1.8V supply voltage and was simulated using Mentor graphics.

1. Introduction

Most of the real world signals are analog in nature. They are difficult to store and even more difficult to process. So it becomes necessary to convert them into digital form. An analog to digital converter converts any analog signals into digital data, which makes it easier to store and process as well as more accurate and reliable. These ADCs vary in speed and performance. Flash ADC which is also called parallel ADC is best suited for high speed low resolution applications. It consists of a resistor ladder network, a comparator and an encoder circuit. For an N bit Flash ADC there are 2^N resistors and 2^N-1 comparators. The resistor ladder provides the reference voltage. The comparator compares the input voltage with the reference voltage. Each comparator produces a '1' when its analog input voltage is higher than the reference voltage applied to it. Otherwise the comparator output is '0'. Output of the comparator is a thermometer code. Thermometer code resembles the output of a thermometer. In thermometer code, a value representing number 'N' has the lowermost N bits as '1', others are '0'. Code consists of sequence of zeros followed by sequence of ones. There cannot be zero in between two ones. The encoder circuit converts the thermometer code to binary. Fig. 1 shows the block diagram of Flash ADC.

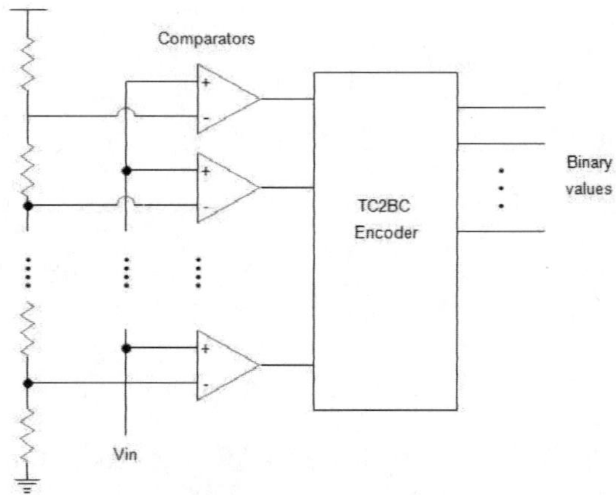

Fig. 1: Block diagram of Flash ADC

Encoder performance is very important for the design of Flash ADC. High speed encoders are used in Flash ADCs for applications such as disk drive read channels and UWB communication receivers involving medium to high resolution and high speed [1]. Normally the thermometer code consists of series of zeroes followed by ones. There should be only one transition. But in some cases zeroes come in between ones or vice versa. This condition is called bubble error. So while designing an encoder care should be taken to reduce this error. The basic building block of Wallace tree encoder is the full adder circuit. For the inputs 011 and 101 the sum and carry outputs are the same. 011 represent the normal thermometer code since there is only one transition and 101 represent the error condition. For both these input conditions output of the full adder remains the same. So even though bubble error exists in the input of Wallace tree encoder the output will not be affected.

Reference [2] gives a Wallace tree encoder with a power dissipation of 1.506nW and a power-delay product of 7.71aJ. This paper proposes a low power Wallace tree encoder by modifying the basic component i.e. full adder circuit. This circuit is having a power dissipation of 939.43pW and a power-delay product of 5.476aJ.

2. Wallace Tree Encoder

The Wallace tree encoder also called as ones counter is built with full adder cells. It counts the number of ones present at the output of comparator and represents them as binary code. Fig. 2 shows a 3 bit Wallace tree encoder with full adder as its basic cell. The full adders count the number of logic "1" at their input to give final binary output. The full adders are arranged to sum the inputs so as to form Wallace tree.

IOP Conf. Series: Materials Science and Engineering **396** (2018) 012042 doi:10.1088/1757-899X/396/1/012042

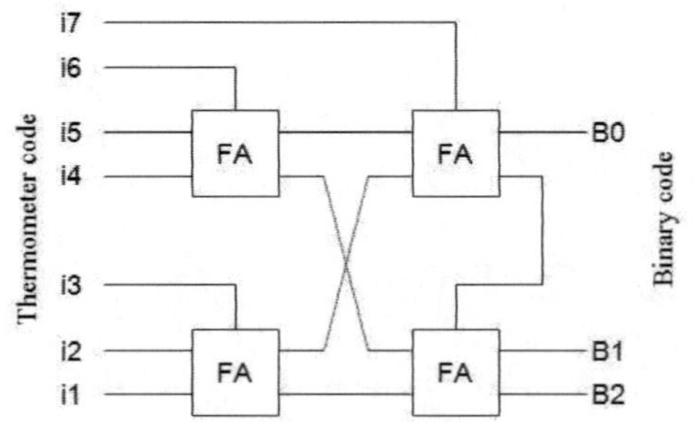

Fig. 2: 3-bit Wallace tree encoder

Even when bubble errors are present, this encoder converts output of comparator into correct binary output. Another advantage of this encoder is flexibility such that suitable topology can be selected according to speed and power requirements [5]-[7]. The number of full adder cells needed to implement an N bit encoder is given by the equation,

$$X_N = \sum_{i=1}^{N}(i-1).2^\wedge(N-i) \quad (1)$$

3. Wallace Tree Encoder Using 28T Full Adder

Reference [2] gives a 28 transistor full adder circuit as shown in fig. 3. The equations for sum and carry output is given as,

$$Sum = A\ xor\ B\ xor\ Cin \quad (2)$$

$$Cout = A.B + Cin\ (A\ xor\ B) \quad (3)$$

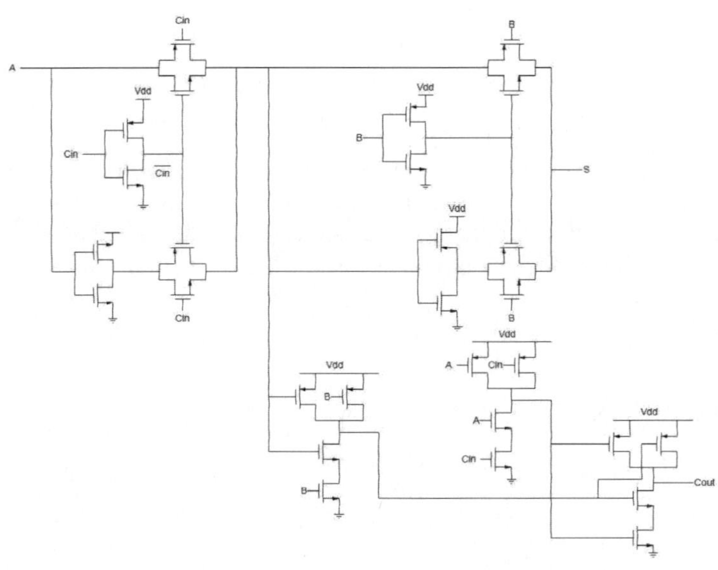

Fig. 3: Conventional 28T full adder

The conventional CMOS full adder shows robustness against voltage scaling and transistor sizing. The disadvantages are high input capacitance and requirement of buffers [3]. The schematic of Wallace tree encoder built using this 28T full adder is as shown in fig. 4.

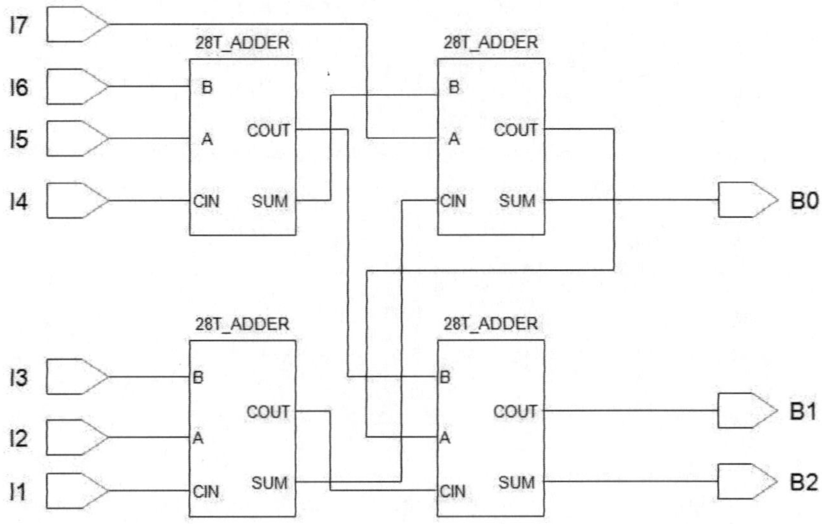

Fig. 4: Schematic of Wallace tree encoder using 28T full adder

The above circuit is simulated using Mentor graphics simulation tool. It gives a full swing output with a supply voltage of 1.8V. The simulation results show that this encoder circuit is having a delay of 5.12 nS and power dissipation of 1.506nW. In order to design a low power Flash ADC it is necessary to reduce the power dissipation of the encoder circuit. By changing the internal full adder circuit of this encoder a remarkable change in the power dissipation is observed.

4.Wallace Tree Encoder Using 14T Full Adder

A new full adder circuit using 14 transistors is proposed. Although the device count is very low, the circuit has full voltage swing in all nodes [4]. More than one logic style is employed for its implementation. It uses Ex-OR/ Ex-NOR module and transmission gate logic. 14 transistor full adder circuit is shown in fig. 5

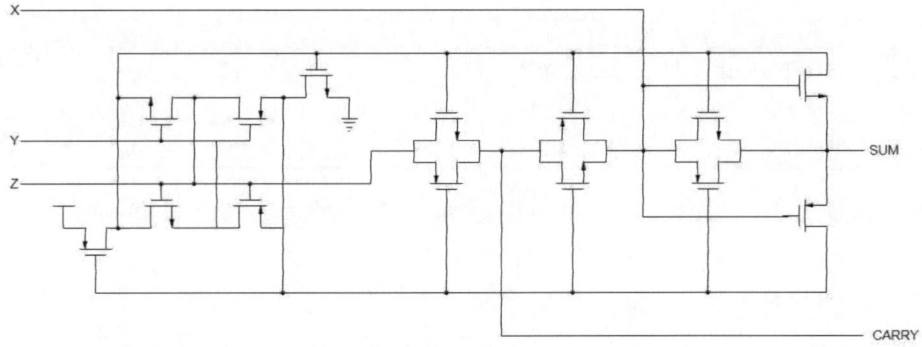

Fig. 5: Proposed 14T full adder

IOP Conf. Series: Materials Science and Engineering **396** (2018) 012042 doi:10.1088/1757-899X/396/1/012042

A low power encoder is designed using this full adder circuit and simulated using Mentor graphics tool. Fig. 6 shows the schematic of the encoder.

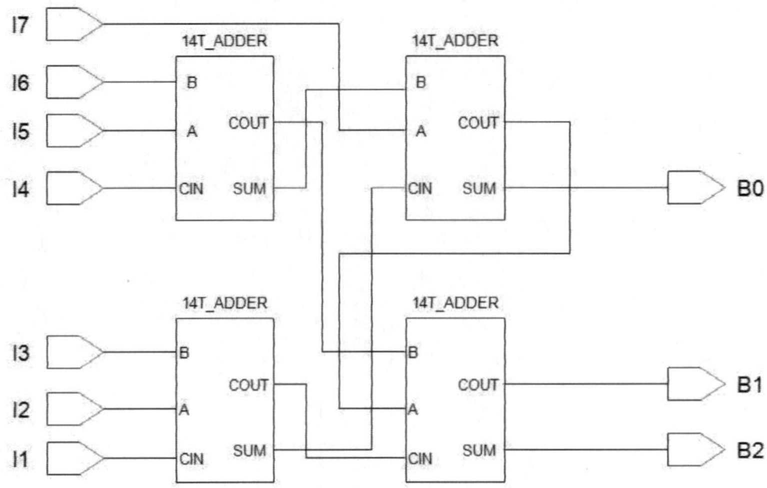

Fig. 6: Schematic of Wallace tree encoder using 14T full adder

This circuit is simulated using mentor graphics simulation tool in 180nm technology. It is having a delay of 5.83nS . Even though the delay of this circuit is slightly greater than that of the circuit shown in fig. 4 its power dissipation is very low. It is observed to be 939.43pW. Hence the overall power-delay product of this circuit is comparatively low.

5.Power Consumption And Delay Analysis

Wallace tree encoder circuit given in reference [2] which is constructed using 28T full adder is simulated using mentor graphics and results are compared with that of proposed encoder. The comparison is tabulated.

Table. 1 Comparison of power, delay and power delay product

Wallace tree encoder	Power Dissipation	Delay (nS)	PDP (aJ)
Using 28T full adder	1.506nW	5.12nS	7.71aJ
Using 14T full adder	939.43pW	5.83nS	5.47aJ

6.Conclusion

Thermometer code to binary code conversion in Flash ADC is done using suitable encoder circuits. Wallace tree encoder is capable of reducing all the bubble errors in the thermometer code and it provides a proper binary output. This paper proposes an efficient Wallace tree encoder which

IOP Conf. Series: Materials Science and Engineering **396** (2018) 012042 doi:10.1088/1757-899X/396/1/012042

dissipates less power. It has a power dissipation of 939.43pW which is comparatively less than previous designs. Hence the proposed encoder is well suitable for the design of low power Flash ADC.

References

[1] He Tang, Hui Zhao, Xin Wang, Yuha Cheng, "Capacitive interpolated Flash ADC design technique", *International SoC design Conference, November 2010.*

[2] Thottempudi Pardhu, Manusha S, Katakam Sirisha, "A low power Flash ADC with Wallace tree encoder", *Eleventh International Conference on Wireless and Optical communications Networks, September 2014.*

[3] Partha bhattacharya, Vinay Kumar, Anup Dandapat, "Performance analysis of a low power high speed hybrid 1-bit full adder circuit", *IEEE Transactions on Very Large Scale integration Systems, September 2014.*

[4] Mark Vesterbacka, "A 14-Transistor CMOS full adder with full voltage swing nodes", *IEEE workshop on Signal Processing Systems, October 1999.*

[5] Ankush Chunn, Rakesh Kumar Sarin, "Comparison of thermometer to binary encoders for Flash ADC", *Annual IEEE India Conference (INDICON), December 2013.*

[6] Syed Masood Ali, Rabin Raut, "Digital encoders for high speed Flash ADCs: Modelling and comparison", *IEEE North-East Workshop on Conference, June 2006.*

[7] Bui Van Hieu, Seunghyun Beak, "Thermometer-to-binary encoder with bubble error correction circuit for Flash Analog to Digital converter (FADC)", *Third International Conference on Communications and Electronics, August 2010.*

IOP Conf. Series: Materials Science and Engineering **396** (2018) 012043 doi:10.1088/1757-899X/396/1/012043

A Review on Autonomous Relay For Millimeter Wave Wireless Communications

GREESHMA SOSA THOMAS
M-Tech Student, Dept.of ECE,
St.Joseph's College Of Engineering and Technology, Palai
E-mail: greesht94@gmail.com

SARATH CHANDRAN C
Assistant Professor, Dept of ECE
St.Joseph's College Of Engineering and Technology, Palai
E-mail: sarath@sjcetpalai.ac.in

Abstract. Millimeter-wave wireless communication is an uprising technology. Millimeter-wave has large bandwidth and short wavelength. Eventhough it is advantageous, it gets easily blocked by obstacles. As a result, effective communication does not take place. Relays are used widely to solve this problem. Unmanned Autonomous Vehicles (UAV), such as drones, allows the mobile relays in real applications. It is difficult for relay to find its position quickly and accurately. Here, a method AutoRelay specialized for mmwave communication have been reviewed. In AutoRelay, the link qualities of mmwave beams are sampled while moving. Based on real-time sampling, UAV adjusts its path to reach the optimal location by using compressive sensing theory which increases accuracy and saves time.

1. Introduction

Mm wave communication operates at 3 to 300 GHz and its available bandwidth is upto multiple GHz[2]. It provides multi gigabit transmittability for wireless information shower[3]. Because of its short wavelength, multiple antennas can be consolidated in a tiny space. But the communication range of mm wave is shorter than existing wireless technologies as it attenuates more severely in air. Also, it is easily blocked by obstacles[4]. It does not have sufficient power to go through obstacles[5].

To compensate these disadvantages, relays are essential. In simple terms, relays are interconnection between source and destination. they receive and re transmit signals. They do not need wired connections to the network. With the development of Unmanned Aerial Vehicles (UAVs), mobile relays have become possible in real applications.

IOP Conf. Series: Materials Science and Engineering **396** (2018) 012043 doi:10.1088/1757-899X/396/1/012043

It is challenging for a UAV to find its optimal location accurately and quickly. Measuring the link qualities between transmitter and receiver in all possible space and selecting the optimal one is an accurate method. But it takes long time.

Thus, an autonomous relay (AutoRelay) measures real time link qualities while moving. UAV automatically adjusts its path to approach the optimal relay location quickly by taking advantage of compressive sensing and online estimation of link qualities in the whole candidate space.

AutoRelay has two advantages:

- i) AutoRelay can estimate the optimal location accurately by combining the mmWave model, real-time sampling, and compressive sensing theory.
- ii) AutoRelay can quickly arrive at the relay location by making use of online update and continual approaching mechanism.

2. AutoRelay

Relays are common solutions to maintain the connection between disconnected transmitter and receiver. At the begining, works were concentrated on static relays [6]. Static relays can work for a long term because of its stable power supply. However, it required a large amount of redundant relays to increase the connectivity of such networks. Then, the cooperative strategies and capacity theorems were analyzed[7]. Signal alignment along with its degrees of freedom for MIMO relay channels with clustered pairwise exchange was also studied[8]. With the development of UAVs, mobile relays took charge[9]. The performance of UAV based wireless relay communications were analyzed[10]. Mobile relays were used to offer high quality LTE services[11]. But these relays does not fit the relay problem to ensure the quick search of optimal location.

2.1. System Model

The basic system includes one transmitter (Tx) and one receiver (Rx). The transmitter transmits signal to the Receiver, but due to its shorter wavelength, it is blocked by obstacles.

2.1.1. Space Model A 3D space is set up by taking the mmWave transmitter as the origin o with coordinates (0, 0, 0) and the mmWave receiver as the point c with coordinates (n_1, n_2, n_3). The optimal relay location is somewhere in this 3D space.

A 3D matrix Q is created to represent the link qualities where (n_1, n_2, n_3) are the scales of the three dimensions. The space is split into (n_1, n_2, n_3) small cubes and link quality in one certain cube is assumed to be same. The value of each cube Q (i, j, k) of the matrix denotes the link quality of mmWave transmission if the relay is set in (i, j, k).

2.1.2. Link Quality Model In order to measure the link quality for each location,a model is provided for mmWave transmission by considering the large-scale path loss, the shadowing loss and the non-shadowing loss.

Both the source side and the destination side should be considered for link quality evaluation. The obvious method to describe the link quality is the received transmission quality from certain points. The transmission quality at the destination point received from the relay point is equivalent to that at the relay point received from the destination point. So, we consider the receiving qualities at the relay point from both the source point and the destination point as factors in representation of link quality metrics.

Let S_o be the sending quality at the origin o, S_c be the sending quality at the receiver c, R_o(i, j, k) be the receiving quality from point o at point (i, j, k) and R_c(i, j, k) be the receiving quality from point c at point (i, j, k).

The link quality Q (i, j, k) is described as the product of the receiving qualities R_o (i, j, k) at point (i, j, k) from origin point o and R_c (i, j, k) at point (i, j, k) from receiver c, according to data transmission model. The product of these two values is taken as the metric to achieve the optimal transmission quality for the relay.

2.2. Mobile Relay System

The mobile relay aims to find the optimal location with the best link quality. A strategy of path choosing is developed for the relay and a method to update the quality matrix Q in an online manner. The mobile relay uses the data transmission model and flies along the path according to the strategy in the space with the initialization of the quality matrix Q.

The space model is based on the transmitter o (0, 0, 0) and the receiver c (n_1, n_2, n_3). A strategy is designed for the relay to find the optimal point $P(P_x, P_y, P_z)$ so that the link quality $Q(Q_x, Q_y, Q_z)$ is maximized, given the start point of mobile relay $S(S_x, S_y, S_z)$ and the initial quality matrix Q. Because noise, multi-path, dynamics, and other factors cannot be perfectly modeled, the link quality calculated by the theoretical model contains errors. Hence, the online measurement and updates are necessary to correct the theoretical results. However, measuring the link quality at all locations of candidate space costs too much time. So the AutoRelay only samples at a few locations and estimates other locations using compressive sensing.

2.2.1. Optimal Location Finding It is assumed that the points o, c, s are known initially. These points are obtained by the mobile using the camera to observe the location of the transmitter and the receiver. Then the corresponding 3D space is setup by taking the location of transmitter as the origin o (0, 0, 0) and the location of the receiver as $c(n_1, n_2, n_3)$. Also, the mobile relay start is setup at the initial point $S(S_x, Sy, S_z)$.

Then, the initial values for the 3D quality matrix Q are set. The quality matrix is initialized taking only the large-scale path loss into account, as shadowing and non-shadowing losses are relatively less as compared to large scale path loss. Any form of

IOP Conf. Series: Materials Science and Engineering **396** (2018) 012043 doi:10.1088/1757-899X/396/1/012043

the distortion on the transmission link quality is corrected in the matrix update using the sample real data.As a result, the initialization does not consider the distortion or the attenuation. Optimal location is selected and the relay flies towards that location.

Samples are taken for updating the link quality matrix. Signals are sent by both the transmitter and the receiver to the relay to acquire the link quality of each point. The data contained in the signals includes the information about the sender which includes the signal strength S and the transmission start time. Signal loss along the path is obtained. After getting the receiving quality R_o from the transmitter point o and R_c from the receiver point c, the link quality value for the current location is calculated. The procedure ends if the current location is identified as the optimal one. Else, the relay moves to the next location.

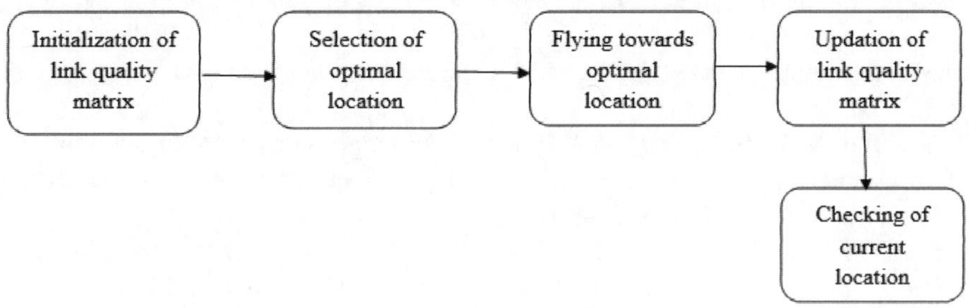

Figure 1. Procedure to find optimal location

3. Conclusions

In the simplest scenario, a communication system consists of a transmitter and a receiver. Effective communication takes place only when the messages send by the transmitter reaches the receiver and the receiver is successful in decoding the message. If the distance between the transmitter and receiver is greater than their transmission range, communication does not takes place. Relays become an important factor in this scenario.

The method reviewed is used to bypass the obstacles or prolong the communication range to compensate for the disconnection between transmitter and receiver. An Auto Relay method which samples the real-time link qualites and keep updating the matrix is designed to quickly determine the optimal location so as to tackle the challenges of accurate location determination.

References

[1] Linghe Kong, Linsheng Ye, Fan Wu, Meixia Tao, Guihai Chen, Athanasios V. Vasilakos, Autonomous Relay for Millimeter-Wave Wireless Communications, *IEEE Journal on Selected Areas in Communication*, 2017

IOP Conf. Series: Materials Science and Engineering **396** (2018) 012043 doi:10.1088/1757-899X/396/1/012043

[2] Zhouyue Pi and Farooq Khan. , An Introduction to Millimeter-Wave Mobile Broadband Systems.,*IEEECommunications Magazine* ,2011.

[3] T. S. Rappaport, R. W. Heath, R. C. Daniels, and J. N. Murdock ,. Millimeter wave wireless communications,,2015

[4] S. Geng, J. Kivinen, X. Zhao, and P. Vainikainen , Millimeter-wave propagation channel characterization for short-range wireless communications. , *IEEE Transactions on Vehicular Technology*,2009; 58(1):313

[5] Y. Zhu, Z. Zhang, Z. Marzi, C. Nelson, U. Madhow, B. Y. Zhao, and H. Zheng. ,Demystifying 60GHz outdoor picocells ,*Synth. Lectures Data Manage.* , 2014

[6] G. Kramer, M. Gastpar, and P. Gupta ,Cooperative strategies and capacity theorems for relay networks,*IEEE Transactions on Information Theory*,2005; 51(9):30373063

[7] K. Liu and M. Tao, Generalized signal alignment: On the achievable DoF for multi-user MIMO two-way relay channels,*IEEE Transactions on Information Theory* , 2014;61(6):33653386

[8] R. Wang, X. Yuan, and M. Tao ,. Degrees of freedom of MIMO multiway relay channel with clustered pairwise exchange,*IEEE Journal on Selected Areas in Communications*,2015;33(2):337351, 2015

[9] M. F. Pinkney, D. Hampel, and S. DiPierro . Unmanned aerial vehicle (UAV) communications relay,*In IEEE MILCOM*, 1996

[10] P. Zhan, K. Yu, and A. L. Swindlehurst Wireless relay communications with unmanned aerial vehicles: Performance and optimization,*International Journal of Computer Applications*,IEEE Transactions on Aerospace Electronic Systems

[11] L. Chen, Y. Huang, F. Xie, Y. Gao, L. Chu, H. He, Y. Li, F. Liang, and Y. Yuan, Mobile relay in LTE-advanced systems,*IEEE Communications Magazine*,51(11):144151, 2013..

[12] D. L. Donoho, Compressed sensing,*IEEE Transactions on Information Theory*, 52(4):12891306, 2006.

[13] FCC , Millimeter wave Propagation: Spectrum Management Implications,, *Office of Eng. and Tech., Bulletin no. 70, July*, 1997

A Method for the Investigation of Bearing Vibration Based on Spectrogram Image Comparison

Arun P[1], Madhukumar S[2], Careena P[3]

[1&2] Department of Electronics and Communication Engineering,
St. Joseph's College of Engineering and Technology, Palai-686 579
[3]Department of Electronics & Communication Engineering,
Amal Jyothi College of Engineering, Kerala, 686518, India

E-mail: [1]arunpeethambaran@yahoo.com

Abstract. The possibility of analyzing the condition of bearing systems from the vibrations generated during their operation, by means of signal processing methods, has been of extensive research, over last few years. As, vibration signal is highly non-stationary, time as well as frequency domain features cannot designate its behavior well. Though, Spectrogram is a time-frequency domain feature extraction method, its analysis is tedious and maybe, subjective. In the proposed method, the spectrogram images of the normal vibration data is compared with that of the contextual vibration, using Peak Signal-To-Noise Ratio (PSNR). It has postulated that, the pattern similarity between the contextual spectrogram and baseline is little when the bearing is faulty. The PSNR between the spectrogram image of normal bearing vibration data and the baseline is different from those between the baseline and vibration data corresponding to Inner Race Failure (IRF), Roller Element Defect (RED) and Outer Race Failure (ORF).The PSNR analogous to the vibrations picked up from normal and faulty bearings vary with a P value of 4.58445×10^{-20}. The method can discriminate faulty bearings with, 96.77% sensitivity and 100% specificity.

1. Introduction

Bearings are the prominent components in most of the machines. They set up free rotational or linear movement, by lessening friction. Bearing may turn broken because of substantial stacking, lacking grease and inadequate fixing. A few investigations have expressed that the significant reason for failure of machines is because of bearing problem [1-2]. Sudden failure of the bearing may crush up different parts of the machines additionally and may raise downtime costs. Regular strategies for observing bearing wellbeing incorporate motor current analysis, wear debris analysis, noise monitoring, temperature monitoring, vibration monitoring, chemical analysis, Laser displacement measurement etc [3-4].

The misalignment of the bearing produces vibration and noise. Signal processing techniques are the most vital part of the systems intended for the automated analysis and interpretation of bearing vibration. Feature extraction techniques used in these automated systems generally belong to time domain, spectral domain [5-6] or Time-Frequency (TF) domain. Peak/peak to peak and Root Mean Square (RMS) amplitude of vibration, skewness, kurtosis, crest factor, measures of Central Tendency (CT) , impulse factor, shape factor and clearance factor[7] are few of such time domain features usually employed.

K. Czarnecki [8] suggested that energy density estimated via ordinary spectrogram or Instantaneous Frequency Rate Spectrogram (IFRS) is a reliable feature as far as mechanical vibrations are concerned. R. Klein [9] demonstrated that faults in turbofan engine can be detected from the spectrogram of its vibrations, computed via Wigner-Ville, wavelets or Short Time Fourier transform (STFT). A. Belsak and J. Prezelj [10] used spectrogram of the vibrations produced in gear units for monitoring their condition. S. K. Yadav and P. K. Kalra [11], demonstrated that the spectrogram of vibration has the potential to detect faults in single cylinder four stroke IC engine. Griffaton et.al [12] suggested that damaged bearings in highly sophisticated systems like aircraft engines can be identified just via simple visual inspection of spectrograms of vibration data. Yu et.al [13] employed a modified version of the ordinary spectrogram, termed as 'adaptive gaussian chirplet spectrogram', for analyzing the vibrations in rolling bearings. In any case, manual understanding of spectrogram is repetitive as it relies upon the mastery of the onlooker. In addition, the subjectivity characteristic in its manual judgment is a mind boggling issue. Image processing methods can be employed for the automated evaluation and comparative interpretation of spectrograms. As a base to such efforts Li et.al [14], had used the images of spectrum of vibration data, for diagnosing faults in bearings. Hua et.al [15] used Quaternion invariable moment to account for color features of the vibration spectrograms. To interpret the spectrogram Klein et.al [16] solely utilized the ridge information.

In the spectrogram image, the variation of pseudo colors or grey levels in a particular column reflects how different the spectral power of a particular frequency component is, between distinct epochs. Thus, the variation of pseudo color or grey level in each column accounts for the variation of spectral power of a particular frequency component over time ie. the non-stationarity of the signal. Similarly, the variation of pseudo color or grey level in each raw account for the relative distribution of spectral power over different frequency components within a single epoch itself. Quaternion invariable moment of the spectrogram image [15] does not seem to be sufficient to characterize the non-stationarity behavior of vibration data. Likewise, the ridges [16] alone cannot express continuous variation in the pseudo colors or grey levels in the spectrogram image which in turn reflects the time varying spectral features of the vibration data.

A method for condition monitoring and fault diagnosis of bearing system using pattern matching of spectrogram images is proposed in this paper. The spectrogram images of the baseline or reference vibration data which belongs to a normal bearing and contextual vibration data, whose fundamental bearing state is unknown, are compared for their mutual similarity as far as the variation in pseudo color or grey level with the help of Peak Signal To Noise Ratio (PSNR) .It is postulated that when the bearing is not healthy, the grey level distribution of the spectrogram image of its vibration could be entirely diverse from that of the baseline image. In this context, the PSNR between them would be comparatively less. Rest of the paper is organized as follows. The details of vibration data used in the analysis and mathematical formulation of spectrogram images and PSNR are furnished in section 2.The statistical significance of PSNR between spectrogram images of base line and contextual vibration data to yield an accurate classification of faulty and normal bearings is analyzed in section 3.

2. Methodology

The method proposed in this paper, the spectrogram image of the contextual vibration has to be compared with a baseline spectrogram image of the vibration data from a healthy bearing, in terms of similarity in color or grey level distribution to identify whether the status of the bearing from which the contextual vibration is recorded is healthy or not. The bearing vibration data used for developing the automated method is availed from Prognostics Center of Excellence, National Aeronautics and Space Administration (NASA) [17].

The vibration information utilized in this paper comprises two conditions of bearing ie; healthy and faulty. Three kinds of faults (healthy, Inner Race Failure (IRF), Roller Element Defect (RED) and Outer Race Failure (ORF).) are mainly considered. Every vibration sample is of one second span containing 20480 samples. The aggregate data collection contains 120 recordings,, 30 recordings, from each fault group. That implies total data set comprises 30 recordings from healthy bearings and 90 recordings from faulty bearings.

To make the base line spectrogram image more consistent, the spectrograms of 8 consecutive vibrations, recorded at 10 minutes interval, from a healthy bearing, are averaged before color mapping. The spectrogram is converted to image or visual form through color or grey scale mapping. For a vibration signal 'x(n)' sampled at a rate of 'Fs', split in to 'Q' segments of length 'M' samples each, given M<N [18], the STFT of a segment of length 'M',

$$X_i(k) = \sum_{l=0}^{M-1} w(l)x(l)e^{\frac{-j2\pi kl}{M}}, 1 \leq i \leq Q \qquad (1)$$

IOP Conf. Series: Materials Science and Engineering **396** (2018) 012044 doi:10.1088/1757-899X/396/1/012044

where $w(l)$ is the window function. The power spectrum of each segment,

$$P_i(k)=\frac{1}{\sum_{k=0}^{M-1}|w(k)|^2}|X_i(k)|^2 \tag{2}$$

Discrete spectrogram 'S' of the vibration data,

$$S=\begin{bmatrix} P_0(0) & P_0(1) & P_0(2) & \cdots\cdots & P_0(Q\text{-}1) \\ P_1(0) & P_1(1) & P_1(2) & \cdots\cdots & P_1(Q\text{-}1) \\ P_2(0) & P_2(1) & P_2(2) & \cdots\cdots & P_2(Q\text{-}1) \\ \vdots & \vdots & \vdots & \cdots\cdots & \vdots \\ P_{M-1}(0) & P_{M-1}(1) & P_{M-1}(2) & \cdots\cdots & P_{M-1}(Q\text{-}1) \end{bmatrix} \tag{3}$$

As vibration data is highly non-stationary, Hamming window is used for computing the spectrogram in this paper. Number of samples in each segment is set to 4550 following the usual practice of setting M=N/4.5. The total number of samples, 'N' in each vibration recording is 20480 and the sampling frequency, Fs is equal to 20 KHz. The segments have 50% overlap. The similarity between the spectrogram image of the contextual vibration data and baseline spectrogram image is quantified using PSNR. The PSNR for an image of size M*N between the base line spectrogram image 'X' and the spectrogram image 'Y' of the contextual vibration data is computed as [19],

$$PSNR=20\log\left(\frac{Max(X)}{\sqrt{\frac{1}{MN}\sum_{i=1}^{M}\sum_{j=1}^{N}[X(i,j)-Y(i,j)]^2}}\right) \tag{4}$$

The computation of spectrogram, color mapping, comparative evaluation of the spectrogram images with PSNR and the statistical evaluation of the PSNR values are done in Matlab®. The statistical significance of PSNR between the spectrogram images of the contextual vibration data and baseline spectrogram image is verified for their capability to differentiate healthy, IRF, ORF and RED via Kruskal-Wallis. The flow diagram of the processes involved in the proposed method is illustrated in fig.1.

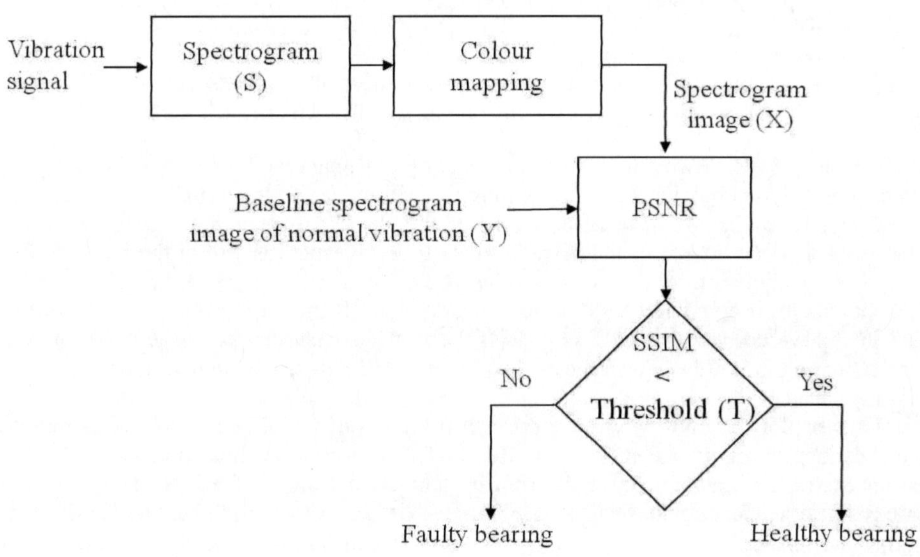

Figure 1. Block diagram of the processes involved in the proposed framework.

IOP Conf. Series: Materials Science and Engineering **396** (2018) 012044 doi:10.1088/1757-899X/396/1/012044

3. Results and Discussions

The baseline spectrogram image of the vibration data is shown in fig. 2 (a). Spectrogram images of the vibration data analogous to healthy, IRF, ORF and RED bearings are furnished in fig. 2 (b) – (e). The color distribution of the spectrogram images corresponding to the faulty bearings is completely different from the baseline in fig. 2(a). It can be noted that the baseline spectrogram image (fig. 2 (a).) is similar to the spectrogram image of healthy class (fig. 2(b)) than the spectrogram image of the fault classes (fig. 2 (b) – (e)).This is a strong indication that the color distribution of spectrogram images is a trustworthy feature to distinguish normal and faulty bearings.

From table 1, the PSNR between spectrogram images of vibration data analogous to healthy, IRF, ORF and RED and the baseline are 39.38±2.52, 32.06±5.82, 30.40±2.45 and 23.87±0.95, respectively. It can be noted that, the PSNR between baseline and spectrogram images of vibration data corresponding to faulty bearings are significantly less than PSNR between baseline and spectrogram images of vibration data corresponding to healthy bearings. As far as numerical values of PSNR between baseline and spectrogram images of vibration data corresponding to ORF is concerned, they are confined to a narrow range compared to numerical values of the PSNR between baseline and spectrogram images of vibration data corresponding to other classes.

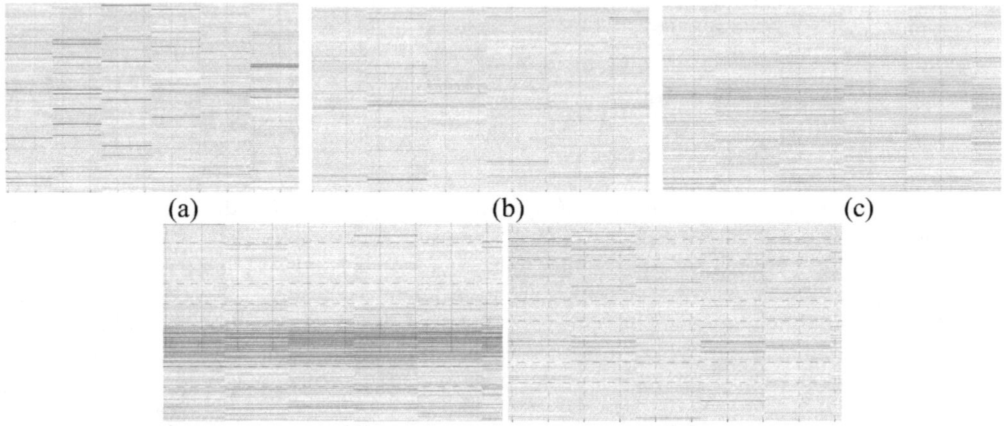
(a) (b) (c)
(d) (e)

Figure 2. Spectrogram images of bearing vibrations of normal and faulty classes (a) base line spectrogram image (b) healthy bearing (b) IRF (c) ORF (d) RED

The ranges of the PSNR between the base line spectrogram image and a spectrogram image of the vibration data of different bearing states are as follows; healthy:32.10-46.16, IRF: 16.93 - 39.51, ORF: 21.18-25.32 and RED: 25.34-34.80, respectively. If the PSNR between the base line spectrogram image and a spectrogram image of the vibration data to be analysed is within the range of 25.34 to 32.10, it can be concluded that the bearing is either healthy or RED. For PSNR values between 34.80 and 39.51 then the bearing could be either healthy or of IRF. If the PSNR values lies between 21.18 and 25.32 then it would be either ORF or IRF. If the PSNR is specifically above 46.16, it is a clear sign that the bearing is healthy. This reveals that, from the PSNR values, faulty and healthy bearings can be distinguished easily.

The Box-Whisker plot of PSNR between spectrogram images of vibration data corresponding to four classes of bearings, healthy, IRF, ORF and RED and the baseline is illustrated in fig. 3. The box whisker plot states the assumptions made from the numerical values of PSNR. The box of PSNR between baseline and spectrogram images of vibration data corresponding to healthy bearings is lying sufficiently apart spatially from the boxes of PSNR between baseline and spectrogram images of vibration data corresponding to faulty bearings. Similarly the box of ORF is lying sufficiently apart spatially from the boxes of other class of vibration data.

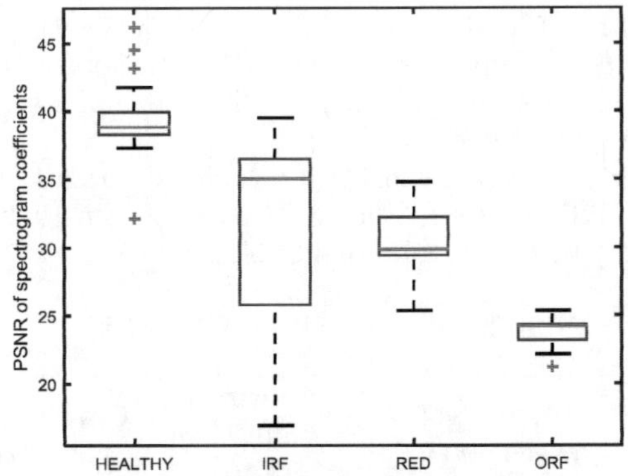

Figure 3. Box-Whisker plot of PSNR between spectrogram images of vibration corresponding to healthy, IRF, ORF and RED and the base

Table 1. Numerical values of PSNR between spectrogram images of vibration data corresponding to four classes of bearings, healthy, IRF, ORF and RED and the baseline

Sl No:	Healthy	IRF	RED	ORF
1	39.41	38.59	29.81	24.28
2	39.62	35.11	29.71	24.29
3	39.95	36.36	29.94	23.28
4	43.13	35.3	29.63	24.29
5	37.68	39.51	30.33	25.23
6	38.68	38.28	29.75	24.32
7	39.46	36.5	30.88	22.32
8	46.16	38.4	31.65	24.31
9	39.6	35.4	33.3	23.21
10	38.3	36.58	33.69	23.18
11	39.97	36.5	34.8	24.29
12	43.17	35.66	34.1	25.32
13	39.02	35.83	32.73	24.28
14	38.61	34.95	32.11	24.39
15	38.82	34.45	33.24	24.48
16	38.74	37.17	33.68	24.23
17	38.87	34.67	32.22	24.18
18	38.31	36.24	31.73	24.17
19	44.53	29.51	30.05	24.21
20	37.47	28.79	29.22	23.21
21	38.18	27.03	29.75	23.18
22	41.75	26.86	29.54	23.13
23	32.1	25.62	29.44	23.11
24	37.31	25.78	29.53	23.11
25	38.74	25.39	28.27	25.13
26	39.21	25.03	27.67	24.18
27	40.21	25.43	27.43	24.42
28	37.72	25.04	26.21	25.12
29	38.73	25.02	26.17	22.14
30	37.91	16.93	25.34	21.18

The statistical significance of PSNR to distinguish healthy, IRF, ORF and RED is tested via Kruskal-Wallis. ANOVA table of the test is furnished in table 2. The Chi-Square value (H) obtained from the Kruskal-Wallis test is 93.16, which is greater than the critical value 12.838. The Chi-Square is outside the critical region and it shows that the PSNR values corresponding to at least one of the bearing state could be different from others. From box whisker it has already been inferred that the class which differ from others in terms of PSNR is ORF. The PSNR corresponding to the vibrations acquire from normal and faulty bearings vary with a 'P' value of $4.58445x10^{-20}$.

Table 2. Kruskal-Wallis ANOVA table

Source	Sum of Squares (SS)	Degree of Freedom (DF)	Mean Square (MS)	Chi-Square value (H)	Probability value (p)
Colu mns	112721.3	3	37573.8	93.16	$4.58445 \text{ x}10^{-20}$
Error	31262.7	116	269.5		
Total	112721.3	3	37573.8		

4. Conclusions

A signal processing method for the fault identification of bearing systems, based on pattern matching of spectrogram images of the contextual vibration data with spectrogram images of the baseline or reference vibration data which goes to a normal bearing by PSNR was proposed in this paper. It was observed that, if the PSNR between the spectrogram image of the contextual vibration and baseline is exclusively beyond 46.16, it is apparent that the bearing is healthy. The PSNR analogous to the vibrations attained from normal and faulty bearings vary with a 'P' value of 4.58445x10-20.The suggested method is capable of detecting faulty bearing with 96.77% sensitivity and 100% specificity. Through the proposed method, subjectivity in analysis of spectrogram has been completely avoided.

Acknowledgment

The authors would like to thank Prognostics Center of Excellence, National Aeronautics and Space Administration (NASA) and the Center for Intelligent Maintenance System, University of Cincinnati, for providing the experimental data.

References

[1] G. Z. Li, Y. Jiang, C. Hu and Z. Peng, Recent progress on decoupling diagnosis of hybrid failures in gear transmission systems using vibration sensor signal: a review, Measurement, Vol.90,pp.4-19,Aug.2016.

[2] Z. Li, Y. Jiang, X. Wang and Z. Peng, Multi-mode separation and nonlinear feature extraction of hybrid gear failures in coal cutters using adaptive nonstationary vibration analysis, Nonlinear Dynamics,Vol.84.,pp. 295–310, April 2016.

[3] Y. Jiang and H. Zhu,The nonlinear dynamics response of cracked gear system in a coal cutter taking environmental multi-frequency excitation forces into consideration, Nonlinear Dynamics, Vol. 84, pp. 203–222,April 2016.

[4] M.S. Safizadeh and S.K. Latifi, Using multi-sensor data fusion for vibration fault diagnosis of rolling element bearings by accelerometer and load cell, Information fusion, Vol.18, pp.1–8, July 2014.

[5] G. Carson, A. J. Mulholland, A. Nordon, A. Gachagan and G. Hayward, Theoretical analysis of ultrasonic vibration spectra from multiple particle-plate impacts, in IEEE Transactions on Ultrasonics, Ferroelectrics, and Frequency Control, vol. 56, pp. 1034-1041, May 2009.

[6] M. Amar, I. Gondal and C. Wilson, Vibration spectrum imaging: a novel bearing fault classification approach, in IEEE Transactions on Industrial Electronics, Vol. 62, pp. 494-502, Jan. 2015.

[7] Tuncay Karacay and Nizami Akturk, Experimental diagnostics of ball bearings using statistical and spectral methods, Tribology International, Vol. 42, pp. 836–843, June 2009.
[8] K.Czarnecki, The instantaneous frequency rate spectrogram, Mechanical systems and signal processing, Vol. 66,pp. 361-373, Jan.2016.
[9] R. Klein, A method for anomaly detection for non-stationary vibration signatures, Annual conference of the prognostics and health management society, Vol. 4, pp.1-7, Oct. 2013.
[10] A. Belssak and J. Prezelj, Analysis of vibrations and noise to determine the condition of gear units, Advances in Vibration Analysis Research, April 2011.Online [Available at], http://www.intechopen.com/books/advances-in-vibration-analysis-research/analysis-of-vibrations-and-noise-to-determine-the-condition-of-gear-units.
[11] S. K. Yadav and P. K. Kalra, Automatic fault diagnosis of internal combustion engine based on spectrogram and artificial neural network, International conference on robotics, control and manufacturing technology, pp. 101-107, Jan. 2010.
[12] J. Griffaton, J. Picheral and A. Tenenhaus, Enhanced visual analysis of aircraft engines based on Spectrograms, Proceedings of the leuven international conference on noise and vibration engineering, pp.2809-2822, April 2015.
[13] H. Yu, Q. Guo, J. Hu and A Xu, Rolling bearings fault diagnosis based on adaptive gaussian chirplet spectrogram and independent component analysis, Advances in Natural Computation, Vol. 4221, pp. 321-330, Sept. 2006.
[14] W. Li, M. Qiu, Z. Zhu, B. Wu and G. Zhou, Bearing fault diagnosis based on spectrum images of vibration signals, Measurement Science and Technology, Vol. 27, pp. 035005, Feb. 2016.
[15] L. Hua, Y. Qiang, J. Gu, L. Chen, X. Zhang and H. Zhu, Mechanical fault diagnosis using color image recognition of vibration spectrogram based on quaternion invariable moment, Mathematical problems in engineering,Vol. 2015 , pp. 11, July 2015.
[16] R. Klein, E. Masad, E. Rudyk and I. Winkler, Bearing diagnostics using image processing methods, Mechanical systems and signal processing, Vol. 45, Issue 1, pp. 105-113,March 2014.
[17] J. Lee, H. Qiu, G. Yu, and J. Lin, Rexnord Technical Services: Bearing Data Set. Moffett Field, CA: IMS, Univ. Cincinnati. NASA Ames Prognostics Data Repository, NASA Ames, 2007.
[18] A.V. Oppenheim and R.W. Schafer, Discrete-Time Signal Processing, Prentice-Hall, Englewood Cliffs, NJ, pp. 713-718, 1989.
[19] J. Joseph and R. Periyasamy, A fully customized enhancement scheme for controlling brightness error and contrast in magnetic resonance images, Biomedical Signal Processing and Control, vol. 39, pp. 271-283, Jan.2018

IOP Conf. Series: Materials Science and Engineering **396** (2018) 012045 doi:10.1088/1757-899X/396/1/012045

A Noise Reduced Hard Thresholding method for Noisy Sparse Reconstruction

Sabarinath G
Dept. of Electronics and Communication Engineering
St. Joseph's College of Engineering & Technology, Palai, Kerala
E-mail: sabari@sjcetpalai.ac.in

Babita R Jose
Division of Electronics
CUSAT, Kochi, Kerala

Abstract. The problem of noisy sparse signal recovery has been studied in this paper. The generalized expectation maximization (GEM) algorithm provides solutions over state of the art convex relaxation methods for sparse signal recovery by considering the noisy measurements as missing data. However, in the case of images, artifacts in the reconstructed signal renders it to have low quality, especially based on qualitative performance metrics and so are unsuitable for further application specific processing. The problem may be reduced significantly by minimizing the total variation in the signal considering the constraints created by the statistics of noise in the reconstructed image. A modified iterative GEM for low noise reconstruction is presented and analyzed. Even though time consuming, simulations and numerical analysis show significantly higher performance metrics, especially PSNR and Structural Similarity, which imply reconstructed images may be suitable for further processing.

1. Introduction

The popular Shannon/Nyquist sampling theorem requires that a signal needs to be sampled at approximately twice the highest frequency components for accurate reconstruction. In certain applications like digital imaging and video, the Nyquist rate tends to be very high and hence, the number of samples acquired per seconds also increases. Storage and transmission of such high volume data becomes difficult, leading to the requirement of compression. Increasing the sampling rate also turns out to be expensive in terms of its computational complexity. The method of Compressive Sensing overcome these disadvantages wherein accurate reconstruction of the signal is possible even if the sampling frequency is less than twice the maximum frequency. Signals are acquired compressed, using non adaptive linear projection methods and are reconstructed using an optimization techniques like convex relaxation, greedy pursuit and probabilistic models [1–3].

Among various thresholding methods for the reconstruction of sparse signals, the GEM hard thresholding [4] uses the principle of noise variance. The sparse signal and the noise variance are found out by the GEM algorithm and the signal is reconstructed by probabilistic model with L_0 - norm. The reconstructed images using GEM algorithm is observed to exhibit low quality in terms of qualitative metrics, rendering such reconstructed images unsuitable for further application specific processing. This paper aims to increase the overall reconstructed image quality especially in the higher quantization levels by reducing noise effects in the reconstructed signal based on multi-grid noise reduction algorithms [5]. Performance based on Structural Similarity Index [12] is also reported.

2. Background Theories

2.1. The GEM Sparse Recovery Algorithm

Compressive sensing theory enables a sparse or compressible signal to be acquired with far fewer measurements than what is required as per the Nyquist/Shannon sampling theory. A signal is generally modeled as

$$y = Ax + b \tag{1}$$

where $y = [y_1, y_2, ...y_n]^T$, A is the NXM sensing matrix, x is the unknown sparse vector of dimension $MX1$ containing r non zero elements (rM), b is the additive white Gaussian noise with zero mean and covariance matrix σ^2. Here the noise variance is assumed to be unknown. In this model there are two unknown parameters x and σ^2. Then the set of unknown parameters can be written as

$$\Theta = (x, \sigma^2) \in \Theta_r \tag{2}$$

where Θ_r is the parameter space. Parameter space means it is the set of all possible combinations of values for all the different parameters in the model. This can be written as

$$\Theta_r = S_r x(\theta, +\alpha) \tag{3}$$

where S_r is the sparse signal parameter space and r is the sparsity level. Here, assuming that the sparsity level is known and the noise variance σ^2 is positive, the elements of y are quantized into code word b, represented as $[b_1, b_2 b_N]^T$, where b_i indicate the quantization interval.

$$y_i \in D(b_i) = [l(b_i), u(b_i)] = [l_i, u_i], l_i < u_i, i = 1, 2N \tag{4}$$

where l_i and u_i are the lower and upper boundaries of the quantization interval. Our aim is to estimate the parameters θ from the quantized data b and the unobserved or the missing data y. For finding the missing data the concept of GEM algorithm was developed [4]. First the joint distribution of the observed data b and the missing data y given the unknown parameters θ are solved for [4]. Then the marginal log likelihood function of θ is obtained by integrating y from the joint distribution of b and y.

$$L(\theta) = ln(P_{b/\theta}(b/\theta)) \tag{5}$$

For the computation of $L(\theta)$ the noise variance σ^2 is positive and set the parameter space accordingly in (3). The maximum likelihood (ML) estimate of θ is found out by maximizing L(). The exact ML estimate requires a combinatorial search and it is difficult task.

Solutions are obtained in two steps: expectation (E) - step and maximization (M) - step. In the expectation step, the missing data are estimated and in maximization step the likelihood function is increased under the assumption that the missing data is known. Assuming that the parameter estimate $^(p) = (x^(p), ()^(p))$ is known , where p indicate the iteration index, the E-step can be found out by finding the expected complete data log likelihood function of (5).

$$Q(/^(p)) = E_y/b, [ln_y, b/(y, b/)/b, ^(p)] = -1/2Nln(2\pi\sigma^2) - E_y/b, [(y - H_s)^T(y - H_s)/b, ^(p)]/(2\sigma^2) \quad (6)$$

Expectation step reduces the Bayesian minimum mean square error and variance using the mean and variance of the truncated pdf [6]. Similarly the M-step increases the expected complete data log likelihood function. Here finding the new parameter estimate by the equation

$$x^{(p+1)} = Tr(s^{(p)} + (1/c^2)H^T(\hat{y}^{(p)} - H_s)) \quad (7)$$

c being the step size coefficient which should satisfy the inequality

$$c \geqslant \rho H \quad (8)$$

where ρH denotes the largest singular value of H and T_r in (7) is the hard thresholding operator. So we get the new parameter $(p+1) = (s(p+1), ()(p+1))$. The iteration between E and M is repeated until two consecutive sparse signal estimates x^p and $x^{(p+1)}$ do not differ significantly. The signal is reconstructed by l_0 - norm which counts the number of non-zero entries in s. This optimization can recover a sparse signal exactly. However, the PSNR of such recovered signal turns out to be low, rendering the signal not suitable for further processing.

2.2. Noise Reduction Methods

The ROF Model [8] considers reducing the total variation or integral of the absolute gradient, thereby reducing unwanted detail preserving edge informations. Given an input signal x, the total variation denoising aims to find an approximate y, with smaller total variation than x, but at the same time close to that of x. Sum of square errors being a measure of closeness, the total variation denoising problem reduces to minimizing the function

$$E(x, y) = \lambda V(y) \quad (9)$$

where λ is termed the regularization parameter and plays a critical role in the denoising algorithm. Differentiating this equation with respect to y a Lengrange equation is arrived at which can be solved for to obtain a solution for y. Various approaches for arriving at a solution has been reported in literature [5–7].

3. The Proposed Model

The sparse signal obtained from the GEM Hard Threshold may be modeled as in 10 to contain an unknown noise η.

$$Y(x, y) = X(x, y) + \eta(x, y) \quad (10)$$

The aim is to compute x of the nonlinear equation 11 with different LaGrange multipliers λ, so as to satisfy the constrained condition in 12

$$L_\lambda(x) = -\bigtriangledown(\frac{\bigtriangledown x}{\sqrt{(\bigtriangledown x^2 + \in)}}) + \lambda(x-y) = 0 \tag{11}$$

$$F(x) = \frac{1}{2}(\parallel x - y \parallel^2 - \sigma^2) = 0 \tag{12}$$

where σ is the variance between the noisy image and the original image and λ is the regularization parameter.

The algorithm is iterative; by first computing weigted λ to further, solve the linear equation (11) to obtain the unique solution $x\lambda_k$. The steps continue until no significant difference in values of x is observed. Numerical examples in the subsequent sections prove this solution to produce higher PSNR to recovered sparse signals, and also higher values of Structural Similarity Index, thereby rendering them useful to further processing.

4. Simulation Results and Discussions

Two dimensional signals have been considered to test the method with a peak signal-to-noise ratio (PSNR) and Structural Similarity Index (SSIM) as performance metrices. Images of size $256X256$ were considered and testing was performed on an intel dual core machine with 2Gb RAM and $Matlab^{(R)}2011b$. The two unknown parameters i.e; sparse signal and the noise variance was computed by the proposed Low-Noise GEM algorithm. The smoothening parameter are kept fixed; upon achieving the residue to the constrained function, both the diffusion equation and the Lagrange multiplier are stopped. The signal was then reconstructed using l_0 norm technique and empirical Bayesian signal estimate. Comparisons on PSNR and SSIM indexes of the proposed algorithm against the state of the art GEM hard thresholding is reported and discussed.

Table 1: PSNR and SSIM values of Bird Image

Quantization Level	PSNR GEM	SSIM GEM	PSNR Proposed	SSIM Proposed
3	55.2	0.39	76.52	0.56
4	56.5	0.43	76.78	0.61
8	58.3	0.48	78.02	0.69
16	60.4	0.53	79.37	0.73

The Bird image in Figure 1 shows (a) the original image (b) sampled and reconstructed using the GEM hard threshold and (c) the proposed low-noise hard threshold solving for the Lagrange Multiplier in 100 iterations. Atificats introduced in (b) is significantly reduced in (c), as visible to the naked eye. Significant improvement in qualitative metrices is also observed. However, some noise with the Salt and Pepper nature is observed in lower quantization levels. This effect is significantly reduced at higher quantization levels as seen in Figure 2(b). Again, since the number of divisions to solve the non-linear equations are lower, a faster convergence is observed

IOP Conf. Series: Materials Science and Engineering **396** (2018) 012045 doi:10.1088/1757-899X/396/1/012045

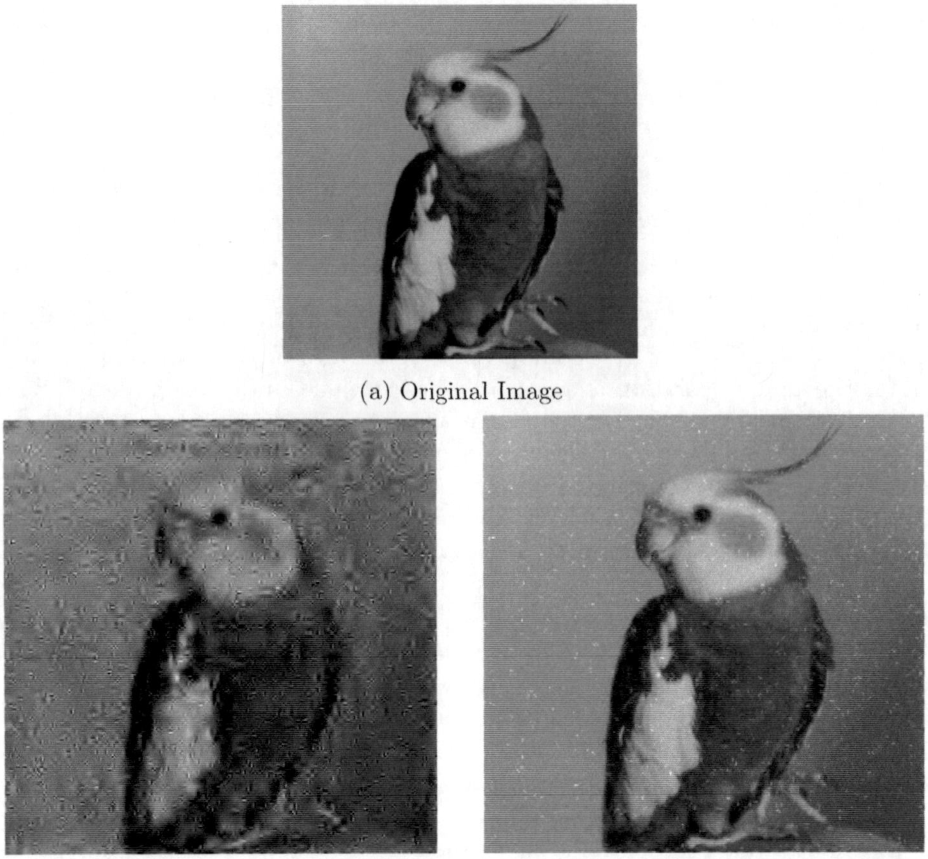

(a) Original Image

(b) GEM Reconstructed (c) Proposed lownoise reconstruction

Figure 1: Bird Image

(a) GEM Reconstructed (b) Lownoise Reconstructed

Figure 2: Cameraman Image

at lower bins. This might also enable faster and easy processing in the event of system on chip developments for standalone systems.

Figure 3 show results pertaining to simulations on complex valued biomedical images. It

Table 2: PSNR and SSIM values of Cameraman

Quantization Level	PSNR GEM	SSIM GEM	PSNR Proposed	SSIM Proposed
3	54.8	0.54	72.39	0.64
4	55.5	0.61	73.88	0.73
8	58.0	0.69	75.87	0.81
16	60.4	0.75	77.03	0.89

is observed that results correspond to observations as in natural images. Such significant improvemet renders compressed sense reconstructed bio-medical feasible for clinical applications.

It may be observed as in Table that at higher quantization levels like 16 and more, the GEM algorithm provides equally good PSNRs as with the proposed method. However, at lower quantization levels, the proposed method provides considerably better PSNRs, make them usable for further processing.

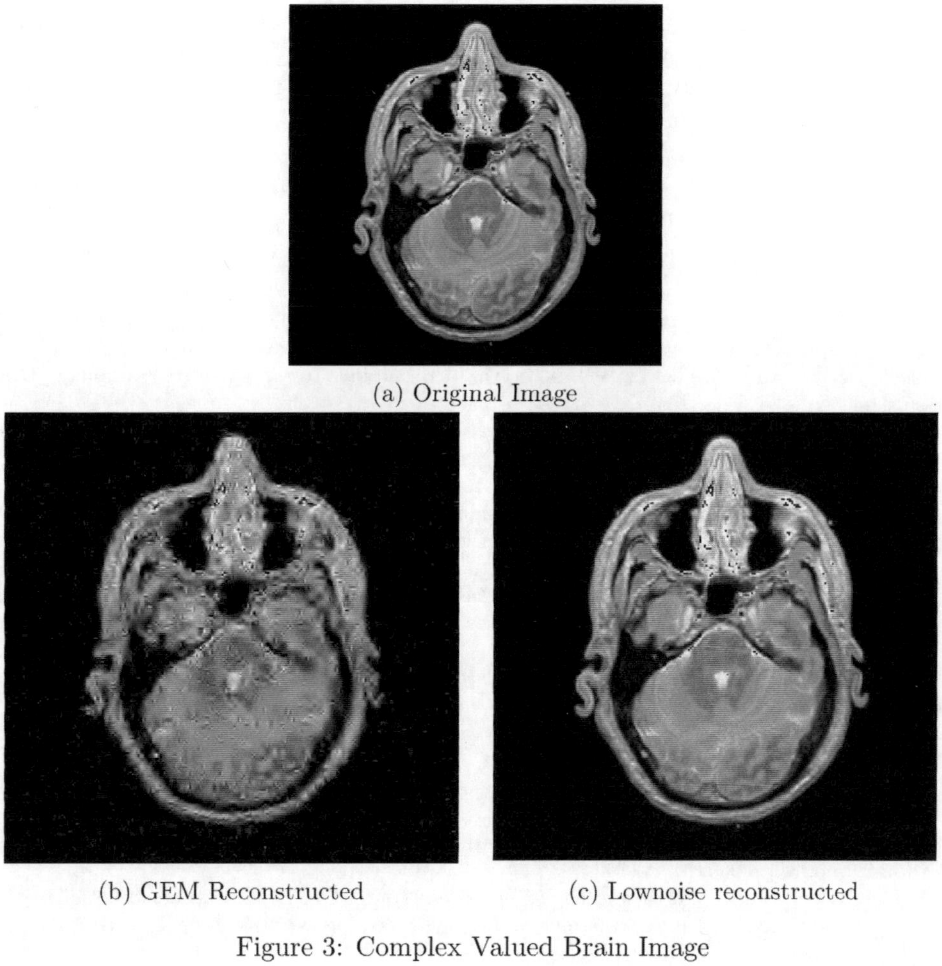

(a) Original Image

(b) GEM Reconstructed (c) Lownoise reconstructed

Figure 3: Complex Valued Brain Image

Table 3: PSNR and SSIM values of Brain

Quantization Level	PSNR GEM	SSIM GEM	PSNR Proposed	SSIM Proposed
3	62.8	0.58	72.78	0.67
4	63.5	0.62	74.64	0.70
8	65.2	0.71	76.44	0.79
16	67.4	0.79	77.55	0.87

5. Conclusions

The issue of accurate recovery of noise corrupted sparse signal has been reviewed in this paper. The generalized expectation-maximization (GEM) hard thresholding reconstruction algorithm which automates noise parameter estimation from quantized measurements, reconstructs noisy sparse signals, however, with artifacts. This might render the reconstructed images not performing up to expectations in further application specific processing. This problem, as reported, can be reduced by minimizing the total variance in the reconstructed signal. An iterative Low-Noise GEM Hard Threshold, which minimizes the variance in the signal has been proposed. Simulations on both natural and biomedical images show that increased PSNR and signal quality can be achieved at comparable computational complexity, thus paving way for better Compressive Sensing applications to be explored.

References

[1] Richard G. Baraniuk., Compressive sensing, IEEE signal processing magazine , July 2007.

[2] Emmanuel J. Candes and Michael B. Wakin., An Introduction To Compressive Sampling, IEEE signal processing magazine , March 2008.

[3] Emmanuel J. Candes and Terence Tao., Decoding by Linear Programming, IEEE transactions on information theory, vol. 51, no. 12, december 2005.

[4] Kun Qiu and Aleksander Dogandzic ., Sparse Signal Reconstruction from Quantized Noisy Measurements via GEM Hard Thresholding, IEEE transactions on signal processing, Volume 60-No.5, May2012.

[5] Jianping, Zhang, K. E Chen and Boyu, An iterative lagrange multiplier method for constrained Total-Variation based Image Denoising, siam j. numer.anal, Vol. 50, No.3, pp.983-1003, May3, 2012.

[6] N. L. Johnson and S. Kotz., Continuous Univariate Distributions, 2nd ed. ed. New York: Wiley, vol. 1, 1994.

[7] R. H. Chan and K. Chen, A multilevel algorithm for simultaneously denoising and deblurring Images, SIAM J. Sci. Comput., pp. 10431063, 2010.

[8] Rudin, L. I.; Osher, S.; Fatemi, E. (1992). "Nonlinear total variation based noise removal algorithms". Physica D 60: 259268. doi:10.1016/0167-2789(92)90242-f

[9] Chambolle, A. (2004). "An algorithm for total variation minimization and applications". Journal of Mathematical Imaging and Vision 20: 8997

[10] A. Zymnis, S. Boyd, and E. Cands, Compressed sensing with quantized measurements, IEEE Signal Process. Lett., vol. 17, pp. 149152, Feb. 2010 .

[11] T. Blumensath and M. E. Davies, Iterative hard thresholding for compressed sensing, Appl. Comput. Harmon. Anal., vol. 27, no. 3, pp.265274, 2009.

[12] Zhou, W., A. C. Bovik, H. R. Sheikh, and E. P. Simoncelli. "Image Qualifty Assessment: From Error Visibility to Structural Similarity." IEEE Transactions on Image Processing. Vol. 13, Issue 4, April 2004, pp. 600612.

IOP Conf. Series: Materials Science and Engineering **396** (2018) 012046 doi:10.1088/1757-899X/396/1/012046

Simulation and Control of MPPT Based Hybrid System with Boost Converter

J Barsana Banu[1], M Balasingh Moses[2] and S Rajarajacholan[3]

[1]Research Scholar, University College of Engineering, BIT Campus, Tiruchirappalli, Tamilnadu, India.
[2]Assistant Professor, University College of Engineering, BIT Campus, Tiruchirappalli, Tamilnadu, India.
[3]Teaching Fellow, University College of Engineering, Anna University, Ariyalur Campus, Tamilnadu, India.
E-mail: barsanajamal@gmail.com

Abstract. A new topology of the hybrid (solar plus wind) power system with enhanced boost converter is proposed in this paper for standalone applications. The proposed hybrid power system comprises of solar panel and permanent magnet synchronous generator (PMSG) based wind turbine followed by rectifier and enhanced boost converter with resistive load. These natural sources deliver the load independently or concurrently depending on the requirement and accessibility of sources. The efficiency of the hybrid power system is increased by maximum power point operation. The most usually used Perturb and Observe (P&O) MPPT technique extracts the maximum power at the certain point. The complete structure was modelled and simulated in MATLAB/ SIMULINK package with variable solar irradiation and wind speed. The outcomes obtainable demonstrate the confirmation of the projected hybrid power system.

1. Introduction

Numerous inadequately occupied remote territories, stand-alone power system is necessary due to the low-power requirements, and it is too expensive to stretch out main power spread line to the remote territories. For instance, wind or solar power scheme can be used as individual or combined in nature to meet out the power requirement in such areas. Regardless, the characteristics of the renewable sources in general are dissimilar than the traditional power sources. The power production from these sources belongs to atmosphere conditions and is sporadic and/or fluctuating. Hence, hybrid power system is needed to fulfill the continuous load requirements.Among the various structures of hybrid system, photovoltaic and wind power systems are complimentary behavior to provide best outcomes. These natural sources are inconsistent due to its discontinuous availability. However, the system dependability can be developed by combining these two sources as a hybrid system as described in [1-3]. This hybrid system can be worked as either standalone or grid-connected operation depending upon the applications. The hybrid system is additional reliable for standalone system when PV-Wind energy system associated with battery storage is reported in [4-6]. The performance of PV–Wind–Diesel–Battery hybrid configuration is analyzed in [7]. Hybrid micro grid systems (wind/hydro/solar) likewise, created with MPPT strategies [8, 9] and the distinctive technique for controlling the wind turbine speed in a micro grid is depicted in [10] yet it needs an excellent speed sensor. The broad review of the different optimum sizing procedures and optimization criteria of the hybrid power system in perspective of small PV, wind, hydro and storage devices is reported in [11]. The power output developed from RES such as PV and wind power system is unreliable because of changing

IOP Conf. Series: Materials Science and Engineering **396** (2018) 012046 doi:10.1088/1757-899X/396/1/012046

every time to time. MPPT controller can be used to get the maximum power obtainable from the two natural sources. Various MPPT procedures are given in the literature. Among them perturb and observe MPPT technique is widely utilized because of simple structure. The paper [12] comprises of solar and DFIG driven wind hybrid system with boost converter. This system requires three distinct controllers such as MPPT controller utilized alone for PV; DFIG needed rotor and grid side controller with special converter and control circuit to regulate the voltage and rotor speed. An topology of PV-wind PMSG hybrid generator is proposed [13] with two low price controllers like MPPT and hysteresis controllers, which are working independently to regulate the duty cycle of the DC-DC converter and the inverter for extracting the most extreme power from the solar and wind power system. These papers [12] and [13] required two additional controllers. During the power scarcity by the hybrid system, a battery bank is needed to satisfy the load demand as suggested in [14] still it needs a complicated vector control method in addition with solar and wind MPPT controllers.

By using fuzzy rules, duty cycle of the boost converter is generated to track the maximum power for both solar and wind power system individually. Hence, the reliability of the HPS can be enhanced by the closed-loop system is reported in [15]. In the paper [16], PV-wind, energy has been coordinated together utilizing the combination of CUK-SEPIC converters, CUK converter is operated by the solar cell and SEPIC converter works under wind turbine. Therefore, this methodology has suggested rather than the conventional converter techniques to reduce current harmonics. The soft switching dc-dc converter discussed in [17] is incorporated with wind-PV hybrid configuration is proposed in this paper. The simulation of wind- PV with common MPPT controller is utilized to control the power devices of the enhanced boost converter to extract MPP are done in MATLAB. Results are investigated by varying contribution parameters such as irradiation level, temperature for solar source and wind speed for wind turbine. The modeling of solar, wind sources and PMSG are described in detail. The simulation results reveal the confirmation of the proposed hybrid power system suitable for standalone applications.

2. The proposed hybrid configuration

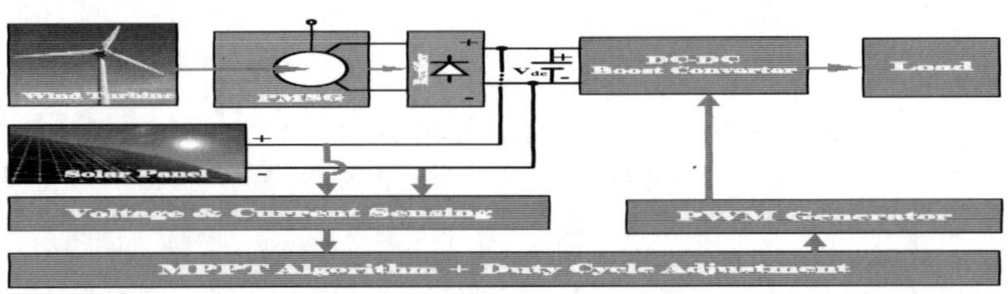

Figure 1. Outline configuration of the proposed hybrid power system

Figure.1 illustrates the configuration of the proposed hybrid power configuration. The proposed hybrid power system contains solar panel with parallel and series connected cells, Permanent magnet synchronous generator(PMSG) driven wind turbine, Rectifier, DC-Dc boost converter and MPPT controller. Rectifier receives AC input from PMSG via wind turbine depending upon wind speed, pitch angle, and converts this AC in to DC. Then the output from the solar panel and rectifier is synchronously connected to boost converter to boost the voltage gotten from the rectifier. MPPT controller sense voltage and current from the rectifier and control the duty ratio of the switch used in the proposed boost converter. The operation technique proposed in this study is gone for reducing the utilization of fossil fuel and attractive the use of renewable energy sources (RES).

3. PV module modeling

A solar cell is made up of the semiconductor device, which transfer solar energy into electrical energy. Solar array is made up of the number of parallel and serious connected solar cells. A generic identical circuit as shown in figure.2 represents the solar module consists of the single diode with series R_s and parallel resistance R_p.

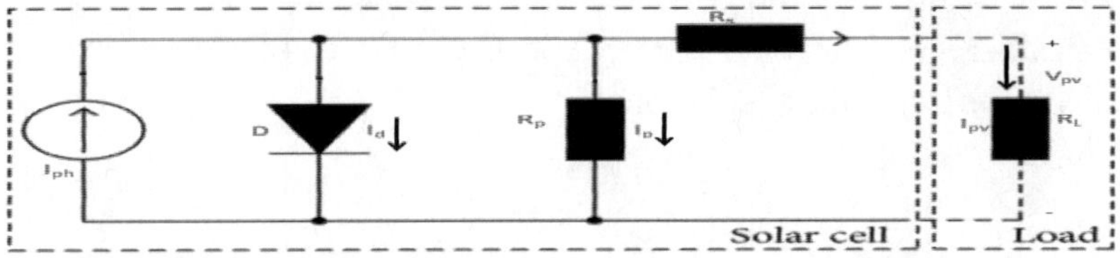

Figure 2. Equivalent circuit of PV module

The four variables required for this model are the two input variables, sunlight based radiation E_G (W/m^2) and surrounding temperature T_a (°C), and the two output terminal variables, PV cell current I_{pv} (A) and voltage V_{pv} (V). According to KCL, the PV cell current (I_{pv}) is expressed by.

$$I_{pv} = I_{ph} - I_d - I_p \tag{1}$$

Where, I_{ph} is the Photo current; I_d is the Diode current; I_p is the Shunt current.

$$I_{pv} = [I_{sc} + K_i(T - T_r)]\frac{S}{100} - I_0\left[e^{q\frac{(V_{pv}+I_{pv}R_s)}{AKT}}\right] - I_p \tag{2}$$

Where I_{sc} is the Short circuit current at T_r; K_i is the temperature coefficient of short circuit current; S is the Global solar radiation in mW/cm^2. The module immersion current (I_o) changes with cell temperature, which is communicated as,

$$I_0 = I_{satr}\left(\frac{T}{T_r}\right)^3 e^{\frac{qE_G}{KA}\left(\frac{1}{T_r}-\frac{1}{T}\right)} \tag{3}$$

Where, E_G is the Band gap energy; T_r is the Reference temperature = 301.18 K; T is the Cell temperature. The module's reverse immersion current (I_{satr}) and the Shunt current I_p is given by.

$$I_{satr} = \frac{I_{sc}}{e^{\left(qV_{oc}/N_{ss}AKT\right)}} \tag{4}$$

$$I_p = \left(\frac{V_{pv} - I_{pv}R_s}{R_p}\right) \tag{5}$$

The output current obtained from the PV module (I_{pv}) is given by.

$$I_{pv} = N_{pp}I_{ph} - N_{pp}I_0 e^{\left[\frac{q(V_{pv}+I_{pv}R_s)}{N_{ss}AKT}\right]} - 1 \tag{6}$$

Where, N_{pp} is the Quantity of parallel cells; N_{ss} is the Quantity of series cells, q = electron charge (1.6 × 10−19 C); V_{oc} = open-circuit voltage (V) of the PV board; k = Boltzmann's constant with the quality 1.3805 × 10−23 J=K; A = ideality variable with the worth 1.2; I_o = PV module immersion current (An); The power extracts from the PV array can be estimated by the subsequent equation.

$$P = N_{pp}I_{ph}V_{pv} - N_{pp}I_0V_{pv}\left(e^{K_0\left(\frac{V_{pv}}{N_{ss}}+I_{pv}R_s\right)} - 1\right) - \frac{V_{pv}}{R_p}\left(\frac{V_{pv}}{N_{ss}} + IR_s\right) \tag{7}$$

The temperature (T) of a PV cell is spoken by.

$$T = 3.12 + 0.25S + 0.899T_a - 1.3w_s + 273 \tag{8}$$

Where, T_a is the ambient temperature; W_s is the Wind speed; S is the solar radiation. In the scientific model, number of parallel and serious connected cells are N_{ss} = 36 and N_{pp} =2.

4. Wind turbine model

The overall characteristic of the wind turbine is described by the non-dimensional curve i.e. coefficient of performance (Cp) versus tip-speed ratio λ. It is given by equation (9)

$$C_P(\lambda)=0.043-0.108\lambda+0.146\lambda^2-0.0602\lambda^3+0.0104\lambda^4-0.0006\lambda^5 \quad (9)$$

The tip-speed ratio is specified by the expression:

$$\lambda=\frac{R\varpi_m}{V_w} \quad (10)$$

Where, R is the rotor radius of the wind turbine in m, ϖm is the rotor angular speed in rad/sec. and V_w is the wind velocity in m/s. The power output of wind turbine P_t is measured utilizing equation (3) as:

$$P_t=0.5Cp(\lambda)AV_W^3 \quad (11), \qquad Pt=Tt\varpi_m \quad (12)$$

Where, A is the rotor swept area of wind turbine. Combining equations (9), (11) and (12), the expression for torque T_t created by the wind turbine is expressed as:

$$T_t=0.5rAR\frac{Cp(\lambda)}{\lambda}V_W^2 \quad (13)$$

The power obtained from the wind is more extreme at maximum C_p. This happens at a characterized estimation of the tip speed proportion. Consequently, for every wind speed; there is an ideal rotor speed where greatest power is extricated from the air. In this way, if the wind velocity is thought to be steady, the estimation of C_p relies upon the wind turbine rotor speed. In this manner, by controlling the rotor speed, the power output of the turbine is controlled.

5. PMSG modeling

WECS can be utilized together synchronous and induction generators depending upon the requirement. Changeable speed permanent magnet synchronous generator (PMSG) is broadly utilized as a part of wind energy systems as it has maximum efficiency, weight less, simple maintenance, easy control and no requirement for magnetizing and reactive current. The gearbox present in the changeable speed wind turbine creates additional cost and maintenance. Utilizing direct driven PMSG enhances reliability as well as reductions weight in the nacelle. The PMSG is modeled using d-q synchronous reference frame. The voltage equations are described below

$$V_d=-R_si_d-L_d\frac{di_d}{dt}+\omega L_qi_q \quad (14), \qquad V_q=-R_si_q-L_q\frac{di_q}{dt}-\omega L_di_d+\omega\lambda_m \quad (15)$$

The electronic torque is specified as

$$T_e=1.5\rho\{\lambda i_q+(L_d-L_q)i_di_q\} \quad (16)$$

$$\begin{bmatrix}V_{qs}\\V_{ds}\\V_{qr}\\V_{dr}\end{bmatrix}=\begin{bmatrix}R_s+\rho L_s & 0 & PL_m & 0\\ 0 & R_s+\rho L_s & 0 & PL_m\\ PL_m & -\omega_rL_m & R_r+\rho L_r & -\omega_rL_r\\ \omega_rL_m & PL_m & \omega_rL_r & R_r+\rho L_r\end{bmatrix}\begin{bmatrix}i_{qs}\\i_{ds}\\i_{qr}\\i_{dr}\end{bmatrix} \quad (17)$$

Where, L_d and L_q are direct and quadrature axis inductances, i_d and i_q are straight and quadrature axis, currents, V_d and V_q are direct and quadrature axis voltages, ω_r is the rotor angular velocity, λ is amplitude of flux induced and p is number of set of poles. If there is a squirrel cage induction generator (SCIG), dynamic modeling is done by stationary d-q reference frame accompanying with the following expression. The expressions of stator side are

IOP Conf. Series: Materials Science and Engineering **396** (2018) 012046 doi:10.1088/1757-899X/396/1/012046

$$\lambda_{ds} = L_s i_{ds} + L_m i_{dr} \quad (18), \qquad \lambda_{qs} = L_s i_{qs} + L_m i_{dr} \quad (19)$$

$$L_s = L_{ls} + L_m \quad (20), \qquad L_r = L_{lr} + L_m \quad (21)$$

$$V_{ds} = R_s i_{ds} + \frac{d}{dt}\lambda_{ds} \quad (22), \qquad V_{qs} = R_s i_{qs} + \frac{d}{dt}\lambda_{qs} \quad (23)$$

The expressions of rotor side are

$$\lambda_{dr} = L_r i_{dr} + L_m i_{ds} \quad (24), \qquad \lambda_{qr} = L_r i_{qr} + L_m i_{qs} \quad (25)$$

$$V_{dr} = R_r i_{dr} + \frac{d}{dt}\lambda_{dr} + \omega_r \lambda_{qr} \quad (26), \qquad V_{qr} = R_r i_{qr} + \frac{d}{dt}\lambda_{qr} - \omega_r \lambda_{dr} \quad (27)$$

The equations of the gap flux linkages are

$$\lambda_{dm} = L_m (i_{ds} + i_{dr}) \quad (28), \qquad \lambda_{qr} = L_m (i_{qr} + i_{qs}) \quad (29)$$

Where, R_s and R_r are stator and rotor winding resistances, L_m is the magnetizing inductance, L_{ls} and L_{lr} are stator and rotor leakage inductances, ω_r is the angular speed of the electrical rotor, i_d and i_q are straight and quadrature axis currents, V_d and V_q are direct and quadrature axis voltages, λd and λq are direct and quadrature axis fluxes. The power output and the turbine torque (T_t) in expression of rotational speed can be gotten by substituting (17) in (16)

$$P_W = 0.5\rho A C_P(\lambda, \beta)\left(\frac{R\omega_{opt}}{\lambda_{opt}}\right)^3 \quad (30)$$

$$T_t = 0.5\rho A C_P(\lambda, \beta)\left(\frac{R}{\lambda_{opt}}\right)^3 \omega_{opt} \quad (31)$$

The power coefficient (Cp) is a nonlinear function communicated by the fitting condition is given by

$$C_p = 0.5176\left(\frac{116}{\lambda_i} - 0.4\beta - 5\right)e^{\frac{-21}{\lambda_1}} + 0.0068\lambda \quad (32)$$

With,

$$\frac{1}{\lambda_i} = \frac{1}{\lambda + 0.08\beta} - \frac{0.035}{\beta^3 + 1} \quad (33)$$

6. Three-phase diode rectifier

The 3Φ Full bridge rectifier can be coupled straight to the three-phase source to rectify the generated output voltage from the PMSG. The output dc voltage across the rectifier is

$$V_{dc} = \left(3\sqrt{2}/\pi\right)V_L \quad (34)$$

Where V_{dc} is output DC voltage across the rectifier and V_L is the input ac voltage from PMSG.

7. DC-DC boost converter

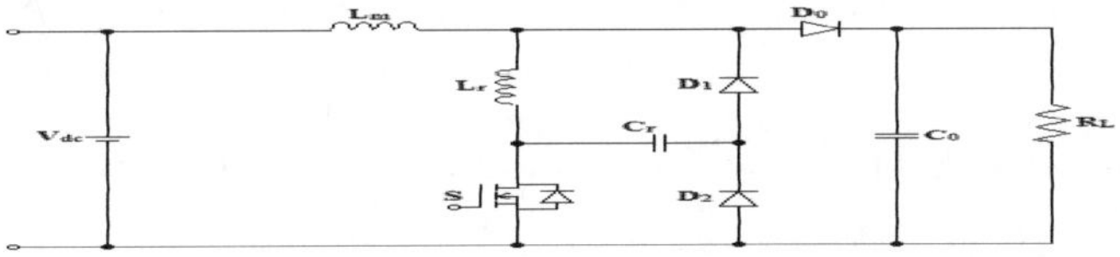

Figure 3. Boost converter used in the hybrid system

IOP Conf. Series: Materials Science and Engineering **396** (2018) 012046 doi:10.1088/1757-899X/396/1/012046

Figure.3 shows the recently proposed improved dc-dc converter with LCD snubber configuration for battery applications [17]. The circuit [17] is introduced to attain high efficiency and reduces switching losses. The unregulated DC voltage from the diode rectifier is known as input to the developed DC-DC boost converter shown in fig.3. The rectifier output voltage will change due to variations in wind speed. Improved DC-DC boost converter is needed to obtain the regulated dc voltage output as well as to realize the MPPT operation. A single power switch S is needed, which reduces the cost and simplifies the control over the system. For ease, a resistance load R_L is linked directly to the output from the boost converter to consume the power generated by the WECS. The output voltage of the boost converter is given by

$$\frac{V_0}{V_{dc}} = \frac{1}{1-D} = \frac{I_{dc}}{I_0} \qquad (35)$$

The control algorithm uses only the dc-link voltage V_{dc} (t) and the dc-current I_{dc} (t) measurements to adjust directly the duty cycle D.

8. MPPT implementation

To enhance the efficiency of the hybrid power system MPPT is implemented to shift the working voltage nearest to the most extreme power under varying weather conditions. Among the various configurations of MPPT, Perturb and observe algorithm is applied in this paper as shown in figure.4. This is simple and admired method. The procedure is given below.

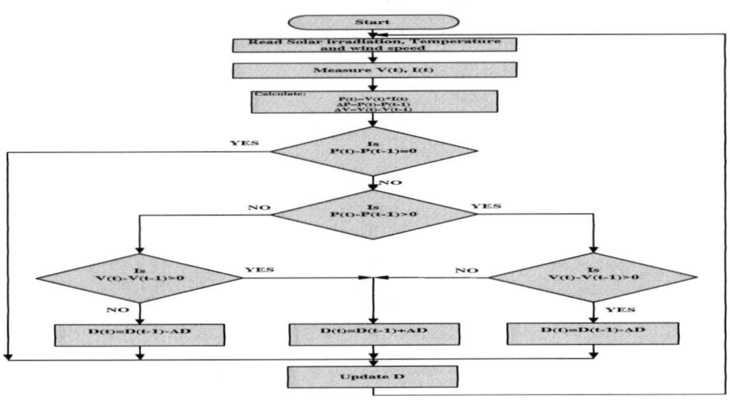

Figure 4. The flowchart of perturb and observe (P&O) MPPT technique

Step1: Read the input parameters of the hybrid system such as solar irradiation, ambient temperature and wind speed.

Step2: Measure the solar panel voltage and current, which are same as that of the diode rectifier voltage and current.

Step 3: The voltage V (t) and I (t) are given as input to the MPPT controller and compute the Power, P (t) = V (t)*I (t).

Step 4: After the power, P (t) is obtained and perturbing the duty ratio of the boost converter switch (increment or decrement) to vary the operating point of the system (perturb V (t) and I (t)) and the difference between Present and previous values of power is computed by $\Delta P = P(t) - P(t-1)$.

Step 5: Then the power increases with voltage increment; the operating point of the system is shifting towards the most extreme power point. Hence, the size of the perturbation is positive or left constant. The mathematical expressions are,

$\Delta P/\Delta V>0$; $D(t) = D(t-1) + \Delta D$ and $\Delta P/\Delta V<0$; $D(t) = D(t-1) - \Delta D$

Step 6: Next the power decreases with voltage decrement; the operating point of the system is shifting away from the most extreme power point. Hence, the size of the perturbation is return back and pushes the system to be operating in the reverse MPP.

Step 7: The MPPT controller generates the desired duty ratio for the boost converter.

Step 8: This procedure is frequent awaiting the MPP is achieved.

9. Results and discussions

A solar-wind hybrid power system with Common MPPT control has been simulated with MATLAB environment, and the results are presented here. The solar panel and wind turbine parameters for 20 m/s wind speed and 1500 W/m^2 irradiation are appeared in the table 1. As per solar irradiation and temperature level, the dc voltage obtained from PV panel changes time to time similarly as per variation in the wind speed and pitch angle, the ac output voltage obtained from wind turbine via the permanent magnet synchronous generator fluctuates as shown in Figure.5.

Figure. 6 shows the simulated results of the hybrid system at the wind speed of 20m/s at 1500 W/m^2 and pitch angle =0°. Figure. 6. (A) and Figure. 7. (A) Demonstrates the voltage and power obtained from the rectifier side. Up to 0.52 sec the system receive energy from wind turbine and after 0.52 sec the system collects energy from solar panel. Figure 6.(B) and Figure 7.(B) illustrates the voltage and power acquired from the dc-dc boost converter. It clearly shows that voltage doubles at every second with respect to rectifier voltage and power losses slightly than the rectified power. From this, we reach the efficiency of the boost converter by 93%.

Table.1.Parameters of hybrid system

Source	Parameters	Values
Wind turbine	Input(Wind speed)	20 m/s
	Maximum Power	1600 W
	Pitch angle	0
	Base wind speed	10 m/s
Solar Panel	Input(Irradiation and Temperature)	1500 W/m^2
	Maximum power	400 W
	Number of series cells	36
	Number of parallel cells	2
Boost Converter	Input voltage	80-180 V
	Output voltage	160-360 V
	Output power	380-1500W
	Switching frequency	50 KHZ

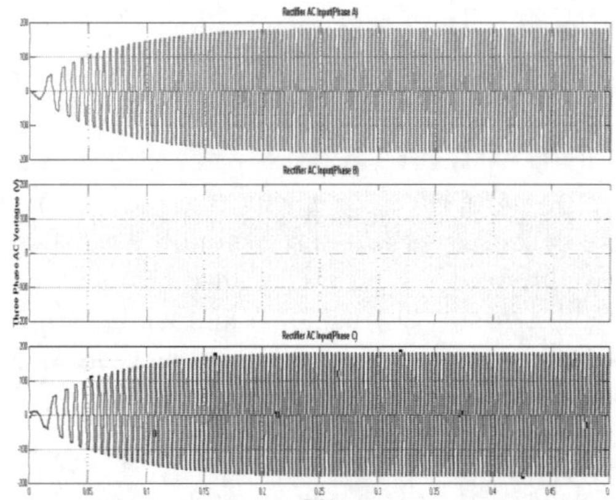

Figure 5. Three Phase AC Voltage generated by PMSG.

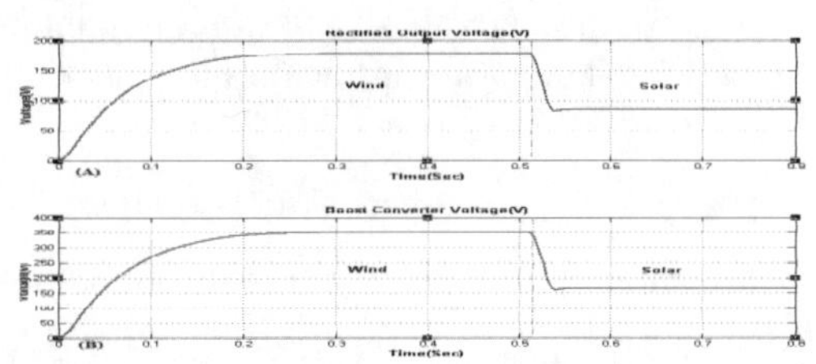

Figure 6. Simulated voltage results for wind speed 20m/s at 1500 W/m^2 and pitch angle =0° (A) Rectified DC output Voltage (B)

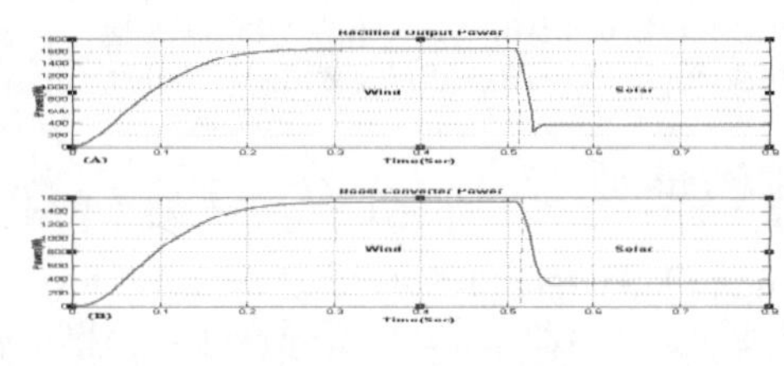

Figure 7. Simulated power results for wind speed 20m/s at 1500 W/m^2 and pitch angle =0° (A) Rectified output power

Boost converter voltage. (B) Boost converter power.

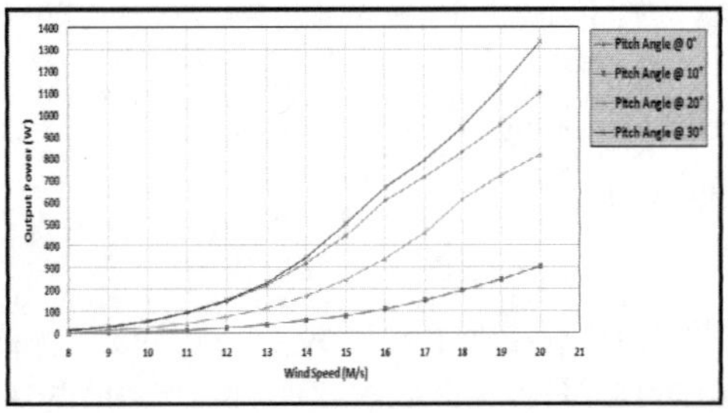

Figure 8. Maximum-power tracking concerning the assortment in the wind speeds at the different pitch angle.

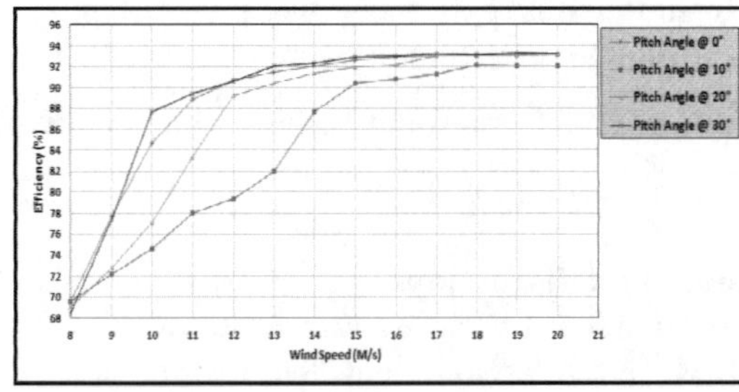

Figure 9. Tracking efficiency concerning the assortment in the wind speeds at different pitch angle.

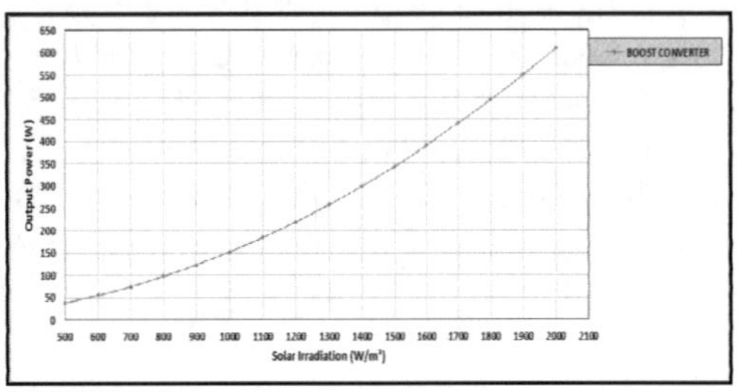

Figure 10. Maximum-power tracking concerning the assortment in the solar irradiation at constant temperature and fixed wind speed.

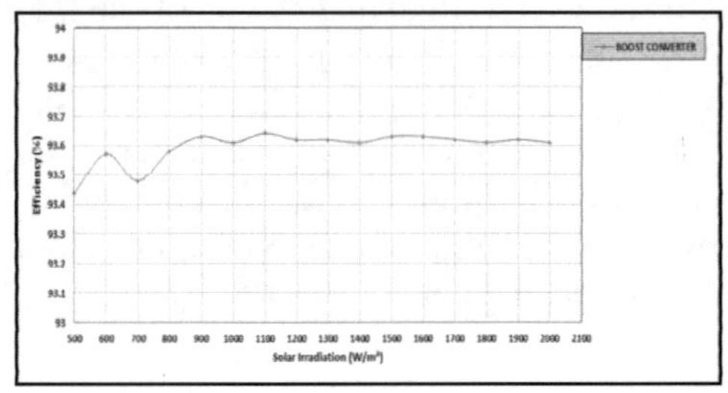

Figure 11. Tracking efficiency concerning the assortment in the solar irradiation at constant temperature and fixed wind speed.

Figure. 8 and figure. 9 depicted the maximum-power tracking and tracking efficiency of the hybrid system concerning the assortment in the wind speed at the different pitch angle. It is observed that the proposed hybrid system tracks maximum power at a pitch angle 30°. Figure. 10 and figure. 11 shows the maximum power tracking and tracking efficiency of the hybrid system in regard to the assortment in the solar irradiation at constant temperature and fixed wind speed and saw that the power increases with an irradiation level.

10. Conclusion

The MPPT based solar-wind hybrid energy system with boost converter is studied in this paper. To increase the conversion efficacy of the suggested standalone hybrid system, an MPPT algorithm is utilized. Perturb and observe algorithm is used to provide the required duty ratio to control the boost converter with unpredictable weather conditions. A detailed analysis of solar, wind and PMSG modeling has been done in this work. The proposed system is simulated in MATLAB/ SIMULINK environment under varying solar irradiation, wind speed and temperature and these results showed that the conversion efficiency reaches around 93%. According to the overall discussion, it could be inferred that the proposed hybrid system is the suitable choice for standalone applications.

11. References

[1] Subhadeep B and Shantanu A 2015 PV-wind hybrid power option for a low wind topography *Energy Conversion and Management.* **89** pp 942–954.

[2] Chih-Lung S, Tsair-Chun L, Siou-Hong Z and Su-Wen W 2015 Voltage Gain Derivation Based on Energy-Balanced Criterion for a Novel Hybrid-Input PV-Wind Power Conversion System *Hindawi Publishing Corporation Mathematical Problems in Engineering.* http://dx.doi.org/10.1155/2015/623161(2015).

[3] Bhave A G 1999 Hybrid solar-wind domestic power generating system-a case study *Renewable Energy.* **17** pp 355-358.

[4] Naveen kumar R and Baskaran J 2014 Energy Management system for Hybrid RES with Hybrid Cascaded Multilevel inverter *International Journal of Electrical and Computer Engineering (IJECE).* **4** pp.24-30.

[5] Cheng-Wei C, Chien-Yao L, Kun-Hung C and Yaow-Ming C 2015 Modeling and Controller Design of a Semiisolated Multiinput Converter for a Hybrid PV/Wind Power Charger System *IEEE Transactions on Power Electronics.* **30** pp. 4843-4853.

[6] Bogdan S. borowy and Ziyad M. salameh 1996 Methodology for Optimally Sizing the Combination of a Battery Bank and PV Array in a Wind/PV Hybrid System *IEEE Transactions on Energy Conversion.* **11** pp. 367-375.

[7] Fazia B, Kamal M, Said D and Omar Behar 2015 Feasibility study and energy conversion analysis of stand-alone hybrid renewable energy system *Energy Conversion and Management.* **105** pp. 471–479.

[8] Binayak B, Shiva raj P, Kyung-Tae L and Sung-Hoon A 2014 Mathematical Modeling of Hybrid Renewable Energy System: A Review on Small Hydro-Solar-Wind Power Generation *International Journal of Precision Engineering and Manufacturing-Green Technology.***1** pp. 157-173.

[9] Szeidert I, Filip I and Prostean O 2016 Issues regarding the modelling and simulation of hybrid micro grid systems *IOP Conf. Series: Materials Science and Engineering.***106** pp. 1-5.

[10] Rashid Al B, Mohammad A and Tapas M 2015 Speed control of synchronous machine by changing duty cycle of DC/DC buck converter *AIMS Energy.* **3** pp. 728-739.

[11] Binayak B, Kyung-Tae L, Gil-Yong L, Young-Man C and Sung-Hoon A 2015 Optimization of Hybrid Renewable Energy Power Systems: A Review *International Journal of Precision Engineering and Manufacturing-Green Technology.* **2** pp. 99-112.

[12] Rajesh K, Kulkarni A.D and Ananthapadmanabha T 2015 Modeling and Simulation of Solar PV and DFIG Based Wind Hybrid System *Procedia Technology.* **21** pp. 667 – 675.

[13] Rajan singaravel M.M and Arul daniel S 2015 MPPT With Single DC–DC Converter and Inverter for Grid-Connected Hybrid Wind-Driven PMSG–PV System *IEEE Transactions on Industrial Electronics.* **62** pp. 4849-57.

[14] Nahidul hoque S, Norhafizan A, Imtiaz ahmed C and Zahari T 2015 Technical Study of a Standalone Photovoltaic–Wind Energy Based Hybrid Power Supply Systems for Island Electrification in Malaysia, DOI:10.1371/journal.pone.0130678 June 29, 2015.

[15] Bogaraj T, Kanakaraj J and Chelladurai J 2015 Modeling and simulation of stand-alone hybrid power system with fuzzy MPPT for remote load application *Archives of Electrical Engineering.* **64** pp. 487-504.

[16] Sajib C Design and Analysis of a Hybrid Solar-Wind Energy System Using CUK & SEPIC Converters for Grid Connected Inverter Application *Journal of Electrical Engineering.* pp.1-7.

[17] Barsana banu J , Balasingh moses M and Rajarajacholan S A non isolated bidirectional DC-DC converter with LCD snubber *Revista tecnica de la facultad de ingenieria universidad del zulia.* **39** pp.131-143.

IOP Conf. Series: Materials Science and Engineering **396** (2018) 012047 doi:10.1088/1757-899X/396/1/012047

Review of Solar PV Powered Water Pumping System Using Induction Motor Drive

Akhila V T[1] and Arun S[2]

[1]PG Scholar, Department of Electrical Engineering,
Amal Jyothi College of Engineering, Kottayam, India
[2]Assistant Professor, Department of Electrical Engineering,
Amal Jyothi College of Engineering, Kottayam, India

[1] akhilavt01@gmail.com [2]sarun@amaljyothi.ac.in

Abstract. This paper presents the review of the Solar Photovoltaic (SPV) array fed water pumping system using an Induction Motor Drive (IMD). The penetration of renewable energy powered water pumping systems in industrial and domestic applications receiving wide attention nowadays. The introduction of a three-phase induction motor brings an improved solution to the commercial water pumping system. SPV array fed water pumping system using IMD mainly consist two stages; first stage extracts the maximum power from the solar Photovoltaic (PV) array by restraining the duty ratio of the DC-DC boost converter. In the second stage, a controller is employed to control the switching pulses of the Voltage Source Inverter (VSI). Different control methods are available for regulating IMD like Scalar (V/f control), Direct Torque Control (DTC), Vector control, and DTC use Space Vector PWM (SVPWM) control.

1. Introduction

Contemporarily more than 900 million people in various countries do not have potable water available for consumption. In isolated and rural areas the water supply only comes from the rain or distant rivers. In such places, the conventional pumping system is indecorous due to the unavailability of electric power. The use of SPV system is one of the perfect remedies for this problem [1]. This sort of energy source is cheaper and can be used to work for several years without the need for maintenance. Such systems are not new and are already used for more than three decades [2].

GDP of agriculture in India is around 16% and it plays an important role in Indian economy. Approximately 69 percentage of the population in India living in rustic areas and the sole source of their income is from agriculture and cognate activities. The water is the prerequisite for agriculture and the main water resource used for the agricultural purpose are rivers, canals, wells, and monsoons. Around 64% of cultivated land in India mostly depends on monsoons [2]. Irrigation is important to lessen the dependence on monsoons in India. In rural areas, diesel pump sets are used for irrigation but its maintenance cost is high and creates pollution. Motor pump set is used for water supply in those places where grid electricity exists. Grid connection is not available in all part of India, especially in remote areas. In such cases, stand-alone water pumping system is employed [1]. Negligible maintenance cost and pollution free operation are the main advantages of solar pumping system but installation cost is high for such systems. The lifespan of solar PV water pumping system is observed nearly around 25-30 years [1].

Different variety of motor-pump set configurations is used in the solar pumping system. The most conventionally employed pump type is the centrifugal pump. Solar PV shallow water pumping using single-stage centrifugal pumps are suitable for low head applications, while the multistage centrifugal pump is more suitable for PV subterranean water pumping and surface water pumping with high heads applications. Other pump types such as progressive cavity pumps and piston pumps have also been utilized. Both DC and AC motors are used for solar PV pumping applications. DC motor needs frequent maintenance hence AC motors are preferred in most of the pumping applications. Permanent magnet synchronous motors are suitable for the submersible application. Most of the pumping system uses induction motor because it is mechanically simple; having rugged construction and reliable operation. It has low cost, high efficiency, low maintenance than the DC motors and can be used for hazardous and contaminated areas [3].

2. Solar Water Pumping System

Block schematic of solar PV water pumping system is shown in figure 1. Solar panel drives the three-phase induction motor using DC-DC and DC-AC converters. The capital cost of the solar power plant is high hence its main aim is to acquire peak power from the installed plant. MPPT algorithms are used to gather maximum energy from the plants. A comparison study of different MPPT methods is described in [6]-[8]. Perturb and Observe is the most recurrently used MPPT algorithm particularly for low priced implementation. DC-DC converters such as a buck, boost, and buck-boost are used for solar pumping applications. The converters like buck and boost have the fewer number of components but the MPPT region is limited to such kind of converters. The buck-boost converters have boundless MPPT region and it also gives ripple free input-output current [4] [5].

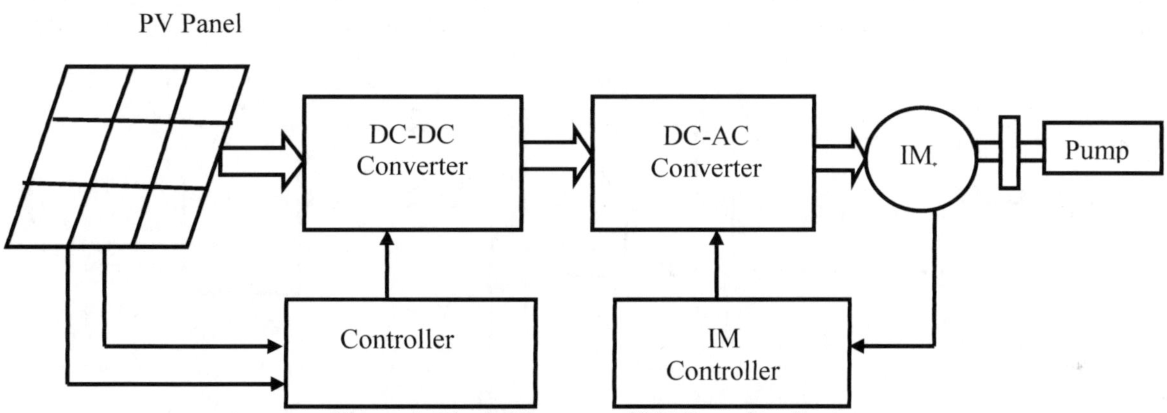

Figure 1 Block Diagram of PV Pumping System.

2.1. Design of PV Array

An induction motor of 2.5 kW is selected for the proposed system. If the losses of motor and pump are neglected, the capacity of the PV array should be equivalent to the motor capacity. In this case, a PV array is selected to be of 2.7 kW.

2.2. Design of Voltage Source Inverter

The apparent power rating of VSI is given as,

$$S_{VSI} = \sqrt{P^2 + Q^2} \tag{1}$$

The RMS current through a VSI is given as,

$$I_{VSI} = \frac{KW * 10^3}{\sqrt{3}\, V_m} \tag{2}$$

The DC bus voltage of VSI is estimated as,

$$V_{DC} = \frac{2V_{LL}\sqrt{2}}{\sqrt{3}} \tag{3}$$

IOP Conf. Series: Materials Science and Engineering **396** (2018) 012047 doi:10.1088/1757-899X/396/1/012047

Open loop V/f control can be used for low power water pumping systems.

2.3. Design of Centrifugal Water Pump

The Centrifugal water pump is designed using its torque-speed relationship as,

$$T_p = Kw^2 \tag{4}$$

Where K is the proportionality constant, w is the angular velocity and T_p is the torque.

3. Control Methods of IMD for SPV Water Pumping System

3.1. Scalar Control (V/f Control)

A simple V/f approach is used to control IMD in [8] [9] but this approach suffers from DC link voltage instability. However, V/f control is simple, easy to implement and cost-effective. The direct torque control and vector control approaches are complicated and they require extra current sensors for implementation. In V/f control, only DC bus voltage, PV current, and voltage are sensed. The system tracks the maximum power point by varying the modulation frequency so that the IMD is able to extract the maximum power from the solar PV array at sustained torque for various solar insolations. By utilizing V/f control, the starting behavior of the IMD is improved even IMD is started with lower solar irradiation level. Hence, it is likely to pump water from morning to till the evening. The starting current of the induction motor connected to the fixed voltage AC mains is approximately 5 to 6 times of full load current [3]. Thus, large numbers of solar modules are required to start the motor without any control. A tranquil starting of the induction motor is feasible by using V/f control without drawing high starting current. Figure 2 shows the block schematic of scalar control.

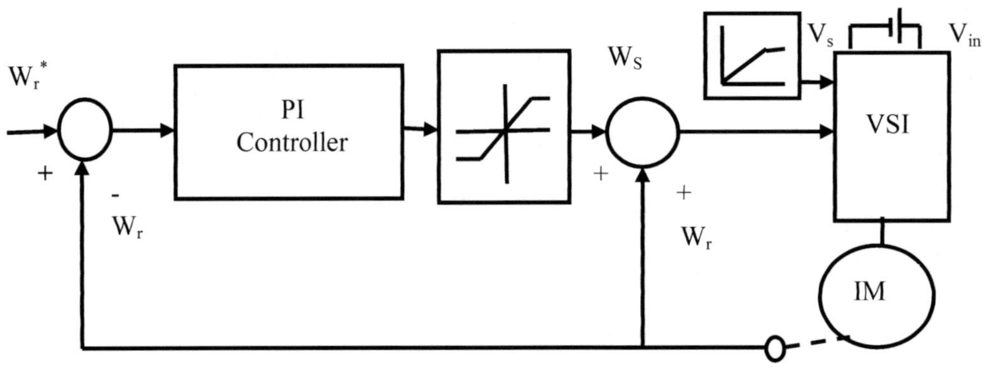

Figure 2 Scalar Control of Induction Motor.

A disadvantage of scalar control is that the torque developed is load dependent as it is not controlled directly. Also, the transient response of such a control is not rapid due to the predefined switching pattern of the inverter. However, if there is a contiguous block to the rotor rotation, it will lead to heating of the motor in spite of the functioning of the over current control loop. By means of a speed/position sensor, the problem relating to the blocked rotor and the load-dependent speed can be overcome [8]. However, this will add to the system cost, size, and complexity.

3.2. Vector Control

The scalar control of IMD gives a sluggish response and it can avoid using vector control [10]. This control is also known as the "field oriented control", "flux oriented control" or "indirect torque control". The field-oriented control performs the Clarke-Park transformation, which converts three phase vectors to a two-dimensional rotating reference frame (d-q). The "d" component stands for the flux producing component and the "q" component symbolize the torque producing component. These two decoupled components can be individually controlled by passing through separate PI controllers. The outputs of the PI controllers are transformed back to the three-dimensional stationary reference plane using the inverse of the Clarke-Park transformation.

Depending on the measurement method, the vector control is grouped into two subcategories: direct and indirect vector control . Indirect vector control, the flux measurement is done by employing flux sensing coils or the Hall devices. It will cause additional hardware cost and in addition, measurement is not highly accurate. Hence, this method is not a very good control technique. The most common method is indirect vector control. In this method, the flux angle is measured indirectly but is estimated from the equivalent circuit model and from measurements of the rotor speed, the stator current, and the voltage. One common technique for determining the rotor flux is based on the slip relation. This needs the measurement of the rotor position and the stator current. With current and position sensors, this method performs reasonably well over the full speed range.

3.3. *Direct Torque Control (DTC)*

The difference between the traditional vector control and the DTC is that it has no fixed switching pattern. The inverter switched by the DTC according to the load needs. Due to the elimination of the fixed switching pattern (characteristic of the vector and the scalar control), the DTC response is extremely fast during the instant load changes. Decoupled control of torque and flux, the absence of mechanical transducers, having low computational time, reduced parameter sensitivity, and very simple control scheme are the main features of DTC [11]. In simple DTC control current regulator, PWM pulse generation, PI control of flux and torque, and coordinate transformation are not required. In this method, one nonzero and two zero voltage vectors of the inverter are selected for the operation. The vectors are selected based on the instantaneous errors in torque and stator flux magnitude. The block diagram of DTC control is shown in figure.3.

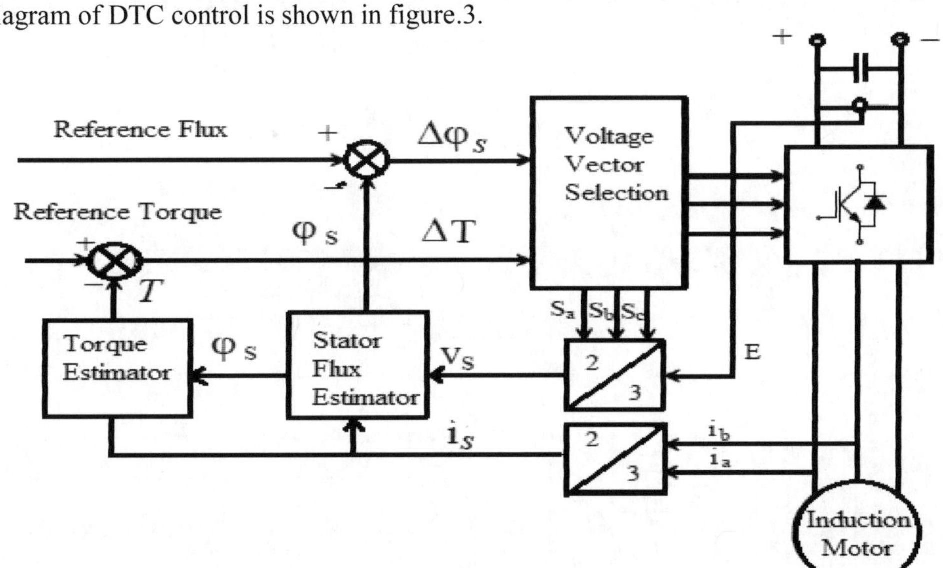

Figure 3 Block Diagram of Simple DTC Control.

3.3.1. DTC Using SVPWM. As compared to the Sinusoidal Pulse Width Modulation (SPWM) space vector control has the advantages like reduced switching frequency, low current ripples, and good utilization of DC bus voltages [12]. Based on instantaneous current and voltage measurements it is possible to calculate the voltage required to drive the current output torque and stator flux to the necessitated values within fixed time duration. This computed voltage then synthesized using space vector modulation. The instantaneous stator flux and output torque can be estimated by using the applied voltage and motor current. The estimated values are used to calculate the required change in output torque and stator flux. The back EMF can be determined from the stator flux and current. The voltage required to achieve the changes in stator current can be calculated knowing the back EMF of the motor and the changes in the torques corresponds to the changes in stator current. The block schematic of SVPWM is shown in figure 4.

IOP Conf. Series: Materials Science and Engineering **396** (2018) 012047 doi:10.1088/1757-899X/396/1/012047

SVPWM [13][14]. Space vector modulation method is suitable for variable speed applications. The triggering pulses for the inverter switches are developed using this control strategy. This method gives the higher magnitude of the fundamental output voltage and lowers harmonic distortion as compared to SPWM technique. The steps involved in SVPWM method is given below.

STEP 1: Co-ordinate Transformation:

Consider a three-phase two level Voltage Source Inverter (VSI) as shown in figure 5. The voltage e_a, e_b and e_c are the output voltages applied to the winding of the motor. S_1 through S_6 are the six power switches which are controlled by a, b, c, a', b', and c' gating signals. There are eight possible combinations of on and off states for the three upper power switches.

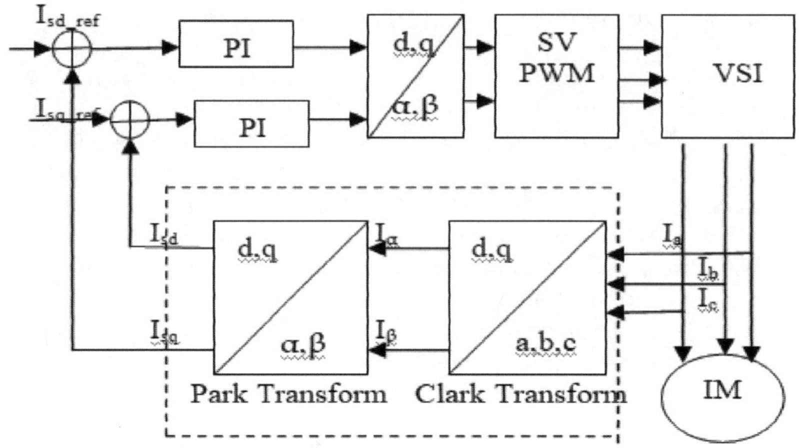

Figure 4 SVPWM Control of IMD.

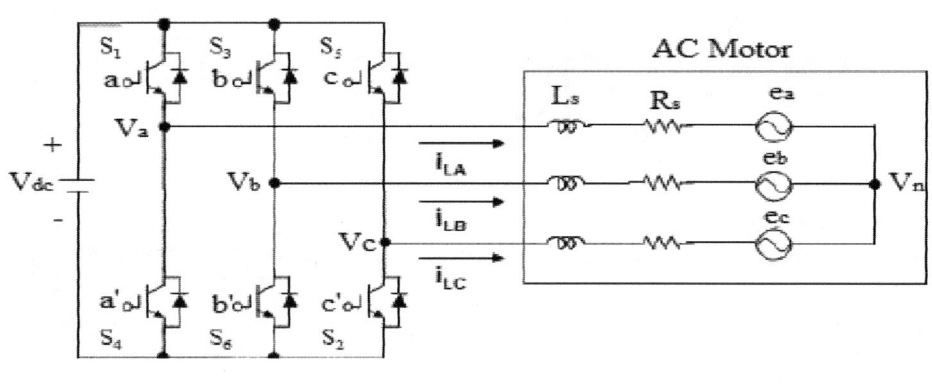

Figure 5 Circuit Diagram of VSI.

The relationship between the switching variable vector [a, b, c], phase voltages [Va, Vb, Vc] and the line-to-line voltage vector [V_{ab}, Vbc, Vca] is given in the following Table 1. The phase voltages corresponding to the eight combinations of switching pattern can be mapped into the d-q plane by using d-q transformation as shown in Table II. This mapping results in six non-zero vectors and two zero vectors. The d-q transformation can be applied to the reference a, b, and c voltages to obtain the reference Vout in the d-q plane as shown in figure 6.

STEP 2: Sector Determination and Time Period Calculation:

It is necessary to know in which sector the output voltage is located to determine the switching time periods and switching sequence. By judging the value of the phase angle of space vector Vout, the sector number can be determined. The output voltage Vout can be in any one of sector 1 to sector 6 and it is approximated by switching between the two non-zero basic vectors that border the sector of

IOP Conf. Series: Materials Science and Engineering **396** (2018) 012047 doi:10.1088/1757-899X/396/1/012047

the current output voltage. If Vout is in Sector1, it can be approximated by switching the inverter between states V1 and V2 for a time period of T1 and T2 respectively. Because the sum of T2 and T1 should be less than or equal to Tpwm, the inverter should remain in T0 or T7 for the rest of the period.
STEP 3: Switching Pattern Generation:
The symmetric space vector PWM switching pattern is a better and conventionally used switching pattern. As the name indicates, it is symmetric pulsation and both zero vectors are applied in this method and it is shown in figure 7.

Table 1 Switching States and Output Voltage of VSI.

A	b	c	V_a	V_b	V_c	V_{ab}	V_{bc}	V_{ca}
0	0	0	0	0	0	0	0	0
1	0	0	2/3	-1/3	-1/3	1	0	-1
1	1	0	1/3	1/3	-2/3	0	1	-1
0	1	0	-1/3	2/3	-1/3	-1	1	0
0	1	1	-2/3	1/3	1/3	-1	0	1
0	0	1	-1/3	-1/3	2/3	0	-1	1
1	0	1	1/3	-2/3	1/3	1	-1	0
1	1	1	0	0	0	0	0	0

Table 2 Switching Voltages in d-q plane.

a	b	c	V_q	V_d	V_{dq}
0	0	0	0	0	$V_0=0$
0	0	1	$-1/3V_{dc}$	$1/\sqrt{3}V_{dc}$	$V_1=2/3V_{dc}$
0	1	0	$-1/3V_{dc}$	$-1/\sqrt{3}V_{dc}$	$V_2=2/3V_{dc}$
0	1	1	$-2/3V_{dc}$	0	$V_3=2/3V_{dc}$
1	0	0	$2/3V_{dc}$	0	$V_4=2/3V_{dc}$
1	0	1	$1/3V_{dc}$	$1/\sqrt{3}V_{dc}$	$V_5=2/3V_{dc}$
1	1	0	$1/3V_{dc}$	$-1/\sqrt{3}V_{dc}$	$V_6=2/3V_{dc}$
1	1	1	0	0	$V_7=0$

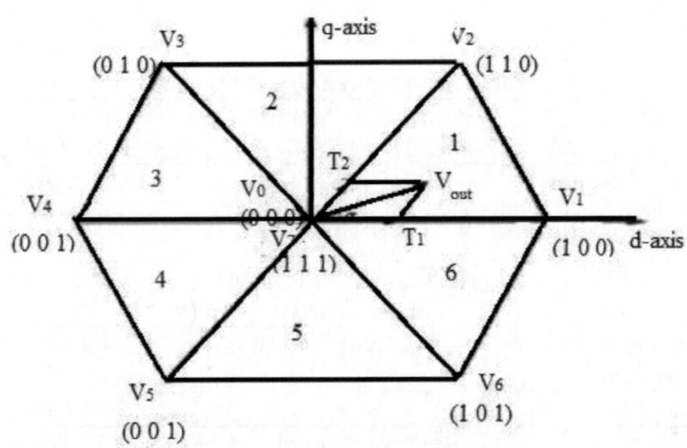

Figure 6 Space Vector Diagram.

4. Comparison

The induction motor control methods can be classified in to three; scalar control, vector control and direct torque control. The scalar control methods have the advantages like simple implementation, sensorless operation and simple control strategy .It is also known as V/f control. The scalar controlled induction motor provides sluggish response that is the main short coming of this method and it more suitable for low performance drives.

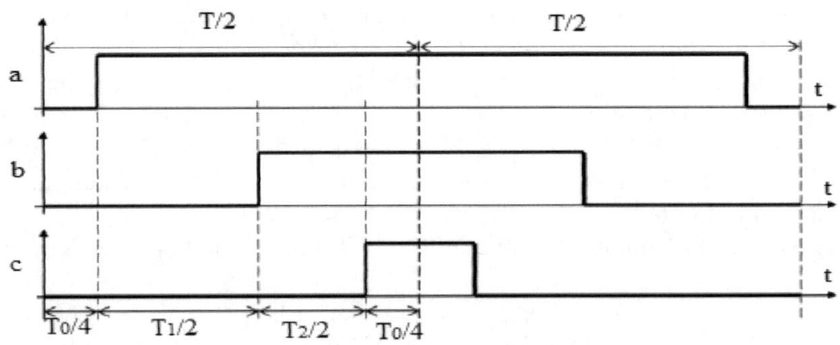

Figure 7 Symmetrical Switching Patterns.

The vector controlled method which is also known as field-oriented control (FOC) provides decoupled control of flux and torque. These methods have the control similar to that of separately excited DC motor. The presence of reference frame transformation makes the implementation of FOC is complex as compared with other methods. DTC or direct self-control (DSF) is a sensorless type control and by using this control it is possible to obtain good dynamic control without any mechanical transducers on the machine shaft. The conventional DTC is preferable in the high power range applications. Compared to FOC, DTC does not require any PWM signal generator, coordinate transformation, and current regulator. DTC has the advantages of the simplicity of operation, good torque control in steady state and transient operating conditions and it is less sensitive to the parameter variation in comparison with FOC. In DTC based scheme the control of torque and flux at very low speed is difficult. In conventional DTC method, direct torque control can be achieved but the direct current control is absent [15].

4.1. Comparison of DTC methods

Table 3 Comparison of DTC Methods

Properties	Simple DTC Control	DTC using SVPWM
Current Distortion	High	Low
Torque Ripple	High	Low
Switching Frequency	Change with motor speed	Constant
Position or Speed Sensor	Not required	Not required
Torque Response	Fast	Fast
Control Method	Relatively simple	Very complex

5. Conclusion

Different control methods of the induction motor are reviewed. Scalar control of IMD provides a sluggish response but its implementation is simple compared to others. The decoupled control of flux and torque is the main attractiveness of FOC. Both V/f control and vector control have fixed switching pattern while DTC using space vector control have varying switching pattern and it provides simple direct control of torque. In DTC method, the ripples in flux, torque, and current are high which can be reduced by employing SVPWM. Scalar control methods are more preferred for a low-performance drive and vector control is preferred for variable frequency application. The DTC method is suitable for high power range applications. The detailed comparison of DTC scheme is described in this paper.

References

[1] Chunting .M, Correa M. B. R, and J. O. P. Pinto 2010 *Proc. IFEC* The IEEE international future energy challenge-Request for proposals **45** 1-24.

[2] J. V. M. Caracas, L. F. M. Teixeira, G. de C. Farias, and A. L. De Ribeiro 2014 *IEEE Conf. APEC* Implementation of a High-Efficiency and Low-Cost Converter for a Photovoltaic Water Pumping Syste **50** 1048-2334.

[3] Utkarsh Sharma, Shailendra .K and B. Singh 2016 *IEEE* Solar Array Fed Water Pumping System using Induction Motor Drive 978-1-4673-8587.

[4] Trishan Esram and Patrick Chapman. L 2007 *IEEE Transactions on Energy Conversion EC* Comparison of photovoltaic array maximum power point tracking techniques **22** 439.

[5] Subudhi and R. Pradhan 2013 *IEEE Trans. Sustain. Energy* A comparative study on maximum power point tracking technique for photovoltaic power systems **4** 89-98.

[6] M. Calavial , 1. M. Periel , 1. F. Sanz, and J. SaBan 2010 *Proc.int.Conl Renewable Energy Power Quality* Comparison of MPPT strategies for solar panels 22-25.

[7] Garrigos,Blanes, Carrascoa, and J. Ejea 2007 *Renew. Energy.* Real time estimation of PV modules characteristics and its application to maximum power point operation **32** 1059-1076.

[8] D.Sun, K.H. Koh, B .G. Yu and M. Matsui 2009 *IEEE Trans. Fndus. Elect.* Fuzzy-Logic Based V/f Control of an Induction Motor for a DC Grid PowerLevelling System Using Flywheel Energy Storage Equipment **56** 3161-3168.

[9] S.R.Bhat, A. Pittet and B. S. Sonde 1987 *IEEE Trans. On fnd.App. I* Performance Optimization of Induction Motor-Pump System Using Photovoltaic Energy Source **23** 995-1000.

[10] Saurabh Shukla and Bhim Singh 2016 *IEEE* Solar PV Array Fed Speed Sensorless Vector Control of Induction Motor Drive for Water Pumping 978-15090-4530.

[11] D. Casadei, G. Serra, A.Tani and Luca Zarri 2006 *Bulletin of the Polish Academy of Science Technical Science* Assessment of direct torque control for induction motor drives **54**.

[12] B.Singh, A.K. Mishra and R.Kumar 2014 *IEEE Trans. Industrial Applications 52 3949-3957.*

[13] Iqbal, A 2008 *IE journel-EL* Analysis of space vector pulse width modulation for a five phase voltage source inverter 8-15.

[14] Syed Mahamood Ali, Dr. Mohammad Haseeb Khan and Dr. Arshia Azam,2015, *International Journal of Science, Engineering and Technology* Control of Induction Motor Drive Using Space Vector PWM 2395-4752.

[15] D.Casadei,F.Profumo,G.Serra and A. Tani 2002 *IEEE Trans. On Power Elec.* FOC and DTC: Two Viable Schemes For Induction Motor Torque Control 779-787

Nano tuned conducting polymer for plastic solar cell

Ishwar Naik[1], RajashekharBhajantri[2], VinayakBhat[3], B.S Patil[4]

[1]Asst.Professor in Physics, Govt. Arts& Science College, Karwar,Karnataka.
[2]Professor in Physics, Karnatak University Dharwad, Karnataka.
[3]Asst. Professor in Physics, Dr. A V BaligaCollege of Arts and Science,Kumta, Karnataka.
[4]Asst. Professor in Physics,Govt. First Grade College, Dharwad,karnataka
E-mail id:iknaik@rediffmail.com

Abstract: Nano technology coupled with the plastic electronics has boosted the research field of plastic solar cells. Usually semiconducting nanocrystals have higher electron affinityto be used as electron acceptors and the conjugated polymers act as electron donors with the preferable energy gap of 1eV-3eV. Presently, we have doped TiO_2nano particles (particle size < 100nm) acting asthe N type semiconductor onto P-type organic semiconductor Poly (3-hexyl thiophene-2,5-diyl)P3HT.Embedded nano particles form multi P-N junctionsinside the polymer matrix forming a mixed junction photoactive blend. P3HT is tuned by adding TiO_2nano particles at 3:1, 1:1 and 1:3 weight ratio in Bromo -Benzene (BB) as the solvent and the photoactive films are prepared on glass plates by solution cast method. The absorption curve gets broadened indicating the reduction in the band gap (energy difference between HOMO & LUMO). Among the samples, 1:3 composite has the broad spectral sensitivity for absorption and can be considered to be the best photoactive blend for constructing the solar cell. Absorption by the blend can be further enhanced by either dye sensitization or by exploiting plasmon resonance without increasing the thickness of the film. It is planned to construct the solar cell using optimized1:3 blend.

Key words:P3HT, HOMO, LUMO

1.Introduction

Solar energy is the abundant natural source of energy to be used effectively to solve the problem of energy crisis.The first generation silicon based cells and the second generation thin film solar cells are not cost effective. They have the drawback of infrastructure and installation complication also.Basically solar cell is a specialized semiconductor P-N junction diode which can convert solar energy into electrical energy. Usual solar cells are made up of inorganic P and N type semiconductors like Silicon and Germanium. Although these cells have enough efficiency, they are not affordable to common people. Scientists are on their way to search for a cost effective,efficient solar cell and organic or the plastic solar cells have emerged to be the promising in this direction. Unlike inorganic or conventional solar cells,these organic solar cells are made up of organic P and N type materials.Based on the constructional hierarchy polymeric solar cells are classified as single layer cells, Bilayer cells, Bulk heterojunction and Plasmonic cells. These constructional configurations are aimed at achieving more efficiency. But achieving higher efficiency is a global challenge for researchers. The use of nano technology has laid a strong hope in organic electronics. Usually

IOP Conf. Series: Materials Science and Engineering **396** (2018) 012048 doi:10.1088/1757-899X/396/1/012048

TiO_2doped P type conducting polymers will act as an efficient mixed junction photoactive material for solar cells.The nano particles property of large surface to volume ratio is effective in creating P-N junction of large interfaces. In the present work we have prepared mixed junction photo active blend of TiO_2 doped Poly (3-hexyl thiophene-2,5-diyl)P3HT.

2.Experimental

2.1 Materials

The Poly (3-hexyl thiophene-2,5-diyl)P3HT, and TiO_2nano particles are procured from Sigma Aldrich Corporation. The solvent Bromo Benzene is purchased from Rankem Chemicals. These chemicals are used as received without further purification. The chemical structure and specifications are indicated below.

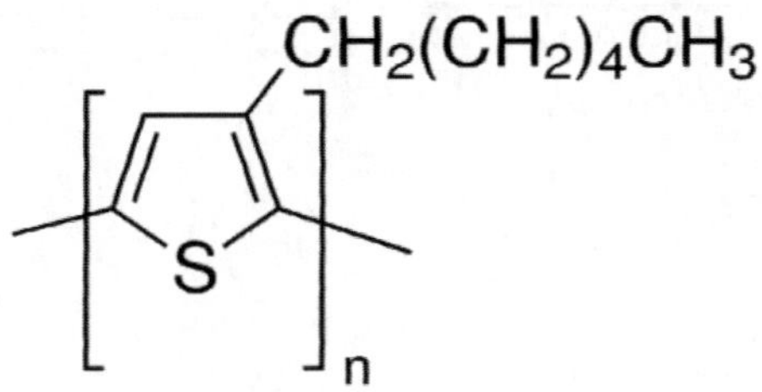

P3HT

P type polymer used – P3HT, Mol wt- average mol wt is15000 – 45000 ,HOMO-5eV, LUMO 3eV

Nano powder used -TiO_2 ,Particle size is less than 100nm

2.2 Processing Methodology

20 mg of P3HT is dissolved in 100cc of Bromo benzene resulting in 0.2mg /cc concentration. The solution is magnetically stirred for 48hourscontinuously at room temperature to ensure the uniform solution. TiO_2nano particles are directly added to the solution at 3:1,1:1 1:3 weight ratios as indicated in the table1. The mixture is again magnetically stirred for 6 to 7 hours to ensure optimum dispersion of nanoparticles in the polymer matrix. Doped solutions are transferred to petriplates and dried in an oven.

Table1. Sample composition

P 3HT	TiO_2	Sample state
4mg (20cc)	------	pure
3mg (15cc)	1mg	3:1
2mg (10cc)	2mg	1:1
1mg (5cc)	3mg	1:3

IOP Conf. Series: Materials Science and Engineering **396** (2018) 012048 doi:10.1088/1757-899X/396/1/012048

3. Result and discussion

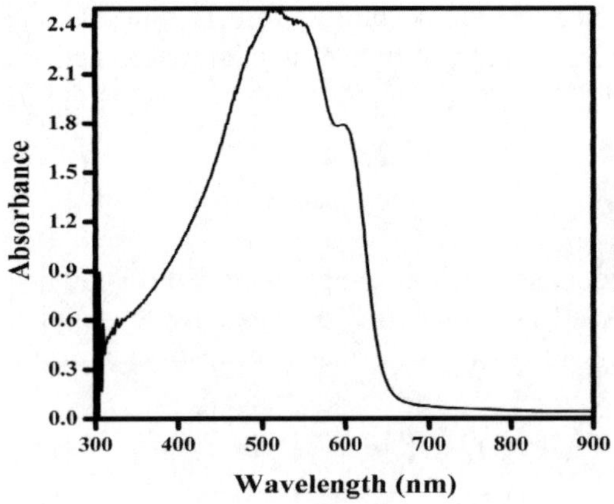

Figure 1.

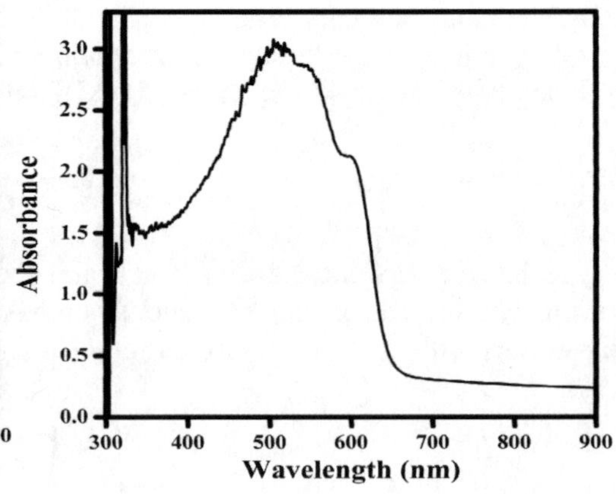

Figure 2.

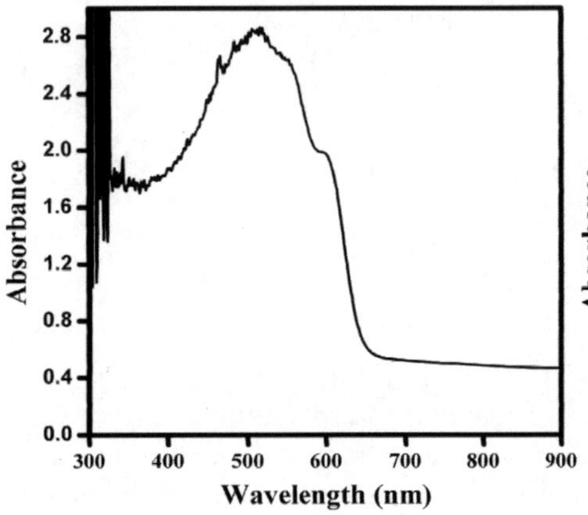

Figure 3.

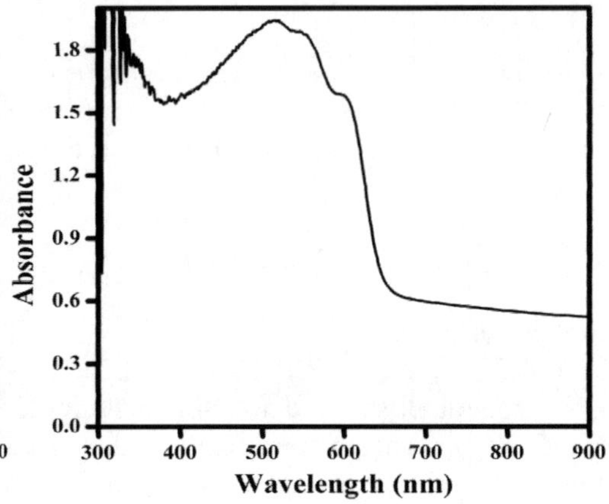

Figure 4.

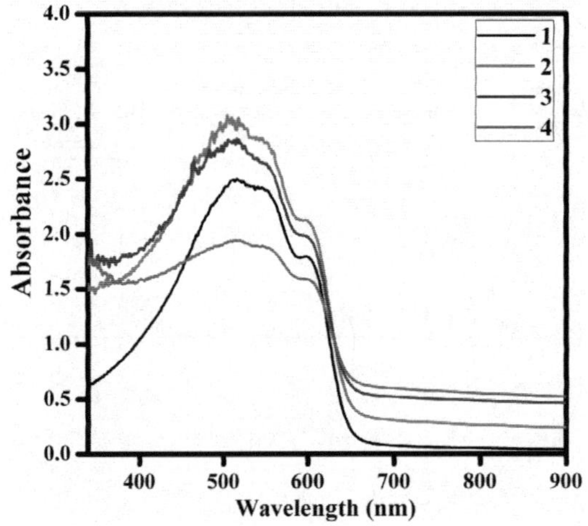

Figure 5.

IOP Conf. Series: Materials Science and Engineering **396** (2018) 012048 doi:10.1088/1757-899X/396/1/012048

Pure Poly (3-hexyl thiophene-2,5-diyl)P3HT is a conjugated polymer with alternate double bonds. It is the most widely used P type polymer in the construction of organic electronic devices. Optimum absorption in the visible region and good transport property of the polymer has made it a novel polymer to be used in Plastic electronic devices, Thefilm has absorption extending from 300 nm to 650nm having λ_{max} between 520nm to 550nm and withonset of absorption 650nm (Fig1). TiO_2nanoparticles arepromising material in a variety of applications like dyesensitized solar cells, photocatalysts and organic photovoltaics.TiO_2nanoparticleshave good dispersionproperty in the polymer matrix and their insertion in the polymer can improve the photocurrentgeneration,that can increase the device efficiency. $TiO_2$is the most promising photoanode materialdue to its chemical stability,ease control of shape and size, low cost and abundance[4]. Fig2, Fig3, Fig4 are the spectra for 3:1,1:1 and 1:3 P3HT:TiO_2 nanocomposite samples.Addition of TiO_2nano particles to P3HThastuned the absorption spectrum of P3HT. With increase in the doping level, the absorption in the UV region is enhanced followed by the slight reduction in absorption in the visible range as indicated by the overlay of spectra in Fig5. But there is increased absorption in the region beyond the visible range so that there is net broadening in the absorption spectra.

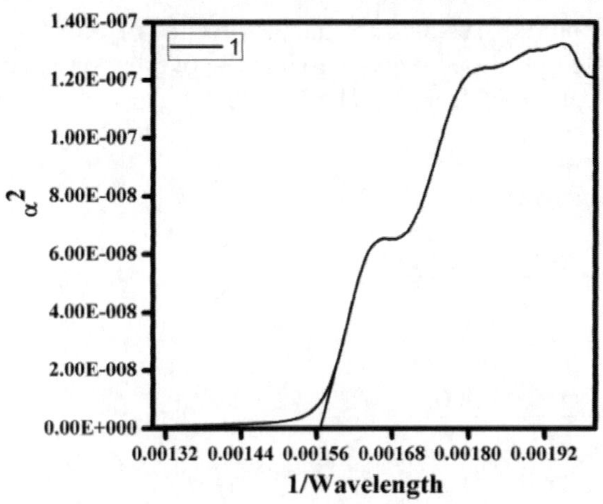

Figure 6

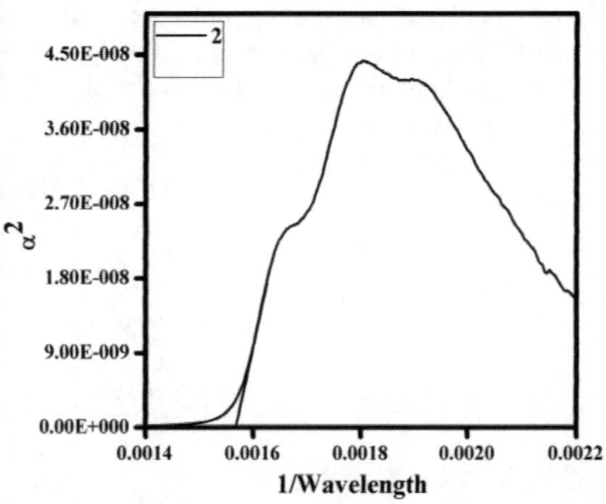

Figure 7

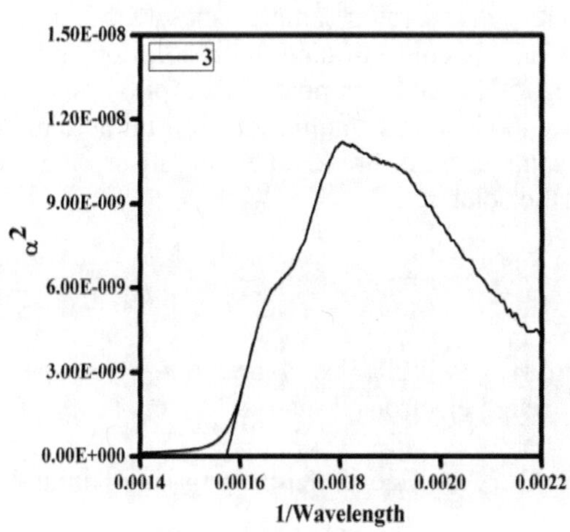

Figure 8

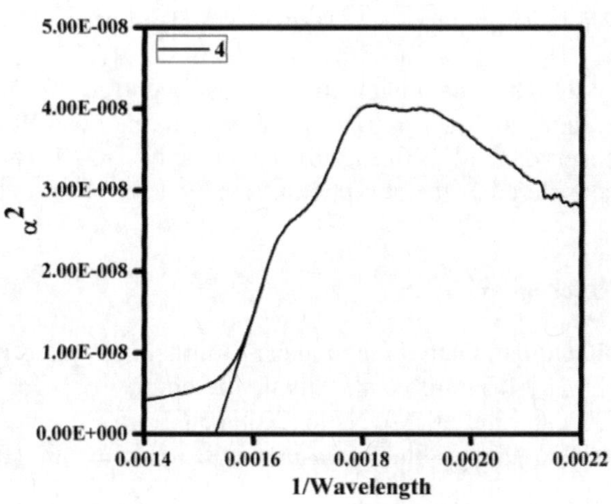

Figure 9

IOP Conf. Series: Materials Science and Engineering **396** (2018) 012048 doi:10.1088/1757-899X/396/1/012048

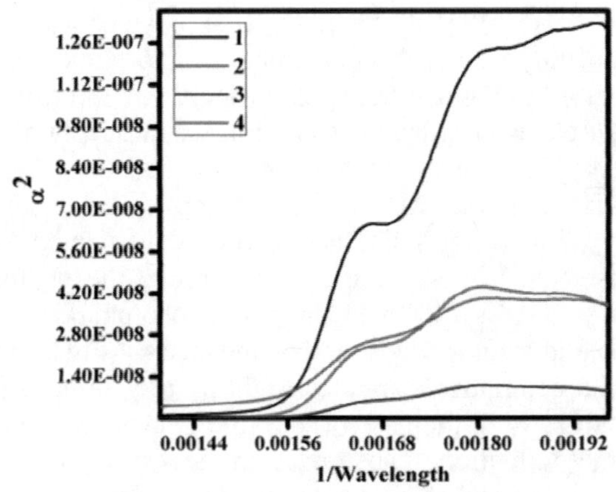

Figure 10

We have calculated the band gap of coated films of pure P3HT,3:1, 1:1 and 1:3 blends of P3HT and TiO_2,usingTauc's plot as shown in Fig6, Fig7, Fig8, Fig9 respectively. Fig10 is the overlay of Tauc's plot. The calculated band gaps are tabulated in the following Table2. The least band gap is observed for 1:3 P3HT:TiO_2 sample.

Table 2. Band gap values.

Sample	Band gap in eV
Pure P	1.941
3:1 P:N	1.938
1:1 P:N	1.941
1:3 P:N	1.900

4. Conclusion

We have prepared 3:1, 1:1 and 1:3 blend of P3HT:TiO_2 glass coated samples and characterized them by UV –Visible spectroscopy. Addition of TiO_2 to P3HT is going to tune the absorption spectrum of P3HT.The maximum tuning has occurred in 1:3 P3HT:TiO_2 sample. Absorption spectrum is broadened covering UV,visible region and even IR region.The maximum reduction in the band gap is observed for 1:3blend of P3HT and TiO_2.This sample with broad spectral absorption can be considered as the best photoactive blend to construct the solar cell.

References

[1] Quiqan Qiao, Lianyong et al Journal of applied Physics, 98, (2005) 094906.
[2] Ritesh Tipnis et. al. Polymer based materials for printed electrons, Material Matters (2008),3,4,92
[3] Yang,J et.al. Asc Nano (2011),5,62.
[4]Hee –Sang Shim, Seok-In Na, Sang Hoon nam,Hyo-Jin Ahn,Hae Jin Kim,Dong –Yu Kimand Won-Bae Kim. Applied Physics Letters 92,1831079, (2008).

Load harmonics extraction based decoupled control of grid connected solar photovoltaic system

Fossy Mary Chacko[1*], Jayan M V[2] and Prince A[3]

[1*]Research Scholar, [2]Professor and [3]Associate Professor, Department of Electrical Engineering, Rajiv Gandhi Institute of Technology, Kottayam, Kerala, India
E-mail*: marynbiji@gmail.com

Abstract. The global energy shortage and the requirement for energy systems that are sustainable has led to the widespread penetration of Renewable Energy Sources (RES). The intermittent characteristics of these Renewable Energy Sources make it a daunting task to integrate renewable energy, especially solar energy resources into the electric grid. The increasing number of Distributed Generation (DG) systems along with the existing non-linear loads create several issues due to variable output power and further worsens power quality problems like overvoltage, voltage fluctuations and harmonics. This paper describes an attempt on the development of a control strategy based on decoupled control with load harmonics extraction for the mitigation of power quality issues in grid integrated Solar Photovoltaic (SPV) systems. The performance of the grid integrated SPV system with linear and non-linear loads is investigated by simulation using MATLAB/Simulink. The effectiveness of the developed control strategy is validated by comparing the performance with an uncompensated system consisting of grid with linear and non-linear loads.

Keywords: Distributed Generation, Power Quality, Solar Photovoltaic (SPV)

1. Introduction

The global energy crisis and the need for sustainable energy systems has paved way for the development of power supply structures that are based mainly on renewable resources. The high penetration of RES requires new strategies for the operation and grid management in order to maintain or even to improve the electric power supply reliability and quality [1]. This paper focuses on the SPV system which is now gaining increased importance as a Renewable Energy (RE) source owing to advantages such as inexhaustibleness, the absence of fuel cost and harmful emissions, reduced maintenance and less wear and tear. The intermittent nature of RES makes it a challenging task to integrate renewable energy, especially solar energy resources into the power grid [2]. The increasing number of SPV systems imposes several issues [3] like variable output power, power quality problems, islanding, storage and compliance to standards. The integration of Photo Voltaic (PV) systems in low voltage distribution systems has led to the introduction of power quality problems like overvoltage, voltage fluctuations and harmonics. Moreover, the output power being fed to the grid depends on the existing conditions of insolation and temperature. This situation combined with the existing non-linear loads at the Point of Common Coupling (PCC) further deteriorates the power quality. There exist several solutions to the power quality problems like traditional passive filters, active filters and hybrid filters [4]. Distribution STATic COMpensators (DSTATCOMs) based on Voltage

IOP Conf. Series: Materials Science and Engineering **396** (2018) 012049 doi:10.1088/1757-899X/396/1/012049

Source Inverter (VSI) have become popular to reduce the problems associated with the ac distribution system, such as reactive power compensation, load balancing, and harmonic current elimination [5]. In this work, a control strategy based on decoupled control with load harmonics extraction for the PV inverter is developed for the amelioration of power quality issues in grid integrated SPV systems. The effectiveness of the proposed control strategy is validated by comparing the performance with an uncompensated system. The paper is organized as follows. Firstly, Section 2 briefly describes the grid connected PV system configuration. The modeling of PV array and the detailed design of the various components of the grid connected PV system is presented in Section 3. The proposed load harmonics extraction based decoupled control strategy is illustrated in Section 4. Section 5 presents the simulation results. Finally, the conclusions are given in Section 6.

2. System Description

PV systems can be mainly categorized as Stand-alone, Grid-connected and Hybrid PV systems [6]. A stand-alone system is the one which is not connected to the power grid. The PV systems integrated to the grid are called grid-connected PV systems. Hybrid PV systems may be of stand-alone or grid-connected type, but have at least one other power source other than PV. The major difference between the grid-connected and stand-alone systems comes from the energy storage feature. While the stand-alone PV systems usually have a provision for energy storage, the grid-connected PV systems have none. This research work focuses on the grid integrated PV System. The Block Diagram representation of a typical grid connected PV System is shown in figure 1.

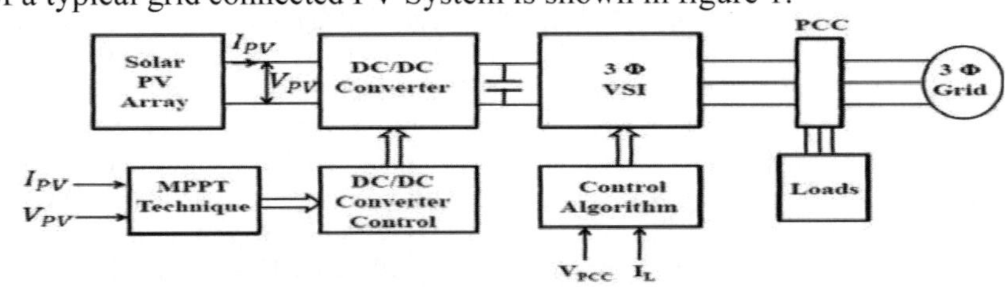

Figure 1. Grid connected PV system configuration

The grid interfaced solar PV generating system consists of a Solar PV Array, DC-DC boost conversion stage and a current controlled VSI interfaced to three-phase ac grid [7]. The output power from the Solar PV array is fed to a DC-DC boost converter stage in order to boost the PV array voltage to the required DC bus voltage level of VSI. Thus the active power generated by the SPV array is fed to the common DC link. The intermittent nature of the solar energy resource necessitates the use of Maximum Power Point Tracking (MPPT) [8]. In this paper, the Perturb and Observe (P & O) method is adopted to extract maximum power from the SPV array under different conditions of insolation.

3. Design of grid-connected SPV system

The system consisting of PV array and VSI is interfaced with a relatively large ac system, i.e. a stiff utility grid of 415 V. The operating frequency is dictated by the ac system and is fairly constant. The design of the various stages of the grid integrated SPV system [9] is given as follows:

3.1. SPV array design

The modelling of PV Array is based on the Single Diode Model (SDM) of PV cell [10]. The Light generated current for any operating conditions of the PV array is related to the photocurrent at Standard Test Conditions (STC) as: $I_{PV} = \left[I_{SC,n} + K_I(T - T_n)\right]\left(\frac{G}{G_n}\right)$ (1)

where, $I_{SC,n}$= light generated current at nominal temperature and irradiance, K_I = short circuit current temperature co-efficient, T = temperature of the p-n junction in Kelvin, T_n= nominal temperature = 273 K, G = solar insolation in W/m^2, G_n = solar insolation at STC= 1000 W/m^2

For a SPV array with N_S modules in series and N_P modules in parallel, the output current equation is defined as: $I = I_{PV}\, N_P - N_P I_O \left[exp\left(\frac{(V+R_S I)}{N_S V_t a C}\right) - 1\right] - \left(\frac{V+R_S I}{R_P}\right)$ (2)

where, I_O = reverse saturation current, V_t= thermal voltage of solar cell, a = Diode ideality factor, C = Number of cells connected in series in a module, R_S = equivalent series resistance of the array, R_P = equivalent parallel resistance of the array. The parameters of PV array model used for simulation are given in table 1. The SPV array is designed to generate a maximum power of 25 kW with input voltage to the boost converter V_{boost} in the range of 500 V. Here, the open circuit voltage of module V_{oc} is 32.9 V, and current at maximum power point I_{mp} is 7.61 A. The number of PV modules to be connected in series and parallel are given by:

$$N_s = \frac{V_{boost}}{V_{oc}} = 15.19 \text{ or } 16 \text{ modules and } N_p = \frac{P_{max}/N_s V_{oc}}{I_{mp}} = 6.24 \text{ or } 7 \text{ modules} \quad (3)$$

3.2. Design and selection of DC link voltage V_{dc}

The minimum value of DC bus voltage for VSI must be selected such that it is twice the value of peak phase voltage at the PCC: $V_{dc} = \frac{2\sqrt{2}V_L}{\sqrt{3}m} = 677.7\,V$ (4)
where, V_L = Line Voltage of the grid, m = Modulation index= 1. The DC link voltage is selected as 700 V.

3.3. Design of VSI DC Link Capacitance

The VSI DC Link Capacitance value is determined as: $C_{dc} = \frac{P_{dc}/V_{dc}}{2\omega V_{dc,ripple}} = 4062\,\mu F$ (5). Where, P_{dc} is the power available at the dc link and ω is the angular frequency. The % Ripple in dc link voltage $V_{dc,ripple}$ is taken as 2% of V_{dc}. The value of VSI DC link capacitance is selected as 4000 μF.

3.4. Design of Coupling Inductances

The value of the coupling ac inductor is computed as: $L_f = \frac{\sqrt{3}\,mV_{dc}}{12hf_s\Delta i} = 7.86\,mH \cong 8\,mH$ (6)
Where, h =overloading factor = 1.2, f_s =Switching frequency, Δi = Ripple current which is assumed to be 3% of the peak current.

3.5. Design of Boost Converter

A two stage grid connected PV system is designed here. The intermediate dc-dc converter is in the unidirectional boost topology that boosts the output voltage from the Solar PV array V_{PV} and feeds the active power generated to the DC Bus of the VSI at a voltage V_{dc}. The boost converter consists of the energy storage reactor L_b, switch S, diode D and output capacitor C_b. The boost converter voltage relation is given by : $V_{dc} = V_{PV}/1-D$ (7). The energy storage reactor L_b is designed for a switch Duty Ratio of 0.25, switching frequency f_{sw} of 10 kHz and input current ripple of 3% as: $L_b = \frac{V_{dc}D(1-D)}{2f_{sw}\Delta i_{in}}$ (8). The DC link capacitance C_{dc} serves as the output capacitor C_b of the boost converter stage. The boost converter is linked to the PV system with a filter capacitor C_1 for decreasing the ripple caused by the high frequency switching of switch S. The designed parameters of grid-connected SPV system used for simulation are given in table 2.

Table 1. PV Array parameters.

Array Parameters	
R_S: 0.23 Ω	a:1.3
R_p: 600 Ω	$I_{SC,n}$: 8.21 A
V_t: 0.0257	K_I: 3.2 x 10^{-3}
C:54	T_n: 273 K

Table 2. Parameters of grid-connected SPV System.

Grid-connected SPV system Parameters	
N_S: 16	L_f:8 mH
N_P:7	f_s: 10 kHz
V_{dc}: 700 V	L_b:4.59 mH
C_{dc}: 4000 μF	C_b:4000 μF

IOP Conf. Series: Materials Science and Engineering **396** (2018) 012049 doi:10.1088/1757-899X/396/1/012049

4. Proposed Control Strategy: Load Harmonics Extraction Based Decoupled Control

The flexibility and reliability of the grid integrated SPV system depends on the control algorithm employed for the switching of the PV inverter. The control scheme is based on the concepts of instantaneous power in the synchronously rotating d-q reference frame. The rotating reference frame is employed in this work because it offers higher accuracy compared to stationary frame based techniques. The PV inverter can be modeled [11] as follows in the synchronously rotating reference frame (d-q frame). Here, the inverter output emfs are represented as V_{invd}, V_{invq} and the grid voltages are represented as V_{gd}, V_{gq}.

$$p\begin{bmatrix} i_d \\ i_q \end{bmatrix} = \begin{bmatrix} \frac{-R_s}{L_s} & \omega \\ -\omega & \frac{-R_s}{L_s} \end{bmatrix} \begin{bmatrix} i_d \\ i_q \end{bmatrix} + \frac{1}{L_s} \begin{bmatrix} V_{invd} - V_{gd} \\ V_{invq} - V_{gq} \end{bmatrix} \tag{9}$$

The new co-ordinate system is defined with the d-axis always coincident with the instantaneous grid voltage vector. Thus in the d-q frame, the q-axis component of the grid or source voltage is zero i.e. $V_{gq} = 0$.

$$p\begin{bmatrix} i_d \\ i_q \end{bmatrix} = \begin{bmatrix} \frac{-R_s}{L_s} & \omega \\ -\omega & \frac{-R_s}{L_s} \end{bmatrix} \begin{bmatrix} i_d \\ i_q \end{bmatrix} + \frac{1}{L_s} \begin{bmatrix} V_{invd} - V_{gd} \\ V_{invq} \end{bmatrix} \tag{10}$$

4.1. Decoupled control of i_d and i_q

Inspection of equation (10) leads directly to a rule that will provide decoupled control of i_d and i_q. The inverter voltage vector is controlled as follows:

$$V_{invd} = L_s(x_1 - \omega i_q) + V_{gd}$$
$$V_{invq} = L_s(x_2 + \omega i_d) \tag{11}$$

Substitution of equation (11) in equation (10) gives rise to the following relation:

$$p\begin{bmatrix} i_d \\ i_q \end{bmatrix} = \begin{bmatrix} \frac{-R_s}{L_s} & 0 \\ 0 & \frac{-R_s}{L_s} \end{bmatrix} \begin{bmatrix} i_d \\ i_q \end{bmatrix} + \begin{bmatrix} x_1 \\ x_2 \end{bmatrix} \tag{12}$$

Equation (12) shows that i_d and i_q respond to x_1 and x_2 respectively through a simple first order transfer function with no cross coupling. The control rule of inverter voltage equations is completed by defining feedback loops of Proportional and Integral (PI) compensators PI_1 (with proportional gain K_1 and integral gain K_2) and PI_2 (with proportional gain K_3 and integral gain K_4) as follows:

$$x_1 = \left(K_1 + \frac{K_2}{s}\right)\left(i_{dref} - i_d\right) \tag{13}$$

$$x_2 = \left(K_3 + \frac{K_4}{s}\right)\left(i_{qref} - i_q\right) \tag{14}$$

4.2. Detection Algorithm based on Load Harmonics Extraction

One of the important aspects in the control scheme design is the detection algorithm used for determining the compensating current and voltage references. The effectiveness of the VSI for compensation depends on the method implemented to generate the reference template. The compensating current reference for the PV inverter consists of d and q axis components i_{dref} and i_{qref}. The compensating current reference template is derived from the load currents. The load currents are sensed and transformed from abc to dq frame, namely i_{ld}, i_{lq}. The d-axis component of compensating current reference for the PV inverter consists of two parts: (i) The d-axis component of load current harmonics i_{ldh} which is extracted as follows. The fundamental d-axis component of load current i_{ldf} is first extracted using a Low Pass Filter. Then, the d-axis component of load current harmonics i_{ldh} is obtained by subtracting this fundamental component from the d-axis component of load current i_{ld}. (ii) The current necessary for regulating the dc bus voltage to the desired level. This current is denoted as Δi_d and its amplitude is determined by the dc bus voltage regulation loop. Thus, the real power can be regulated by varying i_{dref} in response to the error in the dc link voltage via PI compensator PI₃ (with proportional gain K_5 and integral gain K_6). The q-axis component of compensating current reference for the PV inverter consists of the q-axis component of the load current i_{lq}. Thus, the PV inverter is designed to inject the entire reactive power requirement of the load. The compensating current reference template for the PV inverter is thus given by:

$$i_{dref} = i_{ldh} + \Delta i_d = i_{ldh} + \left(K_5 + \frac{K_6}{s}\right)(V_{dcref} - V_{dc}) \tag{15}$$

Where, V_{dcref} is the reference value of dc link voltage.

$$i_{qref} = i_{lq} \tag{16}$$

The compensating current references are then compared with the actual inverter currents as shown in figure 2. The inverter output voltages can be expressed in terms of the dc link voltage and switching function matrix as: $\begin{bmatrix} V_{invd} \\ V_{invq} \end{bmatrix} = \frac{1}{2} maV_{dc} \begin{bmatrix} S_d \\ S_q \end{bmatrix}$ (17). For modulation index m and turns ratio a of 1, equation (17) becomes: $\begin{bmatrix} V_{invd} \\ V_{invq} \end{bmatrix} = \frac{V_{dc}}{2} \begin{bmatrix} S_d \\ S_q \end{bmatrix}$ (18). Thus, $\begin{bmatrix} S_d \\ S_q \end{bmatrix} = \frac{2}{V_{dc}} \begin{bmatrix} V_{invd} \\ V_{invq} \end{bmatrix}$ (19).

The switching function matrix in dq frame is transformed back to abc frame. The modulating signals S_{abc} are compared with triangular carrier signals (10 kHz) to obtain the PWM gating pulses for switches S_1 to S_6 of the PV inverter.

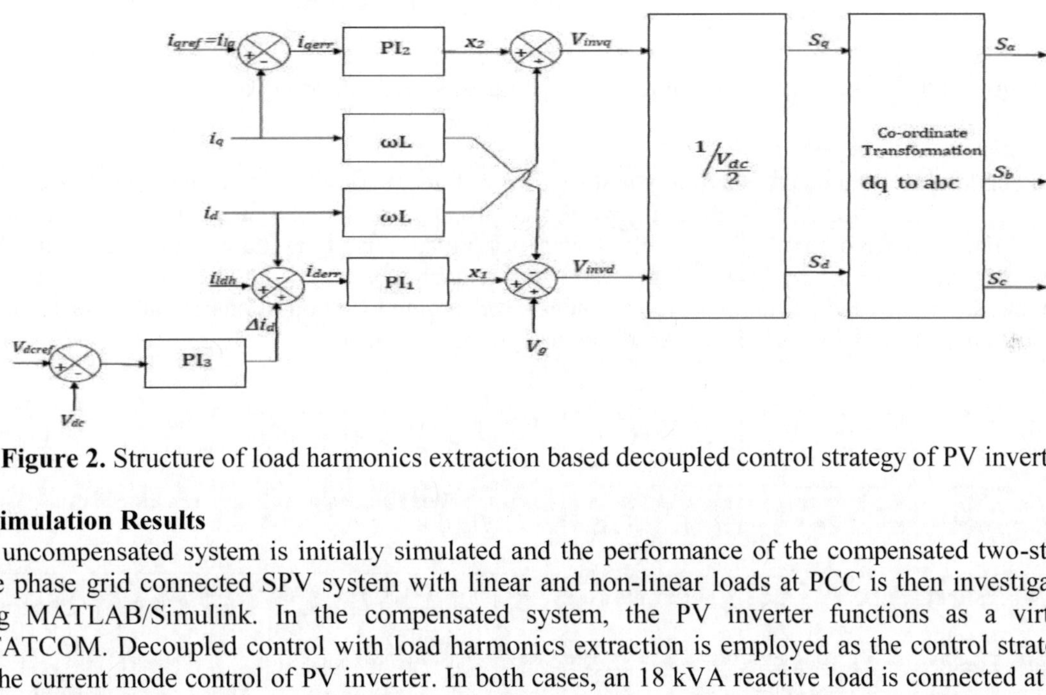

Figure 2. Structure of load harmonics extraction based decoupled control strategy of PV inverter

5. Simulation Results

The uncompensated system is initially simulated and the performance of the compensated two-stage three phase grid connected SPV system with linear and non-linear loads at PCC is then investigated using MATLAB/Simulink. In the compensated system, the PV inverter functions as a virtual DSTATCOM. Decoupled control with load harmonics extraction is employed as the control strategy for the current mode control of PV inverter. In both cases, an 18 kVA reactive load is connected at the PCC from 0 to 0.25 s. From 0.25 s to 0.5 s, a three phase 14 kVA non-linear diode rectifier load and 6 kVA reactive load are connected at the PCC.

5.1. Uncompensated System

The uncompensated system consists of three phase utility grid with linear and non-linear loads at the PCC. Figure 3 depicts the grid and the load voltage and current waveforms under the changing load conditions. It can be seen that distorted load currents cause distortion in the grid side currents. Also, the grid power factor is lagging. Figure 4 depicts the grid and the load real and reactive power. It can be observed that the entire real and reactive power requirement of the load is met by the utility grid.

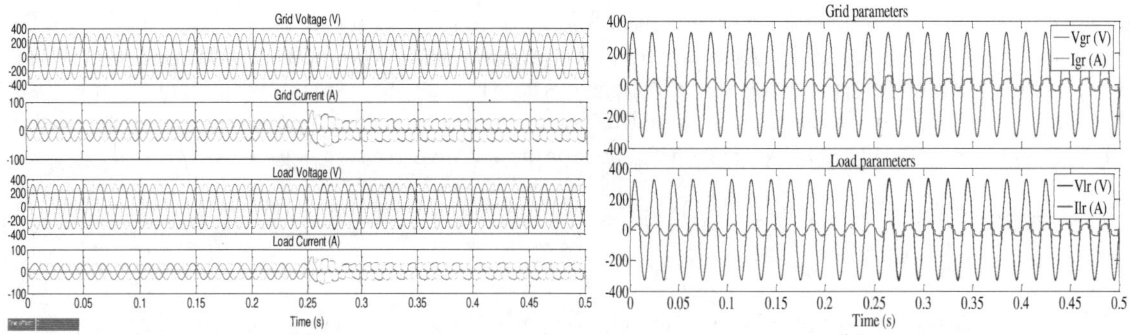

(a) (b)

Figure 3. Response of uncompensated system under changing load conditions

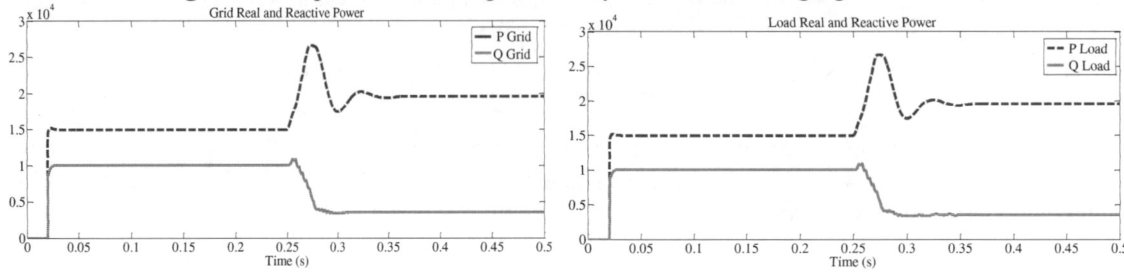

(a) (b)

Figure 4. Real and reactive power of uncompensated system under changing load conditions

5.2. Compensated System

The compensated system consists of a two-stage 25 kW three phase grid connected SPV System with linear and non-linear loads at the PCC. Decoupled control with load harmonics extraction is employed as the control scheme for the PV inverter. The performance is investigated under changing load conditions with constant irradiance of 400 W/m^2 and temperature of 25° C. Figure 5 depicts the grid and the load voltage and current waveforms under the changing load conditions. It can be seen that the grid side currents remain sinusoidal even if the load currents are distorted.

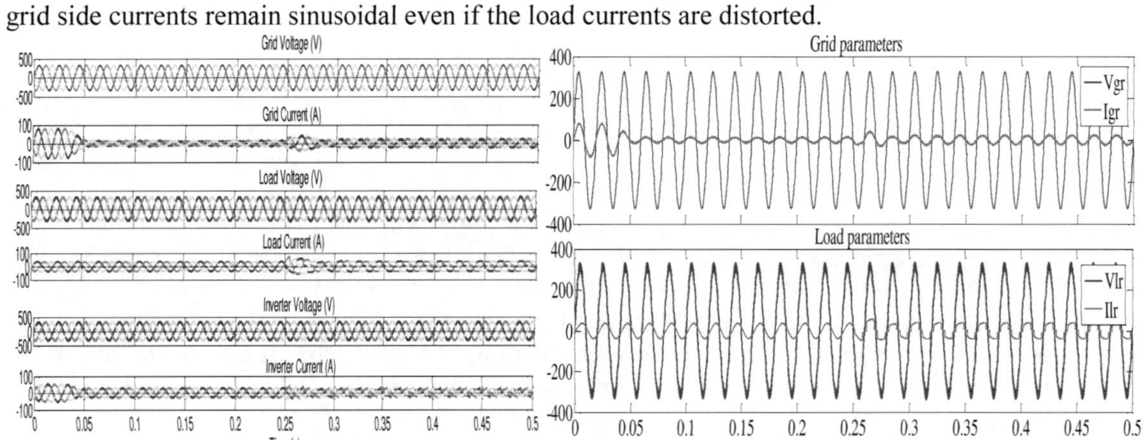

(a) (b)

Figure 5. Response of compensated SPV system under changing load conditions

Figure 6 depicts the grid, the load and the inverter real and reactive power. It can be observed that the entire reactive power requirement of the load is met by the PV inverter. Hence, the reactive power burden of the grid reduces completely and the grid power factor improves to unity. The PV inverter also injects the available real power from the SPV system at the given conditions of temperature and irradiance to the PCC. Hence the SPV system supplies a portion of the real power requirement of the load and the grid has to supply only the remaining real power. Figure 7 shows the boost converter output voltage i.e. the dc link voltage which is regulated to the reference value V_{dcref} of 700 V by 0.8% even under changing load conditions.

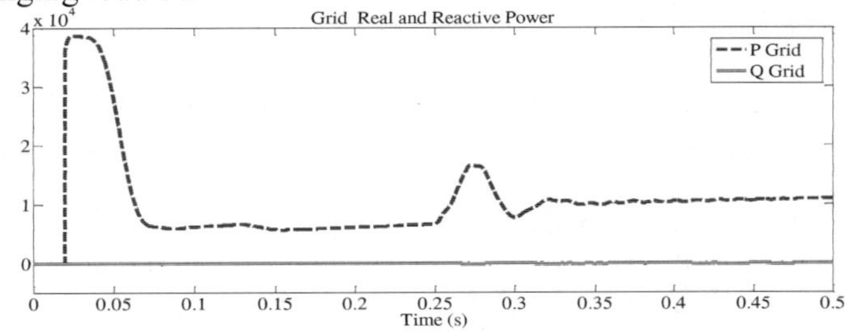

(a)

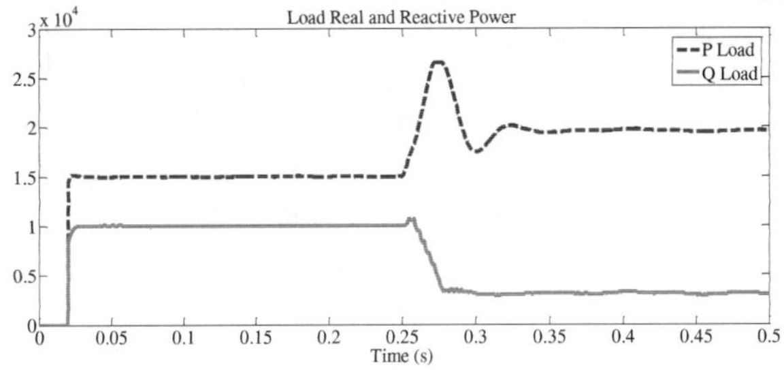

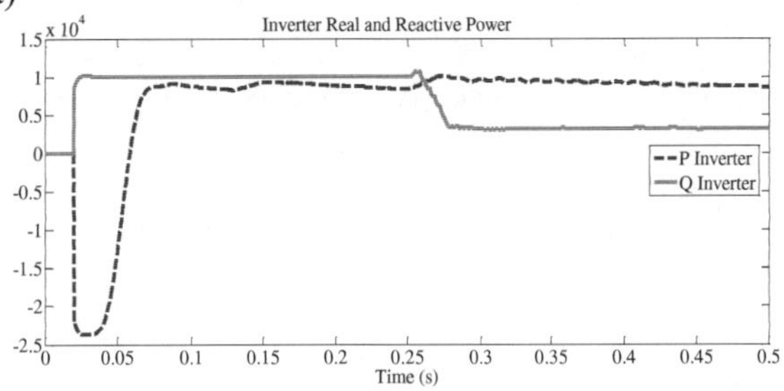

(b) (c)

Figure 6. Real and reactive power of compensated SPV system under changing load conditions

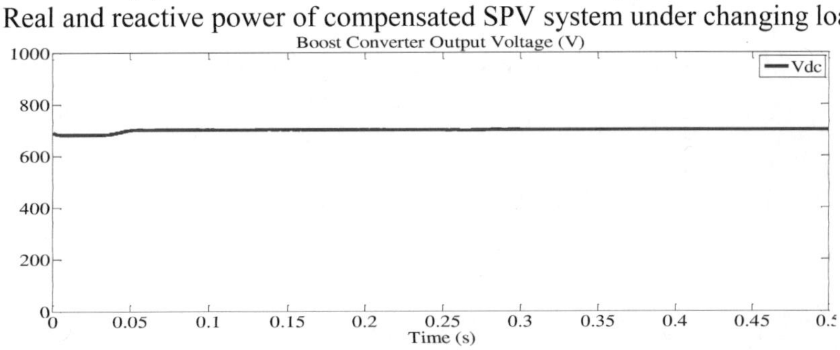

Figure 7. Voltage at the dc link of SPV system under changing load conditions

Table 3. THD Comparison.

	Uncompensated System		Compensated System with SPV-DSTATCOM	
THD (%)	Reactive Load	Reactive and Diode Rectifier Load	Reactive Load	Reactive and Diode Rectifier Load
Grid Voltage	0%	0%	0%	0%
Grid Current	0%	22.07%	1.46%	5.41%
Load Voltage	0%	0.97%	1.73%	1.67%
Load Current	0%	22.07%	0%	22.43%

IOP Conf. Series: Materials Science and Engineering **396** (2018) 012049 doi:10.1088/1757-899X/396/1/012049

Based on the simulation results, a comparison of THD values of the grid and the load voltages and currents for the uncompensated system and compensated system with SPV-DSTATCOM is given in table 3. It can be seen that the implementation of decoupled control approach with load harmonics extraction for the grid integrated SPV system causes significant reduction in the THD values of grid current even in the presence of highly non-linear loads.

6. Conclusion

In recent years, new challenges and opportunities in power quality management have emerged due to the increased penetration of DG systems based on RES and the proliferation of power electronic based non-linear loads. In an uncompensated system, the presence of non-linear diode rectifier loads at the PCC causes the grid currents to become distorted and the grid has to supply the entire real and reactive demand of the load. Also, the grid p.f. is lagging. In this work, a control strategy based on decoupled control with load harmonics extraction is developed for the grid integrated SPV system. It can be seen that the PV inverter functions as a virtual DSTATCOM and the grid current remains distortion free even with highly non-linear loads connected at the PCC.

The simulation results show that the implementation of decoupled control provides independent control of real and reactive power and regulation of the dc link voltage to the desired level. The PV inverter supplies the entire reactive power requirement of the load hence reducing the reactive burden of the grid completely and the grid p.f. is improved to unity. According to the available generated PV power at the given conditions of temperature and irradiance, the PV system provides a portion of the real power demanded by the load. Thus, the grid has to supply only the remaining real power required by the load. However, this system suffers from the drawback that the response time of the controller is more. A modification on the control strategy may give faster response under changing conditions of temperature, irradiance and load.

References

[1] Chidurala A, Saha T K and Mithulananthan N 2016 Harmonic impact of high penetration photovoltaic system on unbalanced distribution networks - learning from an urban photovoltaic network *IET Renew. Power Gener.* **10 (4)** 485–94

[2] Blaabjerg F, Teodorescu R, Liserre M and Timbus A V 2006 Overview of control and grid synchronization for distributed power generation systems *IEEE Trans. Ind. Electron.* **53 (5)** 1398–409

[3] Eltawil M A and Zhao Z 2010 Grid-connected photovoltaic power systems: technical and potential problems- a review *Renewable and Sustainable Energy Reviews* **14** 112-29

[4] Akagi H 1996 New trends in active filters for power conditioning *IEEE Trans. Ind. Appl.* **32** 1312–22

[5] Singh B, Chandra A and Solanki J 2009 A comparison of control algorithms for DSTATCOM *IEEE Trans. Ind. Electron.* **56 (7)** 2738–45

[6] Kroposki B and DeBlasio R 2000 Technologies for the new millennium: photovoltaics as a distributed resource *Proc. IEEE Power Engineering Society Summer Meeting (Seattle, WA)* vol 3 pp 1798–801

[7] Blaabjerg F, Chen Z and Kjaer S B 2004 Power electronics as efficient interface in dispersed power generation systems *IEEE Trans. Power Electron.* **19 (5)** 1184–94

[8] Esram T and Chapman P L 2007 Comparison of photovoltaic array maximum power point tracking techniques *IEEE Trans. Energy Convers.* **22 (2)** 439–49

[9] Agarwal R K, Hussain I and Singh B 2016 LMF-based control algorithm for single stage three-phase grid integrated solar PV system *IEEE Trans. Sust. Energy* **7 (4)** 1379–87

[10] Villalva M G, Gazoli J R and Filho E R 2009 Comprehensive approach to modeling and simulation of photovoltaic arrays *IEEE Trans. Power Electron.* **24 (5)** 1198–208

[11] Schauder C and Mehta H 1993 Vector analysis and control of advanced static VAR compensators *IEE Proc. C- Generation, Transmission and Distribution* vol 140 pp 299–306

Soiling of photovoltaic modules- Review

Anitha Shaju[1], Rani Chacko[2]

[1] PG Scholar, Electrical and Electronics Engineering, Amal Jyothi College of Engineering, Kottayam, Kerala, India
[2] Assistant Professor, Electrical and Electronics Engineering, Amal Jyothi College of Engineering, Kottayam, Kerala, India

E-mail: anithaabin09@gmail.com[1], chacko.rani@gmail.com[2]

Abstract. The atmosphere is contaminated with certain minute particles of size 500μm. The deposition of these dust particles over the panel surfaces is referred as Soiling of Photovoltaic (PV) panels. The particulate matters originating from agricultural activity as well as construction works, air pollution, pollen, bird droppings and growth of lichens are the main sources of soiling on the PV panels. It is influenced by location-based meteorological conditions like rainfall, wind speed, and humidity, tilt angle, the sources of soiling, and the texture of cover glasses. The soiling is one among the major losses in the PV performance. Owing to this reason, the PV panels should be cleaned at regular intervals. There are manual and self- cleaning methods. The cleaning by manpower is most common. The self-cleaning comprises of active as well as passive cleaning techniques. This paper reviews about different methodologies used to study the soiling of PV modules and the different factors affecting soiling.

1. Introduction

Nowadays, the solar energy is the fastest growing renewable energy in residential, commercial, agricultural, and industrial applications. The photovoltaic (PV) system converts the solar power obtained from the sun to electricity through the semiconductor materials like silicon and cadmium telluride. The solar energy is abundant, clean and infinite energy[2]. The main advantage of solar energy is that it provides green energy with no carbon emission. But the solar energy is intermittent and varying depending on the environmental and climatic conditions. Besides these, there are many other factors affecting the PV performances like the photovoltaic technology used, tilt angle, array mismatch, distribution losses, inverter efficiency, soiling of PV panels, etc.

The deposition of dust and other minute particles over the PV modules is termed as Soiling[1]. It is one of the major loss factors affecting the PV performance. The soiling loss is the power loss resulting from the dust, snow, dirt, and other particles that covers the PV module surface. The various sources of soiling are pollution, bird droppings, the growth of lichens, particulate matter arises due to the wind, agricultural activities, construction works, pedestrian and vehicular movements[3].

Section II describes the different factors affecting the soiling of PV panels. In section III, the various methodologies used in soiling study at different locations are given in detail. Section IV describes the

IOP Conf. Series: Materials Science and Engineering **396** (2018) 012050 doi:10.1088/1757-899X/396/1/012050

tilt angle optimization and PV potential assessment. Section V describes the cleaning aspects of PV panels.

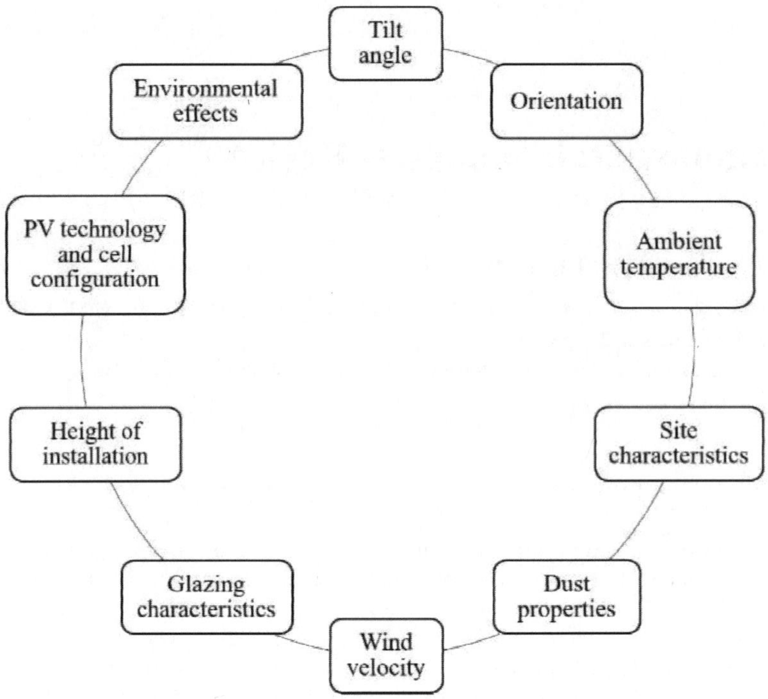

Figure 1. Factors affecting soiling of PV panels [4][5][6].

2. Factors affecting soiling

The soiling is a phenomenon influenced by location-based meteorological factors and climatic conditions, tilt angle, the sources [1]. Soiling may lead to partial shading of cells as the surface is covered with dust and thereby optical losses may occur. The different factors affecting soiling of PV panels is given in figure 1. It includes PV installation factors and environmental factors. The installation factors that affect soiling are height of installation, tilt angle, orientation, glazing characteristics, and PV technology used. The different environmental factors are wind speed, dust properties, ambient temperature, rainfall, humidity, and vegetation.

2.1. Tilt angle and orientation

The dust particles may settle more on horizontal surfaces than the tilted surfaces [7]. The larger dust particles will roll down from top to bottom of a tilted PV module. Hence, the effect of gravitational force in dust accumulation increases as the tilt angle increases. The PV modules that are facing towards the wind will get the more influence in dust deposition compared to the PV modules kept far away from the direction of the wind. Hence, the orientation of PV panels is an important factor influencing the cleaning of panels by wind movement.

2.2. Ambient temperature, Pressure and humidity

Dust, humidity, and wind are the factors depending on the sun's radiation and the environment. At high ambient temperature and low relative humidity, wind carries the dust easily. The water vapour will condense and form water drops at low temperature [3]. High RH increases the dirt adhesion and forms a sticky surface over the solar panels [8].This led to adhesive panel surface which attracts more particles from surroundings. When the airspeed is higher and there is low pressure, less soil accumulation occurs and vice-versa. The humidity changes with the irradiance in a non-linear manner and irradiance itself causes little variations in open circuit voltage and causes large variations in short-circuit current [3].

2.3. Site characteristics
The site characteristics involve the characteristic features of the location for PV panel installation. The area can be a metropolitan city or a village. The vegetation, natural habitats, pollution rate, pedestrian and vehicular traffic varies with location. Based on these variations, the soiling of solar panels vary as the dust pollutant types differ with geographic location.

2.4. Dust properties
The different properties of dust include chemical, biological, electro-static and physical properties. The dust deposition will be different for different dust properties. The physical properties of dust includes its size, shape, and weight. The small dust particles settle down more than that of the larger dust particles on the PV panel surface [2]. The different electrostatic property includes neutral, positively charged and negatively charged particles. Generally, the charged particles cause more dust deposition than the neutral particles because the like charges attract and unlike charges repels.

2.5. Wind velocity
There are two effects of wind speeds on the dust settlement. In [5], it is claiming that the high wind velocities result in high soiling rate than the low wind speeds. But in other papers, the high wind velocity will clean the panels against soiling if it is oriented towards the wind movement. It is seen that the chances of dust settlement are more at light wind speeds. For the wind with high amount of airborne dust, the possibility of dust accumulation is more [3].

2.6. Glazing characteristics
The dust accumulation on PV panels varies with the property of panel surface. The surface texture and additional coating on the PV panels comprise the surface property. For plane surfaces, the dust accumulation is lesser. The textured panels have rough and irregular surfaces increase the soiling on panel surface [2]. A protective layer of glass on the PV module surface is less affected by dust deposition.

2.7. Environmental effects
The environmental effects on soiling arise from the variations in atmospheric and climatic conditions. The air borne dust concentration, probability of dust storm, the occurrence of dew, rainfall, volcanic eruptions, etc. contribute to soiling of PV panels.

2.8. PV technology and cell configuration
The soiling affects the different PV technologies differently based on the spectral transmittance. The PV technologies include monocrystalline PV panels and polycrystalline PV panels. The performance of PV panels varies with its technology.

3. Methodologies for soiling study
The methodology of soiling studies comprises of natural and artificial soiling. The natural soiling study is done by keeping the PV panels exposed to sun for a certain period of time and allows the soiling of panels from the surrounding dust particles. In artificial soiling, the dust particles are artificially prepared and spread on the panels using the technologies like spray gun. In both the cases, the effect of soiling on the electrical parameters (voltage, current, Fill Factor, efficiency, etc.) and optical parameters (transmittance, reflectance, etc.) of solar panels are studied and analysed.

3.1. Natural Soiling

The natural soiling of PV panels is the dust accumulation over solar panels when it is exposed to sun. The main sources of natural soiling are the dust particles present in the atmosphere due to air pollution, bird droppings, agriculture, construction works, pedestrian and vehicular movements. The effect of soiling on PV performance is studied through natural soiling by measuring the electrical and optical parameters of the PV panels.

3.1.1 Measurement of electrical parameter. In Garcia [1], a long-term soiling study of about 30 years was conducted in a moderate sub-tropical climatic region in Italy. They installed 28 silicon PV panels in 1985 and pull apart in 2014 without cleaning. They used manual cleaning and high-pressure water sprayer for cleaning the soiled panels. The figure 2 shows the I-V characteristics of PV panels before and after cleaning. It is observed that the power output increased after the cleaning of the PV panels.

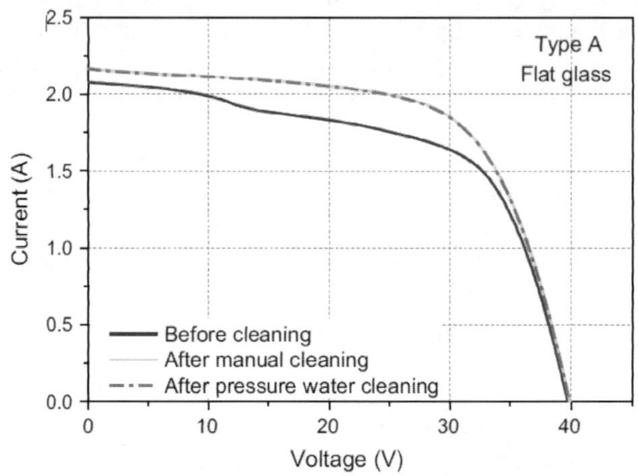

Figure 2. I-V characteristics of the PV panel [1].

Another study conducted by Jose Cano et. al [9], the influence of tilt angle on dust deposition over the panels is investigated. As the short circuit current is proportional to the sun's radiation reaching the solar panels, the measurement of I_{sc} can analyse the effect of tilt angle on soiling loss. The experimental setup for the study is given in figure 3. The figure 4 inferred that the soiling losses decreases with increase in the tilt angle of solar panels.

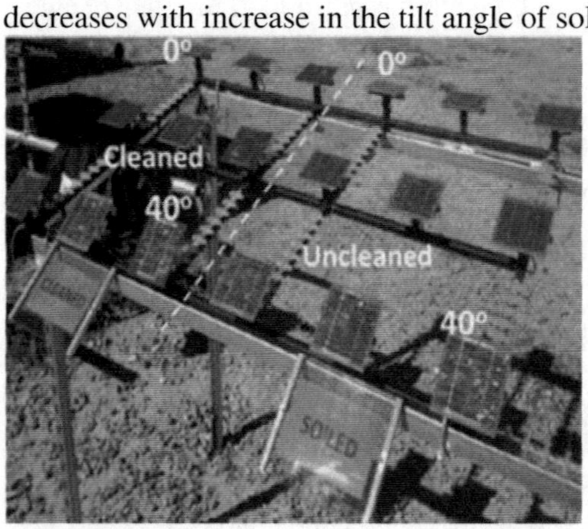

Figure 3. Experimental set-up for studying the effect of tilt angle on soiling [9].

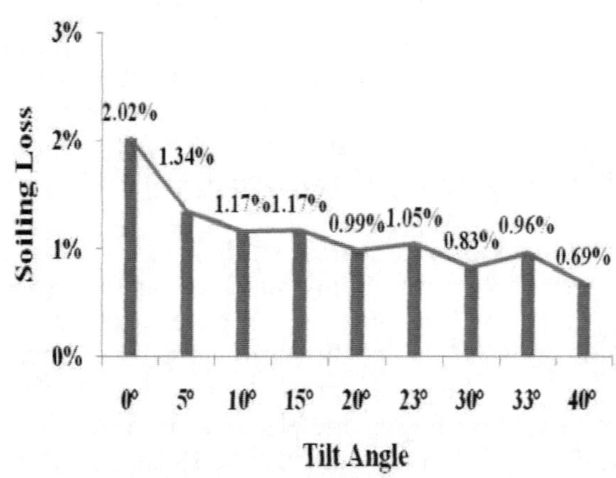

Figure 4. Plot between average soiling loss and tilt angle [9].

The setup consists of 18 PV modules which are calibrated. The modules are divided into two groups of nine PV panels, tilted at nine different angles. One group of panels are cleaned periodically and others kept uncleansed. They calculated the soiling loss at different tilt angle by measuring the daily insolation received by cleaned panels (DI_{cm}) and uncleansed panels (DI_{sm}). The soiling loss is calculated from equation (1).

$$\text{Soiling loss (\%)} = [(DI_{cm}-DI_{sm}) / DIcm] \times 100 \quad (1)$$

King et. al [10] done an improved work of [9] where the experimental setup is named as soiling stations. The soiling stations monitor the soiling losses by comparing the PV output of soiled PV panels to the PV output of cleaned panels [12]. A soiling station is shown in figure 5. The modified experimental setup consists of modules with two half cells at different tilt angles. In each half cells, one sensor was kept clean and others are uncleansed throughout the study. The Data Acquisition System (DAS) collects the Isc values and other required data at an interval of one minute. The soiling stations were installed in five different locations in the USA.

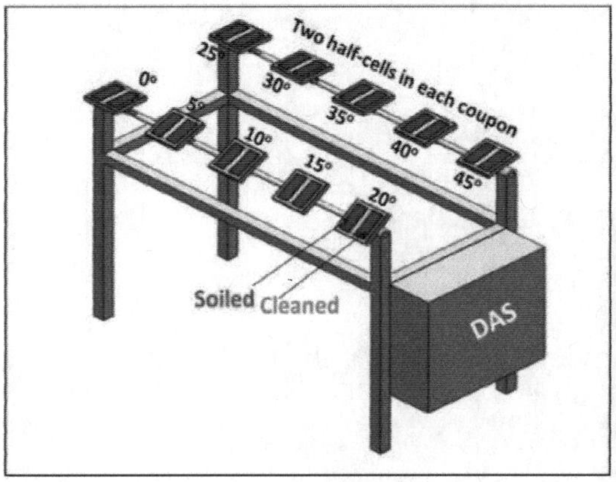

Figure 5. Pictorial representation of a Soiling station [10].

Figure 6. Experimental set-up for studying the effect of tilt angle and air pollution [11].

The air pollution is tremendously increasing in big cities. In order to study the effect of air pollution and tilt angle on solar panel performance, an experimental study was done in Iran [11]. The figure 6 shows the experimental setup in which the panels are tilted at 0°, 23°, 29°, 35°, 42°. The PV output energy is measured in the first month of every season. The results shows that the panels tilted at 29° produced more energy and owing to the effect of pollution, the output reduced up to 60%.

In another study, the effect of tilt angle on soiling is studied in [13] by using the mini-modules (replica of commercially available PV modules) at different tilt angles. The data of voltage, current are measured using multimeters at regular intervals of time. The results reveal that the cleaning of panels is greatly influenced by the rain event and wind speed depending on the tilt angles.

In Micheli et. al [14], the soiling stations were installed in 20 different locations in the USA. A time series data of these locations during a year is plotted with soiling ratio and rainfall against the month of the year. The soiling ratio is the ratio of short circuit current of soiled panels to the short-circuit current of the cleaned solar panels.

In [15], the effect of different PV technologies on electrical parameters is studied. Three different types of PV panels is used in the study. The PV panels are amorphous silicon, monocrystalline silicon and polycrystalline silicon panels. The parameters like irradiation, electrical energy output, and accumulated dust are measured during a period of five months. A mathematical model is developed by relating the dust accumulated over the PV panels and reduced electrical output of PV panels. The dust accumulated is measured using Dust-fall jar methods from Pollution Control Department.

3.1.2 Measurement of optical parameters. The optical parameters like transmittance of cover glasses used in solar panels is an important factor while considering the PV performance since the irradiation from the sun has to pass through the cover glass before it is converted into electricity.

In Gholami [16], a different methodology with a time period of 70 days was adopted for finding the factors affecting dust settlement and its influence in transmittance of the cover glasses. The figure 7 shows the experimental arrangements. The surface density of dust is calculated from the ratio of the weight of dust accumulated to the glass surface area. The amount of dust accumulated is the difference between the mass of glass samples before and after the experiment. The effect of wind direction is taken into account while doing the experiment. The wind swept away the dust deposited over the glass panels facing towards the wind. This will reduce the surface density of dust. In the paper, a correlation equation between transmission coefficient and surface density of dust is specified but there is no evidence for this equation.

Figure 7. Arrangement of glasses for finding the factors affecting dust settlement [16].

A study conducted in Belgium[17] analyses how the transmittance get affected the PV performances. For that, they have done the experiments for natural as well as artificial dust accumulation over the PV modules. In the experiment for natural soiling, the normal glass samples are installed as two sets (one is shielded from rain and other is unshielded from rain) which are tilted at different angles. It is observed that the transmittance is reduced for the panels shielded from the rainfall. The physical properties of dust and its cleaning aspects were also studied show that demineralized water reduces the transmittance loss percentage. The study infer that the transmittance increases with increase in tilt angles.

Elminir et. al [18] conducted a similar study to analyse the reduction in transmittance due to soiling. The experimental setup consists of 100 glass plates installed at different tilt angles in different orientations. They also studied the reduction in power output in monocrystalline solar cells. The experimental study shows that at smaller tilt angles, larger will be the reduction in transmittance. The effect of angle of incidence on soiling is studied with five different PV technologies like monocrystalline silicon, polycrystalline silicon, a-Si, CdTe, CIGs technologies [19]. They have compared the angle of incidence effect with soiling density and found that the angle of incidence effect is inversely proportional to the soiling density. A similar study is conducted in different locations in John [20]. They compared the dust deposition on a-Si, monocrystalline, multicrystalline, and CIGs modules in a hot and dry climate by measuring the spectral reflectance and quantum efficiency.

A natural soiling study done in Kuwait [21] studies the property of dust accumulated over the PV panels and the reduction in transmittance in detail. The dust samples were collected during one month in 2010 and 2011. The composition of materials contained in the collected dust was determined using X-ray diffraction (XRD) analysis. The spectrometer arrangement is used to find the transmittance of each sample. The study reveals that the dust accumulation adversely affects the regions with lesser wavelength.

3.2. Artificial Soiling

The artificial soiling of PV panels is done to analyse the effect of soiling on the electrical and optical parameters of PV panels. The dust samples will be prepared in the laboratory and applied over the PV panels in a confined environment. The artificial soiling study is a fast process when compared with natural soiling.

3.2.1 Measurement of electrical parameters. In 1993, El. Shobokshy and F. Hussein [22] done the artificial soiling on PV cells. The dusting is done in the laboratory from the artificially created dust. The standard deviation and mean size of dust particles were verified. They studied the I-V characteristics of the cells in every tests. The deposition density of dust over the cell surface is calculated and its effect on the different electrical parameters like voltage, current, Fill Factor (FF) were also verified. The result says that the parameters are greatly affected by dust settlement and observed a reduction of 84% in power output for 250 g/m^2 dust deposition.

The natural soiling and its evaluation is a time consuming and expensive process. The soiling loss estimation is affected by the site-specific factors like soil type, geographical location, and the climatic conditions [23]. A work done in India gives a solution for this problem. The methodology involves the dust collection from six interested locations inside the country and the soiling is done in the laboratory. A soil suspension is prepared and deposited over a borosilicate solar glass substrate using spray gun technology. The difference in weight of soiled and clean glass substrate gives a parameter called Soil Gravimetric Density (SGD). The constituents of dust particles is determined by XRD elemental analysis of dust. The quantum efficiency loss, short-circuit current density and spectral transmittance were calculated. The study showed that the soiling loss increases with the SGD. The study concluded with the comparison of dust samples collected from different locations.

While many type of research and studies concentrated on the influence of dust on PV performances, Z Ahmed et. al [24] find the effect of various dust pollutant types on the electrical parameters like voltage, current, Fill Factor (FF), etc. The soiling experiment is conducted in a laboratory and dust density is calculated. The electrical parameters are compared with dust density for various dust types. The I_{sc} and power output has the same effect on dust accumulation. But the change in the open circuit voltage is not affected by deposition of dust till the dust particles block the light completely.

3.2.2 Measurement of optical parameter. Burton and King [25][26] have done the artificial soiling experiment in a laboratory. The Arizona dust, a soot mixture, a mixture of Arizona dust and soot mixture were the dust types used in the experimental study. The dust is prepared and sprayed to a glass substrate using a spray gun. The quantum efficiency and transmission of light through the samples is measured using UV/vis spectroscopy. The UV/vis spectroscopy shows that the soot mixture blocks more amount of light to fall on the substrate than Arizona dust. In a work done by S Boppana [27], the dust is prepared and sprayed using a spray gun with laser pointer. They measured the quantum efficiency, soiling density, and reflectance of the polycrystalline and monocrystalline silicon mini-modules and the results are obtained by comparing the measurements between the two PV technologies. The paper claims that the reflectance loss can be used a measure of soiling density.

IOP Conf. Series: Materials Science and Engineering **396** (2018) 012050 doi:10.1088/1757-899X/396/1/012050

Appels [17] also done an artificial soiling experiment using white sand, clay, and cement and measured the transmittance using a spectrometer. They found a relation between the dust accumulation and power output by conducting artificial dust deposition on two types of PV panels exposed to the atmosphere. The results show that the decrease in power output and the decrease in transmittance of glass is approximately equal.

Kathy Brown et. al [28] analyses the PV performance in the desert regions considering the soiling and humidity. Normally, desert regions are characterized by high temperature and low Relative Humidity (RH). A laboratory experiment for the dust deposition on different glass samples is done and its reflectance is measured. The paper compares the optical performance of glass samples covered with hydrophilic anti-soiling coatings to that of uncoated samples.

4. Tilt angle optimization and PV potential assessment

The tilt angle should be such that to obtain maximum output from the PV panels. The tilt angle can be fixed or solar trackers. The solar tracker is expensive. So fixed tilt angle is very common. It is proven that the maximum PV output in a particular location is obtained when the fixed tilt angle is equal to the latitude angle of that location. The PV potential depends on the various environmental factors and weather conditions. So a location-based study is relevant for the maximization of PV potential. The tilt angle optimization is a methodology adopted for enhancing the PV potential.

In [29], tilt angle optimization and potential assessment are done by considering the solar radiation data of Aligarh city. The monthly average of daily solar radiation is estimated. From that optimization of the tilt angle is carried out by Liu Jordan model. The paper [30] says that there are various types of potentials namely, theoretical, technical, economic and implementation. For the land occupation of the PV generator, some terms such as Packing Factor (PF), spacing factor, performance ratio, ground cover ratio and occupation factor and inter-array spacing. The inter-array spacing between the solar panels is an important factor to avoid the shading effect. The ratio of the total area of PV array to the actual area of land occupied by all the PV array is known as packing factor. In [30], the photovoltaic energy potential and power potential is derived by considering the packing factor. This ensures the maximum of land area utilization for PV installation with maximization of PV potential.

5. Cleaning aspects of PV panels

The PV modules should be cleaned frequently for obtaining maximum PV performance. The manual cleaning and self-cleaning are the two methods of panel cleaning. The manual cleaning is found to be more efficient for better PV performances [1]. The cleaning by manpower is the conventional and effective manual cleaning method. Through manual cleaning, the effective removal of hard soiling sources like bird droppings is possible [5]. The problems with manual cleaning are high labour cost, abrasive damage of panels with improper handling [31]. Also, manual cleaning is not suitable for PV installations at large heights and off-shores. There are active and passive methods of cleaning the PV panels.

The PV panels should be kept at optimum tilt angles to obtain not only the best performance but also to get the cleaning effect by rainfall. Heavy precipitation will keep the PV panels clean. However, the light rain will wash away the dust partially and dust will be deposited on the bottom side of the PV panels. It is found that a small rainfall (less than 1mm) is enough for the reduction of soiling losses and continue to increase immediately after the rain [9]. The water cleaning through high-pressure water sprayer, mechanical cleaning by using wipers, use of electrodynamic screens and hydrophilic or hydrophobic surfaces are the other methods for mitigation of dirt settlement. The active methods of self-cleaning include cleaning with water, mechanical cleaning, and electrodynamic coatings. Some of the passive methods of self-cleaning are natural cleaning by rain, super hydrophobic and hydrophilic surfaces on the PV panels. The robots for cleaning of PV panels against soiling are advanced methods which are under research. The cleaning with robots are advantageous because of its features like fast response, low power consumption, stable operation, and more reliable than other cleaning techniques [5].

6. Conclusion

Dust causes severe performance degradation of PV modules. Both the environmental factors and installation factors contribute to the soiling of PV panels. It blocks the solar irradiation falling on solar panels, reduces the transmittance of cover glasses, and introduces a partial shading effect on the PV panels. The paper discusses the different methodologies adopted for soiling study. The soiling can be analysed from natural and artificial soiling. In natural soiling, the PV panels are directly exposed to sunlight and allowed to soiling from the surrounding dust particles. The artificial soiling is carried out in laboratories with dust particles prepared artificially. The studies analyse the effect of electrical parameters (voltage, current, Fill Factor), optical parameters (transmittance, soiling density), and other environmental factors like tilt angle, wind speed, rainfall, humidity on soiling. The cleaning of solar panels are expensive and different techniques for cleaning are emerging day by day to enhance the PV performance.

7. Future work

A future work on the optimization of tilt angle of the panels to maximize the PV potential by reducing the effect of soiling will be done. The enhancement of self-cleaning capability of panels against soiling will be analysed.

References

[1] J. Lopez-garcia, A. Pozza, and T. Sample, "Long-term soiling of silicon PV modules in a moderate subtropical climate," Sol. Energy, vol. 130, pp. 174–183, 2016.

[2] F. Mohd, S. Mekhilef, M. Seyedmahmoudian, and B. Horan, "Dust as an unalterable deteriorative factor affecting PV panel â€TM s ef fi ciency : Why and how," Renew. Sustain. Energy Rev., vol. 65, pp. 1267–1278, 2016.

[3] S. Mekhilef, R. Saidur, and M. Kamalisarvestani, "Effect of dust , humidity and air velocity on efficiency of photovoltaic cells," Renew. Sustain. Energy Rev., vol. 16, no. 5, pp. 2920–2925, 2012.

[4] M. Mani and R. Pillai, "Impact of dust on solar photovoltaic (PV) performance : Research status , challenges and recommendations," Renew. Sustain. Energy Rev., vol. 14, no. 9, pp. 3124–3131, 2010.

[5] W. Juzaili, H. Abdul, S. Shaari, and Z. Salam, "Performance degradation of photovoltaic power system : Review on mitigation methods," Renew. Sustain. Energy Rev., vol. 67, pp. 876–891, 2017.

[6] J. J. John, "Characterization of Soiling Loss on Photovoltaic Modules , and Development of a Novel Cleaning System," p. 146, 2015.

[7] M. R. Maghami, H. Hizam, C. Gomes, M. A. Radzi, M. I. Rezadad, and S. Hajighorbani, "Power loss due to soiling on solar panel: A review," Renew. Sustain. Energy Rev., vol. 59, pp. 1307–1316, 2016.

[8] A. Sayyah, M. N. Horenstein, and M. K. Mazumder, "Energy yield loss caused by dust deposition on photovoltaic panels," Sol. Energy, vol. 107, pp. 576–604, 2014.

[9] J. Cano, J. J. John, S. Tatapudi, and G. Tamizhmani, "Effect of tilt angle on soiling of photovoltaic modules," 2014 IEEE 40th Photovolt. Spec. Conf. PVSC 2014, pp. 3174–3176, 2014.

[10] B. King, G. Tamizhmani, S. Tatapudi, V. Rajasekar, and S. Boppana, "Regional soiling stations for PV: Design, calibration and installation," 2015 IEEE 42nd Photovolt. Spec. Conf. PVSC 2015, pp. 1–4, 2015.

[11] S. Farhangi and M. S. Zabihi, "The effect of tilt angle , air pollution on performance of photovoltaic systems in Tehran," vol. 24, pp. 459–468, 2001.

[12] M. Gostein and B. Stueve, "Accurately Measuring PV Soiling Losses Using Module Power Measurements," pp. 2–4, 2013.

[13] M. A. Afridi et al., "Determining the Effect of Soiling and Dirt Particles at Various Tilt

Angles of Photovoltaic Modules To cite this version :," 2017.
[14] L. Micheli, D. Ruth, M. G. Deceglie, M. Muller, D. Ruth, and M. G. Deceglie, "Time Series Analysis of Photovoltaic Soiling Station Data : Time Series Analysis of Photovoltaic Soiling Station Data : Leonardo Micheli," no. September, 2017.
[15] N. Ketjoy and M. Konyu, "Study of Dust Effect on Photovoltaic Module for Photovoltaic Power Plant," Energy Procedia, vol. 52, pp. 431–437, 2014.
[16] A. Gholami, A. Saboonchi, and A. A. Alemrajabi, "Experimental study of factors affecting dust accumulation and their effects on the transmission coefficient of glass for solar applications," Renew. Energy, vol. 112, pp. 466–473, 2017.
[17] R. Appels, B. Lefevre, B. Herteleer, and H. Goverde, "Effect of soiling on photovoltaic modules," Sol. Energy, vol. 96, pp. 283–291, 2013.
[18] H. K. Elminir, A. E. Ghitas, R. H. Hamid, F. El-hussainy, M. M. Beheary, and K. M. Abdel-moneim, "Effect of dust on the transparent cover of solar collectors," vol. 47, pp. 3192–3203, 2006.
[19] V. Rajasekar, S. Boppana, and G. Tamizhmani, "Angle of incidence effect on five soiled modules from five different PV technologies," 2015 IEEE 42nd Photovolt. Spec. Conf. PVSC 2015, 2015.
[20] J. J. John, V. Rajasekar, S. Boppana, S. Chattopadhyay, A. Kottantharayil, and G. Tamizhmani, "Quantification and Modeling of Spectral and Angular Losses of Naturally Soiled PV Modules," IEEE J. Photovoltaics, vol. 5, no. 6, pp. 1727–1734, 2015.
[21] H. A. and R. G. Hassan Qasem*, Thomas R. Betts, Harald Müllejans, "Dust-induced shading on photovoltaic modules," Prog. PHOTOVOLTAICS Res. Appl., pp. 218–226, 2014.
[22] M. S. El-Shobokshy and F. M. Hussein, "Degradation of photovoltaic cell performance due to dust deposition on to its surface," Renew. Energy, vol. 3, no. 6–7, pp. 585–590, 1993.
[23] J. J. John, S. Warade, G. Tamizhmani, and A. Kottantharayil, "Study of soiling loss on photovoltaic modules with artificially deposited dust of different gravimetric densities and compositions collected from different locations in India," IEEE J. Photovoltaics, vol. 6, no. 1, pp. 236–243, 2016.
[24] Z. Ahmed, H. A. Kazem, K. Sopian, M. A. Al-goul, and H. Alawadhi, "Effect of dust pollutant type on photovoltaic performance," Renew. Sustain. Energy Rev., vol. 41, pp. 735–744, 2015.
[25] P. D. Burton and B. H. King, "Artificial soiling of photovoltaic module surfaces using traceable soil components," Conf. Rec. IEEE Photovolt. Spec. Conf., pp. 1542–1545, 2013.
[26] P. D. Burton and B. H. King, "Application and characterization of an artificial grime for photovoltaic soiling studies," IEEE J. Photovoltaics, vol. 4, no. 1, pp. 299–303, 2014.
[27] S. Boppana, V. Rajasekar, and G. Tamizhmani, "Working towards the development of a standardized artificial soiling method," 2015 IEEE 42nd Photovolt. Spec. Conf. PVSC 2015, 2015.
[28] K. Brown, T. Narum, and N. Jing, "Soiling test methods and their use in predicting performance of photovoltaic modules in soiling environments," Conf. Rec. IEEE Photovolt. Spec. Conf., pp. 1881–1885, 2012.
[29] B. Jamil, A. T. Siddiqui, and N. Akhtar, "Engineering Science and Technology , an International Journal Estimation of solar radiation and optimum tilt angles for south-facing surfaces in Humid Subtropical Climatic Region of India," Eng. Sci. Technol. an Int. J., 2016.
[30] N. Martín-chivelet, "Photovoltaic potential and land-use estimation methodology," Energy, vol. 94, pp. 233–242, 2016.
[31] T. Sarver, A. Al-Qaraghuli, and L. L. Kazmerski, "A comprehensive review of the impact of dust on the use of solar energy: History, investigations, results, literature, and mitigation approaches," Renew. Sustain. Energy Rev., vol. 22, pp. 698–733, 2013.

Evolution of wind energy conversion system and recent challenges in grid integration: A review

Tomson Thomas[1*], **Prince A.**[2]

[1*]Research Scholar, [2]Associate Professor, Department of Electrical Engineering, Rajiv Gandhi Institute of Technology, Kottayam, Kerala, India 686501.

E-mail: tomsonpala@gmail.com

Abstract. The global demand for electrical energy is expanding twice as quick as the demand for primary energy. Highly reliable and fine quality power are demanded in the present-day power system scenario. Distributed generation including wind turbines and solar photovoltaic systems are significant in the perception of green energy. Electricity generation with renewable power generators like wind and solar are preferred to avoid CO_2 emissions. In the sustainable alternatives, wind energy is considered as one among the fastest evolving energy resources. The percentage of renewable energy in the worldwide generation increases day by day. In this paper, evolution of various wind energy conversion systems (WECSs) with their benefits and demerits are addressed. The increasing entry of power electronic converter (PEC) based renewable energy is transforming power system dynamics with new stability concerns. Several challenges like power quality issues, low voltage ride through (LVRT) capability, fault ride through (FRT) capability, power oscillations, primary frequency regulation (PFR), virtual inertia support etc in power system with the inclusion of wind energy are also discussed.

Keywords: Renewable energy, wind energy conversion system, maximum power point tracking, primary frequency regulation

1. Introduction

The contribution of renewable energy in the electricity generation is extremely significant due to the truth that they are bountiful in nature and inoffensive to the environment. The progress of WECS is very momentous in the inclusion of renewable energy in power system. Figure 1 shows that total worldwide installed capacity of wind energy in December 2017 is 539,581MW [1]. According to Global Wind Energy Council (GWEC), top five countries had greater than 20GW of established capacity including China (188GW), United States of America (89GW), Germany (56GW), India (33GW) and Spain (23GW). In India, 19% of total electricity generation is with renewable energy having a capacity of 63GW. Figure 2 shows the cumulative installed wind energy in India [2]. An installed capacity of 175GW using renewable energy is targeted by the Government of India by 2022 which includes 100GW of solar power, 60GW of wind power, 10GW of bio-power and 5GW of small hydro-power capacity [3].

The inclusion of sustainable energy in power system introduce various technical confrontations like voltage stability issues, flicker, harmonic distortion etc. Nowadays, the electrical power generation from renewables like wind, is having more importance because of environmental perspectives and shortage of traditional fossil fuel based energy sources.

The presence of PEC in renewable power generators has utmost difference with conventional

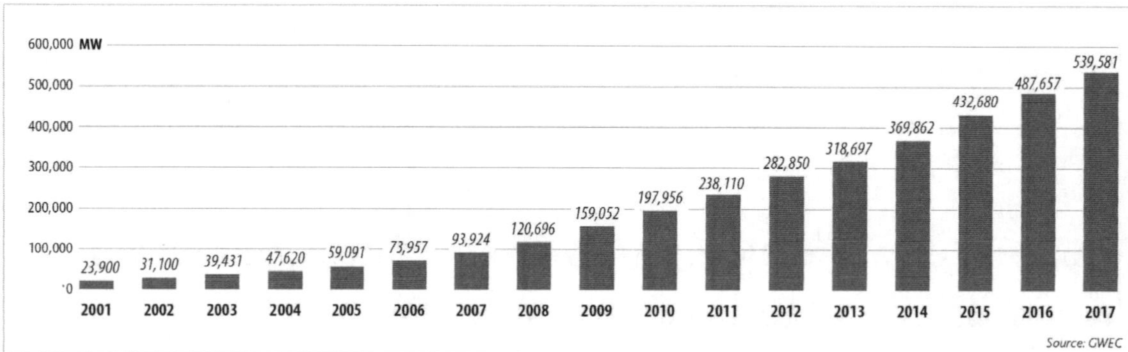

Figure 1. *Wind energy installed across the world [1]*

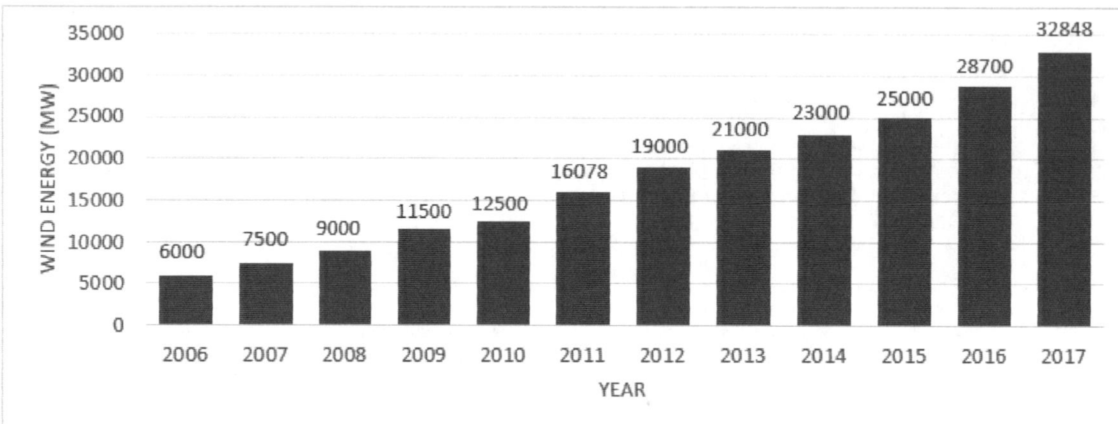

Figure 2. *Wind energy installed in India*

synchronous generators. The active and reactive power flow control, maximum power point tracking (MPPT) and power quality are the considerable aspects to be analysed in the grid connected operation. The various challenges of wind energy systems connected to the power system are LVRT capability, FRT capability, frequency stability, virtual inertia support etc.

This paper concentrates on the evolution and recent challenges in grid integration of WECSs. Basic principles of wind energy is addressed in Section 2. Various categories of WECSs with their fundamental principles are elaborated in Section 3. In Section 4, various challenges incorporated with the grid integration of WECSs are investigated. Conclusions and research aspects are addressed in Section 5.

2. Basic principles of wind energy

Kinetic energy present in the wind is transformed into mechanical energy by wind turbines. Kinetic energy of the wind rely on air density and wind velocity. The wind power developed by the turbine is stated by

$$P_w = \frac{1}{2}\rho A V^3 C_p(\lambda, \beta) \tag{1}$$

where C_p is the power coefficient, ρ is the air density ($1.225 kg/m^3$ at STP), A is the blade area of turbine in m^2, V is the velocity of wind in m/sec, β is pitch angle of the blade, λ is the tip speed ratio (TSR).

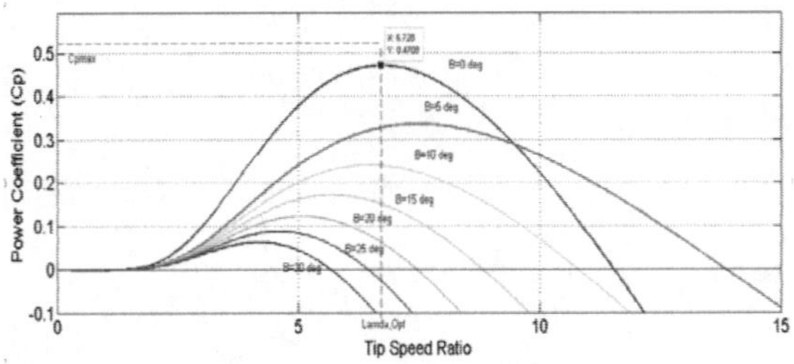

Figure 3. *Power coefficient Vs TSR for different pitch angles [4]*

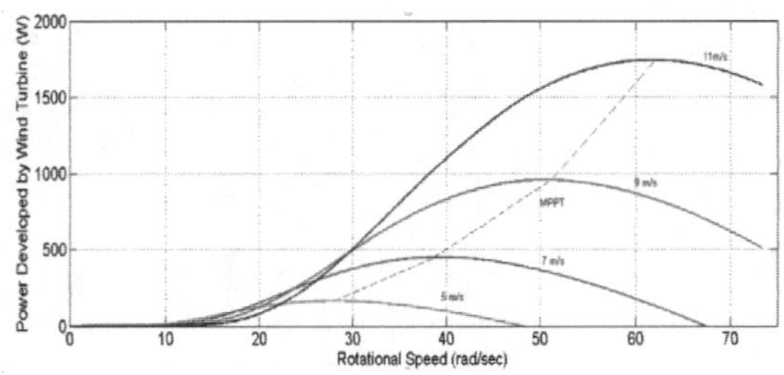

Figure 4. *Maximum power point tracking of wind turbine[4]*

The power coefficient, C_p gives the ratio of wind turbine kinetic energy to mechanical energy. No wind turbine can transform greater than 59.3% of wind kinetic energy into mechanical energy by rotating a rotor as per Betz Law. The power coefficient depends on the blade pitch angle, β and TSR, λ. TSR is the ratio of turbine blade linear speed to the wind velocity. Tip speed ratio is given by,

$$\lambda = \frac{R\omega}{V} \tag{2}$$

where, R is the radius of turbine rotor in metre and ω is the rotational speed in rad/sec. The power coefficient is given by,

$$C_p(\lambda, \beta) = \frac{1}{2}\left(\frac{98}{\lambda_i} - 0.4\beta - 5\right)exp\left(\frac{-16.5}{\lambda_i}\right) \tag{3}$$

where,

$$\frac{1}{\lambda_i} = \frac{1}{\lambda + 0.089} - \frac{0.035}{\beta^3 + 1} \tag{4}$$

The variation of power coefficient corresponding to TSR for different blade pitch angles is shown in Figure 3 [4]. The power developed and the maximum power point (MPP) are different for various wind speeds for a wind turbine as indicated in figure 4 [4]. The maximum power can be tracked by controlling generator speed for varied wind velocities.

3. Evolution of wind energy conversion system

In this section, various types in the evolution of WECS are discussed. Wind has been used for sailing of ships and pumping water. Earlier, wind turbines were utilized for grinding or milling grain and they were called windmills. Electricity was generated initially using wind turbines at Denmark in 1890. WECSs are classified into fixed and variable speed systems.

3.1. Fixed speed wind energy systems

Fixed speed WECSs consist of a grid connected self-excited induction generator which is started as a motor using soft starter [5]. It can run as a generator in grid connected mode and the reactive power requirement is supported by shunt capacitors. This is treated as fixed speed WECS due to the reason that variation in shaft speed and operating slip of induction generator are not so much with the change in wind velocity. The conversion efficiency is very less for fixed speed WECSs.

3.2. Variable speed wind energy systems

Variable speed WECSs are incorporated with a PEC to operate at MPP for varying wind speeds. Variable speed systems are classified as systems with fully rated converter (FRC) and reduced rated converter (RRC). Fully rated converter based system is normally employed with permanent magnet synchronous

IOP Conf. Series: Materials Science and Engineering **396** (2018) 012051 doi:10.1088/1757-899X/396/1/012051

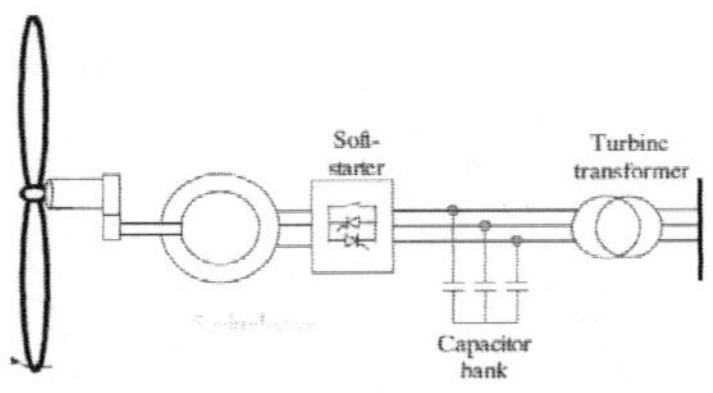

Figure 5. *Configuration of fixed speed wind energy conversion system*

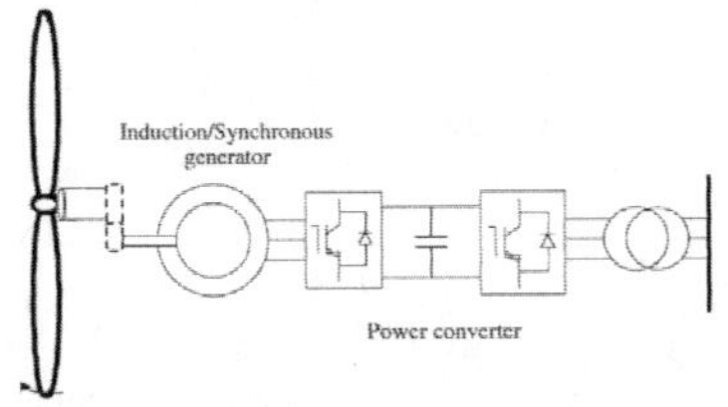

Figure 6. *Configuration of fully rated converter based wind energy system*

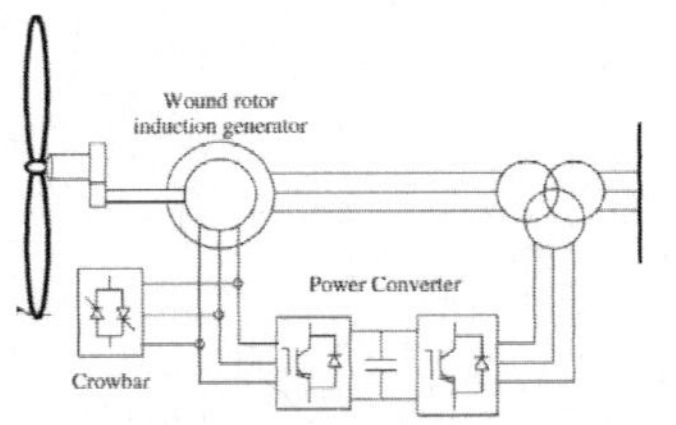

Figure 7. *Configuration of DFIG based wind energy system*

generator (PMSG) for recent wind energy extractions. The generator speed is dissociated by using power electronic converters. The output will be variable voltage and variable frequency, subjected to the wind velocity. The major demerit of WECS using PMSG is that it requires a PEC with a rating equal to the generator.

WECSs with reduced rated converter includes doubly fed induction generator (DFIG) which is extensively used in large scale applications. The main benefit of DFIG is its aptness to run as generator in sub synchronous and super synchronous speeds. Its PEC rating is only one third of the generator rating. Controls are possible through rotor side converter (RSC) and grid side converter (GSC). The operating principle of DFIG is based on injecting three phase currents of variable amplitude and frequency into the rotor to allow the operation under different rotor speeds to achieve peak wind power extraction with optimum value of power coefficient. In sub-synchronous mode of operation, active power flow will be from stator to utility grid and from utility grid to the rotor. But in super synchronous operation, the real power flows from stator and rotor to the utility grid.

4. Recent challenges in wind energy conversion systems

Wind power plants (WPPs) are operated with PECs whereas conventional generating stations with synchronous generators are straight away connected to the grid. Wind turbine input is uncontrollable and it fluctuates stochastically. Wind resource assessment studies are very relevant in finding the optimum location to extract maximum wind power.

Extraction of maximum power for a location at different wind speeds by controlling the generator speed corresponding to the optimum TSR is a challenging task. MPPT technique in a PMSG based WECS connected to the network using optimum speed and wind speed with a fast controller is established in [6]. Attempt has been made with different methods of MPPT algorithm using fixed and variable step size for a small-scale WECS and are compared with fuzzy logic and the look-up table indirect methods in [7]. An adaptive MPPT algorithm for fast tracing of MPP under fast varying wind conditions for a WECS is discussed [8]. A new MPPT approach for WECS with PMSG depending on rectified dc voltage and current is discussed in [9]. A rapid tracking MPPT algorithm using matrix converter for a WECS is analysed in [10]. In [11], a comparison of WECS with a maximum electrical power tracker (MEPT) and maximum mechanical power tracker (MMPT) for the extraction of maximum electrical power and maximum mechanical power respectively. An MPPT strategy for WECS with DFIG using a Lyapunov-based analysis is discussed in [12]. Sensorless control of a grid interfaced DFIG based WECS for MPPT capability and power smoothening using a rotor position estimation algorithm is discussed in [13].

In fixed speed WECSs, fluctuations in wind are directly conveyed to the alterations in output power and it may produce flicker. In unfixed speed systems flicker will not be present due to decoupling of rotor and grid. An exhaustive control of a WECS is elaborated with a controller enhancing power quality using DC link capacitor and power converter in the absence of wind in [14]. Power quality troubles due to the grid connected WECS are addressed using IEC61400 and a STATCOM with battery energy storage system (BESS) is adopted to enhance the power quality in [15]. A STATCOM and BESS control approach is employed for enhancing power quality and stability of a fixed speed WECS in [16]. A TakagiSugeno fuzzy model based linear quadratic regulator is employed for the mitigation of power quality issues of a wind-hybrid power system with BESS in [17].

If the percentage of variable speed wind turbines in the system is high and they disconnect at relatively small voltage reductions and a voltage drop may occur. Therefore wind turbines should have a FRT capability to endure during voltage drops for certain durations without tripping. Fixed speed WECSs contribute to the network fault currents. DFIG based wind generators will lend to the fault currents but it may be disconnected from the network. WECSs with FRC do not donate to system fault current. Fault ride-through capability means WECSs should have an ability to remain connected to the network during fault and must contribute system stability. Crowbar protection converts DFIG to a squirrel cage induction generator for improving FRT capability. A DC-link controllable fault current limiter in RSC is suggested for improving the FRT capacity which abolishes the demerits of AC crow bar protection in [18]. LVRT means the proficiency of the generator to be connected to the network during a voltage sag. Resonant controllers are proposed for asymmetrical voltage sags in DFIG to enhance LVRT competence in [19]. Improving FRT capability of a DFIG with superconducting magnetic energy storage fault current limiter and RSC control is discussed in [20]. A control approach for the betterment of LVRT capability of DFIG using passive and active compensators are discussed in [21]. A virtual impedance control method for enhancing high voltage ride through (HVRT) capability of DFIG during voltage swell is analysed in [22]. A genetic algorithm based LVRT control method for DFIG connected to the network is addressed in [23].

Voltage on a transmission network is determined chiefly by the interaction of reactive power flows. Fixed speed induction generators absorb reactive power to maintain their magnetic field. Voltage of the network can be supported by using shunt capacitors to reduce the reactive power drawn from the network. Variable speed WECSs can control the reactive power to improve the voltage of power system. Reactive power and voltage control can be achieved by using reactive power compensators such as static var compensator (SVC) and static synchronous compensator (STATCOM). A direct torque and reactive power control of grid connected DFIG with a matrix converter is discussed in [24]. In [25], enhancement of transient voltage stability of grid tied DFIG based WECS by controlling superconducting fault current limiter and reactive power is discussed. A changeable reactive current to voltage idea of a DFIG for voltage regulation of a WPP is addressed in [26].

Reactive power control of DFIG based WPPs for the damping of power system oscillations is analysed in [27]. Power system oscillation damping can be attained by controlling the active power of BESS and reactive power modulation of line side converter in [28]. Control of WECS with DFIG and power system stabilizer for power oscillation stabilization under uncertainities in power system is addressed in [29]. Design of controllers for power oscillation damping of WPPs with DFIG are discussed in [30]. Various aspects on sub-synchronous oscillation in series compensated power system with WPPs are discussed in [31,32].

Because of the presence of PEC, the generator inertia is decoupled from the grid. Virtual inertia is needed for the WECS to take part in the frequency regulation. Frequency support can be furnished by a WECS to the power system by increasing the power at the instant of requirement. It can be attained by deloading the wind turbines from the MPP and hence by creating a margin for the frequency control. Primary frequency regulation using variable droop for deloaded WECSs are discussed in [33]. Virtual inertia technique and optimized power point tracking for a DFIG based WECS are analyzed in [34].

5. Conclusions

In this paper, a comprehensive review is done on the evolution of WECS. The various challenges of WECSs connected to the power system such as MPPT, power quality issues, FRT capability, LVRT capability, reactive power support, voltage stability, power oscillation damping, primary frequency regulation and virtual inertia support are discussed. Design of better controllers are required for the

WECSs to enhance the dynamic behaviour in power system.

6. Acknowledgments

This work is supported by Centre for Engineering Research and Development (CERD), Kerala, India.

7. References

[1] 'Global Wind Statistics 2017', http://gwec.net, accessed 15 February 2018.

[2] 'Installed Capacity in India 2017', http://www.cea.nic.in, accessed 15 January 2018.

[3] 'Ministry of New and Renewable Energy', http://mnre.gov.in, accessed 15 February 2018.

[4] T. Thomas and E. P. Cheriyan, Wind energy system for a laboratory scale micro-grid," in Electrical, Electronics and Computer Science (SCEECS), 2012, IEEE Students' Conference on, pp. 15, March 2012.

[5] Olimpo Anaya-Lara, Janaka B. Ekanayake, Phill Cartwright, Nick Jenkins, Michael Hughes, "Wind Energy Generation: Modelling and Control", Wiley 2009.

[6] R. M. Linus and P. Damodharan, "Maximum power point tracking method using a modified perturb and observe algorithm for grid connected wind energy conversion systems," in IET Renewable Power Generation, vol. 9, no. 6, pp. 682-689, 8 2015.

[7] H. Liu, F. Locment and M. Sechilariu, "Experimental analysis of impact of maximum power point tracking methods on energy efficiency for small-scale wind energy conversion system," in IET Renewable Power Generation, vol. 11, no. 2, pp. 389-397, 2 8 2017.

[8] J. Hussain and M. K. Mishra, "Adaptive Maximum Power Point Tracking Control Algorithm for Wind Energy Conversion Systems," in IEEE Transactions on Energy Conversion, vol. 31, no. 2, pp. 697-705, June 2016.

[9] Y. Xia, K. H. Ahmed and B. W. Williams, "A New Maximum Power Point Tracking Technique for Permanent Magnet Synchronous Generator Based Wind Energy Conversion System," in IEEE Transactions on Power Electronics, vol. 26, no. 12, pp. 3609-3620, Dec. 2011.

[10] V. Agarwal, R. K. Aggarwal, P. Patidar and C. Patki, "A Novel Scheme for Rapid Tracking of Maximum Power Point in Wind Energy Generation Systems," in IEEE Transactions on Energy Conversion, vol. 25, no. 1, pp. 228-236, March 2010.

[11] H. Fathabadi, "Novel Maximum Electrical and Mechanical Power Tracking Controllers for Wind Energy Conversion Systems," in IEEE Journal of Emerging and Selected Topics in Power Electronics, vol. 5, no. 4, pp. 1739-1745, Dec. 2017.

[12] F. Fateh, W. N. White and D. Gruenbacher, "A Maximum Power Tracking Technique for Grid-Connected DFIG-Based Wind Turbines," in IEEE Journal of Emerging and Selected Topics in Power Electronics, vol. 3, no. 4, pp. 957-966, Dec. 2015.

[13] N. K. S. Naidu and B. Singh, "Grid-Interfaced DFIG-Based Variable Speed Wind Energy Conversion System With Power Smoothening," in IEEE Transactions on Sustainable Energy, vol. 8, no. 1, pp. 51-58, Jan. 2017.

[14] A. S. Bubshait, A. Mortezaei, M. G. Simes and T. D. C. Busarello, "Power Quality Enhancement for a Grid Connected Wind Turbine Energy System," in IEEE Transactions on Industry Applications, vol. 53, no. 3, pp. 2495-2505, May-June 2017.

[15] S. W. Mohod and M. V. Aware, "A STATCOM-Control Scheme for Grid Connected Wind Energy System for Power Quality Improvement," in IEEE Systems Journal, vol. 4, no. 3, pp. 346-352, Sept. 2010.

[16] A. Arulampalam, M. Barnes, N. Jenkins and J. B. Ekanayake, "Power quality and stability improvement of a wind farm using STATCOM supported with hybrid battery energy storage," in IEE

Proceedings - Generation, Transmission and Distribution, vol. 153, no. 6, pp. 701-710, November 2006.

[17] H. S. Ko and J. Jatskevich, "Power Quality Control of Wind-Hybrid Power Generation System Using Fuzzy-LQR Controller," in IEEE Transactions on Energy Conversion, vol. 22, no. 2, pp. 516-527, June 2007.

[18] A. Jalilian, S. B. Naderi, M. Negnevitsky, M. Tarafdar Hagh and K. M. Muttaqi, "Controllable DC-link fault current limiter augmentation with DC chopper to improve fault ride-through of DFIG," in IET Renewable Power Generation, vol. 11, no. 2, pp. 313-324, 2 8 2017.

[19] V. F. Mendes, C. V. de Sousa, W. Hofmann and S. R. Silva, "Doubly-fed induction generator ride-through fault capability using resonant controllers for asymmetrical voltage sags," in IET Renewable Power Generation, vol. 9, no. 7, pp. 783-791, 9 2015.

[20] X. Y. Xiao, R. H. Yang, X. Y. Chen, Z. X. Zheng and C. S. Li, "Enhancing fault ride-through capability of DFIG with modified SMES-FCL and RSC control," in IET Generation, Transmission and Distribution, vol. 12, no. 1, pp. 258-266, 1 2 2018.

[21] M. Rahimi and M. Parniani, "Efficient control scheme of wind turbines with doubly fed induction generators for low-voltage ride-through capability enhancement," in IET Renewable Power Generation, vol. 4, no. 3, pp. 242-252, May 2010.

[22] Z. Xie, X. Zhang, X. Zhang, S. Yang and L. Wang, "Improved Ride-Through Control of DFIG During Grid Voltage Swell," in IEEE Transactions on Industrial Electronics, vol. 62, no. 6, pp. 3584-3594, June 2015.

[23] T. D. Vrionis, X. I. Koutiva and N. A. Vovos, "A Genetic Algorithm-Based Low Voltage Ride-Through Control Strategy for Grid Connected Doubly Fed Induction Wind Generators," in IEEE Transactions on Power Systems, vol. 29, no. 3, pp. 1325-1334, May 2014.

[24] S. Mondal and D. Kastha, "Improved Direct Torque and Reactive Power Control of a Matrix-Converter-Fed Grid-Connected Doubly Fed Induction Generator," in IEEE Transactions on Industrial Electronics, vol. 62, no. 12, pp. 7590-7598, Dec. 2015.

[25] R. Ou, X. Y. Xiao, Z. C. Zou, Y. Zhang and Y. H. Wang, "Cooperative Control of SFCL and Reactive Power for Improving the Transient Voltage Stability of Grid-Connected Wind Farm With DFIGs," in IEEE Transactions on Applied Superconductivity, vol. 26, no. 7, pp. 1-6, Oct. 2016.

[26] J. Kim, E. Muljadi, J. W. Park and Y. C. Kang, "Flexible IQV Scheme of a DFIG for Rapid Voltage Regulation of a Wind Power Plant," in IEEE Transactions on Industrial Electronics, vol. 64, no. 11, pp. 8832-8842, Nov. 2017.

[27] M. Edrah, K. L. Lo and O. Anaya-Lara, "Reactive power control of DFIG wind turbines for power oscillation damping under a wide range of operating conditions," in IET Generation, Transmission and Distribution, vol. 10, no. 15, pp. 3777-3785, 11 17 2016.

[28] G. Xu, L. Xu and J. Morrow, "Power oscillation damping using wind turbines with energy storage systems," in IET Renewable Power Generation, vol. 7, no. 5, pp. 449-457, Sept. 2013.

[29] T. Surinkaew and I. Ngamroo, "Coordinated Robust Control of DFIG Wind Turbine and PSS for Stabilization of Power Oscillations Considering System Uncertainties," in IEEE Transactions on Sustainable Energy, vol. 5, no. 3, pp. 823-833, July 2014.

[30] K. Elkington and M. Ghandhari, "Non-linear power oscillation damping controllers for doubly fed induction generators in wind farms," in IET Renewable Power Generation, vol. 7, no. 2, pp. 172-179, March 2013.

[31] C. Gao, H. Liu, H. Jiang, Y. Li and X. Tang, "Research on the sub-synchronous oscillation in wind power connected to series compensated power system and its influencing factors," in CES Transactions on Electrical Machines and Systems, vol. 1, no. 3, pp. 334-340, September 2017.

[32] A. E. Leon and J. A. Solsona, "Sub-Synchronous Interaction Damping Control for DFIG Wind Turbines," in IEEE Transactions on Power Systems, vol. 30, no. 1, pp. 419-428, Jan. 2015.

[33] K. V. Vidyanandan and N. Senroy, "Primary frequency regulation by deloaded wind turbines using variable droop," in IEEE Transactions on Power Systems, vol. 28, no. 2, pp. 837-846, May 2013.

[34] D. Ochoa and S. Martinez, "Fast-Frequency Response Provided by DFIG-Wind Turbines and its Impact on the Grid," in IEEE Transactions on Power Systems, vol. 32, no. 5, pp. 4002-4011, Sept. 2017.

IOP Conf. Series: Materials Science and Engineering **396** (2018) 012052 doi:10.1088/1757-899X/396/1/012052

Modular battery charge equalization circuit based on state of charge for electric vehicle application

Ginu Ann George[1*], Prince A.[2], Jayan M.V.[3]

[1]Research Scholar, [2]Associate Professor, [3]Professor

Department of Electrical Engineering, Rajiv Gandhi Institute of Technology, Kottayam, Kerala, India
E-mail*:ginuanngeorge@gmail.com

Abstract. Electric Vehicle (EV)/ Hybrid Electric Vehicle (HEV) enabled by high-efficiency electric motor and controller, provide the means for a clean, efficient and environmental friendly transportation system. Portable rechargeable batteries are the essential basic component in EVs. In EV/HEV system a large number of cells should be connected in series to supply enough energy, while the voltage balancing problem between cells leads to poor performance. This paper attempts modular battery charge equalization based on state of charge (SOC) under MATLAB Simulink environment.

Keywords: Electric Vehicle/ Hybrid Electric Vehicle (EV/HEV), State of Charge (SOC), Battery Management System (BMS).

1. Introduction

Advanced vehicle technologies alternative to the conventional IC engine vehicles are referred to as alternative vehicles. Electric vehicle (EV), hybrid electric vehicle (HEV), plug-in hybrid vehicle (PHEV) and fuel cell electric vehicle (FCEV) are the examples of alternative vehicles. The main characteristic of EVs/HEVs is that the electrical energy would be efficiently utilized to power electric motors for propulsion. The battery is a fundamental component of electric vehicles, which represent a step forward towards sustainable mobility [1]. Rechargeable batteries such as Lithium ion batteries are widely used in automobile industry because of its high energy density, high power density, long cycling life etc.,

The battery management system (BMS) is a critical component of electric and hybrid electric vehicles [2]. The nominal voltage of a lead-acid cell is 2V whereas lithium-ion cell is about 3.65 V and it varies based on cell type. Applications, such as telecommunication systems, uninterruptible power supplies (UPSs), and off-grid solar energy systems require voltages ranging from 12 V nominal to 300 V or more with dozens of stacked cell in series-parallel modules. Electric Vehicle applications often use 300 V-400 V strings. The energy storage device in an EV, consist of number of cells connected in series string to deliver required voltage. The characteristics of individual battery cells could be different in a long string, due to natural dispersion during manufacturing, chemical degradation and ambient temperature variation during charging and discharging [3-4]. As a result, the battery voltages through the entire string are not uniform during the charging and discharging cycle.

The state of charge (SOC) represents the present capacity of the battery. If the SOCs of stacked batteries are different, it lead to overcharge or over discharge condition which in turn reduce the capacity of the cell and pack. Sophisticated battery users are aware that SOC balance is necessary, particularly for lithium-ion types.

Lithium-ion batteries pose a significant safety risk if over-charged. If the voltage across a single cell becomes higher than 4.3 V, a flammable metallic lithium is formed. If the cell voltage is less than 2.5 V, metallic copper as well as a short circuit is formed, leads to heating and even a catastrophic failure. In both charge and discharge modes, a certain magnitude of current creates unequal voltage drop across cells if their electrochemical properties are not completely matched, and this is very likely to happen in the real world. Therefore, it is absolutely necessary to employ a cell equalizer, which is far more feasible than developing manufacturing technology, in order to eliminate the cell unbalancing.

Cells with higher capacities reach full capacity before the others, and cells with smaller capacities tend to discharge quickly as shown in figure 1. Cell unbalance problem tends to be worse through continuous charging and discharging cycles.

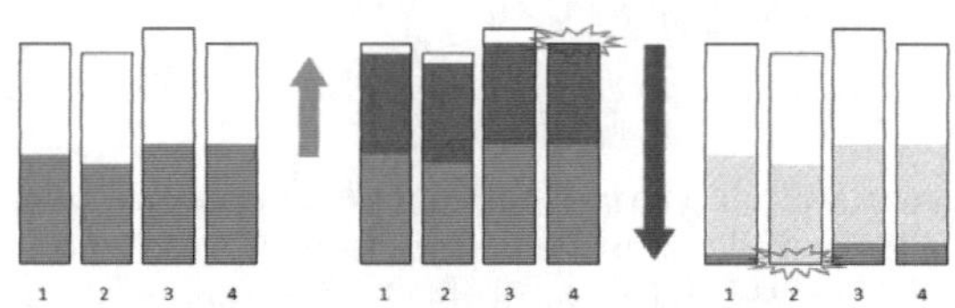

Figure 1.Over-charged and over-discharged condition due to variations in SOC

In this paper, two balancing methods, using an external charger and switched capacitor are analyzed based on the simulation works by considering the SOC of the modular batteries in MATLAB Simulink environment. Section II describes the typical battery balancing method and switched capacitor (SC) method, section III presents the necessity of SOC based balancing method, section IV deals with the simulation works and results of two balancing methods by considering two cases of SOC and cell voltage and section V gives the conclusion.

2. Battery Balancing Method

The available energy stored in a battery cell is E_{avai}=qV, which states that both charge and voltage need to be balanced in a series string to maximize the output of a pack. The simplest strategy adopted for charging a series connected string of cells is to monitor the cell voltages and discontinue charging when one of the cells (strongest cell) reaches the voltage limit for individual cells [5]. Extended charging is another option where charging is continued even after the strongest cell has reached its capacity to bring the weaker cells up to capacity as shown in figure 2. Cell balancing circuit consisting of three cells B1, B2 and B3 connected in series along with S1-S10 switches. The first task is to read the battery voltages. This is done by switches S1, S2 and S3. There is a dedicated path to each battery with the switches so as to route the charging current to the identified weakest battery.

The simplest protection during discharge of a pack is to shut down when the first cell (weakest cell) reaches the minimum voltage limit. But this method of charging or discharging prevents the effective utilization of battery pack as a whole. Therefore various balancing methods are adopted in considerations of energy consumption and transformation. They can be divided into dissipative balancing circuits and non-dissipative balancing circuits. Dissipative balancing circuits violate the future trend of energy-saving, although it is practical with simple topologic and convenient control strategy.

IOP Conf. Series: Materials Science and Engineering **396** (2018) 012052 doi:10.1088/1757-899X/396/1/012052

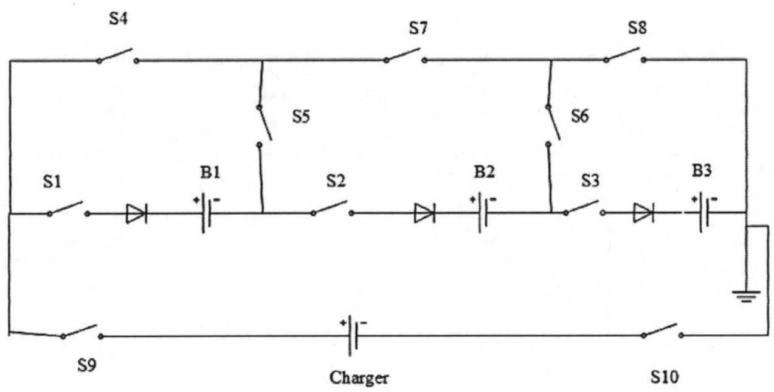

Figure 2. Cell balancing circuit using an external charger.

One of the non-dissipative method is switched capacitor method [6-8]. The basic SC equalization circuit is shown in figure 3. There are two sets of switches. Set one (S1, S3 and S5) connects the switched capacitors to all of the positive battery terminals, while set two (S2, S4 and S6) connects the switched capacitors to all of the negative battery terminals.

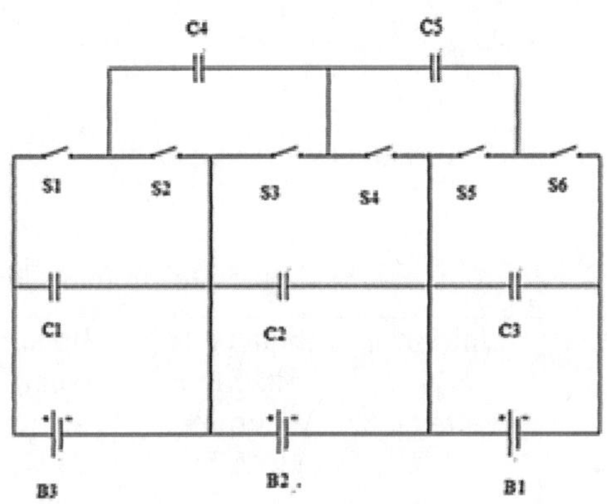

Figure 3. Basic switched capacitor balancing circuit

The two sets are switched alternately. The switched capacitor voltages do not change much, but are switched at a high enough frequency to move charge between adjacent batteries. The direction of current flow will depend on the relative voltages of the adjacent batteries and may vary throughout the pack.

3. SOC based charge equalization and its necessity.

SOC is the amount of capacity that remains after discharge from a top-of-charge condition. If the string is treated as a two terminal device while being charged and discharged, the individual cells will have different states of charge [9]. Because of this different SOC some cells will be chronically overcharged, undercharged or over-discharged. Over many cycles, this tends toward capacity decrease for the cell and the pack. If this unbalance remains in the pack, several cells are underutilized, causing the reduced cycle life for the entire battery pack. Therefore, keeping the SOCs of the cells at a uniform level thereby voltages is critical for enhancing the battery life.

4. Simulation Results

In this paper, balancing circuit using an external charger and switched capacitor methods are simulated using MATLAB Simulink environment. Lithium-ion battery parameters used for simulation are listed as shown in table 1.

IOP Conf. Series: Materials Science and Engineering **396** (2018) 012052 doi:10.1088/1757-899X/396/1/012052

Table 1. Lithium-ion battery parameters

Parameters	Values
Nominal Voltage (V)	11
Rated Capacity (Ah)	42
Initial SOC (%)	100
Maximum Capacity (Ah)	44
Fully Charged Voltage (V)	12
Nominal Discharge Current (A)	3.5
Internal Resistance (ohm)	0.002

In each circuit two cases are considered. In the first case SOCs of all the cells are equal and 100% but the cell voltages are different. In the second case both SOC and cell voltages are different.

4.1. Case I: Same SOC and Different Voltage

In this case all the batteries have same SOC in both circuits and different voltages. The values are listed in table 2.

Table 2. Condition: Same SOC and Different Voltage

Battery	SOC	Initial cell voltage	Final cell voltage in external charger circuit	Final cell voltage in switched capacitor circuit
B1	100	10.8	12.02	11.89
B2	100	11	12.02	11.89
B2	100	11.5	12.02	11.89

If SOCs are equal, all the cells attain balanced voltage irrespective of its equalization method. But in external charger balancing circuit, there is a possibility of overcharging. In table 2, set condition is same SOC and different cell voltages, and the voltage to be balanced is 12V. But the balanced cell voltage of external charger is 12.02V slightly above the preset value of 12V. This one of the draw backs of external charger method. Figure 4 and figure 5 represent the plot corresponding to the same SOC and different voltage condition.

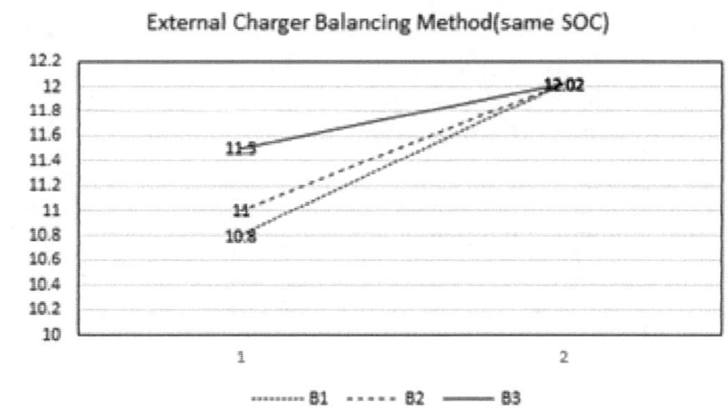

Figure 4.Case1: Balanced cell voltages of external charger method

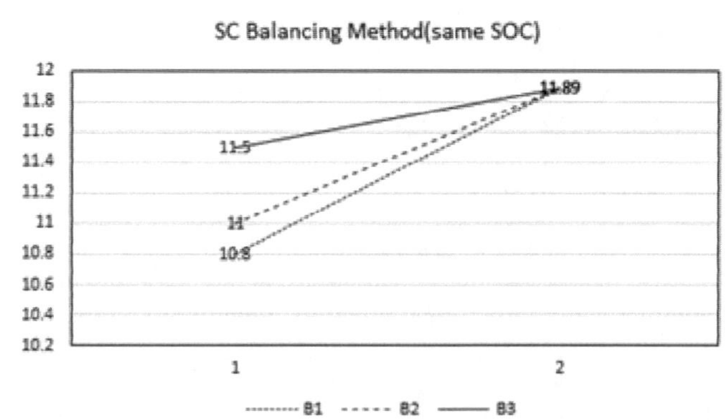

Figure 5.Case 1: Balanced cell voltages of switched capacitor method.

In figure 5 it is clear that all batteries are balanced and closely settled to the set value of 12V without overcharge.

4.2. Case II: Different SOC and Different Voltage

In this case all the batteries have same SOC in both circuits and different voltages. The values are listed in table 3

Table 3. Condition: Different SOC and Different Voltage

Battery	SOC	Initial cell voltage	Final cell voltage in external charger circuit	Final cell voltage in switched capacitor circuit
B1	90	10.8	12.45	11.90
B2	97	11	12.01	11.91
B2	100	11.5	12.02	11.90

This case describes the behavior of external circuit and switched capacitor circuit under different SOC. The plots corresponding to that are shown in figure 6 and figure 7. In different SOC condition cell voltage of external charger circuit is not balanced and all are exceed the set value of 12V. Over discharge may also be happened in the similar way during discharging cycle. In SC method all the cell voltages are balanced and within the set limit of 12V.

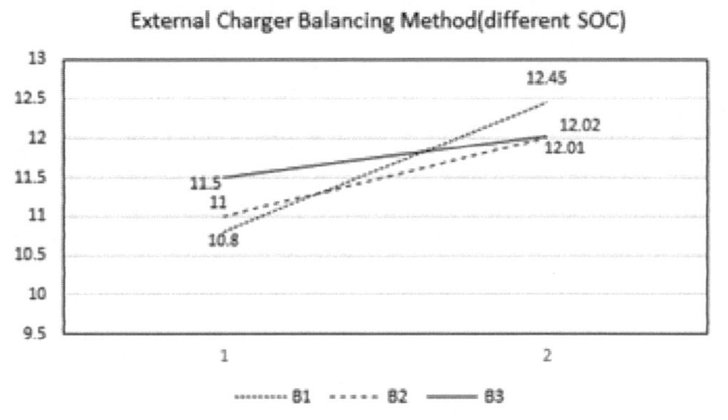

Figure 6. Case 2: Cell voltages of external charger method

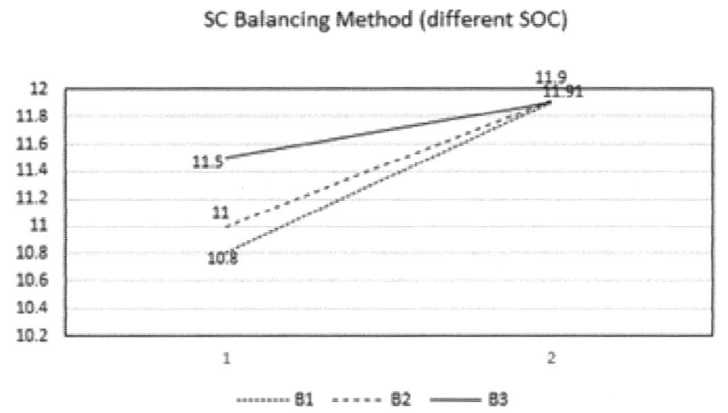

Figure 7. Case 2: Balanced cell voltages of SC method

Thus by using non-dissipative cell balancing method (Active balancing method) like SC method overcharge and over discharge can be eliminated

5. Conclusion and Future Scope

An external charger and switched-capacitor equalization methods are analyzed through MATLAB simulation. Plots of the balanced cell voltages are also shown. The primary advantage of an external charger method is its simplicity. The main limitation of this technique is over charge condition, increased number of switches and low efficiency operation. Active method of cell balancing is very essential component in BMS especially in high power applications such as EV/HEV applications [10-11]. The switched capacitor method discussed in this paper requires a long equalization time compared to low efficient external charger method. To solve the deficiency, utilization of a high-performance control strategy is essential. Fuzzy control is beneficial to reduce balancing time, while Proportional Integral (PI) control is an effective method to improve balancing accuracy.

References

[1] M. A. Hannan, F. A. Azidin, and A. Mohamed, "Hybrid electric vehicles and their challenges: A review," Renewable and Sustainable Energy ReViews, vol. 29, pp. 135-150, Jan. 2014

[2] H. Maleki and J. Howard, “Effects of overdischarge on performance and thermal stability of a Li-ion cell,” J. Power Sources, vol. 160, no. 2, pp. 1395–1402, Oct. 2006.

[3] Karnjanapiboon, C.; Jirasereeamornkul, K.; Monyakul, V., "High efficiency battery management system for serially connected battery string," Industrial Electronics, 2009. ISIE 2009. IEEE International Symposium on , vol., no., pp.1504,1509, 5-8 July 2009 doi: 10.1109/ISIE.2009.5214247.

[4] Nasser H. Kutkut, Deepakraj M, Donald W. Novotny, “Charge Equalization for Series Connected Battery Strings”, IEEE transactions on industry applications, Vol. 31, No. 3, May- June 1995.

[5] B. S. Sagar, B. P. Divakar, K. S. Venkatesh Prasad, "Series battery equalization using sequential difference algorithms and communications”, IEEE International conference on advances in electronics computers and communications, pp. 1-6, 2014.

[6] Moon-Young Kim, Chol-Ho Kim, Jun-Ho Kim and Gun-Woo Moon, "A Chain Structure of Switched Capacitor for Improved Cell Balancing Speed of Lithium-Ion Batteries," iEEE Trans. ind. Electron. , vol. 61, no. 8, pp. 3989-3999, Aug. 2014.

[7] A. Baughman and M. Ferdowsi, “Double-Tiered Switched-Capacitor Battery Charge Equalization Technique,” IEEE Transactions on Industrial Electronics, vol. 55, June 2008.

[8] A. Baughman and M. Ferdowsi, “Double-Tiered Capacitive Shuttling Method for Balancing Series-Connected Batteries,” in Vehicle Power and Propulsion, 2005 IEEE Conference, pp. 109–113, Sept. 2005.

[9] Xu, J.; Li, S.; Mi, C.; Chen, Z.; Cao, B. SOC Based Battery Cell Balancing with a Novel Topology and Reduced Component Count. Energies 2013, 6, 2726-2740.

[10] Kyung-Min Lee, Yoo-Chae Chung, Chang-Hyeon Sung, and Bongkoo Kang, "Active Cell Balancing of Li-Ion Batteries Using LC Series Resonant Circuit," IEEE Trans. Ind. Electron. , vol. 62, no. 9, pp. 5491-5501, Sep.2015.

[11] J. W. Kimball, B. T. Kuhn, and P. T. Krein, “Increase performance of battery packs by active equalization,” in Proc. 3rd IEEE VPPC, 2007, pp. 323–327.

IOP Conf. Series: Materials Science and Engineering **396** (2018) 012053 doi:10.1088/1757-899X/396/1/012053

LES OF THERMALLY STRATIFIED TURBULENT CHANNEL FLOW AT LOW PRANDTL NUMBER

Waquar Raza[1], Ajiv Alam Khan[1], Syed Mohd Yahya[1,*]

Department of Mechanical Engineering, Aligarh Muslim University, Aligarh-202002, India

Corresponding Author: *smyahya@zhcet.ac.in

Abstract. In this paper Large Eddy Simulation (LES) is carried out for turbulent channel flow to assess the role the Prandtl and Richardson number on the turbulent heat transport. Fully developed turbulent channel flow configuration is used in the present study with top and bottom walls kept at constant temperature of T_h and T_c. The Reynolds number calculated on the basis of friction velocity and channel half width are kept as 180, the molecular Prandtl number is 0.1 and Richardson number are 0, 20, 40, 60, 120 and 240. Wall Adaptive Local Eddy viscosity (WALE) model is used for modelling sub-grid stress tensor and Dynamic thermal model is used for modelling sub-grid heat flux. Mean velocity and temperature statistics as well as its variance, turbulent heat flux, turbulent Prandtl number are carried out and the role of Prandtl and Richardson number are examined. Also the instantaneous flow structures and temperature contours are visualized in order to understand the inherent physics of of streaks and vortices.

1. Introduction

1.1. Background and Motivation

Most of the flows appearing in everyday life and engineering applications are turbulent. In some case turbulence is favourable, in other not. Whatever the case is, a high level of knowledge about the physics of turbulence and related phenomena is necessary. Turbulent mixed convection is a mode of heat transfer, which combines both natural and forced convection. It is frequently encountered in variety of engineering applications, e.g. nuclear reactor cooling systems, turbine blades, cooling of electronic equipment, heat exchangers, solar panels etc. as well as in the nature, e.g. dynamic of oceanic and atmospheric circulations. As a consequence of density or temperature gradients, the buoyancy force arise, changing the structure and intensity of both mean and turbulence fields. Effects of buoyancy enter the Navier-Stokes equation through a body force. Temperature acts as an active scalar so that the momentum and energy transport are closely coupled. Because the density variations directly depend on temperature, it is possible to use only the temperature differences to represent the buoyancy force. This approach is known as the Boussinesq approximation. Many studies pertaining to experimental and computational work of heat transfer between a solid wall and the adjacent turbulent fluid flow at various Reynolds and Prandtl numbers were carried out by researchers. Results concerning the effects of Richardson number on the heat transfer in stably stratified wall-bounded flows were reported in literature for Prandtl number of 0.71. As per the knowledge of author none of the researcher have studied the effect of Richardson number on the heat transfer at low Prandtl number. The basic mechanism and fluctuation characteristics of the temperature and its effect on thermal transport need to be investigated in detail.

1.2 Literature Review

Thermally stratified channel flows are difficult to simulate in the laboratory because of technical issue of maintaining the walls at higher temperatures, the influence of sidewalls, etc. Thus, the available experimental studies are very few and at low Reynolds number. Fukui, et al.[1] experimentally studied the stratification effects on thermal transport process in the atmospheric surface layer or in the wall region. They reached a fully developed stratified condition in turbulent flow between horizontal parallel plates maintained at different temperatures for both stable and unstable situation and provided elaborative data on stratification effects. Most laboratory experiments on stratified wall turbulence have been made for other configurations such as open channels or boundary layers. All these experiments have shown that when the class level is stable, and it is increased, the skin friction and the heat transmission decreases and suppressed the turbulence as the potential energy toll is required to yield larger mean shear. Kim & Moin [2] were the first who carried out a DNS for Prandtl numbers *Pr* = 0.1, 0.7 and 2.0 at a moderate Reynolds number of $\mathrm{Re}_\tau = 180$. They assumed generation of heat within the fluid and removal of heat at both walls. They examined the computed scalar fields for structures and found that the temperature fields were highly correlated with the streamwise velocity. They observed that streaky structures in the wall region were also present in the temperature field with the same non-dimensional mean spacing, almost independent of the molecular Prandtl numbers. Kasagi & Ohtusubo[3] used a constant heat flux at both walls at Prandtl numbers of 0.71 and 0.025 at $\mathrm{Re}_\tau = 150$ and allowed the temperature to vary with time and position. They observed that as the Prandtl number becomes very low, the temperature variance and the turbulent heat fluxes are markedly decreased with their peak values moving away from the wall. They also showed that the turbulent Prandtl number at *Pr*=0.025 is much larger that for the moderate Prandtl number fluids.

The earliest studies of homogeneous shear flows with stratification was performed by Rohr[4] who showed, that at a threshold gradient Richardson number (i.e, 0.25), turbulence neither grows nor decays. However, at low Richardson numbers, the turbulent kinetic energy increases & at higher values, the turbulence suppressed. Iida & Kasagi[5] performed DNS of turbulent channel flow under unstable density stratification. They reported that at increased Grashoff numbers the large-scale buoyant thermal plumes diminish the quasi-coherent streamwise vortices which result in increased bulk mean velocity and a decreased turbulent friction coefficient. Consequently, the vertical fluid motion of thermal plumes drastically changes the transport mechanism of the Reynolds shear stress. In addition, found that the thermal plumes are spatially aligned in the direction of flow, and the low-speed streaks and vortical structures are concentrated in the region where the thermal plume begins to rise.

1.3 Problem description

The main phenomenon we want to study is the effect of Prandtl number on the turbulent transport of scalar filed under moderate to high stratification. We consider a flow in a plane channel with constant pressure gradient driven flow in streamwise direction, periodic boundary conditions in streamwise as well as in spanwise direction and no slip condition at the wall. Top wall is heated and bottom wall is cooled. We chose a large enough domain which does not alter the characteristic of flow through boundary conditions having size $L_x = 4\pi\delta$, $L_y = \frac{4}{3}\pi\delta$, $L_z = 2\delta$, in the x, y, and z direction respectively. We studied for molecular Prandtl number of 0.1 with friction Richardson number of 0, 20, 40, 60, 120 and 240. Shear Reynolds number based on friction velocity and channel half width is fixed as 180 for all the simulations [6,7].

IOP Conf. Series: Materials Science and Engineering **396** (2018) 012053 doi:10.1088/1757-899X/396/1/012053

2. Governing equations and numerical methodology

An incompressible, homogenous, Newtonian fluid with constant thermo-physical properties is considered. Assuming constant kinematic viscosity and a reference fluid density ρ_{ref} at a reference temperature $T = T_{ref}$. The equations of motion can be written in the following dimensional form in terms primitive variables of velocity u_i, pressure p , and temperature T. These fields are solution of the equations of conservation of mass, momentum and energy. They can be written in the following dimensional form[8]

$$\frac{\partial u_i}{\partial x_i} = 0 \tag{2.1}$$

$$\frac{\partial u_i}{\partial t} + u_j \frac{\partial u_i}{\partial x_j} = -\frac{1}{\rho_{ref}} \frac{\partial p}{\partial x_i} + \frac{1}{\rho_{ref}} \frac{\partial(\tau_{ij})}{\partial x_j} + B_T \delta_{i3} + F\delta_{i1} \tag{2.2}$$

With $\tau_{ij} = \mu\left(\frac{\partial u_j}{\partial x_i} + \frac{\partial u_i}{\partial x_j}\right)$. These partial differential equations (PDEs) (equation (2.1) and equation (2.2)) represent the conservation of mass and momentum respectively, applied to an infinitesimal region of space along with a constitutive relation called Stokes Law which relates the velocity gradients to the stresses in a fluid particle. In thermally coupled flow problems, the governing equations are fully coupled. First, as density is temperature sensitive, temperature variations may lead to density gradients. This can result in buoyancy forces due to gravitational effects. In equation (2.2), B_T is the thermally buoyancy force which is equal to $g(\rho - \rho_{ref})/\rho_{ref}$, where g is the gravitational acceleration. F is a constant pressure gradient needed to maintain the flow in the stream-wise direction. In non-isothermal flow situation, the fluid density ρ and viscosity μ generally depends upon temperature and pressure. The variations of density are highly important because of the production of buoyancy forces. The buoyancy forces appear automatically when an equation of state is inserted to the equation of motion. For the cases that density as coincident with the reference value ρ_{ref} . Expanding a Taylor series for $\rho(T)$ at constant pressure p and keeping only first two terms of the series, we can write

$$\rho(T) = \rho_{ref} - \rho_{ref}\beta(T - T_{ref}) \tag{2.3}$$

Where $\beta = -\frac{1}{\rho}\left(\frac{\partial \rho}{\partial T}\right)_p$ is the isobaric coefficient of thermal expansion, evaluated at the reference temperature T_{ref} .

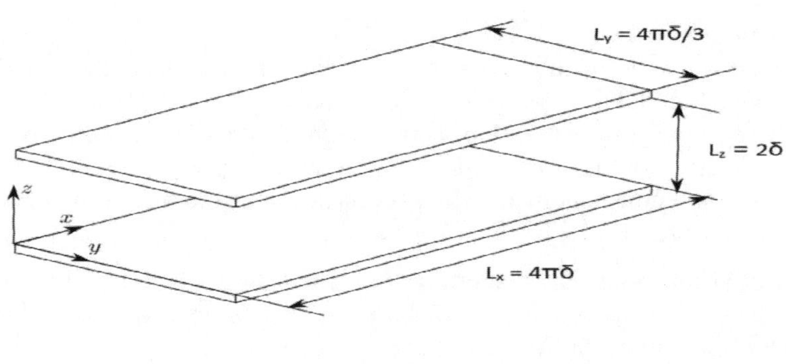

Figure 2.1. Schematics of a channel flow with the adopted coordinate system

IOP Conf. Series: Materials Science and Engineering **396** (2018) 012053 doi:10.1088/1757-899X/396/1/012053

3. Validation

A schematic sketch of the flow configuration is shown in figure 3.1. The distance between the two-parallel walls of the channel is 2δ where δ is referred to as the channel half-height. Validation is presented in subsequent figures. Mean velocity profile non-dimensionalized by friction velocity is shown in fig. 3.3. Also shown is the mean velocity profiles from other references of DNS and LES database as given by Armenio & Sarkar[9]. As can be seen in small differences are present when comparing solutions with each other. A maximum deviation of 2% occur at center of channel while comparing with DNS database of Kawamura [10]. The present neutral profiles collapse perfectly with those calculated by Armenio & Sarkar [9].The profiles are symmetric about channel's vertical center plane and thus are only presented for half of channel height. The small discrepancies seen in the velocity profiles are most likely a result of numerical error, and a consequence of different numerical methods being employed.

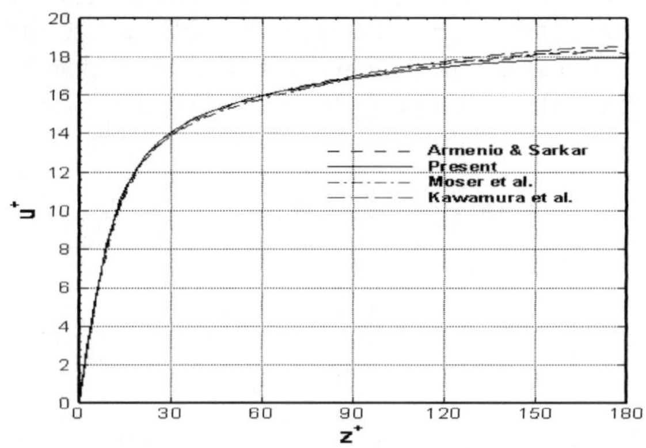

Figure 3.1. Mean velocity profile u_τ as a function of wall normal height

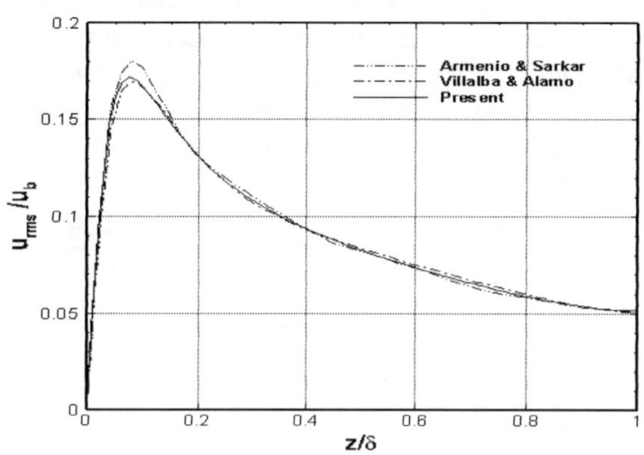

Figure 3.2. RMS of fluctuations in stream-wise direction non-dimensionalized by friction velocity

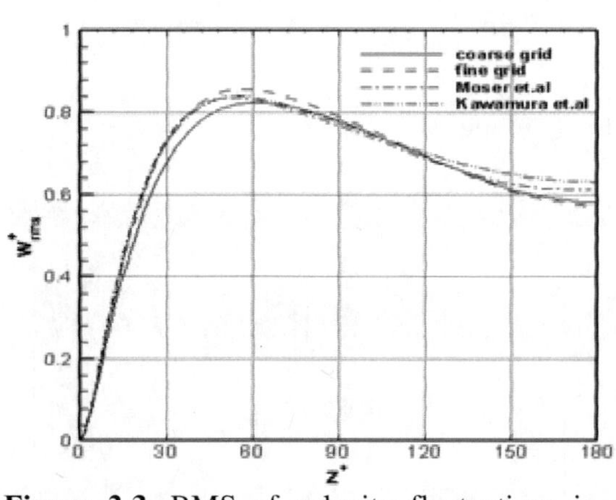

Figure 3.3. RMS of velocity fluctuations in wall-normal direction

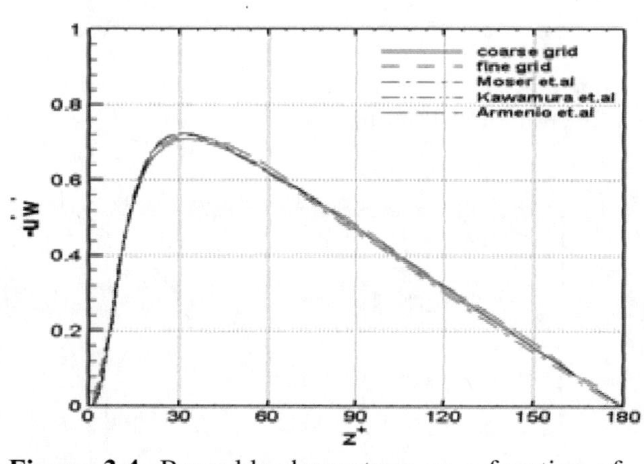

Figure 3.4. Reynolds shear stress as a function of wall normal height

IOP Conf. Series: Materials Science and Engineering **396** (2018) 012053 doi:10.1088/1757-899X/396/1/012053

4. Results and discussions

In this section, we present detailed numerical results of the considered test problems in order to check the accuracy and durability of the chosen algorithm. This is the case of large-eddy simulation (LES) of turbulent channel flow at $\boldsymbol{Re_\tau = 180}$, and $\boldsymbol{Pr = 0.1}$. The main purpose of this study is to examine how the turbulent structures are going to change in a stably stratified channel flow for different Prandtl number under temperature stratification. This is done by analyzing turbulence statistics, heat fluxes, relevant non-dimensional quantities and flow visualization to examine the structure of streaks and vortices.

4.1 Effect of stable stratification on the mean flow

In figure 4.1 the mean stream wise velocity profiles, u^+ are shown as a function of the dimensionless wall-normal coordinate z^+. As in the previous investigations, of Armenio & Sarkar [9], Garcia-Villalba & del Alamo [11], stable stratification resulted in a systematic decrease in both friction coefficient and Nusselt number. Since the driving force is pressure gradient which is held constant for all simulations, the wall shear stress and therefore the slope of the mean velocity profile in the case of stable stratification should be invariant compared to those of the non-buoyant case (solid line). It can also be seen that mean velocity increases compared to the non-buoyant simulation and so does the mass flow rate.

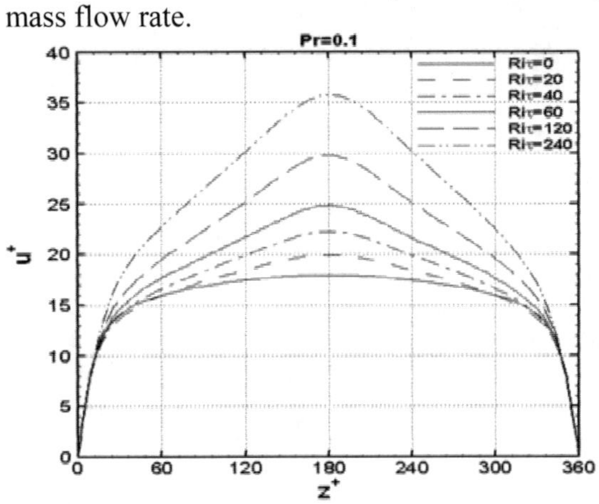

Figure 4.1. Comparison of mean velocity profile a function of wall-normal height

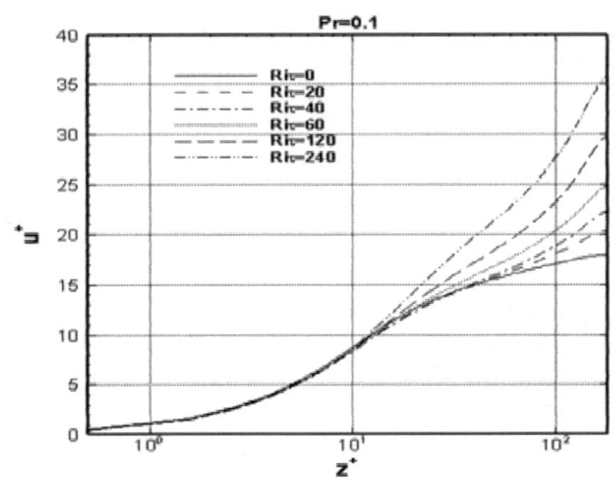

Figure 4.2. Comparison of mean streamwise velocity in logscale

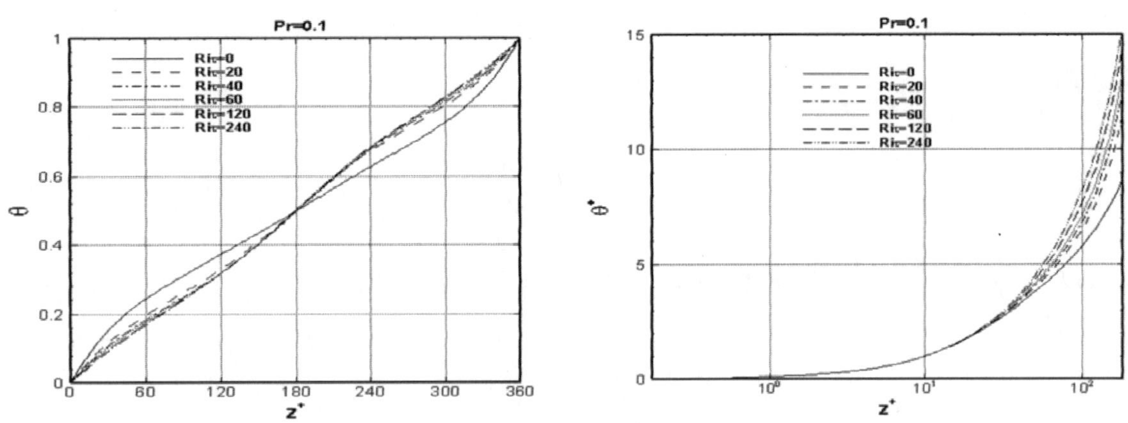

Figure 4.3. Comparison of mean temperature profile as a function of wall-normal height

IOP Conf. Series: Materials Science and Engineering **396** (2018) 012053 doi:10.1088/1757-899X/396/1/012053

4.2 Effect of stratification on Reynolds stress

The root mean square (rms) of fluctuations (Fig. 4.4-4.5) in streamwise velocity field and Reynolds shear stress distribution for different stratification is analyzed here. The scaling factor chosen here is bulk velocity and in the subsequent section we will see that the second order statistics of turbulence for this simulation condition is better described by bulk velocity rather than shear velocity, also Amenio & Sarkar [9] has chosen similar type of scaling factor. We have seen a kink type formation at the core which is nothing but symbolize the IGW at the centre and fluctuations get suppressed on both the walls. Same is the case with wall shear stress.

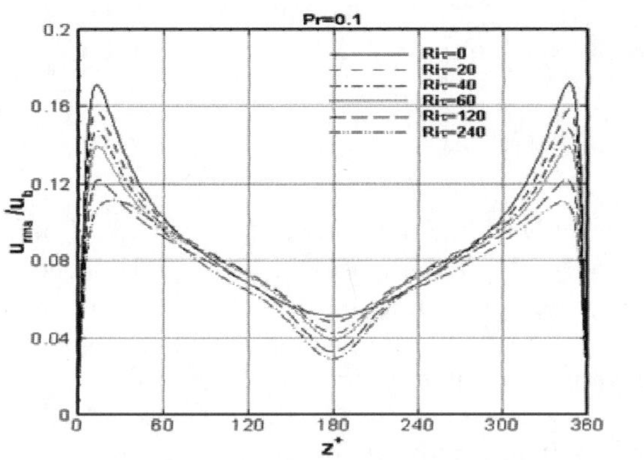

Figure 4.4. Comparison of streamwise fluctuation of velocity field upon stable stratification

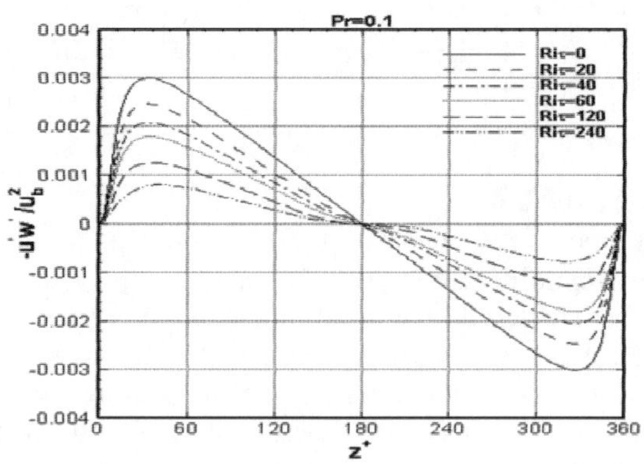

Figure 4.5. Comparison of wall shear stress upon stable stratification

4.3 Visualization of temperature contour

The temperature contour is plotted after a sufficient time interval so that the fluid properties reach a statistical equilibrium (Fig.4.6). The plane of contour is taken at half of channel i.e. mid y-plane [12]. As stable stratification is imposed we see internal gravity waves at the core of the channel. As we increase the stratification relaminarization is observed at the hot wall and the major portion of heat transfer is through molecular diffusion rather than turbulent heat flux. In Fig.4.7 iso-surfaces of non-dimensional temperature is shown. We can see clearly from figure that there exist some turbulent mixing at the core of the channel which further get dampened at higher Richardson number confirms the laminarizing behaviour of overall flow.

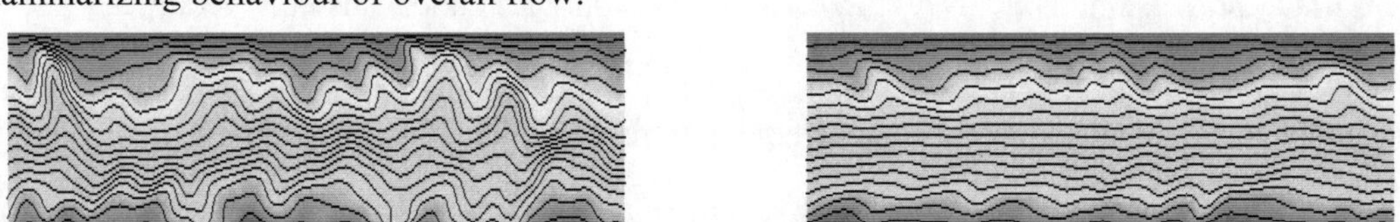

$Ri_\tau = 20$ $Ri_\tau = 240$

Figure 4.6. Temperature contours for Pr = 0.1 at different stratification at mid y-plane

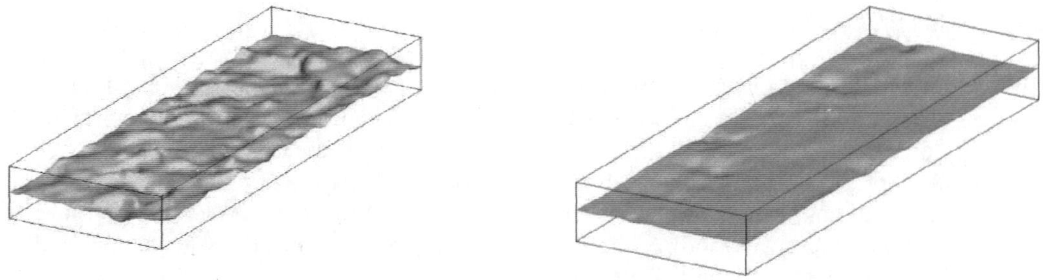

$Ri_\tau = 20$ $Ri_\tau = 240$

Figure 4.7. Iso-surfaces of non-dimensional temperature $\theta = 0.5$ for Pr = 0.1

5. Conclusions

A Large eddy simulation (LES) of a turbulent channel flow with thermal buoyancy was carried out. The Reynolds number (Re_τ) based on the friction velocity and half-channel height is 180 and the molecular Prandtl number is taken as 0.1. We have seen a kink type formation (IGW) at the centre and fluctuations are suppressed on both the walls. As we increase the stratification relaminarization is observed at the hot wall and the major portion of heat transfer is through molecular diffusion rather than turbulent heat flux. There exists some turbulent mixing at the core of the channel which further get dampened at higher Richardson number confirms the laminarizing behaviour of overall flow.

References

[1] **Fukui, K., Nakajima, M. & Ueda, H**., 1983. A laboratory experiment on momentum and heat transfer in stratified surface layer. Quaterly Journal of Royal Metrological Society, 109, 661.

[2] **Kim, J. & Moin, P**., 1989. Transport of passive scalars in a turbulent channel flow, In: Andre, J.-C. et al. (Eds), Turbulent Shear Flows. *Springer,* Volume 6, pp. 85-96.

[3] **Kasagi, N. & Ohtusubo, Y**., 1993. Direct numerical simulation of low Prandtl number thermal field in a turbulent channel flow, In: Durst, F. et al. (Eds). *Turbulent Shear Flows, Springer, Berlin,* Volume 8, pp. 97-119.

[4] **Rohr J. J.,** 1989. Growth and decay of turbulence in a stably stratified shear flow. Journal of Fluid Mechanics, 200, 563-594.

[5] **Iida, O. & Kasagi, N.,** 1997. Direct numerical simulation of unstably stratified turbulent channel flow. Journal of Heat Transfer, 119(1), 53-61.

[6] **Yahya S. M., Anwer S. F., and Sanghi S.**, 2013. Phenomenological and statistical analyses of turbulence in forced convection with temperature-dependent viscosity under non-Boussinesq condition. Eur. Phys. J. E 36, 120.

[7] **Yahya S. M., Anwer S. F., and Sanghi S.,** 2012 A conservative pressure correction method on collocated grid for low mach number flows," World J. of Mech. 5, 253.

[8] **Yahya S. M., Anwer S. F., and Sanghi S.,** 2015, Turbulent forced convective flow in an anisothermal channel" Int. J. Thermal Sciences, 2015, 88, 84-95.

[9] **Armenio, S. & Sarkar, S**., 2002. An investigation of stably stratified turbulent channel flow using lage-eddy simulation. *Journal of Fluid Mechanics* , pp. 459 1-42.

[10] **Abe, H., Kawamura, H. & Matsura, Y**., 2004. Surface heat flux fluctuations in a turbulent channel flow upto Re(tau)=1020 with Pr=0.025 and 0.071. *International Journal of Heat anf Fluid Flow,* Volume 25, pp. 404-419.

[11] **Garcia-Villalba, M. & del Alamo, J. C**., 2011. Turbulence modification by stable stratification in channel flow. *Physics of Fluids* , p. 23 (4) 045104.

[12] **Sagaut, P**., 2006. *Large Eddy Simulation for incompressible Flows.* s.l.:Springer.

IOP Conf. Series: Materials Science and Engineering **396** (2018) 012054 doi:10.1088/1757-899X/396/1/012054

Estimation of in-plane thermal conductivity of copper clad board by inverse analysis using artificial neural networks

K Niju Mohammed[1] and Samarjeet Chanda[2]

[1]Indian Institute of Technology Madras, Chennai 600036, India
[2] Indian Institute of Technology Kanpur, Kanpur, India

nijumk25@gmail.com

Abstract. This study presents a methodology to estimate the in-plane thermal conductivity of copper clad board used in electronic applications. Experiments are performed in vacuum environment and an inverse heat conduction problem (IHCP) is solved employing artificial neural networks to estimate the in-plane thermal conductivity. A comparison of estimates of thermal conductivity as obtained by solving the inverse problem using back propagation artificial neural networks trained using two algorithms namely Levenberg-Marquardt and Scaled Conjugate Gradient are presented.

1. Introduction

In electronic applications copper clad boards are used as substrate material for mounting electronic components. The copper clad boards are etched to make printed circuit boards. These boards are in-plane isotropic in nature. The objective of the present study is to estimate the in-plane thermal conductivity $(k_x = k_y = k)$ of copper clad boards by solving an inverse heat conduction problem (IHCP) using steady state temperature measurements. The inverse analysis is performed using backpropagation artificial neural networks (ANNs) employing both Levenberg-Marquardt and Scaled Conjugate Gradient algorithms.

There are different techniques proposed by researchers in the literature which are implemented to solve a variety of inverse problems. Beck [1] proposed a nonlinear parameter estimation technique to estimate the thermal transport properties such as thermal conductivity and volumetric heat capacity of solids simultaneously under transient conditions. Garnier et al. [2] introduced a novel technique to carry out surface temperature measurements on the test specimen to estimate thermal properties of composites. The advantages of parameter estimation techniques in terms of more information and faster results in comparison with other steady state techniques such as guarded hot plate or transient methods such as line source method, flash method etc. were brought out. Sawaf and Ozisik [3] performed inverse analysis for the simultaneous estimation of the principal thermal conductivities of orthotropic materials using conjugate gradient method and Levenberg-Marquardt method. Mejias et al. [4] carried out a comparative study on non-linear parameter estimation techniques like conjugate gradient method and Levenberg-Marquardt method for estimating thermal conductivities of an orthotropic solid by solving an inverse problem. Gobbé et al. [5] conducted an experimental study for the measurement of thermal conductivity of multilayer orthotropic media characterized by isotropic behaviour along planes parallel to layers.

Artificial neural networks have now been used extensively for solving inverse heat transfer problems. Krejsa et al. [6] presented the advantage and limitations of various back propagation algorithms that are implemented to obtain the successful solution of inverse heat conduction problems using ANNs. Cortés et al. [7] harnessed the capability of ANN for solving the inverse problem to retrieve the heat source generation term of the hot plate in a guarded hot plate apparatus. Chanda et al. [8] presented a novel methodology for the simultaneous estimation of the thermal conductivities along principal directions of an anisotropic honeycomb composite by solving an inverse heat conduction problem using ANN trained with Levenberg-Marquardt algorithm.

The literature review shows the immense capabilities of inverse methodology for estimating the thermal conductivities. Also ANNs have been successfully employed to solve inverse heat transfer problems. These facts motivate to undertake the present study. Hence, the present study involves numerical simulation of direct problem, temperature measurements using of steady state experiments and estimation of in-plane thermal conductivity using artificial neural networks trained using Levenberg-Marquardt and scaled conjugate gradient algorithms.

2. Direct problem

The direct problem is formulated to simulate the exact experimental conditions numerically and to obtain the required temperature distribution. The copper clad test board is modeled as a homogenous and in-plane isotropic board with $k_x = k_y = k$, where the in-plane thermal conductivity need to be determined and out-plane thermal conductivity k_z is already known. Commercial software Ansys Fluent is employed for solving the direct problem. The governing equation for the direct problem is given by equation (1).

$$k_x \frac{\partial^2 T}{\partial x^2} + k_y \frac{\partial^2 T}{\partial y^2} + k_z \frac{\partial^2 T}{\partial z^2} = 0 \quad (1)$$

where k_x, k_y and k_z are the thermal conductivities in x, y and z directions. The direct problem geometry consists of a square board of dimensions $180 \times 180 \times 1.60$ mm with a foil heater fixed at the centre of the top surface of the board. The whole assembly is mounted on a copper frame which serves as a heat sink. The direct problem geometry is depicted in figure 1.

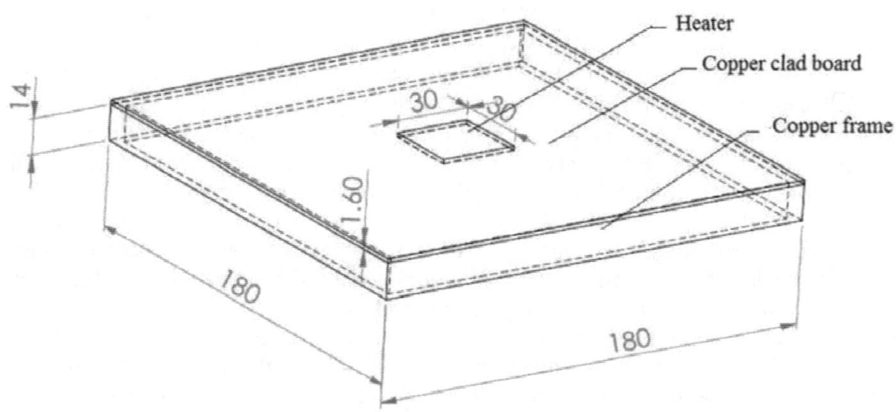

Figure 1. Geometry for direct problem simulation

2.1. Meshing of direct problem geometry

The direct problem geometry is discretised into quadrilateral cells. Regular auto meshing with proximity option is used to mesh the geometry. A grid independence study is conducted to obtain the optimum number of cells to save the computational time while not affecting the accuracy of the solution. The results of the grid independence study are shown in figure 2. A mesh with 0.24 million nodes yielded grid independent results, hence it was used for the present study.

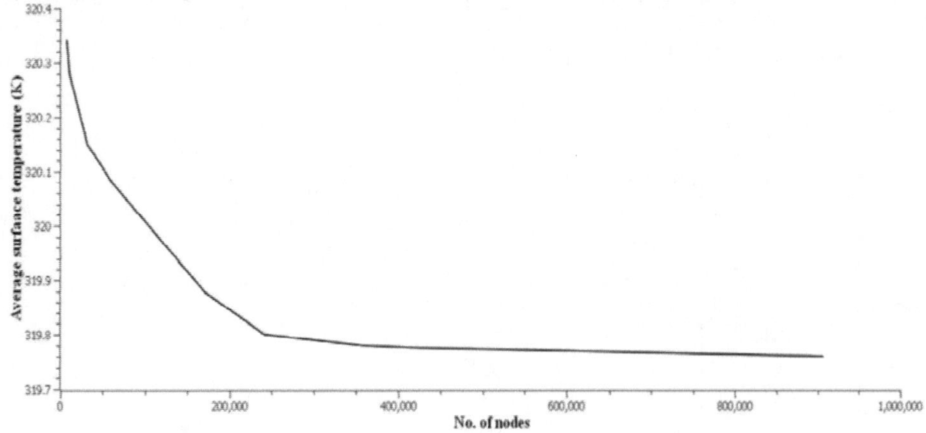

Figure 2. Grid independence study

2.2. Boundary conditions

The following boundary conditions are imposed on the direct problem,

- A heat flux of 1111 W/m^2 is applied to heater domain.
- Isothermal condition (298 K) is imposed on copper frame.
- All exposed surfaces of copper clad board except the surface in contact with the copper frame are given adiabatic condition.

2.3. Direct problem solution

The governing heat conduction equation is solved using Ansys Fluent. The convergence criterion is said to be reached when the temperature residuals fall below 10^{-9}.

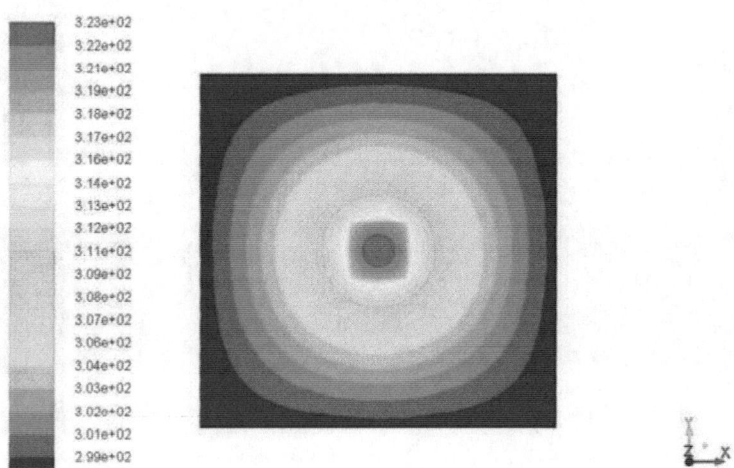

Figure 3. Temperature contour for the simulation case

The results for a sample simulation case with $k_x = k_y = k$ = 10W/mK and k_z = 0.34W/mK is shown in figure 3. As seen in figure 3, the temperature contour assumes a circular profile implying equal heat transfer in the in-plane direction which corroborates well with the laws of heat transfer. The temperature is maximum near the heater and appreciable variations of temperatures are observed along both X and Y directions to a certain distance after which appreciable temperature gradients do not exist as depicted in figure 4. These temperature gradients hold the key to the successful estimation of thermal conductivity.

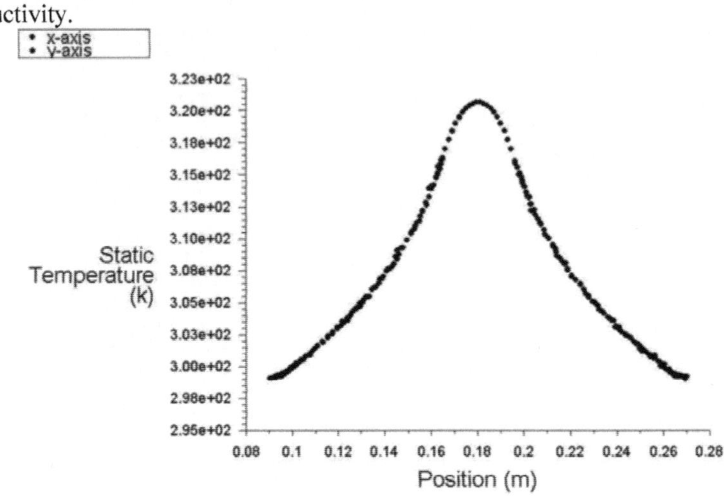

Figure 4. Temperature profile along X and Y axis of test sample

3. Experimental setup and methodology

The photograph of the entire experimental setup is depicted in figure 5. It consists of a vacuum chamber that develops a vacuum level of 10^{-5} mbar with the help of vacuum pumps.

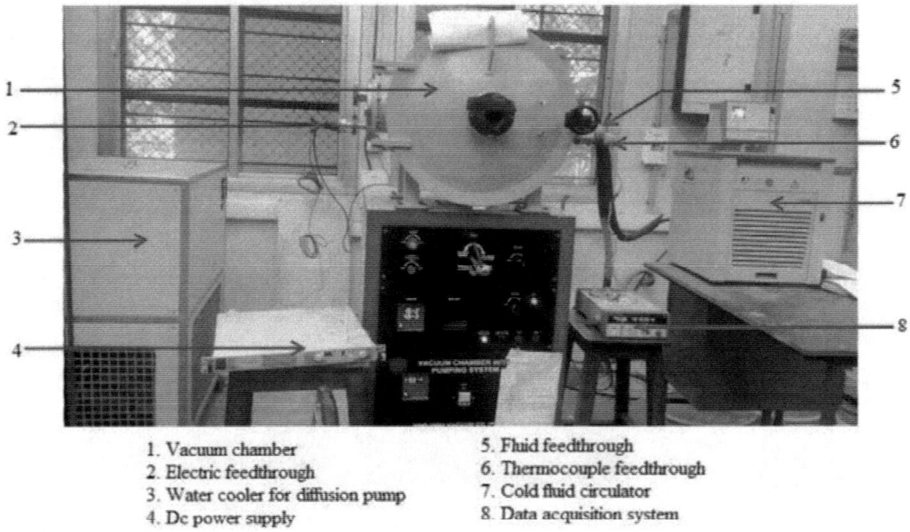

Figure 5. Photograph of the experimental setup

The vacuum is maintained to eliminate convection losses. A square copper frame along with an aluminium cold plate serves as heat sink. A constant temperature water bath is used to circulate cold fluid to aluminium plate to maintain it at a constant low temperature.

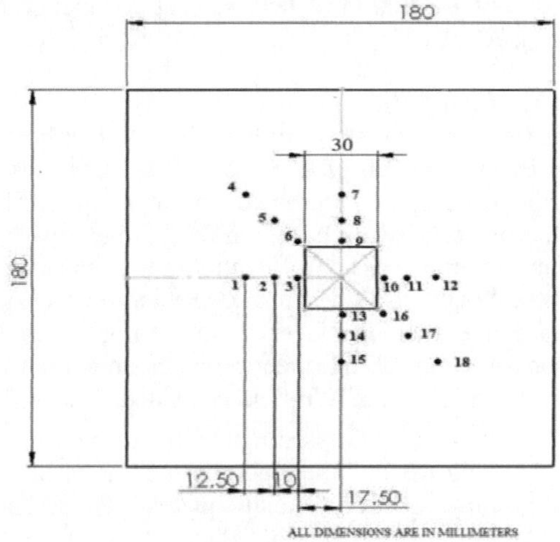

Figure 6. Locations of thermocouples in the test sample

T type thermocouples are fixed at the locations on the top surface of the test as shown in figure 6. A foil type heater is placed at the middle of the board (test sample). A low emissivity sheet is used to cover all the exposed surfaces of the board to minimize radiation losses along with a multilayer insulation cover. This entire test assembly is placed in the vacuum chamber. A vacuum level of 10^{-5} mbar is maintained in the chamber. The foil heater is energized to 1W and the sink is kept at 25 $^{\circ}$C. The steady state is deemed to be achieved when the temperature sensor readings do not change by $\pm 0.1^{\circ}$ C for a time period of one hour. These steady state temperatures are used for the estimation of thermal conductivity of copper clad board by solving the inverse problem using ANN. Details of the experimental set up are provided in study by Chanda et al. [8]. More details are omitted for the sake of brevity.

An uncertainty analysis is carried out to determine the uncertainty in both measured and derived quantities. For the direct measurements like temperature, voltage, current and vacuum pressure the least count of the instruments are taken as uncertainty. Uncertainty in the derived quantities i.e. power is evaluated using equation (2) as given by Venkateshan [9].

$$\sigma_P = \sqrt{\left(\frac{\partial P}{\partial V}\sigma_V\right)^2 + \left(\frac{\partial P}{\partial I}\sigma_I\right)^2} \tag{2}$$

where σ is the uncertainty in measurement, P is the power, V is the voltage and I is the current. The results of the uncertainty analysis are given in table 1.

Table 1. Uncertainties in measured and derived quantities

Sl. no.	Quantity measured	Uncertainty
1	Temperature	±0.5 K
2	Voltage	±0.001V
3	Current	±0.0001A
4	Power	0.102%
5	Vacuum pressure	10^{-6}mbar

IOP Conf. Series: Materials Science and Engineering **396** (2018) 012054 doi:10.1088/1757-899X/396/1/012054

4. Inverse methodology and artificial neural network

The inverse methodology involves the solution of inverse heat conduction problem to obtain the in plane thermal conductivity using ANN where the experimental temperatures are fed as inputs to the ANN.

Artificial neural networks are computational structures used to simulate nonlinear and complex relationships between input and output data. By mimicking the human brain ANNs can identify the correlated patterns between inputs and outputs. Therefore ANNs trained for sample set of inputs and outputs can be used to predict the outputs when presented with input data set in the similar range.

In the present study artificial neural network based on feed forward back propagation principle is implemented. Temperature data at nine pre assigned locations are used as inputs and in-plane thermal conductivity as output of the neural network. A neuron independence study is undertaken to determine the number of neurons required in the hidden layer of the artificial neural network to accurately simulate the problem under consideration. Using direct problem simulations, 300 different values of $k_{xx} = k_{yy} = k$ lying in the range from 0.5 to 15 W/mK is generated. Out of these, 210 samples (70%) are used for training and 90 samples (30%) for testing the neural network. The number of hidden neurons is varied from 3 to 12. Using performance metrics such as mean relative error (MRE) and correlation coefficient (R^2) as given by equation (3) and equation (4), the optimum number of hidden neurons is determined.

$$MRE = \frac{1}{m}\sum_{i=1}^{m}\left|\frac{k_{actual} - k_{retrieved}}{k_{actual}}\right| \quad (3)$$

$$R^2 = 1 - \left[\frac{\sum_{i=1}^{m}(k_{actual} - k_{retrieved})^2}{\sum_{i=1}^{m}(k_{actual} - \bar{k}_{actual})^2}\right] \quad (4)$$

Results of the neuron independence study are given in table 2.

Table 2. Neuron independent study results

Sl. no.	No. of hidden neurons	MRE	R^2
1	3	0.5786	0.9458
2	5	0.5728	0.9332
3	7	0.6098	0.9371
4	9	0.8461	0.9387
5	**10**	**0.4985**	**0.9568**
6	12	0.9134	0.9012

A network with 10 neurons in the hidden layer is found to have lowest MRE and highest R^2 values of 0.4985 and 0.9568 respectively and is used for the estimation of in-plane thermal conductivity.

5. Results of the estimation of thermal conductivity of copper clad board

As stated earlier, the test specimen is an in-plane isotropic copper clad board of dimensions 180×180×1.6 mm. Steady state experiments are performed on the test sample instrumented with T-type thermocouples placed at pre-assigned locations. These steady state temperatures are used as an input to the inverse model. The training range of ANN and the search range used in this problem for performing parameter estimation has been limited to $2 \leq k \leq 15$ W/mK. The artificial neural network used is trained with a data set comprising of 210 samples obtained by simulating the direct problem in the range as stated earlier, with temperatures as inputs and thermal conductivities as outputs.

Artificial neural network used in this study is trained using the feed forward backpropagation technique. The algorithms employed to train the ANN in the current study are Levenberg - Marquardt algorithm and scaled conjugate gradient algorithm with back propagation of errors. Levenberg - Marquardt algorithm is one of the robust algorithms implemented by many researchers because of its efficacy in retrieving the parameters of interest from inverse problems. Scaled conjugate gradient algorithm utilises less computational resources. The tan sigmoid transfer function is applied to all the neurons in the hidden layer and pure-linear to the neurons in the output layer. A single hidden layer network with 10 neurons is used in the network. The neural networks are trained for 1000 epochs. The results of the estimated values of thermal conductivity as obtained from ANNs trained using two different algorithms are listed in table 3.

Table 3. Retrieved values of thermal conductivity

Sl. no.	Algorithm	Thermal conductivity(W/mK)
1	Levenberg - Marquardt	10.08
2	Scaled conjugate gradient	9.60

Figure 7 shows the parity plot with error bars depicting the comparison of measured temperatures with those obtained from direct problem simulation using the retrieved value of thermal conductivity using Levenberg - Marquardt algorithm and scaled conjugate gradient algorithm. The error bar corresponding to maximum measurement error in temperature i.e. 0.5 K

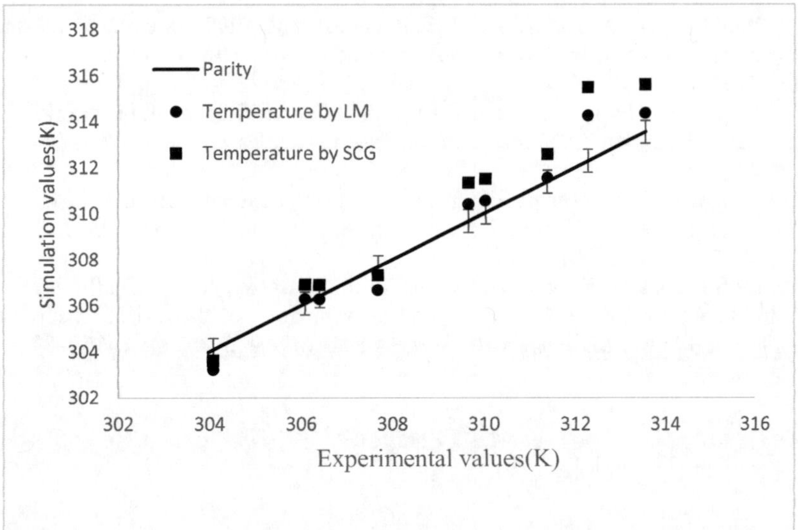

Figure 7. Parity plot of measured and simulated temperatures of copper clad board

As evident from figure 7, the temperature values obtained from simulation and measurement agrees well with each other. The Levenberg - Marquardt algorithm gives a slightly better approximation than scaled conjugate gradient algorithm. The bias free plot ensures that the governing physics has been captured well in the model. The thermal conductivity of the copper clad board estimated by using this technique agrees well with that reported by Azar and Graebner [10]. Therefore the present study using inverse methodology employing ANN is able to retrieve the in-plane thermal conductivity of copper clad board successfully.

6. Conclusion

The present study focused on estimating the in-plane thermal conductivity of copper clad board by employing synergistic combination of direct problem solution, experiments and inverse methodology employing artificial neural networks using Levenberg - Marquardt algorithm and scaled conjugate gradient algorithm. Two ANN models trained using different algorithms namely Levenberg - Marquardt and Scale Conjugate Gradient with 210 sample data obtained from direct problem simulations were used to retrieve thermal conductivity from experimental temperature data. Both the ANNs were found to predict the thermal conductivity values with good accuracy. It was also seen that, ANN trained using Levenberg Marquardt algorithm out performed that trained using Scale Conjugate Gradient algorithm slightly in terms of predicted estimates.

References

[1] Beck J 1966 Transient determination of thermal properties *Nucl. Eng. Des.* **3** 373–381.

[2] Garnier B, Delaunay D and Beck J V. 1992 Estimation of thermal properties of composite materials without instrumentation inside the samples *Int. J. Thermophys.* **13** 1097–111.

[3] Sawaf B and Özisik M N 1995 Determining the constant thermal conductivities of orthotropic materials by inverse analysis *Int. Commun. Heat Mass Transf.* **22** 201–11.

[4] Mejias, M., H. Orlande and M O 1999 A comparison of different parameter estimation techniques for the identification of thermal conductivity components of orthotropic solids *3rd Int. Conf. Inv. Prob. Eng.* 1–8

[5] Gobbé C, Iserna S and Ladevie B 2004 Hot strip method: application to thermal characterisation of orthotropic media *Int. J. Therm. Sci.* **43** 951–958.

[6] Krejsa J, Woodbury K A, Ratliff J D and Raudensky M 1999 Assessment of strategies and potential for neural networks in the inverse heat conduction problem *Inv.Prob. Eng* **7**, Taylor and Francis, pp. 197-213.

[7] Cortés O, Urquiza G, Hernandez J, and Cruz M A 2007. Artificial neural networks for inverse heat transfer problems *Robotics and Automotive Mechanics Conf. CERMA IEEE*, pp.198-201.

[8] Chanda S, Balaji C, Venkateshan S P, Ambirajan A and Ramakrishnan V 2013 Simultaneous Estimation of Principal Thermal Conductivities of an Anisotropic Composite Medium: An Inverse Analysis *J. Heat Transfer* **135** 21301.

[9] Venkateshan S P 2008 *Mechanical Measurements* (Ane Books, New Delhi, India)

[10] Azar K and Graebner J E 1994 Experimental determination of thermal conductivity of printed wiring boards *Twelfth Annu. IEEE Semicond. Therm. Meas. Manag. Symp. Proc.* 169–82

IOP Conf. Series: Materials Science and Engineering **396** (2018) 012055 doi:10.1088/1757-899X/396/1/012055

Development of Tool to Design Piston Bowl Considering Spray Parameters to Reduce Emissions

Sajil N Sreedharan[1], Rajiv Krishnan[2a]
[1]Assistant Professor, Department of Mechanical Engineering, SJCET, Palai, India
[2]Assistant Professor, Department of Mechanical Engineering, SJCET, Palai, India

[a]Email: krirajiv@gmail.com

Abstract: The piston bowl design is one of the most important factors that affect the air/fuel mixing, combustion and pollutant formation process in a diesel engine. The piston bowl dimensions such as the bowl radius, hitting plane etc... have an effect on combustion process in-cylinder mixing. This paper aims the development of a tool that can generate the bowl profiles by combining spray parameters. Initially the spray parameters were validated against the engine test data .The bowl geometries generated with same bowl volumes as of the base bowl geometry.1D thermodynamic simulation software Diesel RK is used to predict the performance and emission characteristics of the bowls at full load and part load. The performance parameters such as rated power, Indicated mean effective pressure were compared for all the bowl geometries. Also the emission parameters such as soot and NO_x were compared. Best bowl with increased performance and least emissions are selected for the particular engine specification. Increase in performance as well as reduction in soot has been observed with the new bowl geometry.
Keywords: Piston Bowl, Diesel RK, Spray Hitting Plane, Re-Entrancy

1. Introduction

The enormous increase in the use of fossil energy sources has caused increased pollution of the air we breathe. Exhaust emissions by the vehicles have a significant contribution to the overall pollution. Diesel engine, due to its high compression ratio and un-throttled mode of operation it has been a power source for heavy duty vehicles & stationary applications in terms of fuel economy. But problems of weight, noise, odour & cost have limited its applications. If diesel engines can meet future emission standards, they offer a viable means for meeting the future fuel economy goals. Among several engine design parameters, piston bowl design is one of the important factor that affect the air/fuel mixing and the subsequent combustion, pollutant formation processes. The bowl geometry & dimensions such as pip region, bowl lip area and torroidal radius have major effect on in-cylinder mixing, combustion process and engine out emissions.

1.1 *Effect of Piston Configuration*

To achieve a better combustion with less pollutant emissions in D.I. engines, the method which is widely used and proved effective is the accelerating of fuel/air mixing to improve the combustion in the cylinder and reduce combustion period. The NO_x emission can be reduced by retarding the injection timing without facing a considerable increase in exhaust pollutants and fuel consumption. The air motion in the chamber with contracted throat is stronger than that of without a throat.

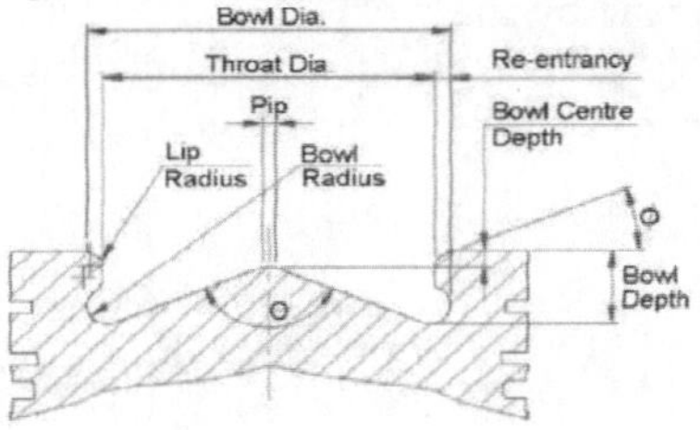

Figure 1.Piston Bowl Configuration

1.2 .*Effect of Cavity Size*

For the same volume if the cavity wall distance is varied, spray spreading areas do not differ much. But a too shallow bowl will create a long floor jet which is undesirable and too deep bowls with fewer diameters will cause much fuel to move to the cavity outside. Moreover too short distances will cause interference between the sprays injected from two consecutive holes on the cavity wall. Too long wall distance creates wider unused space between two consecutive sprays. The larger radius of the bottom corner reduces the amount of air near the side wall and the air distribution in the chamber changes and results in poor balance between air and fuel.

1.3. *Effect of Lip* Shape

Sharp edge at the cavity entrance and the lip corner are generally avoided to reduce thermal stresses. In the cavity with round lip the spray impinges and then follows smoothly along the lip and energy loss due to impingement is small.

1.4 *Effect of Bottom Corner Radius*

With a sharp bottom corner, the energy loss is very high due to the large change in direction. This creates a smaller wall jet diameter which affects fuel air mixing. But too much bottom corner diameter will reduce the amount of air near the side wall and the air distribution changes and results in poor balance between air and fuel. Thus an adequate value of the bottom corner radius is preferable to optimize the fuel dispersion and air-fuel distribution.

1.5 *Effect of Impinging Position*

Spray impingement position changes significantly from start to the end of spray impingement and the impinging position on the cavity wall .this will have a prominent effect on fuel distribution and combustion. The position where the spray impinges on the cavity wall has a significant effect on the fuel distribution.

1.6 *Effect of Throat Diameter of the Piston Bowl*

A sharp squish lip shape can reduce the smoke without increase NO_x emissions. But its smoke reduction capability depends on the interaction with other bowl geometry parameters. It was found out that the bowl with larger throat diameter and limited pip height has the best air-fuel mixing than the narrow throated bowls.

1.7 *Fuel Injection Parameters*

In a diesel engine, the fuel is injected into the cylinder at the end of compression stroke. The spray has to be injected finely and precisely. So that it could reach the combustion chamber to form air-fuel mixture. In addition, spray formation and injection process are the two important mechanisms for in-cylinder air/fuel mixing, which controls the combustion and emissions.

1.8 Spray *Angle*

Spray angle is the angle made by the spray core. It is influenced by the injection pressure, fuel density, nozzle hole diameter etc… Spray angle affects the fuel dispersion. Higher values of spray angle will increase the mixing rate inside the cylinder. Normal value of spray angle is between 8-14 degrees.

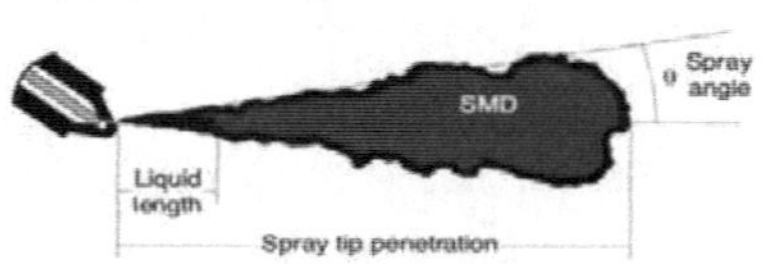

Figure 2. Spray Parameters

IOP Conf. Series: Materials Science and Engineering **396** (2018) 012055 doi:10.1088/1757-899X/396/1/012055

1.9. *Spray Penetration Length*

Spray penetration length is the length at which it penetrates into the combustion chamber. It is influenced by nozzle hole diameter, injection pressure, spray angle, fuel density etc… Spray penetration is most important in characterization of spray formation within a combustion chamber. The penetration length must be neither too long nor too short for efficient in-cylinder combustion. If too long, impingement could occur. This will result in wetting of the combustion chamber wall and piston crown. Consequently, formation of soot and wastage of fuel will be enhanced. If the length is too short, mixing efficiency and good combustion is compromised.

1.10 *Spray Breakup*

Spray breakup is the disintegration of the spray into fine droplets. The disintegration of liquid jets is described by two mechanisms. The first mechanism is the breakup of the intact liquid core into droplets called primary breakup. The second mechanism is the breakup of droplets into smaller ones, which is called secondary breakup. Both breakup length and droplet size are dependent on the properties of the liquid and the surrounding gas. Primary breakup is the most important mechanism in diesel injection systems. It determines the size of droplets that separate from liquid core.

1.11 *Spray Breakup Time*

Spray breakup time is defined as the time for primary breakup. This duration is directly affects the ignition delay. The normal delay period is about 1-4 degrees after the commencement of injection. Therefore spray breakup time is relevant for computing the ignition delay in diesel engine .K-H (Kelvin Helmholtz) breakup model is commonly used for calculating breakup time.

$$t_b = 4.351\frac{\rho_f * d_{nh}}{C_D^2 * \sqrt{\rho_a * \Delta P}}$$

Where, t_b is the breakup time, ρ_f is the fuel density, d_{nh} is nozzle diameter, C_D is the coefficient of discharge, ρ_a is ambient density, ΔP is the differential pressure.

2. Literature Review

Y. Zhu et. al. [1] conducted a computational study on effect of re-entrant lip shape and torroidal radii of piston bowl on HSDI diesel engine's performance and emissions. Three piston bowls were selected having different torroidal radius and lip shapes designs with the same lip area and pip inclination. Engine modeling used is computational software KIVA3V. The results have shown that the large torroidal radius bowl produces better combustion than the small torroidal radius bowl.

Jongyoon Lee et. al. [2] conducted a computational study on bowl shape design optimization for engine-out PM reduction in heavy duty diesel engine. Different piston bowls with different bowl diameter, lip height, pip heights were considered for the study. As the bowl diameter is increased, soot emissions are decreased and its reduction rate.

Lu Lin et. al. [3] conducted a numerical investigation on effects of combustion chamber geometry on in-cylinder air motion and performance in DI diesel engine. Three different shapes of diesel engine combustion chamber are selected (Chamber A is a central projection type, chamber B is a shallow type, and chamber C is a patalloid type). A stronger air motion is induced in chamber A, which makes the combustion faster compared to chamber C and B. This is because that the strong air motion in chamber A increases the premixed combustion. This concludes that central projected type piston bowl is much efficient to meet the performance and emissions.

Rahman M. Montajir et. al. [4] conducted an experiment on fuel spray behavior in a small DI diesel engine and the effects of combustion chamber geometry. The fuel spray development is visualized from two directions in an actual engine with a transparent cylinder and piston arrangement. The place where spray impinges, the distance from impinging wall to nozzle tip, and the shape of the chamber entrance and bottom were varied to determine their effects on the fuel spray development in the combustion chamber. In this study the spray spreading area, the equivalent wall jet diameter, and the spray path length were

IOP Conf. Series: Materials Science and Engineering **396** (2018) 012055 doi:10.1088/1757-899X/396/1/012055

considered as the characteristic parameters. A maximum spreading area and good balance is obtained for impingement just on the lip corner. Introduction of a bottom corner radius helps disperse the fuel accumulated at the bottom corner and the spray volume increases. Larger values of the bottom corner radius produces long floor jets and the air and fuel balance deteriorates due to the decreases in the amount of air near the side wall.

S. Pasupathy Venkateswaran et. al. [5] investigated Effects of the Re-Entrant Bowl Geometry on a DI Turbocharged Diesel Engine Performance and Emissions using CFD. Three different bowl geometries, namely, bowl 1 (baseline), bowl 2, and bowl 3 were selected and investigated. All the other relevant parameters, namely, compression ratio, maximum diameter of the bowl, squish clearance and injection rate were kept constant. There was a large reduction in soot (46% of baseline) and increase in NOx (15% of baseline). It can be concluded that bowl 3 is the best trade-off between performance and emissions. The swirl and TKE levels of bowl 3 are much higher than baseline bowl. So bowl 3 with more reentrancy produces better combustion and hence better engine performance than the baseline bowl. Bowl 3 is effective in lowering ISFC and soot emission.

Saravanan Subramanian et. al. [7] conducted piston bowl optimization for single cylinder diesel engine using CFD. Naturally aspirated engine having flat type combustion chamber was selected for the study. The in-cylinder flow are measured numerically on different planes at various crank angles to see the effect of air motion caused by the tangential swirl groves created on the piston bowl. Five different piston bowls (Hemispherical combustion chamber, hemispherical combustion chamber with scoop, torroidal combustion chamber, shallow combustion chamber, hemispherical combustion chamber groove) were considered for the study to see the swirl ratio during suction and compression stroke. The CFD results have supported to identify the piston shape which is good in terms of in-cylinder motion with the following order: TCC scoop >SCC scoop>HCC scoop> HCC groove>HCC. It is concluded that the torroidal combustion chamber can enhance the air-fuel mixing rate by improving the in cylinder turbulence

Jaeman Lim et. al. [9] conducted a study on the Effects of Spray Angle and Piston Bowl Shape on Diesel Engine Soot Emissions Using 3-D CFD Simulation. 2 piston bowls (A & B) were considered for the study by varying the spray angles. In order to predict the performance and emissions of the engine, a flamelet combustion model with the kinetic chemical mechanism for NOx and soot was used. Piston A had the typical shape of a conventional HSDI diesel engine. Piston B was the modified piston, with the same volume of piston A. In the shape of piston A, performance was improved and soot emission was reduced for the 70° spray angle where distribution of the fuel vapor was more homogeneous. Based on the results of the piston A, new piston shape was designed as piston B. For spray angle 100°, IMEP was improved by 3.0% and soot was reduced to 41% as compared to the case of piston A than spray angle 70° at 1,400 rpm. IMEP and soot emission of piston B were better than any other condition in piston A at 2,500 rpm. This study shows the relationship between the piston bowl and spray angle such that spray angle should be more enough to enhance fuel dispersion.

It has been well understood from the literature that piston bowl geometry strongly affects the performance and emissions in a CI engine. The main concern regarding the diesel engines are the emissions. So it has to be controlled within the cylinder. Many studies have been done based on spray –bowl interactions using CFD as well as experimental setups. In order to perform these, the piston bowl geometry should be either designed by trial and error and by CAD models. It would take much time to develop the required bowls. So it will be quite logical to develop a code that could generate the piston bowl geometry considering spray parameters upon user inputs.

3. Engine specifications

The engine specifications and input data were provided by a leading testing and research agency in India. The test engine is a 4 cylinder inline turbocharged DI diesel engine. However the engine faced problem of late injection in order to reduce NOx emissions. At the same time the soot emissions were prominent. Since the engine is being stationary the exhaust after treatments could not be installed due to its bulkiness. In order to reduce the soot emissions the piston bowl geometry should modified in such a way that the emissions can be controlled in-cylinder itself.

IOP Conf. Series: Materials Science and Engineering **396** (2018) 012055 doi:10.1088/1757-899X/396/1/012055

Table 1: Engine Specifications and spray parameters

Number of cylinders	4
Swept volume	4.7 litre
Rated speed	1500 rpm
Compression ratio	16.9
Nozzle holes	5
Spray cone angle(deg)	69
Length of nozzle(mm)	1.2
Spray penetration length	.01883m
Maximum injection pressure	633bar
Nozzle opening pressure	230bar
Injection duration	26deg CA
Spray angle	10deg
Start of injection	4deg aTDC

3.1. *Spray Parameters*

The injection duration is 26 degrees and it is clear from the test data the engine is a retarded injection of about 4deg after TDC. Late injection is provided by the manufacturer to reduce the NO_x emissions. At the same time this particular engine produces significant amount of soot. The following table shows the injection parameters of the test engine at maximum injection pressure (633bar) i.e. at 763 CA.

4. Base bowl specifications

The detailed sketch of the base bowl geometry is shown in figure 3. The geometry is having a bowl radius of 8.6mm.

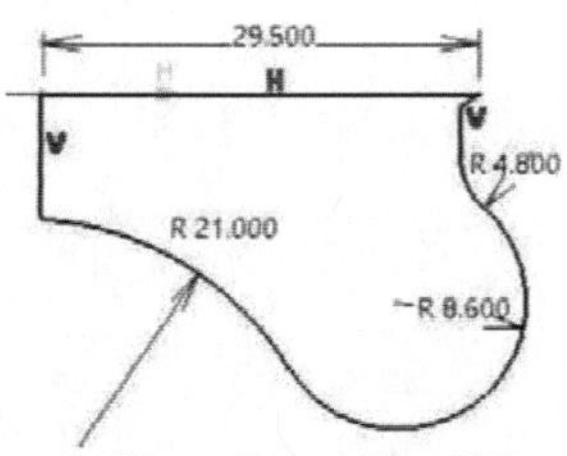

Figure 3. Base Bowl Specifications

The base bowl faced following drawbacks, Lack of reentrancy, Bowl is having more soot emissions, Hitting plane 66 %(undesirable), Spray cone angle 69deg. The objective is to modify the base piston geometry of the engine having the same bowl volume (55.3cc). The new bowl should reduce soot emissions and increase the performance.

The new bowl geometry should include the following features: It was recommended to design bowl with flat floor region instead of curved shape, Hitting plane should be 44%, Spray cone angle should be increased to 74 deg, and Bowl volume should be maintained as 55.3cc in order to preserve the constant compression ratio of 16.9.

5. MATLAB Tool for Piston Bowl Geometry Generation

The tool for generating the bowl geometry is carried out using MATLAB. The tool can generate different types of bowl profiles. The inputs required for working of this code is spray Hitting plane (44% from piston crown), Bowl radius, center offset, Floor angle, Pip height, Lip radius, center offset. The process used for generation of bowl profile is as follows:

The bowl profile is divided into 4 curves, 2 lines: one in pip region, floor angle, 2 curves: one in bowl region and lip region

IOP Conf. Series: Materials Science and Engineering **396** (2018) 012055 doi:10.1088/1757-899X/396/1/012055

6. Bowls generated using MATLAB Tool

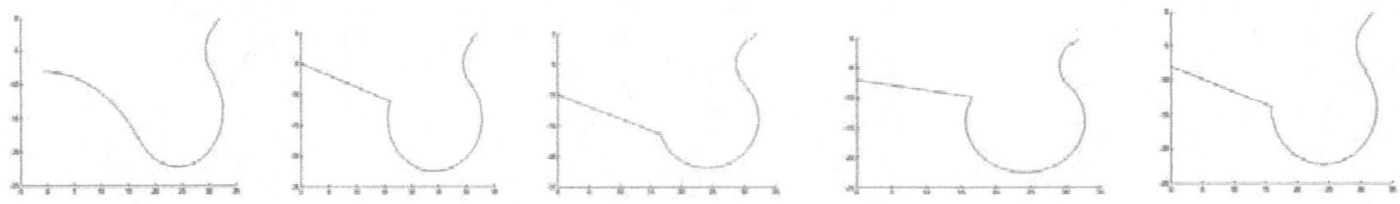

Base bowl Bowl 1 Bowl 2 Bowl 3 Bowl 4

Figure 4. Generated Bowls

7.Results and Discussions

The 5 bowls are simulated using the diesel RK to find the performance and emissions. Simulation is carried out at both full load and part for all the set of bowls. Cylinder pressure data, performance and emission data are plotted for the 4 bowls in comparison with the base bowl

7.1 *Cylinder Pressure v/s Crank Angle*

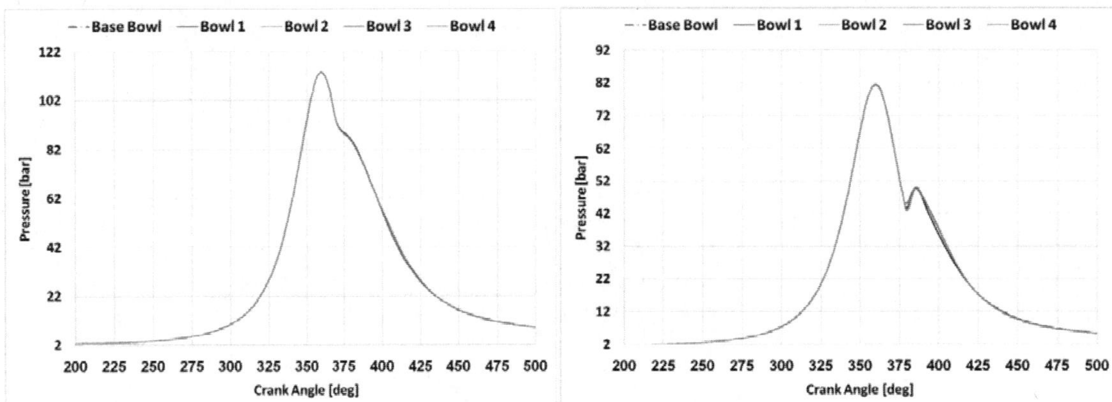

Figure 5. Cylinder pressure v/s Crank Angle (Full load, Part Load)

The cylinder pressure data for part load condition has been plotted, the values showed a good agreement with each other. At the diffusion combustion stage a slight deviation has been observed for bowl 3. But the deviation is considerable since the premixed phase is an exact match.

7.2 *Performance Parameters*

The above graph of IMEP shows bowl 4 is having a higher value of 16.30 bar with an increase of 1.05% from the base bowl at full load. At part load the bowl 1 and bowl 4 has showed the higher values with an increase of 2.08% from the bas bowl. At the full load IMEP for bowl 1 is less compared to the bowl 4. Therefore considering IMEP values, bowl 4 shows better result than all the other bowls

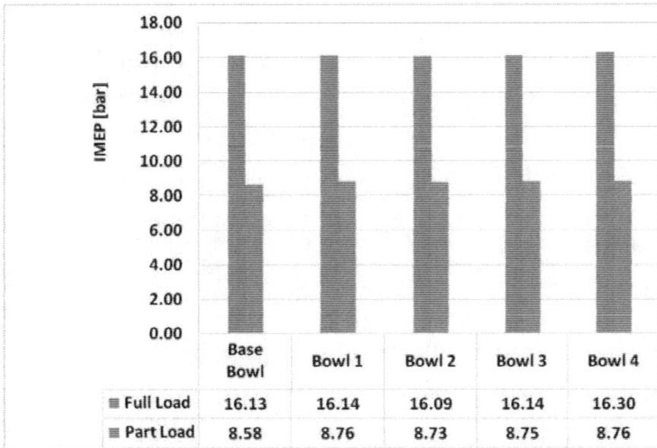

Figure 6. IMEP Data

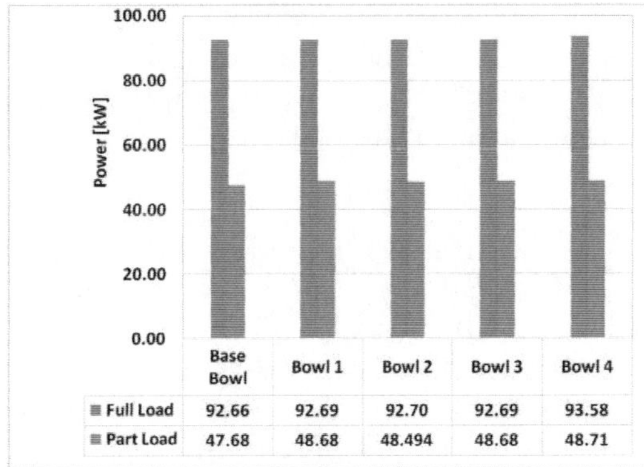

Figure 7. Power Data

The graph for rated power at full and part load for all set of bowls has been shown in figure 8. Bowl 4 shows the highest values for rated power for full load as well as part load. At full load there is an increase of 1.01% compared with the base bowl. While at part load bowl 4 has showed 2.16% increase in its rated power.

7.3 *Emission Parameters*

The soot and NO_x are plotted for the 4 bowls to analyze the emission trends. The comparisons are made with the base bowl geometry to find the bowl having least emissions.

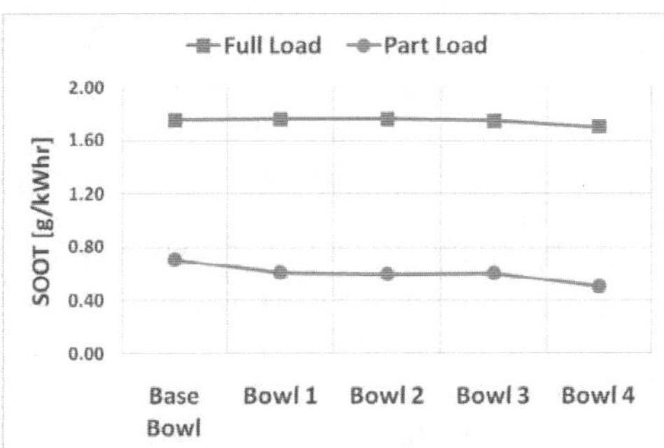

Figure 8. Soot Emissions

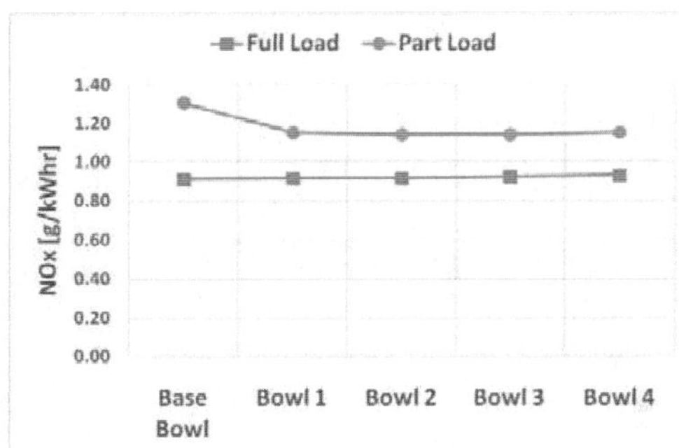

Figure 9. NO_x Emissions

Figure 8 shows the soot values for full load and part loads. It is observed that bowl 4 showing the least values for soot emissions among all set of bowls both at full load and part load. There is a decrease of 2.81% of soot at full load for the bowl 4 when compared with the base bowl. Also at part load, bowl 4 shows a reduction of 27.14% of soot when compared with the base bowl. Bowl 4 has shown a significant reduction in soot emission at part load conditions. Since the engine operates mostly on part load conditions, this reduction will have a positive control over the engine emissions.

8. Conclusions

The MATLAB code has been developed to generated bowl geometries upon user input conditions. The program could generate the bowl profiles within the same bowl volume. Different bowl geometries has been generated and 4 of them (Bowl1, Bowl2, Bowl3, Bowl 4) were selected based on the bowl volume and hitting plane for further analysis and compared with the base bowl geometry.

A significant change has been observed when the floor angle has changed from 138 degree to 148 degree. It had been a suggestion from the engine testing authority to increase the spray cone angle from 69 degree to 74 degree. The hitting plane target of 44% is achieved by increasing the spray cone angle. Bowl 4 showed a significant effect on performance and emissions while satisfying these conditions. On comparison with the base bowl geometry the bowl radius has been reduced by 3% and increase in floor angle has increased the reentrancy.

References

[1] Y. Zhu, H. Zhao, D. A. Melas and N. Ladommatos *Computational Study of the Effects of the Re-entrant Lip Shape and Torroidal Radii of Piston Bowl on a HSDI Diesel Engine's Performance and Emissions* 2004-01-0118
[2] Jongyoon Lee, Sangyul Lee, Jungho Kim, and Duksang Kim *Bowl Shape Design Optimization for Engine-Out PM Reduction in Heavy Duty Diesel Engine*, 2015-01-0789
[3] Lu Lin, Duan Shulin, Xiao Jin, Wu Jinxiang and Gao Xiaohong *Effects of Combustion Chamber Geometry on In-Cylinder Air Motion and Performance in DI Diesel Engine* 2000-01-0510
[4] Rahman M. Montajir, H. Tsunemoto and H. Ishitani *Fuel Spray Behavior in a Small DI Diesel Engine: Effect of Combustion Chamber Geometry* 2000-01-0946
[5] S. Pasupath Venkateswaran and G. Nagarajan *Effects of the Re-Entrant Bowl Geometry on a DI TurbochargedDiesel Engine Performance and Emissions—A CFD Approach* Journal *of Engineering for Gas Turbines and Power* December 2010, Vol. 132 / 122803
[6] Helgi Skuli Fridriksson, Martin Tuner, Oivind Andersson, and Bengt Sunden *Effect of Piston Bowl Shape and Swirl Ratio on Engine Heat Transfer in a Light-Duty Diesel Engine* 2014-01-1141
[7] Saravanan Subramanian, Balamurugan Rathinam, I J Isaac JoshuaRamesh,and Kandaswamy Annamalai *Piston Bowl Optimization for Single Cylinder Diesel Engine Using CFD* 2016-28-0107
[8] Rushikesh Dakhore ,Naresh G. Gandhi, Nitin Gokhale, Yogesh Aghav, M N Kumar Dattatray and B. Hulwan *Effect of Piston Cavity Geometry on Combustion, Emission and Performance of a Medium Duty DI Diesel Engine* 2015-26-0198
[9] Jaeman Lim and Kyoungdoug Min *The Effects of Spray Angle and Piston Bowl Shape on Diesel Engine Soot Emissions Using 3-D CFD Simulation* 2005-01-2117
[10] Vinod Karthik Rajamani and Avnish Dhongde *Parametric Analysis of Piston Bowl Geometry and Injection Nozzle Configuration using 3D CFD and DoE* 2012-01-0700
[11] Muzaffar Ali,Quazi Shakti Kumar Singh Escorts Limited and Mangeshkumar *Jadhao Effect of Piston Bowl Shape, Swirl Ratio and Spray Angle on Combustion and Emission in Off Road Diesel Engine* 2015-26-0142
[12] B.V.V.S.U. Prasad, C.S. Sharma, T.N.C. Anand and R.V. Ravikrishna *High swirl-inducing piston bowls in small diesel engines for emission reduction Applied Energy*, 88 (2011) 2355–2367
[13] Abdul Gafoor C.P, Rajesh Gupta *Numerical investigation of piston bowl geometry and swirl ratio on emission from diesel engines, Energy Conversion and Management* 101 (2015) 541–551
[14] Helgi Skuli Fridriksson, Martin Tuner, Oivind Andersson, and Bengt Sunden *Effect of Piston Bowl Shape and Swirl Ratio on Engine Heat Transfer in a Light-Duty Diesel Engine* 2014-01-1141
[15] Abdul Gafoor C.P and Rajesh Gupta *Numerical investigation of piston bowl geometry and swirl ratio on emission from diesel engines, Energy Conversion and Management* 101 (2015) 541–551
[16] F. Dos Santos and L. Le Moyne *Spray Atomization Models in Engine Applications, from Correlations to Direct Numerical Simulations* Vol. 66 (2011), No. 5, pp. 801-822 DOI: 10.2516/ogst/2011116
[17] Pundlik R. Ghodke A and Jiwak G. Suryawanshi *Investigation Of Diesel Engine For Low ExhaustEmissions With Different Combustion Chambers* Year 2015, Vol. 19, No. 6, pp. 2013-2024
[18] Luis Bravo and Chol-Bum *Kweon A Review on Liquid Spray Models for Diesel Engine Computational Analysis* ARL-TR-6932 May 2014
[19] Karaaslan S and Erman C *Numerical Investigation of Ethanol Fuel Blends in Engine Performance Characteristics by Using Diesel-Rk Software Journal of Engineering Sciences 2013.*

Numerical investigations on a proton exchange membrane fuel cell of active area 50 cm^2

Tino Joe Tenson[1], Rajesh Baby[2*]

[1]Graduate Student, [2*]Associate Professor,
Dept. of Mechanical Engg., St. Joseph's College of Engineering and Technology, Palai, Kerala, India
E-mail*: rbaby55@gmail.com

Abstract. The present study focusses on the results of a numerical work carried out on a proton exchange membrane fuel cell with an active area of the bipolar plate as 50 cm^2. The investigations are employed on various operating conditions such as different flow rates of hydrogen and oxygen, cell voltage and temperature. Assumptions made in the study are steady state, single phase, laminar, compressible ideal gasses, the by-product water in liquid phase, transfer of heat in the cell only through conduction mode and so on. The landing to channel ratio employed in the present study is 1:1. The results obtained from the numerical investigations are validated with the help of an experimental study on a 50 cm^2 serpentine channel of a bipolar plate in a proton exchange membrane fuel cell with same operating parameters. Results show that mass flow rates and temperature have a significant effect on the performance of the fuel cell.

Keywords: Proton exchange membrane fuel cells, active area, serpentine channel

1. Introduction

Fuel cells are direct energy conversion devices which convert the chemical energy of the fuel into electrical energy. The PEMFCs (Proton Exchange Membrane Fuel Cells) are considered promising power sources and command attraction now- a- days due to advantages such as low operating temperature (50-90°C), high efficiency (60-70%), low emissions and quick start up. Main application areas of PEMFCs are automobile, stationary and portable sector. A number of studies have been done on the numerical investigations on PEMFCs.

The numerical approach has become an effective tool to analyse the performance of the fuel cells. Cheddie et al. [1] do the three-dimensional modelling of a high temperature PEMFC. This model gives all transport and polarization phenomena. The model reveals that the oxygen depletion happens in the catalyst area under the ribs. Velocities of inlet gases and conductivity of membrane are numerically studied by Sezgin et al. [2]. They develop a three-dimensional, isothermal and steady state model. At sufficiently high flow rates for air, this model gives best results for mass transport region. At an inlet hydrogen gas velocity of 0.133 m/s and an inlet air velocity of 1.3 m/s, the fuel cell gives results closely matching with experimental data. Meng et al. [3] develop a three-dimensional numerical model to study current distribution and performance of the cell under a current density boundary condition. Their results show that long channel direction gives rise to a difference in the electronic phase potential in the bipolar plate.

The numerical model of both parallel and serpentine flow field is done by Iranzo et al. [4]. The modelling results are calibrated with the help of exchange current density. Water in the channels produces a blockage effect that is difficult to model with this numerical analysis. An automotive application-based modelling of PEMFC is carried out by Haddad et al. [5]. Model responds to both low and high dynamics of power demand and the results show that membrane humidity has a considerable effect on the cell's efficiency. Wong et al. [6] conduct a theoretical investigation of relative humidity control in PEMFC. Authors conclude that peak power of a PEMFC depends on the combination of anode and cathode humidification factors. Cooling flow field performance of a PEMFC is assessed by Asghari et al. [7]. They studied the thermal management of the cell. Results show that free convection from stack faces has a negligible role in heat removal, but the inlet and outlet manifolds of reactant gases have an influence on the temperature distribution within the bipolar plates. Yu and Jung [8] conduct thermal management strategy for a PEMFC system with a large active cell area. Automobile applications require PEMFCs with the large active area, which produces high temperature, so this model shows proper thermal management to handle the high temperature. A review of thermal management issues in a PEMFC stack is conducted by Kandlikar and Lu [9]. Authors review the basic heat transfer mechanisms at PEMFC at the component level.

Cozzolino et al. [10] analyse the theoretical and experimental investigations on thermal management of a PEMFC stack. Micro cogeneration applications have been investigated through experimental and numerical investigations. Results show that electrical efficiency decreases as the current intensity rises, due to higher ohmic losses and that thermal efficiency always increases from partial to full loads. Huang et al. [11] show the interaction between heat and water transport in PEMFCs. Authors also address the economic factor of the cell due to the use of platinum metal. Results show the influence of porosity gradient and enhancement of performance of the cell. Jang et al. [12] conduct both experimental and numerical investigations on spiral flow field channels. In this study, authors focus their attention on spiral channel design which produces secondary vortices, that can improve the heat and mass transfer characteristics and reduce the large pressure thereby enhancing the fuel cell performance. Due to the presence of secondary vortices in the spiral channel flow field design, the current density of spiral flow field channel is 11.9% higher than the conventional serpentine channel. Roshandel et al. [13], conduct a numerical study on a bio inspired pattern for fuel cells. Authors suggest a new bipolar design inspired from the existing biological fluid flow patterns in a plant leaf. From the results obtained, they find that bio inspired flow channel design shows 26% higher power density than the serpentine channel and 56% more than parallel flow design. But this power density shown by bio inspired flow channel design does not consider the liquid water formation, temperature variations and other minor losses. As a result, there will be a decrease in the power density of the bio-inspired flow channel when these factors are taken into account.

Damain, et al. [14] present a numerical analysis of channels with a design based on the leaf like flow field design. Numerical investigations are conducted at 9 cm^2 active area fuel cell. To find the leaf like arrangements of the flow field design, authors investigate thirty most common trees in Mexico region and 20 leaves from each tree are photographed. From these observations, they statistically identify two most repetitive configurations. They find that the increase in the inlet humidity causes an increase in the current density. The best fuel cell performance is obtained from flow field channel with two levels of sub-branches. They observe that increasing the number of sub branches gives more uniform gas distribution, hence it will enhance more uniform current density production.

The main contributions of the current work are:

1) Numerical investigations on the serpentine flow field (Landing to Channel ratio, L:C=1:1) of bipolar plates with area 50 cm^2.

2) Mesh sensitivity analysis and validation of numerical results with experimental results.

3) Effect of electrical performance of the fuel cell with respect to various operating conditions such as flow rates, cell voltage and temperature.

2. Numerical modelling

Numerical modelling involves three major steps. The first step is the creation of fuel cell geometry. Solidworks 2017 is used for this purpose. The second step is the creation of mesh, ICEM meshing module available in ANSYS Fluent 15.0 is used. The last step is the analysis part. A block diagram depicting this sequence is shown in figure 1.

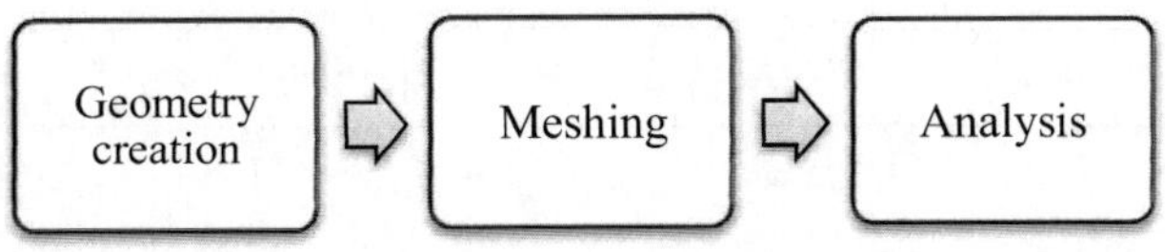

Figure 1. Flow diagram of modelling.

The assumptions made in the numerical modelling are: steady state, single phase, laminar flow, gases are compressible ideal gasses, the by-product water is in liquid phase and heat transfer inside the cell is only through conduction. The physical phenomena occurring in a fuel cell can be represented by solutions of conservation equations like mass, momentum, energy, species and current transport. In addition, equations that deal specifically with a phenomenon in a fuel cell may be used where applicable such as Darcy's equation for fluid flow in porous media, Fick's law of diffusion, Stefan-Maxwell equation for multispecies diffusion, Faraday's law, Butler-Volmer equation, Ohms law and so on.

2.1. Governing equations

The continuity equation is

$$\frac{\partial(\varepsilon\rho)}{\partial t} + \nabla\cdot(\varepsilon\rho u) = 0 \qquad (1)$$

For the two-phase mixture in the gas channel, the Maxwell-Stefan mass transport equation is used

$$F_i = \sum_{i\neq j}\zeta_{i,j}x_j(u_i - u_j) \qquad (2)$$

where, F_i is the driving force on i, at a given T and p, $F_i = -\frac{RT}{x_i}\frac{dx_i}{dz}$, $\zeta_{i,j}$ is the friction coefficient between i and j, x_j is mole fraction of j. u is velocity.

For the multiphase mixture flow in the GDL, Darcy's Law is used for momentum transport modelling

$$\frac{\partial(\rho u)}{\partial t} + \nabla\cdot(\rho uu) = -\nabla P + \nabla\cdot(\nabla\mu u) - \frac{\mu}{K}(\varepsilon u) \qquad (3)$$

The species conservation equation for the multiphase mixture is

$$\varepsilon\frac{\partial}{\partial t}(\rho C^\alpha) + \nabla\cdot(\gamma_\alpha \rho u C^\alpha) = \nabla\cdot(\varepsilon\rho D\nabla C^\alpha) + \nabla\cdot\left[\varepsilon\sum_k\left[\rho_k s_k D_k^\alpha(\nabla C_k^\alpha - \nabla C^\alpha\right]\right] - \nabla\cdot\left[\sum_k C_k^\alpha j_k\right] \qquad (4)$$

Boundary and initial conditions at the interface between the membrane and the cathode are:

$$u(x=0) = u_{in},\ v(x=0) = 0,\ \frac{\partial C^{H2}}{\partial x}(x=L) = C_{in}^{H2},\ C^{H2O}(x=L) = \frac{\rho_{g,sat}^{H2O} RH_{in}^{H2}}{\rho_g} \qquad (5)$$

The dimensions employed for this work are shown in table 1.

Table 1. Dimensions of the fuel cell model for 50 cm^2 active area

Component Name	Width (mm)	Height (mm)	Thickness (mm)
Current collector	100	100	10
Gas Diffusion Layer	100	100	0.3
Catalyst Layer	100	100	0.06
Membrane	100	100	0.15

Model is shown in figure 2.

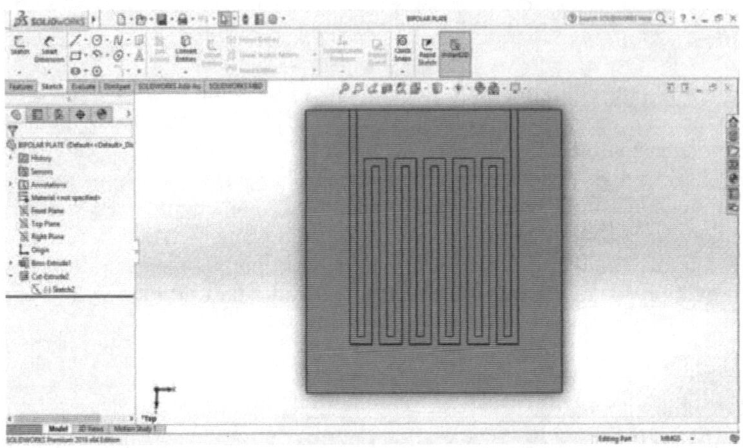

Figure 2. Bipolar plate, L:C ratio = 1:1 (50 cm^2 active area).

A Dell Optiplex PC with the i7 processor having 8 GB ram and graphics card of 4 GB is used for the numerical analysis. CPU time for 1 simulation is 8 hours.

3. Mesh sensitivity analysis

The present model is discretized into 10,22,547 elements, 17,18,652 elements and 20,51,432 elements and the results are evaluated and are shown in table 2.

Table 2. Current density at various no. of discretized elements

No. of discretized element	Current density(A/cm^2) @ 0.4 V
10,22,547	1.0121
17,18,652	1.0186
20,51,432	1.0188

From table 2 it is observed that the results predicted with the numerical code are grid independent after 17,18,652 elements.

4. Validation

The numerical investigations of the PEMFC with an active area of the bipolar plate 50 cm^2 are carried out to study how close the predicted results match with experimental work. Experimental work conducted by the same research group is used for validation purpose [15]. Figure 3 shows the comparison of power density against cell voltage for experimental and numerical work. The numerical results of power density obtained for 50 cm^2 PEMFC is quite close to the experimental work as shown in figure 3. The percentage deviation between experimental and numerical results are shown in table 3. The deviation between the predicted results and that of the experimental work for the power density is between 4.70 -14.20 %. It is found that the numerical results achieved by CFD technique resemble the experimental work. Thus, the CFD code ANSYS Fluent 15.0 can be extended further with confidence to study the fuel cell characteristics at other operating conditions.

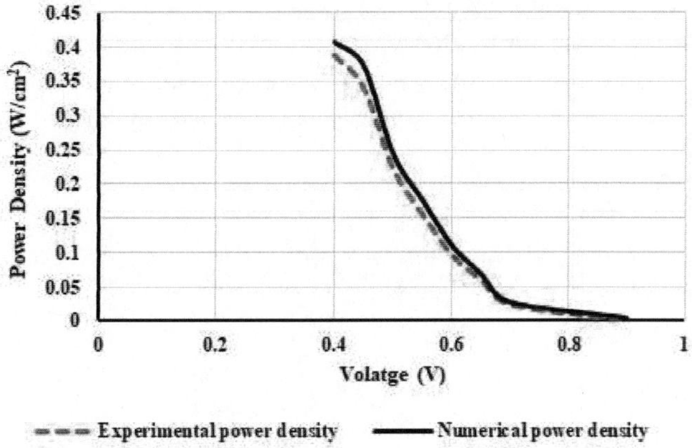

Figure 3. Validation of power density of 50 cm^2 PEMFC model

Moreover, the cost and time required for machining a scaled up PEMFC for experimental studies are quite high, and hence numerical study with CFD technique gives an opportunity to optimize qualitatively the flow channel design of scaled up PEMFCs.

Table 3. Percentage deviation between experimental and numerical work

Voltage (V)	Power Density (W/cm^2) Experimental	Power Density (W/cm^2) Numerical	Percentage deviation (%)
0.40	0.3891	0.4074	4.70
0.45	0.3416	0.3729	9.16
0.50	0.2243	0.2456	9.49
0.55	0.1557	0.1757	12.85
0.60	0.0966	0.1100	13.91
0.65	0.0608	0.0691	13.67
0.70	0.0246	0.0281	14.20

5. Results and Discussion

Results obtained from the work are shown below. The various input parameters used in the present investigations are shown in table 4.

From figure 4 it is clear that maximum power output obtained is between the voltage range of 0.4 and 0.6 V. From numerical investigation it is observed that power output has a significant influence on the increase in flow rates of hydrogen and oxygen. Maximum power is obtained when the hydrogen flows at 600 and oxygen at 1200 SCCM (Standard Cubic Centimetre per Minute) respectively and its value is 23.79 W at 0.5 V. Meanwhile the maximum power obtained at 300 and 450 SCCM of hydrogen and oxygen respectively is 22.57 W at 0.5 V. Increased mass flow rate improves the mass transport between the anode and cathode side which results in the enhancement of reaction rate and hence both reactions at anode and cathode side increase and give better current

IOP Conf. Series: Materials Science and Engineering **396** (2018) 012056 doi:10.1088/1757-899X/396/1/012056

density. But an increase in oxygen supply beyond 1200 SCCM will significantly degrade the performance of the cell because of the formation of water flooding.

Table 4. Input parameters used in the present study

Varying Parameter	Values
Hydrogen flow rate	300, 600 SCCM
Oxygen flow rate	450, 1200 SCCM
Temperature variation	50, 60, 65, 70°C
Voltage variation	0.40, 0.42, 0.45, 0.48, 0.50, 0.52, 0.55, 0.57, 0.59, 0.60, 0.63, 0.65, 0.68, 0.70, 0.72, 0.75, 0.78, 0.80, 0.83, 0.85, 0.88, 0.90, 0.92, 0.95 V

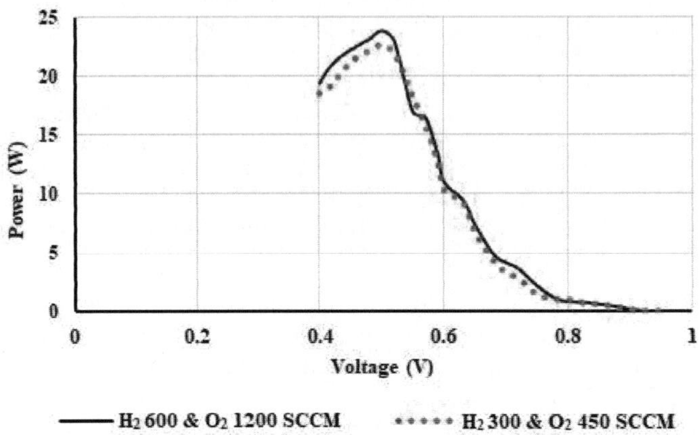

Figure 4. Variation of power at various flow rates of hydrogen and oxygen

After finding the better flow rates of fuel and oxidant, next results show the best operating temperature of the cell under 600 and 1200 SCCM of hydrogen and oxygen. Following are the four temperatures taken into account: 50, 60, 65, 70°C. Results are shown in figure 5. From the figure 5, it is clear that maximum power is obtained at 65°C and its value is 24.49 W at 0.5 V. Further increase in temperature reduces the performance of the cell. It is observed that 70°C gives a performance level lower than at 50°C and 60°C. Two temperature regions which gives better results are 60 and 65°C.

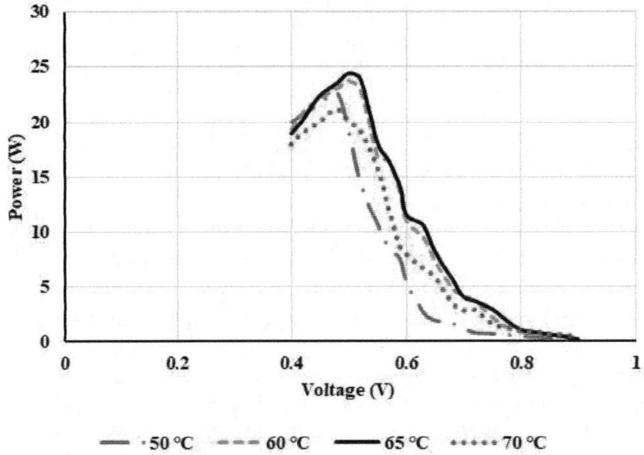

Figure 5. Variation of power at various temperatures

Temperature is acting as a catalyst in the reaction, since further increase in the temperature value also improves the rate of reaction which in turn improves the product formation rate. Since water is one of the by-products of this reaction between anode and cathode side, water formation rate will increase with increase in temperature value. So, it is observed that temperatures above 65°C will improve the rate of reaction and hence the water accumulation rate will increase. This water accumulation degrades the performance of the cell.

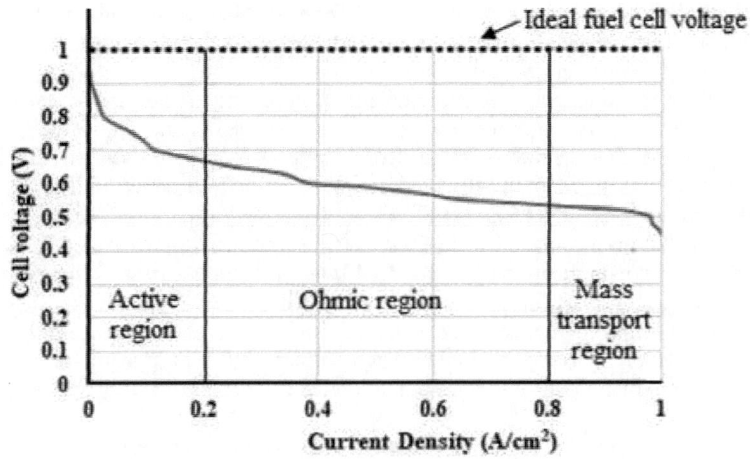

Figure 6. Fuel cell performance

Figure 6 shows the performance of the cell at 600 and 1200 SCCM of hydrogen and oxygen respectively with temperature at 65°C. Fuel cell operates between 0.2 and 0.8 A/cm^2 range which gives voltage in the range of 0.52 to 0.67 V. Regions of operation of a fuel cell is shown in figure 6. In activation region current density becomes very low which is due to the presence of activation losses. Activation losses arise because of the losses due to electrochemical reaction. Ohmic region is the region in which fuel cell operates. Reduction in voltage in this region is due to ohmic losses because of ionic and electronic conduction. Last is the mass transport region, in which further reduction in voltage occurs due to the losses arising during mass transport and these losses are known as concentration losses.

6. Summary and conclusions

In the present work, numerical studies are conducted on a 50 cm^2 active area PEMFC. The conclusions made from this investigation are as follows:

- Various flow rates of hydrogen and oxygen are used. The flow rates 600 SCCM of hydrogen and 1200 SCCM of oxygen give the best results.
- Numerical investigations establish the influence of temperature and its significance in the performance of the cell. 65°C gives the best performance, further increase in temperature reduces the performance of the cell. This is because of the water flooding the cell.
- Performance of a fuel cell can be assessed by current density – voltage curve, i.e. the polarisation curve of the fuel cell. Activation region, ohmic region and mass transport region are the three operating regions of the fuel cell. Among these, the major operating region of the fuel cell is the ohmic region. Figure 6 reveals the ohmic region is between 0.2 and 0.8 A/cm^2 with cell voltage between 0.52 and 0.67 V.

The key factors affecting the performance of PEMFC are flow rates of hydrogen and oxygen, temperature and cell voltage. This study can be extended further to predict the performance of the cell when it is a part of a more complex energy conversion system.

Acknowledgments
The authors are grateful for the financial support by APJ Abdul Kalam Technological University through Research Seed Money scheme, Centre for Engineering Research and Development (Project no. KTU/RESEARCH 2/2743/2017) and St. Joseph's college of Engineering and Technology, Palai.

References
[1] Denver F. Cheddie, Norman D.H Munroe 2006 Three dimensional modeling of high temperature PEM fuel cells *Journal of Power Sources* **160** pp 215–23
[2] Berna Sezgin, Dilara Gulcin Caglayan, Yilser Devrim, Thomas Steenberg, Inci Eroglu 2016 Modelling and sensitivity analysis of high temperature PEM fuel cells by using Comsol Multiphysics *International Journal of Hydrogen Energy* pp 1–9
[3] H. Meng and C Y Wang 2004 Multidimensional modelling of polymer electrolyte fuel cells under a current density boundary condition pp 1–8
[4] Alfredo Iranzo, Miguel Munoz, Felipe Rosa, Javer Pino 2010 Numerical model for the performance prediction of a PEM fuel cell. Model results and experimental validation *International Journal of Hydrogen Energy* **35** pp 11533–50
[5] Ahmad Haddad, Marc Mannah, Hasan Bazzi 2015 Nonlinear time variant model of the PEM type fuel cell for automotive applications *Simulation Modelling Practice and Theory* **51** pp 31–44
[6] K.H. Wong, K.H. Loo, Y.M. Lai, Siew-Chong Tan, Chi K. Tse 2011 A theoretical study of inlet relative humidity control in PEM fuel cell *International Journal of Hydrogen Energy* **36** pp 11871–85
[7] Saeed Asghari, Hooman Akhgar, Bagher Faghih Imani 2011 Design of thermal management subsystem for a 5kW polymer electrolyte membrane fuel cell system *Journal of Power Sources* **196** pp 3141–48
[8] Sangseok Yu and Dohoy Jung 2008 Thermal management strategy for a proton exchange membrane fuel cell system with a large active cell area *Renewable Energy* **33** pp 2540–48
[9] Satish G. Kandlikar and Zijie Lu 2009 Thermal management issues in PEMFC stack – A brief review of current status *Applied Thermal Engineering* **29** pp 1276–80
[10] R. Cozzolino, S.P. Cicconardi, E. Galloni, M. Minutillo, A. Perna 2011 Theorteical and experimental investigations on thermal management of a PEMFC stack *International Journal of Hydrogen Energy* **36** pp 8030–37
[11] Yu-Xian Huang, Chin-Hsiang Cheng, Xiao-Dong Wang, Jiin-Yuh Jang 2010 Effects of porosity gradient in gas diffusion layers on performance of proton exchange membrane fuel cells *Energy* **35** pp 4786–94
[12] Jiin- Yuh Jang, Chin-Hsiang Cheng, Wang-Ting Liao, Yu-Xian Huang, Ying-Chi Tsai 2012 Experimental and numerical study of proton exchange membrane fuel cell with spiral flow channels *Applied Energy* **99** pp 67–79
[13] Roshandel R, Arabi F, Moghaddam GK 2012 Simulation of an innovative flow-field design based on a bio inspired pattern for PEM fuel cells *Renewable Energy* **41** pp 86–95
[14] Cesar E. Damian-Ascencio, Adriana Saldana-Robles, Abel Hernandez-Guerrero, Sergio Cano-Andrade 2017 Numerical modeling of a proton exchange membrane fuel cell with tree-like flow field channels based on an entropy generation analysis *Energy* **133** pp 306–16
[15] Antony Alex, Austin George, Bonny Mathew, Bijo Thomas, Rajesh Baby 2018 Experimental investigations on 49 cm^2 serpentine bipolar plate of a proton exchange membrane fuel cell *Proc. of 11th Int. conf. on thermal engineering: theory and applications* to be held at Doha, Qatar during February 25–28, 2018

IOP Conf. Series: Materials Science and Engineering **396** (2018) 012057 doi:10.1088/1757-899X/396/1/012057

Estimation of characteristic length of a horizontal cylindrical annulus for heat transfer through conduction using Rayleigh number

Ramesh Krishnan S [1] and Narayanan NamboothiriVN[2]
[1]Assistant Professor in Mechanical Engineering,Government Rajiv Gandhi Institute of Technology,Kottayam, Kerala, India 686501rameshs.krishnan@gmail.com
[2]Associate Professor in Mechanical Engineering, School of Engineering, Cochin University of Science and Technology, Kochi, Kerala, India 682022
nnamboothirivn@gmail.com

Abstract: Heat transfer from or to a horizontal cylinder has got substantial significance in many engineering applications, such as in heating and cooling devices, steam pipelines, heat exchangers etc. Coaxial cylinder technique may be used for the analysis of heat transferred by conduction through a concentric cylindrical annulus where the two surfaces are maintained at a constant temperature. The annular space is filled with the fluid sample whose thermal conductivity needs to be measured. If a heat flux is imposed on the sample from the inner to the outer cylinder, it propagates through the fluid in a radial direction under steady state condition. Insulating the terminal ends ensures heat flow primarily in radial direction and prevents heat loss at the ends. The focus is on the use of dimensionless numbers used in multi-domain coupled conduction-convection flow problems so as to find out the optimum gap thickness in between the two concentric cylinders at which convection will just onset and upto which conduction is the only heat transfer mode. The values of these numbers can be used as guidelines in finding the optimum value from which an experimental set up for determination of thermal conductivity of various fluids based on co-axial cylinder technique can be designed. it can be later extended to analyse thermal conductivity of nanofluids as well.

1. Introduction:

The mode of conductionheat transfer involves energy transfer within a mediumwith no motion of the medium as a whole. Conduction takes place as a result of the temperature gradient existing in a solid or stationary fluid medium. The direction of conductive heat flow is in the direction of decreasing temperature because a higher temperature is associated with a higher molecular energy and molecular movement. Energy transfer from the more active to the less active molecules occurs when the neighbouring molecules collide with each other. If the heat transfer coefficient is to be maximised, the dominating parameter is the thermal conductivity of the fluid.

In the coaxial cylinder method, a very thin layer of homogeneous fluid having a uniform thermal conductivity which is to be determined, is enclosed between two coaxial cylinders of infinite length. If r_i and r_o are respectively the external radius of the inner cylinder and internal radius of the outer cylinder which are at steady state temperatures of T_i and T_o, then the heat transfer through conduction Q in cylindrical annulus for the stagnant fluid is given by $Q = 2\pi Lk\,(Ti - To) / \ln\,[ro/ri]$. From this equation, thermal conductivity of the fluid can be determined.

2. Mechanism of natural convection in enclosures:

When a horizontal layer of fluid is heatedfrom below, the difference in fluid density between the different layers of fluid increases as we move away from the heating surface. Although it may appear that the fluid would be unstable and that convective motion would set in immediately, the fluid will in fact remain in a quiescent state for a range of values of temperature difference between the bottom and top surfaces. Only when the temperature difference surpasses a critical value which is dependent on the layer thickness and also on the properties of the fluid, the motionless quiescent state becomes unstable and thermal convection will set in.[1,2]

The physical model of natural convection in the annular space between concentric cylinders is as shown in figure 1. The fluid movement is under the combined effects of gravity g and the temperature difference $T_i - T_o$ between the two surfaces.

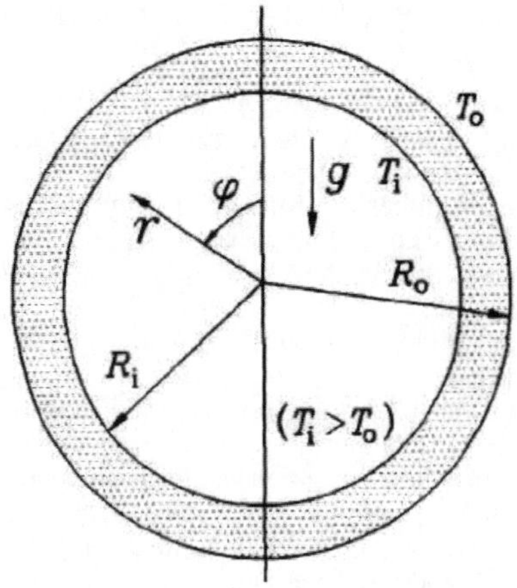

Figure 1. The physical model of natural convection in annular space between concentric cylinders [3]

3. Rayleigh's model:

In this model, a thin layer of fluid is restrained completely between twoinfinite flat plates so that there is no air gap. A thin layer means that length of fluid layer is very large when compared to the thickness. The fluid is to be heated in such a way that the temperature gradient in the fluid remains uniform (spatial invariance) and steady (temporal invariance) [4]. Under such a condition, the graph of temperature Vs height is a straight line when the fluid is at rest (figure 2).

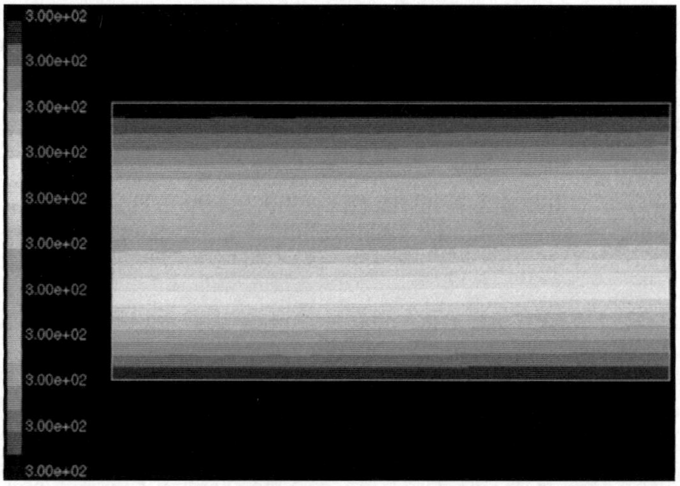

Figure 2. Linear evolution of temperature when convection is absent

The variation of temperature along a radius will change with length as soon as the effect of convection sets in. The larger the variation more is the convection and the smaller is the conduction. Higher the departure from the linear temperature profile more is the fluid layer getting susceptible to instability [2]. With a further increase in the temperature of the bottom plane, the complexity of the structure will become more [4]; the turbulent flow will become disordered as shown in figure 3.

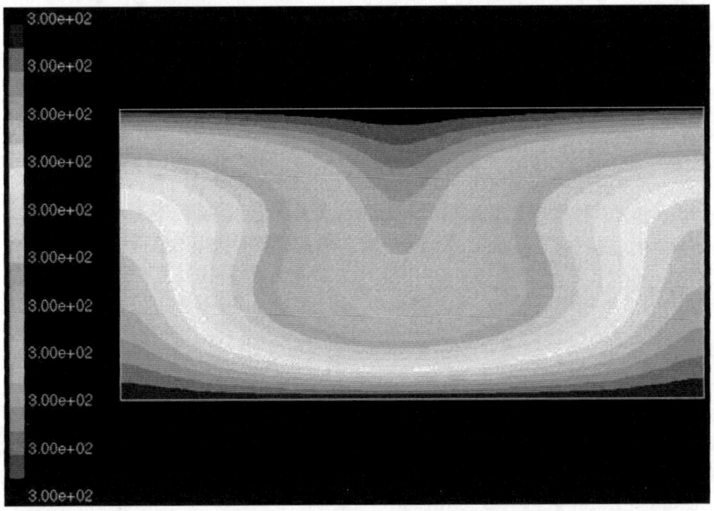

Figure 3. Nonlinear evolution of temperature with increased convection

The statement of Rayleigh Bénard problem is based on the hydrodynamic equations in Boussinesq approximation. The fluid is assumed to be incompressible i.e., density ρ is independent of pressure and is linearly dependent on temperature T i.e., $\rho - \rho_0 = - \rho_0 \alpha (T - T_0)$ where ρ_0 is the density at temperature $T = T_0$. Also, the temperature distribution is a linear function of vertical coordinate z (height) i.e., $T_s = T_i - \lambda z$, where Ti is a constant [5].

This approximation is valid for temperature difference ratio defined as the difference in temperatures of the two surfaces to the outer surface temperature $(T_i - T_o)/ T_o$ upto 0.2 to predict heat transfer rates with reasonable accuracy [6].

4. Governing equations:

The flow and heat transfer in a Boussinesq fluid are governed bycontinuity equation, momentum equation and energy equation.

4.1. Continuity Equation:

$$\partial\rho/\partial t + \nabla.(\rho V) = 0 \quad (1)$$

In this case, it can be written as

$$\nabla.V = 0 \quad (2)$$

4.2. Momentum Equation:

The forces which are acting on a moving fluid element are the bodyforces which act directly on the volumetric mass of the fluid element (such as the gravitational force), and the surface forces which act directly on the surface of fluid element. If gravity alone is the body force acting on the fluid element, the momentum equation may be written as

$$\rho(dV/dt) = - \nabla p + \nabla.\tau ij + \rho g \quad (3)$$

where, V is the velocity vector, t is the time, ρ is the fluid density, p is the static pressure, g is the acceleration due to gravity. If ρ_0 is the density of the surrounding fluid (reference density), the density difference $\Delta\rho = \rho - \rho_0$ causes the buoyancy force which is acting in the direction of body force gravity. The statements of Boussinesq approximation are valid only when the condition $\Delta\rho/\rho_0 \ll 1$ is satisfied. Since the density variation is due to change in temperature alone,

$$\Delta\rho = \rho - \rho 0 = -\rho 0\beta(T - T0) \quad (4)$$

IOP Conf. Series: Materials Science and Engineering **396** (2018) 012057 doi:10.1088/1757-899X/396/1/012057

where β is thermal expansion coefficient.
Since p is the actual static pressure in any given point and p_h is the hydrostatic pressure, the pressure difference (p_m = p−p_h) is the change in pressure that arises through the fluid motion. Hence the momentum equation takes the form

$$dv/dt = (-1/\rho 0) \bigtriangledown pm + (1/\rho 0) \bigtriangledown .\tau ij - g\beta(T - T0) \quad (5)$$

4.3. Energy Equation:

$$\rho Cp[(\partial T/\partial t) + v. \bigtriangledown T] = k \bigtriangledown^2 T \quad (6)$$

whereC_p is the specific heat capacity, k is the thermal conductivity, and T is the temperature. The characteristic length of the system d greatly impacts the Rayleigh-Be´nard cell structure and pattern.The heat transfer within the fluid can be expressed in terms of the dimensionless numbers, Nusselt number and Rayleigh number [7,8,9,10].

5. Dimensionless numbers:

Grashoff's number which isthe ratio of buoyancy force to viscous force in fluids helps to predict whether natural convection is dominant. A very low value of Prandtl number (Pr<<1), signifies that rate of thermal diffusion is dominant than therate ofmomentum diffusion. This means the fluid will move with diffusion of thermal energy more rapidly than its flow and conduction is dominant. 0.24 is a critical value of Prandtl number for the annular geometry. For Pr ~ 0, there is a uniform distribution of the like-rotating cells in the upper and the lower parts of annulus. The instability of conduction regime is first formed in the region near φ = π/2 [9]. A unit value for Nusselt number indicates that the heat transfer is only through conduction, but the exchange is both through conduction and convection if Nu >1. The steady convection rolls begin to appear at this stage.

6.Significance of Rayleigh number:

The Rayleigh number is defined as the product of Grashof number andPrandtl number i.e., Ra = Gr Pr. The convective flow will onset if buoyancy force, which results from temperature gradient and density gradient, exceeds the dissipative forces of viscous drag and heat diffusion. To understand better the influence of these forces on the onset of convection, a non-dimensional number Ra, called the Rayleigh number is used, which is proportional to buoyant force divided by the product of the viscous drag and the rate of heat diffusion.

$$Ra = g\beta\Delta T\, d^3 / \nu\alpha$$

where g is acceleration due to gravity, β is coefficient of thermal (volumetric) expansion of the fluid, T is temperature difference between top and bottom planes, d is the distance between the two planes or fluid thickness (characteristic length), ν is kinematic viscosity and α is thermal diffusivity of the fluid.

At small values of Ra (Ra ≤ 1000), the Nusselt number Nu is equal to one and there is little increase in heat transfer over that due to conduction alone. However, as Ra increases, several flow regimes have been found to occur, resulting in a significant increase in the Nusselt number.

Sparrow et al who investigated to find the effect of boundary conditions on the thermal instability of a horizontal fluid layer obtained the critical Rayleigh number corresponding to linear initial temperature profile which is presented graphically in figure 4 when the lower surface is at a fixed temperature. The figure has two curves: the dashed curve corresponds to a free upper surface and is referred to the left-hand ordinate scale; the solid curve corresponds to a rigid upper surface and is referred to the right-hand ordinate scale.

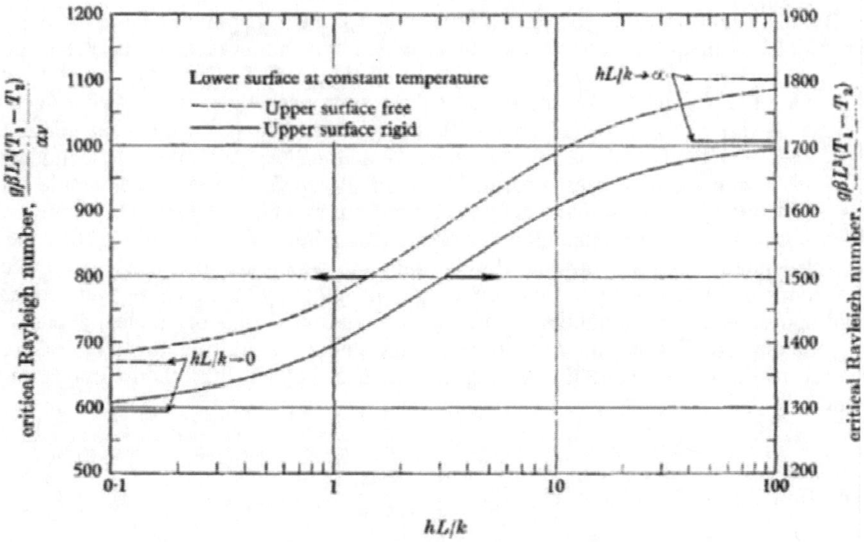

Figure 4. The Rayleigh numbers marking the onset of instability when the lower surface is at a fixed temperature [2]

The horizontal lines adjacent to the extremities of the curves provide the Rayleigh numbers for the limiting cases of hL/k → 0 and hL/k → ∞. The cases designated by hL/k → 0 and hL/k → ∞ respectively correspond to fixed heat flux and to fixed temperature at the upper surface.

Below the critical Rayleigh number, the Nusselt number is one. Once the critical Rayleigh number is exceeded, it grows. In this case Nusselt number may be defined as Nu = hd/k where h is heat transfer coefficient and d is characteristic length of the system.

Hence the mode of heat transfer can be ascertained by calculating the Rayleigh Number and checking whether the critical value has been exceeded or not. When the Rayleigh number is below the critical value for that fluid, heat transfer is primary in the form of conduction; when it exceeds the critical value, heat transfer is primarily in the form of convection. Radiation is not a significant heat transfer mechanism if a material with low emissivity is chosen.

For the case of prescribed temperature at both boundaries and a rigid upper surface, Pellew and Southwell find a critical Rayleigh number of 1707.8. In the Bénard configuration (fixed top surface), at the critical Rayleigh number of 1708, the instability sets in and convection cells appear [2].

7. Conclusion and problem formulation:

This study has examined the problem of heat transfer through a concentriccylindrical annulus and the procedure for determination of thickness of air gap (characteristic length) in which conduction is the heat transfer mode and convection will just set in. The findings of this study may be summarized as given below:

1. The fluid in the annular gap is stable for a range of values of temperature differences between the two surfaces. Convection does not set in as soon as a temperature difference is developed. Only when the temperature difference exceeds a critical value which in turn depends on the layer thickness and the fluid properties, the quiescent state breaks down and thermal convection sets in.
2. The Boussinesq approximation is valid for a temperature difference ratio of upto 0.2.
3. The values of Grashoff number, Prandtl number, Nusselt Number and Rayleigh Number may be used for determining the characteristic length. For the annular geometry, 0.24 is the critical value of Prandtl number at which convective instability sets in. When Nusselt number is unity, heat transfer is only due to conduction. If the temperatures of both boundaries are prescribed and the upper surface is

rigid, the critical Rayleigh number at which convection cells will appear and instability sets in is 1708.

By analysing the values of the different dimensionless parameters as mentioned above, the optimum thickness of the cylindrical annulus at which convection just onsets or the thickness upto which conduction is the heat transfer mode can be determined. The experimental set up whose dimensions are to be arrived at on the basis of the analysis can be used for finding out the thermal conductivity of various fluids including nanofluids. For this, coaxial cylinder technique can be used which is a steady-state method used for the measurement of heat exchanged by conduction between two concentric cylindrical surfaces separated by a small gap filled with fluid sample, each of these surfaces maintained at a constant temperature.The important prerequisites to apply the coaxial cylindrical technique is that the annular liquid layer must be sufficiently thin so that the convection currents are negligible and that the heat flow through the sample is more nearly one-dimensional.If a heat flux is imposed on the sample from the inner to the outer cylinder, it propagates radially through the liquid under steady state condition.Insulation at the terminal ends of the device ensures heat flow primarily in radial direction and heat losses at the ends. Measuring the temperature difference between the two cylinders using thermocouples and also the heat flux, thermal conductivity can be determined.

References:

[1] Narasimhan A., (1999), Rayleigh-Benard Convection: Physics of a Widespread Phenomenon, *Resonance: Journal of Science Educ*ation, 4(6),pp 82-90.
[2] E. M. Sparrow, R. J . Goldstein and V. K .Jonsson, Thermal instability in a horizontal fluid layer: effect of boundary conditions and non-linear temperature profile, *Journal of Fluid Mechanics*, Volume 18, Issue 4 April 1964 ,pp 513-28.
[3] Joo-SikYoo, Natural convection in a narrow horizontal cylindrical annulus: Pr≤ 0.3, *International Journal of Heat and Mass Transfer* 41 (1998),pp 3055-73.
[4] Amir Faghri, Yuwen Zhang and John R. Howell, *Advanced Heat and Mass Transfer*, Global Digital Press 2010, pp 555-71.
[5] *Rayleigh-Bénard Convection: Structures and Dynamics*- Alexander V. Getling, World ScientificPublishing Company Pvt. Ltd., Singapore.
[6] D. N. Mahony, R. Kumar and E. H. Bishop, Numerical Investigation of Variable Property Effects on Laminar Natural Convection of Gases between Two Horizontal Isothermal Concentric Cylinders, *Journal of Heat Transfer* November 1986, Vol. 108,pp 783-89.
[7] Mark P. Dyko and KambizVafai, Three-dimensional natural convective states in a narrow-gap horizontal annulus, *Journal of Fluid Mechanics* (2001), vol. 445, pp 1-36.
[8] Gregory J. Sheard and Martin P. King, Horizontal convection: Effect of aspect ratio on Rayleigh number scaling and stability, *Applied Mathematical Modelling*, Volume 35, Issue 4, April 2011,pp 1647-55.
[9] A.Bazylak, N. Djilali, and D. Sinton, Natural Convection in an Enclosure with Distributed Heat Sources, *Numerical Heat Transfer*, Taylor & Francis Group Part A,2006pp 655–67
[10] J.F. Wendt (ed.), Governing Equations of Fluid Dynamics, *Computational Fluid Dynamics*, 3rd ed., J.D. Anderson, Springer-Verlag Berlin Heidelberg 2009.

IOP Conf. Series: Materials Science and Engineering **396** (2018) 012058 doi:10.1088/1757-899X/396/1/012058

Enhancement of thermal performance of a loop heat pipe using alumina – water nanofluid : An experimental investigation

Akshay J[1], Aswin J Joseph[1], Jobin Jose[2] and Rajesh Baby[3*]

[1]Under Graduate Students, [2]Graduate Student, [3]Associate Professor

Department of Mechanical Engineering, St. Joseph's College of Engineering and Technology, Palai, Kerala, India.

[*]E – mail : rbaby55@gmail.com

Abstract : The thermal performance of a loop heat pipe (LHP) with alumina – water nanofluid is experimentally investigated. The experiments are carried out in a LHP, in which the setup consists of condenser, evaporator, two transport lines (vapour and liquid lines), rotameter, power supply, thermocouples. The experiments are conducted for heat input ranging from 30 – 310 W using deionised water (baseline) and alumina – water nanofluid. Alumina nanoparticles with 2% mass concentration and water are used as working fluids. The experimental results indicate that the nanofluid can enhance the thermal performance of loop heat pipe and lower the evaporator temperature by 12 % from the baseline case. The experimental results prove the capability of loop heat pipe with alumina – water nanofluid for various applications.

Key words : Loop heat pipe, nanofluid, thermal performance

1. Introduction

The loop heat pipe (LHP) is a device that operates on a multi phase fluid-flow cycle, maintained by a capillary medium in the evaporator. It is a heat transfer device which can transport a large amount of heat over a long distance with negligible temperature difference. In steady state operation these heat pipes have high thermal conductance. Thus LHPs are also known as super thermal conductors. Their advantages include electricity free operation, reliability and ability to work even with negligible temperature difference and transfer heat over long distances with small pressure drop [1,2].

The performance of an LHP is greatly affected by characteristics of working fluid in the heat pipe. Gunnasegaran *et al.* [3,4] conducted experimental and numerical investigations on LHPs with alumina and silica as nanofluids. It was found that the thermal resistance decreases when using silica-water nanofluid. The LHP charged with alumina-water nanofluid yielded lower wall temperature difference between evaporator and condenser and steady state is attained faster. Zhenping *et al.* [5] experimentally investigated the influence of a nanofluid on the heat transfer characteristics of a miniature loop heat pipe. It was concluded that performance improvement of the LHP using the nanofluid resulted in the reduction of the contact angle, the enhancement of boiling heat transfer and a deposited nanoparticle coating on the boiling surface. The thermal performance of a miniature loop heat pipe was investigated with graphene nanofluid by Trijo *et al.* [6]. The results showed an increase in the thermal conductivity and thermal efficiency of the device. The thermal performance of a loop heat pipe had been experimentally investigated by many researchers. Pauken *et al.* [7] experimentally studied the thermal performance of LHP with two working fluids i.e., ammonia and propylene. The constant conductance performance of the LHP was found to be 170 W/K with ammonia and 44 W/K for propylene. The thermal performance of an LHP with acetone as working fluid was experimentally investigated by Riehl *et al.* [8]. It was concluded that the system could present reliable performance in all situations with power level as low as 1 W. Also higher operating temperatures could be obtained when evaporator is above the condenser. Zhao *et al.* [9] also studied the thermal performance of a double compensation chamber LHP for anti icing system in aircrafts. Results showed an increase in surface temperature of the aircraft wing with increase of heat load on the evaporator.

The application of nanofluid in LHP is a challenging area of research, which is still in its initial stages. Also only a few researchers have considered nanofluids as the working fluid in LHP. The present study deals with the experimental investigations of LHP with alumina nanoparticles. Heat input ranges from 30 to 310 W in steps of 40 W. The experiments are conducted in an LHP which is of new design. The thermal performance of the LHP using nanofluid is compared with that of deionised water (baseline case).

2. Nanofluid preparation

In the present study, the working fluids used are alumina – water nanofluid and deionised water. The nanofluid is prepared by using a two-step method. Deionised water is taken as the base fluid for the preparation of nanofluid. The alumina nanoparticles having an average size of 13 nm and density 3.95 g/cm^3 are used to prepare the nanofluid with 2 % mass concentration . The particle mass concentration of the nanofluid is calculated using the equation (1) [10]

$$\text{Percentage mass concentration} = W_n/W_{bf} * 100\ \% \quad (1)$$

Where,

W_n = Amount of nanofluid in gram

W_{bf} = Amount of base fluid in gram

3. Experimental setup

The LHP has six parts: evaporator, compensation chamber, wick, condenser, liquid and vapour line. All parts of the heat pipe are made with copper. The figure 1 shows the schematic diagram of LHP indicating the thermocouple positions used in the present study. Figure 2 shows the photograph of compensation chamber and the wick which is placed inside the evaporator. The detailed specifications of the loop heat pipe are given in the table 1.

The evaporator area is same as that of the area of heater top surface. In order to reduce contact resistance between evaporator and heater, a flat evaporator is employed. The wick with a porosity of 63 % is used. Wick is used to produce the capillary force which is required to overcome pressure losses in the LHP. The compensation chamber acts as the reservoir of working fluid. After placing the wick inside the evaporator the compensation chamber and evaporator are brazed together.

The schematic diagram of experimental setup LHP is shown in figure 3. It consists of LHP, thermocouples, rotameter, pump, chilling unit, voltmeter, ammeter, data logger, heater and computer. The heater block consists of a copper block fitted with two cartridge heaters. Arrangement of heaters are made such that the evaporator surface is maintained at uniform temperature. The heaters are connected to a dimmerstat to give different heat inputs which are measured using voltmeter and ammeter.

A thick layer of glass wool insulation is given at the heater so that heat loss can be minimised. The chilling unit is used to maintain the cooling water at constant temperature. Rotameter and pump are used to measure and maintain constant cooling water supply to condenser. Cooling water enters the condenser at bottom left corner and leaves at top right corner. The uncertainty in measurement of temperature, voltage and current are $\pm$ 0.2 ^{o}C, $\pm$ 0.1 V and $\pm$ 0.01 A respectively. Fifteen calibrated T-Type thermocouples are used to measure temperatures at different locations of the heat pipe assembly. The thermocouples are fixed to the outer surface of heat pipe.

Table 1. Specification of LHP

Specification	Dimension/material
Evaporator	
Dimension (mm)	L45 × W45 × H22
Compensation chamber	
Dimension (mm)	L45 × W38 × H22
Material	Copper
Condensor	
Dimension (mm)	L315 × W96 × H60
Material	Acrylic
Vapour line	
Outlet diameter (mm)	13
Inlet diameter (mm)	11
Length (mm)	490
Liquid line	
Outlet diameter (mm)	13
Inlet diameter (mm)	11
Length (mm)	490
Pump	
Performance (litres/hour)	40

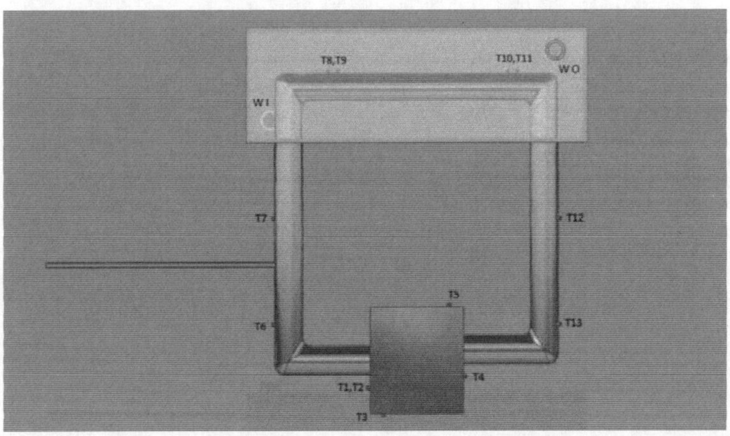

Figure 1. Schematic diagram of the loop heat pipe indicating the thermocouple positions

IOP Conf. Series: Materials Science and Engineering **396** (2018) 012058 doi:10.1088/1757-899X/396/1/012058

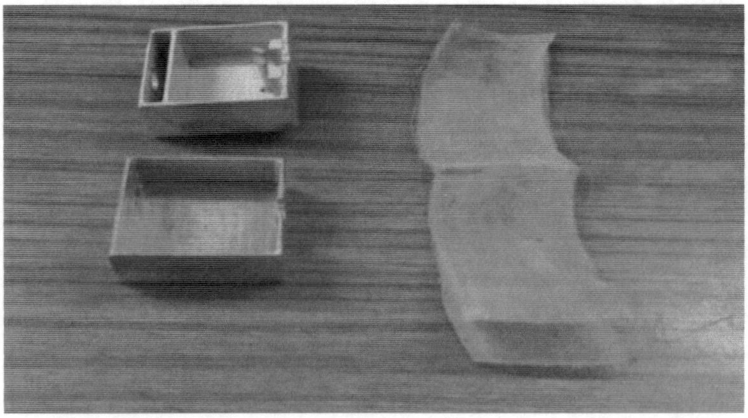

Figure 2. A photograph of compensation chamber and wick used in the present study

The joint of thermocouple is strengthened with fast curing epoxy compounds (M-seal and araldite). T_1 and T_2 measure the heater temperature and interface temperature between heater and evaporator. Two thermocouples (T_3 and T_4) measure the temperature of evaporator: one on vapour line side and one on liquid line side. T_5 and T_9 give the vapour temperatures at evaporator and condenser. Vapour line (T_6 and T_7) and liquid line (T_{12} and T_{13}) have two thermocouples each. The average condenser temperature is measured by T_8, T_{10} and T_{11}. Cooling water inlet and outlet temperatures are measured using W_i and W_0. The computer connected to data logger stores all the data which are used for data analysis.

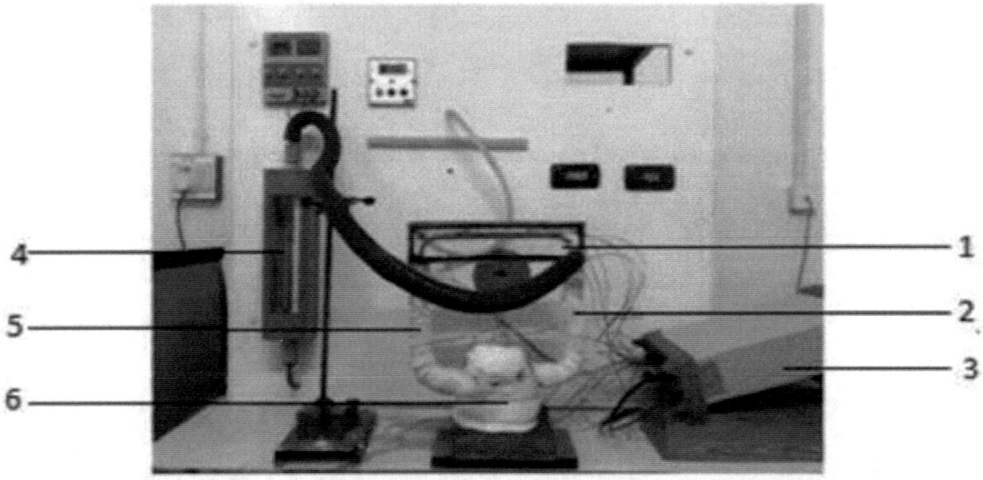

1. Condensor 2. Liquid line 3. Data acquisition system 4. Rotameter 5. Vapour line 6. Evaporator

Figure 3. Experimental setup

4. Experimental procedure

The loop heat pipe is tested with deionised water and alumina–water nanofluid to study its thermal performance for the heat load range of 30 - 310 W. The LHP is fabricated and working fluid is filled.

Then evaporator is connected to heater block and the whole assembly is covered with a thick layer of glass wool insulation. For all the experiments LHP is kept in vertical orientation i.e. condenser is above the evaporator. The condenser is connected to cooling water inlet and outlet. The cooling water with a constant flow rate of 40 litres per hour is allowed to flow through the condenser with a temperature of 20 ^{0}C. The thermocouples are connected to the data logger. Heater wires are connected to the dimmerstat and power supply is switched on. An initial heat load of 30 W is applied to the evaporator base and the heat pipe is allowed to reach steady state. Then heat input is increased in steps of 40 W. The above procedure is repeated until the heat load becomes 310 W. The thermocouples measure the temperature of various locations in every 5 s and the values are noted. After the experiment, the heat input is reduced to zero and waited till the loop heat pipe is allowed attain the room temperature. Later cooling water supply is cut off and loop heat pipe is removed from the experimental set up.

5. Results and discussions

5.1. Effect of heat input on evaporator temperature

Figure 4 plots the evaporator temperature versus heat loads. When heat load is applied to the evaporator base, the temperature increases and vapour is formed in the vapour line. In this experiment the reservoir is maintained at ambient temperature. The LHP shows a very good response to the change in the heat load from 30 to 310 W and reaches steady state within a short time period.

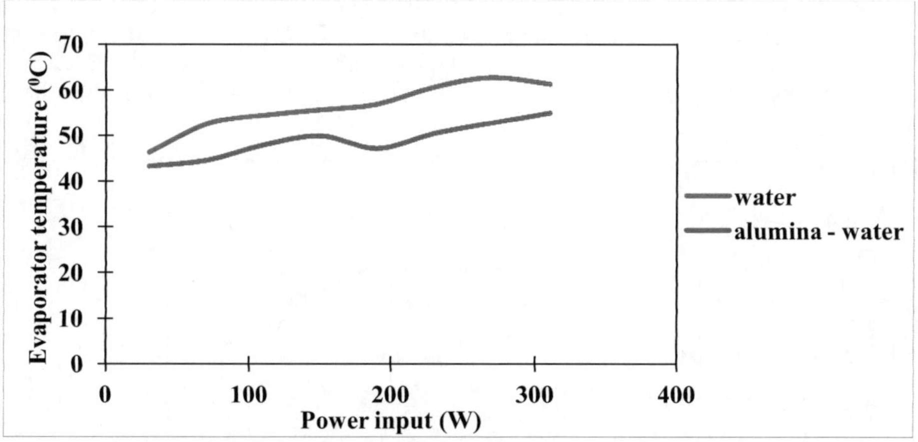

Figure 4. Evaporator temperature versus applied heat load

It can be seen that LHP is able to maintain the evaporator temperature within a range of 43.28 °C to 54.98 °C and 46.302 °C to 61.35 °C for alumina–H_2O nanofluid and deionised water respectively. The LHP exhibits very efficient control over the operating temperature of the evaporator for the entire range of input power. The temperature at the evaporator of the LHP using alumina – H_2O nanofluid is lower by 12 % than that of the deionised water, since alumina – H_2O nanofluid has higher thermal conductivity and high heat transfer rate.

5.2. Temperature analysis

The transient temperature distribution of LHP with deionised water and alumina–H_2O nanofluid at heat load of 30 W is shown in figure 5 and figure 6 respectively. Here temperatures considered are average values of evaporator (T_e), vapour line (T_v), condenser (T_c) and liquid line (T_l) temperatures.

IOP Conf. Series: Materials Science and Engineering **396** (2018) 012058 doi:10.1088/1757-899X/396/1/012058

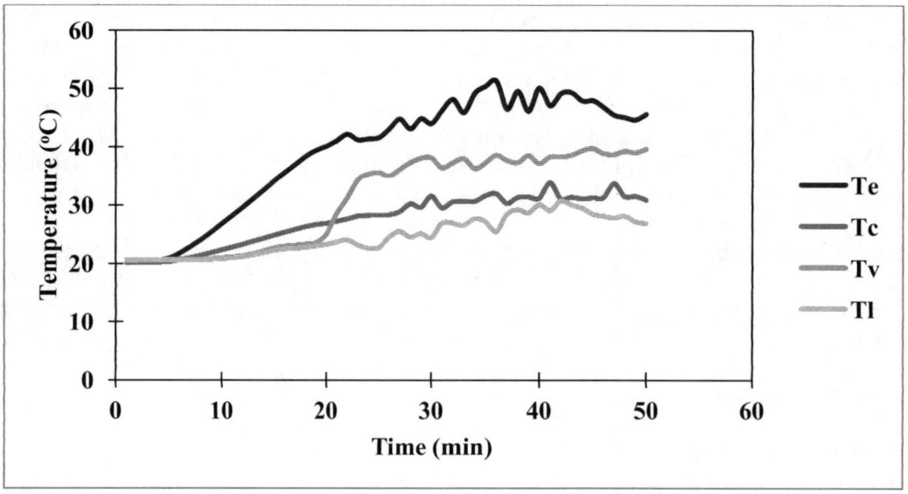

Figure 5. Transient temperature distribution of LHP using deionised water at Q = 30 W

It can be seen from figure 5 that the temperatures of all points increase gradually until 23 min and become almost steady thereafter. It is also observed from figure 6 that the temperatures at all points have been reduced due to the enhanced heat removal in the LHP by higher thermal conductivity and convective heat transfer of alumina–H_2O nanofluid. As a result, the higher thermal characteristics of the nanofluid have proved its potential as alternative for conventional deionised water in the LHP.

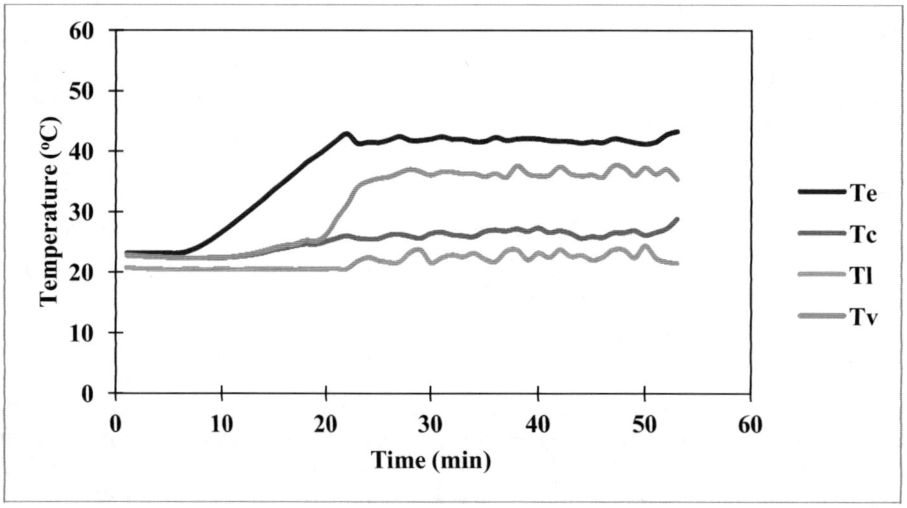

Figure 6. Transient temperature distribution of LHP using alumina–H_2O nanofluid at Q = 30W.

6. Conclusion

A loop heat pipe is experimentally studied to determine its thermal performance for a heat load range of 30 – 310 W in steps of 40 W. The use of alumina–water nanofluid in LHP enhances the thermal performance. The transient temperature distributions are also obtained from experiment. The LHP using alumina–H_2O nanofluid yields lower evaporator temperature and attains its steady state faster than LHP using deionised water. The evaporator temperature of LHP reduces by 12 % when nanofluid is used. The experimental results show that nanofluid can enhance the thermal performance of the loop heat pipe. Thus loop heat pipes using nanofluids are highly promising in the thermal management of electronic devices.

References

[1] Faghri A 1995 Heat Pipe Science and Technology (Washington : Taylor & Francis).
[2] Shafahi M, Bianco V, Vafai K and Manca O 2010 An investigation of the thermal performance of cylindrical heat pipes using nanofluids *Int. J. Heat Mass Transfer* **53** pp. 376 – 383.
[3] Gunnasegaran P, Abdullah M Z and Shuaib N H 2013 Influence of nanofluid on heat transfer in a loop heat pipe *International Communications in Heat and Mass Transfer* **47** pp. 82 – 91.
[4] Gunnasegaran P, Abdullah M Z and Yusoff M Z 2014 Effect of alumina-water nanofluid concentration on heat transfer in a loop heat pipe *International conference on advances in Manufacturing and Materials Engineering* **5** pp. 137 – 146.
[5] Zhenping Wan, Deng J, Li B, Xu Y, Wang X and Tang Y 2014 Thermal performance of a miniature loop heat pipe using water-copper nanofluid *Applied Thermal Engineering* **78** pp. 712 -719.
[6] Trijo Tharayil, Lazarus Godson Asirvatham, Vysakh Ravindran and Somchai Wongwises 2016 Thermal performance of miniature loop heat pipe with graphene – water nanofluid *International Journal of Heat and Mass Transfer* **93** pp. 957 - 968.
[7] Pauken M and Rodriguez J I 2000 Performance characterisation and model verification of a loop heat pipe *Society of Automotive Engineer* Paper No. 2000-01-0108.
[8] Riehl R R and Dutra T 2005 Development of an experimental loop heat pipe for application in future space missions *Applied Thermal Engineering* **25** pp. 101 - 112.
[9] Zhao Y, Chang S, Yang B and Zhang W 2017 Experimental study on the thermal performance of loop heat pipe for the aircraft anti-icing system *International Journal of Heat and Mass Transfer* **111** pp. 795 - 803.
[10] Naphon P, Thongkum D and Assadamongkol P 2009 Heat pipe efficiency enhancement with refrigerant–nanoparticles mixtures *Energy Conversion and Management* **50** pp. 772 – 776.

IOP Conf. Series: Materials Science and Engineering **396** (2018) 012059 doi:10.1088/1757-899X/396/1/012059

Performance studies on serpentine flow channel of a proton exchange membrane fuel cell

Antony Alex[1], Austin George[1], Bijo Thomas[1], Bonny Mathew Antony[1] and Rajesh Baby[2*]

[1]Under Graduate Students, [2*]Associate Professor,
Dept. of Mechanical Engg., St. Joseph's College of Engineering and Technology, Palai, Kerala, India
E-mail*: rbaby55@gmail.com

Abstract. The results of the experimental investigations on the performance of a Proton Exchange Membrane Fuel Cell (PEMFC) with serpentine flow field is reported in the present work. The serpentine flow channel considered in the present work has an active area of 50 cm^2 and has a landing to channel ratio of 1:1. The power densities at various operating temperatures are determined experimentally. The testing is done on the PEMFC for three different temperatures: 50, 60 and 70ºC. The output power at a particular cell voltage has been found to increase as the temperature increases. The maximum power obtained is 23.01 W at 0.48 V when the cell operating temperature is 70ºC. In order to characterize the nafion membrane, a scanning electron microscope analysis is also performed and the results are reported in the present paper.

1. Introduction

Today, the world is mostly dependent on fossil fuels for meeting its energy demands. Fossil fuels are non-renewable and their use has serious detrimental effects on the earth's environment. Among the alternative sources of energy fuel cells are very promising. Fuel cells are electrochemical devices, which can produce electricity by electrochemical reactions between the hydrogen fuel and oxygen or another oxidizing agent. The various types of fuel cells are proton exchange membrane fuel cells, molten carbonate fuel cells, direct methanol fuel cells, solid oxide fuel cells, phosphoric acid fuel cells and alkaline fuel cells. The PEMFCs offer greater advantages over others due to their lower operating temperature, less running cost and increased lifetime.

The various components of a PEMFC are a) end plates b) current collector plates c) polymer electrolyte membrane d) gas diffusion layer and e) bipolar plates. Unlike the battery, fuel cells need a continuous supply of fuel and oxidizer for the production of electricity. Higher initial investment and the complications involved in the storage of hydrogen are the major drawbacks for the implementation of fuel cells for power generation.

IOP Conf. Series: Materials Science and Engineering **396** (2018) 012059 doi:10.1088/1757-899X/396/1/012059

Following are the chemical reactions involved in the operation of PEMFC:

At the anode: $H_2 \longrightarrow 2H^+ + 2e^-$ (1)

At the cathode: $(½)\ O_2 + 2H^+ + 2e^- \longrightarrow H_2O$ (2)

Cell reaction: $H_2 + (½)\ O_2 \longrightarrow H_2O$ (3)

2. Literature Review

Several numerical as well as experimental investigations have been done on various flow fields of PEMFCs. Still, a lot of work needs to be done on fuel cells for their useful applications in our day to day lives. Wang et al. [1] investigated various operating features like pressures and temperatures, anode and cathode humidification. They also carried out numerical studies and verified that their numerical results with reference to their experimental outcomes. They reported that, with proper humidification the performance of a PEMFC would increase with the increasing operating temperature. In their work Carton and Olabi [2], studied the various flow fields such as serpentine, maze and parallel types for low temperature PEMFCs. Various experiments were carried out on these three flow channels in order to determine the reaction of the fuel cell to output current, voltage, power and also their efficiency. It was found that the serpentine flow channel exhibited better performance characteristics and polarization curve compared to other two flow fields.

Karthikeyan et al. [3] observed that the power density of PEMFC decreased with increase in cross-sectional area. They also found that the performance of a PEMFC stack depended on the temperature management and the distribution of fuel in the flow channel. Kanezaki et al. [4] focused their attention towards the effective water removal in a PEMFC. They conducted numerical studies on the influence of cross-leakage flow through a PEMFC. It was reported that, in a PEMFC with serpentine flow channel, cross-leakage through the electrode reduced the concentration over potential and helped to effectively remove the water formed in the fuel cell. Chang and Hung [5] studied the influence of channel depth and flow rates on the performance of a PEMFC. They reported that to maintain necessary pressure in order to push the reactant into the flow field and to have a sufficient water balance, an appropriate flow rate was required for shallow channels. Ararimeh and Sastry [6] studied the effect of baffles on various flow channels. They reported that when baffles are added to various flow field designs, the performance of the PEMFC improved.

Rostami et al. [7] numerically studied the effect of bend sizes in serpentine flow fields of PEMFC. They carried out various numerical studies by changing the bend sizes from 0.8 mm to 1.2 mm. They found that 1.2 mm bend size serpentine flow channel had better electrochemical properties due to better catalyst contact area and also due to uniform temperature distribution. Khazaee and Sabadbafan [8] investigated the influence of flow channel design and relative humidity of a rectangular PEMFC with active area of 24.8 cm^2. They established the effect of water management in the performance of a PEMFC. They developed a three-dimensional PEMFC of various serpentine flow fields and studied the flow patterns of the inlet gases on the fuel cell performance. They also found that the flow channel design had much influence over the performance of a PEMFC compared to the directional properties. A serpentine geometry having three tapered channels and a diverging channel at the cathode side was proposed by Wang et al. [9]. They reported that the design would enhance the oxygen transport rate within the channel and thus the current density.

A three-dimensional computational model with various sizes, shapes and patterns of bipolar plates with serpentine flow channel was developed by Boddu et al. [10]. They reported that square bends had higher pressure drop than curvilinear bends. Also, it is seen that more uniform distribution of inlet gases occurred when the cross-section area was decreased. Teccani and Zuliani [11] studied the influence of flow field on the performance of a PEMFC. They compared the results of serpentine and parallel flow fields. The results showed that the flow field having a higher number of serpentine step had better performance in terms of power density.

IOP Conf. Series: Materials Science and Engineering **396** (2018) 012059 doi:10.1088/1757-899X/396/1/012059

3. Experimental Setup

The experimental setup of a fuel cell test station has mainly five parts: a computer, temperature control system, flow rate control system, humidifier and an electronic load system. The computer stores the various values of current, voltage, power density and inlet flow rate of hydrogen and oxygen. The temperature control system is used to vary the temperature of the fuel cell. A calibrated thermocouple is inserted into the graphite plates, in order to measure the temperature of the fuel cell. The flow rate control system has hydrogen, oxygen and nitrogen stored in special cylinders for supply to the fuel cell. High purity hydrogen is supplied to the anode side of the PEMFC and medical grade oxygen is supplied at the cathode side. Nitrogen is used so as to flush the excess water and residual gases formed in the fuel cell during the reaction. If the hydrogen and oxygen are supplied without proper humidification the membrane electrode assembly (MEA) may suffer damages. So a humidifier is used to humidify the inlet hydrogen and oxygen supplied to the fuel cell. The electronic load system is employed to obtain the various current and voltage loads. The load system has a 600 W maximum power output, 500 V maximum voltage and 120 A maximum current. A photograph of the fuel cell test station is shown in figure 1.

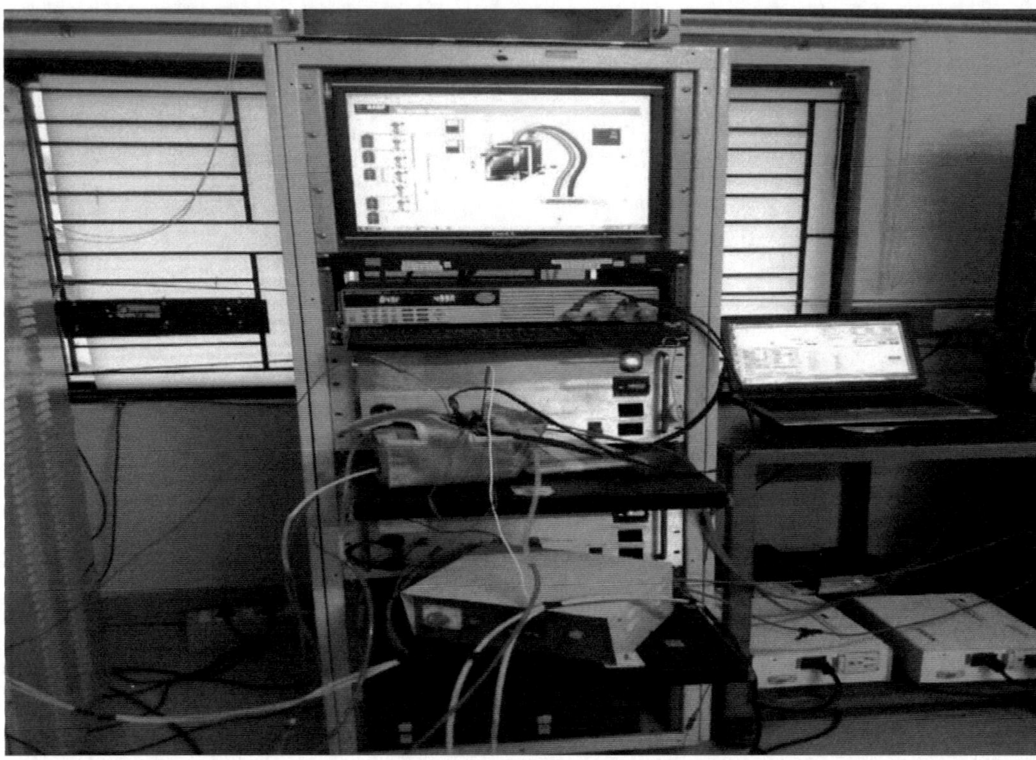

Figure 1. A photograph of the fuel cell testing station used in the present study

The flow channel is made of graphite plate with a total area of 100 cm^2. The active area of the fuel cell selected for the present experimental work is 50 cm^2. The landing to channel ratio considered is 1:1. The diagram of the serpentine flow channel used is shown in figure 2.

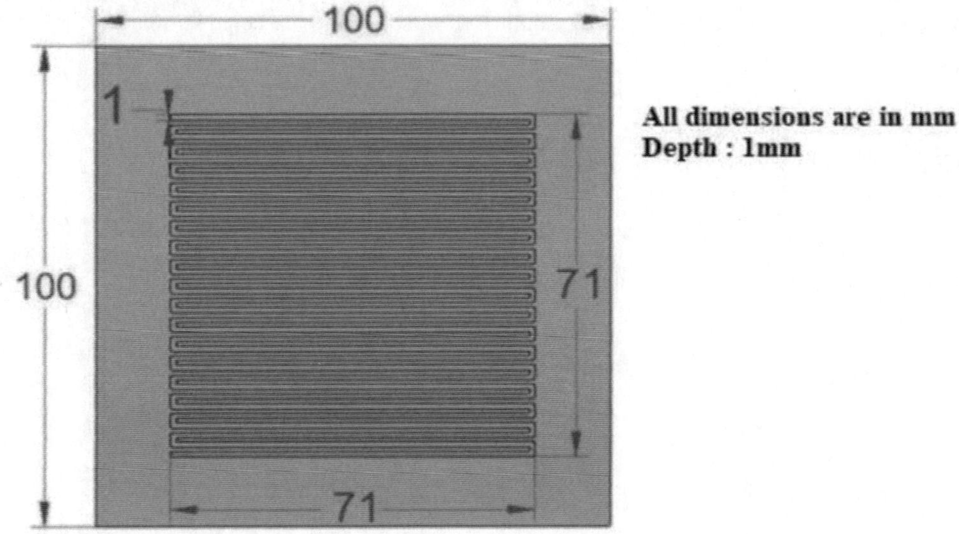

Figure 2. Design of the 50cm^2 bipolar plate.

The fuel cell has a membrane electrode assembly through which the exchange of proton takes place. Nafion membrane is used as the binder to hold the gas diffusion electrodes (GDE) together. A scanning electron microscope image of the Nafion membrane used in the present study is shown in figure 3.

Figure 3. A 850x magnified image of the Nafion membrane.

3.1 Experimental Procedure

The values of current, voltage, and power of the fuel cell are recorded during the testing process. Oxygen and hydrogen gases are humidified using a humidifier at various temperatures. Calibrated thermocouples and heating coils are used to measure and vary the temperature respectively. The voltage developed is recorded for a particular current. A computing device is used to record current density, power density and the power developed, with the help of a testing apparatus. The flow rate,

IOP Conf. Series: Materials Science and Engineering **396** (2018) 012059 doi:10.1088/1757-899X/396/1/012059

line pressure, line temperature, humidity values and supply temperatures of hydrogen and oxygen gases respectively are recorded throughout the experiment. An electronically varying load testing apparatus is set to 1 V and 50 A parameter limits. Relevant readings are noted for temperatures of 50, 60 and 70ºC.

4. Results and Discussion

4.1 Effect of temperature on current density

Polarization curves are plotted with cell voltage along the Y-axis and current density along the X-axis for temperatures of 50, 60 and 70ºC. The polarization curve at 70ºC is shown in figure 4. From the figure it can observed that the maximum current density of 0.96 A/cm^2 is obtained for a cell voltage of 0.48 V. The cell potential decreases as current density increases.

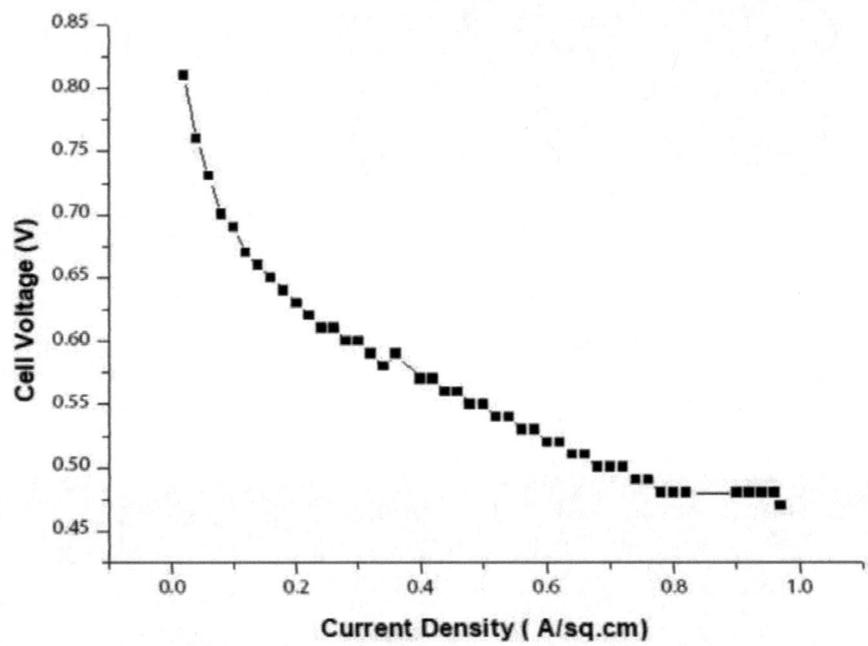

Figure 4. Polarization curve at 70ºC.

The polarization curve at 60ºC is plotted and is shown in figure 5. From the graph it can be inferred that the maximum current density of 0.88 A/cm^2 is attained for a cell voltage of 0.48 V. There is a slight decrease in the maximum current density with temperature.

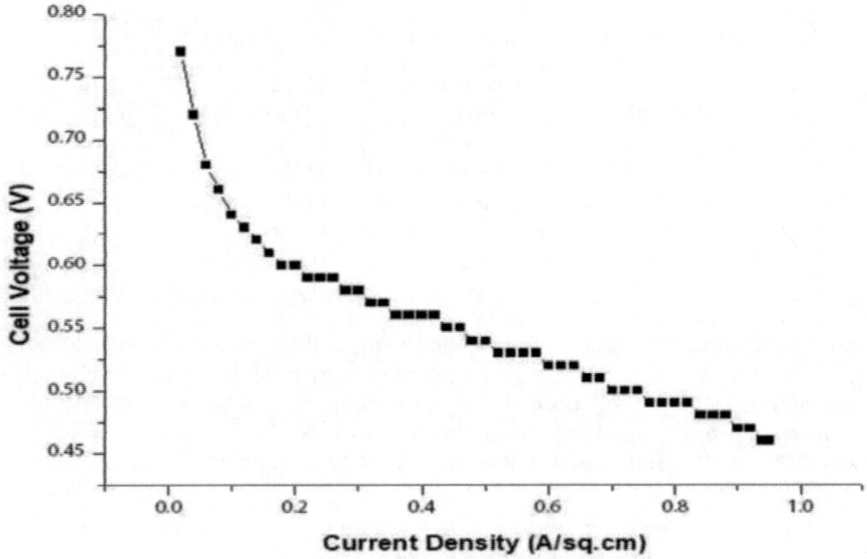

Figure 5. Polarization curve at 60°C.

The polarization curve at 50°C is shown in figure 6. From the graph it can be seen that the maximum current density of 0.80 A/cm^2 is attained for a cell voltage of 0.48 V. From the figures 4 to 6, it is clear that current density increases with increase in temperature for a particular cell voltage.

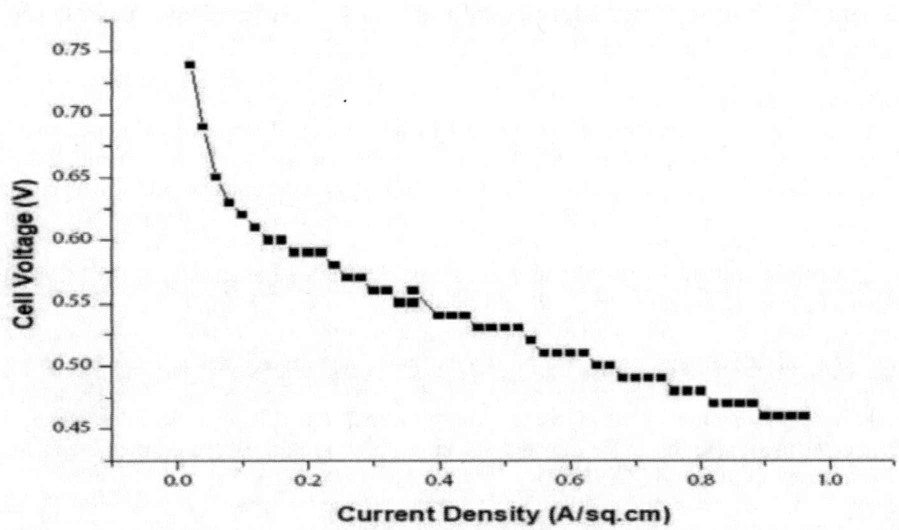

Figure 6. Polarization curve at 50°

4.2 Effect of temperature on power

The variation of power with cell voltage for the three temperatures are given in table 1. From the table it is inferred that the power decreases as the cell voltage increases.

Table 1. Variation of power at different temperatures

Sl. No	Cell Voltage(V)	Power at 50°C (W)	Power at 60°C (W)	Power at 70°C (W)
1	0.47	20.65	21.60	22.88
2	**0.48**	**19.18**	**21.09**	**23.01**
3	0.50	16.47	18.47	17.97
4	0.55	9.87	12.62	13.72
5	0.60	4.77	5.97	8.97
6	0.65	1.91	3.21	5.16

Operating range of the fuel cell is the ohmic region and the voltage varies between 0.40 and 0.50 V. From the table 1 it is observed that the peak power output values occur between these voltages. The maximum power output of 23.01 W is obtained at 0.48 V and 70°C. At 0.48 V and 50°C the power output is 19.18 W. The deviation when temperature rises from 50 to 70°C is 20%. Similarly, deviation between power at 50°C and 60°C is 10%. Better performance is due to the increase in the rate of reaction when temperature is on the increase.

5. Conclusions

In the present work, experimental investigations are conducted on a 50 cm^2 active area PEMFC. The main conclusions obtained from this investigation are given below.

- The power output at three different temperatures and cell voltages are obtained. From the present study it is seen that for a particular voltage as the temperature increases power level also increases.
- From the cell voltage and power, it can be inferred that as the temperature is increased from 50 to 70°C the power output increases by 20%. Increase in the temperature will improve the rate of chemical reaction. So, at 70°C electrochemical reaction and its rate will be high when compared with those at lower temperatures and therefore gives the maximum power value.

The insights developed from the present experimental studies may be used for further improvement in the design of proton exchange membrane fuel cells.

Acknowledgments

The authors are grateful for the financial support by APJ Abdul Kalam Technological University through Research Seed Money scheme, Centre for Engineering Research and Development (Project no. KTU/RESEARCH 2/2743/2017) and St. Joseph's college of Engineering and Technology, Palai.

References

[1] Wang Lin, Husar Attila, Zhou Tianhong, Liu Hongtan 2003 A parametric study of PEM fuel cell performances *Int J Hydrogen Energy* **28** pp 1263-72

[2] Carton J G, Olabi A G 2010 Design of experiment study of the parameters that affect performance of three flow plate configurations of a proton exchange membrane fuel cell *Energy* **35** pp 2769-806

[3] Karthikeyan P, Velmurugan P, George Abby Joseph, Ram Kumar R, Vasanth R J 2014 Experimental investigation on scaling and stacking up of proton exchange membrane fuel cells *J Hydrogen Energy* **39** pp 11186-95

[4] Kanezaki Toshihiko, Li Xianguo, Baschuk J J 2006 Cross-leakage flow between adjacent flow channels in PEM fuel cells *J Power Sources* **162** pp 415-25

[5] Dyi-Huey Chang, Jung-Chung Hung, 2012 Effects of channel depths and anode flow rates on the performance of miniature proton exchange membrane fuel cells, *Journal of Applied Science and Engineering* pp 273-280

[6] Aiyejina Ararimeh, Sastry M K S 2012 PEMFC flow channel geometry optimization: a review. *Journal of Fuel Cell Science and Technology* 9:011011.1-011011.24

[7] Rostami L, Nejad P M, Vatani A 2016 A numerical investigation of serpentine flow channel with different bend sizes in polymer electrolyte membrane fuel cells *Energy* **97** pp 400-10

[8] Khazaee I, Sabadbafan 2016 Effect of humidity content and direction of the flow of reactant gases on water management in the 4-serpentine and 1-serpentine flow channel in a PEM (proton exchange membrane) fuel cell *H. Energy* **101** pp 252-65

[9] Wang X, Huang Y, Cheng Ch, Jang J, Lee D, Yan W 2010 An inverse geometry design problem for optimization of single serpentine flow field of PEM fuel cell *Int J Hydrogen Energy* **35** pp 4247-57

[10] Boddu R, Kumar Marupakula U, Summers B, Majumdar P 2009 Development of bipolar plates with different flow channel configurations for fuel cells *J Power Sources* **189** pp 1083-92

[11] Taccani R, Zuliani Nicola 2011 Effect of flow field design on performances of high temperature PEM fuel cells: experimental analysis *Int J Hydrogen Energy* **36** pp 10282-87

IOP Conf. Series: Materials Science and Engineering **396** (2018) 012060 doi:10.1088/1757-899X/396/1/012060

Recent advances in loop heat pipes : A review

Jobin Jose[1] and Rajesh Baby[2*]

[1] Graduate Student, [2] Associate Professor
Department of Mechanical Engineering, St. Joseph's College of Engineering and Technology, Palai, Kerala, India.

[*]E – mail : rbaby55@gmail.com

Abstract : Among the various technologies available today heat pipes can be considered as an efficient heat transfer device. A heat pipe is a device which can transfer heat though long distances. Heat pipes (HPs) have gained importance in recent years, mainly in the field of cooling electronic components, since they need heat removal from limited volumes to the environment. Small HPs are widely used in electronic applications, which are normally limited by the compact structure and dimensions of the electronic device. Among small heat pipes, loop heat pipes (LHPs) are usually preferred due to high efficiency and compact size. Nano fluids also have attracted attention due to their superior heat transfer properties in recent years. The lower thermal conductivity of conventional working fluids puts a limit on heat transfer. The thermal performance of heat pipes may be improved by using nano fluids as working fluids. This review presents an investigation on the design, operating principle, advantages and applications of loop heat pipes. Furthermore, emphasis is given on the effect of the nano fluid in the thermal performance of LHPs.

Key words : Loop heat pipe, nano fluid, thermal performance

1. Introduction

The loop heat pipe (LHP) is a device that operates on a multi phase fluid-flow cycle, maintained by a capillary medium in the evaporator. It is a heat transfer device which can transport a large amount of heat over a long distance with negligible temperature difference. It utilizes boiling and condensation to transfer heat, and the surface tension force developed at the liquid/vapor interface on the evaporator wick to maintain the flow circulation. The schematic of loop heat pipe is shown in figure 1. The operation of a LHP may be summarised as follows. At start-up, the liquid load is sufficient to fill the condenser and the liquid and vapour lines and there is sufficient liquid in the evaporator and compensation chamber to saturate the wick. When a heat load is applied to the evaporator, fluid evaporates from the surface of wick. Since the wick has considerable thermal resistance, the temperature and pressure within the compensation chamber will be less than that in the evaporator. The capillary forces in the wick obstruct the flow of the vapour from evaporator to compensation chamber. When the pressure difference between compensation chamber and evaporator rises, the liquid is displaced from the vapour line and the condenser and it enters the compensation chamber.

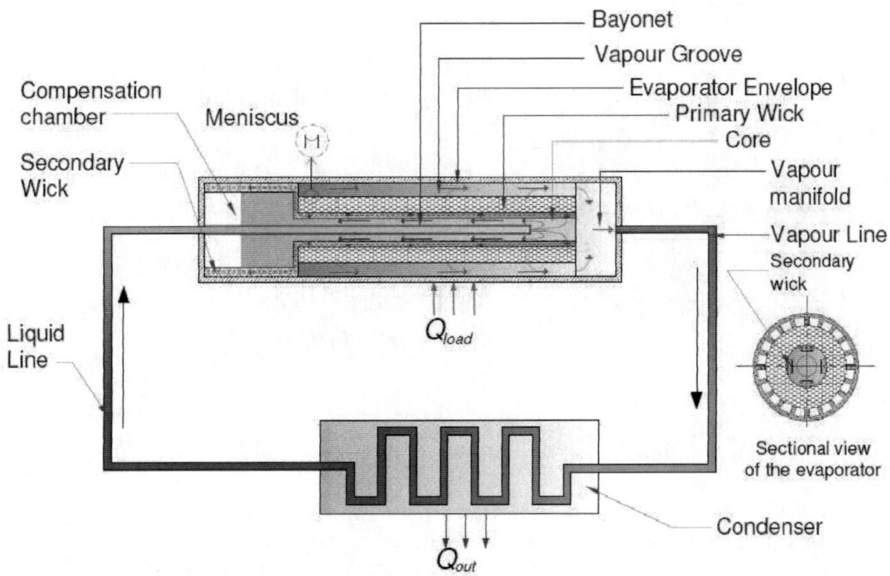

Figure 1. Schematic of loop heat pipe [1]

1.1. Advantages of loop heat pipes
Along with the features of a heat pipe, the LHP has the following additional advantages:
- High heat flux capability
- Ability to transfer energy over long distances
- Ability to operate over a range of environments
- Wick is not provided in the transport lines
- Chance of entrainment is very low

1.2. Applications
LHPs can be used in thermoelectric generators. Thus it avoids the use of power consuming fan and results in noise free operations. They are also used in LED street lights for heat dissipation. In solar water heaters the LHP is considered as an integral component due to its better performance. In 1989, LHP was first adopted for space application in a Russian spacecraft. Even now they are being used for space applications. In the thermal management of electronic components also LHPS are widely used.

2. Literature review
A pie-chart demonstrating the literature explored for the present review is shown in figure 2. This paper is an attempt to review recent advances in loop heat pipes with special emphasis on nano based loop heat pipes as of 2017. A total of 32 papers are taken for this study. Out of these seventeen papers focus on experimental, eight on numerical, three on combined experimental and numerical investigations, and four explaining various reviews of loop heat pipes.

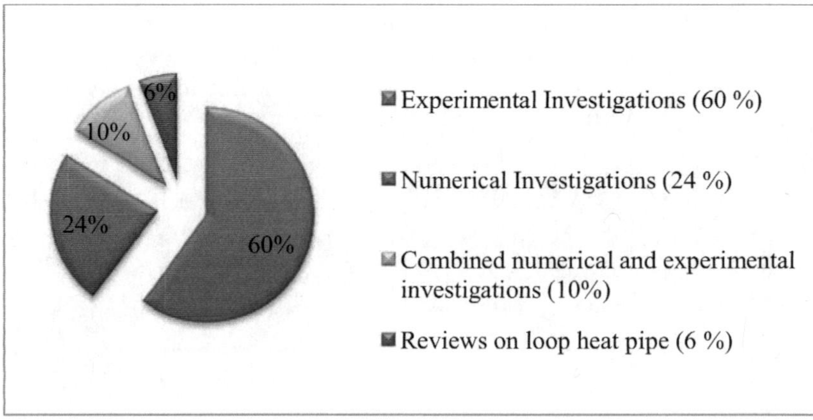

Figure 2. Chart demonstrating the literature

2.1. Experimental Investigations

The thermal performance of a loop heat pipe has been experimentally investigated by many researchers. Pauken *et al.* [2] experimentally studied the heat transfer characteristics of LHP with two working fluids i.e., ammonia and propylene. The constant conductance performance of the LHP was found to be 170 W/K with ammonia and 44 W/K for propylene. The thermal performance of an LHP with acetone as working fluid was experimentally studied by Riehl *et al.* [3]. Results showed that the system could present reliable performance in all situations with power level as low as 1 W. Also higher operating temperatures could be obtained when evaporator was above the condenser. Bai *et al.* [4] developed a model for start-up process on the basis of node network method. These nodes were mainly classified into two, namely fluid nodes and wall. Zhao *et al.* [5] introduced a solar water heater based on LHP technology. It resulted in the reduction of fossil fuel consumption and carbon emissions associated with hot water production and supply. Vlassov *et al.* [6] developed a considerably precise condenser and transient nodes in the compensation chamber and evaporator. It was found that his model was able to predict transient behaviour of LHP within 3 - 4^{0}C of accuracy. Guiping *et al.* [7] conducted a fundamental study of a dual compensation chamber. The experimental results showed that a V-shaped operational curve was obtained for heat loads above 50 W. Thermal behaviour of a cryogenic LHP was experimentally studied by Philippe *et al.* [8] for space applications. Results indicated that a small heating load applied to the cold reservoir allowed maintenance of a constant sub cooling and the system behaved as a capillary loop. Leonid *et al.* [9] developed a LHP for cooling of high power electronic components. The experimental results suggested that when high power dissipation levels were available low conductivity wick should be adopted for LHP applications.

Huang *et al.* [10] categorised the start-up phenomenon of LHP into four modes i.e., oscillation, failure, normal and overshoot. In the failure mode, evaporator temperature couldn't reach a steady state instead it continued to rise. In oscillation mode, evaporator temperature increased due to the continued heat input and vapourisation of the fluid. Zhang *et al.* [11] developed a solar photovoltaic heat pump system for generating hot water. After the experimental investigations the thermal, electrical and overall efficiency of the module were 40%, 10% and 50% respectively. The system's overall performance coefficient was 8.7. Shen *et al.* [12] experimentally studied the application of self rewetting fluid to LHP. It was concluded that total thermal resistance reduced by 60% and heat transfer characteristics improved. A LHP using flat evaporator with biporous wick was investigated by Chen *et al.* [13]. The wick was made from nickel powder. The LHP showed a very quick response to variable heat load and operated steadily without temperature oscillation. Zhao *et al.* [14] developed a LHP with liquid guiding holes and the performance characteristics were analysed. The results showed that LHP could start successfully with heat loads from 70 W to 220 W. The thermal performance of a miniature LHP was investigated with graphene nano fluid by Trijo *et al.* [15]. The results showed that thermal conductivity and thermal efficiency of the LHP increased. Zhao *et al.* [16] also studied the thermal performance of a double compensation chamber LHP for anti icing system in aircrafts. The results indicated that surface temperature of the aircraft wing increased with increase of heat load on the evaporator. Zhenping *et al.* [17] experimentally investigated the influence of a copper nano fluid on the heat transfer characteristics of a miniature loop heat pipe. It was concluded that performance enhancement of the loop heat pipe using the nano fluid resulted in the reduction of the contact angle, improvement in boiling heat transfer and deposited nanoparticle coat on the boiling surface.

2.2. Numerical investigations

Alekey *et al.* [18] conducted 3D numerical investigation of flow and conjugate heat and mass transfer in a model TacSat-4 satellite LHP evaporator. The simulation results showed that evaporation mainly occurred at the vapour groove corners near the evaporator. In order to determine thermodynamic behaviour of LHP a steady state model was developed by Benjamin *et al.* [19]. Simulation results showed the major effect of evaporation coefficient and wick conductivity on the LHP evaporator temperature and its temperature field. Esarte *et al.* [20] developed a new model for cooling high power emitting diodes through loop heat pipes. It was based on the steady state energy balance equations for each component. Several parameters such as tube radius and length and their effects were also studied. Mariya *et al.* [21] developed a 3D model of heat and mass transfer in the flat evaporator of an LHP. EFD Lab software package was used for numerical simulation. The simulation results were analysed for getting heat transfer dependences of heat load for different orientations of the evaporator. Hui *et al.* [22] numerically investigated the steady state model of an anti-gravity LHP. It was found that anti-gravity length and condensation length influenced the circulation of working fluid. The thermal characteristics of LHP evaporator were analysed by Masahito *et al.* [23]. It was done by using a 3D pore network model with a dispersed pore size wick. It was inferred that optimum shape can be achieved by varying the three phase contact line. Masahito *et al.* [24] also developed a mathematical model to study the heat transfer characteristics of evaporator in LHP with a small gap between case and wick. Also the effect of this gap on the thermal properties was discussed. After the numerical investigation it was concluded that micro gap was an efficient approach for improving thermal performance. Riehl [25] studied the effect of nickel - water nano fluid in a LHP. Result showed that heat transfer coefficient was lower at the evaporator side and operating temperatures were higher throughout the LHP.

2.3. Combined experimental and numerical investigations

Gunnasegaran *et al.* [26, 27] conducted experimental and numerical investigations on LHPs with alumina and silica as nano fluids. It was found that the thermal resistance decreased when using silica - water nano fluid. The LHP charged with alumina - water nano fluid yielded lower wall temperature difference between evaporator and condenser and steady state was attained faster. Nookarajua *et al.* [28] experimentally and numerically investigated the enhancement in thermal performance of a copper sintered wick heat pipe with deionised water. From the results it could be concluded that as heat input increased the efficiency value decreased. The investigations on loop heat pipes with nano fluids can be summarized as shown in the table 1.

Table 1. Investigations on LHPs with nanofluids

Authors	Parameter analysed	Nanoparticle	Conclusions
Trijo Tharayil et al. [15]	Thermal conductivity	Graphene	The thermal conductivity and thermal efficiency increased. Very low concentrations of nanofluid gave improvement in heat transfer.
Zhenping et al. [17]	Contact angle	Copper	Results from the reduction of the contact angle, the enhancement of boiling heat transfer. Water recirculation rate of nano fluid is more than that of pure water during boiling.
Reihl [25]	Heat transfer coefficient	Nickel	Heat transfer coefficient is lower at the evaporator side and operating temperatures are higher throughout the LHP.
Gunnasegaran et al. [26]	Thermal resistance	Silica	Thermal resistance of LHP decreased. Therefore, the presence of nanoparticles could greatly enhance the cooling of LHP.
Gunnasegaran et al. [27]	Thermal resistance	Alumina	It reaches steady state faster than pure water. Thermal resistance of LHP decreased.

2.4. Reviews on loop heat pipes

Zhangyuan *et al.* [29] conducted a review on solar water heating application of LHP. This review also discussed the problems that still existed in the solar water heaters and identified new research subjects to improve the performance of solar water heaters. Stephane *et al.* [30] focused the review on parametric analysis of LHP operation. It was shown that LHP operation was mainly affected by the boiling and capillary limitations. The LHP performance was characterized by their thermal resistance and maximum heat transport capability. A review on the development of cryogenic LHPs was presented by Lizhan *et al.* [31]. This work contributed to a better understanding of the development and operating characteristics of cryogenic LHPs. Maydanic *et al.* [32] conducted a review on the development of LHPs using flat evaporators. It was concluded that stainless steel – nickel – ammonia was the most efficient combination for 40 – 70 °C and Copper – copper – water combination was the most efficient for 70 – 100 °C.

2.5. Application domains of loop heat pipes

Loop heat pipes can be used for a wide variety of applications like aircraft anti – icing system, cooling of electronic components, solar water heating etc. The specific applications and corresponding literature are listed in table 2.

Table 2. Applications of LHP

Sl. no	Literature	Applications
1	[3], [6], [8], [16], [30]	Cryogenics, future space missions, aircraft anti-icing system
2	[5], [11], [28]	Solar water heating, photovoltaics
3	[9], [17], [20], [25]	Cooling and thermal management of electronic components

3. Conclusions

This paper reviews recent findings related to loop heat pipes and factors affecting their performance. It is observed that thermal efficiency and thermal resistance are two important factors which determine the thermal performance of a loop heat pipe. Nano fluid based LHPs form a recent area of investigation with promising results. The thermal resistance of loop heat pipes reduces with increase in nano particle concentration. Thermal efficiency of loop heat pipes can be improved by using nano fluids. There are only limited investigations on loop heat pipes with nano fluids. The three dimensional numerical investigations on loop heat pipes are scarce. Also a few researchers have considered combined experimental and numerical investigations. Various parameters, such as wick materials, structure factors of evaporator and heat conduction through the tube wall, on heat transfer performance and fluid flow properties are yet to be analysed in full detail. Furthermore, theoretical and experimental investigations are required to determine the thermal characteristics of nano fluid based loop heat pipes.

References

[1] Adoni A A, Ambirajan A, Jasvanth V S, Kumar D and Dutta P 2010 Theoretical Studies of Hard filling in Loop Heat Pipes *Journal of Thermophysics and Heat Transfer* **24** pp. 173 - 183.
[2] Pauken M and Rodriguez J I 2000 Performance characterisation and model verification of a loop heat pipe *Society of Automotive Engineers* Paper No. 2000 - 01 - 0108.
[3] Riehl R R and Dutra T 2005 Development of an experimental loop heat pipe for application in future space missions *Applied Thermal Engineering* **25** pp. 101 - 112.
[4] Bai L, Lin G and Wen D 2000 Modelling and analysis of start-up of a loop heat pipe *Applied Thermal Engineering* **30** pp. 2778 - 2787.
[5] Zhao X, Wang Z and Tang Q 2010 Theoretical investigation of the performance of a novel loop heat pipe solar water heating system for use in Beijing *Applied Thermal Engineering* **30** pp.

2526 - 2536.

[6] Vlassov S V and Riehl R R 2005 Modelling of a loop heat pipe for ground and space conditions *SAE International Paper* Paper No. 2005 - 01 - 2935.

[7] Guiping Lin, Nan Li, Lizhan Bai and Dongsheng Wen 2010 Experimental investigation of a dual compensation chamber loop heat pipe *International Journal of Heat and Mass Transfer* **53** pp. 3231 – 3240.

[8] Philippe Gully, Qing Mo, Tao Yan, Peter Seyfert, Laurent Guillemet, Pierre Thibault and Jingtao Liang 2011 Thermal behaviour of a cryogenic loop heat pipe for space application *Cryogenics* **51** pp. 420 - 428.

[9] Leonid Vasiliev, David Lossouarn, Cyril Romestant, Alain Alexandre, Yves Bertin, Yauheni Piatsiushyk and Vladimir Romanenkov 2009 Loop heat pipe for cooling of high-power electronic components *International Journal of Heat and Mass Transfer* **52** pp. 301 - 308.

[10] Huang B J, Huang H H and Liang T L 2009 System dynamics model and start-up behaviour of loop heat pipe *Applied Thermal Engineering* **29** pp. 2999 - 3005.

[11] Zhang X, Zhao X, Xu J and Yu X 2013 Characterization of a solar photovoltaic/loop-heat-pipe heat pump water heating system *Applied Energy* **102** pp. 229 - 1245.

[12] Chen B B, Liu W, Liu Z C, Li H and Yang J G 2012 Experimental investigation of loop heat pipe with flat evaporator using biporous wick *Applied Thermal Engineering* **42** pp. 34 - 40.

[13] Shen-Chun Wu 2015 Study of self-rewetting fluid applied to loop heat pipe *International Journal of Thermal Sciences* **98** pp. 374 - 380.

[14] Zhao Y, Chang S, Zhang W and Yang B 2016 Experimental research on thermal characteristics of Loop heat pipe with liquid guiding holes *Applied Thermal Engineering* **101** pp. 231 - 238.

[15] Trijo Tharayil, Lazarus Godson Asirvatham, Vysakh Ravindran and Somchai Wongwises 2016 Thermal performance of miniature loop heat pipe with graphene–water nanofluid *International Journal of Heat and Mass Transfer* **93** pp. 957 - 968.

[16] Zhao Y, Chang S, Yang B and Zhang W 2017 Experimental study on the thermal performance of loop heat pipe for the aircraft anti-icing system *International Journal of Heat and Mass Transfer* **111** pp. 795 - 803.

[17] Wan Z, Deng J, Li B, Xu Y, Wang X and Tang Y 2014 Thermal performance of a miniature loop heat pipe using water-copper nanofluid *Applied Thermal Engineering* **78** pp. 712 - 719.

[18] Aleksey A Pozhilov, Dmitri Zaitsev K, Evgueni Smirnov M and Aleksander Smirnovsky A 2017 Numerical simulation of heat and mass transfer in a 3D model of a loop heat pipe evaporator *St. Petersburg Polytechnical University Journal: Physics and Mathematics* **3** pp. 210 - 217.

[19] Benjamin Siedel, Valerie Sartre and Frederic Lefevre 2013 Numerical investigation of the thermohydraulic behaviour of a complete loop heat pipe *Applied Thermal Engineering* **61** pp. 541 - 553.

[20] Esarte J, Bernardini A, Blanco J M and Sancibrian R 2016 Optimizing the design for a two-phase cooling loop heat pipe: Part A: Numerical model, validation and application to a case study *Applied Thermal Engineering* **99** pp. 892 - 904.

[21] Mariya A Chernysheva and Yury F Maydanik 2012 3D - model for heat and mass transfer simulation in Flat evaporator of copper-water loop heat pipe *Applied Thermal Engineering* **34** pp. 124 - 134.

[22] Hui Li, Yong Tang and Rui Zhou 2017 Numerical analysis on heat transfer of a complete anti – gravity loop-shaped heat pipe *International Journal of Heat and Mass Transfer* **109** pp. 824 - 834.

[23] Masahito Nishikawara, Hosei Nagano and Marc Prat 2017 Numerical study on heat-transfer characteristics of loop heat pipe evaporator using three-dimensional pore network model *Applied Thermal Engineering* **126** pp 1098 - 1106.

[24] Masahito Nishikawara and Hosei Nagano 2016 Numerical simulation of capillary evaporator

with microgap in a loop heat pipe *International Journal of Thermal Sciences* **102** pp. 39 - 46.

[25] Riehl R R 2012 Heat transport behaviour of a miniature loop heat pipe using water – nickel nanofluid *Heat Pipe Science and Technology* **1** pp. 83 - 96.

[26] Gunnasegaran P, Abdullah M Z and Shuaib N H 2013 Influence of nanofluid on heat transfer in a loop heat pipe *International Communications in Heat and Mass Transfer* **47** pp. 82 - 91.

[27] Gunnasegaran P, Abdullah M Z and Yusoff M Z 2014 Effect of alumina-water nanofluid Concentration on heat transfer in a loop heat pipe *International conference on advances in Manufacturing and Materials Engineering* **5** pp. 137 - 146.

[28] Nookarajua B Ch, Kurma Raob P S V and Naga Saradac S 2017 Experimental and Numerical investigation on enhancement in thermal characteristics of Sintered Copper wick Heat pipe using deionized water as fluid Proceedings *5th International Conference of Materials Processing and Characterization* **4** pp. 1321 - 1331.

[29] Zhangyuan Wang and Wansheng Yang 2014 A review on loop heat pipe for use in solar water heating *Energy and Buildings* **79** pp. 143 - 154.

[30] Stephane Launay, Valerie Sartre and Jocelyn Bonjour 2007 Parametric analysis of loop heat pipe operation: a literature review *International Journal of Thermal Sciences* **46** pp. 621 - 636.

[31] Lizhan Bai,Lianpei Zhang, Guiping Lin, Jiang He and Dongsheng Wen 2015 Development of cryogenic loop heat pipes : A review and comparative analysis *Applied Thermal Engineering* **89** pp. 180 - 191.

[32] Maydanik Y F, Chernysheva M A and Pastukhov V G 2014 Review: Loop heat pipes with flat Evaporators *Applied Thermal Engineering* **67** pp. 294 - 307.

IOP Conf. Series: Materials Science and Engineering **396** (2018) 012061 doi:10.1088/1757-899X/396/1/012061

Thermal performance of multilayer insulation: A review

Sutheesh P M [1*], Alex Chollackal[2]

[1] Graduate Student
Department of Mechanical Engineering, St. Joseph's College of Engineering and Technology, Palai, Kerala, India

[2] Assistant Professor
Department of Mechanical Engineering, St. Joseph's College of Engineering and Technology, Palai, Kerala, India

*E – mail: sutheeshpm@gmail.com

Abstract: A multilayer insulation (MLI) is a passive thermal protection system used in cryogenics and space exploration programs as a thermal insulator. It is very thin and light weight. Due to less weight and higher thermal performance of MLI, it found application in space programs to store cryogenic liquid propellant. The effective thermal conductivity of MLI is in the order of 10^{-5} W/mK. The prediction of heat transfer in MLI is very complex due to the anisotropic conductivity and combination of radiation, gas conduction and solid conduction modes of heat transfer. This work is an attempt to gather some significant research outcome in MLI performance prediction and the factors to be considered while modeling the MLI.

Key words: multilayer insulation, thermal protection system, thermal control

1. Introduction

MLI was introduced in 1950's. It consist of multiple foils made of Kapton or Mylar coated with a highly reflective metal placed parallel to each other and low thermal conductive spacers are arranged in between the foils to avoid direct contact with the foils[1]. Radiation, solid conduction and gas conduction are the significant modes of heat transfer in MLI. Multilayer insulations are capable of maintaining hundreds of temperature gradient across a thin insulation. The effective thickness of a typical MLI is within few millimeters. MLI found application in storage of cryogenic fluids storage used for space missions as propellant. It is also used in MRI scanning systems to produce high intensity magnetic fields with the help of superconductivity. It also known as a passive thermal control system used in satellites to maintain the temperature of electronic equipment within the working temperature limit.

The radiation heat from the outer space strikes on the first reflective layer where a part is reflected back to the environment, and the remaining radiation energy heats up the first layer of spacer. As the temperature of the layer increases, solid conduction, gas conduction and radiation takes place through the spacer material to the next foil. Thus the second foil temperature will increase. The second foil reflects some radiation back to the first foil and the remaining energy transfers to the third foil. This process continues up to the bottom layer.

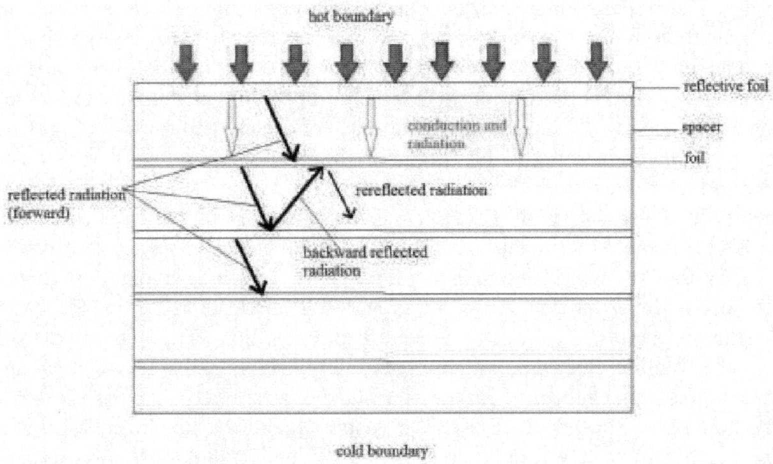

Figure 1. Schematic of MLI

2. Developments in MLI

In the area of multilayer thermal insulation, large quantity of investigation result is generated over the last few decades. Experimental, numerical as well as combined work is being carried out with probable statistics as given in Figure 1. This work is an attempt to review the research progress in the performance of multilayer thermal insulation.

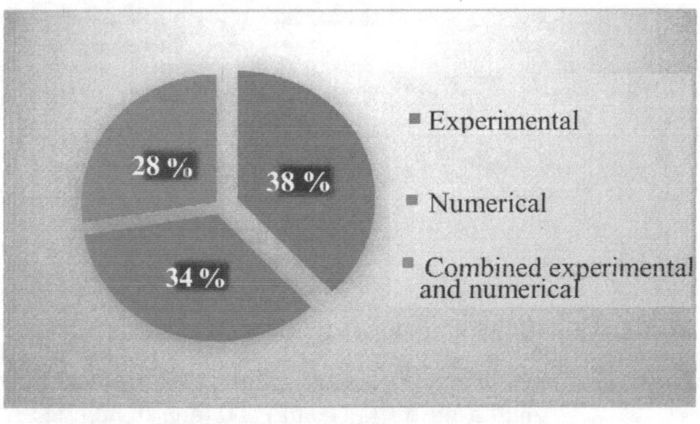

Figure 2. Chart demonstrating the literature

2.1 Experimental investigations

Bapat *et al* [3] conducted an experimental investigation in different combinations of spacer and shield material and compared with numerical model. The result shows that the combination of aluminized Mylar of 12μm thick and glass fabric of 76.2μm thick is the best for MLI. Wei *et al* [8] experimentally investigated the effect of structure and shape of perforated MLI blanket. A new calorimeter with spherical top and bottom is compared with a cylindrical calorimeter and found that 13% excess heat flux is transferred in new calorimeter. It is due to the "edge effect" formed by deformation of MLI. Thus it is concluded that shape and structure will affect the performance of MLI. A steady state experimental investigation is carried out by Sun *et al* [9] to measure the effect of interstitial gas

IOP Conf. Series: Materials Science and Engineering **396** (2018) 012061 doi:10.1088/1757-899X/396/1/012061

pressure (degraded vacuum). In this experiment argon, air, nitrogen and carbon-dioxide are used as the interstitial gas. The experiment is carried out at different heat transfer condition ranging from free molecular to continuum regime with gas pressure ranges from 10^{-3}-10^{5} Pa by using a nitrogen boil-off instrument. The temperature is also varied from 77K to 300K. The energy accommodation coefficient (EAC) introduced by Kogan is the energy exchange between gas and surface which depends on the factors such as surface roughness, gas properties etc. It is calculated for the test gas. Result reveals that the apparent thermal conductivity will be higher at the hot boundary region than the cold boundary and this increase with increase in magnitude of vacuum.

Wesley [13] tested several MLI specimens to find the optimal layer density in Kennedy space Centre, NASA. But the experimental result shows an increase in layer density than the numerical prediction. It is due to the predominance of radiation which is avoided in the model. It is also suggested to use lower layer density for mass and heat load optimization. Heat pipes (HP) are also used in the thermal control system of spacecraft and satellites. They are used to dissipate the heat generated in the satellite electronic components [12]. Taig *et al* [14] used a phase change Material along with heat pipe to enhance the heat transfer rate from the internal components of satellites.

Thomas *et al* [17] used a calorimeter cryostat which is capable of measuring MLI performance at different temperatures. The low temperature cryogenic fluid (LHe) is passed through the test cylinder and the specimen wrapped around it. Thermal performance of high vacuum multilayer insulation (HVMLI) reduces due the leakage of fluid stored in the tank (hydrogen gas is considered). Small amount of nitrogen and oxygen present in between the layers affect the effectiveness of insulation. Getters are used in such case to absorb the gas present in HVMLI. Jian *et al* [23] compared a composite getter of CuO & C with usually used materials to absorb H_2 gas. It is found that CuO & C getter is a cost effective way of absorbing H_2 gas.

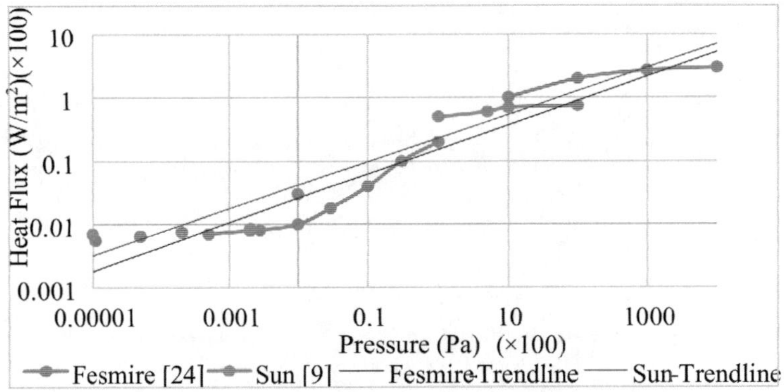

Figure 3. Comparison of heat flux variation with temperature (Sun's [9] and Fesmire's [24] data) and their trend lines.

Fesmire [24] developed a layered composite with aerogel for insulation of non-vacuum application such as space vehicles, propulsion test stand and launch pad and for cryogenic fluid transportation in Kennedy space center, NASA. Johnson *et al* [29] tested the transmissivity of MLI at very low temperature (2K) by using a laser based measuring system. The laser system directly measure the transmittance of MLI and emmitance of aluminium single layer foil. It is found that there is no transmittance of energy through the foil corresponds to 2K. A cylindrical calorimeter is used by Fesmire *et al* [30] to test six different type of MLI. Conclusion is that different combination of MLI is available and the selection of best MLI depends on the application and design of MLI. Table 1 shows a brief idea of experimental investigations in MLI.

Table 1. Experimental investigators in MLI

Author	Parameter analyzed	MLI material	Comments
Bapat [3]	Number of layers and layer density	Few indigenous materials	Thermal conductivity of MLI increases with increase in layer number
Wei [8]	Shape and structure	Double-face aluminum-coated terylene film with fiber paper spacer	Edge effect caused by shape and size leads to reduction in performance of MLI
Sun [9]	Degradation of air	Aluminized Mylar foil with vegetal super-thin fibrous spacer	The effective thermal conductivity increases with temperature and vacuum level
Johnson [13]	Optimal layer density	~	Low layer density will be suitable for thermal and mechanical load optimization.
Thomas [17]	Performance at different temperatures	Double aluminized Mylar with Polyester spacer	Heat load is calculated at different time intervals
Wang [23]	Getters	~	CuO & C is the cost effective getter
Johnson [29]	Transmissivity of MLI	Aluminium single layer foil	Three is no energy transmission through MLI at 2K

2.2 Numerical investigations

Bapat *et al* [2] formulated a numerical model with combined conduction, radiation and gas conduction to predict the performance of MLI with double aluminized Mylar reflective foil and glass fabric spacer. The gas conduction increased with increase in layer density due to the increase in effective thermal conductivity of the insulation.

McIntosh [4] developed a general equation for optimizing heat flux through MLI using a computer program. He used helium as the interstitial gas having a pressure of 1.33E-5 Pa. The radiation heat transfer though MLI is given by

$$Q/A = \sigma\,(T_w^4 - T_c^4)/(1/\varepsilon_w + 1/\varepsilon_c - 1)$$

Where, σ = Stephan-Boltzmann constant = 5.675E(-8) W/m$^2\cdot$K^4

T_w = temperature of the warm surface, K

T_c = temperature of the cold surface, K

ε_w & ε_c are the warm and cold surface emissivities

The gas conduction equation is given by

$$H = C_1\ P\ \alpha\,(T_w - T_c)$$

$$k_g = C_1\,P\,\alpha$$

Where, k_g = gas conduction, W/m$^2\cdot$K

P = gas pressure, Pa

$C_1 = [(\gamma + 1)/(\gamma - 1)].[R/8\pi.M.T]^{1/2}$

α= accommodation coefficient

$\gamma = C_p/C_v$

R = gas constant, 8.31441 J/mol·K

M = molecular weight of gas, kg/mol

T = temperature of vacuum gauge, normally 300 K

The solid conduction equation is given by

$$k_s = C_2\, f\, k / \Delta X$$

k_s = the solid conductivity per unit thickness, W/m^2·K

C_2 = an empirical constant

f = relative density of the separator compared to solid material

k = separator material thermal conductivity, W/mK

ΔX = actual thickness of separator between reflectors, m

These equations are capable of predicting the performance of MLI with a variety of materials by using a computer application like Microsoft Excel. The result concluded that higher layer density near the warm layer can create better performance.

Alifanov *et al* [10] calculated the thermal and radiative properties such as thermal conductance and emissivity by measuring the temperature and heat flux by inverse problem method. Mavromatidis *et al* [11] studied the effect of air gaps between wall and MLI. The test MLI is placed between two wooden walls maintaining a small gap. It is found that the thermal resistance is maximum when the air gap is 3cm. Gongan *et al* [16] designed a 3D model of lightweight thermal protection system (TPS). A corrugated core sandwiched panel is generated using ANSYS and is subjected to mechanical and thermal loads by using ANSYS parametric design language (APDL) codes. The APDL along with GCMMA algorithm is used for the optimization of minimum weight TPS. 37% lighter TPS is obtained after optimization. But the mechanical loads such as transverse load and in-plane mechanical compressive loads are not considered in the optimization procedure.

Table 2. Numerical investigators in MLI

Author	Model	Comments
Bapat [2]	Combined conduction, radiation and gas conduction	The gas conduction will increases with increase in layer density
McIntosh [4]	Conduction, radiation and gas conduction	The equation can be used for optimization by using a computer program
Alifanov [10]	Inverse problem method	Calculated the properties of MLI
Gongan [16]	3D model of lightweight thermal protection system	37% lighter TPS is created through optimization by APDL & GCMMA algorithm
Tigwu [20]	Conduction, radiation and gas conduction	Calculated 2D temperature distribution inside MLI. Increase in number of layer will increase the performance of MLI up to a limit.
Dye [22]	Discrete spacer	Low area to length ratio spacer reduces conduction heat transfer
Zhan[25]	Quasi steady state	Compared MLI and foam insulation
Xie [26]	Transient heat transfer	MLI with phase change material will increase its performance

Tingwu *et al* [20] numerically investigated the influence of factors such as emissivity, thickness, density layout, location and number of the foils, the density of materials and the emissivity of foils. It has been found that the best performance is achieved at a particular number of layers (18 layers). If the number of layer increases the overall conductivity of the insulation increases due to the domination of highly conductive metallic foils. Dye *et al* [22] designed a wrapped MLI with new discrete spacers having low area to length ratio to reduce the conduction heat transfer.

Zhan *et al* [25] compared the thermal performance of form and MLI in a cryogenic liquid hydrogen tank. Quasi steady state model is used for the analysis process. MLI performance is higher than form at higher temperature where the radiation heat transfer is predominant. The performance of MLI can be improved by introducing the phase change material at the middle layer [26]. But the performance enhancement fails if the phase change material is at the bottom of the insulation.

Paola *et al* [27] developed a design criterion for increasing the effectiveness of multilayered wall insulation for building insulation. Table 2 shows a brief idea of numerical investigation in MLI.

2.3 Combined experimental and numerical investigations

Many researchers used combined numerical simulation and experimental works to verify and improve the accuracy in predicting the performance of MLI.

Krishnaprakas *et al* [5] compared the experimental data of heat flux with four empirical models such as conductance (h-model), effective emittance (ε_{eff}-model), Conduction-radiation (CR-model) and Cunnington and Tien (CT-model). It has been concluded that the CT-model can be used to predict the heat flux accurately. Markus *et al* [6] studied the thermal performance of MLI with porous and fibrous materials which are capable of absorbing, emitting and scattering the radiation. Ohmori [7] introduced a dimensionless contact pressure, P^* in wrapped MLI which is independent of specific weight of foil and radius of cylinder. There is high dependence of MLI performance with the contact pressure.

Table 3. Combined experimental and Numerical investigators in MLI

Author	Numerical	Experimental	Comments
Krishnaprakas [5]	Thermal performance prediction by different models	Heat flux is determined	Cunnington and Tien (CT) model can estimates heat flux accurately
Huang [15]	Finite volume method & inverse problem method	Thermal conductivity is measured	Slight variation between experimental and numerical data
Haim [19]	Conductivity of nitrogen	Thermal conductivity of spacer particle	Contribution of spacer particle in heat transfer is very low

Huang *et al* [15] generated a theoretical model with FVM for analyzing the behavior of high temperature MLI by using inverse problem method. The thermal conductivity of the material is experimentally calculated. The theoretical model shows 4% deviation from the experimental thermal conductivity. Takeshi *et al* [18] developed a new type of small discrete non-interlayer-contact spacer MLI (NICSMLI) made of polyetheretherketone (PEEK) for SPICA satellite. A/L (area to length ratio) of spacer is 10^{-5}m. Haim *et al* [19] tested an annular MLI having stainless steel coaxial foils. 50μm spherical zirconia particle is used as the spacer material which is dispersed in between foils with nitrogen. The overall conductivity of annular MLI shows that the contribution of heat transfer through the particle is very low at vacuum condition.

Nast *et al* [21] tested MLI in large Dewar tanks. A simulation model is presented to measure the cryogenic fluid boil off and thermal conductivity of MLI. The test results are used for designing large propellant tanks for NASA. Wang *et al* [28] tested MLI with Dacron net and fiber cloth as spacer material with different layer density. The heat transfer rate will be comparatively large in fiber cloth than Dacron net due to the higher contact area in the cloth. Table 3 shows significant works of combined experimental and numerical investigation in MLI. Also, Figure 4 portraits the comparison of Nast [21] and Wang [28] data for variation of heat flux with temperature

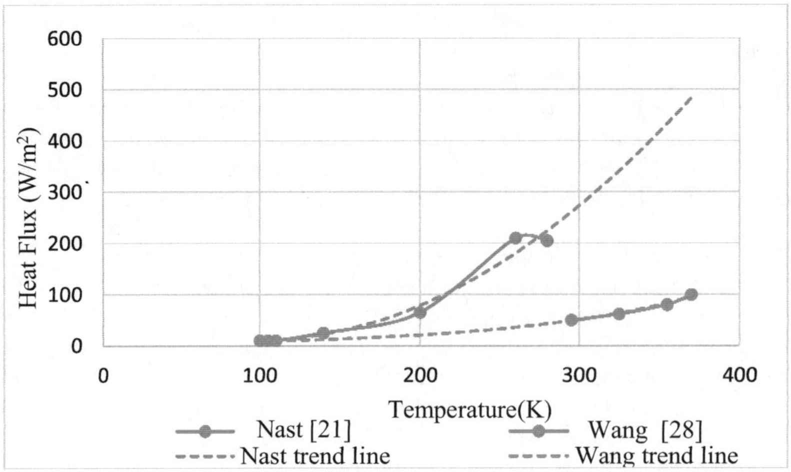

Figure 4. Heat flux variation with temperature (comparison of Nast and Wang data)

3. Conclusions

This review has provided concise summary of investigations on MLI. Based on the studies, MLI is an important insulating material in the field of cryogenic and space industry. The performance prediction of MLI with minimum error is a crucial step in the design of TPS for spacecraft. The performance of MLI is affected by many factors such as contact pressure, boundary temperature, interstitial gas and its pressure etc. Numerical models and simulation software are helpful in analyzing MLI. The experimental method does not match with the exact working condition of MLI in space, but the experimental data can be used to predict the performance of MLI. The use of MLI helps to reduce energy consumption for pumping of heat and heat removal.

4. References

[1] Meseguer J, Perez G and Sanz A 2012 *Spacecraft thermal control* (Cambridge : woodhead publishing limited) chapter **7** pp 111-118

[2] Bapat S L, Narayankhedkar K G and Lukose T R 1990 Experimental investigations of multilayer Insulation *Cryogenics* 1990 Vol **30**

[3] Bapat S L, Narayankhedkar K G and Lukose T R 1990 Performance prediction of multilayer insulation *Cryogenics* Vol **30**

[4] McIntosh G E Layer by layer mli calculation using a separated mode equation *Cryogenic Technical SeJVices Inc*. pp 1683-1690

[5] Krishnaprakas C K, Badari N K and Pradip D 2000 Heat transfer correlations for multilayer insulation systems *Cryogenics* Vol **40** pp 431-435

[6] Markus S, Edgar R and Raymond Viskanta 2004 Studies on high-temperature multilayer thermal insulations *IJHMT* Volume **47** Issues 6–7 pp 1305-1312

[7] Ohmori T 2006 Thermal performance of multilayer insulation around a horizontal cylinder. *Cryogenics* **45** pp725–732

[8] Wei W, Xiangdong L, Rongshun W and Yang L 2009 Effects of structure and shape on thermal performance of Perforated Multi-Layer Insulation Blankets *App Therm Engg* **29** pp 1264–1266

[9] P J Sun, J Y Wu, P Zhang, L Xu and M L Jiang 2009 Experimental study of the influences of degraded vacuum on multilayer insulation blankets *Cryogenics* **49** pp 719–726

[10] O M Alifanov, A V Nenarokomov and V M Gonzalez 2009 Study of multilayer thermal

insulation by inverse problems method *ActaAstronautica* **65** pp 1284–1291

[11] L E Mavromatidis, A Bykalyuk, Mohamed E M, P Michel and M Santamouris 2012 Numerical estimation of air gap's influence on the insulating performance of multilayer thermal insulation, *Building and Environment* **49** pp 227-237

[12] Philippe G, Qing M, Tao Y, Peter S, Laurent G, Pierre T and Jingtao L 2011 Thermal behavior of a cryogenic loop heat pipe for space application *Cryogenics* **51** pp 420–428

[13] Wesley J 2012 Thermal analysis of low layer density multilayer insulation test results *AIP Conference Proceedings pp* 1434- 1519

[14] Taig Y, Bum S, Jang J and Juhun R 2013 Numerical study of the spacecraft thermal control hardware combining solid–liquid phase change material and a heat pipe *Aerospace Sci and Tech* **27** pp 10–16

[15] Huang C and Zhang Y 2014 Calculation of High-temperature Insulation Parameters and Heat Transfer Behaviors of Multilayer Insulation by Inverse Problems Method *Chines journal of aeronautics* **27** pp 10-16

[16] Gongnan X, Qi W, Bengt and Weihong Z 2013 Thermomechanical optimization of lightweight thermal protection system under aerodynamic heating *Appl Therm Engg* **59** pp 425-434

[17] Thomas F and Christoph H 2014 A calorimeter for measurements of multilayer insulation at variable cold temperature *25th International Cryo Engg Conference and the International Cryogenic Materials Conference. Physics Procedia* **67** pp 1062 – 1067

[18] Takeshi M, Ryuta H, Hiroyuki S, Masanori S and Tomoyuki H 2014 Development of a new multi-layer insulation blanket with non-interlayer-contact spacer for space cryogenic mission *Cryogenics*

[19] Y Haim, Y Weiss and R Letan 2014 Effect of spacers on the thermal performance of an annular multi-layer Insulation *Applied Thermal Engineering* **65** pp 418-421

[20] Tingwu J, Ruiping Z, BengtSunden and Gongnan X 2014 Investigation on thermal performance of high temperature multilayer insulations for hypersonic vehicles under aerodynamic heating condition *Applied Thermal Engineering* **70** pp 957-965

[21] Nast T C, Frank D J and Feller J 2014 Multilayer insulation considerations for large propellant tanks *Cryogenics*

[22] Dye S A, Tyler P N, Mills G L and Kopelove A B 2014 Wrapped multilayer insulation design and testing *Cryogenics* **64** pp 100–104

[23] Jian W, Ying Z, Wei W, Shujun C and Rongshun W 2016 A new cost effective composite getter for application in high-vacuum-multilayer insulation tank *Vaccum*

[24] J Fesmire 2015 Layered composite thermal insulation system for nonvacuum cryogenic Applications *Cryogenics*

[25] Zhan L, Yanzhong L, Fushou X and Ke Z 2016 Thermal performance of foam/MLI for cryogenic liquid hydrogen tank during the ascent and on orbit perio *Appl Therm Engg* **98** pp 430–439

[26] Tao Xie, Ya-Ling He and Zi Xiang Tong 2016 Analysis of insulation performance of multilayer thermal insulation doped with phase change material *IJHMT* **102** pp 934–943

[27] Paola G, Claudia G, Luca E and Francesco A 2016 Design criteria for improving insulation effectiveness of multilayer walls *IJHMT* **103** pp 349–359

[28] Wang, Huang Y, Li P, Sun P, Chen Z and Wu J 2016 Optimization of variable density multilayer insulation for cryogenic application and experimental validation *Cryogenics* **80** pp 154–163

[29] Johnson W L, Van Dresar N T, Chato D J and Demers J R 2017 Transmissivity Testing of Multilayer Insulation at Cryogenic Temperatures *Cryogenics*

[30] Fesmirea J E and Johnsonb W L 2018 Cylindrical cryogenic calorimeter testing of six types of multilayer insulation Systems, *Cryogenics* **89** pp 58–75

IOP Conf. Series: Materials Science and Engineering **396** (2018) 012062 doi:10.1088/1757-899X/396/1/012062

Pyrolysis process to produce fuel from different types of plastic – a review

Anandhu Vijayakumar[1*], Jilse Sebastian[2]

[1] Graduate Student
Mechanical Engineering Department, St. Joseph's College of Engineering and Technology, Palai, Kerala, India
[2] Assistant Professor
Mechanical Engineering Department, St. Joseph's College of Engineering and Technology, Palai, Kerala, India

[*]E – mail: anandhuvijaykumar@gmail.com

Abstract: Fast exhaustion of oil resources and increase in energy demand have focused the researchers to find alternate ways to produce high quality oils that could replace fossil fuels. The idea of waste to energy recovery is one of the promising techniques for managing the waste plastic. Waste plastics are attractive for energy conversion because of their high heat of combustion and bulk availability. Exponential rate of increase in plastic production happens in every year due to the wide range of plastic appliances in domestic as well as industrial purposes. The drastic increase in the plastics production naturally lead to large amount of plastic waste that endangers the environment because of their disposal problems. The conversion of plastic to high quality liquid oil through pyrolysis process is highly advisable as the oil produced has high calorific value than that of commercial fuel. This paper describes commonly used verities of plastics and potential of pyrolysis process to produce fuel using them.

Key words: Plastic waste, Pyrolysis, Fuel production, Energy recovery

1. Introduction

Plastics are organic compounds having long chained hydrocarbon synthesized from petroleum products. Because of its own special features plastic had acquired wide popularity in short time. Plastic production and consumption rate increased exponentially due to its low cost, non-degradable nature, easy availability and management, wide range of usage and application. According to the estimate given by APME (Association of Plastic Manufacturers Europe), the global production of plastic has crossed 280 million tons in 2011 and it is increasing exponentially [1]. The increasing demand of plastic products also increases the accumulation of plastic waste that endangers the environment because of their disposal problems [2]. The rising plastics demand also led to the exhaustion of non-renewable crude oil as plastics are petroleum-based material. To minimize the adverse environmental impacts of plastic waste, many organizations are implemented with plastic waste management systems, for control/reduction of plastic waste formation.

Various methods of waste plastic treatment can be categorized as follows [3, 4]:

1) Plastic scrap can be re-introduced in the heating cycle of plastic processing line itself to increase production (primary method);
2) Mechanical recycling (secondary methods): mechanical (physical) treatment to re-extrude, process and convert plastic solid waste to new plastic products blended with virgin polymers to reduce the overall production cost;
3) Chemical (tertiary) methods: chemical alteration in the polymer structure of plastic material can be performed through chemical or thermo-chemical means to utilize it as a monomer feedstock in industrial recycling loops and;
4) Energy recovery: which entails recovery of steam, heat and electricity from waste plastic through combustion.

2. Plastic for fuel production

Recycling of plastic is difficult and costly because of the restrictions on contamination of water and labor intensive segregation of different plastics before recycle which is labor intensive [5]. Segregation of different plastic materials are essential since they are made of different resin compound for difference in transparency and color. Dyed or pigmented plastics have a lower market value too. Clearly transparent plastics can be easily dyed to transform into new products, have greater flexibility and are mostly desirable by the manufacturers [6, 7]. Recycling plastic is energy intensive too. As there is an alarming depletion of energy sources, means of energy recovery from plastic waste is a good option. Pyrolysis is a suitable method for energy recovery from plastic waste and is one of the finest techniques for the conversion of mass to energy with liquid and gaseous products with high energy values [8]. Figure 1 represents the processes involved in the pyrolysis of plastic.

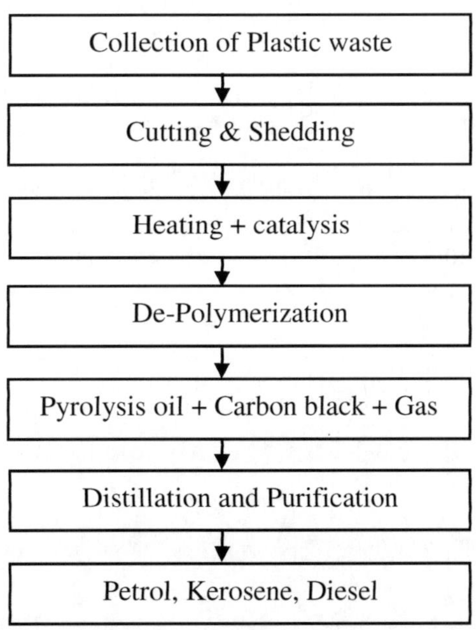

Figure 1. Flow chart of plastic pyrolysis process

Pyrolysis or thermal cracking involves thermal degradation of long chain polymer molecules into less complex smaller molecules. The process takes place in the absence of oxygen at increased pressure and temperature for a short duration. Pyrolysis process is proposed by many researchers since the process is able to produce large quantity of liquid oil up to 80 wt% at temperatures around 500°C [9]. The process parameters can be altered to generate products based on personal preferences. Hence pyrolysis is often referred as a flexible process. The liquid oil produced is of high quality as it can be used in multiple applications without any upgradation or treatment [8]. The gaseous fuel produced as the byproduct of pyrolysis, can be reused to compensate the energy requirement of the pyrolysis plant as the gaseous fuel produced is of high calorific value [12]. Pyrolysis is mostly employed over common recycling processes since handling is much easier and flexible. Moreover, pyrolysis does not require intense sorting process and hence it is less labor intensive process.

Many published research papers are available on the potential of pyrolysis processes on various types of plastics for liquid fuel production. The present review comprehends the properties and use of various categories of plastics and description of pyrolysis process for fuel production for each category.

3. Thermal pyrolysis process

The commonly used plastics can be classified into six different types of plastics such as Polyethylene terephthalate (PET), High-density polyethylene (HDPE), Polyvinyl chloride (PVC), Low-density polyethylene (LDPE), Polypropylene (PP) and Polystyrene (PS). The pyrolysis process results of each of these are described in the following sections.

3.1. Polyethylene terephthalate

PET plastics are mainly used for packaging of food products, fruit juice containers, mineral water bottles, soft drink bottles etc. PET plastics have other applications in the production of electrical insulation, printing sheets, magnetic tapes, X-ray and other photographic films [11]. The extensive applications of PET plastics contribute huge amount of plastic waste that endangers the environment because of their disposal problems. Recycling PET waste is the current method for reducing PET plastic waste. Large use of containers causes fast accumulation of PET plastics which requires high frequency of collections and hence, increases the transport costs. Cepeliogullar and Putun [10] have conducted the pyrolysis of PET to produce liquid oil using fixed-bed reactor at 500°C. In the experiment, the rate of heating was 10°C/min. Nitrogen gas was used as the sweeping gas. The gaseous product obtained was 76.9 wt % and the liquid oil obtained was 23.1wt%. The pyrolysis oil had acidic characteristic which is unfavorable to use in I.C. engines directly due to its corrosiveness that deteriorates the fuel quality [13]. Benzoic acid was a general sublime but needed a serious attention if running in industrial scale [20, 21]. Another liquid oil with slightly higher yield at the same operating temperature and heating rate was found by Fakhrhoseini and Dastanian [15]. The liquid yield was found to be 39.89 wt%, gaseous was 52.13 wt% while solid residue was 8.98 wt%.PET pyrolysis resulted in the generation of liquid oil in the ranges of 23 – 40 wt% while gaseous was in the ranges of 52 – 77 wt%. The polymer structure of PET is shown in Figure 1, together with its most likely thermal cleavage mechanisms.

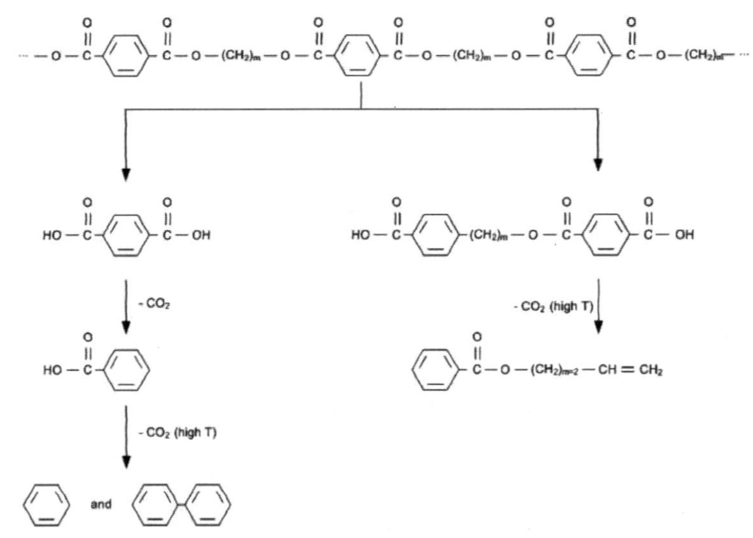

Figure 2. Chemical reaction involved in pyrolysis [35].

3.2. High-density polyethylene

HDPE is a long linear polymer chain having high strength properties. The features contributing to the strength are high degree of crystallinity and low branching. There are wide range of applications for HDPE. It is typically used for the production of milk bottles, detergent bottles, oil containers, toys and many more. It is the third largest plastic type found in MSW (Municipal Solid Waste) [14, 16].

Ahmad et al. [17] conducted the pyrolysis of HDPE using a steel reactor at a temperature range of 300–400°C. The fluidizing medium used was nitrogen. 80.88 wt% of liquid was obtained and the total conversion was happened to be at 350°C. 33.05 wt% solid residue was obtained at 300°C which found to be large. Thermal pyrolysis study of HDPE was conducted by Kumar and Singh using a semi-batch reactor at higher temperature of 400 – 550°C. The highest liquid yield and gaseous product (79.08 wt% and 24.75 wt% respectively) was obtained at a temperature of 550°C. Also a dark brownish oil was obtained. Beltrán et al. [20] have also conduct pyrolysis of HDPE plastics at 550°C. The liquid oil yield obtained was 84.7 wt% and gaseous product was found to be 16.3 wt%. This results show that higher liquid oil yield is possible from HDPE at higher temperature. However, too high temperature would reduce the liquid oil yield and would increase the gaseous product yield as the process may go beyondthe maximum thermal degradation point of HDPE. This result was proved by the experiment on HDPE pyrolysis in a fluidized bed reactor at 650°C by Mastral et al [21]. The experiments provided a liquid oil production of around 68.5 wt% and 31.5 wt% of gaseous product. So it can be concluded that at high temperature above 550°C, the liquid was cracked to gaseous state.

3.3. Polyvinyl chloride

PVC is exceptional from other thermoplastics such as polyethylene, polystyrene and polypropylene and it is manufactured from the mixture of chlorine (57%) and carbon (43%) [22]. Due to the chlorine content, PVC have high fire resistance, and is suitable for electrical insulation. The works done on the PVC pyrolysis are very less due to the hydrogen chloride (HCl) fumes that is released when heated at high temperature. Jin et al. [23] conduct pyrolysis of PVC in a batch reactor at temperature range of 220 – 520°C and with a heating rate of 10°C/min. Liquid oil obtained was of range 0.45 wt% to 12.79 wt%. 58.2 wt% of HCl was also obtained from the experiment.

3.4. Low-density polyethylene

LDPE has lower tensile strength and hardness. This is because of the weaker intermolecular force of LDPE. In comparison with HDPE, LDPE has more branching. LDPE waste is the second largest plastic waste in MSW after PP [24]. Bagri and Williams [25] conducted LDPE pyrolysis in fixed-bed reactor at a temperature of 500°C. The experiment was conducted for a duration of 20 min and nitrogen gas was used as fluidizing medium. The liquid yield obtained was 95 wt% with low gas yield and negligible amount of char. 93.1 wt% of liquid oil was obtained when the same experiment was conducted in a batch reactor at a temperature of 550°C by Marcilla et al. [24]. From the research conducted by Koizumi et al. [26], 75.6 wt% of liquid oil was obtained in a batch reactor at 400°C. Aguado et al. [27] have obtained 74.7 wt% of liquid oil when using the same type of reactor at a temperature of 450°C. Onwudili et al. [28] conducted pyrolysis of LDPE at a temperature of 425°C in a pressurized batch reactor (0.8 – 4.3 MPa). The liquid oil yield obtained was 89.5%, gas obtained was 10 wt% and char was 0.5 wt%.

3.5. Polypropylene

PP have high chemical as well as heat resistance properties. It is a saturated polymer having linear hydrocarbon chain. It does not melt below than 160°C. PP has high hardness and rigidity which makes it preferable for plastic industry. Unlike HDPE, PP has a lower density. It is the largest contributor of plastics in MSW [22]. The diversified applications of PP include making of flowerpot, office folders, car bumpers, buckets, carpets, various furniture, storage boxes etc. Many researchers have studied the process of pyrolysis of PP by altering the parameters to optimize the liquid oil yield. One experiment conducted by Ahmad et al. [17] on PP pyrolysis was done at 250 – 400°C temperature. From the experiments it was found that 69.82 wt% of liquid oil was achieved at temperature of 300°C. Pyrolysis of PP was conducted by Sakata et al. [29] on PP at temperature of 380°C and was obtained a liquid yield of 80.1 wt%, gas yield of 6.6 wt% and obtained a 13.3 wt% solid residue. Fakhrhoseini and Dastanian [15] conducted PP pyrolysis at 500°C and obtained 82.12 wt% of liquid yield. The liquid yield was reduced when the temperature is more than 500°C. Demirbas [30] conducted pyrolysis of PP at 740°C in a batch reactor and 48.8 wt% liquid, 49.6 wt% gas and 1.6 wt% char yield were obtained. This experiment inferred a reduction in liquid yield with increasing temperature.

3.6. Polystyrene
PS is synthesized from the liquid petrochemical and is the polymer of styrene monomers. It has compounds of long hydrocarbon chain with phenyl group attached to every carbon atom. It is a colorless compound, but can be colored with suitable colorants. Due to its reasonable durability, strength and lightness, PS is used in variety of applications including in food packaging, electronics, construction, medical, appliances, toys and many more. As the range of applications is higher, large waste accumulation is there. PS cannot be included in the roadside recycling program as the recycling bins only include glasses, papers, cans, and other light plastics. People won't throw the foam food packaging into plastics recycle bin but will be put into the general bin. So recycling is challenging in the case of PS. The only way for the recovery of PS waste is pyrolysis process which converted it into high quality liquid oil. Pyrolysis of PS have been studied by many researchers. Of them Onwudili et al. [28] have conducted pyrolysis process in a pressurized autoclave reactor. The experimental duration was one hour and was conducted at a temperature of 300°C-500°C. The heating rate used was 10°C and the pressure applied was in the range of 0.31MPa-1.6MPa. From the experiment, the oil yield obtained was around 97.0 wt% at a temperature of 425°C. Maximum gas production was only 2.5 wt%. Liu et al. [31] also obtained high yield of liquid oil in their experiments. They used fluidized bed reactor at temperature of 450 – 700°C. 98.7 wt% of liquid oil was obtained at a temperature of 600°C. The liquid oil production was also high at lower temperatures of about 450°C. However, the experiments done by Demirbas [30] on PS resulted a decreased liquid oil production. The production was 89.5 wt% at 581°C in a batch reactor.

3.7. Mixed plastics
Unlike recycling, pyrolysis does not require a keen sorting of different plastics. It is one of the most meritorious point of pyrolysis of plastics. Most plastics are not compatible with each other and hence cannot be processed together during recycling. For example, a slight presence of PVC in PET recycle stream will degrade the whole PET resin by becoming yellowish and brittle [32]. Schlesselmann et al. [33] investigated the pyrolysis of mixed plastic wastes by collecting German households which consist 75% of polyolefin (mixture of PE and PP) and 25% PS.

Similar liquid oil production was found in another study conducted by Demirbas [30]. The experiment was conducted on the polyolefin and PS mixture collected from landfill which was approximately 46.6 wt% of total plastic waste. The respective yields of gas and solid were reported as 35 wt% and 2.2 wt%. The composition of oil showed the presence of about 4 ppm chlorine. Donaj et al. [34] conducted a study on pyrolysis of polyolefin mixed plastics. The mixed plastics consist 75 wt% LDPE, 30 wt% HDPE and 24 wt% PP. The experiments were conducted at temperatures of 650°C and 730°C in a fluidized bed reactor. In this experiment the liquid obtained was more at lower temperature of 650°C which was quantified around 48 wt%.

4. Conclusions
This review provides a concise summary of different types of plastics and the potential of pyrolysis for fuel production. The summary of pyrolysis results of the various category plastics can be consolidated as in Table 1.

Table 1. Summary of studies on plastic pyrolysis

Reference	Types of plastic	Reactor	Process Parameters			Products		
			Temperature	Pressure	Duration (min)	Oil (wt%)	Gas (wt%)	Solid (wt%)
[11]	PET	Fixed bed	500	–	–	23.1	76.9	0
[15]	PET	–	500	1 atm	–	38.89	52.13	8.98
[14]	HDPE	Horizontal steel	350	–	30	80.88	17.24	1.88
[16]	HDPE	Semi-batch	400	1 atm	–	82	16	2
[17]	HDPE	Batch	450	–	60	74.5	5.8	19.7
[20]	HDPE	Fluidized bed	500	–	60	85	10	5
[21]	HDPE	Batch	550	–	–	84.7	16.3	0
[19]	HDPE	Fluidized bed	650	–	20–25	68.5	31.5	0
[22]	PVC	Fixed bed	500	–	–	12.3	87.7	0
[24]	PVC	Vacuum batch	520	2 kPa	–	12.79	0.34	28.13
[25]	LDPE	Pressurized batch	425	0.8–4.3 MPa	60	89.5	10	0.5
[31]	LDPE	–	500	1 atm	–	80.41	19.43	0.16
[28]	LDPE	Fixed bed	500	–	20	95	5	0
[26]	LDPE	Batch	550	–	–	93.1	14.6	0
[29]	LDPE	Fluidized bed	600	1 atm	–	51.0	24.2	0
[22]	PP	Horizontal steel	300	–	30	69.82	28.84	1.34
[17]	PP	Batch	380	1 atm	–	80.1	6.6	13.3
[31]	PS	Semi-batch	400	1 atm	–	90	6	4

Based on the studies, pyrolysis process is approved as a potential method to generate energy from plastic waste. It resulted in the production of valuable liquid oil, gaseous fuel and char instead of waste accumulation in landfills and hence contribute heavily to environmental pollution. For mini-scale batch pyrolysis process, liquid fuel is preferable both in the view of fuel handing and its storage. The plastic category PS found to give maximum liquid yield whereas PET category, the least. Pyrolysis process is not advisable for PVC category as it produce HCl fumes which is highly toxic. But pyrolysis is a reliable and sustainable method to find a solution for the problem of plastic waste accumulation in the country which is increasing tremendously day by day. Waste management can be effectively done by the process of pyrolysis of plastic wastes. The process is economical and cost effective too. Moreover, the proper use of the advantages of the pyrolysis process can reduce the dependence over the conventional energy sources like fossil fuels. Thus the rise in the energy demand can be reduced.

5. References

[1] *An analysis of European plastics production, demand and waste data-Belgium* (Association of Plastic Manufacturers Europe) pp 1–32
[2] *Common wastes and materials US* 2014 (U.S. Environmental Protection Agency)
[3] Zannikos F, Kalligeros S, Anastopoulos G and Lois E 2013 Converting biomass and waste plastic to solid fuel briquettes *J Renew Energy* **9**
[4] Kukreja R2009 Advantages and disadvantages of recycling *Conserve Energy Future*
[5] Masanet E, Auer R, Tsuda D, Barillot T and Baynes A 2002 An assessment and prioritization of

design for recycling guidelines for plastic components in electronics and the environment *IEEE international symposium* pp 5 – 10
[6] Heikkinen JM, Hordijk JC, de Jong W and Spliethoff H 2004 Thermogravimetry as a tool to classify waste components to be used for energy generation *J Anal Appl Pyrol* **71** pp 883–900
[7] Ahmad I, Ismail Khan M, Ishaq M, Khan H, Gul K and Ahmad W 2013 Catalytic efficiency of some novel nanostructured heterogeneous solid catalysts in pyrolysis of HDPE *Polym Degrad Stab* **98** pp 2512–19
[8] Bridgwater AV 2012 Review of fast pyrolysis of biomass and product upgrading biomass *Bioenergy* **38** pp 68–94
[9] Abnisa F and Wan Daud WMA 2014 A review on co-pyrolysis of biomass: an optional technique to obtain a high-grade pyrolysis oil *Energy Convers Manage* **87** pp71–85.
[10] Çepeliog˘ullar Ö, Pütün AE 2013 Thermal and kinetic behaviors of biomass and plastic wastes in co-pyrolysis. Energy Convers Manage **75** pp 263–70.
[11] Fakhrhoseini SM and Dastanian M 2013 Predicting pyrolysis products of PE, PP, and PET using NRTL activity coefficient model *Hindawi Publishing Corporation* pp 1–5
[12] Cepeliogullar O and Putun AE 2013 Utilization of two different types of plastic wastes from daily and industrial life. In: ed Ozdemir C, Sahinkaya S, Kalipci E and Oden (MK *ICOEST Cappadocia)* pp 1–13
[13] Wan Ho M 2015 *Waste plastics into fuel oil* (UK Institute of Science in Society)
[14] Shioya M, Kawanishi T, Shiratori N, Wakao H, Sugiyama E and Ibe H 2005 Development of waste plastics liquefaction technology, *feedstock recycling in Japan* ed Muller-Hagedorn M and Bockhorn H (Feedstock recycling of plastics, Germany) pp 19–42
[15] Fakhrhoseini SM and Dastanian M 2013 Predicting pyrolysis products of PE, PP, and PET using NRTL activity coefficient model (Hindawi Publishing Corporation) pp 1–5
[16] Michael P A 2010 Plastic waste total in MSW *Society of the Plastic Industry*
[17] Ahmad I, Khan MI, Khan H, Ishaq M, Tariq R and Gul K 2014 Pyrolysis study of polypropylene and polyethylene into premium oil products *Int J Green Energy* **12** pp 663–71.
[18] Kumar S and Singh RK 2011 Recovery of hydrocarbon liquid from waste high density polyethylene by thermal pyrolysis. Braz J Chem Eng **28** pp 659–67.
[19] Boundy B, Diegel SW, Wright L and Davis SC 2011 *Biomass energy data book* (US Oak Ridge National Laboratory) **4**th ed
[20] Marcilla A, Beltrán MI and Navarro R 2009 Thermal and catalytic pyrolysis of polyethylene over HZSM5 and HUSY zeolites in a batch reactor under dynamic conditions. *Appl Catal B Environ* **86** pp 78–86
[21] Mastral F, Esperanza E, Garcia P and Juste M 2001 Pyrolysis of high-density polyethylene in a fluidized bed reactor Influence of the temperature and residence time *J Anal Appl Pyrol* **63** pp 1–15
[22] *Polyvinyl chloride (PVC)* 2015 British Plastics Federation London
[23] Miranda R, Jin Y, Roy C and Vasile C 1998 Vacuum pyrolysis of PVC kinetic study *Polym Degrad Stab* **64** pp 127–44
[24] López A, de Marco I, Caballero BM, Laresgoiti MF and Adrados A 2011 Dechlorination of fuels in pyrolysis of PVC containing plastic wastes *Fuel Process Technol* **92** pp 253–60
[25] Bagri R and Williams PT 2001 Catalytic pyrolysis of polyethylene. J Anal Appl Pyrol **63** pp 29–41
[26] Uddin MA, Koizumi K, Murata K and Sakata Y 1996 Thermal and catalytic degradation of structurally different types of polyethylene into fuel oil *Polym Degrad Stab* **56** pp 37–44
[27] Aguado J, Serrano DP, San Miguel G, Castro MC and Madrid S 2007 Feedstock recycling of polyethylene in a two-step thermo-catalytic reaction system *J Anal Appl Pyrol* **79** pp 415–23
[28] Onwudili JA, Insura N and Williams PT 2009 Composition of products from the pyrolysis of polyethylene and polystyrene in a closed batch reactor: effects of temperature and residence time *J Anal Appl Pyrol* **86** pp 293–303

IOP Conf. Series: Materials Science and Engineering **396** (2018) 012062 doi:10.1088/1757-899X/396/1/012062

[29] Sakata Y, Uddin MA and Muto A 1999 Degradation of polyethylene and polypropylene into fuel oil by using solid acid and non-solid acid catalysts *J Anal Appl Pyrol* **51** pp 135–55
[30] Demirbas A 2004 Pyrolysis of municipal plastic wastes for recovery of gasolinerange hydrocarbons *J Anal Appl Pyrol* **72** pp 97–102
[31] Liu Y, Qian J and Wang J 1999 Pyrolysis of polystyrene waste in a fluidized-bed reactor to obtain styrene monomer and gasoline fraction *Fuel Process Technol* **63** pp 45–55
[32] Hopewell J, Dvorak R and Kosior E 2009 Plastics recycling: challenges and opportunities *Philos Trans R Soc London B Biol Sci* pp 2115–26
[33] Kaminsky W, Schlesselmann B and Simon CM 1996 Thermal degradation of mixed plastic waste to aromatics and gas *Polym Degrad Stab* **53** pp 189–97
[34] Donaj PJ, Kaminsky W, Buzeto F, Yang W 2012 Pyrolysis of polyolefins for increasing the yield of monomers' recovery *Waste Manage* **32** pp 840–6
[35] Anke Brems , Jan Baeyens , Carlo Vandecasteele and Raf Dewil Polymeric cracking of waste polyethylene terephthalate to chemicals and energy *Journal of the Air & Waste Management Association* **61** pp 721-731

IOP Conf. Series: Materials Science and Engineering **396** (2018) 012063 doi:10.1088/1757-899X/396/1/012063

Development of benchmark for simulation of Mode I time based fracture

A V Gopal, A Jamil

Department of Mechanical Engineering,
Z.H.C.E.T, Aligarh Muslim University, Aligarh-202002 (U.P.), India

azharjamil@zhcet.ac.in

Abstract. Time based fracture is a material failure phenomenon in industrial components which are under stress over a long period of time. In this work an attempt is made to numerically model this time based fracture corresponding to industrial components made of P92 steel thereby providing solutions to structural integrity using the commercial code of ABAQUS®. As such benchmark problems in pure Mode I, *viz.* a compact test Specimen *C(T)* was modelled using static and viscoelastic analysis and fracture parameter *J* & *C** were evaluated respectively. The results were compared with experimental results from literature revealing a good agreement between them, thereby validating the *FE* model. The knowledge gained from this quest is proposed to be further extended to industrial components prone to time based fracture like pressure vessels and pipes.

1. Introduction

Time based fracture mechanics is based on the study of fracture mechanics in which time is also considered as a variable. Time based fracture incorporate nonlinear, time-dependent material behaviour. In these cases the classical approach based on linear elastic fracture mechanics and elastic plastic fracture mechanics which assume the quasi-static, rate independent deformation is not adequate. This time based fracture study is a class of dynamic fracture mechanics which is based on rapid crack propagation. The crack initiation can occur either by quasi-static or rapid application of load. Structural components such as fuselage of rockets, oil carrying pipelines, beams and columns and other structural components are made up of material medium which may have internal flaws. Even with the most precise manufacturing methods there can be some material flaw which might be present in components. With passage of time even if these components are subjected to loads which are static in nature, material creep occurs thus causing deformation in the components during these deformation there is always a possibility that the material flaw present in these components causes crack propagation in the components and fracture occurs, thus causing failure of these components.

Experimental analysis and Analytical solutions for complex Industrial components is a cumbersome task. Thus it is much more feasible to first develop a numerical solution for Simple standard specimen for which analytical or experimental solutions can be found and then these results are applied on complex industrial pipes.

2. Literature Review

Among the basic works, J.R. Rice [1] gave the definition of a path independent integral which gave the same value around a 2-dimensional strain field for every path around a notched region.

This helped in evaluation of complicated stress concertation problems. A.Zahoor *et al*[2] evaluated *J*-Integral for a circumferentially cracked pipe. B.Moran *et al* [3] evaluated the crack tip integrals and calculated point wise release rates along a curvilinear three dimensional crack front. D.M. Finch *et al* [4] used ABAQUS® technique to evaluate interactions between applied & typical residual stress distribution effects in various welded Joints. N.P.O'Dowd *et al* [5] provided definitions of two parametrs *J* & *Q* which provided stress distribution over a large & small areas near crack tip. Y.Lei, *et al* [6] gave definition of path independent *J*-Integral in a residual stress field. Y.Lei [7]evaluated the value of *J*-integral for a crack in a welding residual stress field using the modified *J* approach & ABAQUS® version 6.11 & above.

R. Ehlers *et al* [8] developed a finite element program to analyse the deformation of creeping body. A. Saxena *et al* [9] discussed developments in methods for characterizing creep crack growth in metals. P. A. O'Connell *et al* [10] studied relationship between slow crack propagation & tensile creep behaviour in polyethylene. Frederic dubious *et al* [11] studied the effects of viscoelastic characteristics, on the creep-crack growth process. N. Le *et al* [12] conducted experiments to study the creep & fatigue growth behaviour. M. Yatomi *et al* [13] presented a numerical study of creep crack growth in a fracture mechanics specimen. Masataka *et al* [14] studied virtual methods of predicting creep crack growth *CCG*, using finite element analysis, are implemented in a compact tension specimen, *C(T)*.Masato Yamamoto *et al* [15] studied effects of constraints on creep crack propagation. M. Yatomi *et al* [16] proposed a modified *NSW* Model for plane stress and plane strain conditions. J.P. Tan *et al* [17] analysed the effects of constraints in axially cracked pipelines. F. Bassi *et al* [18] performed High temperature assessment of crack like defects in power plant components. Y. Lei [19] performed validation of contour integral functions. M. Saber [20] presented a doctoral thesis by conducting experimental and finite element studies of creep and creep crack growth in P91 and P92.

3. Methodology

The methodology involved is straight forward and is represented in two separate modules corresponding to static *J* -Integral and time dependent C^* Integral. *J*-Integral which is a path independent integral giving a measure of strain energy release rate *SERR* was primarily evaluated by the numerical implementation in ABAQUS® and was compared with the analytical solutions using Equation 1 taken from [21].

$$K_I = \frac{PY}{BW^{1/2}} \tag{1}$$

where, *P* is remotely applied load, *B* and *W* are the thickness and width of the specimen respectively, as seen in Figure 1. *Y* is shape parameter defined by Equation 2

$$Y = \frac{2+h}{(1-h)^{3/2}}[0.886 + 4.64h - 13.32h^2 + 14.72h^3 - 5.60h^4] \tag{2}$$

Such that

$$h = \frac{a}{W} \tag{3}$$

Where, *a* represents the crack length.

The methodology involved in the numerical modelling and validation for the static *J* -Integral and time dependent C^* Integral is represented in Figures 2.

IOP Conf. Series: Materials Science and Engineering **396** (2018) 012063 doi:10.1088/1757-899X/396/1/012063

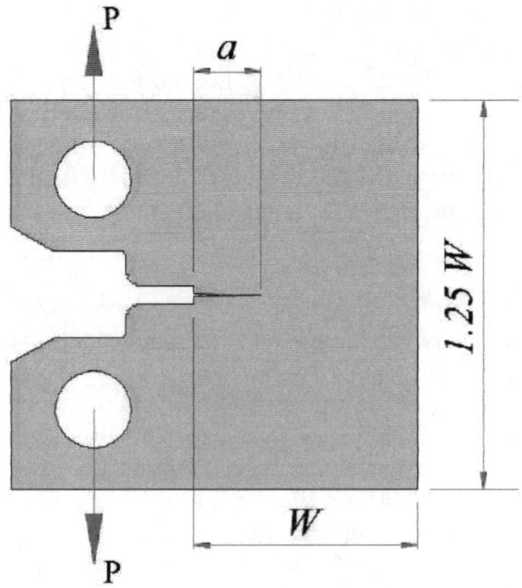

Figure 1. Compact Specimen C(T).

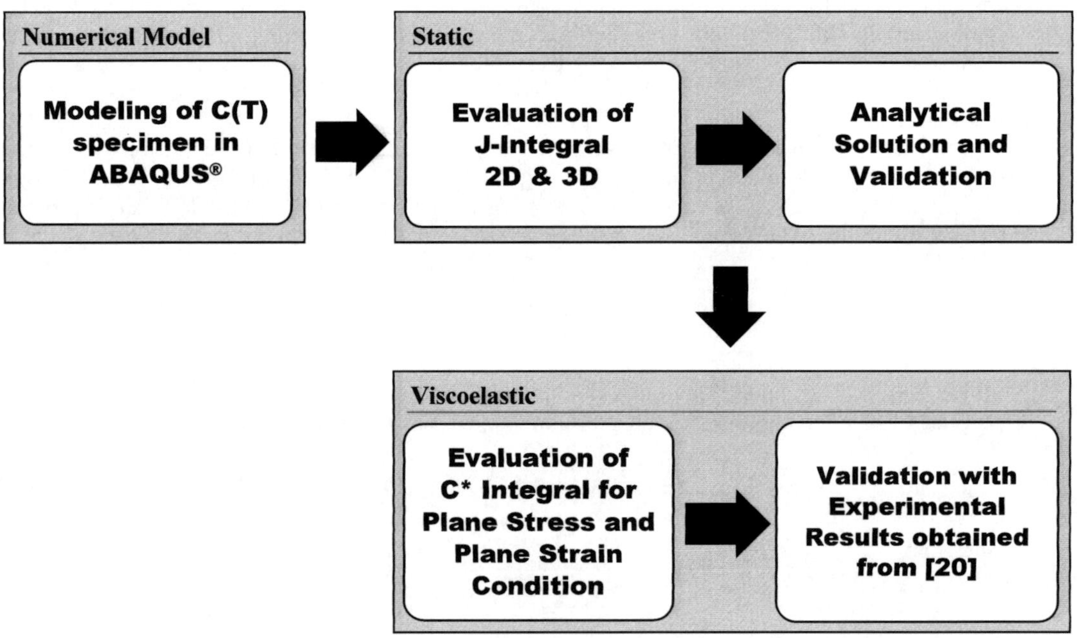

Figure 2. Schematics of the methodology involved.

4. Modelling and Simulation

Standard C(T) specimen was taken into consideration following the works of [20]. The dimension of which are enlisted in Table 1. The material taken into consideration was P92 steel, which is mostly used in the manufacturing of pressure vessels, the properties of which are represented in Table 2. This was implemented in ABAQUS® both in two dimensions having 4 noded CPS4R and CPE4R elements corresponding to plane stress and plane strain conditions and three dimensions using 8 noded C3D8R elements.

Table 1. Geometry for C(T) Specimen.

Specimen Geometrical Parameters		
Width	*W (mm)*	32
Height	*H (mm)*	38.4
Breadth	*B (mm)*	16
Crack Length	*a (mm)*	10
Load	*P(N)*	100

Table 2. Properties of P92 Steel.

Mechanical and Creep Properties of P92 Steel at 650°C		
Modulus of Elasticity	*E (GPa)*	125
Power Law Multiplier	*A (MPa-n/h)*	3.77 E-19
Power-law creep stress exponent	*n*	6.71
Poisson Ratio	*v*	0.3

Mesh refinement was introduced at the crack tip and mesh biasing algorithm was implemented in order to relax the element sizes at regions of less importance and to minimize computational costs as shown in the Figures 3 and 4. The cracked region was defined as a discontinuity with 'Seam definitions' and q-vectors defining the direction of crack propagation. In two dimensions the discontinuity was in the form of a line and a point representing the crack whereas in three dimensions the discontinuity was represented by a plane and the crack front by a line. For the *J*-Integral analysis a quasi-static step was introduced and the strain energy release rate *SERR* was thereby evaluated. Viscoelastic analysis was employed for the evaluation of time based C^* Integral using the mentioned properties listed in Table 2. Grid independence studies, were further performed using mesh sizes varying from 0.025 to 0.2mm in order to optimise the computational costs and the reliability of the results so obtained. Based on this a mesh size of 0.1mm was thereby implemented at the crack tip.

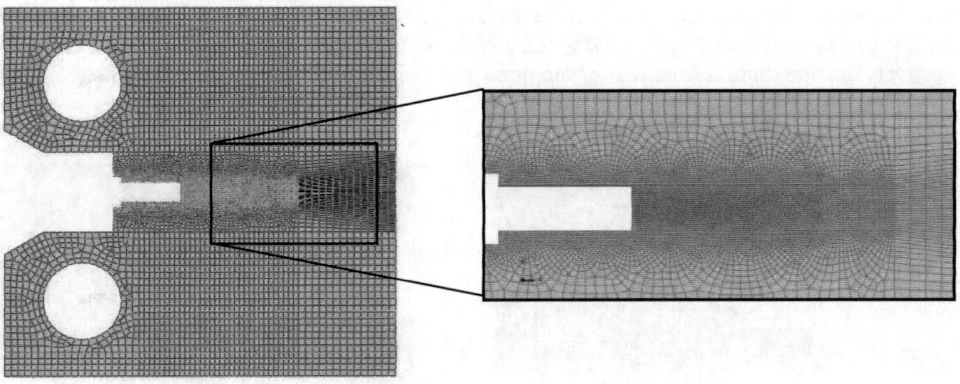

Figure 3. Modelling for two dimensional C(T) Specimen.

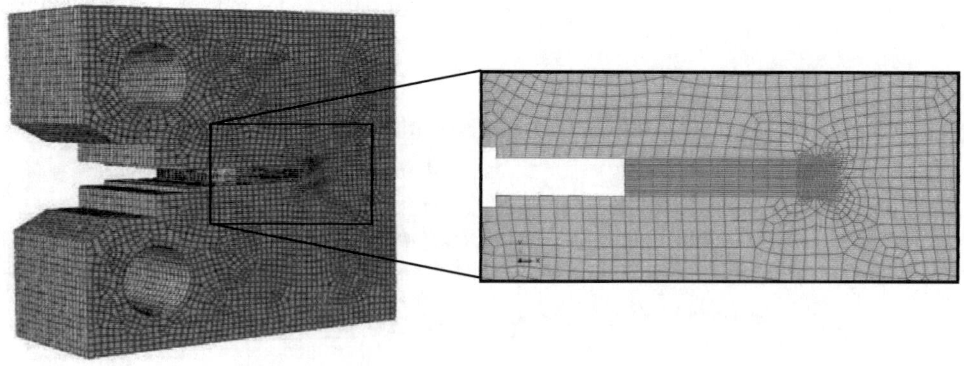

Figure 4. Modelling for Three dimensional C(T) Specimen.

5. Results and Discussion

5.1. Grid independence study

The results obtained for different grit sizes were compared with the result obtained from analytical solution corresponding to plane strain and plane stress conditions respectively, as represented in Table 3. These results let to the selection of 0.1mm as the proper mesh size at the crack tip and was therefore implemented in the model.

Table 3. Comparison between values of J-Integral all values in (N/mm).

S. No.	Mesh Size	Plane Stress % Error	Plane Strain % Error
1	0.025	3.16	1.269
2	0.05	3.06	1.169
3	0.1	2.856	0.968
4	0.15	3.00	1.10
5	0.2	3.038	1.13

5.2. J-Integral

Stress distribution as obtained from static analysis are shown in Figures 6, 7, and 8. Revealing a maximum principal stress of 199.72 MPa and 197.91 MPa for plane stress and plane strain conditions respectively and the same is reported for the three dimensional analysis to be 161.6 MPa.

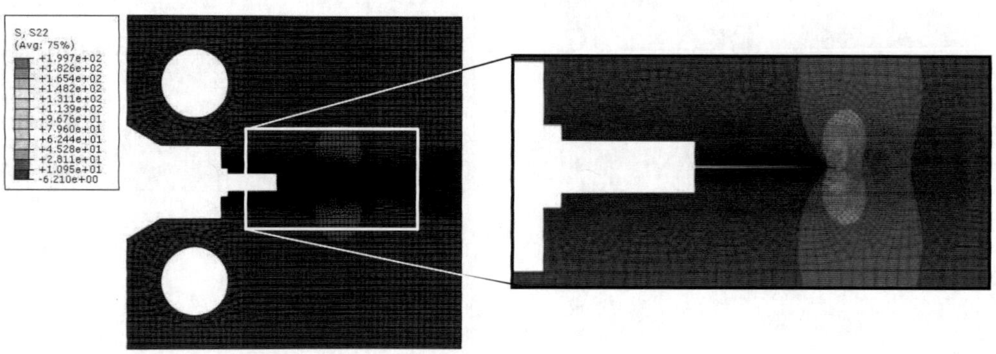

Figure 5. Stress Distribution in two dimensional C(T) Specimen in J-Integral analysis for Plane Stress Condition.

IOP Conf. Series: Materials Science and Engineering **396** (2018) 012063 doi:10.1088/1757-899X/396/1/012063

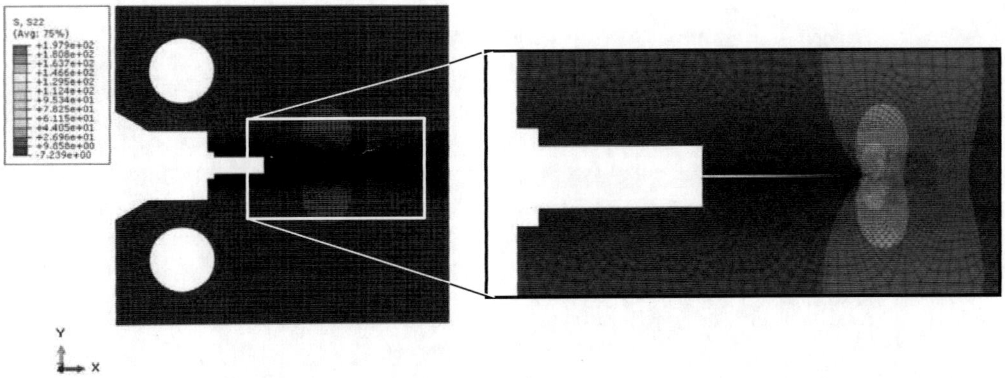

Figure 6. Stress Distribution in two dimensional C(T) Specimen in J-Integral analysis for Plane Strain Condition.

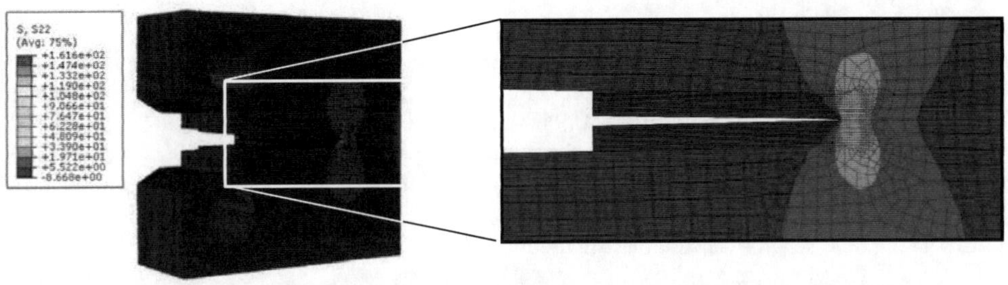

Figure 7. Stress Distribution in three dimensional C(T) Specimen in J-Integral analysis.

J-Integral values obtained by the simulation of two dimensional analysis are presented in Table 3 corresponding to both plane stress and plane strain conditions. As expected the model is stiffer in plane stain conditions than plane stress conditions and as a result reveal lesser amount of strain energy release rate and the percent error when compared with the analytical solution involving Equation 1 was appreciably below 3%. For the three dimensional analysis a variation of the strain energy release rates was observed along the width of the specimen as seen in Figure 8.

Table 4. J-Integral Values for two dimensional analysis.

S. No.	J – Integral (N/mm)				
	Plane Stress	Plane Strain	3D Stress		Analytical
			Min.	Max.	
1	0.0003385	0.0003023	0.0002194	0.0003548	0.0002994

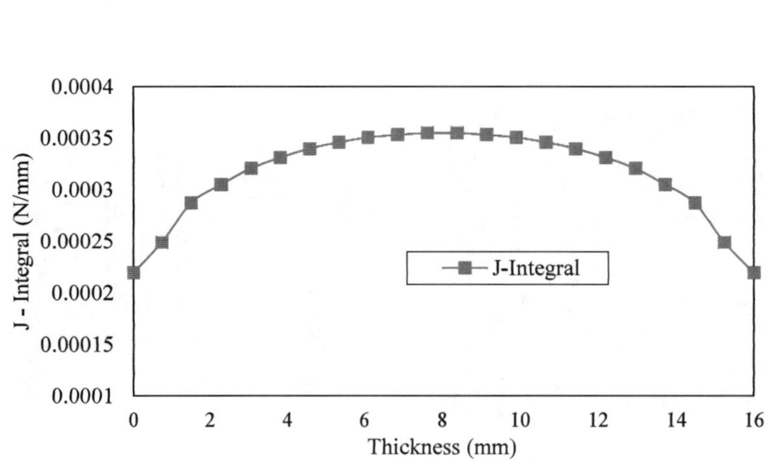

Figure 8. Variation of J-Integral with respect to thickness.

5.3. *C*-Integral*

The results obtained from viscoelastic analysis reveal the development of maximum principal stress measuring 77.73 MPa, 146.1 MPa and a maximum of 133.3MPa for plane stress, plane strain and 3D stress conditions respectively. The corresponding strain values in principal direction were observed to be 5.878E-3, 1.310E-3 and a maximum of 0.94E-3 for plane stress, plane strain and 3D stress conditions respectively. The illustration of the development of strain can be seen in Figure 9(*a-c*) corresponding to time increments from 1E-3 hours to 1000 hours and the actual variation of stress and creep strain with respect to time is seen in Figure 10(*a, b*) respectively, revealing a close resemblance of bulk material conditions *i.e.* plane strain of the actual *3D* model.

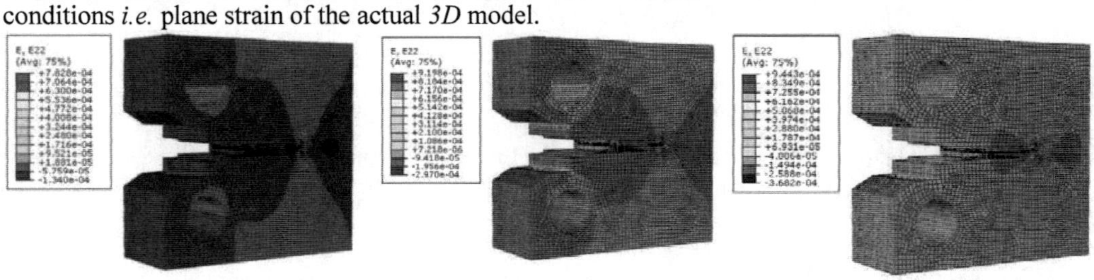

(*a*) Time *t* = *1E-3* hours (*b*) Time *t* = *494.8* hours (*c*) Time *t* = *1000* hours

Figure 9. Variation of strain with respect to time.

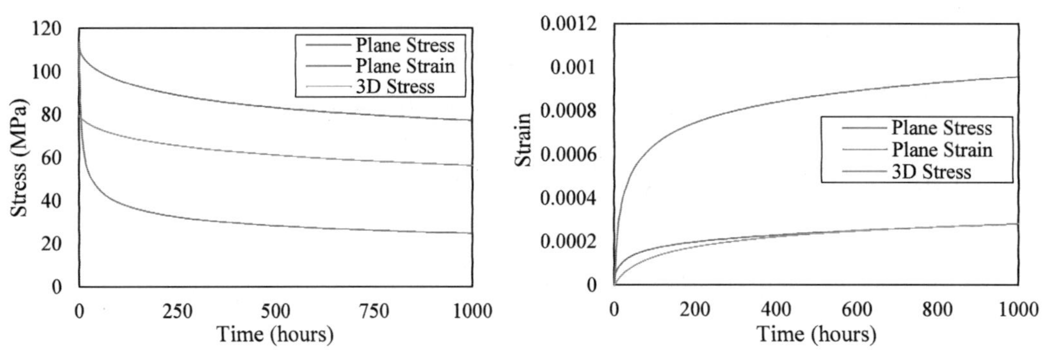

(*a*) Stress variation with time (*b*) Strain variation with time

Figure 10. Stress & Strain Variation with time for Plane Stress and Plane Strain Condition

IOP Conf. Series: Materials Science and Engineering **396** (2018) 012063 doi:10.1088/1757-899X/396/1/012063

Further, variation of C^* with respect to time was evaluated for plane stress, plane strain and 3D stress condition. The energy release rate with respect to time was observed to be of higher magnitude in the starting region and saturates progressively in an asymptotic manner as seen in Figure 11. The results were further compared with the experimental works of Saber [20] and a good agreement between the two can be seen in Figure 12.

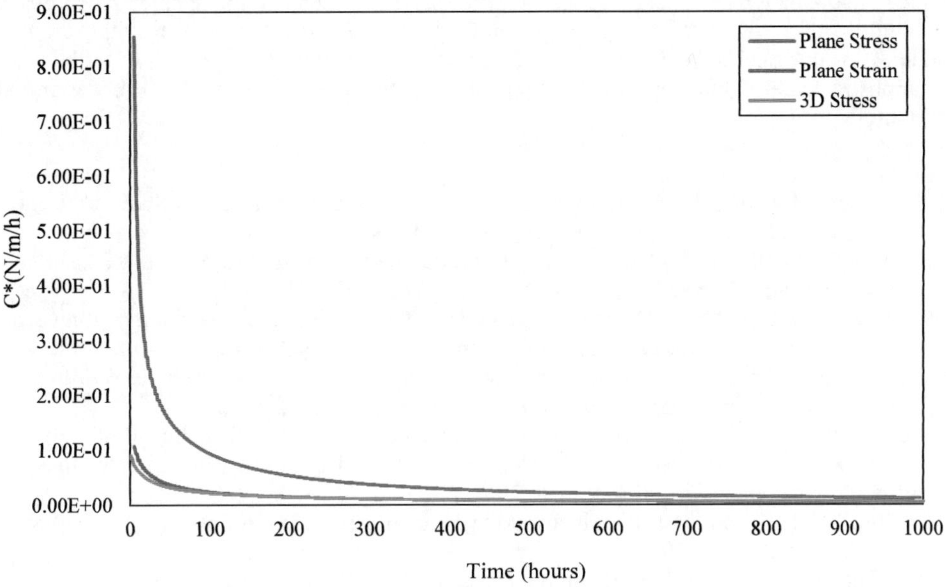

Figure 11. C variation with time*

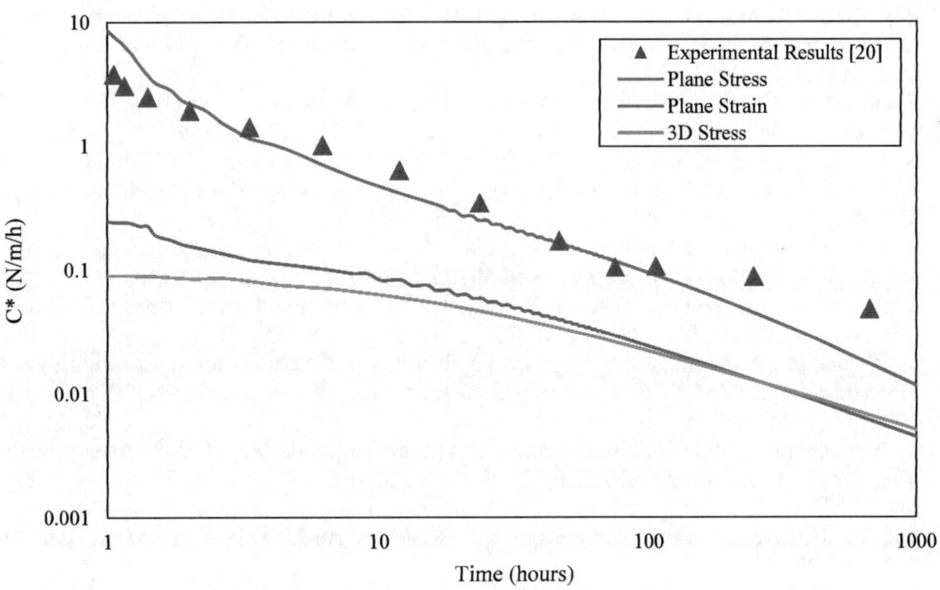

Figure 12. Validation of Results for C with respect to time*

IOP Conf. Series: Materials Science and Engineering **396** (2018) 012063 doi:10.1088/1757-899X/396/1/012063

6. Conclusions

Numerical simulation of time based fracture was successfully implemented in the commercial code of ABAQUS® with the development of a methodology to model time based fracture in industrial components. The results obtained from numerical analysis of the *C(T)* specimen were within idealistic *2D* plane stress & *2D* plane strain conditions confirming the realistic nature of models and the dominance of bulk materials conditions. Viscoelastic analysis is sensitive to the creep properties of the material and therefore must be obtained very carefully. Both J and C^* integral results were in good agreement with analytical and experimental results, thereby validating the approach. In further research these simplistic material models will be applied to pipes which are used for underground oil transportation.

References

[1] J. R. Rice, "A Path Independent Integral and the Approximate Analysis of Strain Concentration by Notches and Cracks," *J. Appl. Mech.*, vol. 35, no. 2, p. 379, 1968.

[2] A. Zahoor, "A Plastic Fracture Mechanics Prediction of Fracture Instability in a Circumferentially Cracked Pipe," vol. 103, no. 80, pp. 352–358, 1981.

[3] B. Moran and C. F. Shih, "Crack tip and associated domain integrals from momentum and energy balance," *Eng. Fract. Mech.*, vol. 27, no. 6, pp. 615–642, 1987.

[4] D. M. Finch, "Effects of welding residual stresses on significance of defects in various types of welded joint-II," *Eng. Fract. Mech.*, vol. 42, no. 3, pp. 479–500, 1992.

[5] O. N.P. and S. C.F., "Two-parameter fracture mechanics: theory and applications," 1992.

[6] Y. Lei, N. P. O 'dowd, and G. A. Webster, "Fracture mechanics analysis of a crack in a residual stress field," *Int. J. Fract.*, vol. 106, pp. 195–216, 2000.

[7] Y. Lei, "J calculation for a crack in a welding residual stress field following a fe welding simulation," vol. 7, no. 2014, 2015.

[8] "6_finite_element_creep_deformation_Ehlers.pdf." .

[9] A. Saxena and J. D. Landes, "Characterization of creep crack growth in metals," *Adv. Fract. Res.*, 1984.

[10] P. A. Oconnell, M. J. Bonner, R. A. Duckett, and I. M. Ward, "The Relationship between Slow Crack-Propagation and Tensile Creep-Behavior in Polyethylene," *Polymer (Guildf).*, vol. 36, no. 12, pp. 2355–2362, 1995.

[11] F. Dubois, C. Chazal, and C. Petit, "A Finite Element Analysis of Creep-Crack Growth in Viscoelastic Media," *Mech. TimeDependent Mater.*, vol. 2, no. 3, pp. 269–286, 1999.

[12] N. Le Mat Hamata and I. a. Shibli, "Creep crack growth of seam-welded P22 and P91 pipes with artificial defects. Part II. Data analysis," *Int. J. Press. Vessel. Pip.*, vol. 78, no. 11–12, pp. 827–835, 2001.

[13] N. P. O. M. Yatomi K.M. Nikbin, "Creep crack growth prediction using a damage based approach," *Int. J. Press. Vessel. Pip.*, vol. 80, pp. 573–583, 2003.

[14] M. Yatomi, C. M. Davies, and K. M. Nikbin, "Creep crack growth simulations in 316H stainless steel," *Eng. Fract. Mech.*, vol. 75, no. 18, pp. 5140–5150, 2008.

[15] M. Yamamoto, N. Miura, and T. Ogata, "Applicability of C* parameter in assessing Type IV creep cracking in Mod. 9Cr-1Mo steel welded joint," *Eng. Fract. Mech.*, vol. 77, no. 15, pp. 3022–3034, 2010.

[16] M. Yatomi and M. Tabuchi, "Issues relating to numerical modelling of creep crack growth," *Eng. Fract. Mech.*, vol. 77, no. 15, pp. 3043–3052, 2010.

[17] J. P. Tan, S. T. Tu, G. Z. Wang, and F. Z. Xuan, "Effect and mechanism of out-of-plane constraint on creep crack growth behavior of a Cr-Mo-V steel," *Eng. Fract. Mech.*, vol. 99, pp. 324–334, 2013.

[18] F. Bassi, S. Foletti, and A. Lo Conte, "Creep fatigue crack growth and fracture mechanisms of T/P91 power plant steel," *Mater. High Temp.*, vol. 32, no. 3, pp. 250–255, 2015.

[19] Y. Lei, "Validation of contour integral functions (J and C(t)) in ABAQUS v6.11-v6.14 for

combined mechanical and residual stresses," *Procedia Struct. Integr.*, vol. 2, no. August, pp. 2566–2574, 2016.
[20] M. Saber, "Experimental and finite element studies of creep and creep crack growth in P91 and P92 weldments," University of Nottingham, 2011.
[21] T. L. Anderson, *Fracture Mechanics: Fundamentals and Applications*, vol. 58, no. 1. 2012.

IOP Conf. Series: Materials Science and Engineering **396** (2018) 012064 doi:10.1088/1757-899X/396/1/012064

Influence of aspect ratio on the linear and non-linear steady state forced vibration response of cylindrical open shells

Arshad Hussain Khan[×], Mohd. Taha Parvez, and Rehan Ahmed

Department of Mechanical Engineering, ZHCET, AMU Aligarh, 202002, India

Corresponding Author: [×] khanahamu@gmail.com

Abstract. The paper deals with the linear and non-linear steady state periodic response estimation of isotropic cylindrical open shells. The displacement field of first order shear deformation theory has been employed and the analysis is based on finite element method. C^0 continuous, eight-noded serendipity quadrilateral shear flexible element with five nodal degrees of freedom has been used for the analysis. The geometric non-linearity is included in the analysis using von Kármán's assumption for small strains and moderately large deflection. The governing equations of motion have been obtained in the time domain and the nonlinear periodic responses are obtained using shooting technique. The entire non-linear steady state frequency response curve consisting of stable and unstable regimes has been obtained employing shooting technique along with arc length and pseudo-arc length continuation schemes. The present method does not involve any apriori assumption on the participating modes. The comparison of linear and non-linear frequency response curves reveals large differences between the two. Frequency response curve, response history and the phase plane plots have been obtained, for different aspect ratio (L/b), to explore the linear and non-linear forced vibration characteristics of cylindrical open shells. The asymmetric nature of the phase plane plots reveal significant higher harmonic contributions. With the increase in aspect ratio, thin open shells reveal increased peak amplitude in both linear and nonlinear analysis. The presence of secondary peak in the nonlinear frequency response for greater aspect ratio is due to modal interaction between first and higher modes.

1. Introduction

Shells are among the most common structural elements used in many engineering structures, including pressure vessels, pipes, submarine hulls, wings and fuselages of airplanes, exteriors of rockets, missiles as well as integral parts of machines where they are subjected to dynamic loading and hence their vibratory response analysis is important for efficient, reliable and failure-proof designs. Thin cylindrical shells are used, particularly in the aerospace application, as it undergoes buckling/vibration either due to static/dynamic axial load or external pressure and sometimes both. Obtaining an effective design presents a difficult and challenging problem, hence a detailed analysis is required of a shell structure. The analysis of a shell is a challenging task, as it resists the load applied largely by its curvature, that is by just changing the curvature of the shell keeping the thickness and the material same, a totally different load carrying capacity can be obtained for the shell. Hence, parameters such as curvature, thickness and the boundary conditions play a crucial role in determining the behaviour of the shell. The peculiarities of shell structural behaviour, the difficulties of analysis, and the wide use of shell structures have instigated a large research effort in shell analysis. A shell is said to be geometrically nonlinear if the strain-displacement relations are nonlinear. This paper deals with the linear and geometrically nonlinear forced vibration analysis of isotropic cylindrical open shells.

Structural members such as shells often undergo large amplitude vibrations where the amplitudes are generally of the order of their thickness. The dynamic behaviour obtained based on linear strain-displacement relations generally provide a conservative estimate and may be treated as the first approximation to the actual behaviour. The forced vibration characteristics obtained by considering the geometric non-linearity predicts a closer approximation of the actual dynamic behaviour. The combined effect of geometric nonlinearity and middle surface curvature for shell panels is expected to significantly alter the dynamic behaviour as compared to linear analysis. The need for non-linear steady state forced vibration response arises from the requirement of safe, optimal and efficient design of structural components which are in the form of cylindrical shells.

Most of the research carried out are on free vibration characteristics of cylindrical shells whereas research on forced vibrations analysis of isotropic cylindrical shells are very limited. Alijani & Amabili [1] and Qatu, et al. [2], have carried out extensive literature reviews on the nonlinear vibrations of shells. Liew, et al. [3] presented a review article with bibliography documents, focussed on the developments in the vibration analysis of thin, moderately thick, and thick shallow shells. The nonlinear free flexural vibration behaviour of the isotropic/laminated orthotropic noncircular rings using the finite element approach and Newmark time marching scheme was investigated by Patel et al. [4-5] and reported softening nonlinearity. Kurylov and Amabili [6] analysed the non-linear forced vibration response of simply supported circular cylindrical shells. Alijani and Amabili [7] analysed the nonlinear forced vibrations of laminated circular cylindrical panels. Khan and Patel [8-9] carried out non-linear forced vibration analysis of bimodular plates and cylindrical panels based on modified shooting technique. The effect of boundary conditions on the nonlinear forced vibration response has been analyzed by Amabili [10]. The linear as well as geometrically non-linear steady state periodic response of cylindrical shell panels has been carried out by Khan, et al [11] by employing the displacement field of first order shear deformation theory based on finite element method. C^0 continuous, eight-noded serendipity quadrilateral shear flexible element with five nodal degrees of freedom has been used in the analysis.

From the literature review, it can be noted that most of the study on the non-linear steady state periodic response of cylindrical shells are based on methods requiring apriori assumptions on the participating mode. Further, it is observed that the convergence to a steady state solution is very slow by employing the direct integration approach. In addition, the steady state response cannot be ascertained in the direct time integration approaches. In this paper the nonlinear steady state periodic response of open cylindrical shell is analysed using the first order shear deformation theory. The governing equation of motion is solved using shooting method coupled with New Mark time marching scheme. The unstable portion of the non-linear frequency response curves are obtained using arc-length and pseudo-arc length continuation methods. A detailed parametric study is conducted to study the influence of aspect ratio (L/b) on the linear and nonlinear dynamic behaviour of isotropic cylindrical open shell. The differences between the linear and nonlinear frequency response curves have been analyzed. The phase plane plots and the steady state response history reveal unequal positive/negative half cycle time. The secondary peak in the frequency response of shell with L/b =2 is due to interaction of first and higher modes.

2. Governing Equations and Solution Procedure

The geometry and coordinate system of an isotropic open cylindrical shell are shown in Figure 1. The coordinates x, y and z are along the meridional, circumferential and radial/thickness directions, respectively. The length, width, thickness, included angle and radius of the shell are L, b, h, ϕ and r respectively.

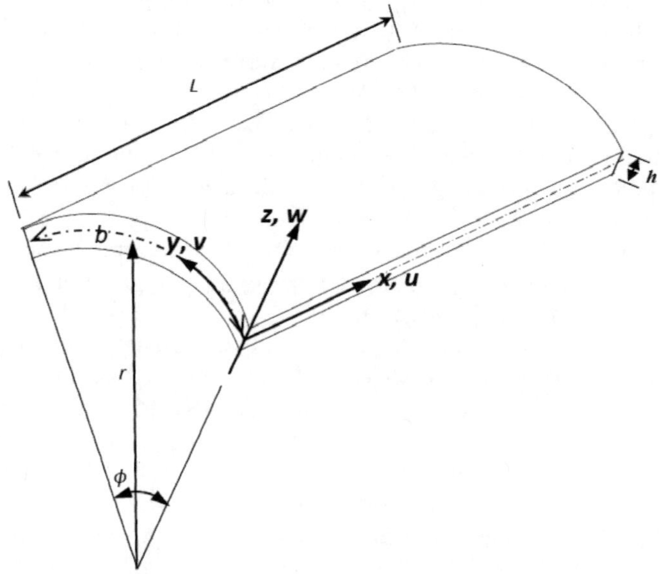

Fig. 1 Geometry and co-ordinate system of open cylindrical shell

The displacement field (u, v, w) at a point(x, y, z) is expressed as a function of middle surface displacements u_0, v_0, w_0 and the independent rotations θ_x and θ_y of the meridional and hoop sections, respectively, using first-order shear deformation theory as:

$$\begin{aligned} u(x,y,z,t) &= u_0(x,y,t) + z\theta_x(x,y,t) \\ v(x,y,z,t) &= v_0(x,y,t) + z\theta_y(x,y,t) \\ w(x,y,z,t) &= w_0(x,y,t) \end{aligned} \tag{1}$$

Based on Sander's shell theory, strain field in terms of mid-surface deformation variables can be written as:

$$\{\varepsilon\} = \{\varepsilon_{xx} \quad \varepsilon_{yy} \quad \gamma_{xy} \quad \gamma_{xz} \quad \gamma_{yz}\}^T = \begin{Bmatrix} \varepsilon_P^L \\ 0 \end{Bmatrix} + \begin{Bmatrix} z\varepsilon_b \\ \varepsilon_s \end{Bmatrix} + \begin{Bmatrix} \varepsilon_P^{NL} \\ 0 \end{Bmatrix} \tag{2}$$

where, ε_P^L, ε_b, ε_s and ε_P^{NL} representing linear mid-surface membrane, bending, transverse shear and nonlinear mid-surface membrane strain vectors, respectively, are defined as:

$$\varepsilon_p^L = \begin{Bmatrix} u'_{0,x} \\ v'_{0,y} + \dfrac{w_0}{r} \\ u'_{0,y} + v'_{0,x} \end{Bmatrix}, \varepsilon_b = \begin{Bmatrix} \theta'_{x,x} \\ \theta'_{y,y} \\ \theta'_{x,y} + \theta'_{y,x} \end{Bmatrix}, \varepsilon_s = \begin{Bmatrix} \theta_x + w'_{0,x} \\ \theta_y + w'_{0,y} - \dfrac{v_0}{r} \end{Bmatrix}, \varepsilon_p^{NL} = \begin{Bmatrix} \dfrac{1}{2}\left(w'_{0,x}\right)^2 \\ \dfrac{1}{2}\left(w'_{0,y} - \dfrac{v_0}{r}\right) \\ w'_{0,x}\left(w'_{0,y} - \dfrac{v_0}{r}\right) \end{Bmatrix}, \tag{3}$$

The membrane stress resultant$\bar{N} = \{N_{xx} \quad N_{yy} \quad N_{xy}\}^T$, moment resultant $\bar{M} = \{M_{xx} \quad M \quad M_{xy}\}^T$ and transverse shear stress resultant $\bar{Q} = \{Q_{xz} \quad Q_{yz}\}^T$vectors are related to the membrane $\varepsilon_p = \varepsilon_P^L + \varepsilon_P^{NL}$, bending ε_b and transverse shear ε_s strain vectors through the constitutive relation as.

$$\begin{Bmatrix} \bar{N} \\ \bar{M} \end{Bmatrix} = \begin{bmatrix} A & B \\ B & D \end{bmatrix} \begin{Bmatrix} \varepsilon_p \\ \varepsilon_b \end{Bmatrix}, \quad \bar{Q} = E\varepsilon_s \tag{4}$$

where $B_{ij} = 0$ and the nonzero elements of A_{ij} , D_{ij} $(i, j = 1,2,6)$ and E_{ij} $(i, j = 4,5)$ for an isotropic shell are

$$A_{11} = A_{22} = \frac{Eh}{1-\nu^2}, \quad A_{21} = A_{12} = \frac{\nu Eh}{1-\nu^2}, \quad A_{66} = \frac{Eh}{2(1+\nu)}$$

$$D_{11} = D_{22} = \frac{Eh^3}{12(1-\nu^2)}, \quad D_{21} = D_{12} = \frac{\nu Eh^3}{12(1-\nu^2)}, \quad D_{66} = \frac{Eh^3}{24(1+\nu)}, \tag{5}$$

$$E_{44} = E_{55} = \frac{Eh}{2(1+\nu)}$$

where [A], [D] and [B] are extensional, bending, and bending-extensional coupling stiffness coefficient matrices, respectively. [E] is the shear stiffness matrix.

The total potential energy functional U consisting of strain energy and potential of the uniformly distributed transverse load is given by:

$$U = \frac{1}{2}\iint \left(d^T K d + \frac{1}{3} d^T K_1 d + \frac{1}{6} d^T K_2 d\right) dxdy - \iint F w_0 dxdy \tag{6}$$

where [K] represents linear stiffness matrix, $[K_1]$ and $[K_2]$ are representing quadratic and cubic non-linear stiffness matrices, respectively.

The kinetic energy of the shell is given by:

$$T = \frac{1}{2}\iint \left(\rho h(\dot{u}_0^2 + \dot{v}_0^2 + \dot{w}_0^2) + \frac{\rho h^3}{12}(\dot{\theta}_x^2 + \dot{\theta}_y^2)\right) dxdy \tag{7}$$

where ρ is the mass density of shell. Dot over the variables denotes the derivative with respect to time.

Using a C^0 continuous, eight-noded serendipity quadrilateral shear flexible shell element with five nodal degrees of freedom $u_{0i}, v_{0i}, w_{0i}, \theta_{xi}, \theta_{yi}$, the field variables are interpolated in terms of their nodal values and shape functions as:

$$\left(u_0, v_0, w_0, \theta_x, \theta_y\right) = \sum_{i=1}^{8} N_i^0 (u_{0i}, v_{0i}, w_{0i}, \theta_{xi}, \theta_{yi}) \tag{8}$$

Employing Hamilton's principle, using standard finite element assembly procedure and considering dissipative forces, the governing equation of motion can be written as:

$$[M]\{\ddot{\delta}\} + [C]\{\dot{\delta}\} + [K + (1/2)K_1(\delta) + (1/3)K_2(\delta)]\{\delta\} = \{F\} \tag{9}$$

where the damping matrix is defined based on Rayleigh proportional damping model and the damping matrix [C] is taken as

$$[C] = \hat{\alpha}[M] + \beta[K] \tag{10}$$

The governing equations are solved using the procedure described in authors work [8-9] and are not presented here for the sake of brevity. The solution is started at frequency far away from resonance using New-Mark time marching and modified shooting method. At the frequency near bifurcation points where the modified shooting method fails to give converged solution the frequency response curve is continued using arc-legth/pseudo-arc length continuation schemes.

IOP Conf. Series: Materials Science and Engineering **396** (2018) 012064 doi:10.1088/1757-899X/396/1/012064

3. Results and Discussion

The influence of aspect ratio on the linear and nonlinear forced vibration response of isotropic cylindrical open shell subjected to uniformly distributed harmonic force with forcing frequency varied in the vicinity of first fundamental frequency has been investigated. The material of the shell is assumed to be Aluminium with Young's Modulous 70 GPa, Density 2778 kg/m^3 and poisons ratio 0.3. Based on mesh convergence a 10x10 mesh size is used to discretize the shell. The boundary condition choosen for the analysis is all edges simply supported.

The nonlinear frequency response curves for both positive and negative half cycle are presented in Fig. 2. It can be inferred from the positive/negative half cycle frequency response curves (Fig. 2 a) that the peak amplitude increase with the increase in aspect ratio (L/b). The nonlinear response amplitude for negative half cycle is significantly greater for L/b=1, 2 as compared to L/b=0.5 indicating that the difference in positive and negative half cycle stress amplitude is greater in the GNL analysis compared to the GL(Fig. 2 b) analysis. The peak amplitude also increases with the increase in aspect ratio for both GNL and GL. The percentage difference between GNL and GL increases with the increase in aspect ratio.

The comparison of peak amplitude in the linear and nonlinear analysis for the positive half cycle reveals that peak amplitude in linear analysis is significantly larger compared to non-linear analysis, with the peak amplitude in the linear analysis being 1.09, 3.02 and 3.77 times the peak amplitude obtained from non-linear analysis for L/b = 0.5, 1 and 2, respectively. The peak amplitude in the linear analysis being 1.06, 2.13 and 2.98 times the peak amplitude obtained from non-linear analysis for L/b = 0.5, 1 and 2, respectively for the negative half cycle. From the positive half cycle GNL response curve it can be observed that the peak amplitude for L/b = 1 and 2 are 2.37 and 4.53 times the peak value for L/b = 0.5 respectively. In the negative half cycle GNL response curve, the peak value for L/b = 1 and 2 are 3.27 and 5.58 times the peak value for L/b = 0.5 respectively.

The steady state response history in the linear analysis (Fig. 3 (b)) reveal equal positive/negative half cycle amplitude with equal time in tension and compression. In contrast the non-linear response history (Fig. 3 (a)) depicts greater negative half cycle time indicating that the panel is in inward motion for greater portion of the periodic cycle. The negative half cycle time is 1.12, 1.29 and 1.10 times the positive half cycle L/b = 0.5, 1 and 2, respectively.

The phase plane plots (Fig. 4 (a) & (b)) corresponding to peak amplitude in the frequency response curves depict asymmetry for the nonlinear case which is due to significant higher harmonic contributions. The secondary peak in the nonlinear response curve for shell with L/b =2, is due to interaction of first and higher modes.

4. Conclusion

The linear as well as nonlinear forced vibration response of isotropic cylindrical open shells have been obtained in time-domain using modified shooting technique and continuation schemes. The complete frequency response curves consisting of stable and unstable regimes have been obtained. It is apt to make a mention here that the direct time integration method fails to predict the unstable regimes. A detailed analysis has been carried out to analyze the influence of aspect ratio on the linear and nonlinear dynamic response of open cylindrical shells. Based on the analysis it is found that with the increase in aspect ratio the peak amplitude increases with the increase in aspect ratio for both linear and nonlinear analysis. It is concluded that the peak amplitude for linear analysis is considerable greater than that for nonlinear analysis and this difference increases with the increase in aspect ratio. It can also be concluded that the negative half cycle amplitude as well as cycle time is greater than the corresponding positive half cycle values with the nonlinear analysis. The nonlinear phase plane plots depict greater higher harmonic contributions. The secondary peak in the frequency response curves of L/b=2 is due to modal interaction between first and third mode.

IOP Conf. Series: Materials Science and Engineering **396** (2018) 012064 doi:10.1088/1757-899X/396/1/012064

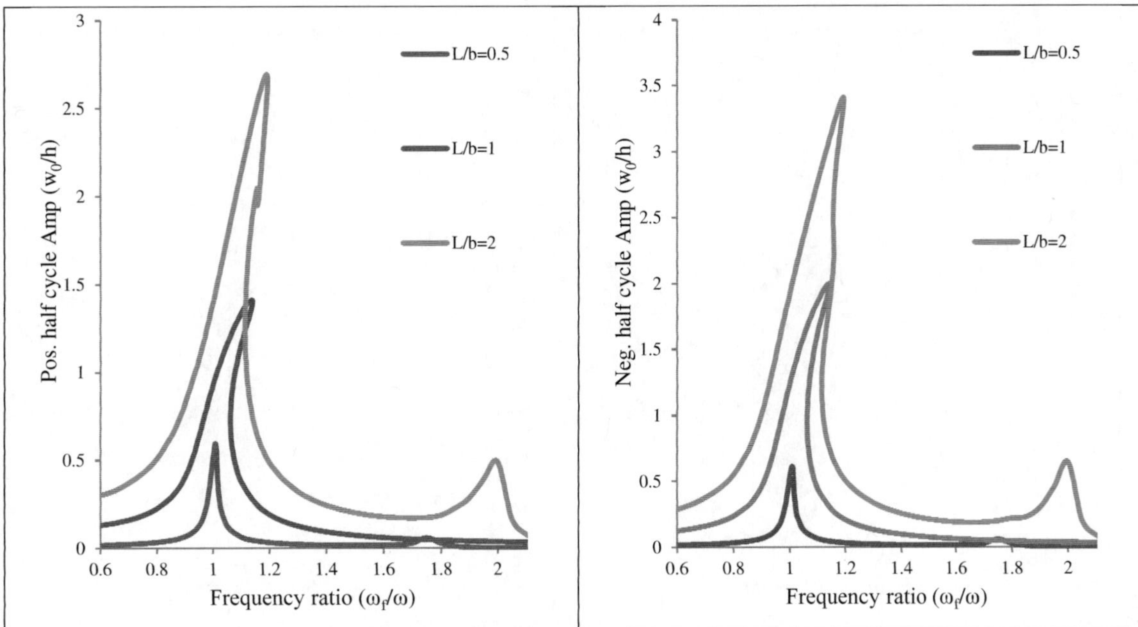

(a) Geometrically Non-linear frequency response

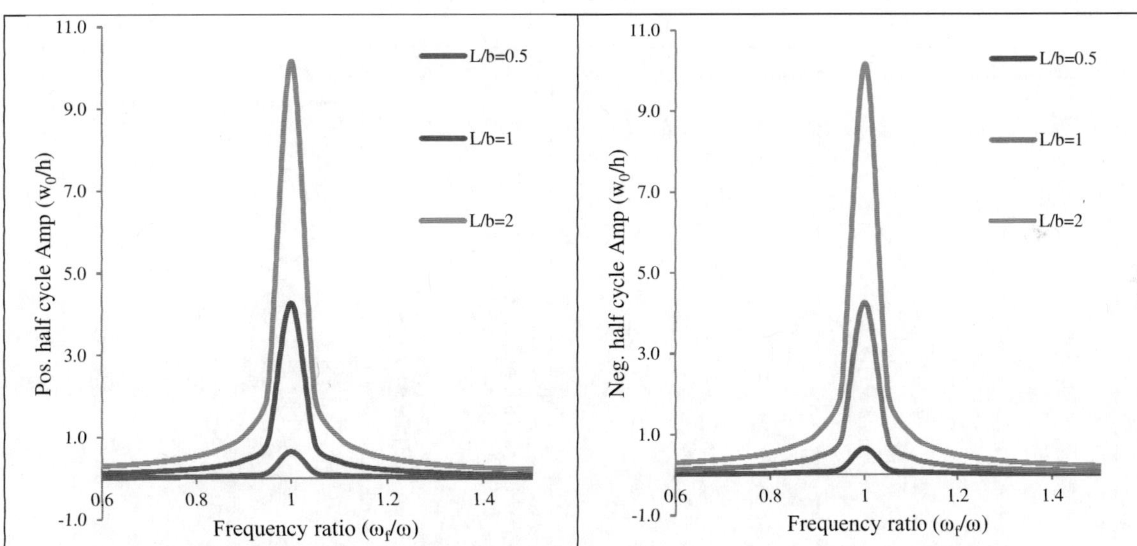

(b) Geometrically linear frequency response

Fig. 2 Steady state forced vibration frequency response curves (a) Non-Linear, (b) linear of cylindrical open shell. (SSSS, b/h=100, b/r=0.1, Load=5 kPa, ξ=0.010)

IOP Conf. Series: Materials Science and Engineering **396** (2018) 012064 doi:10.1088/1757-899X/396/1/012064

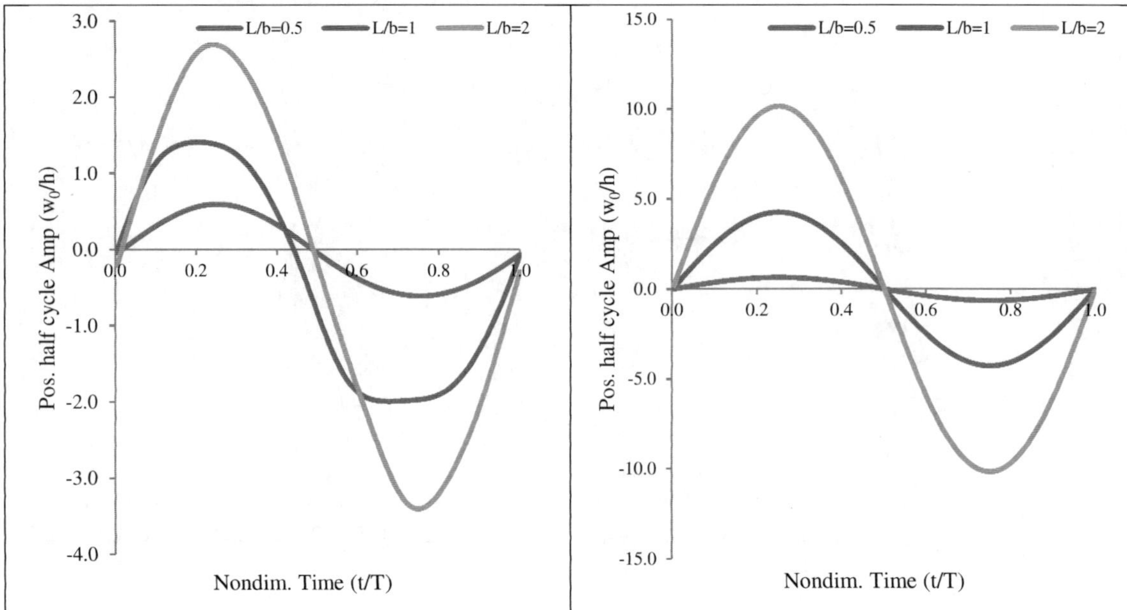

(a)Non-linear response	(b)Linear response

Fig. 3	Non-Linear and linear steady state response history.

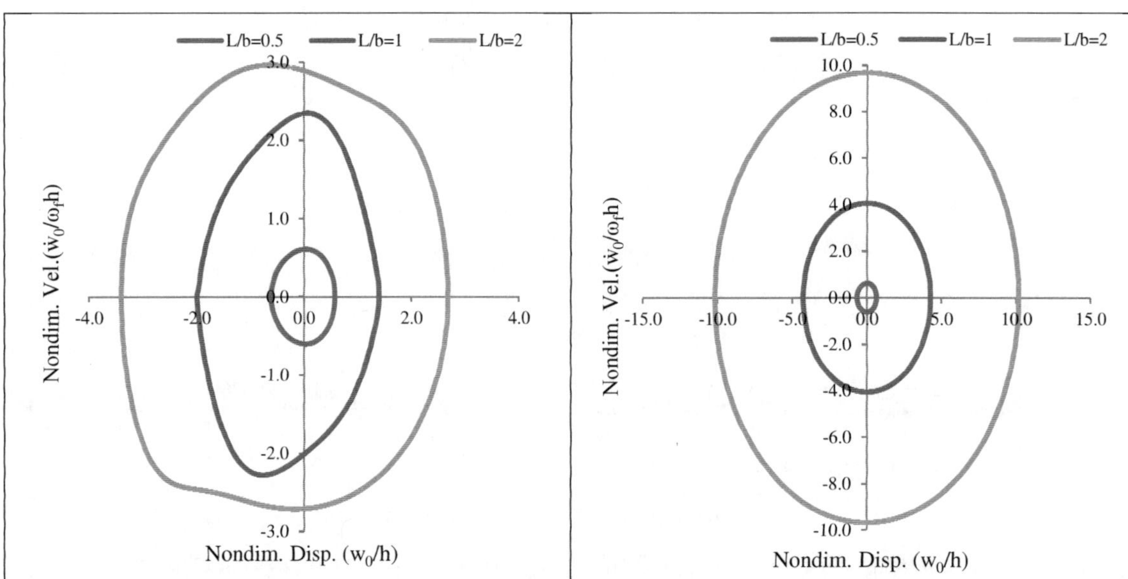

(a)Non-linear response	(b)Linear response

Fig. 4	Phase-plane plots at forcing frequencies.

IOP Conf. Series: Materials Science and Engineering **396** (2018) 012064 doi:10.1088/1757-899X/396/1/012064

References

1. Alijani F, Amabili M. Non-linear vibrations of shells: A literature review from 2003 to 2013. *International Journal of Non-Linear Mechanics* (2013) **58:** 233-257.
2. Qatu MS, Sullivan RS, Wang W. Recent Research advances on the dynamic analysis of composite shells: 2000-2009, Composite Structures (2010) **93**: 14-31.
3. Liew KM, Lim CW, Kitipornchai C. Vibration of Shallow Shells: A review with bibliography. *Applied Mechanics Review* (1997) **50**(8): 431-444.
4. Patel BP, Ganapathi M, Makhecha DP, Gupta SS. On the nonlinear free flexural vibrations problem for thin elastic rings. *Journal of the Aeronautical Society of India* (2002) 54: 298–309.
5. Patel BP, Ganapathi M, Makhecha DP, Shah P. Large amplitude free flexural vibration of rings using finite element approach. *International Journal of Nonlinear Mechanics* (2003) 37: 911–921.
6. Kurylov Y, Amabili M. Nonlinear vibrations of clamped-free circular cylindrical shells. Journal of Sound and Vibration (2011) **330:** 5363-5381.
7. Alijani F, Amabili M. Nonlinear vibrations of thick laminated circular cylindrical panels. Composite Structures (2013) **96:** 643-660.
8. Khan AH, Patel BP. Periodic response of bimodular laminated composite cylindrical panels with and without geometric nonlinearity. International Journal of Non-Linear Mechanics (2014) **67:** 209-217.
9. Khan AH, Patel BP. Nonlinear forced vibration response of bimodular laminated composite plates. Composite Structures (2014) **108:** 534-537.
10. Amabili M, Kurylov Y. Non-Linear Vibrations of Circular Cylindrical Shells with Different Boundary Conditions. *Proceedings of the ASME 2009 International Mechanical Engineering Congress & Exposition IMECE2009*, Lake Buena Vista, Fluorida, USA, November.
11. Khan AH, Parvez MT, Yasin MY. Non-Linear Forced Vibration of Cylindrical Shell Panels. Advanced Science Engineering & Medicine (2018) **10**(3): 496-499.

IOP Conf. Series: Materials Science and Engineering **396** (2018) 012065 doi:10.1088/1757-899X/396/1/012065

Experimental and numerical investigations on failures due to plasticity in thin walled structures

A Jamil, A H Khan, S Ali

Department of Mechanical Engineering,
Z.H.C.E.T, Aligarh Muslim University, Aligarh-202002 (U.P.), India

azharjamil@zhcet.ac.in

Abstract. Plasticity is one of the prominent cause of failures in metallic materials and with the continuous efforts of mankind in exploring domains of high uncertainty, a thorough understanding of structural behaviour under extreme conditions is quite necessary. Particularly, in the case of thin walled structures, where geometrical and physical nonlinearities are quite significant and require sophisticated techniques for failsafe design. In this work, a numerical model was developed in the commercial finite element code of ABAQUS®, after experimentation of 6150 cold rolled steel used in thin walled pressure vessels. For this, proper experimentation quest was carried out based on the ASTM E8/E8M standard and dog-bone specimen were made with both planar and circular cross-sections. Ramberg-Osgood plasticity equation was employed for the smoothening of the test data along with the extrapolation of stresses at higher strain level. A tabular plasticity model was thereby made from the smoothened and extrapolated stress-strain experimental data, and implemented in the material definition module of ABAQUS®. Both plane stress CPS4R and plane strain CPE4R elements were employed in planar dog-bone specimen and C3D8R elements were used for three dimensional planar and circular dog-bone specimen. The results from the simulation were in excellent agreement with the experimental data with the plane stress analysis slightly underestimating, the plane strain analysis slightly overestimating the load and the three dimensional analysis lying in between the two.

Keywords Tensile test, plasticity, Ramburg Osgood plasticity model, FEA

1. Introduction

Cold rolled steel is widely used in industry for storage and transportation of liquids and gases and often these are susceptible to failure due to unavoidable peak loading in the lifetime of these structures. The test has performed to find out the point of failure in industrial components that occurs due to the internal forces such as internal fluid pressure into the thin walled pressure vessels. It occurs due to the hoops stress into components consequently plastic deformation initiated into the surfaces of components. Previous studies have performed tensile test for determining the ductility of materials and determine the plasticity criterion according to simulation analysis. Investigations on necking and post-necking behaviour of uniaxial stretched tensile specimen were performed by Lars *et.al.* [1], using an enhanced *2D* plane stress finite element model. Tensile tests were carried by Jain *et.al.* [2] on aluminium specimen for the determination of ductile fracture criterion. Weng *et.al.* [3], worked on ductile materials and presented standards, using finite element analysis. Experimental and numerical investigations were done by Eduardo *et.al.* [4] inorder to study the mechanical behaviour under tension of SAE 1045 steel sheet specimens. In the area of shear failure, Komori [5], implemented ductile failure criterion on finite element nodes. Fracture criterion of various materials along with failure analysis was done by Yang *et al.* [6]. *FEA* modelling of deformation behaviour of steel specimen was done by Banerjee *et al.* [7], under uniaxial tension test. Prediction of strains and deformation of model were compared with experiment data. The role of imperfection of necking behaviour in *FE* modelling results of uniaxial test was discussed. Chen [8], worked on tensile tests of 7075 aluminium obtaining true stress-strain values along with finite element simulation analysis. In the analysis various ductile fracture criteria were investigated revealing different levels of accuracy in the simulation. Edgar *et al.* [9], analysed the tensile test of mild steel using finite element method using a computer program called ABAQUS/CAE®. The results have been as the stress-strain graph. It is understood that the mild steel follows the Hooke's Law *i.e.*, stress is directly proportional to strain. After the linear region in the graph, there occurs necking on the sample and finally it breaks. Ahmed *et al.* [10], investigated the different stresses of thin walled pressure vessel. Equations of static equilibrium were used to determine the normal stresses and circumferential or hoop stresses.

2. Methodology

In this work, a tensile test of 6150 cold rolled steel was performed used in thin walled pressure vessels which is widely used in industry for storage and transportation of liquids and gases. For this proper experimentation quest was carried out based on the ASTM E8/E8M standard dog-bone specimen of both planar and circular cross-sections and determine the plasticity criterion according to a simulation analysis, as seen in Figure 1.

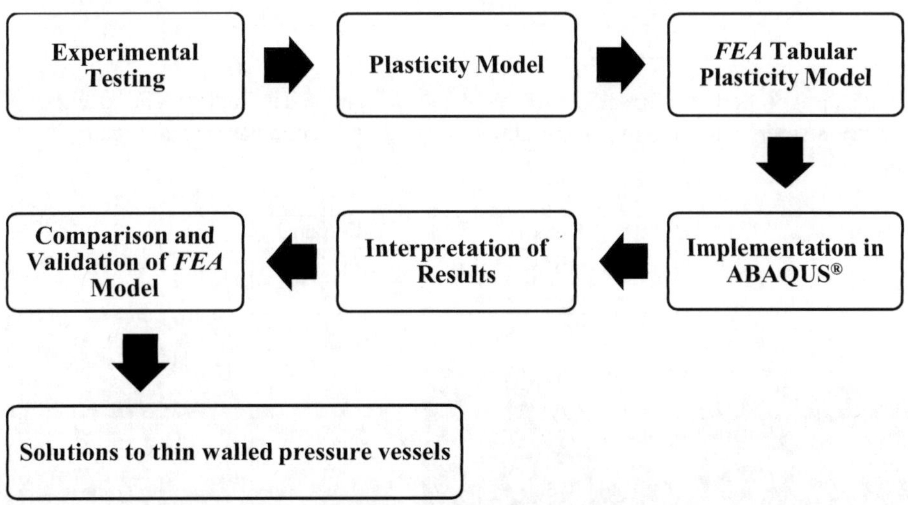

Figure 1. Systematic flow chart of methodology

3. Experimental Testing

Tensile tests were performed in accordance with ASTM E8/E8M for both rectangular and circular section specimen made up of 6150 cold rolled steel. The dog-bone specimen were designed and fabricated in the in-house workshop facility, the schematics of which is shown in Figure 2(a) and one of the fabricated specimen can be seen in Figure 2(b). Final dimensions of which are tabulated in Table 1. Experimental tests were finally performed on the in house universal testing machine (UTM) having a load capacity of 50kN as shown in Figure 3.

IOP Conf. Series: Materials Science and Engineering **396** (2018) 012065 doi:10.1088/1757-899X/396/1/012065

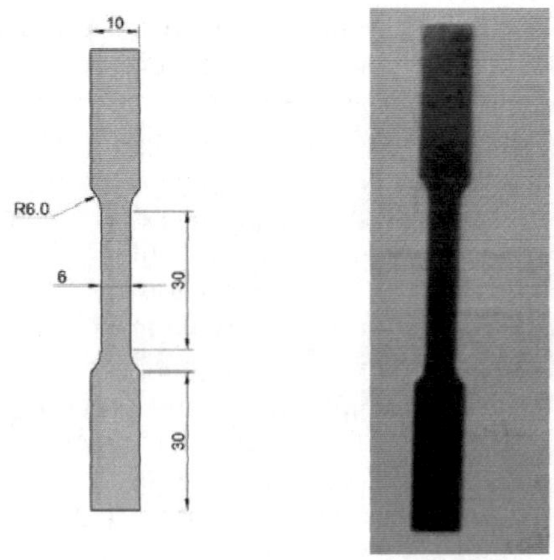

(a) Schematics of specimen (b) Fabrication of test specimen

Figure 2. Dog bone rectangular specimen based on ASTM/E8 standard

Table 1. Dimension of specimen

S. No.	Descriptions	Dimension (mm)
1	Gauge length	30
2	Gauge width	6
3	Grip length	30
4	Thickness	1

The mechanical properties of the materials so obtained are enlisted in Table 2. The specimen showed the typical stress-strain behaviour and post-ultimate strength it fractured as seen in Figure 4.

Table 2. Mechanical properties of 6150 cold rolled steel

Density	7.85 g/cm3
Young's Modulus	210 Gpa
Poison ratio	0.3

Figure 3. Universal testing machine

Figure 4. Fracture specimen

4. Development of the Plasticity Model

The data so obtained from the tests was averaged and was utilised for the plotting of load *vs.* extension. For the planar rectangular section the typical loads *vs.* deformation phenomenon is shown in Figure 5(a) whereas the loads *vs.* deformation for circular cross section is represented in Figure 5(b). From these the value of engineering stress-strain and true stress-strain was evaluated using the basic formulation of stress and strain represented in Equations 1 and 2, as seen in Figures 5(c) and 5(d) respectively for specimen with rectangular and circular cross-sections.

$$\varepsilon = \ln(1 + e) \tag{1}$$

$$\sigma' = \sigma(1 + e) \tag{2}$$

Where ε is known as true strain, 'e' is engineering strain, σ' is true stress and σ is engineering stress. This was further utilised for the calculation of plastic strain using the standard analytical relation represented in Equation 3.

$$\varepsilon_p = \varepsilon - \frac{\sigma}{E} \tag{3}$$

4.1 Ramberg Osgood Plasticity Model

Ramberg Osgood plasticity model was implemented for the smoothening of the test data along with the extrapolation of stresses at higher strain level in accordance with Equation 4. This relationship is exponential and is used to describe the plastic strain in a material. It is especially useful for metals that harden with plastic deformation, showing a smooth elastic-plastic transition.

$$\frac{\sigma}{\sigma_0} = a\left(\frac{\varepsilon}{\varepsilon_0}\right)^n \tag{4}$$

σ_0 and ε_0 are true stress and logarithmic strain corresponding to 0.002 plastic strain and

a and n are constants of the power law.

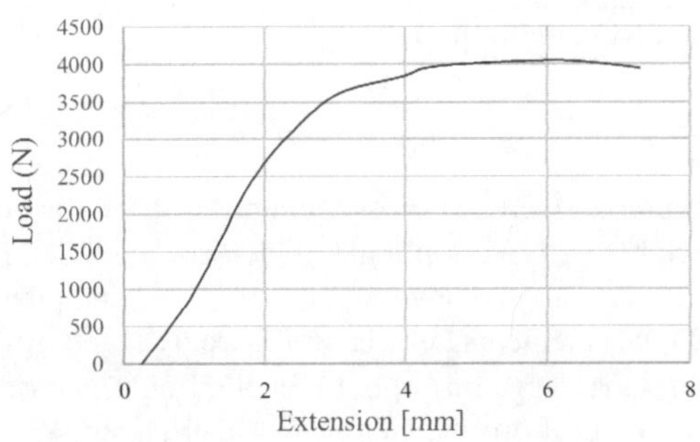

(a) Load vs. Extension curve for rectangular specimen

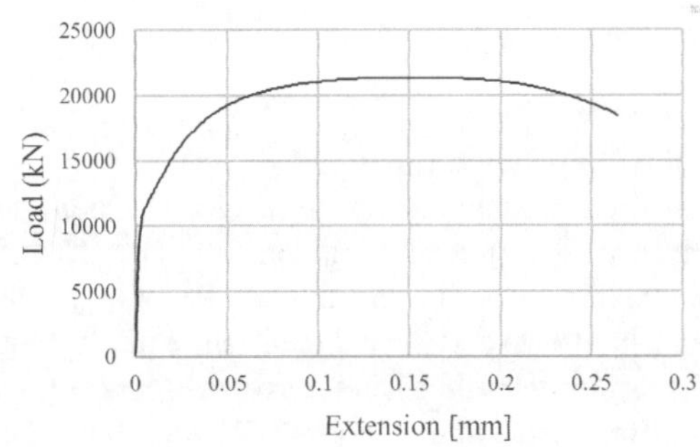

(b) Load vs. Extension curve for circular specimen

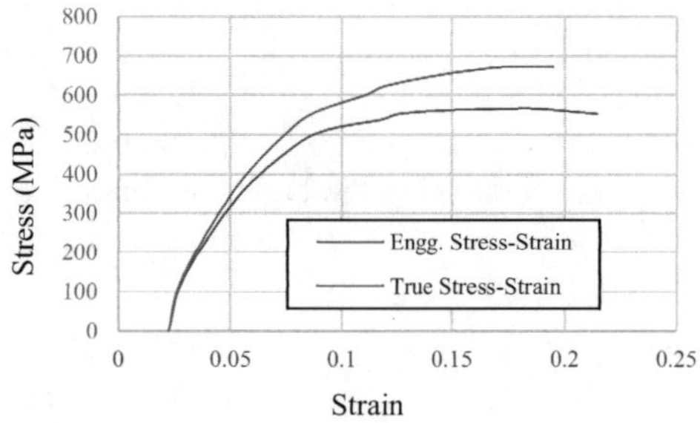

(c) Stress strain curve for rectangular specimen

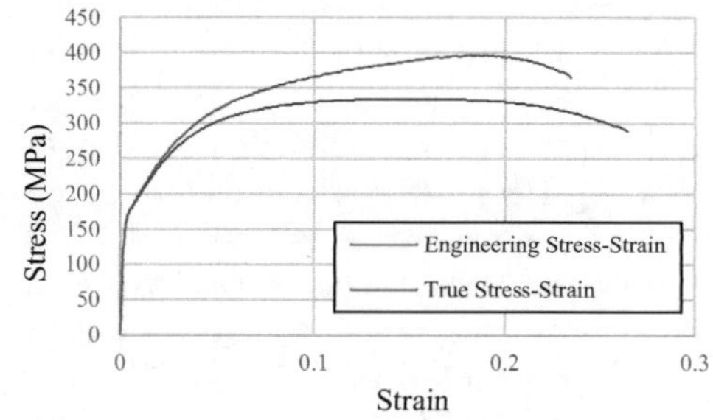

(d) Stress strain curve for circular specimen

Figure 5. Load-Extension and Stresses-Strains graphs

Finally the plastic model was developed after the extrapolations, represented in Table 3, and was finally implemented in the material definitions of ABAQUS®.

Table 3. Final plasticity model

Yield Stress [MPa]	Plastic Strain
1	0
136.2499876	0.0001
361.4609244	0.395023082
502.8731463	0.538420172
558.2141003	0.58376253
600.6772615	0.61560292
624.1550544	0.632254219
648.3509722	0.648771894
666.133737	0.660523157
672.8329214	0.664868961
671.8650519	0.664243779

5. Modelling and Simulation

A numerical model of dog bone specimen of both rectangular and circular cross section was developed in the commercial finite element code of ABAQUS®. Rectangular cross-sectional model was modelled in both two dimensions (with 4 noded CPS4R and CPE4R elements corresponding to plane stress and plane strain conditions) and three dimensions using 8 noded C3D8R elements, as seen in Figure 6 (a) and 6(b). However, on the other hand, circular cross-sectional dog-bone specimen were modelled only in three dimensions using the 8 noded C3D8R elements, as evident from Figure 6(c). Material model so development was inserted in property module along with the value of modulus of elasticity and poison ratio. In interaction module, two coupling corresponding to two end condition of specimen. On one end was perfectly clamped with zero degree of freedom whereas the upper end is provided displacement load of 5mm.

IOP Conf. Series: Materials Science and Engineering **396** (2018) 012065 doi:10.1088/1757-899X/396/1/012065

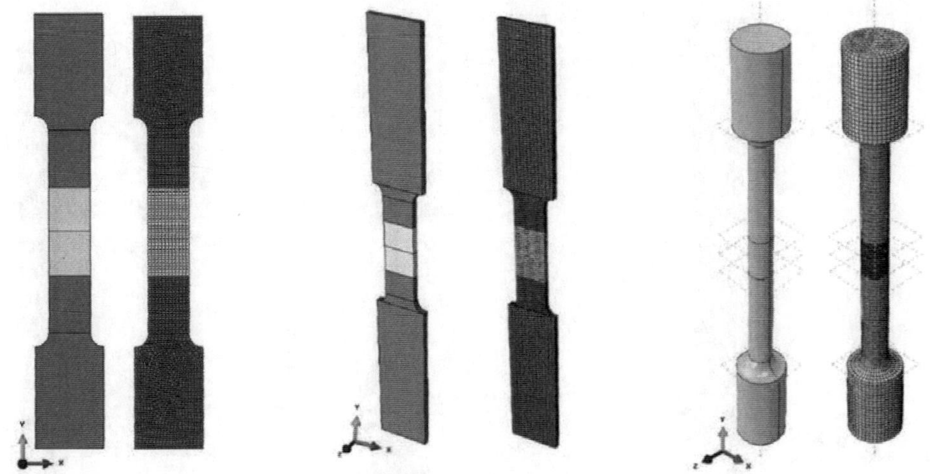

(a) Two dimensional model of rectangular section specimen *(b) Three dimensional model of rectangular section specimen* *(c) Three dimensional model of circular section specimen*

Figure 6. Modelling of rectangular and circular section tensile specimen.

6. Results and Discussion

The simulation of rectangular dog bone reveal a maximum value of Von Mises stress equal to 876.0 MPa at the mid-section in plane stress conditions and 1239.0 MPa for plane strain condition at the end of 5mm displacement load, as seen in Figure 7(a) and (b) respectively. In three dimensional analysis the maximum attained level of Von-Mises stress was 1285.0 MPa corresponding to the same 5mm displacement loading, as seen in Figure 7(c). The corresponding strain levels were recorded to be 0.657 and 0.815 for plane stress and plane strain conditions as seen in Figures 8(a) and (b) respectively and 0.991 for the three dimensional analysis, depicted in Figure 8(c).

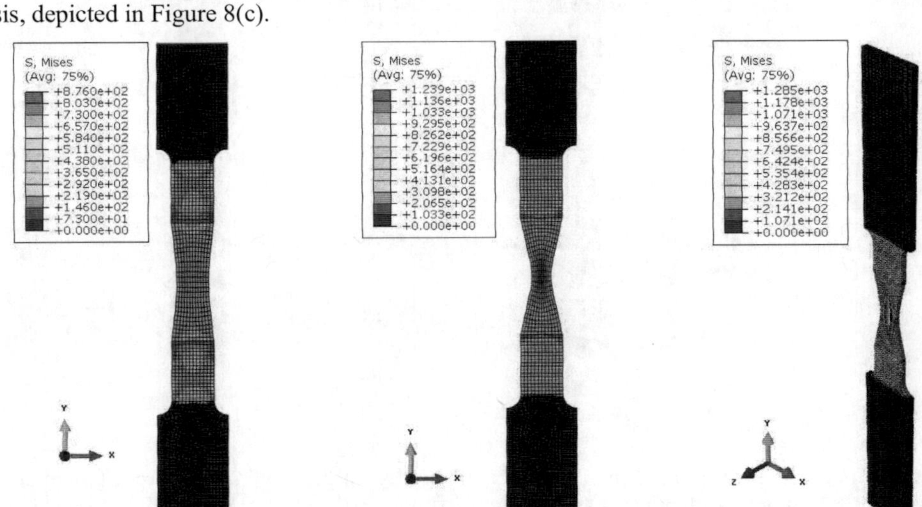

(a) Plane Stress conditions *(b) Plane Strain conditions* *(c) 3D Specimen*

Figure 7. Von Mises Stress distribution of rectangular cross section tensile specimen.

IOP Conf. Series: Materials Science and Engineering **396** (2018) 012065 doi:10.1088/1757-899X/396/1/012065

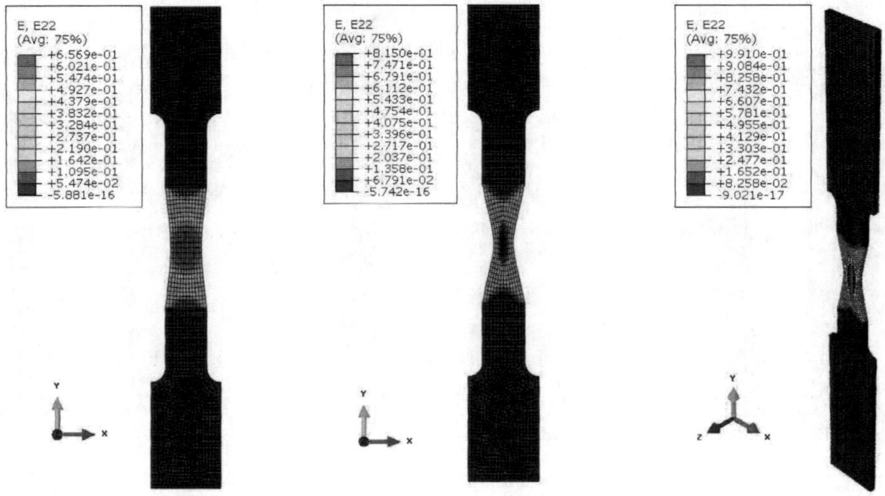

(a) Plane Stress conditions *(b) Plane Strain conditions* *(c) 3D Specimen*

Figure 8. Strain distribution in principal direction of rectangular cross section tensile specimen.

For the circular cross section specimen, the maximum value of Von Misses stress was observed to be 1952.0 MPa and a corresponding strain of 3.66 in the principal direction corresponding to the same 5.0mm displacement load.

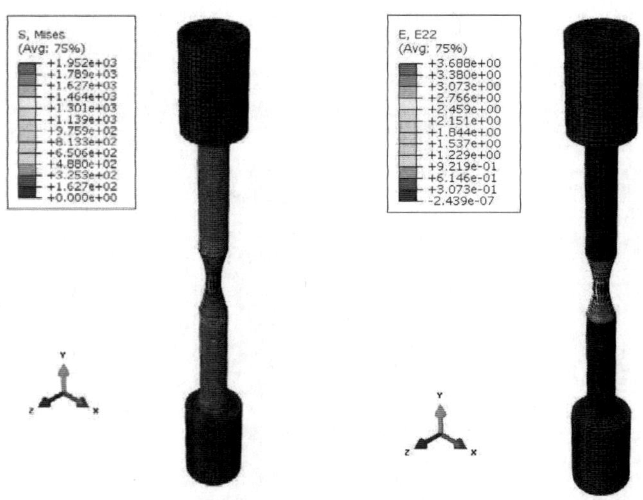

***(a)** Von Mises stress distribution* ***(b)** Strain in principal direction*

Figure 9. Stress-strain contours of circular cross section tensile specimen.

The results obtained from *FEA* model were compared with the experimental results and showed an excellent agreement with each other as can be seen in Figures 10 and 11 corresponding to rectangular and circular sections respectively. The results reveal close resemblance of the experimental data with plane stress and three dimensional analysis. This is quite expected for thin sections are more inclined towards plane stress conditions. Plane strain expresses a stiffer section owing to bulk material conditions of the interior

IOP Conf. Series: Materials Science and Engineering **396** (2018) 012065 doi:10.1088/1757-899X/396/1/012065

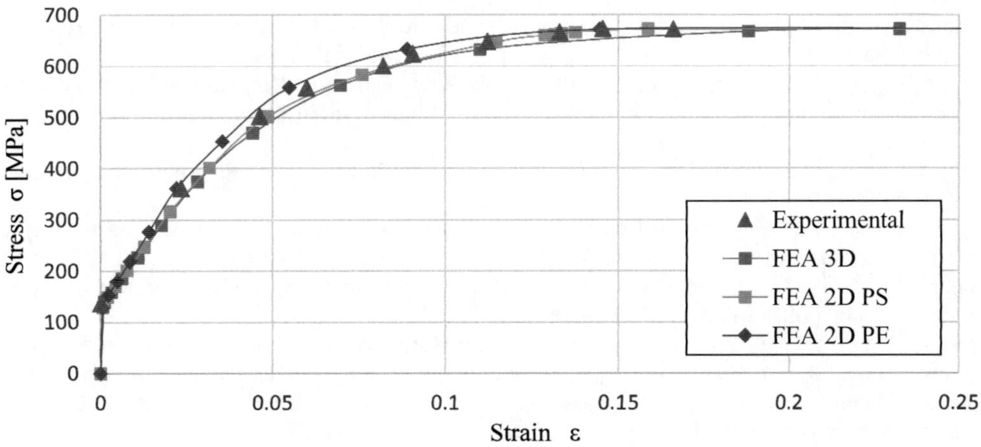

Figure 10. Stress vs. Strain graph for rectangular specimen

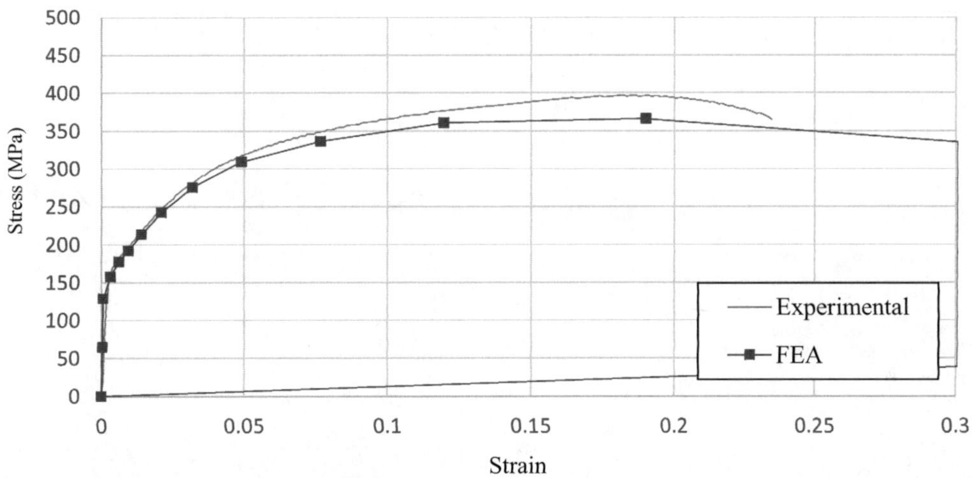

Figure 11. Stress vs. Strain graph for circular specimen

7. Conclusion

The material model was made with the help of the experimental test data. This model was implemented in commercial *FEA* code of ABAQUS® and simulating the experimental quest. Various conclusions can be drawn from work and with the validation of the approach, it can be extended to industrial components of high geometric complexities. The results from rectangular section reveal the closeness of the plane stress conditions to the experimental results which were also in accordance to the three dimensional analysis. Plane strain conditions in such a case slightly overestimates the stress distributions and thin sections are therefore best modelled in plane stress conditions when three dimensional analysis is quite challenging or not possible. The methodology is therefore proposed in the simulation of plasticity in industrial components like thin walled pressure vessels where hoop stress plays an important role in design considerations.

8. References

[1] Mikkelsen, Lars Pilgaard. "Necking in rectangular tensile bars approximated by a 2-D gradient dependent plasticity model." *European Journal of Mechanics-A/Solids* 18.5 (1999): 805-818.
[2] Jain, M., J. Allin, J., D. J. Lloyd, D.J., 1999. Fracture limit prediction using ductile fracture criteria for forming of an automotive aluminum sheet. International Journal of Mechanical Sciences 41, 1273-1288.
[3] Weng, T.L., Sun, C.T., 2000. A study of fracture criteria for ductile materials. Engineering Failure Analysis 7, 101-12
[4] Cabezas, Eduardo E., and Diego J. Celentano. "Experimental and numerical analysis of the tensile test using sheet specimens." *Finite Elements in Analysis and Design* 40.5-6 (2004): 555-575.
[5] Komori, K., 2005. Ductile fracture criteria for simulating shear by node separation method. Theoretical and Applied Fracture Mechanics 43, 101–114
[6] Banerjee, Dilip. "Finite Element Modeling of Deformation Behavior of Steel Specimens under Various Loading Scenarios Dilip Banerjee, Mark Iadicola, Adam Creuziger, and Timothy Foecke."
[7] Yang, X., Chen, C., Hu, Y., Wang, C., 2005. Damage analysis and fracture criteria for piezoelectric ceramics. International Journal of NonLinear Mechanics 40, 1204 -1213.
[8] Analysis and experiment of 7075 aluminum alloy tensile test.et al (2014).
[9] Jose, Edgar, and Tom Anto. "Analysis of Tensile Test of Mild Steel using Finite Element Method." *International journal of innovation in engineerinsg and technology* 5.4 (2015): 247-251.
[10] Ibrahim, Ahmed, Yeong Ryu, and Mir Saidpour. "Stress analysis of thin-walled pressure vessels." *Modern Mechanical Engineering* 5.1 (2015): 1-9.

IOP Conf. Series: Materials Science and Engineering **396** (2018) 012066 doi:10.1088/1757-899X/396/1/012066

Ergonomic optimisation of machining operations of power transmission shaft

Ajith James[1], **Tom Zacharia**[2]

[1.] PG scholar, Department of Mechanical, St. Joseph college of Engineering and Technology, Palai, Kottayam, Kerala, India

[2.] Assistant Professor, Department of Mechanical, St. Joseph college of Engineering and Technology, Palai, Kottayam, Kerala, India

Abstract: Workstation efficiency and operator satisfaction are two main factors which influence the overall productivity of a production unit. Considering ergonomic factors while designing a workstation will help to develop a better environment for working and will help to avoid unsafe working conditions. This paper describes a study conducted in a machining station to find out the various factors which may result in serious health issues to the operator and reduction of productivity. According to the study various modifications are made by considering ergonomic factors and results show that efficiency and operator satisfaction have been improved after implementing the recommendations.

1. Introduction

1.1 Ergonomics

This is an interdisciplinary scientific approach of applying human physiological, psychological and bio-mechanical limitations, capabilities and characteristics to the design of human- machine interfaces and workstations. Workstations and operators are the key factors in a manufacturing unit. Design of the workstations will directly influence the productivity of the operators. It is therefore critical that they are designed to be as ergonomic, flexible and efficient as possible. Design of a workstation should suit a number of persons, and it is to be designed according to the anthropometric data of the region. It is a known fact that the ergonomically ideal layout of workstations and workplaces not only stimulates the efficient manufacture of the product, but also has direct benefits with respect to reducing the psychological demands placed on employees.

2. Literature Review

The first paper reviewed as part of the literature review was "Ergonomics in the workplace". The paper gives detailed description about the influence of ergonomics in the productivity, health and safety of the workers and work satisfaction. It details the importance of considering anthropometric data while designing the workstation. The work station designed by considering the ergonomic factors will reduce the chance of various musculoskeletal disorders. Various Cumulative Trauma Disorders like Tendon disorders, Neurovascular disorders, and Nerve entrapment disorders are also detailed in this paper [1].

The second paper titled "Fundamentals of ergonomics in theory and practice" presents the historical background and a clear definition of the term ergonomics. It details about the context of ergonomics and modern view of ergonomics as a science. Ergonomics consist of various factors which are networked together to represent the interacting system, which shows the overall view of ergonomic system. As a conclusion it describes the various challenges in implementing ergonomic practices [2].

The third paper referred titled "An ergonomics study on assembly line workstation design" describes an ergonomic study conducted in an automotive component manufacturing unit. Twenty workers were taken as participants of the study, anthropometric data and dimensions of the workstation were taken and checked whether it was suitable for a comfortable working condition [3].

The fourth paper titled "Scientific approaches for the industrial workstations ergonomic design: a review" presents a modern and scientific methods used for ergonomic workstation designs. Now a days computer modelling softwares are widely used for workstation designs. Virtual model of the plant layout is made to check the various ergonomic factors and also it is very helpful for the close observation of the work condition and dimensions [4].

The fifth paper titled "Evaluation of work posture by RULA and REBA: A case study" describes a research study conducted at a small scale industry to check the possible occurrence of various musculoskeletal disorders (MSD). RULA (Rapid Upper Limb Assessment) is a survey method developed for use in ergonomics investigations of workplaces where work related upper limb disorders are reported and REBA (Rapid Entire Body Assessment) is a postural analysis tool sensitive to musculoskeletal risks in a variety of tasks [5].

The sixth paper referred title "RULA: a survey method for the investigation of work-related upper limb disorders" describes the development and application of RULA method for ergonomic analysis. It provides a detailed description about the step by step application of this method and how the corresponding values are given. According to the final value obtained we can find out that whether process will result in any MSDs [6].

3. Methodology

3.1 Research goal

Two main objectives of this research are: **a.** to study and ergonomically optimise the secondary machining operations of power transmission shaft, **b.** study the work practices and check whether any modifications are required to the workstation.

3.2 Sample and data collection

This study was focussed on the workers from an automotive spare parts production company, who operate CNC machines. The people working in the first shift are considered for study. There are both male and female operators in the study group. The main data were obtained through personal interviews and direct observation. Each person was personally interviewed to understand the level of satisfaction in their jobs and also to find out the tasks which were felt difficult for them.

Operator motion study was conducted initially, to check for any inefficient motion or work practices. By repeated observation, incorrect work practices or inefficient workstation conditions were determined. For the selected operations RULA method was used. Those operations which reported 'investigate and implement change' was analysed further and proper modifications were made to avoid MSDs.

3.3 Rapid upper limb assessment (RULA)

RULA is a survey method developed for ergonomic investigation of workplaces where work related upper limb disorders are reported. This method gives a quick assessment of postures of body parts, muscle function and other external loads experienced by the body. This method is simple and efficient to use and does not need any special equipment. A standard coding system is used to represent the level of influence of current postures in the operator health and according to the final result corresponding suggestion are mentioned.

RULA method is applied in three steps, the body segments to be evaluated are divided into two and these are evaluated separately and according scores are calculated. In the last step these two values are combined together to get the final value. In the first stage upper arm, lower arm and wrist are monitored and score is provided for the level of wrist twist by considering the amount of load carried and frequency of work. In the second stage posture score for the neck trunk and legs are calculated and recorded. In the third stage these two scores from stage one and two are combined together to form the grand score.

Grand score shows the level of action required for the particular work. If the grand score is 1 or 2 it indicates that the posture is acceptable, if score is 3 or 4 it shows that further investigation is required and changes may be required, if the score is 5 or 6 investigations and changes are required soon and if the score is 7 or greater investigation and immediate changes are required [6].

4. Motion study

A detailed 'motion study' was conducted and five distinct activities were identified:
They are:

i. *Raw material collection*
The first activity is collection of raw materials for part machining. These may be forgings or semi-finished items from previous machines.
ii. *Loading and unloading*
Part is loaded for subsequent operations, and unloaded once the operation is completed.
iii. Visual inspection and measurement
Inspection is carried out after placing the part on a fixture.
iv. *Oiling and protective covering*
Rust oil is applied to finished parts and some parts are provided with a protective covering.
v. *Transferring of parts*
Parts ready for shipment are transferred to the conveyor on pallet.

According to the detailed analysis of the motion study 4 major issues were identified for modification.

i. *Height of inspection table*
Height was found to be less, resulting in an uncomfortable body posture, to take readings. The angle of vision was more than 70°, whereas ergonomic standards stipulate an angle within ±30°.

Figure 1. Table and height gauge.

Figure 2. Height measuring.

ii. *Manual transfer of parts*
The parts are stacked inside cartons, which weigh 18kg, 30kg, and 41kg.
iii. *Manual chip removal*
Production time is lost due to shut down for manual chip removal. Also, the body posture and effort required were found to be against ergonomic considerations.
iv. *Oiling of finished parts*
The method was found to be unhygienic and inefficient

In order to check the impact of these issues on the health and safety of the operators RULA study was conducted for these issues. According to the survey it was found out that position of height gauge was inadequate. This issue scores 7 in RULA assessment and it implies that investigation and corresponding implementation is required. For manual burr removal and material handling also the RULA score was 7.

IOP Conf. Series: Materials Science and Engineering **396** (2018) 012066 doi:10.1088/1757-899X/396/1/012066

5. Discussions

The study shows that the operators lack in knowledge of ergonomics and manual material handling. Each problems was separately considered and analysed. According to the study, proper modification were made. Firstly the height gauge positioning is considered and a design modification is executed for the table. Anthropometric data of the current location is taken and according to that values the position of the height gauge is changed by providing a special stand for the current table and the height gage is placed on that.

Figure 3. Modified table.

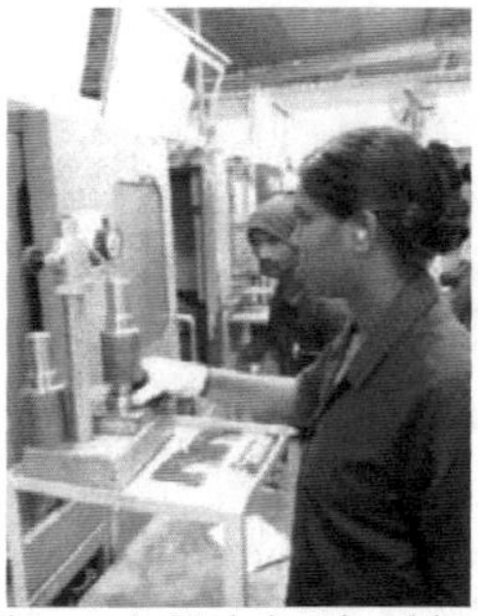

Figure 4. Height checking.

Due to this modification visual angle is corrected and a convenient space is obtained to place the various snap gauges and plug gauges.

Regarding chip removal, it was estimated that this activity reduced plant efficiency as much as 30%. RULA score for removal process also shows that this work practice will result in serious health issues. Establishment of conveyor system is recommended.

RULA score for manual material handling also shows that it is injurious to the health of the operators and there is severe chance for the occurrence of back pain and leg pain. As a result the number of absenteeism will be increased and operator morale will be reduced. It was recommended that a trolley should be provided for material handling purpose.

A proper oiling mechanism has been designed and established so that oil will never get soaked in the cotton gloves. It is designed in such a way that the parts are horizontally placed and the part will be fully immersed in oil.

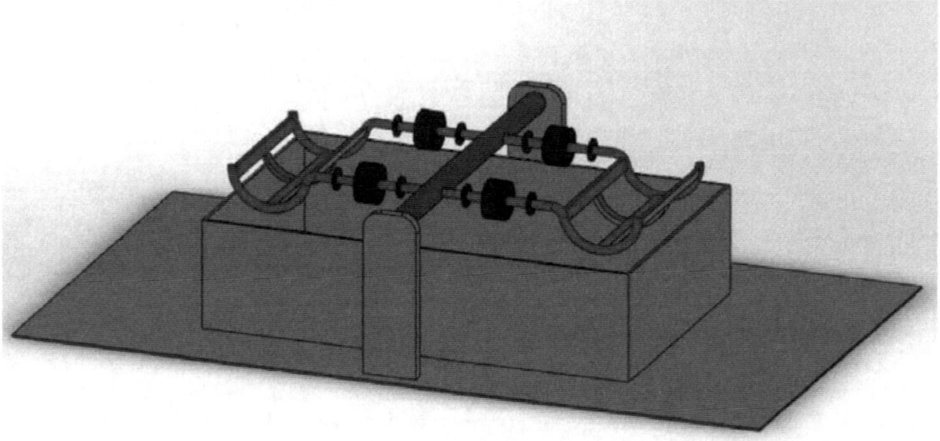

Figure 5. Modified oiling mechanism

IOP Conf. Series: Materials Science and Engineering **396** (2018) 012066 doi:10.1088/1757-899X/396/1/012066

This mechanism works like a 'siso' and a counter weight mechanism is provided for ensuring the part is fully immersed in the oil. When a new part is placed for oiling the previous part which is in the oil will automatically come up and the new part will go down.

6. Conclusion

Evaluation of ergonomic factors in a workstation shows that they give only a moderate consideration to the ergonomic factors. It is observed that even a simple modification can make a lot of improvement in the system. Ergonomic improvements can result in overall improvement in efficiency and employee morale in a production plant.

References

[1] Jeffrey E. Fernandez and Michael Goodman *Ergonomics in Workplace* Exponent health Group

[2] John R Wilson 2000 *Fundamentals of ergonomics in theory and practice* Applied Ergonomics **31** 557-567

[3] Baba Md Deros, Nor Kamaliana Khamis, Ahmad Rasdan Ismail, Haris Jamalddin, Azmi Mat Adam and Sarudin Rosli 2011 *An ergonomic study on assembly line workstation design* American Journal of Applied Sciences **8(11)** 1195-1201

[4] Terry Bossomaier, Agostino Bruzzone, Antonio Cimino, Francesco Longo and Giovanni Mirabelli *Scientific approaches for the industrial workstations ergonomic design: a review*

[5] Ansari N A and Dr. Sheikh M J 2014 *Evaluation of work posture by RULA and REBA: A case Study* **11(4)** 18-23

[6] Lynn McAtamney and Nigel Corlett 1993 *RULA: a survey method for the investigation of work-related upper limb disorders* **24(2)** 91-99

IOP Conf. Series: Materials Science and Engineering **396** (2018) 012067 doi:10.1088/1757-899X/396/1/012067

Design and Simulation of an Automated Road Safety Enhancement System by Testing the Physical Fitness of Driver

ARUN VIJAYAN[1] and ARUN P[2]

[1]M.Tech Scholar, Electronics and Communication Engineering
St. Joseph's College of Engineering And Technology, Palai, Kerala
[2]Assistant Professor, Department of Electronics and Communication Engineering
St. Joseph's College of Engineering And Technology Palai, Kerala

E-mail: arunvijayan111@gmail.com

Abstract. Most of the road mishaps occurred due to the mistakes of driver. The lack of physical fitness is one of the reasons for these kinds of accidents. Alcohol consumption and sleeping during driving also leads to accidents. The biological signals such as ECG, EEG and Heart Rate are considered to be good indicators to check physical fitness. To avoid the road mishaps, a micro-controller based system to sense the above mentioned parameters and to make a decision based on the sensed data is proposed in this paper. It consist of various sensors to detect biological signals, sleeping attitude and presence of alcohol in the blood of the driver. The micro-controller is programmed to make decision based on the sensed data. If the physical fitness of the driver is tested and verified successfully, can drive the vehicle. If not, the system automatically locks the engine and a message will be delivered to the control centre via wireless technology. The design and real time simulation of the proposed system is carried out in Proteus® and the micro controller programming is done in mikroC®.

1. Introduction

The mistake by driver is one of the reasons for major road accidents. The performance of the driver depends upon his current physical condition. This may be because of the driver fatigue and distraction [1]. Alcohol consumption and sleeping during driving also leads to accidents. The biological signals such as ECG, EEG and Heart Rate are considered to be good indicators to check current physical level of the driver [2]. To avoid this mishaps, a micro-controller based system to monitor physical states of driver is proposed in this paper. The proposed system will detect various biological parameters and check whether the physical fitness of driver is good or not. The proposed system will be very useful for elderly car driver or driver with chronic diseases. It ensures safety driving and reduces the chance of road mishaps.

In past decades, many researches have performed to detect the fatigue and drowsiness of driver. Many of them are based on EEG signal bands (α, β and γ bands) [3]. These bands are occurred by the brain activity and it varies with respect to drowsiness or fatigue. For EEG based assessment, an algorithm is developed to assess driver fatigue based on the changes in EEG bands. Then estimated the drowsiness level by independent component analysis (ICA) of EEG and found the optimal locations to place EEG electrodes. Compared four algorithms based on the four EEG Bands [4]. The result showed that a slight decrease of alpha activity and a significant decrease of beta activity were associated with fatigue. The fatigue during driving causes sleepiness.

Previous researchers has proved that the ECG is very helpful to identify the fatigue and sleeping [5]. The ECG signal obtained from the electrode that placed on chest. Then, R-R interval and QRS complex of measured ECG signal will be compared with the clinical one. So the drowsiness can be identified. The fatigue of driver leads to sleeping. So the detection of sleeping along with the fatigue detection is essential. The sleeping can be detected from eye blinking activity and Heart Rate (HR). The eye blinking activity can be sensed by eye blink sensor. While sleeping, the duration of pulse signal obtained from sensor is high and number of blinking is less. Heart Rate and HR variability can calculated from ECG signal by using computing algorithm. The HR during sleeping will be 40-50 bpm (Beats per Minute).

Alcohol impaired driving also leads to road mishaps. Alcohol taken long before driving can still affect the driver, even several hours later. This is because the body can only metabolize alcohol at a fixed rate, no matter how much has consumed. This leads to lack of coordination, poor decision-making and drowsiness. The detection and prevention of alcohol impaired driving achieved by water-cluster-detecting (WCD) breathe sensor [6]. The WCD breathe sensor detects breath by measuring electric currents of positively or negatively charged water clusters in breath that are separated by using an electric field. The WCD breath-alcohol sensor couples the WCD breath sensor with an alcohol sensor and simultaneously detects the electrical signals of both breath and alcohol in the breath.

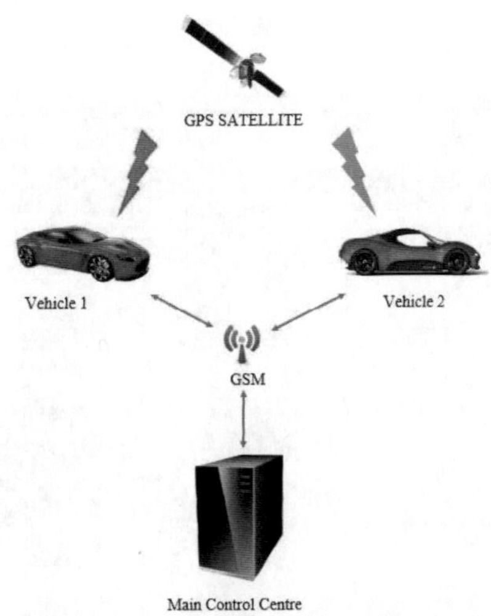

Figure 1. Schematic representation of proposed system

In this paper, a microcontroller based automatic system is proposed to monitor the physical fitness of driver. Here, the vital signals such as EEG, ECG and Eye Blinking are detected using sensor network connected to the microcontroller. These three parameters help to produce more accurate detection of fatigue. An alcohol sensor also used to avoid the alcohol effect on driving. The microcontroller makes decision based on above sensed vital information. The result continuously displayed on the LCD display. If the physical fitness of the driver is tested and verified successfully, then he can start the vehicle else the system automatically locks the engine and a message will be delivered to the main control centre via wireless technology. So, the control station can provide medical aid to drivers. When physical fitness changes dangerously, an audio alert will give through audio buzzer for the attention of driver. The proposed system also consist of a sleep prevention device to awake the driver from sleeping.

The design and real time simulation of the proposed system is carried out in Proteus and the microcontroller programming is done in mikroC. The rest of the paper is organized as follows. Section 2 deals with the overview of the proposed system which consist of schematic diagram, block level representation, details of various blocks, circuit diagram and their process flow diagram. Section 3 details the simulation result and analysis. Section 4 depicts conclusions and future scope followed by references.

2. Overview of Proposed System

Fig 1 shows the schematic representation of proposed system. It consist of Health Monitoring System (HMS), Main Control Centre (MCC), GPS and GSM network. The HMS is placed on vehicle 1 & 2, which monitors the present heath condition of drivers of vehicle 1 & 2 respectively. The HMS also record the current location of vehicle with the help of GPS that helps to identify the location of vehicle by the MCC. All the HMS are connected to MCC via GSM network. When the driver meet with a serious health problems, the HMS automatically send message to MCC. Then the MCC inform the ambulance for medical aid and also inform the police.

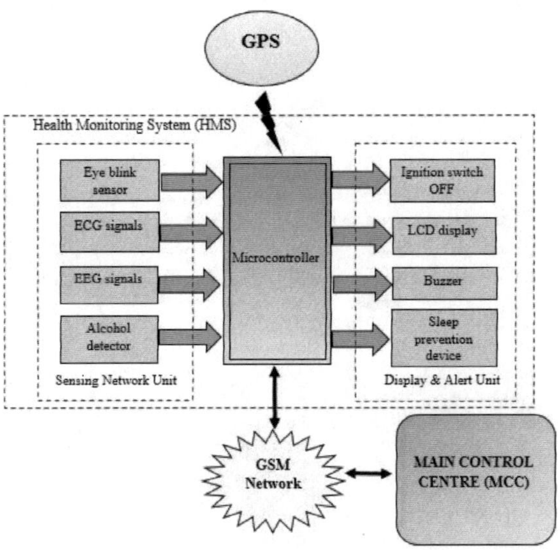

Figure 2. Block diagram of proposed system

The detailed block diagram of the proposed system is given in Fig 2. The major block in this system is HMS. It contains Sensing Network Unit (SNU), Display & Alert Unit (DAU) and micro-controller. As soon as the vehicle is started, the HMS will turn ON. Then micro-controller check for the fitness and alcohol consumption rate of driver by analysing the sensed signals by SNU. The SNU senses biological signals such as EEG, ECG, Eye blinking and Alcohol effects. If the fitness test and alcohol test verified OK, can drive vehicle. Otherwise the ignition switch will be turned OFF. The HMS also monitor the health changes continuously during drive. When a health problem is detected at the time of driving, then DAC is given an alert to drivers. The DAU also consist of an audible alert and a sleep preventing devices in order to avoid sleeping while driving. When a serious health problem is detected by the HMS, an alert message will be sent to the MCC by means of GSM network. The MCC is a central monitoring station under the control of motor vehicle department, which observe the drivers through HMS equipped in each vehicle. The MCC has the contact with hospital and police. So, it can provide medical assistance or any help to drivers, when the emergency message is received from HMS.

2.1. Health Monitoring System (HMS)

The HMS is a micro-controller based system to check physical fitness and alcohol consumption of driver by analysing the biological signals. The biological signal are obtained from driver's body by SNU unit associated with the micro-controller. The fig 2 consist of sensors to detect EEG signal, ECG signal, Eye blinking signal and alcohol effects. Here an innovative non-intrusive sensors for effective detection signals are used [7]. The fitness detection algorithm stored in the micro-controller memory helps to compute the above signals [8]. When the vehicle starts, the HMS begin to operate. Initially it check for the physical fitness level and check if any hazardous level of alcohol is consumed by the driver. If both are ok, then the driver can proceed the driving, otherwise he can't proceed. The HMS continue its operation to monitor the health changes during driving. If any serious health problems detected, the micro-controller send emergency messages to MCC using GSM network. The DAU unit in the HMS give messages and alert to driver with the help of LCD display and audible alert using buzzer. The HMS also consist of a sleep detection and prevention system. The simulation diagram of the HMS is shown in fig 3. The detailed operation of SNU and DAU are given in the following section.

2.1.1. Sensing Network Unit (SNU) The SNU consists a group of sensors. They are EEG and ECG sensors, Eye Blinking detection and Alcohol detection sensors. The EEG sensor consists of different electrodes that placed on the skull which detects the four brain waves (α, β and γ). When the fatigue occurs, the brain waves will changes. The changes can be detected by comparing the sensed brain waves with its clinical waves using comparator. While changes occurs, the output of comparator will be high. Next, the important signal is ECG signal which

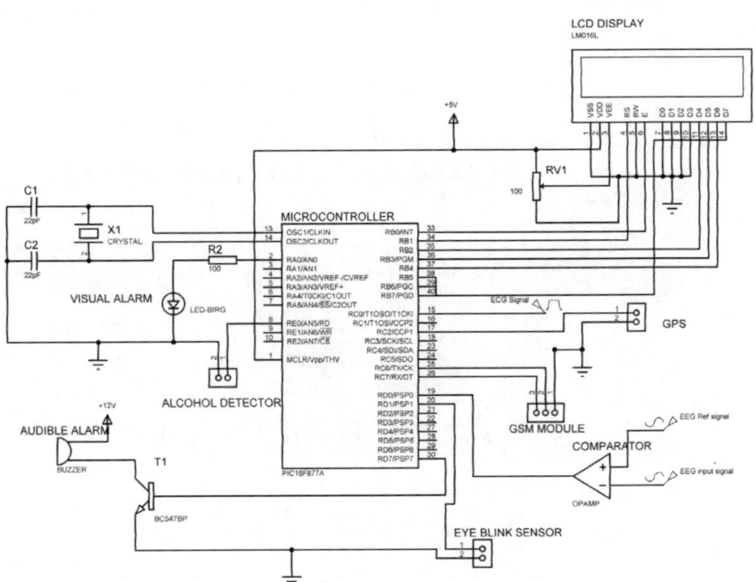

Figure 3. Circuit Diagram of HMS

is detected by the electrode that placed on the chest of the driver [9]. Then the R-R interval and QRS complex in the sensed ECG signal will be identified by algorithm stored in the micro-controller. By counting the QRS per minute, Heart Rate and HR variability can be measured. Next the important sensor is alcohol sensor which is placed in front of driver's face that detects the alcohol effect from the breathing air. It produce a high pulse for the alcohol detection. The last one is, Eye Blinking sensor which produce pulse signal. By computing, no of eye blinking, eye blinking duration and Heart Rate of driver, sleepiness can be detected [10]. For a sleeping person, number of eye blinking is low, the eye blinking duration is high and the HR is in between 40-50 bpm. All the sensors are connected to various ports of micro-controller that are shown in fig 3.

2.1.2. Display and Alert Unit (DAU) The role of DAU are display the alert messages, produce audible alarm, control the engine operation and activate sleep prevention device. In this system an LCD display is used to display the alert messages and computational result of sensed biological signal. This is helpful for drivers to aware about his health condition. Here a buzzer circuit is set up to give audible alarm. When an emergency health problem detected, the buzzer circuit will be activated and give an audible alert. At that time, the health problem is displayed on the LCD. This unit consist of a control circuitry to turn ON/OFF of engine. While detecting an alcohol consumption, the micro-controller will be activate the control circuitry and turn off engine. Sleep prevention during driving is also one of the highlight of the proposed system. Here use a sleep preventive devices. It consists of vibrator and has a contact with the skin of driver. When the sleepiness is detected by the micro-controller, then the device will be activated and produce vibrations. So, the chance to fall into sleep can be avoided.

2.2. GPS and GSM Network

The HMS system consist of a GPS receiver. The HMS system update and store the current location of vehicle by using GPS. This information will be sent to MCC, while the emergency medical aid is requested to the MCC. The HMS and MCC are connected through GSM network. When any health problem has occurred, the HMS automatically send a voice message/text message to MCC. The message consists information of the driver and the details and location of the vehicle.

2.3. Main Control Centre (MCC)

It is a central monitoring system maintained and monitored under the control of motor vehicle department. It consist of microcontroller, host system and communication facilities. It check for the fitness of driver is capable for driving. If the fitness is not good, take action to stop driving

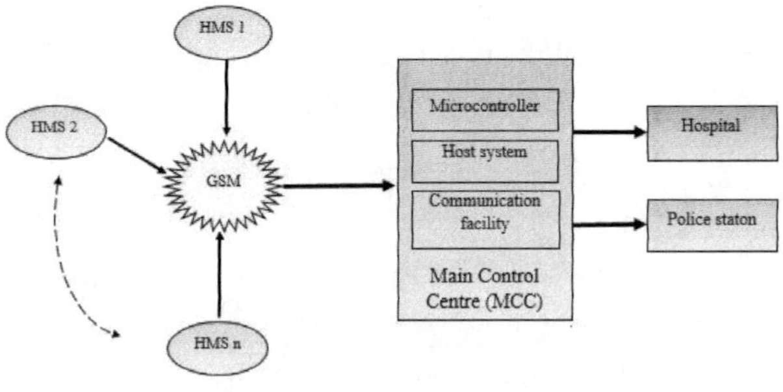

Figure 4. Block diagram of Main Control Centre (MCC)

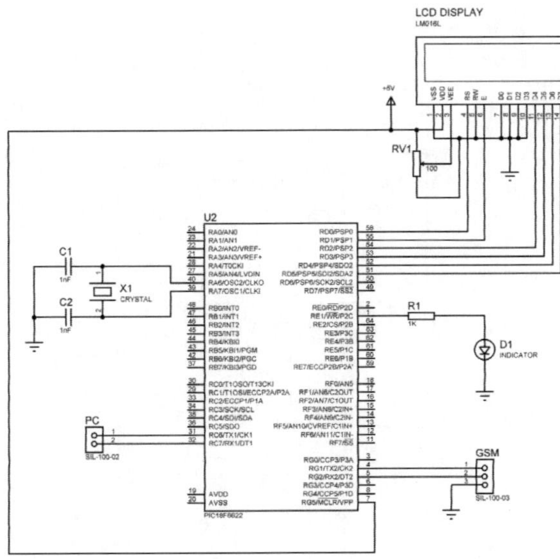

Figure 5. Circuit diagram of MCC

and give medical aid to driver. So, the road mishaps due to physical problems can be avoided. All the operations of MCC are coordinated by the microcontroller. The host system connected to the microcontroller consist of the data base of vehicle. It also provide a software support to the MCC. Communication facilities are used to inform the police station and hospital about the health issues of drivers.

When the physical fitness of driver changes dangerously, an alert from HMS of that vehicle will be sent and it will be receive at the MCC. The message consist of information about health problem, vehicle number, driver details and current location of vehicle. Then the MCC forward this message to the hospital and police station to provide service to driver. Depending on the details of HMS message, the hospital/ police station can be provide their service easily.

The circuit diagram of Main Control Centre (MCC) is shown in fig 5. Here use a PIC microcontroller (PIC 18F6622) that having two serial ports. So the host system and GSM module can be connected to the microcontroller simultaneously. It also consist of an LCD display to display alert messages while receiving any emergency messages from HMS in vehicle.

2.4. Steps Involved in Decision Making of Proposed System

The major steps of decision making involve fitness checking and alcohol detection. The fitness consists of fatigue detection and sleep detection. The sequence of operation that occurred in the proposed system is shown in fig 6. Initially check for an alcohol consumption, then it check for physical fitness if and only if the alcohol detection successfully verified. Next the check for the fitness. If the driver is physically fit, then can start vehicle. In case the physical fitness and alcohol detection failed, the vehicle cant start. The physical fitness ensure the EEG, ECG, Eye Blinking are normal. So the safety can be guaranteed.

3. Simulation Result and Analysis

The design and real-time simulation of the proposed system has been carried out in Proteus. The circuit diagrams of HMS and MCC are simulated and its outputs are verified. In HMS part, different biological signal are produced using signal generator available with Proteus. The GSM module is modelled using virtual terminal. The signal computing decisions for fitness

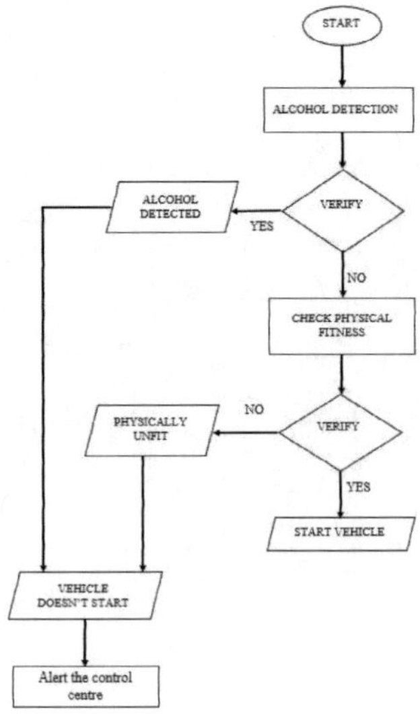

Figure 6. Flow Chart of proposed system

measurement are developed and compiled using microC tool. Then the hex file of computation algorithm is loaded into the micro-controller of the real-time simulation diagram drawn in Proteus. Similarly, the MCC circuits are set up in Proteus and its controlling algorithm is developed and compiled using microC. Then the hex file loaded and result have been verified. The snap shot of Proteus simulation result and result of compilation of microC program is presented in fig 7 and fig 8 respectively.

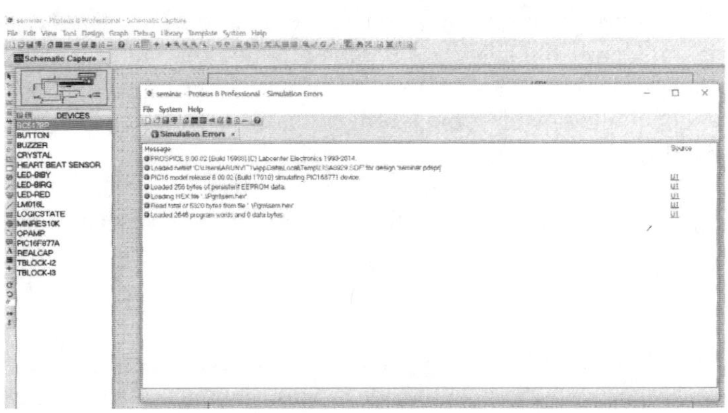

Figure 7. Snap shot of Proteus simulation

4. Conclusions and Future Scope

A microcontroller based system (HMS) to detect fatigue and physical fitness affecting factors such as alcohol consumption, sleepiness, etc. was proposed in this system. Decision has been taken by considering the sensed signals like EEG, ECG, Eye blinking and alcohol effect. If the fitness doesn't meet with threshold value, then the micro-controller takes action to stop driving. In addition to this, the proposed system consist a centralized remote monitoring system (MCC) to check whether the HMS module equipped in vehicles are working normally or not. The

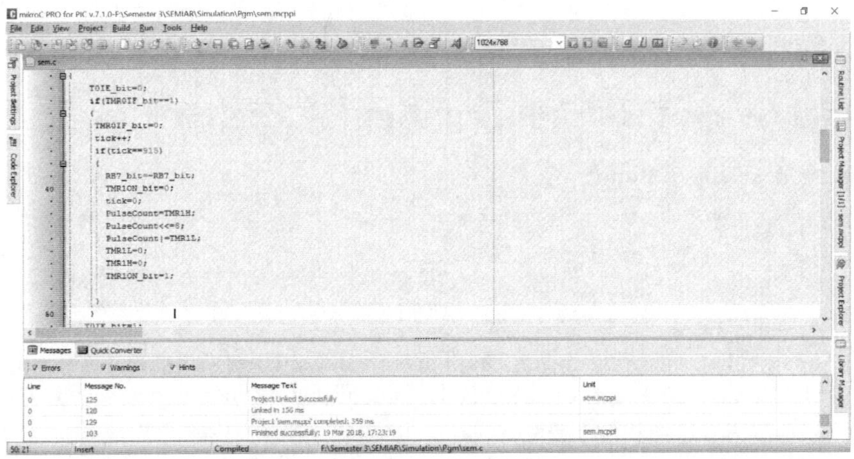

Figure 8. Snap shot of microC compilation result of program

MCC work under the control of motor vehicle department. The design and simulation of all the blocks has been carried out in Proteus and the software code were done in microC. The simulation results were verified and tested for correction. As a future deviation, one can think of the hardware design of this system and its prototype may also has importance.

References

[1] A. H. Taylor and L. Dorn, "Stress, fatigue, health and risk of road traffic accidents among professional drivers: The contribution of physical inactivity", *Annu. Rev. Public Health*,vol. 27, pp. 371391, Aug. 2006

[2] T. Wartzek, B. Eilebrecht, J. Lem, H.-J. Lindner, S. Leonhardt, and M.Walter, "ECG on the road: Robust and unobtrusive estimation of heart rate",*IEEE Trans. Biomed. Eng.*, vol. 58, no. 11, pp. 31123120, Nov. 2011.

[3] C. T. Lin, R. C. Wu, S. F. Liang, W. H. Chao, Y.-J. Chen, and T.-P. Jung, "EEG-based drowsiness estimation for safety driving using independent component analysis,"*IEEE Trans. Circuits Syst. I, Reg. Papers*, vol. 52, no. 12, pp. 27262738, Dec. 2005.

[4] S. K. L. Lal, A. Craiga, P. Boorda, L. Kirkupb, and H. Nguyenc, "Development of an algorithm for an EEG-based driver fatigue countermeasure,",*J. Safety Res.*, vol. 34, pp. 321328, Aug 2003.

[5] T. Wartzek, B. Eilebrecht, J. Lem, H.-J. Lindner, S. Leonhardt, and M.Walter, "ECG on the road: Robust and unobtrusive estimation of heart rate,"*IEEE Trans. Biomed. Eng.*, vol. 58, no. 11, pp. 31123120, Nov. 2011.

[6] Minoru Sakairi, "Water-Cluster-Detecting Breath Sensor and Applications in Cars for Detecting Drunk or Drowsy Driving,"*IEEE Sensors Journal Year: 2012*, Volume: 12, Issue: 5 Pages: 1078 1083

[7] Ye Sun; Xiong (Bill) Yu, "An Innovative Nonintrusive Driver Assistance System for Vital Signal Monitoring",*IEEE Journal of Biomedical and Health Informatics Year: 2014*, Volume: 18, Issue: 6 Pages: 1932 1939

[8] Driver assistant for the detection of drowsiness and alcohol effect Prajnyajit Mohanty; Pallem Siddharth; Kunja Bihari Swain; Rakesh Kumar Patnaik 2017 Third International Conference on Sensing, Signal Processing and Security (ICSSS) Year: 2017 Pages: 279 283

[9] T. Maruyama, M. Makikawa, N. Shiozawa, and Y. Fujiwara, "ECG measurement using capacitive coupling electrodes for man-machine emotional communication" *in Proc. IEEE Int. Conf. Complex Med. Eng.*,2007, pp. 378383.

[10] E. Michail, A. Kokonozi, and I. Chouvarda, "EEG and HRV markers of sleepiness and loss of control during car driving", in *Proc. IEEE 30th Annu. Int. Eng. Med. Biol. Soc. Conf.*,Vancouver, BC, Canada, Aug. 2024, 2008, pp. 25662569.

Bio-active glass synthesis and coating: A review

Athira Thulasidas[1*] and J Babu[1]

1* M Tech student, Department of Mechanical Engineering, St. Joseph's College of Engineering& Technology,Choondacherry,686579, Kerala,India.

1 Department of Mechanical Engineering, St. Joseph's College of Engineering& Technology,Choondacherry,686579, Kerala,India.

*1*Corresponding author email: thushara.athira2991@gmail.com*

Abstract: Implants primarily use Titanium or its alloys. These alloys are biologically inert to living cells. These metal implants has key issues such huge difference elastic moduli of a metal implant and bone tissue which demands a compatible system between these two ,absence stable and reliable chemical compound between implant and bone tissue for biological integration. These two issues can be effectively dealt with intermediate coating of suitable elastic modulus and also biologically active. The aim of the present work is to discuss briefly various synthesis and coating methods of bioactive glass and the key issues in this area.

Keywords: Bioactive glass, Implants, Coating methods, Bio-activity.

1. Introduction

Bioactive glasses are exceptional glasses, react with a living tissues and aids in the formation of biologically-active layer of hydroxyl carbonate apatite (HCA) on outer surface of the implant. This HCA layer has matching structural and chemical properties to the mineral phase in the bone. However, these bioactive glasses elastic modulus is higher than that of human bone but have smaller values of strain-to-failure, fracture toughness and flexural strength [1]. Because of this variation in mechanical properties, these bioactive glasses generally not preferred in load-bearing requirements, where metallic alloys are the best choice for this purpose. Then it was realised that this bio-active glasses can be coated on metallic implants. This will improve Osseo integration of the coated metallic implants, and also prevent from corrosion of implant due to body fluids. This also protects the tissue from the corrosion products [2].

Design criteria for bioactive coatings for metallic implants incudes: (a) Proper selection of heating cycle that do not deteriorate the properties of the both bioactive glass and metal implant; (b) Coefficient of thermal expansion of the bio-active glass and metal and should be nearly same, this reduce the delamination of coating materials. Huge difference of this coefficients cause thermal stresses leads to cracking or delamination of the coating material while fabrication; (c) Heating cycles should result with optimum adhesion at the interface; and (d) Very important is that when coatings come in contact with body fluids must form hydroxyapatite (bioactive) layer. It is the fact that Titanium alloys are very active. This causes the chemical reactions between Titanium implant and bioactive glasses reactions during heating leads to deteriorate adhesion and bioactivity at the interface. Due to this reason coating of titanium implants with bioactive glasses is challenging. Generally bioactive glasses compose of silicate glasses of weight percentage is less than 60%. Coefficient of thermal expansion of these glasses is higher than those of Titanium implants. Coefficient of thermal expansion of bioactive glasses can be minimised by enhancing the SiO_2 content. But increase of SiO_2 content in the glass reduces the bioactivity significantly [3]. Presently there are different methods available to coat bioactive glasses on the Titanium or Titanium alloys substrates which include enamelling, sol-gel, electrophoresis deposition, pulsed laser deposition, plasma spraying, magnetron sputtering, etc. which discussed in detail in the following paragraphs.

2. Synthesis Methods

2.1 Melt Quench

In the method, glass can be formed by mixing the different raw materials thoroughly and then following by solidification using quenching method. This method uses suitable fraction of the ingredients and mixing them to form a batch of glasses. To obtain the glass in uniform powder form the mixture is to be ground to small particles. This mixture is added with a wet medium like acetone and ground for two hours in a ball mill to obtain homogenous sized particles. This resulted mixture is then dried in air to evaporate the acetone content. Mixture of raw materials could also be ground in water before melting if they are not highly hygroscopic. This mixture then placed in platinum/ alumina crucible and placed in a furnace for melting. As alumina diffuses at high temperature cannot be used for glass composition which melt at high temperatures; in this case platinum crucible can be used. Alumina crucible can be the best choice for those glasses, which melt at the low temperatures. To eliminate gaseous composition this mixture is calcined for 2 h, at 500°C temperature. This calcination temperature for aluminosilicate compositions is around 1500°C, and for phosphate and borate compositions it is in the range of 1200–1300°C. Constant melting temperature is to be maintained for at least an hour to obtain homogeneous materials. This homogenous material is then transfer to a preheated graphite mold. Moulds of different shapes can be used to obtain glasses of desirable shapes and sizes [4-9]. This method requires high temperature ranges from 1200-1500°C. The bioactive glass produced by this method possesses less porosity causes defects in tissue integration in-vivo testing.

2.2 Sol-Gel synthesis

Sols are defined as the dispersions of colloidal particles (diameter of 1– 100 nm) in a liquid. A gel is a network of polymeric chains and pores and is interconnected. The polycondensation and hydrolysis of a precursor results interconnected 3-D network of a gel. Growths of the network from a group of discrete colloidal particles also cause the formation of a gel.

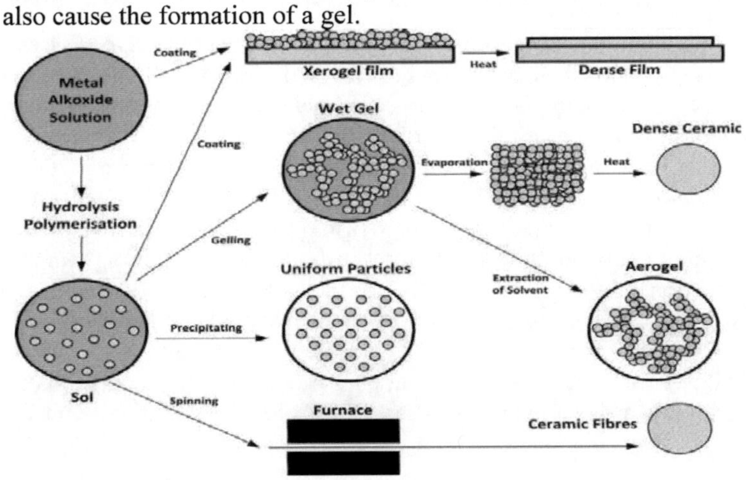

Figure 1.Different stages and routes of the sol-gel processing method

The begining step of this process is proper mixing of all the organometallic precursors or alkoxide. Next step is the use of de-ionized water for hydrolysis of alkoxide. Silicon alkoxide gives silanol groups ($Si(OH)_4$) when hydrolysed by the water. These groups by interacting with each other form Si-O-Si bonds. This leads the formation of silica network (SiO_2) through polycondensation process and water is given out as a by- product.

After the sol is formed, gelation process follows, which is the formation of three-dimensional network by cross-linking and condensation of silica particles and other particles. Gelation results in increase in the viscosity of the sol. Time of gelation depends on the factors like solvent concentration, amount of water used for the hydrolysis and nature of alkoxide group. This time is quite larger for higher molecular weight alkoxi groups and is directly proportional with the hydrolysis water content.

The next step is the aging of this gel, is a process of precipitation and polycondensation of network of the gel. This results in decrease of porosity leads to increase in the strength. The aging process shows a remarkable influence on the physical properties which include, surface area, volume, and density of the gel and finally on the resulting glass. Heating of this resulting gel at higher temperatures eliminates pores and helps in the densification of network [10-13].

This technique is more advantageous because of low temperature requirement and bioglass produced by this technique exhibits more bioactivity during in-vivo testing. Important application of glasses prepared by this method is thin films, where substrate is coated using either dip coating or spin coating.

2.3 Microwave Irradiation
This is a modified sol-gel approach. In this process the resultant obtained was further irradiated at microwave oven, before sintering at high temperatures. This will strengthen gel struts, in order to improve mechanical characteristics.

3. Coating Methods
3.1 Enameling
This method uses the glass powder of particle size < 50 µm for the bioactive glasses coatings. Powders will be suspended in isopropanol then be deposited on Ti or Ti-6Al-4V substrate plates. These plates generally polished with 1 µm diamond and then thoroughly cleaned in acetone and ethanol. These samples then heated in a furnace first by preheating at 600 °C in air then followed by concurrently evacuating the oven to 1.3×10^5 Pa with a heating rate, 40°C/min in the temperature ranges from 700 - 1000 °C. Vacuum to be released on reaching this temperature. The final thickness of the coating generally ranges from 25 to 150 µm. Disadvantages of these methods are when firing temperatures is less than 750 °C results in porous layers which results improper adhesion to the substrate and when firing temperatures greater than 850°C with firing time longer than 1 minute results in the formation of bubbles. Proper control of temperature is necessary. Excellent adhesion can be obtained at firing temperature in the range 800 - 850°C for periods of time less than 1 minute [14]. This is an advantageous method for coating the complicated shapes with varying thickness even thin coating can't crack or delaminate .This method uses a simple technology with low production cost and ease in operation.

3.2 Sol- gel deposition
The main attraction of this method is the better control of chemical composition and needs low temperature for synthesis. In this method the bioactive glass coatings will be given to the substrate by dipping in the solution and removing at different intervals. Then the coatings are to be aged followed by drying at relatively lower temperatures. Last coatings are to be heated to 900◦C to stabilize coating. Different factors influence the solubility and adhesion strength and morphology of the sol–gel coated films. The surface morphology of these films may be administered by processing variables which include use of chemical additives, different precursors and post-heat-treatment [15-17]. Adhesion between

substrates and films can be improved by increasing the roughness of the surface and coating with another oxide. To obtain good adhesion, low thermal stress residue at interfaces and chemical bonding is required [18].The coating presents a higher impedance modulus than the base metal and it acts as a protective layer on the substrate [19].

3.3 Electrophoresis deposition

Electrophoresis deposition (EPD) is a simple coating method with low cost, greater purity and micro structural homogeneity of the coatings on substrates with complex shapes. The process consists of two steps: electrophoresis and deposition. Deposition depends on the congeal of particles. This method is used for the coatings with wider thickness range from less than 1 μm to more than 500 μm. From the beginning of 19th century, the phenomenon of electrophoresis is known, and it has been using tremendously in traditional ceramic for past 40 years [20-21]. When an electric field is applied, the motion of charged particles towards oppositely charged electrode is termed as electrophoresis. Coatings is done by congeal of the particles in the suspension, to form a dense coat on one of the electrodes. An EPD cell is schematically represented in Fig. 2[22]. Important factors include current, voltage, concentration and time are vital to control morphology /composition [23] and thickness. After the bioglass deposit on substrate surface the sample are to be slowly dried for 24 hours in a desiccators at appropriate temperature, and then sintered at 800-1000^{0}C temperature.

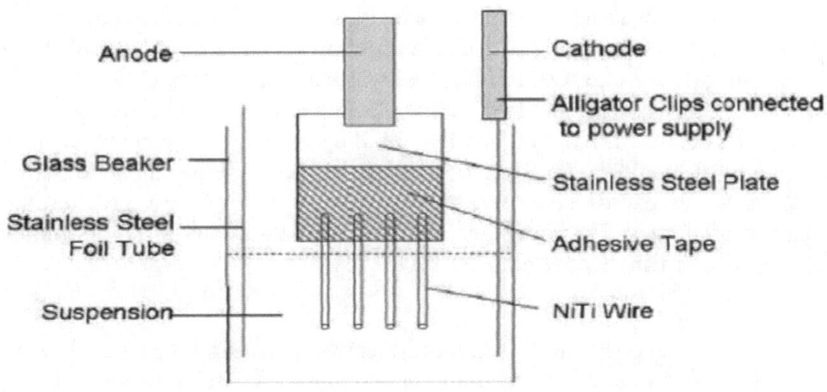

Figure 2. Schematic representation of An EPD cell [22].

3.4 Plasma spraying

This method uses a gas flow (plasma spray) into a chamber by an electric arc. This plasma temperature is about 1000^0-3000^0 K. Now the coating powders are fed into the chamber and are heated up. These particles are accelerated through a nozzle, and sprayed on the substrate. The particles may have speed in the range of 100 and 350ms^{-1}, and so that the flight time is only 10^{-3} of a second. The particle assumed as a lenticular shape and solidifies in 10^{-6} of a second, by releasing their kinetic and thermal energy to the metal plate as they strikes on them. The desired thickness can be achieved by spraying the layers of the material, one top of the other. The thickness is usually about 100 μm. Plasma spraying process is shown in fig.3.

IOP Conf. Series: Materials Science and Engineering **396** (2018) 012068 doi:10.1088/1757-899X/396/1/012068

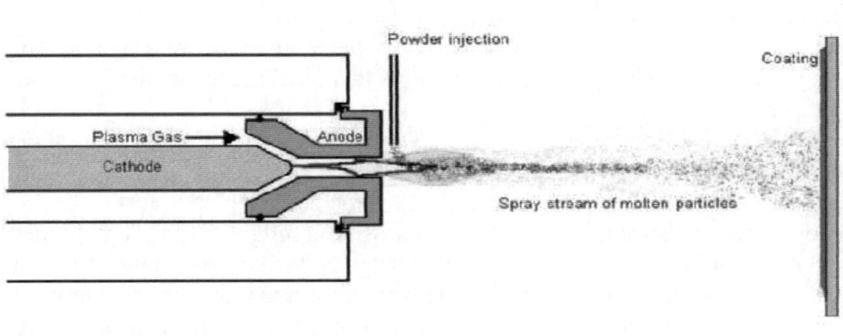

Figure 3. Plasma spraying process

The source powders for spraying contain the main phase of HA. In this method the average content of HA phase is decreased to 76.4%, after plasma spraying. In the coating there are other phases containing calcium: $Ca(OH)_2$, $Ca_4 (PO_4)_2$, $Ca_3 (PO_4)_2$, $Ca_4(PO_4)_2O$, and CaO. This is due to decomposition of HA when exposed to plasma jet. It needs a hydrothermal treatment of a coating to cause a reverse chemical reaction in the coating. For this coatings will be treated with steam at 650°C, the content of the HA phase in the coating can be restored to 94.2% [24]. Research shows ay high content of the HA phase was achieved when coated samples were exposed in distilled water at the temperature of 25°C for 5 to 30 days [24]. Another variation in this coating method is coat layers of other materials (bond coat) of greater adhesion under the main coating for better bonding of the coating. These bond coat layers used are generally finer than that of the main coating and they fit into the pores of the substrate during the spraying. Now the main coating on bond coat layer results in a good adhesion between these two layers will improve the mechanical properties. The advantages of these techniques are minimum modification on the substrate, high deposition speed and less tolerance of size [25].

3.5. Pulsed Laser Deposition (PLD)

This method is widely used to deposit single and multi-layer thin coatings of materials. It works based on pulsed laser ablation. A plasma plume is formed and the coatings after ablation are fed to the substrate to form a coating. Substrates may include metals, alloys, semiconductors, ferroelectrics, dielectrics and superconductors etc. [26]. Deposition of bioglass film on a substrate is shown in Fig. 4. It is provided with a vacuum chamber consisting of a substrate holder attached with a heater and a rotating target. The laser generally used for the ablation is an excimer laser. The rotating targets are made of bioactive glasses. Titanium or Titanium alloy is used as substrate. The pressure is about 10^{-4} Pa.

IOP Conf. Series: Materials Science and Engineering **396** (2018) 012068 doi:10.1088/1757-899X/396/1/012068

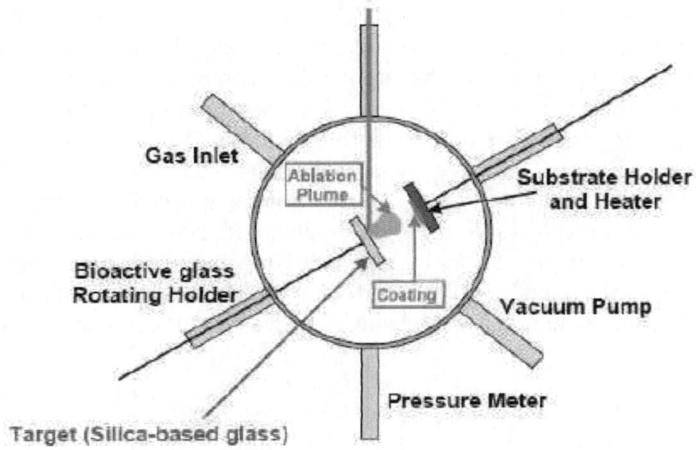

Figure 4. Scheme of PLD experiment [27].

Advantages of this method include: stoichiometric transfer of the target composition without any contamination, growth of particles with high-melting point. In addition the formation of a new compound by the reaction of energetic species in the plasma plumes with the reactive atmosphere, when the ablation of target plate is with low pressure reactive gas [28]. Reactive pulsed laser deposition (RPLD) method is used widely for fine control of the coating properties of bioactive silica-based materials. Typical scratch test results on these coatings showed the mean pressure at which the coating deformations was 1.4 GPa, at the same time that of substrate plate is 9.3 GPa. [29].

3.6. Magnetron sputtering

This method is more flexible method for the improvement of the surface morphology of the metal implants. This method is used to produce well-adhered, dense coatings with adequate composition of elements. In this process, the magnetic field is produced by either electromagnets or permanent magnets or the combination of both. This helps in trapping the electrons. Sputtering of atoms from the cathode to the target is because of magnetically controlled plasma. These high energy ions are speedup towards the cathode and bombard over the target material. Remaining secondary electrons are moved away from the target due to the ions collision. But quality of plasma is maintained by these secondary electrons. This technique is useful in producing low-impurity and high quality coatings at suitable deposition rates [30-35].

Magnetron sputtering can be used for the deposition hard, decorative, wear and corrosion resistance coatings. This method is also useful for coatings with desired electrical or optical properties. Besides, this method can be used to coat calcium phosphate on titanium implants. This deposition method possess good control of coating composition of Ca–P coatings, a better adhesion between the underlying substrate and the coating, and it requires low temperature during the deposition.

4. Conclusion and scope for research

There are various methods of synthesis of the bioactive glass .Among these sol-gel technique is more feasible and economical. The bioglass produced by sol-gel technique is more bioactive and reasonably better mechanical properties. However choice of synthesis method depends on many factors like shear strength, hardness, and wear and corrosion resistance. Similar to synthesis there are many methods

IOP Conf. Series: Materials Science and Engineering **396** (2018) 012068 doi:10.1088/1757-899X/396/1/012068

available for coating implants with bioactive glass powder. But past research shows the limited success of the trials conducted to coat metallic scaffolds with these bioactive glasses. Main reasons may be poor adhesion properties of the coating with the substrate or glass composition degration during the coating process. These Bioactive glasses are generally contains silica content below 60wt%, and the thermal expansion coefficients of these glasses are much higher than those of titanium and its alloys. This is the cause for cracking and low reliability of this system at the metal-glass interface. This can be solved to some extend by increasing the content of SiO_2 while keeping bioactivity. This demands a suitable synthesis process and coat methods to solve the issues mentioned. More studies on multilayer bioactive glass coatings on metal substrates are yet to be conducted to obtain the matching thermal expansion coefficients of coatings and the metal implant. Different pre-treatment of substrates before coating bioactive glasses can be tried for better mechanical and bioactive properties of coated implant.

References

[1] Cao W and Hench L L 1996 Bioactive materials *Ceram. Int.* **22** 493–507
[2] Hench L L and Andersson O 1993 Bioactive glass coating *In An Introduction to Bioceramics,*ed.Hench L L and Wilson J *World Scientific Singapore* 239–260.
[3] Lopez Esteban S, Saiz E, Fujino S, Oku T, Suganuma K and Tomsia A P 2003 Bioactive glass coatings for orthopedic metallic implants *Journal of the European Ceramic Society* **23** 2921–2930
[4] Tilocca A and Cormack A N 2010 Surface Signatures of Bioactivity: MD Simulations of 45S and 65S Silicate Glasses *Langmuir* **26** 545–551.
[5] Zarzycki J 2003 Special Methods of Obtaining Glasses and Amorphous Materials Laboratory of Science of Vitreous Materials *Montpellier, France: University of Montpellier*
[6] Pietrokowsky P 1963 Novel mechanical device for producing rapidly cooled metals and alloys of uniform thickness *Rev SciInstr* **34** 445
[7] Zarzycki I and Naudin F 1967 A study of kinetics of the metastable phase separation in the PbO-B2O3 system by small-angle scattering of X-rays *Phys Chem Glasses* **8** 11–18
[8] Chen H S and Miller C E 1976 Centrifugal spinning of metallic glass filaments *Mat Res Bull* **11**49–54
[9] Topol L E 1974 Formation of new lanthanide oxide glasses by laser spin melting and free-fall cooling *J Non-Cryst Sol* **15** 116–124
[10] Kaur G, Pickrell G, Kimsawatde G, Homa D, Allbee H A and Sriranganathan N *2014* Synthesis, cytotoxicity, and hydroxypatite formation in 27-Tris-SBF for sol-gel based CaO-P2O5-SiO2-B2O3-ZnO bioactive glasses *Sci Rep* 1–14
[11] Oki A, Parveen B, Hossain S, Adeniji S and Donahue H 2004 Preparation and in vitro bioactivity of zinc containing sol-gel–derived bioglass materials *J Biomed Mater Res A* 216–221
[12] Courtheoux L, Lao J, Nedelec J M and Jallot E 2008 Controlled bioactivity in zinc-doped sol-gel-derived binary bioactive glasses *J Phys Chem C* **112** 13663–13667
[13] Li H C, Wang D G, Hu J H and Chen C Z 2013 Crystallization, mechanical properties and in vitro bioactivity of sol–gel derived Na2O–CaO– SiO2–P2O5 glass–ceramics by partial substitution of CaF2 for CaO *J Sol-Gel SciTechnol* **67** 56–65
[14] Pazo A, Saiz E and Tomsia A P 1998 Silicate glass coatings on Ti-based implants *Acta mater.* **46(7)** 2551–2558
[15] Weng W and Baptista J L 1998 Alkoxide route for preparing hydroxyapatite and its coating *Biomaterials* **19(1–3)** 125–135
[16] You C, Oh S and Kim S 2001 Influences of heating condition and substrate-surface roughness on the characteristics of sol–gel-derived hydroxyapatite coatings *J. Sol–Gel Sci. Technol***21(1–3)** 49–54

[17] Weng W, Han G, Du P and Shen G 2002 The effect of citric acid addition on sol–gel preparation of apatite films *Mater. Chem. Phys.* **2(77)** 578–582

[18] Yang Y C and Chang E 2001 Influence of residual stress on bonding strength and fracture of plasma-sprayed hydroxyapatite coatings on Ti6Al4V substrate *Biomaterials* **22** 1827–1836

[19] Garcia C, Cere S and Duran A 2006 Bioactive coatings deposited on Ti alloys *J. Non-Cryst. Solids.* **352(32-35)** 3488–3495

[20] Sarkar P, De D and Rho H 2004 Synthesis and microstructural manipulation of ceramics by electrophoretic deposition *J. Mater. Sci.* **39** 819–823

[21] Zhitomirsky I 1998 Electrophoretic deposition of chemically bonded ceramics in the system CaO–SiO2– P2O5 *J. Mater. Sci. Lett.* **17** 2101–2104

[22] Krause D, Thomas B and Leinenbach C 2006 The electrophoretic deposition of Bioglass® particles on stainless steel and Nitinol substrates *Surf. Coat. Technol.* **200** 4835–4845

[23] Stojanovic D, Jokic B, Veljovic D J 2007 Bioactive glass–apatite composite coating for Ti implant synthesized by electrophoretic deposition *J. Eur. Ceram. Soc.* **27** 1595–1599

[24] Kalita V I, Komlev D I, Komlev V S,Yu A Fedotov and Radyuk A A 2016 Hydroxyapatite-Based Coatings for Intraosteal Implants,ISSN 2075-1133 *Inorganic Materials: Applied Research* Vol. 7 **4** pp. 486–492

[25] Gabbi C, Cacchioli A, Locardi B and Guadagninol E 1995 Bioactive glass coating: physicochemical aspects and biological findings *Biomaterials* **16(7)** 515–520

[26] Chrisey D B and Hubler G K 1994 Pulsed Laser Deposition of Thin Film (*Wiley, New York*)

[27] Borrajo J P, Gonzalez P, Liste S 2005 The role of the thickness and the substrate on the in vitro bioactivity of silica-based glass coatings *Mat. Sci. Eng.* C **25** 187–193

[28] Liste S, Serra J and Gonzale P 2004 The role of the reactive atmosphere in pulsed laser deposition of bioactive glass films *Thin Solid Films* 224–228

[29] Alessio L D, Ferro D, Marotta V, Santagata A, Tegjo R and Zaccagnino M 2001 Laser ablation and deposition of Bioglass® 45S5 films *Appl. Surf. Sci.* **183** 10-17

[30] Shi J Z, Chen C Z, Yu H J and Zhang S J 2008 Application of magnetron sputtering for producing bioactive ceramic coatings on implant materials, *Bull. Mater. Sci,* Vol. 31 **No. 6,** pp. 877–884

[31] Mardarea C C, Mardarea A I , Fernandesa J R F,Joannia E, Pinac S C A , Fernandesc M H V and Correiac R N 2003 Deposition of bioactive glass-ceramic thin-films by RF magnetron sputtering *Journal of the European Ceramic Society* **23** 1027–1030

[32] Wolke J G C, De Groot K and Jansen J A 1998 Dissolution and adhesion behaviour of radio-frequency magnetron-sputtered CaP coatings *J. Mater. Sci.* **33** 3371–3376

[33] Ozeki K,Yuhta T,Fukui Y,Aoki H and Nishimura I 2002 A functionally graded titanium/hydroxyapatite film obtained by sputtering *J. Mater. Sci.—Materials in Medicine* **13** 253–258

[34] Hulshoff J E G, Hayakawa T, Van Dijk K, LeijdekkersGovers A F M, Van der Waerden J P C M and Jansen J A 1997 Mechanical and histological evaluation of Ca-P plasma-spray and magnetron sputter-coated implants in trabecular bone of the goat *J. Biomed. Mater. Res.*, **36** 75–83

[35] Jana C, Nisch W and Grimm G 1995 Production and characterization of thin films of BioveritR-type glasses deposited by RF magnetron-sputtering. *Mater.Clin. Appl.* 257–262

IOP Conf. Series: Materials Science and Engineering **396** (2018) 012069 doi:10.1088/1757-899X/396/1/012069

Experimental and numerical analysis of impact of conical projectile on polyethylene plate

M Junaid Qadri°, Arshad Hussain Khan× and Raisuddin Ansari*

Department of Mechanical Engineering, ZHCET, AMU Aligarh, 202002, India
°junaidqadri472@gmail.com, *raisansari@zhcet.ac.in

Corresponding Author: × khanahamu@gmail.com

Abstract. The impact behaviour of polyethylene plate has been experimentally and numerically investigated. The analysis involves determination of the ballistic impact behaviour of polyethylene plates under dynamic impact of rigid conical nose shape projectile. The experimental investigation has been carried out using pneumatic gun set-up with arrangements for recording the impact velocity and residual velocity after perforation. The experimental investigations, are supplemented with the numerical analysis based on FE code using ABAQUS/EXPLICIT. Mie-Gruneisen equation of state is employed for modelling of the problem. A detailed analysis has been presented to explore the mode of deformation, ballistic limit and the failure characteristics of the polyethylene plate under normal impact of conical projectiles.

1. Introduction

In recent years the use of Polymers in industrial/commercial/household applications has significantly increased owing to low cost, ease of production, excellent anti-corrosion properties and low specific weight. The structural components made up of such polymers are subjected to impact loading during service/use and as such the evaluation of their ballistic performance is of paramount importance. Polyethylene (PE) is a low density material which offers high impact energy absorbing characteristics and found applications in electrical appliances, medical industry, safety equipment, component packaging and house hold application etc. Thus there is need for understanding the mechanism of penetration by a projectile and the associated damage characteristics for the efficient/optimal design of PE structures.

Compared to the metallic targets [1- 3] the ballistic performance evaluation of polymer plates has received little attention [4- 9]. Mannan et al. [3] investigated failure of aluminum beams impacted at smaller velocities with blunt hardened steel projectile. The projectile velocity range choosen for the analysis is 11 m/s to 114 m/s and it was concluded that three different modes of failure occur with increased velocity. The impact behaviour of poly-ethylenes with three different densities and molecular weights has been analyzed by Brough et al. [4]. The impact test based on Charpy test were performed and the fracture surface was analyzed using Scanning Electron Microscopy (SEM) revealing thermal softening for low and medium molecular weights, and less localized deformation for high molecular weight polyethylene. Li et al. [5], carried out Izod impact test and using SEM the impact behaviour was investigated and it was concluded that with the increase in molecular weight the resistance to crack propagation increases for polyethylene. Rafael et al. [6] investigated the quasi-static impact behaviour of HDPE as well as PVC plates using Servo hydraulic Dartec machine and digital image correlation (DIC). The numerical simulation has been done employing LS-DYNA and the rate dependency was modelled using Cowper- Symonds model. Ming-mingxu et al. [7] investigated static as well as dynamic properties of polyethylene. It was concluded from the static and dynamic impact tests that compared to LDPE the extruded UHMWPE exhibit greater yield stress with increased strain to failure. It was also observed that for both LDPE and UHMWPE the strain corresponding to tensile failure decreases with the increase in strain-rate. It was further concluded that at high strain rates, UHMWPE due to its long chain molecular structure with few branches, absorbs more energy compared to LDPE. Afsar et al. [8] analyzed the dynamic impact behaviour of polycarbonate plates. The influence of projectile nose shape on the impact behaviour of thin polycarbonate plates have been analysed experimentally as well as numerically. Mie- Grüneisen equation of state (EOS) has been employed to model the polycarbonate plate in their analysis. The impact behaviour of monolithic polyethylene plate has been analysed by Mohagheghian et al. [9] and the effect of projectile nose shape on energy required for perforation has been investigated. Three varying density polyethylene viz. low density polyethylene (LDPE), high density polyethylene (HDPE) and ultra high molecular weight polyethylene (UHMWPE) were analyzed and dynamic mechanical analysis (DMA), tensile test and shear test for polymer characterization have been carried out. It is concluded that with the increase in strain rate premature failure takes place and UHMWPE offers higher perforation resistance as compared to LDPE and HDPE. Mohagheghian et al. [10], further analyzed the impact perforation of polyethylene-aluminium laminate using quasi static perforation and impact loading. A square plate of 130 mm side with clamped diameter 100 mm is impacted by 12.5 mm diameter projectile of mass 20.2 grams. In quasi static perforation the indenter speed was 1 mm/sec and the impact velocity was varied in the range 50m/s to 150m/s. Instron test machine was used for quasi static perforation and gas gun apparatus was used for impact test. It is observed that the addition of a soft polymer layer significantly improves the perforation resistance of a thin metallic plate.

It is evident from the literature that the impact behaviour of structural components made from polyethylene has not been adequately investigated compared to their metallic counterparts. The excellent impact resistance behaviour of polyethylene can be utilized for the efficient/optimal design of protective equipments if its ballistic performance is understood. In the present analysis the dynamic impact performance of PE plate has been experimentally and numerically investigated. The PE plate has been impacted by a conical nose projectile at varying impact velocities. The ballistic limit which is basically the mean of minimum perforation velocity and maximum bounce back velocity has been obtained. The mechanism of perforation along with the associated deformation for varying impact velocities has been obtained.

2. Experimental procedure

In the present study rigid conical projectile (Fig.1) of 12.8 mm diameter, total length of 41.8 mm and mass 35 g is employed for the dynamic test on pneumatic gun. The projectile is machined from EN31 alloy steel to the required shape and then hardened by heating the specimen at 920°C followed by oil quenching. The hardness of the projectile is measured on Rockwell hardness tester where the hardness is measured in the range of 64-65 HRC. The hardness is reduced to 47-50 HRC by the process of tempering. The camera view and the schematics of the projectile are presented in Fig.1. The diameter of polyethylene plate (Fig. 2) used for dynamic test is 255 mm and thickness is 3.0 mm. Eight holes are made on pitch diameter of 230 mm on the target plate to secure the target plate between two rings on pneumatic gun and effective diameter is 205 mm.

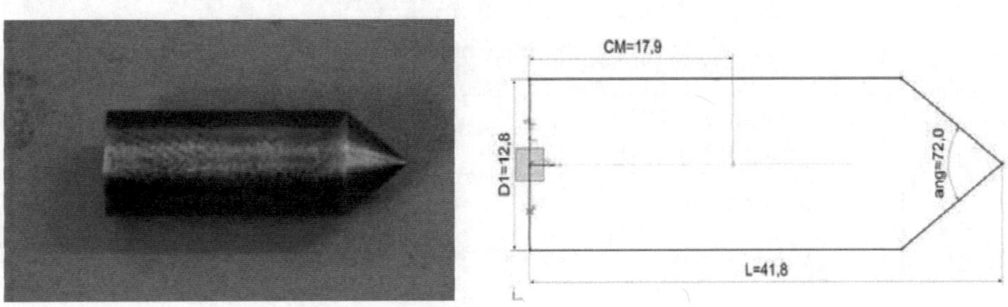

Fig.1 Camera and schematic view of projectile

IOP Conf. Series: Materials Science and Engineering **396** (2018) 012069 doi:10.1088/1757-899X/396/1/012069

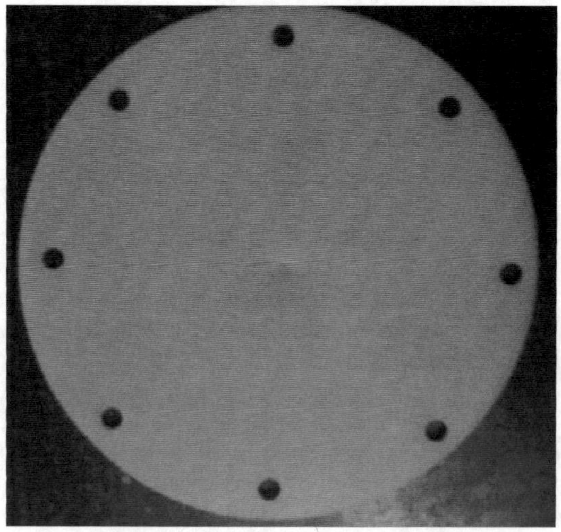

Fig.2 Polyethylene target plate.

2.1. Tensile Test

The mechanical properties of the polyethylene target plate have been obtained from the Tensile test conducted on a Servo_mechanical Universal Testing Machine (UTM). A number of Standard specimens are prepared from 3.0 mm thick polyethylene sheets. The load-deformation curve and stress-strain diagram are presented in Fig.3 and Fig.4 respectively. The properties extracted from the stress-strain curve are tabulated in Table. 1.

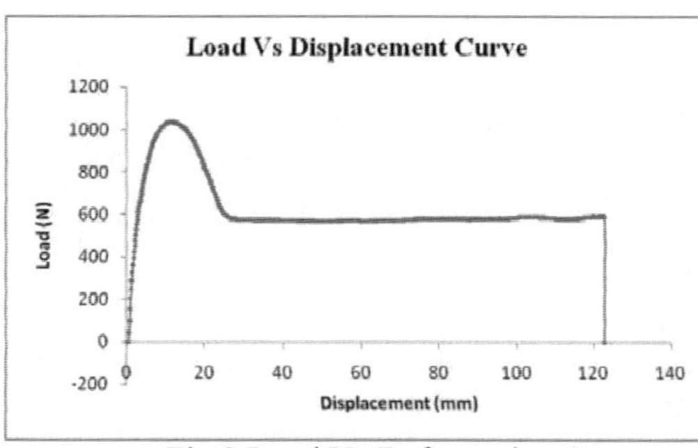

Fig.3 Load Vs Deformation Curve

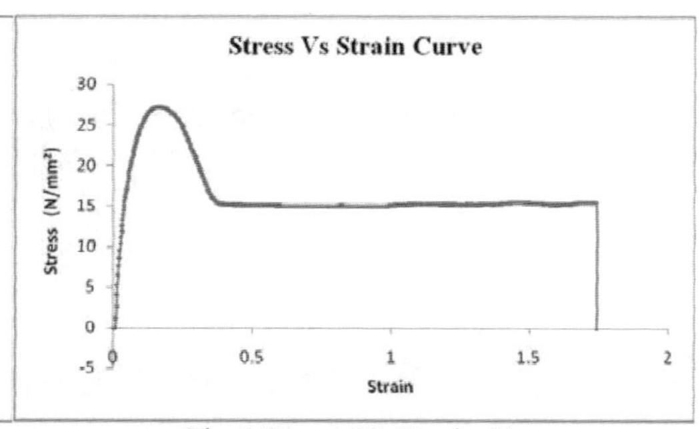

Fig.4 Stress Vs Strain Curve

Table 1 Properties of PE obtained from Stress - Strain curve

Ultimate Tensile Strength (N/mm^2)	**27.00**
Fracture Strength (N/mm^2)	**15.50**
Elongation at Peak (%)	**12.00**
Elongation at Break (%) (Total Elongation)	**350.00**

These properties are required for numerical simulation in ABAQUS/EXPLICIT.

IOP Conf. Series: Materials Science and Engineering **396** (2018) 012069 doi:10.1088/1757-899X/396/1/012069

2.2. Dynamic Impact Test

A pneumatic gun is used to fire the projectiles at different velocities. The gun is designed for firing projectiles with velocities up to 105 m/sec. Impact and residual velocities of the projectile are measured by employing number of light emitting diodes and photo diodes placed in a systematic manner near the exit of the barrel and after the target plate. Figure 5 shows the schematic arrangement of the various components of the experimental setup.

(a) Camera View

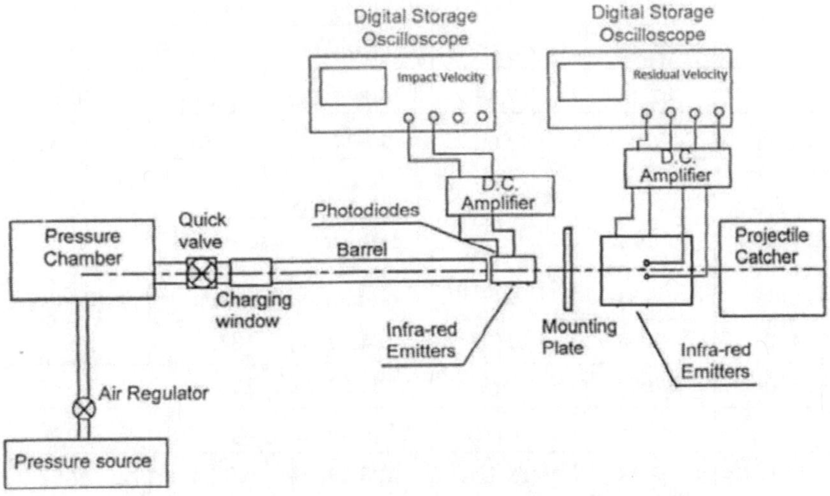

(b) Schematic Diagram

Fig.5 Camera view and Schematic diagram of the experimental set up

3. Numerical Simulation

The finite element analysis is carried out employing ABAQUS CAE. Axi-symmetric analytically rigid projectiles corresponding to the geometry employed in the experiments have been modelled. A reference point corresponding to the center of gravity of has been assigned to the projectile. The modelling of the plate corresponds to axi-symmetric deformable body with shell feature with two partitions. The first partition represents the impact zone where the projectile has actually made contact with the plate and the local deformation around the impact zone was observed using the second partition. The impact of the projectile takes place at the centre of the plate. Mesh convergence analysis has been done and based on the convergence 20 elements along thickness direction for 3.0 mm thick plate has been selected. The element chosen in the analysis is C3D8R element. Mie- Grüneisen equation of state (EOS) with application of linear Us-Up Hugoniot. Has been employed to model the polyethylene plate. Us is the linear shock wave velocity and Up is the particle velocity. It is used in conjunction with Mises plasticity model. The detailed formulation can be referred from the author's previous work [8] and are not discussed here in detail for brevity.

4. Results and Discussion

Dynamic impact test is performed on thin circular Polyethylene plate of diameter 255 mm and thickness 3.0 mm. The impact velocity and residual velocity obtained experimentally have been employed to calculate the velocity drop, impact energy, residual energy and energy absorbed in perforation. The experimental results obtained from the dynamic impact tests are given in Table 2 and the results obtained numerically are provided in Table 3. It can be observed from Table 2 & 3 that the results obtained experimentally from dynamic impact tests compared well with the results obtained numerically.

Table 2 Experimental dynamic impact test result

S.No.	Impact velocity (m/s)	Residual velocity (m/s)	Velocity drop (m/s)	Impact energy (Joule)	Residual energy (Joule)	Absorbed energy (Joule)
1	96.00	85.43	10.57	155.75	123.34	32.41
2	86.00	77.44	8.56	124.99	101.35	23.64
3	75.00	64.36	10.64	95.06	70.00	25.06
4	67.00	50.44	16.56	75.86	43.00	32.87
5	60.00	42.15	17.85	60.84	30.02	30.82
6	42.00	0.00	42.00	29.81	0.00	29.81

Table 3 Numerical results

S.No.	Impact velocity (m/s)	Residual velocity (m/s)	Velocity drop (m/s)	Impact energy (Joule)	Residual energy (Joule)	Absorbed energy (Joule)
1	110	99.64	10.36	204.49	167.78	36.70
2	100	90.58	9.42	169.00	138.66	30.33
3	90	80.23	9.77	136.89	108.78	28.10
4	80	68.88	11.12	108.16	80.18	27.97
5	70	56.45	13.55	82.81	53.85	28.95
6	60	42.00	18.00	60.84	29.81	31.02
7	50	24.24	25.76	42.25	9.93	32.31
8	40	0.00	40.00	27.04	0.00	27.04

The progressive advancement of the conical projectile at different time instants obtained numerically is shown in Figure 6.

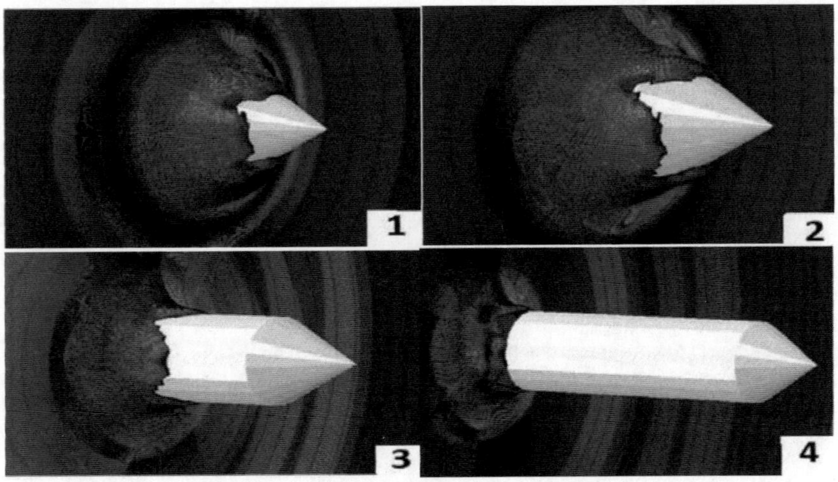

Fig. 6 Progressive advancement of the conical projectile.

The comparison of numerical and experimental perforation reveal similar nature of deformation as can be observed from Figure 7.

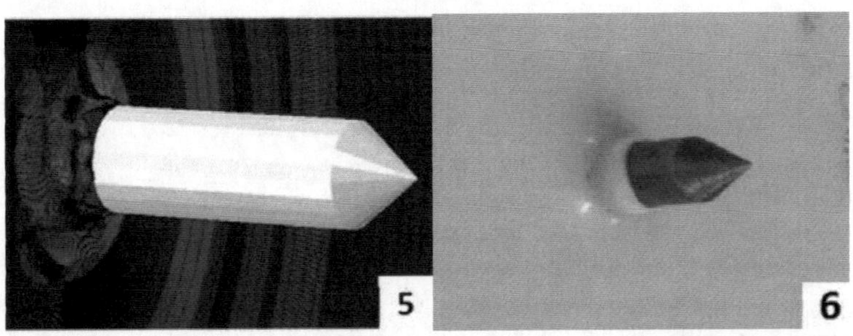

Fig. 7 Comparison of experimental and numerical perforation mechanism

The comparison of residual velocity obtained experimentally and numerically is plotted in Figure.8. It can be seen that the residual velocity of the projectile increases with increase in impact velocity and the numerical results are in excellent agreement with the experimental results. Figure 9 compares the velocity-drop variation with impact velocity revealing decrease in velocity drop with impact velocity. However the velocity drop becomes constant at higher impact velocities.

The experimental as well as numerical impact tests reveal two types of deformation mechanism for polyethylene plate perforated by conical nose shape projectile. First is ductile hole-enlargement and second is a combination of ductile hole-enlargement and petal formation. In ductile hole-enlargement, firstly an indent is made by the tip in the plate which is followed by plate yielding around the nose of the projectile and dishing in rest of the plate. The plate is deformed initially in the shape of projectile nose at the centre. The tip of the projectile later comes out of the plate and a hole formation occurs at the point of contact of plate and projectile and with further deformation this hole expands uniformly causing the flow of material in the direction of the movement of the projectile.

IOP Conf. Series: Materials Science and Engineering **396** (2018) 012069 doi:10.1088/1757-899X/396/1/012069

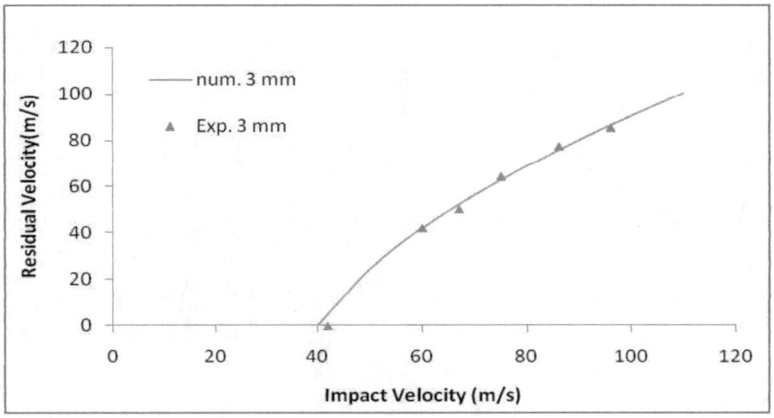

Fig. 8 Variation of residual velocity with impact velocity

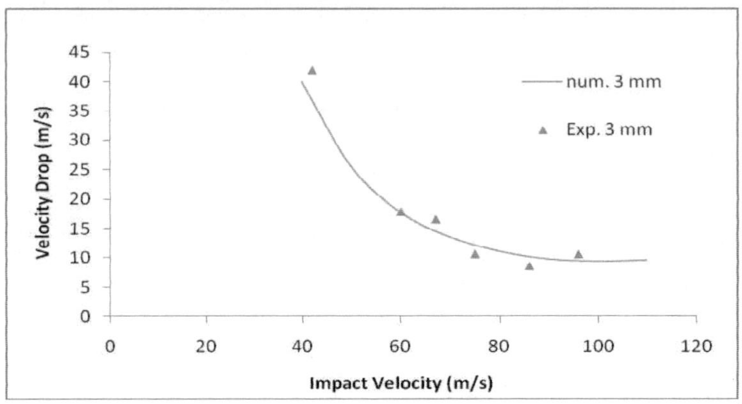

Fig. 9 Experimental and numerical variation of velocity drop with impact velocity

The experimental as well as numerical impact tests reveal two types of deformation mechanism for polyethylene plate perforated by conical nose shape projectile. First is ductile hole-enlargement and second is a combination of ductile hole-enlargement and petal formation. In ductile hole-enlargement, firstly an indent is made by the tip in the plate which is followed by plate yielding around the nose of the projectile and dishing in rest of the plate. The plate is deformed initially in the shape of projectile nose at the centre. The tip of the projectile later comes out of the plate and a hole formation occurs at the point of contact of plate and projectile and with further deformation this hole expands uniformly causing the flow of material in the direction of the movement of the projectile.

5. Conclusion

The dynamic impact behavior of polyethylene plates impacted by conical nose shape projectile has been investigated experimentally and numerically. The mechanical properties of the polyethylene plates have also been obtained from tensile test on UTM. The ballistic performance characteristics in terms of ballistic limit, velocity-drop, absorbed energy and the mechanism of the deformation have been obtained. Conical projectile perforates the target plate by ductile hole enlargement causing the flow of material in the direction of projectile movement. A combination of failure by petal formation and ductile hole enlargement also occurs at higher velocities. Energy absorbed at ballistic velocity and at higher velocity during impact perforation is quite different. It is found that the energy absorbed near ballistic limit is less as compared to the energy absorbed at higher velocities. The experimental results reveal excellent agreement with the numerical results.

IOP Conf. Series: Materials Science and Engineering **396** (2018) 012069 doi:10.1088/1757-899X/396/1/012069

References

[1] Corran RSJ, Shadbolt PJ, Ruiz C. Impact loading of plates-an experimental investigation. Int J Impact Eng (1983)1: 3-22.

[2] Børvik T, Langseth M, Hopperstad OS, Malo KA. Perforation of 12 mm thick steel plates by 20 mm diameter projectiles with flat, hemispherical and conical noses. Int J Impact Eng 27 (2002)19-35.

[3] Mannan MN, Ansari R, Abbas H. Failure of aluminium beams under low velocity impact. International Journal of Impact Engineering 35 (2008) 1201–1212.

[4] Brough I, Haward RN, Healey G, Wood A. Scanning electron micrographs of high density polyethylene fracture surfaces. Polym Guildf 45(2004) 3115-23.

[5] Li B, Gong G, Xie B-H, Yang W, Yang M-B. Influence of molecular weight on impact fracture behavior of injection molded high density polyethylene: scanning electron micrograph observations. J Appl Polym Sci 109(2008)1161-7.

[6] Rafael Traldi Moura, Arild H.Clausen, Egil Fagerholt. Impact on HDPE and PVC plates-Experimental test and numerical simulation. Int J Impact Eng 37 (2010) 580-598.

[7] Ming-mingxu, Mc Shane and William J. Stronge. Static and dynamic properties of semi-crystalline polyethylene. Polymer Science Journal, Polymers 8(2016),8,77; doi:10.3390/polym8040077.

[8] Afsar Husain, Raisuddin Ansari, Arshad Hussain Khan. Experimental and numerical investigation of perforation of thin polycarbonate plate by projectiles of different nose shapes. Latin American journal of solids and structures 14 (2017) 357-372.

[9] Mohagheghian I., Shane G.J.Mc, Stronge WJ. Impact perforation of monolithic polyethylene plate: projectile nose shape dependence. International journal of impact engineering 80 (2015) 162-176.

[10] Mohagheghian I., Shane G.J.Mc, Stronge WJ. Impact perforation of polymer-metal laminates: projectile nose shape sensitivity. International journal of solids and structures 88-89 (2016) 337-353.

IOP Conf. Series: Materials Science and Engineering **396** (2018) 012070 doi:10.1088/1757-899X/396/1/012070

Effect of tool diameter in ECDM process with powder mixed electrolyte

Lijo Paul[1] and Donald Antony[2]

[1]Associate Professor, [2]Graduate Student
Dept. of Mechanical Engg., St. Joseph's College of Engineering and Technology, Palai, Kerala, India
E-mail*: lijo.paul@gmail.com

Abstract. Electro Chemical Discharge Machining (ECDM) combines the basic principles of Electro Chemical Machining (ECM) and Electro Discharge Machining (EDM) using the advantages of both the techniques to machine non-conducting, hard to machine materials like a variety of glasses, ceramics and composites. In the current paper , ECDM is carried out with powder mixed electrolyte with stales steel tool of various diameters. Output responses are measured according to MRR and ROC. From the experimental results smaller diameter tools has shown higher MRR with lower ROC.

Keywords: Electro Chemical Discharge Machining (ECDM), MRR, ROC, Gray relational analysis

1. Introduction

Hybrid machining processes facilitates new applications and opportunities for manufacturing various components which are notable to be produced economically by processes on their own. The main objectives include increasing material removal rate, improving surface integrity, reducing production time, reducing tool wear and extending application areas. By combining different machining actions or phases to be used on the material being removed the technological improvement of machining processes can be achieved. A mechanical conventional single cutting or mechanical abrasive action process can be combined with the respective machining phases of Electro Discharge (ED) in Electro Discharge Machining (EDM) or ECD in ECM. The reason for such a combination and the development of a hybrid machining process is mainly to make use of the combined advantages and to avoid or reduce some adverse effects the constituent processes produce when they are individually applied. As comparing the performance characteristics of a hybrid process with single-phase processes there is considerable difference in terms of accuracy, productivity and surface quality.

Material removal in ECDM process is by machining non-conducting material by removing material through melting or vaporization by electric arcs and sparks. Chemical reaction takes place when a DC current is applied between the electrodes separated by a small distance (0.03 to 0.07mm) and this causes formation of positively charged ionic gas bubbles. These bubbles form a dielectric space around the electrodes. At an optimum voltage the gas bubbles collapses and a current of electrons generated, this results in high temperature spark which erodes the material from the work piece. The material removed gets dissolved in the electrolyte. Electrolyte provides better cooling effect.

2.Literature Review

A. Kulkarni *et al*, [1] the basic ECDM phenomenon is studied and with the help of experimental observations of time-varying current in the circuit concluded the mechanism of ECDM. The basic mechanism of material removal and temperature rise is proposed based on these observations. A high electric field of the order of 107 V/m gets generated across the cathode tip and isolated electrolyte causing an arc discharge within the gas layers covering the tip. That is heat discharge region.
Basak et al. [2] have given analytical model of MRR with the experimental validation, but the basic theory of spark generation proposed by them does not match with the actual situation.
Shilpi Sharma *et.al.* [3] has conducted a review on the effective parameters electrochemical discharge machining. It was found in the researches that applied voltage is the most influential parameter for material removal rate. Electrolyte concentration is the secondary parameter affecting the Tool wear and MRR. For the better performance of ECDM the optimization techniques are required to known the optimized value of parameter. Selection and optimization of parameters is done ECDM has capability to provide better results. Lijo and Somashekhar, [4] had conducted Response Surface Modelling (RSM) of ECDM with pulsating DC to understand the effect of various process parameters on micro hole and micro channel machining for Material Removal Rate (MRR), Tool Wear Rate (TWR), Heat Affected Zone (HAZ) and Radius of Overcut (ROC). From the experiments they found that micro machining was effected by the temperature of the electrolyte. At higher depth MRR decreases due to lower spark which was due to unavailability of electrolyte. Lijo and Somashekhar [5] have also carried out multi objective optimization with grey relational analysis and micro channels have made with optimized values. They have reported that low duty factor during machining can reduce HAZ in micro channels. Lijo *et al*, [6] also reported that higher frequency at low duty factor will increase MRR with reduction in HAZ. Duty factor is found to be a predominant factor affecting the machining characteristics. Ranganayakulu *et al*, [7] developed an ECDM setup for micromachining of acrylic plate.

Despite the considerable amount of research, which has been reported on the theoretical aspects and laboratory experiments, there are many aspects of the process control issues that are yet to be investigated for the achievement of improved productivity of the ECDM process. From the literature survey it is found that ECDM is at research level only and most of the works are experimental. In this investigation experiment is carried out with powder mixed electrolyte. Tool diameters used are 0.52 and 1 mm. The input parameters selected for experiments are tool rotation, voltage (V), graphite powder concentration (%) and duty factor (%). Multi-objective optimization has been carried out using Gray relational analysis. The output parameters selected are Material Removal Rate (MRR) and Radial Over-Cut (ROC). The confirmation tests are conducted based on the best parameter combinations. The effect of tool diameter on output responses are studies in the current paper.

3.Experimental setup

Important components of ECDM machining setup are tool electrode, counter electrode, workpiece, DC source and end sensing part. The performance of ECDM in terms of rate of machining and surface finish is affected by many factors. As far the metal removal mechanism of an ECDM process is concerned, it is necessary to control the gap width, electrolyte flow rate, temperature of the electrolyte, electrolyte concentration and pulse parameters of the pulsed power supply at an optimum level to obtain higher performance in terms of machining accuracy, the surface finish and the metal removal rate. Figure1 shows the ECDM set up.

IOP Conf. Series: Materials Science and Engineering **396** (2018) 012070 doi:10.1088/1757-899X/396/1/012070

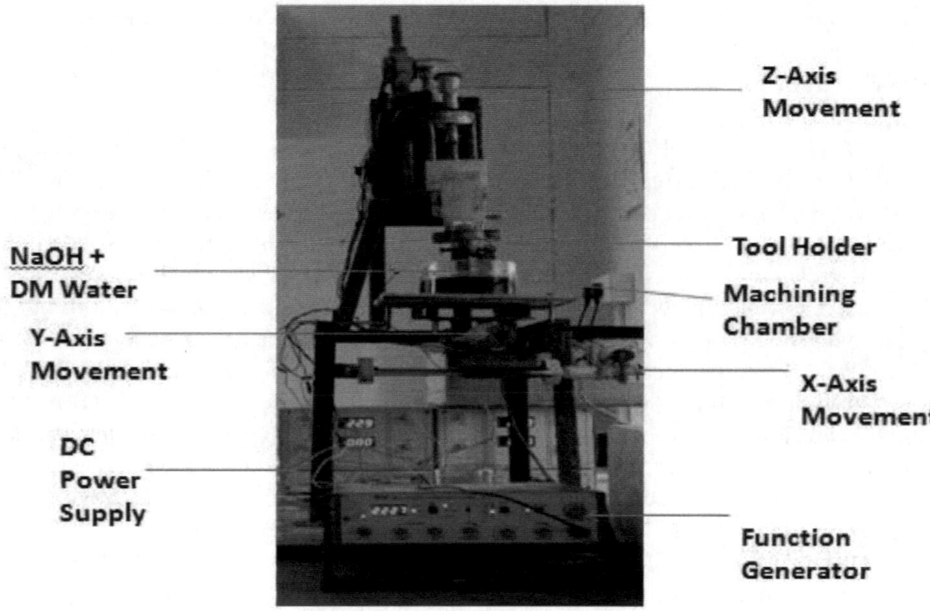

Figure 1 ECDM Setup.

3.1 Tool electrode

Tool will acts as the cathode. At the negative electrode where a high electron potential have been produced via an external voltage source, electrons are pushed out of the electrode, thereby reducing the oxidised species, because the electron energy level inside the electrode(Fermi level) is higher than the energy level and the electrons can lower their energy by occupying this orbital. So the negative electrode will be the one where reduction reaction will take place and thus its called cathode. Tungsten copper alloy of diameter 0.52 mm and 1 mm is used for tool electrode in electro chemical discharge machining process.

3.2 Counter electrode

A rectangular graphite piece of dimension 40*40*10 mm is used as the counter electrode in ECDM process. Graphite is an electrically conductive form of solid carbon. At the positive electrode where low electrons potential have been produced via an external voltage source, electrons are sucked into the electrode leaving behind the reduced species. So the positive electrode will be the one where oxidation reaction will take place and thus it is called anode.

3.3 Workpiece

Borosilicate glass is used as a workpiece material in this ECDM investigation. Borosilicate glass is a type of glass with silica and boron trioxide as the main glass-forming constituents. Borosilicate glasses are known for having very low coefficients of thermal expansion and thermal shock, more so than any other common glass.

Table 1 Properties of the Workpiece

Sl No	Properties/Specifications	Borosilicate Glass
1	Thickness(mm)	0.52
2	Melting point(°C)	820
3	Density(g/cm^2)	2.23
4	Specific heat(J/kg K)	775
5	Thermal conductivity(W/mK)	1.13

IOP Conf. Series: Materials Science and Engineering **396** (2018) 012070 doi:10.1088/1757-899X/396/1/012070

3.4 Tool feeding mechanism

The ECDM set up consists of a 3- axis tool feeding mechanism. The three axis movements are provided with the help of 3 stepper motors. Power is supplied to the stepper motors with the help DC power supply through the power circuit. The rotational motion of the stepper motor is converted into linear movement with the help of worm and worm wheel set up. Here the worm is fixed on to the stepper motor shaft and the worm wheel is fixed on to the framework.

3.5 Machining chamber

Acrylic material was used for the preparation of the chamber. There is also a work holding set up inside this beaker. Workpiece can be easily clamped on to this set up. The chamber can be clamped on to the machining table of the experimental set up using nuts and bolts.

3.6 Electrolyte

Electrolyte concentration is an important factor that influences machining. In this experiment electrolyte is taken as a mixture of NaOH pellets and DM water. The total capacity of the chamber is 650ml. The amount of electrolyte is taken as 500ml where NaOH is of 150 gm and DM water is of 350 ml i.e the electrolyte concentration is 30%.

4. Experimental investigation

Design of experiment is a powerful analysis tool for modelling and analyzing the influence of control factors on performance output. The initial step in the Taguchi model development is to build up an input-output database required for the optimization through the ECDM experiments. In order to have complete knowledge of ECDM process over the range of parameters selected, proper planning of experimentation is essential to reduce the cost and time. Hence an experimental plan based on Taguchi orthogonal array (L9) was chosen. The Taguchi method involves reducing the variation in a process through robust design of experiments. The overall objective of the method is to produce high quality product at low cost to the manufacturer.

5. Design of experiments

The machining tests were conducted on electrochemical discharge machining (ECDM). The experiments with tool rotation were conducted on Borosilicate glass. The process parameters selected for the experiments without tool rotation are voltage (V), graphite powder concentration (%) and duty factor (%). The process parameters selected for the experiments with tool rotation are voltage (V), graphite powder concentration (%), duty factor(%) and RPM.

Voltage is provided to produce spark at the tool tip. MRR increases with increase in voltage. With increase in applied voltage, the machining current in the inter electrode gap increases, the spark rate and bubble formation will increase which leads to enhancement of MRR. For optimal radial overcut, applied voltage have been found out as the most important ECDM parameter. Graphite is used as the powder to be mixed with the electrolyte. Graphite particles are conductive and presence of conductive particles in the electrolyte can reduce the formation of micro cracks and fractures on the machined surface of the workpiece to a certain extend. Duty factor is the ratio of machine on time to the total time. Normally maximum duty factor increases the material removal rate and reduces the radial overcut. Three sets of experiments were planned to be carried out:

- One set of experiments with tool diameter of 1 mm.
- One set of experiments with tool diameter of 0.52 mm and tool rotation.

The experiments were conducted with varying tool diameters in order to obtain holes in the micro-level. The experiments were conducted with and without tool rotation. The machining parameters selected for the experimentation without tool rotation are Voltage(V), Graphite powder

IOP Conf. Series: Materials Science and Engineering **396** (2018) 012070 doi:10.1088/1757-899X/396/1/012070

concentration(%) and Duty factor(%).The machining parameters selected for the experimentation with tool rotation are Voltage(V), Graphite powder concentration(%), Duty factor(%) and tool rotation with three levels. The three levels for each parameters are as follows:

Table 2 Input factors with their levels.

Sl No	Factor/Level	1	2	3
1	Voltage(V)	35	40	45
2	Graphite Conc. (%)	0	0.25	0.5
3	Duty factor (%)	60	70	80
4	Rotation(RPM)	100	300	500

Four factors (voltage, concentration, duty factor, rotation) with three levels are considered. From this input factors experimental design of L9 orthogonal array is constructed with the help of MINITAB software.

Based on the levels of the process parameters selected, the experimental plan is developed using L9 orthogonal array. According to L9 orthogonal array, the experiments are conducted. The experiments are conducted by varying each of the process parameters according to the experimental plan with tool of 1 mm diameter as given in Table 3.

Table 3 Experimental layout plan of L9 orthogonal array.

Sl.No	Voltage(V)	Graphite Conc. (%)	Duty Factor (%)
1	35	0	60
2	35	0.25	70
3	35	0.5	80
4	40	0	70
5	40	0.25	80
6	40	0.5	60
7	45	0	80
8	45	0.25	60
9	45	0.5	70

The experimental details for are given in table 4.

Table 4 Experimental details for L9 experiments.

Sl No	Item	Specifications
1	Tool	Tungsten copper alloy
2	Tool diameter	1 mm, 0.52 mm
3	Counter electrode	Graphite block
4	Dimension	(40*40*10) mm
5	Workpiece	Polypropylene(1.3 mm thick)
6	Electrolyte	NaOH + DM water
7	NaOH pellets	30% of 500 ml(150 gms)
8	DM water	350 ml
9	Powder mixed	Graphite powder
10	Process parameters	Voltage(V),Graphite powder conc(%), Duty factor and tool rotation(tool diameter 0.52mm)

Based on the levels of the process parameters selected, the experimental plan is developed using L9 orthogonal array. A set of nine experiments will be generated with different process parameter combinations. According to L9 orthogonal array, the experiments are conducted. The experiments are conducted by varying each of the process parameters according to the experimental plan. The experimental plan based on the orthogonal array for tool of 0.52 mm is provided in Table 5 below:

Table 5 Experimental layout plan of L9 orthogonal array.

Sl No	Voltage(V)	Graphite Conc.(%)	Duty Factor(%)	Rotation(RPM)
1	35	0	60	100
2	35	0.25	70	300
3	35	0.5	80	500
4	40	0	70	500
5	40	0.25	80	100
6	40	0.5	60	300
7	45	0	80	300
8	45	0.25	60	500
9	45	0.5	70	100

6. Multi-Objective Optimization

Multi-objective optimization is the process in which the optimum levels of different input parameters are found out to provide the best output parameters. Multi-objective optimization can be applied to processes having more than one output. In Multi-objective optimization, the values of output parameters are processed through different steps in order to find out the best input conditions. In this investigation, the output parameters selected are material removal rate (MRR) and radial overcut (ROC). A higher value for MRR and a lower value for ROC is required for the best

IOP Conf. Series: Materials Science and Engineering **396** (2018) 012070 doi:10.1088/1757-899X/396/1/012070

combination. Therefore a higher preference is given for MRR and less preference is given for ROC. The multi-objective optimisation technique used is Gray Relational Analysis

7. Results and Discussions

Multi-objective optimisation has been done on the two output parameters MRR and ROC based on Gray relational analysis. The best parameter combination has been obtained from multi-objective optimisation and confirmation tests have been conducted.

7.1 EXPERIMENTAL ANALYSIS

Material Removal Rate (MRR) and Radial Over-Cut (ROC) have been calculated for each of the experiments based on the experimental observations and microscopic images.

7.2 Experiment set 1 (Tool diameter= 1mm)

Table 6 Experimental observations for experiments with tool diameter = 1mm.

Sl No	Voltage(V)	Graphite Conc(%)	Duty Factor(%)	MRR(mg/min) $*10^{-4}$	ROC(micons)
1	35	0	60	1.34	183
2	35	0.25	70	0.68	156
3	35	0.5	80	0.21	121
4	40	0	70	1.8	96
5	40	0.25	80	1.2	176
6	40	0.5	60	0.93	154
7	45	0	80	0.45	89
8	45	0.25	60	3.98	97
9	45	0.5	70	1.98	110

The experiments were conducted according to the L9 orthogonal array and two outputs MRR and ROC were obtained. Higher MRR and least ROC is preferred in micromachining of holes so that better accuracy is obtained. Higher MRR show that better machining characteristics have been taken place and is required for better results. ROC is a defect that has to be minimised in machining of micro holes. ROC is the deviation of actual hole size from the required hole size. From the above observation table we can find that the maximum MRR obtained is $3.98*10^{-4}$ mg/min and the least ROC obtained is 89 microns.

Based on the experimental observations the signal to noise ratios for MRR and ROC has been generated from MINITAB software. The normalised signal to noise ratios and deviation sequences has been calculated as per the equations in gray relational analysis. Also the gray relational coefficient and gray relational grade has been calculated and ranked according to priority.

Table 7 Ranking for grades in experiment set 1.

Experiment	GRG	Rank
1	0.275	7
2	0.392	2
3	0.287	6
4	0.193	9
5	0.345	5
6	0.382	3
7	0.266	8
8	0.587	1
9	0.372	4

Table 8 Response table of parameters for experiment set 1.

Voltage	0.7342	1.0893	1.2447
Graphite Conc	0.9328	1.1251	1.0103
Duty factor	0.9371	1.0827	1.3341

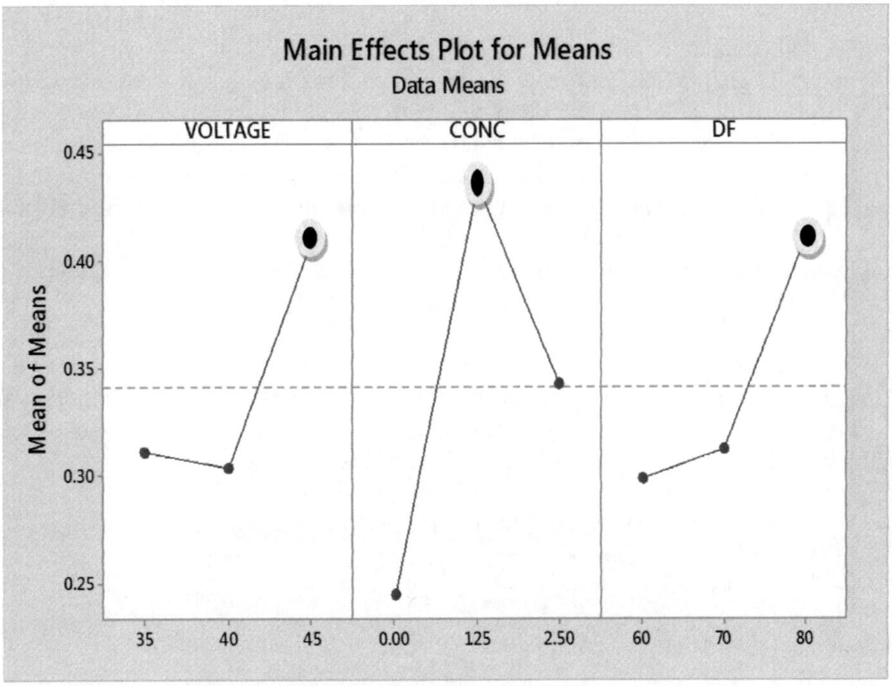

Figure 2 Main effects plot for average (set 1)

(i) Confirmation test

The best parameter combination obtained from multi-objective optimisation of experiment set 1 are:

- Voltage = 45V
- Graphite powder conc. =1.25 gms(0.25%)
- Duty factor = 80%

Maximum MRR obtained from L9 set=$3.98*10^{-4}$ mg/min

Least ROC obtained from L9 set =89 microns

IOP Conf. Series: Materials Science and Engineering **396** (2018) 012070 doi:10.1088/1757-899X/396/1/012070

MRR obtained from confirmation test = $3.85*10^{-4}$ mg/min
ROC obtained from confirmation test = 86 microns

7.3 Experiment set 2(Tool diameter= 0.52mm)

Table 9 Experimental observations for experiments with tool diameter = 0.52mm

Sl No	Voltage (V)	Graphite Conc (%)	Duty Factor(%)	RPM	MRR(mg/min) $*10^{-4}$	ROC(micons)
1	35	0	60	100	1.53	1260
2	35	0.25	70	300	1.10	1372
3	35	0.5	80	500	2.75	910
4	40	0	70	500	1.55	960
5	40	0.25	80	100	3.82	860
6	40	0.5	60	300	2.89	897
7	45	0	80	300	7.30	530
8	45	0.25	60	500	7.50	770
9	45	0.5	70	100	0.81	1453

The experiments were conducted according to the L9 orthogonal array and two outputs MRR and ROC were obtained. Higher MRR and least ROC is preferred in micromachining of holes so that better accuracy is obtained. Higher MRR show that better machining characteristics have been taken place and is required for better results. ROC is a defect that has to be minimised in machining of micro holes. ROC is the deviation of actual hole size from the required hole size. From the above observation table we can find that the maximum MRR obtained is $7.50*10^{-4}$ mg/min and the least ROC obtained is 530 microns. When driven by an external voltage applied to the electrodes, the ions in the electrolyte are attracted to an electrode with the opposite charge and charge transferring reaction can take place. The occurrence of spark depends upon the electrolyte concentration.

Based on the experimental observations the signal to noise ratios for MRR and ROC has been generated from MINITAB software. The normalised signal to noise ratios and deviation sequences has been calculated as per the equations in gray relational analysis. Also the gray relational coefficient and gray relational grade has been calculated and ranked according to priority.

Table 10 Ranking for grades in experiment set 2.

Experiment	GRG	Rank
1	0.214	9
2	0.286	6
3	0.258	8
4	0.299	4
5	0.305	3
6	0.261	7
7	0.298	5
8	0.402	1
9	0.382	2

Table 11 Response table of parameters for experiment set 2.

Voltage	0.8024	0.852	1.054
Graphite conc	0.758	1.0063	0.9412
Duty factor	0.9454	0.9663	0.8774
RPM	0.9013	0.8185	0.9857

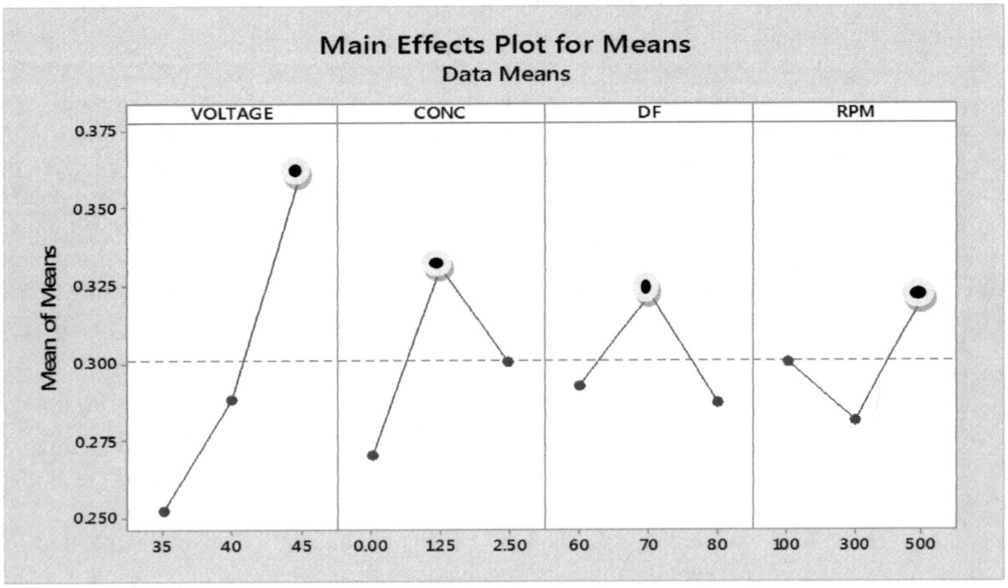

Figure 3 Main effects plot for Average (set 3)

(i) Confirmation test

The best parameter combination obtained from multi-objective optimisation of experiment set 1 are:

- Voltage = 45V
- Graphite powder conc. =1.25 gms(0.25%)
- Duty factor = 70%
- Speed = 500 rpm

Maximum MRR obtained from L9 set=7.50*10^{-4} mg/min
Least ROC obtained from L9 set =530 microns
MRR obtained from confirmation test = 7.56*10^{-4} mg/min
ROC obtained from confirmation test = 528 microns

8. Conclusion

In current study, experiments were conducted with and without tool rotation. Powder Mixed Electro-Chemical Discharge Machining process (PM-ECDM) has been carried out to perform the experiments using Graphite powder as the powder to be mixed in the electrolyte. The process parameters selected for experiments without tool rotation are voltage (V), Graphite powder concentration (%) and Duty factor (%). The process parameters selected for experiments with tool rotation are voltage (V), Graphite powder concentration (%), Duty factor (%) and Rotation (RPM).

Machining was conducted based on L9 orthogonal array. MRR and ROC were taken as the output parameters for both the experiment sets. Multi-objective optimisation using Gray relational analysis was carried out in order to obtain the best parametric combination. Based on the optimised parametric conditions, confirmation tests have been carried out. The best parameter combination obtained for experiments without tool rotation are 45 V, 0.25 % Graphite powder concentration and 80% Duty factor. The best parameter combination obtained for experiments with tool rotation are 45 V, 0.25 % Graphite powder concentration, 70% Duty factor and 500 rpm.

The confirmation test with tool rotation shows a higher MRR value ($7.56*10^{-4}$ mg/min) when compared with the MRR value ($3.85*10^{-4}$ mg/min) of confirmation test without tool rotation. The reason is that; with tool rotation the surface contact between the tool tip and the workpiece is more when compared with that of without tool rotation. Therefore more debris is removed faster while machining. Thus, it can be interpreted that tool rotation provides higher material removal rates. The confirmation test with tool rotation shows a higher ROC value (530 microns) when compared with the ROC value (89 microns) of confirmation test without tool rotation. The reason is that, with tool rotation a larger area comes in contact with the tool due to its rotational motion. This can lead to increase in the hole size from the required hole diameter. The confirmation test shows that the tool rotational speed can be increased beyond 500 rpm since the optimum value for rotation is its maximum value. Thus tool rotation increases both MRR and ROC. Also from the experimental results tool with 0.52 mm diameter has shown higher MRR with lower ROC, due to higher spark rate and tool rotation during machining process.

REFERENCES

1. Kulkarni, R. Sharan and G.K. Lal, 2002 An experimental study of discharge mechanism in electrochemical discharge machining, *International Journal of Machine Tools & Manufacture* **42** pp 1121–27
2. Basak I and Ghosh A 1997 Mechanism of material removal in electrochemical discharge machining: a theoretical model and experimental verification. *J Mater Process Technol* **71**: 350 35934.
3. Shilpi Sharma, Radha Raman Mishra, Veerendra Kumar A N and Rajesha S.2014, Effective Parameters Of Electrochemical Discharge Machining – A Review, *Proceedings Of Sarc-Irf International Conference*,.
4. Lijo, P. and Somashekhar,S.2013. Response Surface Modeling of Micro Holes in Electro Chemical discharge machining process, *Procedia Engineering*:**64**:1395-1404
5. Lijo P. and Somashekhar S. H 2014. Evaluation of Process Parameters of ECDM using Grey Relational Analysis. *Procedia Materials Science* ;**5**:2273-82
6. Lijo P. and Somashekhar S. H. 2014 Effect of process parameters on heat affected zone inmicro. machining of borosilicate glass using μ-ECDM process. *Applied Mechanics and Materials* ;**592**:224-238.
7. Lijo, P. , Somashekhar, S.H. and Ranganayakulu, J. 2014 Experimental investigation and. parametric analysis of electrochemical discharge machining, *Int. J Manufacturing Technology and Management* ;**28**:57-79.

Comparative study of effect of tool materials in EDM process

Lijo Paul[1], Ivin Jose[2*]

[1]Associate Professor, [2*]Assistant Professor,
Dept. of Mechanical Engg., St. Joseph's College of Engineering and Technology, Palai, Kerala, India
E-mail*: lijo.paul@gmail.com

Abstract. Electro Discharge Machine (EDM) has shown tremendous application in machining of various electrically conducting materials precisely and accurately. Micro level precision is attained in EDM with hard alloys of steel. In the current paper EDM is carried out on copper plates with copper and stainless steel tool wires. A comparative study based on the effect of tool materials in machining copper plate is conducted. Stainless steel as tool materials has found to better in terms of higher material removal rate and lower tool wear in EDM process. A multi objective optimisation is carried out t analyse the best parameter combination with each tool material. Microscopic images are taken to study the quality of the micro holes machined in each cases.

Keywords: EDM, grey relational analysis, duty factor

1. Introduction

Going with the improvement of mechanical industry, the requests for compound materials having high hardness, sturdiness and effect resistance are expanding. All things considered, such materials are hard to be machined by customary machining strategies. Consequently, non-conventional machining strategies including electrochemical machining, ultrasonic machining, Electro Discharge Machine (EDM) and so on are connected to machine such hard to machine materials. Moreover, EDM is fit for delivering a fine, exact, erosion and wear safe surface. Customary machining system is regularly in view of the material evacuation utilizing instrument material harder than the work material and can't machine them financially. An EDM depends on the disintegrating impact of an electric spark on both the cathodes utilized. EDM really is a procedure of using the expulsion marvel of electrical-release in dielectric. EDM is a most fundamental non-traditional machining process, where material is evacuated by warm vitality of spark happening by methods for rehashed arrangements of electrical discharges between the little crevice of a terminal and a work piece. EDM is normally utilized for machining of electrically conductive hard metals and combinations in car, aviation and pass on making businesses

2. Literature review

J. Kozak [1] discovered that setting and keeping up a little yet stable hole estimate in PECMM is critical to accomplish a superior dimensional exactness control. The suitable least hole measure, not withstanding, is restricted by entomb cathode crevice conditions, for example, electrolyte bubbling. A

IOP Conf. Series: Materials Science and Engineering **396** (2018) 012071 doi:10.1088/1757-899X/396/1/012071

model created to appraise the base crevice estimate uncovers that a shorter pulse on-time would permit a littler least hole measure without electrolyte bubbling. Test examinations affirmed the hypothetical outcomes, which licenses to suggest numerical models and programming for re-enactment of PECM for useful applications.

Vinod Yadav et al [2] performed numerical simulations to anticipate thermal stresses in HSS workpiece by building up a limited component based code. The outcomes acquired serve to light up the harming way of thermal stress as they create amid EDM. It is observed that after one spark, generous compressive and tractable burdens create in a thin layer around the spark area. It is additionally found that the thermal stress surpass the yield quality of the workpiece generally in a greatly thin zone close to the spark. The results illustrated the damaging nature of thermal stresses during EDM process. W Murray et al [3] determined the effects of debris in EDM. Debris was imagined using SEM and TEM techniques, revealing particle sizes and other properties of debris.

S.L.Chen, et.al [4] stated about Influence of kerosene and distilled water as dielectrics on the electric discharge machining characteristics of Ti-6Al-4V alloy under various machining condtions. Leera Raju et.al [5] carried out an extensive literature study in their journal "A state of the art review on μ-EDM" to give a complete description on μ-EDM process, its requirements, performance and applications. The experimental setups and its subsystems, experimental studies and optimization methods, generated micro feature and various applications are described in this journal.

Wang et.al [6] experimentally investigated a wear-resistant electrode for micro-EDM. The results proved that Cu-ZrB2 composite (copper-zirconium Diboride) coated electrodes have better wear resistance than pure copper electrodes. They also found that it is feasible to use the wear compensation method on the basis of the difference between the wear ratio of matrix and that of coating material to maintain electrode shape precision. S.K.Choudhury et.al [7] conducted experiments using air as the dielectric medium to study the effect of gap voltage, discharge current, pulse-on time, duty factor, air pressure and spindle speed on material removal rate (MRR), surface roughness (Ra) and tool wear rate(TWR).

In the current study an EDM setup is developed and micro holes machined in copper workpiece with different tools. The input variable parameters are Voltage, Frequency and Duty factor. Taguchi method is applied to create an L9 orthogonal array of input variables using the Design of Experiments (DOE). The effect of the variable parameters mentioned above upon machining characteristics such as Material Removal Rate (MRR), and Tool Wear Rate (TWR) is studied. A multi objective optimization of parameters is carried out using Grey Relational Analysis (GRA)

3. Experimental setup

The experimental setup for EDM consists of different subsystems. Fig 1 shows the experimental setup with DC power supply and function generator. The deionized water is used as dielectric and it provides insulation against premature discharging, cools the machined area, and flush away heat, damaged dielectric and the EDM chips and debris

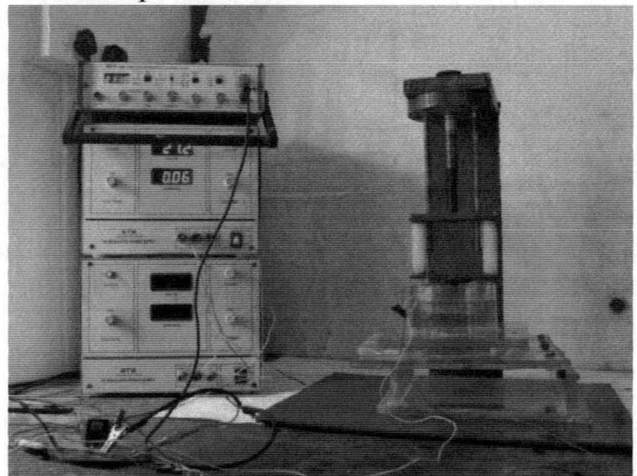

Fig.1 Overall machine setup

The workpiece used is copper and is cut into small square pieces and with a thickness 0.5 mm. Copper of 1 mm diameter and stainless steel tool of 0.5mm diameter are taken as tools.

4. Experimental investigation

The experiment is carried out in order to get maximum material removal rate with minimum tool wear. In this experiment the input parameters such as voltage, frequency and duty factor is varied. In the output of the experiment, the material removal rate and tool wear rate are examined. The maximum material removal rate and minimum tool wear rate can be found at a particular combination of voltage, frequency and duty factor, and can be optimized using Taguchi method. Design of experiment is a powerful analysis tool for modelling and analysing the influence of control factors on performance output. An experimental plan based on Taguchi orthogonal array (L_9) was chosen. Two different workpieces and tool are used during the machining. Two sets of experiments were planned to be carried out:

1. Set of experiments with copper as workpiece and copper rod as tool (diameter of 1mm)

2. Set of experiments with as copper as workpiece and stainless steel as tool (diameter of 0.5 mm.)

The experiments were conducted with varying tool diameters in order to obtain holes in the micro-level. The experiments were conducted with DI water as the dielectric fluid. The machining parameters selected on the basis of trail machining results. The machining parameters selected for the experimentation are Voltage (V), Frequency (KHz), Duty factor (%)s. The three levels for each parameter are as follows:

4.1 L9 Experiments with copper tool and copper work piece

Based on the levels of the process parameters selected, the experimental plan is developed using L9 orthogonal array with output responses is given in Table 1.

Table 1. Experimental layout plan of L9 orthogonal array

EX. NO	VOLTAGE (V)	FREQUENCY (kHz)	DUTYFACTOR (%)	MATERIAL REMOVAL RATE (mg/m)	TOOL WEAR RATE (mg/m)
1	40	2	60	0.575	0.212
2	40	4	70	0.589	0.233
3	40	6	80	0.501	0.201
4	45	2	70	0.616	0.225
5	45	4	80	0.582	0.214
6	45	6	60	0.683	0.298
7	50	2	80	0.654	0.284
8	50	4	60	0.710	0.327
9	50	6	70	0.791	0.341

4.2 L9 Experiments with stainless steel tool and copper workpiece

In experiment set 2 the workpiece used is copper plate of 0.5mm thick and tool used is stainless steel rod of 0.5mm diameter. DI water is used as Di electric fluid. The experimental results obtained are given in Table 2.

Table 2: Experimental observation of experiment set 2

EX NO	VOLTAGE (V)	FREQUENCY (kHz)	DUTYFACTOR (%)	MATERIAL REMOVAL RATE (mg/min)	TOOL WEAR RATE (mg/min)
1	40	2	60	0.541	0.202
2	40	4	70	0.582	0.141
3	40	6	80	0.521.	0.143
4	45	2	70	0.631	0.167
5	45	4	80	0.688	0.181
6	45	6	60	0.701	0.189
7	50	2	80	0.748	0.221
8	50	4	60	0.765	0.269
9	50	6	70	0.795	0.295

5. Results and discussions

Material removal rate (MRR) and Tool wear rate (TWR) have been calculated for each of the experiments based on the experimental observations. In the present work, multi– objective optimization of the EDM process has been carried out by considering maximization of MRR and minimization of TWR as two distinct objectives in both experiments. The Grey relational analysis is used to optimize the machining parameters with multiple performance characteristics such as Material Removal Rate (MRR) and Tool Wear Rate (TWR).

5.1 Experiment set 1

In experiment set 1 the workpiece used is copper plate of 0.5mm thick and tool used is copper rod of 1mm diameter. DI water is used as Di electric fluid. Based on the experimental observations the signal to noise ratios for MRR and ROC has been calculated. The normalised signal to noise ratios and deviation sequences has been calculated as per the equations in grey relational analysis. Also the grey relational coefficient and grey relational grade has been calculated and ranked according to priority in each experimental set.

Table 3: Ranking for grades in experiment set 1

EX.NO	GRG	RANK
1	0.579	3
2	0.515	7
3	0.599	2
4	0.478	9
5	0.576.	4
6	0.524	6
7	0.499	8
8	0.547	5
9	0.733	1

A response table for parameter voltage, frequency and duty factor

Table 4: Response table of parameters for experiment set 1

Voltage	1.556	1.63	1.856
Frequency	1.693	1.578	1.779
Duty factor	1.65	1.726	1.908

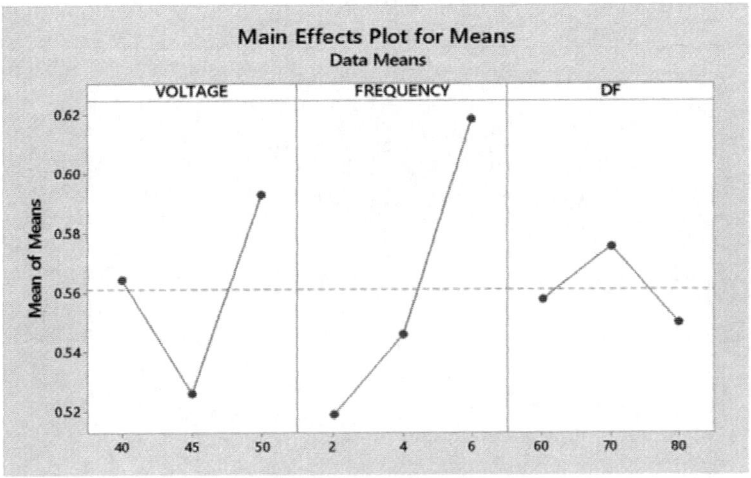

Fig.2 Optimum values for set 1

A confirmation test is conducted based on the best parameter combination obtained from multi-objective optimisation with voltage = 50V, frequency=6KHz, duty factor = 70%. The maximum MRR obtained from the experiment is is 0.791 mg/min with minimum TWR 0.341 mg/min. The MRR obtained from confirmation test is found as 0.797mg/min and TWR of 0.278 mg/min which is in agreement with experimental result.

The ranking table for the experiment with stainless steel as tool is as follows:

Table 5: Ranking for grades in experiment set 2

EX.NO	GRG	RANK
1	0.203	7
2	0.187	8
3	0.167	9
4	0.221	6
5	0.263	5
6	0.278	4
7	0.344	3
8	0.413	2
9	0.5	1

A response table is calculated and is given below.

A confirmation test is conducted with parameter combination of voltage = 50V, frequency=6KHz and duty factor = 70%. The MRR obtained from L9 set= 0.795mg/min TWR obtained from L9 set =0.295mg/min. Also the MRR obtained from confirmation test= 0.801 mg/min TWR obtained from

IOP Conf. Series: Materials Science and Engineering **396** (2018) 012071 doi:10.1088/1757-899X/396/1/012071

confirmation test = 0.241 mg/min. By comparing first two experiments it is noted that the tool wear rate is less in the case of stainless steel tool than the copper tool while machining copper workpiece.

Table 6: Responses of experiment set 2

Voltage	0.768	0.863	0.945
Frequency	0.557	0.651	1.257
Duty Factor	0.894	0.908	0.93

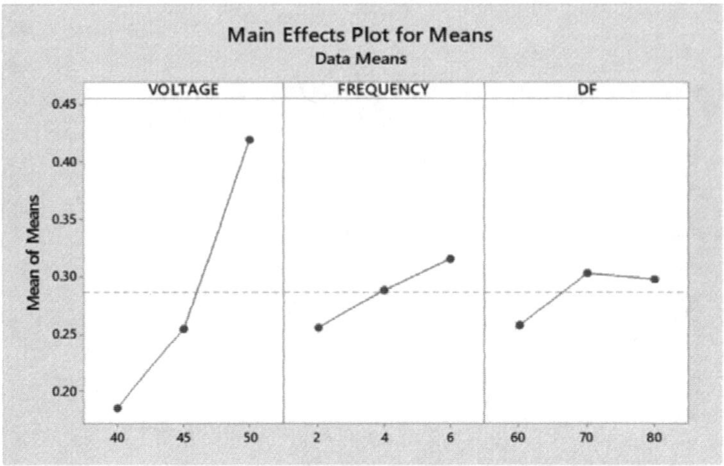

Fig.3 Optimum values of experimental set 2

5.2. Microscopic images

The microscopic images are taken for copper workpiece with stainless steel tool and are given below.

a) Microscopic Images of Workpieces for Voltage 40 V

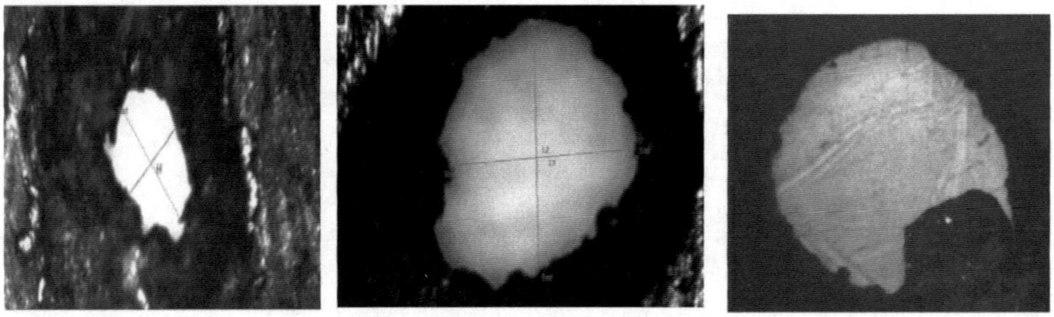

Workpiece 1	Workpiece 2	Workpiece 3
Voltage= 40 V	Voltage= 40 V	Voltage= 40 V
Frequency = 2 kHz	Frequency = 4 kHz	Frequency = 6 kHz
Duty factor= 60	Duty factor= 70	Duty factor= 80

Fig.4 Microscopic Images of samples at 40 V

As the voltage is lower, energy of the spark generated is also less so that the material removal rate is less compared to higher voltages. Because of this the hole is not completely machined in fixed machining time.

b) Microscopic Images Of Workpieces for Voltage 45 V

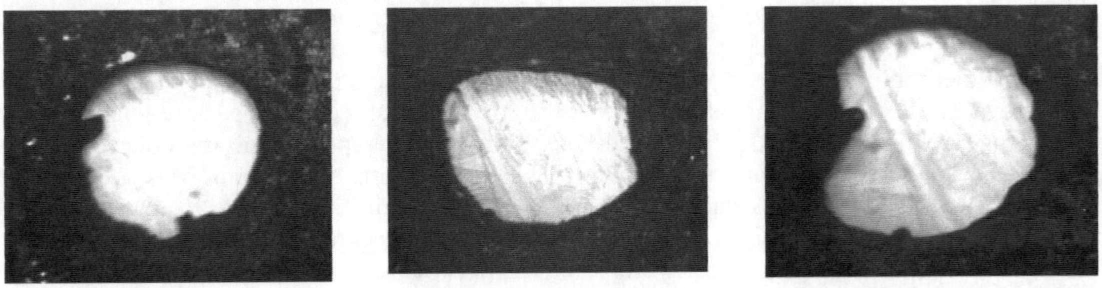

Workpiece 4	Workpiece 5	Workpiece 6
Voltage= 45 V	Voltage= 45 V	Voltage= 45 V
Frequency = 2 kHz	Frequency = 4 kHz	Frequency = 6 kHz
Duty factor= 70	Duty factor= 80	Duty factor= 60

Fig.5 Microscopic Images of samples at 45 V

Here the voltage is increased by 5V and the material removal rate is also gradually increased. This leads to the complete drilling of holes but the hole is irregular and at duty factor 70 micro-hole is much better than others.

c) Microscopic Images Of Workpieces for Voltage 50 V

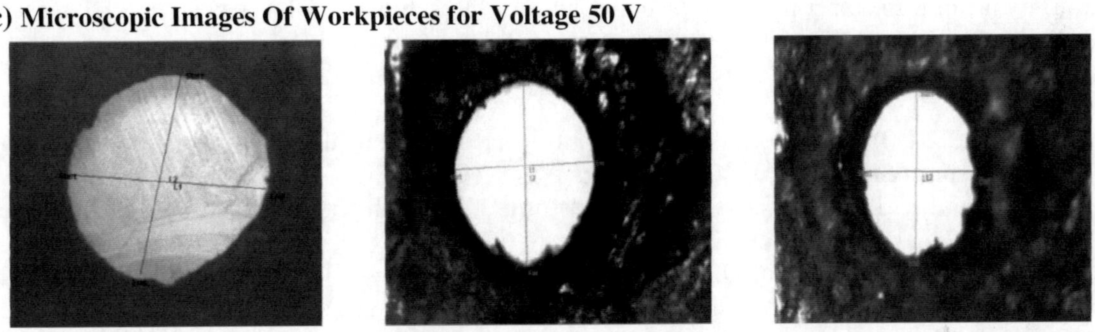

Workpiece 7	Workpiece 8	Workpiece 9
Voltage= 50 V	Voltage= 50 V	Voltage = 50 V
Frequency = 2 kHz	Frequency = 4 kHz	Frequency = 6 kHz
Duty factor= 80	Duty factor= 60	Duty factor =70

Fig.6 Microscopic Images of samples at 50 V

Here the voltage is increased to 50 V and the material removal rate is also gradually increased. This leads to the complete drilling of holes with regular shapes and at duty factor 60 micro-hole is much better than others. From the microscopic images it is concluded that the MRR get increased with higher voltage and regular holes are attained at moderate duty factors. ROC get reduced at lower DF due to more cooling time at lower DF.

6. Conclusion

In this investigation, development of a micro EDM setup for machining micro hole is discussed. Experiments were conducted with different tools on copper workpiece combination. During the experiment the Di-electric medium used is de-ionised water. The process parameters selected for experiments are voltage (V), Frequency (KHz) and Duty factor (%).Machining was conducted based on L9 orthogonal array for experiments. MRR and TWR were taken as the output parameters for all three experiment sets. Multi-objective optimisation using grey relational analysis was carried out in order to obtain the best parametric combination. Based on the optimised parametric conditions, confirmation tests have been carried out. The best parameter combination obtained for experiments done on copper workpiece with copper tool are Voltage of 50 V, 6 KHz Frequency and 70% Duty factor. After that conformation tests are carried out with the optimum parameters obtained and the test results provide MRR = 0.797mg/min and TWR =0.278 mg/min. The best parameter combination obtained for experiments done on copper workpiece with Stainless steel tool are Voltage of 50 V, 6KHz Frequency and 70% Duty factor. Conformation tests are carried out with the optimum parameters obtained and the test results provide MRR = 0.801 mg/min and TWR = 0.241 mg/min. By comparing these two set of experiments it is noted that stainless steel tool is better that copper tool for the micro hole machining in the copper workpiece. The material removal rate is almost similar in two cases and the tool wear rate is less in the case of stainless steel tool compared to copper workpiece. Tool wear is noted that less in case of stainless steel tool than copper tool.

Voltage and Frequency are the energy factors, during the experiments. It is noted that as voltage increases material removal rate also increases. This is due to increase in discharge energy when voltage increases more discharge current will pass through the gap and it causes more material removal. But the voltage is at very high value is not preferable for the machining because in high voltage spark may change to arc so proper machining is not happening. If off time is too less it may affect the removal of debris from the gap and proper machining will not occurs. During the experiments it is noted that for stainless steel workpiece high voltage applied is 60V but the maximum MRR obtained at 55V this may due to at 60V sparking action is not properly done. At 60V arcing may occur. Duty factor is the other parameter; from results it is clear that 70% DF is better. 70% DF means 70% ON Time of pulse and 30% OFF Time.

References

[1] J. Kozak (2000)Thermal models of pulse electrochemical machining, Institute of Manufacturing Technology, Warsaw University of Technology, Vol. 222, 650–663

[2] V. Yadav and K. Jadav,(2002) Thermal modelling of EDM, International Journal of Machine Tools and Manufacturing Vol. 42,877-888.

[3] Murray, J.W., Sun, J., Patil, D.V., Wood, T.A., Clare, A.T. (2015), Physical and Electrical Characteristics of EDM Debris. Journal of Materials Processing Technology 10,10-16.

[4] S L Chen, B H Yan, F Y Huang, (1999) " Influence of kerosene and distilled water as dielectrics on the electric discharge machining characteristics of Ti-6Al-4V", Journal of materials processing technology 87,107-111.

[5] Leera raju, Somashekhar S Hiremath,(2016) A state of art review on micro electro discharge machining, Global colloquium in recent advancement and effectual researches in engineering science and technology Rearest 2016 Procedia Technology 25 (2016)1281.

[6] Weng, F.T.(2001), Micro-hole machining of copper using the electro discharge machining process with a tungsten carbide electrode compared with a copper electrode, International Journal of Advanced Manufacturing Technology, 17(10), pp715-719.

[7] Sourabh K.Saha, S.K. Choudhury "Experimental investigation and empirical modeling of the dry electric discharge machining process" Thesis reports, Department of Mechanical Engineering, Indian Institute of Technology Kanpur, Kanpur, India.

IOP Conf. Series: Materials Science and Engineering **396** (2018) 012072 doi:10.1088/1757-899X/396/1/012072

A FIREFLY ALGORITHM APPROACH FOR MULTI-ROW FACILITY LAYOUT PROBLEM

Akash P Lukose[1], Abyson Scaria[1],Babu George[1]

[1]Department of Mechanical Engineering, St. Joseph's College of Engineering and Technology, Palai, Kerala, India

Abstract: The facility layout problem is an optimization problem that makes efficient layouts. Layout design has a prime importance on the performance of manufacturing systems. The facility layout problem (FLP) helps to reduce the material handling costs by a well organized positioning of facilities. The facility layout problems are generally regarded as NP-Hard problems. The different methodologies for solving FLP's can be exact algorithms, heuristics and meta-heuristics. The exact algorithms are generally used when the number of facilities in the layout is up to 16. For large sized FLP's, heuristics were developed to attain near optimal solution in reasonable computation times. In the recent literatures, there is a great focus on biology-inspired meta-heuristic algorithms to find most suitable solutions for FLP's. Firefly Algorithm is a recently developed Meta-heuristic, which represents some of the traits of the firefly behaviour. This paper focuses on the application of firefly algorithm to multi-row facility layout problem with the goal of reducing total material handling cost. A parameter analysis was done and some general FLP's were solved using the algorithm. A comparative study between Genetic Algorithm and Firefly Algorithm was also performed.

Keywords: Facility layout problem (FLP), NP-Hard, Firefly Algorithm, Genetic Algorithm, Parameter Analysis.

1. Introduction

A facility is an entity that facilitates the performance of any job. Facility can be a manufacturing cell, a work center, a department or a ware house. Facility layout is an arrangement of facilities required for the production of goods or delivery of the services. By considering the various interactions between facilities and material handling system, the facility layout problem helps to make efficient layouts. The facility layout problem (FLP) helps to reduce the material handling costs by determining the most suitable arrangement of facilities within the given space; subject to various constraints such as site plan, the building, service requirements, the departmental area and the decision maker.

Layout problems are usually found in manufacturing systems which is related to the positioning of various facilities in a plant. The arrangement of facilities has specific impact on the net efficiency of operations, reduce manufacturing cost, decrease lead times, and increase productivity. Facility Layout Problem is considered to be significantly complex problem and the most studied problem in facilities planning [1]. The facility layout problem was a typical NP-hard problem, so it is difficult to arrive at the optimal solution using exact methods. This calls for the usage of approximation algorithms for obtaining satisfactory solution.

Global optimization problems were effectively solved by using the swarm intelligence, which was a nature inspired metaheuristic approach. The Firefly algorithm (FA) which is used in this work focus on the flashing pattern of tropical fireflies. This paper focuses on the application of firefly algorithm to multi-row facility layout problem with the goal of reducing total material handling cost.

2. Literature Review

The various objectives taken by the researchers for the facility layout problems include reduction of material handling and curtailment of material handling cost, minimization of material flow and reducing the total distance travel for material and improving the total closeness between facilities. The single row facility layout problem is extensively investigated in the literature with different objectives. But the multi-row facility layout problem is rarely addressed by the researchers.

Liu et al. [2] considered the facility layout problem by using certain logistic flows and stochastic product demand techniques. The rotations were allowed, but the shape of the department was same. Apart from traditional material handling costs, some of the criteria such as the shape ratio factor, the operational performance rating and the Work In Process (WIP) value was considered in this paper. Jannat et al. [3] deals with optimization of multi-objective facility layout problems. The objectives considered include reducing the material handling cost and maximizing the closeness rating score. Chang et al. [4] suggested a mathematical model by considering the production volume and alternative routing to attain the cell formation and cell layout concurrently. A hybrid tabu search algorithm was used to solve it due to the combinatorial nature of this model.

Chung et al. [5] analyze the double row layout problem (DRLP) that includes the positioning of departments or machines on the two sides of a central corridor. Various heuristics were proposed and they were compared with MIP solutions by CPLEX 10.2, in a series of experiments. Alvarenga et al. [6] analyzed the facility layout problems in the situation of manufacturing environment and two procedures based on simulated annealing (SA) and Tabu Search (TS) were presented. For single-row layout problems having numbers of facilities 4, 10 and 11, both of the algorithms produced the optimal solutions. For other problems, they produced better solutions than those given in the literature. The algorithms produced the same results as in the literature for all the multi-row layout problems solved, with the exception of the one with 30 facilities.

Shahin et al. [7] proposed an approach for improving and optimization of the facility layout design. It was an integrated approach of simulation, quality function deployment (QFD), fuzzy analytic hierarchy process (AHP) and multiple criteria decision making (MCDM). Quantitative measure was determined by computer simulation and for qualitative measure analytical hierarchy process (AHP) was used. QFD was implemented to find out weights of criteria and the significance of the alternatives regarding the quantitative and qualitative measures. TOPSIS (Technique for Order of Preference by Similarity to Ideal Solution) method was used for rating the alternatives and selecting the best alternative.

Ohmori et al. [8] addresses Particle swarm optimization (PSO) technique. The proposed method did not use encoding techniques and instead, it searched coordinates of each department so that it can represent any possible layout. The objective of this model was to develop the layout for reducing the total material handling costs between departments.

Miao and Xu [9] described a model by taking account of handling cost and handling time for the facility layout problem. A hybrid algorithm which was formed by the superposition of Tabu Search (TS) and genetic algorithm (GA) was suggested to solve the model.

3. Problem Description

The general facility layout design problem is involved with the positioning of various departments of given areas within given facility. Single row facility layout problem (SRFLP) is concerned with positioning a number of facilities on a line to reduce transportation costs among the facilities. Arranging a number of facilities on different rows to reduce transportation costs among the facilities defines the multiple row facility layout problems (MRFLP).

The material handling cost of a system can be minimized by efficient arrangement of facility layouts. To determine the material handling cost for a possible layout of the system, certain parameters have to be known, such as the handling cost of unit material for unit distance between equipment, the amount of material flow or production volumes among equipment and the rectilinear distance between equipment.

3.1. Assumptions

- '*n*' facilities with the same shape and size to '*n*' distinct predetermined locations.
- Rectilinear distances are considered.
- The facility assigned to each location is one only.
- Each facility is assigned to only one physical location.

3.2. Mathematical Model

Minimize $Z = \sum_{i=1}^{N} \sum_{j=1}^{N} A_{ij}.X_{ij}$

s.t.

$$\sum_{i=1}^{N} X_{ij}=1 \; ; \quad j = 1, 2, 3.......N$$

$$\sum_{j=1}^{N} X_{ij}= 1; \; i = 1, 2, 3........N$$

$$X_{ij} \in (0,1) \; ; \quad i,j = 1,2,3........N$$

where

X_{ij} $\in \{1 , 0\}$
=1 if facility i is assigned to location j,
=0 otherwise.

A_{ij} = The cost of locating facility i at location j
$= f_{ij}. d_{ij}$

f_{ij} = flow cost from facility i to j

d_{ij} = distance from location i to j

Each location is assigned only one facility and each facility is assigned to only one physical location, respectively.

4. Solution Methodologies

The different methodologies for solving FLP's can be exact algorithms, heuristics and meta-heuristics. The exact algorithms are generally used when the number of facilities in the layout is up to 16. For large sized FLP's, heuristics were developed to obtain near optimal solution in reasonable computation times.

4.1. Exact Algorithms

Exact algorithms involve existence of a feasible solution and then the optimal solution was determined with certainty once such exact algorithm is successfully terminated. The main disadvantage of these exact algorithms is that they entail heavy computational requirements when applied even to small size problems.

4.1.1. Problem Specific Algorithms
CRAFT is a helps to improve the facilities of the existing layout. The facility is improved by interchanging two or three departments and attains the optimal floor plan. The inputs for the procedure are: From-To Chart, Distances (determined for a given layout), Cost Matrix and an Initial layout. CRAFT uses a pair wise exchange algorithm, so it may not return the optimal result because the final solution is dependent on the initial layout of the plant.

CORELAP is an acronym for computerized relationship layout planning. A layout is generated by the CORELAP on the basis of total closeness rating (TCR) for each department.

ALDEP is an acronym for automated layout design program. ALDEP is quite similar to CORELAP, but in the case of ties it performs differently. It breaks ties randomly and the first department to enter is also chosen randomly.

4.1.2. Mathematical Techniques
Maximizing or minimizing a linear function subject to linear constraints defines the linear programming problem. The constraints constitute the equalities or inequalities. Constrained optimization is the practical application of linear programming. The objective function and the constraints used in the mathematical expression are linear.

A nonlinear program is similar to a linear program in that it contains an objective function, variable bounds and general constraints. The non linear program differs in the sense that it contains at least one nonlinear function, which can be the objective function or some or all of the constraints.

Dynamic programming is the process of breaking a complex problem into a sequence of simpler problems. The multistage optimization technique is the characteristic feature of it. It helps to form a general framework for solving many problem types.

4.2. Heuristic Algorithms

In order to obtain good (near optimal) solution in a reasonable computational time, heuristic algorithms were developed. A heuristic algorithm is defined by a set of steps for quickly identifying good quality solutions.

Construction algorithms are considered to be the simplest and oldest heuristic approaches. A construction algorithm consists of successive selection and placement of facilities until a complete layout is achieved. But these construction algorithms can be used to provide initial solutions for improvement algorithms.

The improvement methods start with the formation of feasible solution and try to improve it by interchanges of single assignments. The best solution which is produced by the exchange is retained and the process continues till a stopping criterion is reached or until the solution cannot be improved further. The quality of solutions is dependent on the initial layout and the location exchange procedure.

In hybrid algorithms, the solution of Quadratic Assignment Problem is determined by using a combination of two optimal or sub-optimal algorithms. Such combination of algorithms is essential in some cases to improve solution quality. This classification includes certain algorithms, which use the principal of improvement algorithms and construction algorithms.

4.3. Meta-Heuristic Algorithms

The development of meta-heuristic algorithms has high impact on the performance of improvement algorithm and it uses a general strategy like pair-wise exchange heuristic. Genetic Algorithm (GA) works on the theory of Darvin's theory of evolution and the survival of the fittest. By using natural selection and genetic operators, such as crossover, mutation and the selection, the Genetic algorithms guide the search through the solution space. The decision variables or input parameters of the problem are encoded in GA into solution strings of a finite length. Genetic algorithms usually work with the coding, on the other hand traditional optimization techniques work directly with the decision variables or input parameters. Instead of from a single point in the solution space, Genetic algorithms start to search from a population of encoded solutions. The initial population is created by random selection. Genetic Global optimum solutions are created by using genetic operators in GA based on the solutions in the current population. The commonly used genetic operators are selection, crossover and mutation. The evolution process proceeds until certain termination criteria are satisfied by replacing the old population by the newly generated individuals.

Simulated annealing (SA) is a general probabilistic local search algorithm, suggested by Kirkpatrick in 1983, to solve difficult optimization problems. SA is based on the analogy between the annealing of solids and the solving of combinatorial optimization problems. SA is a step-by-step method which could be considered as an improvement of the local optimization algorithm. This process accepts the better solutions. It also accepts worse solutions with a certain probability which is called the probability of accepting. The temperature determines the probability of acceptance. The probability of accepting a worse solution is large at a higher temperature. As the temperature decreases, the probability of accepting a worse solution also decreases as well.

Tabu Search (TS) was suggested by Glover was an efficient method, which uses the local search techniques to obtain near optimal solution to combinatorial optimization problems. It uses a list of prohibited neighbour solutions known as the tabu list, the deterministic local search technique can able to escape local optima. Apart from escaping local optima, the tabu list can also used to avoid cycling by suppressing or penalizing moves which takes the solution in the next iteration, to previously visited points in the solution space and helps to reduce the computational time. A drawback of tabu search was that it cycles the same path in the previously visited solution unless a tabu neighbor exists. In other words, a loop is encountered when the search moves to previously visited solution that has not tabu for the past two iterations.

Kennedy and Eberhart proposed an evolutionary computation technique called Particle Swarm Optimization (PSO). It is commonly used evolutionary computation algorithm. It involves the initialization of a population having random solutions. Then it search for optimal solutions by updating generations. Potential solutions, called particles, are then ''flown'' through the problem space by following the current optimum particles. In particle swarm optimization concept at each step, the velocity of the particle is changed toward its 'pBest' and'gBest' locations.

An optimization algorithm is developed based on the intelligent foraging characteristic of honey bee swarm constitutes the Artificial Bee Colony (ABC) algorithm. The artificial bee colony contains three groups of bees. They are employed bees, onlookers and scouts .The employed bee searches for destination in which food is available. They collect the food and return to its origin. Then they perform waggle dance based on the quantity of food available at the destination. The onlooker bee watches the dance and follows the employed bee depending on the probability of the available food. So, more onlooker bees will follow the employed bee associated with the destination having

IOP Conf. Series: Materials Science and Engineering **396** (2018) 012072 doi:10.1088/1757-899X/396/1/012072

more amount of food. When the food source becomes deserted the employed bee transforms to scout bee. Then it searches for the new food source. The optimization problems are solved by considering the foraging characteristic of honey bee. It divides population into two parts containing the employed bees and onlooker bees. An employed bee searches for the solution in the search space and the amount of food associated with the solution defines the value of objective function associated with the solution.

5. Firefly Algorithm

Firefly Algorithm is a recently developed Meta-heuristic, which represents some of the traits of the firefly behaviour. The flashing pattern of fireflies is used to develop this algorithm. Xin-She Yang developed the Firefly Algorithm (FA) at Cambridge University in 2007.

5.1. Behaviour of Fireflies

The intensity of light at a certain distance r from the light source follows the inverse square law. That is the intensity of the light I goes on decreasing as the distance r increases. Also, the air which absorbs the light becomes weaker with the increase in the distance. When these two factors are combined, majority of the fireflies are visible at a limited distance. This extends to a few hundred meters at night, which is sufficient for the communication of fireflies with each other.

5.2. Concept

Brightness and attractiveness are proportional to each other. Thus while considering any two flashing fireflies, the less bright one will move towards the one which is brighter. Both the brightness and attractiveness decreases as their distance increases. They will move randomly if there is no one brighter than other firefly. The objective function is determined by the brightness of a firefly.

5.3. Light Intensity and Attractiveness

In the firefly algorithm, there are two key aspects: the variation in the intensity of light and formulation of the attractiveness. We can assume that the brightness of a firefly is determined by its attractiveness which in turn is connected with the encoded objective function.

In the simple cases, such as for maximum optimization problems, the brightness I of a firefly for a particular location x could be chosen as I(x) is proportional to f(x). Even so, the attractiveness β is relative, it should be judged by the other fireflies. Thus, it will differ with the distance R_{ij} between firefly i and firefly j. Also light intensity decreases with the distance from its source. Hence light is absorbed by the media and we should allow to vary the attractiveness with the varying degree of absorption. For a stated medium with a fixed light absorption coefficient γ, the light intensity I vary with the distance r as:-

$$I = Io\, e^{-\gamma r2} \tag{1}$$

Since a firefly's attractiveness is proportional to the light intensity seen by adjacent fireflies, we can define the attractiveness β of a firefly as:-

$$\beta = \beta o\, e^{-\gamma r2} \tag{2}$$

5.4. Implementation for Facility Layout Problem

For the implementation of firefly algorithm to the multi- row facility layout problem considering 1 firefly as 1 layout and randomly creating an initial population of n fireflies. The light intensity (I) is treated as the fitness function of the population. The light intensity level helps to determine the movements. Attractiveness is find out by using an unstable distance value. The iterative process is repeated for different fireflies until the maximum generation level is reached. The light intensity values are updated. Thus the fireflies are ranked and the current best is found until the stop condition occurs.

Considering eachfirefly asa single layout and the total number offireflies as thepopulation size respectively; 3 main functions are used to implement firefly algorithm to facility layout problem. Fitness function is used to calculate the fitness of firefly. Ranking function is used to rank the Fireflies according tothe fitnessvalues(sorting).Attractiveness function is to modify the current layout with the best layout. The FA algorithm is implemented on MATLAB (R2014a).

6. Computational Results

6.1. Parameter Analysis

Population size is the total number of fireflies. To find the best population size, a parameter analysis is required. A 3 rows 9 machines layout is considered for calculating the best population size.

Table 1. Parameter Analysis table

Sl. No.	Population Size	Average Best Score
1	10	627
2	20	617.40
3	30	612.80
4	40	617.60
5	50	614.20
6	60	608
7	70	609.80
8	80	610.40
9	90	610.20
10	100	611.40

The populationsize value varies from 10 to 100 fireflies and the results obtained for eachproblems are plotted. From these values,the best population size is determined using the curve fitting techniquein SPSS software. From this graph the best value for population size is estimated as 75 fireflies.

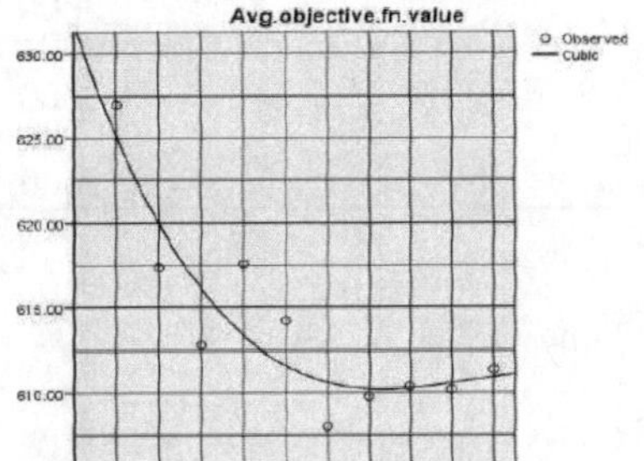

Figure 1. Parameter Analysis result using SPSS

IOP Conf. Series: Materials Science and Engineering **396** (2018) 012072 doi:10.1088/1757-899X/396/1/012072

6.2. Solution for general FLP's

The flow matrix and the cost matrix are randomly generated in the code, and the rectilinear distance is calculated. Using the newly developed FA code for FLP, different facility layout problems are solved.

Total material handling cost = Flow of material * Unit Cost * Distance (3)

The result for 4 rows 20 machines problem is obtained as:-

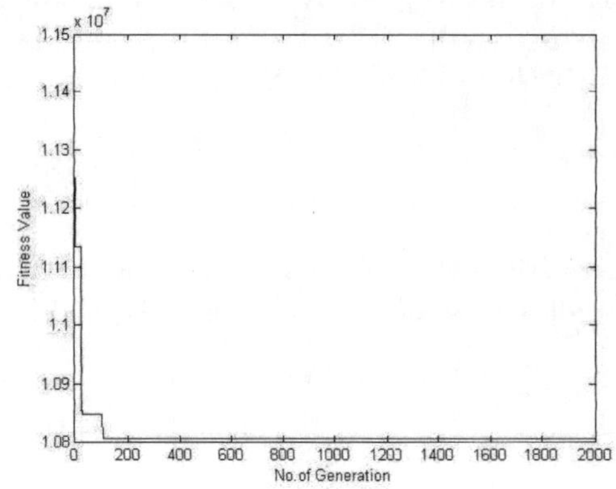

Figure 2. Final result of 4 rows 20 machines general problem

Total material handling cost = $1.081 * 10^7$.

Table 2: Final result for general problems

Sl. No	Number of Rows	Number of Machines	Population Size	Number of Generations	Best Score
1	4	20	75	2000	$1.081*10^7$
2	8	40	75	2000	$7.51*10^7$
3	12	60	75	2000	$2.371*10^8$
4	16	80	75	2000	$5.37*10^8$
5	20	100	75	2000	$9.98*10^8$

6.3. Comparative Study between GA and FA

Comparative study betweenfirefly algorithm and genetic algorithm is done. For this, a 3 rows 9 machines layout is considered and population size is taken as 10.The number of total generations is taken as 2000.

IOP Conf. Series: Materials Science and Engineering **396** (2018) 012072 doi:10.1088/1757-899X/396/1/012072

	1	2	3	4	5	6	7	8	9
1	-	2	6	1	3	7	4	5	8
2	1	-	2	4	5	8	9	7	3
3	5	2	-	1	3	7	4	5	8
4	1	2	6	-	3	7	4	5	8
5	3	2	6	1	-	7	4	5	8
6	7	2	6	1	3	-	4	5	8
7	4	2	6	1	3	7	-	5	8
8	5	2	6	1	3	7	4	-	8
9	8	2	6	1	3	7	4	5	-

Figure 3. Flow matrix for the sample problem

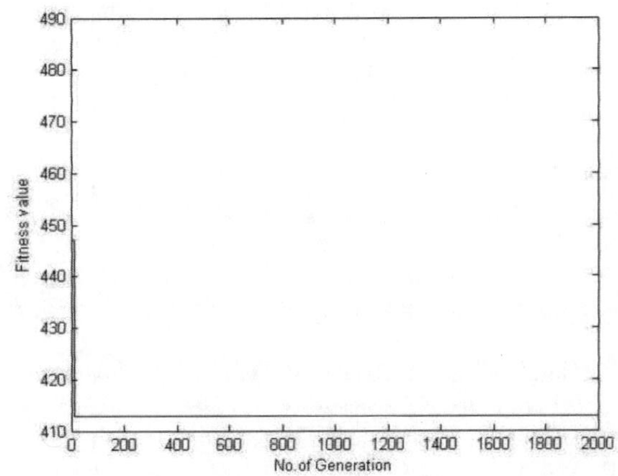

Figure 4. Final result of 3 rows 9 machines using FA

2	4	8
7	6	9
5	3	1

Figure 5. Best layout using FA

Total material handling cost using Firefly Algorithm = 413.

Total material handling cost using Genetic Algorithm is also obtained as 413. When both algorithms are computed for the same layout problem, both FA and GA give good result. So Firefly Algorithm is effective for solving layout problems.

IOP Conf. Series: Materials Science and Engineering **396** (2018) 012072 doi:10.1088/1757-899X/396/1/012072

7. Conclusion

The Facility Layout Problem provides the optimum arrangement of various facilities in the layout.The exact methods can be used to obtain an optimal layout when the number of departments is up to 16. In this work, we use a meta-heuristic approach to attain the optimal facility layout for larger problems. The Firefly algorithm used in this work is a latest meta-heuristic approach.Thecharacteristics of firefly behaviour were used to generate an algorithm for minimizing the objective of total material handling cost.A Multi-Row Facility Layout Problem is considered in the work.

A parameter analysis was done to determine the population size for Firefly Algorithm and was found to be 75. Different problems are solved using the Firefly Algorithm approach for upto 100 facilities in which the number of rows ranges from 4 to 20. A comparative study was also performed with Genetic Algorithm which establishes that Firefly Algorithm gives good results.

The closeness gap between the facilities is another crucial factor for facilitly layout problems which can be considered for future works. As another extension of the work, multiple objectives can be considered for the problem to enhance the system performance. The FLP problem with more number of machines, more number of rows and different number of machines in each row can also be considered as an extension of the work.

8. References

[1] Maricar G. Misola and Bryan B. Navarro "*Optimal Facility Layout Problem Solution Using Genetic Algorithm*", World Academy of Science, InternationalJournal of Mechanical, Aerospace, Industrial, Mechatronic and Manufacturing Engineering, Vol:7, No:8, 2013.

[2]F. Liu, M. Dong, F. Hou and F. Chen, "*Facility Layout Optimization with Stochastic Logistic Flows*", Service Operations and Logistics, and Informatics, SOLI, International Conference, pp. 534 – 539, 2006.

[3] S. Jannat, A. A. Khaled, and S. K. Paul, "*Optimal Solution for Multi-Objective Facility Layout Problem Using Genetic Algorithm*", Proceeding of The International Conference on Industrial Engineering and Operations Management, Dhaka, Bangladesh, 2010.

[4] C. C. Chang, T. H. Wu and S. H. Chung, "*A Novel Approach for Cell Formation and Cell Layout Design in Cellular Manufacturing System*", Management and Service Science, MASS,International Conference, pp. 20 -22, 2009.

[5] Jaewoo Chung and J.M.A. Tanchoco, "*The double row layout problem* ",School of Industrial Engineering ,International Journal of Production Research, Vol. 48, No. 3, 2010, 709–727.

[6] A.G.D. Alvarenga, F.J.N. Gomes And M. Mestria, "*Metaheuristic Methods for a Class of the Facility Layout Problem*", Journal of Intelligent Manufacturing-11, pp. 421-430, 2000.

[7] A. Shahin and M. Poormostafa,"*Facility Layout Simulation and Optimization: anIntegration of Advanced Quality and Decision Making Tools and Techniques* ", Modern Applied Science, Vol. 5, No. 4, pp. 95-111, 2011.

[8] S. Ohmori, K. Yoshimoto and K. Ogawa, "*Solving Facility Layout Problem via ParticleSwarm Optimization*", Third International Joint Conference on Computational Science and Optimization, pp. 409 – 413, 2010.

[9] Z. Miao and K. Xu, "*Research of Multi-Rows Facility Layout Based on Hybrid Algorithm*",International Conference on Information Management, Innovation Management and Industrial Engineering, pp. 553 – 556, 2009.

Green supplier selection using hybrid grey relational analysis with fuzzy logic method

Ashish Vishnu V S[1], Jalumedi Babu[1], Babu George[1]

[1]Department of Mechanical Engineering, St. Joseph's College of Engineering & Technology, Choondacherry, Kerala, 686579 India.

Abstract: Green supplier selection is an integral part of Supply Chain Management in recent times. With increasing concern on environment protection and ecological preservation use of green factors in supplier selection is vital for every organization. The general concept behind green supplier selection is to evaluate suppliers based on certain environmental and cost effective criteria and select the one with best ranking. Hence green supplier selection is a crucial process. Present work proposes the application of a hybrid MCDM method, Grey relational analysis with fuzzy logic for green supplier selection. An example problem is solved by this method and a comparison with the outcome obtained using MOORA and COPRAS method for same problem is presented to verify the proposed method. Results revealed that ranking orders for the first supplier and second supplier were interchanged with grey relational analysis when compared with COPRAS and MOORA methods. But after using fuzzy logics GFRG values of both the supplier obtained are same. This indicates that both the suppliers are equally competitive and there is flexibility to choose any one of them. It indicates that when vagueness is involved, grey- fuzzy logic algorithm can be effectively applied for solving multi-criteria supplier selection problems.

Keywords: Green Supplier selection, Grey Relational Analysis Fuzzy Logic.

1. Introduction;

Supplier assessment and selection is an important strategic decision for minimizing operating costs and enhancing organizational efficiency. Every customer has different expectations from the suppliers. Various companies may have varies organizational and cultural backgrounds, which may also affect the supplier selection process. Therefore, a systematic method is required to determine which criteria are suitable and should be used for assessment of suppliers for an enterprise. At the same time environment protection and ecological preservation is one of the main challenges faced in recent times. And measures are being taken in all possible fields to do so. Until recent times supply chain management was based completely on economic factors but now there is a need for taking the environmental factors too in consideration for supplier selection. Green supplier selection (GSS) is a supplier selection process where equilibrium between the economics and environmental consideration of the suppliers are considered for

IOP Conf. Series: Materials Science and Engineering **396** (2018) 012073 doi:10.1088/1757-899X/396/1/012073

evaluation and selection. There are numerous sources which focuses on the GSS issues, using different strategies include, linear weighting models, data envelopment analysis (DEA), decision models for the final choice-phase, mathematical programming models , cluster analysis, case-based-reasoning systems etc. [1]. It's been seen over a period of time that supplier evaluation has been done keeping the economic aspects into considerations rather than the environmental considerations in the supplier selection [2]. Environmental issues are growing concern and it is the time to consider environmental and sustainable matters. When taking criteria into consideration for supplier selection, environmental factors should also be incorporated in GSS [3-5]. Criteria such as cost, delivery, pollution control, green product, service, quality, environment management, energy and natural resource consumption etc. These criteria are provided with weightage according to their importance, customer requirements and technical requirements.

The study on Green Supplier Chain (GSC) is gaining wide popularity specifically, in multi-objective decision making for green supply chain management (GSCM). AHP method is used in order to solve green supplier selection problems [6]. In order to eliminate or minimize the uncertainty of opinion of experts a hybrid AHP - fuzzy logic was used to solve supplier selection problem [7]. Methods including integration of hybrid fuzzy ANP with Fuzzy PROMETHEE are also used in GSS problem and solution can be obtained by analysis using sensitivity analyses [8]. Green supplier selection is done on the foundation of certain criteria including hazardous substance management (HSM), ANP process was applied to solve problems with that for supplier selection [9]. To solve the uncertainties arising in ANP method for GSCM due to interdependent criteria Grey relation analysis is integrated with ANP [10]. Another method for GSS in the Fuzzy –NGT- VIKOR method where the problem is solved by evaluating the criteria for selection using NGT method and then the substitutes are evaluated using VIKOR and finally the uncertainties are solved by the fuzzy logic [11]. Instead of only focusing on traditional criteria for evaluation human perceptions can also be converted to values and used to solve the supplier selection problem, these values can be considered as criteria and evaluated along with the other criteria using fuzzy TOPSIS [12]. Keeping the general criteria for the GSS the main criteria are senior management commitment, product designs with three R's (recycle, reuse, reduce) and agreement with legal environmental needs and auditing programs these criteria are incorporated for GSS using fuzzy TOPSIS [13]. Fuzzy entropy TOPSIS for green supplier selection was used for selection of supplier in thermal power plant were linguistic ratings were provided for the method for green supplier selection [14]. The integration of Fuzzy AHP and TOPSIS is used to govern selection of green supplier; the subjectivity in the problem solving is eliminated by use of fuzzy logics [15]. Fuzzy integrated VIKOR method was used for supplier selection based on four alternative supplier companies with main criteria being green, purchasing, transportation and recycling etc. for comparison and selection [16]. Another model was based on the framework obtained by the preprocessing of the socio-environmental and economic data of suppliers for use in fuzzy AHP method coupled with multi product lot sizing and multi period supplier selection problem [17]. A combined approach of (DEA-GEP) gene expression programming (GEP) with Data Envelopment Analysis (DEA) was used to select supplier and then the results were compared with results obtained by (DEA- ANN) Artificial Neural Network [18]. A new approach of GSS using SWARA, QFD using WASPAS tool was introduced where customer attitudes in the process of supplier evaluation is considered. Here customer requirements are given priority [19].

Recently an integrated approach using QFD and MCDM method was proposed by Morteza Yazdani et al [20] for GSS using DEMTEL and QFD for customer evaluation by incorporating green factors in the selection. COPRAS and MOORA methods were used to arrange and rank supplier then both results are compared with each other. Present work uses the integration of QFD and an integrated MCDM method grey base fuzzy logic method for solving the same problem in order to eliminate the uncertainties. The ranking obtained by this method is compared with ranks and results obtained by using earlier methods for the same supplier selection problem for the validity of the suggested method. Next section present the GSS using grey based fuzzy logic approach.

2. Methodology for green supplier selection integrated grey-based fuzzy logic

The procedures adopted for selection of a supplier with multiple conflicting criteria are presented in the following nine steps:

Step 1: Identification of criteria and weightage for each criteria.

Step 2: Shortlisting of the green suppliers satisfies the criteria and assigning the indices for different criteria to the suppliers

.

Step 3: Normalization of the data (data preprocessing) applying Eqns. (1)-(3). As most of material properties require either greater-the-better or lower-the-better characteristics, Eqns. (1) and (2) are most widely used.

Step 4: Calculation of the GRC for each criterion.

Step 5: Calculation of the overall GRG value.

Step 6: Fuzzification of the grey relational coefficients for each criteria and the overall GRG applying appropriate fuzzy membership functions.

Step 7: Establishing of fuzzy rules relating to the grey relational coefficients and with GRG.

Step 8: Obtaining the fuzzy output using max-min interface operation.

Step 9: Calculation of GFRG (G_0) using centroid defuzzification approach.

Step 10: Selection of the supplier having the maximum GFRG.

A detailed formula for the calculation of GRG and GFRG is presented elsewhere [21]

3. Illustrative example

To validate accuracy and potentiality of the proposed methodology, the following green supplier selection problems is considered and subsequently solved. The ranking orders obtained using the adopted method is compared with those obtained by the past research work. Illustrative example problem is chosen from Yazdani et al. [20], Green Supplier Selection (GSS) which is applied in planning and logistics section of a typical Iranian dairy industry. They considered ten green suppliers to evaluate on the basis of seven different criteria considering the customer requirements (CRs) and Technical Requirements (TRs) and the chosen criteria are namely Price (P), Green design (GD), Energy and natural resource consumption (ENRC), quality adaptation (QD), Delivery system (DS), Reuse and Recycle rate (RRR) and Production planning (PP).They used integration of DEMTEL and QFD for determining the weights for the criteria. The initial decision matrix is shown in Table.1. They used COPRAS and MOORA methods for ranking the suppliers and to choose the supplier of lowest rank. In variation this method present work uses a grey based Fuzzy logic method for ranking and selection of supplier for the same GSS problem.

Table 1. Initial Decision Matrix for GSS problem along with criteria weight [20]

Weight	.132	.135	.138	.162	.0.90	.223	.120
Supplier	QD	P	ENRC	DS	GD	RRR	PP
A_1	.068	.066	.150	.098	.156	.114	.098
A_2	.078	.076	.108	.136	.082	.171	.105
A_3	.157	.114	.128	.083	.108	.113	.131
A_4	.106	.139	.058	.074	.132	.084	.120
A_5	.103	.187	.125	.176	.074	.064	.057
A_6	.105	.083	.150	.051	.134	.094	.113
A_7	.137	.127	.056	.133	.122	.119	.114
A_8	.100	.082	.086	.060	.062	.109	.093
A_9	.053	.052	.043	.100	.050	.078	.063
A_{10}	.094	.074	.097	.087	.080	.054	.106

In this problem P and ENRC are non-beneficial attributes having lower-the-better characteristic and remaining five are beneficial attributes higher-the better characteristics, normalized values are calculated using normalized data, the deviation sequence, GRCs and overall GRG values are computed. The GRCs and overall GRG values for all the ten suppliers are provided in Table 2. From the Table 2 it can be noted that supplier A_2 is the best supplier whereas A_{10} is the worst supplier.

Table 2. GRCs and GRG values for green suppliers

Suppliers	Grey relational coefficients							GRG	Rank
	QD	P	ENRC	DS	GD	RRR	PP		
A_1	0.3688	0.8282	0.3333	0.4448	1.0000	0.5065	0.5286	0.5449	5
A_2	0.3969	0.7377	0.4515	0.6098	0.4173	1.0000	0.5873	0.6441	1
A_3	1.0000	0.5212	0.3863	0.4019	0.5248	0.5021	1.0000	0.6000	3
A_4	0.5049	0.4369	0.7810	0.3799	0.6883	0.4021	0.7708	0.5391	6
A_5	0.4906	0.3333	0.3948	1.0000	0.3926	0.3535	0.3333	0.4804	9
A_6	0.5000	0.6853	0.3333	0.3333	0.7067	0.4317	0.6727	0.4991	7
A_7	0.7222	0.4737	0.8045	0.5924	0.6092	0.5294	0.6852	0.6214	2
A_8	0.4771	0.6923	0.5544	0.3501	0.3605	0.4855	0.4933	0.4896	8
A_9	0.3333	1.0000	1.0000	0.4513	0.3333	0.3861	0.3524	0.5485	4
A_{10}	0.4522	0.7542	0.4977	0.4125	0.4109	0.3333	0.5968	0.4799	10

The corresponding GFRG values are obtained using Fuzzy logic tool box in MATLAB (R2017b). The grey relational coefficients for QD, P, ENRC, DS, GD, RRR, and PP are the inputs to the Fuzzy logic system. Triangular membership functions are adopted for Fuzzy modeling. Membership functions, like Lowest (LT), Low (L), Medium (M), High (H) and Highest (HT) are utilized to represent the GRCs of inputs. For the related output, the GFRG membership functions are represented as Lowest (LT), Very Low (VL), Medium Low (ML), Low (L), High (H), Medium High (MH), Higher (HR), Medium Higher (MHR) and Highest (HT). These Fuzzy membership functions are shown in Figures 1.

IOP Conf. Series: Materials Science and Engineering **396** (2018) 012073 doi:10.1088/1757-899X/396/1/012073

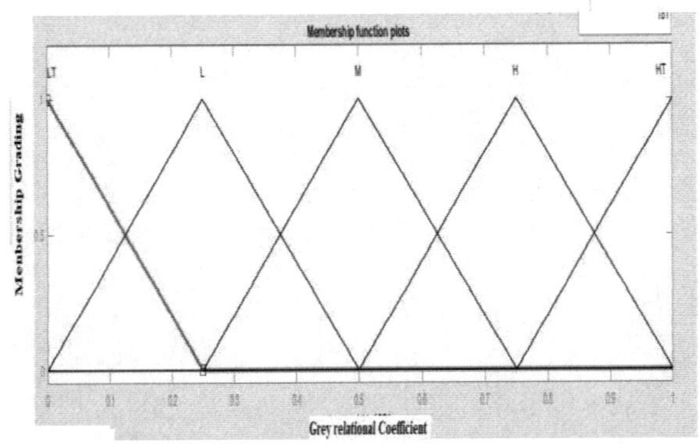

Fig 1. Fuzzy membership functions for QD, P, ENRC, DS, GD, RRR and PP.

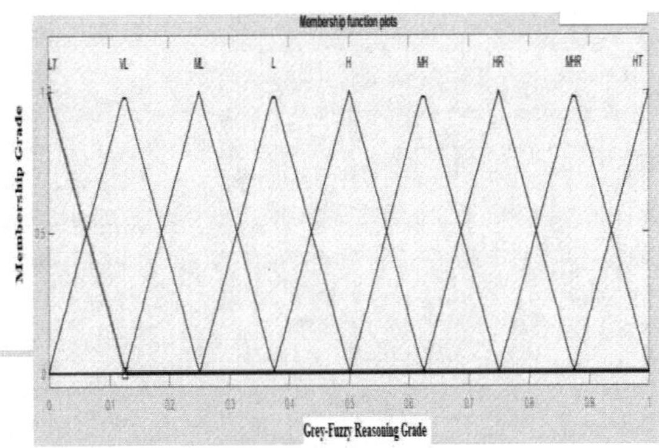

Fig 2. Fuzzy membership functions for grey-fuzzy reasoning grade

The GFRG values are shown in Fig. 2 as shown in the rule viewer in the Fuzzy tool box. In Fig. 3 &4, rows represent the Fuzzy rules and the first five columns represent the grey relational coefficients for QD, P, ENRC, DS, GD, RRR, and PP, and the last column provides the defuzzified GFRG values. The GFRG values obtained for the ten suppliers are given in Table 3.

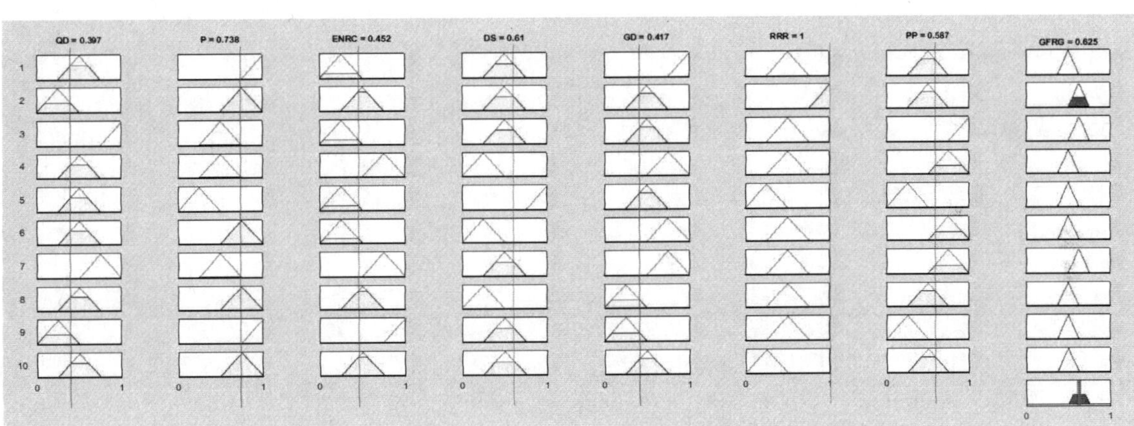

Fig 3. Fuzzy logic rule viewer (for alternative S_2)

IOP Conf. Series: Materials Science and Engineering **396** (2018) 012073 doi:10.1088/1757-899X/396/1/012073

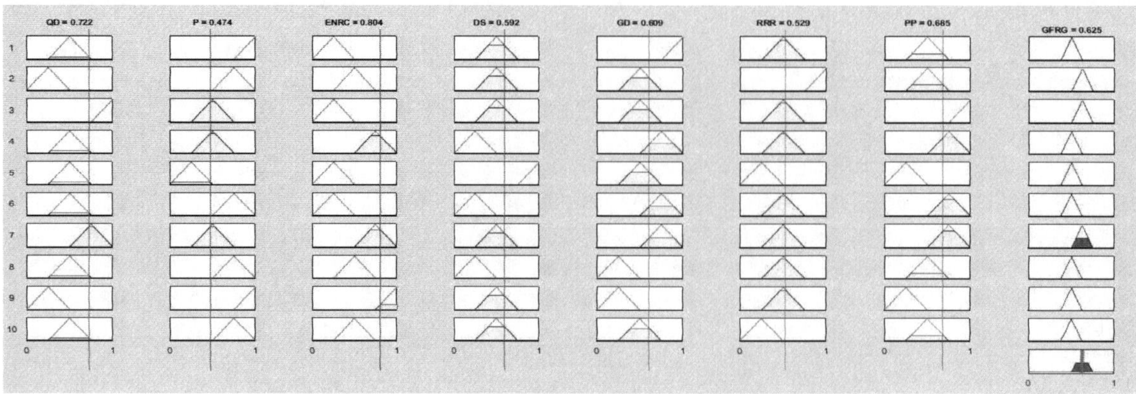

Fig 4. Fuzzy logic rule viewer (for alternative S_7)

Table 3. Grey fuzzy reasoning grade and rank for green suppliers

Supplier	A_1	A_2	A_3	A_4	A_5	A_6	A_7	A_8	A_9	A_{10}
GFRG	0.500	.625	.600	.500	.500	.500	.625	.500	.505	.500
Rank	7.5	1.5	3	7.5	7.5	7.5	1.5	7.5	4	7.5

A comparison of the rank arrangements for suppliers obtained using the proposed method is carried out with those attained by Yazdani et al. [20], using COPRAS and MOORA methods is shown in Table 4. It is fascinating to note that ranking orders for the first and second were interchanged with grey relational analysis when compared with COPRA and MOORA methods. But after using fuzzy logics (grey based fuzzy logic) GFRG for both the suppliers A_1, A_7 is 0.625. This indicates that both the suppliers are equally competitive and there is flexibility to choose any one of them. It is also observed that the GFRG values of all the suppliers except A_1, A_3, A_7 and A_9 shows same value as 0.5. This may be because of close GRG values, which are ranging from 0.4799-0.5449 and hence their corresponding fuzzy values are same. The computed r_s values between the rank orders of the considered methods are depicted in Table 5.

Table 4. Comparison of rankings with different methods for green suppliers

Supplier	COPRAS	MOORA	Grey-relational analysis	Grey-based fuzzy logic
S_1	5	4	5	7.5
S_2	2	2	1	1.5
S_3	4	3	3	3
S_4	6	5	6	7.5
S_5	10	10	9	7.5
S_6	8	8	7	7.5
S_7	1	1	2	1.5
S_8	7	6	8	7.5
S_9	3	7	4	4
S_{10}	9	9	10	7.5

Table 5. Spearman's Rank Correlation Coefficients

Method	COPRAS	MOORA	Grey-relational analysis	Grey-based fuzzy logic
COPRAS	-	0.878	0.939	0.878
MOORA	-	-	0.878	0.764
Grey-relational analysis	-	-	---	0.999

From Table 5, it can be noted that the computed r_s values for the ranking orders of the proposed method with the other considered approaches are above 0.7 showing strong correlation between grey-based fuzzy logic and the other methods.

4. Conclusions

Present work shows an application of grey-based Fuzzy logic for solving green supplier selection problems. It was found that ranking orders for the first and second rankings were interchanged with grey relational analysis when compared with COPRA and MOORA methods. Using fuzzy logics (grey based fuzzy logic) GFRG for both the suppliers A_1, A_7 is 0.625 showing both the suppliers are equally competitive and there is flexibility to choose any one of them. For most of the cases, the Spearman's correlation coefficients between the ranking orders of the previously employed methods and the grey-based fuzzy logic are greater than 0.7, showing strong correlation between these methods. This integrated grey relational - fuzzy logic approach can be suitably adopted for solving other complex supplier selection problems. When GRG values are close, precise GFRG values are same. This may be a limitation. Present study uses triangular membership function in future attempts can be made with other shapes of member ship function to obtain accurate values

5. References

[1] Boer, L., E. Labro, and P. Morlacchi. 2001. "*A Review of Methods Supporting Supplier Selection.*" European Journal of Purchasing and Supply Management 7 (2)

[2] Kumar, A, Jain V, Kumar S. (2014). *A comprehensive environment friendly approach for supplier selection, Omega*, 42(1): 109-123. http://dx.doi.org/10.1016/j.omega.2013.04.003.

[3] Govindan, K., & Sivakumar, R. (2016). *Green supplier selection and order allocation in a low-carbon paper industry: integrated multi-criteria heterogeneous decision-making and multi-objective linear programming approaches. Annals of Operations Research*, 238(1-2), 243-276

[4] Bai, C., & Sarkis, J. (2010*). Green supplier development: analytical evaluation using rough set theory.* Journal of Cleaner Production, 18(12), 1200-1210.

[5] Khaksar, E., Abbasnejad, T., Esmaeili, A., & Tamošaitienė, J. (2016). *The effect of green supply chain management practices on environmental performance and competitive advantage: a case study of the cement industry. Technological and Economic Development of Economy*, 22(2), 293-308.

[6] Louis Y.Y. Lu , C.H. Wu & T.-C. Kuo (2007) *Environmental principles applicable to green supplier evaluation by using multi-objective decision analysis, International Journal of Production Research*, 45:18-19, 4317-4331, DOI: 10.1080/00207540701472694

[7] Lee HI, Kang HY, Hsu CF, Hung HC. (2009). *A green supplier selection model for high-tech industry, Expert Systems with Applications,* 36(4): 791-798. http://dx.doi.org/10.1016/j.eswa.2008.11.052.

[8] Tuzkaya, G, Ozgen A, Ozgen D, Tuzkaya U. (2009*). Environmental performance evaluation of suppliers: A hybrid fuzzy multi-criteria decision approach, International Journal of Environmental Science and Technology*, 6(3): 477- 490. http://dx.doi.org/ 10.1007/BF03326087.

[9] Hsu, C.W, & Hu, AH. (2009). *Applying hazardous substance management to supplier selection using analytic network process, Journal of Cleaner Production*, 17(2): 255-264. http://dx.doi.org/10.1016/j.jclepro.2008.05.004.

[10] Hashemi, S.H, Karimi, A, Tavana, M. (2015*). An integrated green supplier selection approach with analytic network process and improved Grey relational analysis, International Journal of Production Economics*, 159: 178-191

[11] Awasthi, A., Chauhan, S.S., Goyal, S.K. (2010). *A fuzzy multicriteria approach for evaluating environmental performance of suppliers, International Journal of Production Economics*, 126(2): 370-378. http://dx.doi.org/10.1016/j.ijpe.2010.04.029

[12] Shen, L, Olfat, L, Govindan, K, Khodaverdi, R, Diabat, A. (2013). *A fuzzy multi criteria approach for evaluating green supplier's performance in green supply chain with linguistic preferences, Resources, Conservation and Recycling,* 74: 170-179. http://dx.doi.org/ 10.1016/j.resconrec.2012.09.006

[13] Kannan, D, Jabbour, ABLDS, Jabbour CJC. (2014*). Selecting green suppliers based on GSCM practices: Using fuzzy TOPSIS applied to a Brazilian electronics company, European Journal of Operational Research,* 233(2): 432- 447. http://dx.doi.org/10.1016/j.ejor.2013.07.023.

[14] Zhao, H, & Guo, S. *Selecting green supplier of thermal power equipment by using a Hybrid MCDM Method for sustainability, Sustainability* 2014; 6 (1): 217-235. http://dx.doi.org/10.3390/su6010217.

[15] Kannan, D, Khodaverdi R, Olfat L, Jafarian A, Diabat A. (2013). *Integrated fuzzy multi criteria decision making method and multi-objective programming approach for supplier selection and order allocation in a green supply chain, Journal of Cleaner Production*, 47: 355-367. http://dx.doi.org/10.1016/j.jclepro.2013.02.010.

[16] Rostamzadeh, R, Govindan K, Esmaeili A, Sabaghi M. (2015). *Application of fuzzy VIKOR for evaluation of green supply chain management practices, Ecological Indicators*, 49: 188-203. http://dx.doi.org/10.1016/j.ecolind.2014.09.045

[17] Azadnia, AH, Saman, MZM, & Wong, KY. *Sustainable supplier selection and order lot-sizing: an integrated multiobjective decision-making process, International Journal of Production Research* 2015; 53(2): 383-408. http://dx.doi.org/10.1080/00207543.2014.935827.

[18] Fallahpour, A., Amindoust, A., Antuchevičienė, J., & Yazdani, M. (2016). *Nonlinear genetic-based model for supplier selection: a comparative study. Technological and Economic Development of Economy*, 1-18.

[19] Yazdani, M., Hashemkhani Zolfani, S., & Zavadskas, E. K. (2016). *New integration of MCDM methods and QFD in the selection of green suppliers. Journal of Business Economics and Management,* 1-17.

[20] Yazdani M, Chatterjee P, Zavadskas EK, Hashemkhani Zolfani S, *Integrated QFD-MCDM framework for green supplier selection, Journal of Cleaner Production (*2016), doi: 10.1016/ j.jclepro.2016.10.095.

[21] Babu J. James Ajith, Philip J. Chakraborty S, *Application of the grey based fuzzy logic approach for material selection, International journel of material research* (2017); 702-709. Doi: 10.3139/146.111538

IOP Conf. Series: Materials Science and Engineering **396** (2018) 012074 doi:10.1088/1757-899X/396/1/012074

A review on magnetic field assisted micro machining

Renjith R[1] and Lijo Paul[2*]

[1]Post Graduate Student, [2*]Associate Professor,
Dept. of Mechanical Engg., St. Joseph's College of Engineering and Technology,
Palai, Kerala, India, 686579
E-mail*: lijo.paul@gmail.com

Abstract. Micromachining has developed in the last few decades with precision as primary aim. Many conventional and non-conventional processes are developed with precision in the range of micro level. Hybrid machining processes are developed in non-conventional machining area by combining the principles of basic machining process. In the current paper a review is carried out on Magnetic field assisted micro machining with thrust on EDM, ECM and ECDM machining process. The effect magnetic force during the machining has improved the machining output responses in terms on MRR and surface qualities.

Keywords: ECDM, Magnetic Field Assisted Machining, MRR, ECM, EDM

1. Introduction

In the last few decades, the need of advanced materials increased due to their unique metallurgical properties such as Super alloys, Composites, Ceramics. These properties have been developed to meet the demand of extreme applications such as in aerospace, automotive and biomedical industries etc. These advanced materials are harder, tougher, less heat sensitive and more resistant to corrosion and fatigue. Advanced materials are being widely used these days not only in the aerospace but also in defense and automotive. Therefore, the machining of difficult-to-machine materials is a predominant issue in the field of micromachining. Micromachining plays a vital role in miniaturization of parts used in aerospace (cooling holes in jet turbines blades), micro holes in fuel injection system of automotive, micro probes and micro needle for bio-medical application etc. Micromachining has great advantages in machining of hard materials in non-conventional machining as well as in conventional machining such as micro-mechanical milling, chemical etching, and micro-detonation of striking arc machining. In the current paper a literature review is

IOP Conf. Series: Materials Science and Engineering **396** (2018) 012074 doi:10.1088/1757-899X/396/1/012074

conducted focusing on the micro machining process such as Micro-EDM and Micro-ECDM and the enhancement of machining properties by the introduction of a magnetic field.

2. Magnetic Field Assisted EDM

The micro machining processes paved the way for advanced machining operations. The use of magnetic field assisted micro machining was introduced in 2008 [1]. The author has reported that an introduction of magnetic field in the conventional EDM will eliminate limitations and enhance the material properties. The major limitation in conventional EDM process is the settlement of debris between the machining gaps. A rotating magnetic field is introduced in a conventional EDM machine is converted. The rotating magnetic field was created by placing a rotating magnet below the workpiece, which can be rotated about 1200 rpm by the assistance of an electric motor. The workpiece material that used in this experiment was SKD 61 steel. In this research, the stability of the magnetic force-assisted EDM was determined by the analysis of discharge waveforms, and the machining characteristics such as Material Removal Rate (MRR), EWR, and surface integrities were explored. The results reveals that the MRR of the magnetic force-assisted EDM was increased nearly three times compared with that of conventional EDM, and the surface roughness was also finer than that of conventional EDM. However, the EWR was slightly higher than that of conventional EDM. The magnetic force-assisted EDM revealed the potentiality of expelling the debris and improving the machining efficiency. The MRR of the magnetic force-assisted EDM added-on with peak current and pulse duration, and the MRR reached a peak value at 20 A peak current and 350 ms pulse duration. The author recount that surface cracks became more obvious in the conventional EDM than that in the magnetic force-assisted EDM. Since the magnetic force-assisted EDM could facilitate the expulsion of machining debris to diminish the probability of abnormal electrical discharge, so the surface cracks were probably less on the machined surface.

Y.C.Lin and H.S Lee [2] reported that the magnetic field assisted EDM will exaggerate the conventional machining processes. He has optimized the machining parameters in magnetic field assisted EDM with Taguchi method. The effects of magnetic force on EDM machining characteristics were scrutinized in their work. The proposed work was adopted an L18 orthogonal array based on Taguchi method to conduct a series of experiments and statistically evaluated the experimental data by analysis of variance (ANOVA). In order to determine the EDM machining characteristics such as material removal rate (MRR) and surface roughness (SR), the main machining parameters such as machining polarity (P), peak current (Ip), pulse duration (p), high-voltage auxiliary current (IH), no-load voltage (V) and servo reference voltage (Sv) were chosen. The benefits of magnetic force assisted EDM were confirmed from the analysis of discharge waveforms and from the micrograph observation of surface integrity. The experimental result reveals that the magnetic force assisted EDM has a higher MRR, a lower relative electrode wear ratio (REWR), and a smaller SR as compared with standard EDM. In addition, the substantial machining parameters, and the optimal combination levels of machining parameters associated with MRR as well as SR were also discussed. In addition to it the work emphasis on the contribution provided by magnetic force assisted EDM in expelling machining debris attain a high efficiency and high quality of surface integrity to meet the demand of modern industrial applications.

The use of magnetic field in the dry EDM was studied by S. Joshi et al. [3]. They have investigated a hybrid dry EDM process in a pulsating magnetic field for improving process performance. In this cogitation, the pulsating magnetic field is applied tangential to the electric field, for increasing the movement of electrons and degree of ionization in the plasma. Experiments with parametric variations showed that this hybrid approach leads to productivity improvement by 130% and zero tool wear as compared to the dry EDM process without the magnetic field. The improvement in surface quality is illustrated by scanning electron microscopy.

Govindan et al.[4] also modeled the effect of magnetic field in dry EDM with single spark in magnetic field assisted dry EDM. They reported that the Material Removal Rate (MRR) increased about 130% as compared to conventional dry EDM. He found that presence of Lorentz force developed due to the presence of magnetic field. An investigation was made by them over 100 single- discharges in dry and liquid EDM with and without magnetic field with variation in machining parameters such as varying magnetic field, varying voltage, varying current, varying pulse on time & bi-pulse current. They reported that the magnetic field increased the depth and crated diameter about 80% in the case of dry EDM.

The optimization of magnetic field assisted EDM using ACO algorithm was proposed by R. Teimouri and H. Baseri [5]. They have supported the result proposed by Y.C.Lin and H.S Lee [1] in using a rotating magnetic field for better fleshing of the debris from the machining zone. Based on experimental observations two Adaptive Neuro-Fuzzy Inference System (ANFIS) models have been designed to correlate the EDM parameters to MRR and surface roughness (SR). They proposed the continuous ant colony optimization technique to select the best process parameters. Experimental trials divided into three main regimes of low energy, the middle energy and the high energy. Results showed that the CACO technique which used the ANFIS models as objective and constrain functions can successfully optimize the input conditions of the magnetic field assisted rotary EDM process.

A new variety magnetic field assisted EDM was proposed recently by Khan et al. [6]. They mixed a powder of along with the magnetic field. A permanent cylindrical magnet is used in this experiment and the workpiece used is steel grade 760.The electrode used was copper and peak current, pulse-on-time and powder concentration.

Brains et al. [7] conducted experiments in Magnetic field assisted EDM with machining parameters such as magnetic field intensity, peak current, duration of pulse on/off, tool electrode material and SiC% distribution. They adopted the Taguchi method for optimization and the analysis of results shows that increase in magnetic field, increase in peak current and increase in pulse-on-time increases the MRR.

3. Magnetic Assisted ECM

A novel concentrated magnetic field-assisted electrochemical machining (ECM) technology is proposed by L. Tang & W. M. Gan [8], to machine complex cavity with high efficiency and good precision. An ECM clamping apparatus with concentrated magnetic field, periodic magnetic field, and no magnetic field was described. The magnetic field simulation was implemented. Comparing the results of the concentrated magnetic field to the periodic magnetic field, the magnetic field intensity of the former is elevated by 9.8 % than the last-mentioned. Under the experimental conditions of 12 % $NaNO_3$, 14-V voltage, 0.8-MPa electrolyte pressure, temperature 32 °C, cathode feed rate 0.9 mm/min, initial machining gap 0.1 mm, and the S-03 special stainless steel work piece material, the experiments with concentrated magnetic field, periodic magnetic field, and no magnetic field were carried out. The results show that the magnetic field strength was increased by 16.7 % in the concentrated magnetic field than in the periodic magnetic field. Through a sectioning test, the precision in the concentrated magnetic field is increased by 33.3 % compared with no magnetic circuit and increased by 14.8 % compared with the periodic magnetic field.

4. Magnetic Assisted ECDM

C.P .Cheng [9] studied the importance of magnetic field in hybrid micro machining process, Electro Chemical Discharge Machining (ECDM on non-conducting material. They have reported about the lack of electrolyte flow at the narrow gap between the tool and workpiece in conventional ECDM. They have found the formation of a magneto hydrodynamic convection in the magnetic field assisted ECDM and the enhancement of electrolyte circulation. They have reported application of Lorentz force in inducing charged ions to form magneto hydro dynamic force and there by enhance the circulation by forcing electrolyte. They insisted that the magneto hydrodynamic convection induced by the magnetic field plays the key role in the enhancement of electrolyte circulation in the micro hole. They also reported the formation of a gas film by the coalescence of gas bubbles that evolved at tool surface and argues that the gas film quality influence the discharge performance. This gas film quality is the dominant factor that determines the machine quality such as geometric accuracy, material removal rate and surface integrity. The magnetic field was produced by introducing a ring magnet of Nd-Fe-B on the electrode and the work piece used is Pyrex glass. The results reveal that the gas film quality was enhanced and the maximum stability of gas film occurred at a higher voltage.

IOP Conf. Series: Materials Science and Engineering **396** (2018) 012074 doi:10.1088/1757-899X/396/1/012074

5. CONCLUSION

The micromachining process is discussed in the current paper with thrust on non-conventional machining process. The area of consideration includes the magnetic field assistance in micro machining process which includes EDM, ECM and ECDM process. The study shows that the limitations in micro machining can be eliminated by the application of a magnetic field. The results show that the presence of magnetic field enhances the MRR and the presence of Lorentz force enhances the machining operations.

REFERENCE

[1] Yan-Cherng Lin, Ho-Shiun Lee, 2008 Machining characteristics of magnetic force-assisted EDM
[2] Cherng Lina, Yuan-Feng Chena, Der-An Wanga, Ho-Shiun Leeb, 2009, Optimization of machining parameters in magnetic force assisted EDM based on Taguchi method.
[3] S. Joshi , P. Govindan, A. Malshe , K. Rajurkar, 2011, Experimental characterization of dry EDM performed in a pulsating magnetic field
[4] P. Govindan, A. Guptaa, Suhas S. Joshi a, Ajay Malsheb, K.P. Rajurkar, 2013, Single-spark analysis of removal phenomenon in magnetic field assisted dry
[5] R. Teimouri, H. Baseri, 2014, Optimization of magnetic field assisted EDM using the continuous ACO algorithm
[6] R. N. H. R. Ismail, A. A. Khan, A. N. M. Karim, 2017, Improvement of Material Removal Rate and Surface Roughness for Steel Grade 760 Using Magnetic Field Assisted Al2O3 Powder-Mixed EDM
[7] Preetkanwal Singh Bains, Sarabjeet Singh Sidhu & H. S. Payal, 2017, An investigation of magnetic-field-assisted EDM of composites
[8] L. Tang &W. M. Gan, 2013, Experiment and simulation study on concentrated magnetic field-assisted ECM S-03 special stainless steel complex cavity
[9] Chih-Ping Cheng, Kun-LingWu, Chao-Chuang Mai1, Yu-Shan Hsu and Biing-Hwa Yan, 2010, Magnetic field-assisted electrochemical discharge machining.